Angles

180°	90°
Straight Angle	**Right Angle**

45°

Acute Angle

An acute angle measures less than 90°.

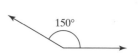

150°

Obtuse Angle

An obtuse angle measures between 90° and 180°.

- Two angles are **complementary** if the sum of their measures is 90°.

- $\angle a$ and $\angle c$ are vertical angles, and $\angle b$ and $\angle d$ are vertical angles.
- The measures of vertical angles are equal.

- Two angles are **supplementary** if the sum of their measures is 180°.

- The sum of the measures of the angles of a triangle is 180°.

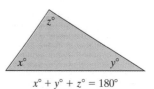

$$x° + y° + z° = 180°$$

Application Formulas

Simple interest = (principal)(rate)(time): $I = Prt$

Distance = (rate)(time): $d = rt$

Linear Equations and Slope

The slope, m, of a line between two distinct points (x_1, y_1) and (x_2, y_2):

$$m = \frac{y_2 - y_1}{x_2 - x_1}, \quad x_2 - x_1 \neq 0$$

Standard form: $ax + by = c$ (a and b are not both zero.)

Horizontal line: $y = k$

Vertical line: $x = k$

Slope-intercept form: $y = mx + b$

Point-slope formula: $y - y_1 = m(x - x_1)$

Properties and Definitions of Exponents

Let a and b ($b \neq 0$) represent real numbers and m and n represent positive integers.

$$b^m b^n = b^{m+n}; \qquad \frac{b^m}{b^n} = b^{m-n}; \qquad (b^m)^n = b^{mn};$$

$$(ab)^m = a^m b^m; \qquad \left(\frac{a}{b}\right)^m = \frac{a^m}{b^m}; \qquad b^0 = 1;$$

$$b^{-n} = \left(\frac{1}{b}\right)^n$$

Pythagorean Theorem

$$a^2 + b^2 = c^2$$

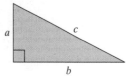

A Wealth of Tools for You
Miller/O'Neill Beginning Algebra

Instructor's Resource Manual   ISBN 0072553995

Written by the authors, the Instructor's Resource Manual (IRM) will help you guide your developmental math students to *success* in algebra.

The IRM contains:

- Classroom Activities**, keyed to the major concepts in each section of the text. The lessons are short (5-10 minutes) and can be assigned to groups or to individuals. The activities can also be provided as worksheets to complete outside the classroom.**
- Technology Connections
  These activities are designed for use with the Internet.
- Student Portfolio Materials
  To reinforce better organization and study habits, the authors have provided materials for a student portfolio. The structure of the portfolio will encourage students to keep well-organized class notes, homework, and definitions of vocabulary terms. Instructors need no prior experience with student portfolios.
- Cooperative Learning
  For instructors who may want to incorporate group work into their classes, the authors provide suggestions for constructing groups, information on group structures, tips for establishing accountability and roles within groups, and more.

Both students and instructors have access to the following elements of the IRM: Classroom Activities, Technology Connections, and Student Portfolio materials. In addition to these elements, instructors have access to answer keys and special notes on collaborative learning.

To review elements of the IRM available to *students* <u>now</u>, see below under "Student Learning Site Passcode." A printed version of the IRM that contains all elements for *instructors* can be requested by contacting your McGraw-Hill Higher Education (MHHE) representative. You can also download a copy from the Instructor Resource Site found on the Online Learning Center at www.mhhe.com/miller_oneill (you will need a passcode available from your MHHE representative).

Online Learning Center

The Online Learning Center (OLC), located at www.mhhe.com/miller_oneill, contains resources for students and instructors. In the OLC you will find:

- Student Learning Site
  In order to access the Student Learning Site, *students use passcodes that appear at the front of new copies of the student edition*.
- Instructor Resource Site
  The Instructor Resource Site is passcode-protected. Instructors can obtain passcodes by contacting their MHHE representatives. Instructors also have access to the Student Learning Site.
- Information Center
  The Information Center provides general information about the text and supplements.

Student Learning Site Passcode

For immediate access to materials available to students on the OLC, including access to elements of the IRM available to students, please refer to the back of this page. We have provided a student passcode and instructions for accessing the *Student Learning Site* of the OLC. Important** Your *Student Learning Site* account is time-sensitive. It will expire approximately eight months following the time you first log-on.

View Key
Resources Online
(See Student Learning
Site Passcode below)

IMPORTANT:

HERE IS YOUR REGISTRATION CODE TO ACCESS

YOUR PREMIUM McGRAW-HILL ONLINE RESOURCES.

For key premium online resources you need THIS CODE to gain access. Once the code is entered, you will be able to use the Web resources for the length of your course.

If your course is using **WebCT** or **Blackboard**, you'll be able to use this code to access the McGraw-Hill content within your instructor's online course.

Access is provided if you have purchased a new book. If the registration code is missing from this book, the registration screen on our Website, and within your WebCT or Blackboard course, will tell you how to obtain your new code.

Registering for McGraw-Hill Online Resources

TO gain access to your McGraw-Hill web resources simply follow the steps below:

(1) USE YOUR WEB BROWSER TO GO TO: **http://www.mhhe.com/miller_oneill**

(2) CLICK ON **FIRST TIME USER**.

(3) ENTER THE REGISTRATION CODE* PRINTED ON THE TEAR-OFF BOOKMARK ON THE RIGHT.

(4) AFTER YOU HAVE ENTERED YOUR REGISTRATION CODE, CLICK **REGISTER**.

(5) FOLLOW THE INSTRUCTIONS TO SET-UP YOUR PERSONAL UserID AND PASSWORD.

(6) WRITE YOUR UserID AND PASSWORD DOWN FOR FUTURE REFERENCE.
KEEP IT IN A SAFE PLACE.

TO GAIN ACCESS to the McGraw-Hill content in your instructor's **WebCT** or **Blackboard** course simply log in to the course with the UserID and Password provided by your instructor. Enter the registration code exactly as it appears in the box to the right when prompted by the system. You will only need to use the code the first time you click on McGraw-Hill content.

Thank you, and welcome to your McGraw-Hill online Resources!

* YOUR REGISTRATION CODE CAN BE USED ONLY ONCE TO ESTABLISH ACCESS. IT IS NOT TRANSFERABLE.

0-07-236371-1 MILLER/O'NEILL, BEGINNING ALGEBRA

MCGRAW-HILL
ONLINE RESOURCES

REGISTRATION CODE

perceiving-45603942

Mc Graw Hill Higher Education

BEGINNING ALGEBRA

ANNOTATED INSTRUCTOR'S EDITION

JULIE MILLER
MOLLY O'NEILL

Daytona Beach Community College

Boston   Burr Ridge, IL   Dubuque, IA   Madison, WI   New York   San Francisco   St. Louis
Bangkok   Bogotá   Caracas   Kuala Lumpur   Lisbon   London   Madrid   Mexico City
Milan   Montreal   New Delhi   Santiago   Seoul   Singapore   Sydney   Taipei   Toronto

BEGINNING ALGEBRA

Published by McGraw-Hill, a business unit of The McGraw-Hill Companies, Inc., 1221 Avenue of the Americas, New York, NY 10020. Copyright © 2004 by The McGraw-Hill Companies, Inc. All rights reserved. No part of this publication may be reproduced or distributed in any form or by any means, or stored in a database or retrieval system, without the prior written consent of The McGraw-Hill Companies, Inc., including, but not limited to, in any network or other electronic storage or transmission, or broadcast for distance learning.

Some ancillaries, including electronic and print components, may not be available to customers outside the United States.

This book is printed on acid-free paper.

1 2 3 4 5 6 7 8 9 0 VNH/VNH 0 9 8 7 6 5 4 3
1 2 3 4 5 6 7 8 9 0 VNH/VNH 0 9 8 7 6 5 4 3

ISBN 0–07–236371–1 (Student Edition)
ISBN 0–07–252561–4 (Annotated Instructor's Edition)

Publisher: *William K. Barter*
Senior sponsoring editor: *David Dietz*
Developmental editor: *Erin Brown*
Executive marketing manager: *Marianne C. P. Rutter*
Senior marketing manager: *Mary K. Kittell*
Lead project manager: *Peggy J. Selle*
Production supervisor: *Sherry L. Kane*
Senior media project manager: *Tammy Juran*
Media technology producer: *Jeff Huettman*
Designer: *K. Wayne Harms*
Cover/interior designer: *Rokusek Design*
Cover image: *David Woodfall/Gettyimages*
Lead photo research coordinator: *Carrie K. Burger*
Photo research: *LouAnn K. Wilson*
Supplement producer: *Brenda A. Ernzen*
Compositor: *TechBooks*
Typeface: *10/12 Times Ten*
Printer: *Von Hoffmann Corporation*

Photo Credits

About the Authors: Photo of Julie Miller by Marc Campbell; Photo of Molly O'Neill by Gail Beckwith. Chapter R: Opener: © Vol. 38/CORBIS; p. 25: © CORBIS website; p. 42: © PhotoDisc website. Chapter 1: Opener: © Vol. 154/CORBIS; p. 68: © Elena Rooraid/Photo Edit; p. 76: © Vol. 44/CORBIS; p. 88: © CORBIS website; p. 106: © David Young-Wolff/Photo Edit. Chapter 2: Opener: © CORBIS website; p. 150: © Vol. 44/CORBIS; p. 159: © Judy Griesedieck/CORBIS; p. 169: © CORBIS website; p. 181: © CORBIS website; p. 195: © Robert Brenner/Photo Edit; p. 207: © Tony Freeman/Photo Edit. Chapter 3: Opener: © CORBIS website; p. 254: © Vol. 26/CORBIS; p. 256: © CORBIS website; p. 293: © CORBIS website; p. 308: © CORBIS website. Chapter 4: Opener: © Vol. 1/PhotoDisc; p. 377: © Michael Newman/Photo Edit; p. 385: © Vol. 4/CORBIS. Chapter 5: Opener: © Vol. 62/CORBIS; p. 398: © Paul Morris/CORBIS; p. 408: © Vol. 56/CORBIS; p. 415: © Michael Newman/Photo Edit; p. 434: © EyeWire/Getty Images website; p. 441 (left): © Vol. 117/CORBIS; p. 441 (right): © Vol. 35/CORBIS; p. 460: © PhotoDisc website. Chapter 6: Opener: © Vol. 188/CORBIS; p. 477: © EyeWire/Getty website; p. 489: © Vol. 107/CORBIS; p. 500: © EyeWire/Getty website; p. 520: © Susan Van Etten/Photo Edit. Chapter 7: Opener: © Vol. 247/CORBIS; p. 550: © Vol. 132/CORBIS; p. 560: © Burke/Triolo Productions/FoodPix/Getty Images; p. 570: © Jeff Greenberg/Photo Edit. Chapter 8: Opener: © Bill Ross/CORBIS; p. 605: © AP/Wide World Photos; p. 620: © Dennis O'Clair/Stone Images/Getty Images; p. 628: © CORBIS website; p. 643: © Vol. 101/CORBIS. Chapter 9: Opener: © Vol. 110/PhotoDisc; p. 679: © PhotoDisc website; p. 697 (left): © Massimo Listri/CORBIS; p. 697 (right): © PhotoDisc/Getty website; p. 719: © Vol. 527/CORBIS; p. 726: © Frank Whitney/Brand X Pictures/PictureQuest.

Library of Congress Cataloging-in-Publication Data

Miller, Julie, 1962–
    Beginning algebra / Julie Miller, Molly O'Neill.—1st ed.
        p.   cm.
    Includes index.
    ISBN 0–07–236371–1
    1. Algebra. I. O'Neill, Molly 1953–. II. Title.

QA152.3 .M55   2004
512—dc21   2002032577
                                            CIP

www.mhhe.com

DEDICATION

To Geoff and Pam, and Joelle and Bob

—Julie Miller

To my parents, Doris and Richard Krajewski

—Molly O'Neill

ABOUT THE AUTHORS

JULIE MILLER

Julie Miller has been a member of the Mathematics Department at Daytona Beach Community College for 14 years where she has taught developmental and upper level courses. Prior to her work at DBCC, Julie worked as a software engineer for General Electric in the area of flight and radar simulation. Julie earned a bachelor of science degree in applied mathematics from Union College in Schenectady, New York and a master of science in mathematics from the University of Florida. In addition to her textbook, Julie has authored several course supplements for college algebra, trigonometry, and precalculus and several short works of fiction and nonfiction for young readers.

MOLLY O'NEILL

Molly O'Neill is also from Daytona Beach Community College where she has taught for 16 years in the Mathematics Department. She has taught a variety of courses from developmental mathematics to calculus. Before she came to Florida, Molly taught as an adjunct instructor at the University of Michigan—Dearborn, Eastern Michigan University, Wayne State University, and Oakland Community College. Molly earned a bachelor of science in mathematics and a master of arts and teaching from Western Michigan University in Kalamazoo, Michigan. Besides her textbook, Molly has authored several course supplements for college algebra, trigonometry, and precalculus and reviewed texts for developmental mathematics.

TABLE OF CONTENTS

chapter **R**

REFERENCE: FRACTIONS, DECIMALS, PERCENTS, GEOMETRY, AND STUDY SKILLS   1

chapter **1**

SET OF REAL NUMBERS   43

chapter **2**

LINEAR EQUATIONS AND INEQUALITIES   121

chapter **3**

POLYNOMIALS AND PROPERTIES OF EXPONENTS 225

chapter **4**

FACTORING POLYNOMIALS 313

chapter **5**

RATIONAL EXPRESSIONS   389

chapter **6**

GRAPHING LINEAR EQUATIONS
IN TWO VARIABLES   465

chapter **9**

FUNCTIONS, COMPLEX NUMBERS, AND QUADRATIC EQUATIONS

665

PREFACE TO THE INSTRUCTOR

Instructors of developmental mathematics encounter a wide variety of challenges. The students under their tutelage often face motivational problems and lack the reading and study skills necessary to succeed. We wrote *Beginning Algebra* to address the challenges.

To help students overcome deficiencies in mathematical language, special "translation" problems and writing exercises are presented throughout the text. To combat motivational problems, we provide applications based on the kinds of everyday facts and information students encounter by watching the news, reading a magazine, or surfing the World Wide Web. A special "Help Yourself" preface in the Student Edition of this text offers students practical suggestions for making the most of the resources available to them. In addition, we devote a section of the text to study skills to promote better study habits. Ultimately, through these and similar features, students will have more positive experiences with mathematics and their chances for success will increase.

In writing *Beginning Algebra*, we were mindful of the standards for introductory college mathematics prepared by the American Mathematical Association of Two-Year Colleges (AMATYC). We constructed applications and examples intending to show the presence of mathematics in a variety of disciplines. We incorporated real-world data for students to create and interpret mathematical models in the form of equations and graphs. Furthermore, applications to geometry are threaded throughout the text and in the homework exercises. We designed classroom activities and writing problems to encourage students to communicate mathematical ideas both orally and in writing. Implementation of these measures is intended to help students become stronger problem-solvers and critical thinkers.

While *Beginning Algebra* offers special features to address the unique challenges of teaching today's developmental mathematics students, it also maintains a solid approach to algebraic instruction. *Beginning Algebra* provides students with plenty of exercises and examples, clear explanations, and the core content needed to build a firm foundation in algebra.

CORE FEATURES

We draw attention to the following features, which we believe set *Beginning Algebra* apart from other beginning algebra texts found in the marketplace.

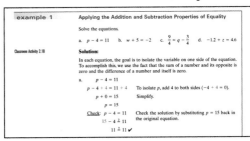

Classroom Activities

References to Classroom Activities are provided in each section of the Annotated Instructor's Edition and appear on the pages near the text and examples to which the activities apply. The activities themselves are included in the Instructor's Resource Manual* and can be used at the discretion of the instructor. With increasing demands on faculty schedules, these ready-made lessons offer a convenient means for both full-time professors and part-time adjuncts to promote active learning in the classroom. The classroom activities can be assigned to groups or to students individually. Alternatively, these worksheets may be provided directly to students for extra practice inside or outside the classroom. In performing these activities in class, students become better conditioned to take responsibility for their learning and thereby show greater interest when asked to participate in classroom discussions.

Language of Mathematics ⬌

Throughout the text, the language of mathematics is emphasized. Special "translation" examples and exercises ask students to convert between English phrases and algebraic statements. Through these exercises and examples, students gain increased familiarity with mathematical symbols and terminology, and in turn develop a fuller understanding of the content.

Concept Retention

To promote long-term concept retention, review problems have been infused in the Practice Exercises so that concepts learned earlier in the chapter are continually reinforced. These exercises appear at the beginning of the Practice Exercises and are labeled in the Annotated Instructor's Edition with section numbers to designate each concept's point of origin.

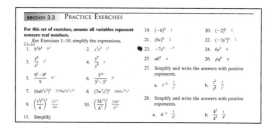

Midchapter Reviews

A set of retention exercises can be found in the middle of each chapter. With these questions, students can review concepts introduced in the first half of the chapter before expanding to new ideas presented in the second half. The Midchapter Reviews also help promote long-term concept retention.

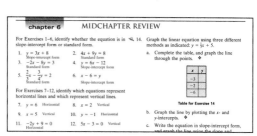

Student Portfolio

The materials that constitute the Student Portfolio, found in the Instructor's Resource Manual, offer students additional opportunity to take responsibility for their learning. Using the portfolio materials, students can maintain an organized notebook that contains classroom lecture notes, homework, tests, test corrections, a calendar, a record of grades, and vocabulary terms.

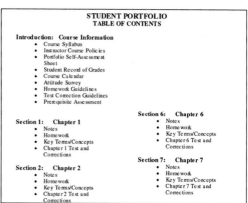

*Available from your McGraw-Hill Higher Education representative, ISBN 0072553995

COVERAGE

Early Graphing

The graphical interpretation of algebra is a critical skill that carries through to upper level mathematics courses, yet students often have difficulty when they encounter graphing. In *Beginning Algebra*, we offer the opportunity for early and repeated exposure to graphing through special sections entitled, Connections to Graphing. Some or all of these sections may be covered at the instructor's discretion to begin building a foundation before a traditional coverage of graphing is presented in Chapter 6. Each Connections to Graphing section is a self-contained unit and may be covered at different points in the text. This offers instructors the flexibility of introducing graphing early, late, or intermittently throughout the course. The Connections to Graphing sections provide an early introduction to such topics as plotting points (Chapter 1, Section 7), graphing linear equations in two variables (Chapter 2, Section 8), and finding x- and y-intercepts (Chapter 4, Section 7); however, these concepts are covered again formally in Chapter 6. Therefore, instructors who prefer to teach graphing as a stand-alone chapter can easily do so.

Calculator Connections and Calculator Exercises are provided as a tool for instructors who incorporate calculator activities in their course. The Calculator Connections are self-contained units found at the end of applicable sections of text and may easily be omitted by instructors who choose not to implement the calculator.

Chapter R

Chapter R is a reference chapter and is provided for students and instructors who want to engage in a brief review of basic mathematics, geometry, or study skills.

Geometry

Special emphasis is placed on geometry in the text. Exercises have been included throughout, to help reinforce students' learning and mastery of geometry skills. In Section 3 of Chapter R, students can review basic concepts in geometry.

Chapter 4 Midchapter Review

The Midchapter Review for Chapter 4, found on page 345, has been specially titled *Factoring Strategy*. It is designed to provide a cumulative review of the methods used to factor different types of polynomials. This review offers an opportunity for students to see a variety of polynomials and to apply the skills they have learned in the first three sections to successfully factor each polynomial.

ACKNOWLEDGMENTS

Preparing a first edition mathematics text is an enormous undertaking that would never have been possible without the creative ideas and constructive feedback offered by many reviewers. We are especially thankful to the following instructors for their valuable feedback and careful review of the manuscript:

Mary Kay Best, *Coastal Bend College*
Connie Buller, *Metropolitan Community College*
Gerald F. Busald, *San Antonio College*
Elizabeth Condon, *Queens Community College*
Pat C. Cook, *Weatherford College*
Jacqueline Coomes, *Eastern Washington University*
Cynthia M. Craig, *Augusta State University*
Andres Delgado, *Orange County Community College*
Irene Durancyzk, *Eastern Michigan University*
Pat Foard, *South Plains College*
Dr. Paul F. Foutz, *Temple College*
Jacqueline B. Giles, *Houston County Community College*
Celeste Hernandez, *Richland College*
Julie Hess, *Grand Rapids Community College*
Kayana Hoagland, *South Puget Sound Community College*
Rosalie Hojegian, *Possaic County Community College*
Lori Holdren, *Manatee Community College*
Sarah Jackman, *Richland College*
Nancy R. Johnson, *Manatee Community College*
Steven Kahn, *Anne Arundel Community College*
Jane Keller, *Metropolitan Community College*
Jeff A. Koleno, *Lorain County Community College*
Mary M. Leeseberg, *Manatee Community College*
Deann Leoni, *Edmonds Community College*
Pamela A. Lipka, *University of Wisconsin*
J. Robert Malena, *Community College of Allegheny County—South Campus*
Maria M. Maspons, *Miami Dade Community College*
Debbie K. Millard, *Florida Community College*
Jean P. Millen, *Georgia Perimeter College*
Cameron Neal, Jr., *Temple College*
Linda Padilla, *Joliet Junior College*
Bernard J. Pina, *Dona Ana Branch Community College—NMSU*
Rita Beth Pruitt, *South Plains College*
Nancy C. Ressler, *Oakton Community College*
Reynaldo Rivera, Jr., *Estrella Mountain Community College*
Mary Romans, *Kent State University*
Fred Safier, *City College of San Francisco*

Ned W. Schillow, *Lehigh Carbon Community College*

Mary Lee Seitz, *Erie Community College*

Cindy Shaber, *Boise State University*

Lisa Sheppard, *Lorain County Community College*

Brian Starr, *National University*

Dr. Bryan Stewart, *Tarrant County College*

Alexis Thurman, *County College of Morris*

Dr. Roy N. Tucker, *Palo Alto College*

Sandra Vrem, *College of the Redwoods*

Thomas L. Wolters, *Muskegon Community College*

Special thanks go to Doris McClellan Lewis and Karyn Anderson for preparing the *Instructor's Solutions Manual* and the *Student's Solutions Manual*, to Cynthia Cruz of College of the Canyons for her appearance in and work on the video series, and to Lauri Semarne for performing an accuracy check of the manuscript.

In addition to the assistance provided by the reviewers, we would also like to thank the many people behind the scenes at McGraw-Hill who have made this project possible. Our sincerest thanks to Erin Brown, David Dietz, and Bill Barter for being patient and kind editors, to Peggy Selle for keeping us on track during production, and to Mary Kittell and Marianne Rutter for their creative ideas promoting all of our efforts. We further appreciate the hard work of Wayne Harms, Carrie Burger, Tammy Juran, and Jeff Huettman.

Finally, we give special thanks to all the students and instructors who use *Beginning Algebra* in their classes.

Julie Miller and Molly O'Neill

PEDAGOGY

Chapter Openers

Each chapter opens with an application relating to topics presented in the chapter. The Chapter Openers also contain website references for **Technology Connections**—Internet activities found in the Instructor's Resource Manual—that further the scope of the application.

8

RADICALS

The area of a triangle can be found if the length of one side (the base) and the corresponding height of the triangle are known.

$$A = \frac{1}{2}bh$$

However, if the height of a triangle is not known, but the lengths of the three sides, a, b, and, c are given, the area of the triangle can be found using Heron's formula:

$$A = \sqrt{s(s - a)(s - b)(s - c)}$$

where

$$s = \frac{1}{2}(a + b + c)$$

Heron's formula and other applications of radicals are studied in this chapter.

The Louvre pyramid, designed by architect I. M. Pei, is a glass structure that serves as the entrance to the Louvre Museum in Paris. Each triangular face is made of glass with dimensions as shown.

The area of each face can be found using Heron's formula:

$$s = \frac{1}{2}(108.5 + 108.5 + 116) = 166.5$$

$$A = \sqrt{166.5(166.5 - 108.5)(166.5 - 108.5)(166.5 - 116)}$$
$$\approx 5318.4$$

The area of each triangular face required approximately 5318.4 ft^2 of glass.

108.5 ft   108.5 ft
116 ft

Concepts

1. Review of Exponential Notation
2. Evaluating Expressions with Exponents
3. Multiplying and Dividing Common Bases
4. Simplifying Expressions with Exponents
5. Applications of Exponents

section
3.1   EXPONENTS: MULTIPLYING AND DIVIDING COMMON BASES

1.   Review of Exponential Notation

Recall that an **exponent** is used to show repeated multiplication of the **base**.

Definition of b^n

Let b represent any real number and n represent a positive integer. Then,

$$b^n = \underbrace{b \cdot b \cdot b \cdot b \dots b}_{n \text{ factors of } b}$$

example 1   **Evaluating Expressions with Exponents**

For each expression, identify the exponent and base. Then evaluate the expression.

a. 6^2 b. $\left(-\frac{1}{2}\right)^3$ c. 0.8^4

Solution:

Expression	Base	Exponent	Result
a. 6^2	6	2	$(6)(6) = 36$
b. $\left(-\frac{1}{2}\right)^3$	$-\frac{1}{2}$	3	$\left(-\frac{1}{2}\right)\left(-\frac{1}{2}\right)\left(-\frac{1}{2}\right) = -\frac{1}{8}$
c. 0.8^4	0.8	4	$(0.8)(0.8)(0.8)(0.8) = 0.4096$

Note that if no exponent is explicitly written for an expression, then the expression has an implied exponent of 1. For example,

$$x = x^1$$
$$y = y^1$$
$$5 = 5^1$$

2.   Evaluating Expressions with Exponents

Recall from Section 1.2 that particular care must be taken when evaluating exponential expressions involving negative numbers. An exponential expression with a negative base is written with parentheses around the base, such as $(-3)^2$.

To evaluate $(-3)^2$, we have: $(-3)^2 = (-3)(-3) = 9$

If no parentheses are present, the expression -3^2, is the *opposite* of 3^2, or equivalently, $-1 \cdot 3^2$.

Hence: $-3^2 = -1(3^2) = -1(3)(3) = -9$

Special Elements

Concepts

A list of important concepts is provided at the beginning of each section. Each concept corresponds to a heading within the section, making it easy for students to locate topics as they study or as they work through homework exercises.

example 6 — Writing an Equation of a Line Through Two Points

Write an equation of the line passing through the points $(-3, 0)$ and $(3, -4)$. Write the answer in slope-intercept form.

Solution:

Given two points on a line, the slope can be found with the slope formula.

$$(-3, 0) \quad \text{and} \quad (3, -4)$$
$$(x_1, y_1) \qquad\qquad (x_2, y_2) \qquad \text{Label the points.}$$

$$m = \frac{y_2 - y_1}{x_2 - x_1} = \frac{(-4) - (0)}{(3) - (-3)} = \frac{-4}{6} = -\frac{2}{3}$$

With the slope $m = -\frac{2}{3}$, the slope-intercept form of the line is: $y = -\frac{2}{3}x + b$.

To find b, substitute the values of x and y from *either* ordered pair and then solve the resulting equation for b. We will use the point $(-3, 0)$.

$$y = -\frac{2}{3}x + b$$

$$0 = -\frac{2}{3}(-3) + b \qquad \text{Substitute } y = 0 \text{ and } x = -3.$$

$$0 = 2 + b \qquad \text{Simplify.}$$

$$-2 = b \qquad \text{Solve for } b.$$

The slope-intercept form is $y = -\frac{2}{3}x - 2$.

A sketch of the line shows that the line passes through the points $(-3, 0)$ and $(3, -4)$ as desired (Figure 6-25).

Figure 6-25

Tip: The value of b could also have been found by using the point $(3, -4)$:

$$y = -\frac{2}{3}x + b$$
$$-4 = -\frac{2}{3}(3) + b$$
$$-4 = -2 + b$$
$$-2 = b$$

7. Writing an Equation of a Line Parallel or Perpendicular to Another Line

example 7 — Writing an Equation of a Line Parallel to Another Line

Write an equation of the line passing through the point $(1, 2)$ and parallel to the line $y = 3x - 4$. Write the answer in slope-intercept form.

Solution:

Figure 6-26 shows the line $y = 3x - 4$ (pictured in black) and a line parallel to it (pictured in blue) that passes through the point $(1, 2)$. The given line, $y = 3x - 4$, is written in slope-intercept form and its slope is easily identified as 3. The line parallel to the given line must also have a slope of 3.

The slope-intercept form of the line we want to find is

$$y = 3x + b$$

Figure 6-26

Tips

Tip boxes appear throughout the text and offer helpful hints and insight.

---

4. Use the addition and subtraction properties of equality to collect the constant terms on the other side of the equation.
5. Use the multiplication and division properties of equality to make the coefficient of the variable term equal to 1.
6. Check your answer.

example 3 — Solving Linear Equations

Classroom Activity 2.4A

a. $\frac{1}{6}x - \frac{2}{3} = \frac{1}{5}x - 1$ b. $\frac{1}{3}(x + 7) - \frac{1}{2}(x + 1) = 4$

Solution:

a.
$$\frac{1}{6}x - \frac{2}{3} = \frac{1}{5}x - 1 \qquad \text{The LCD of 6, 3, and 5 is 30.}$$

$$30\left(\frac{1}{6}x - \frac{2}{3}\right) = 30\left(\frac{1}{5}x - 1\right) \qquad \text{Multiply by the LCD, 30.}$$

$$\frac{\overset{5}{30}}{1} \cdot \frac{1}{6}x - \frac{\overset{10}{30}}{1} \cdot \frac{2}{3} = \frac{\overset{6}{30}}{1} \cdot \frac{1}{5}x - 30(1) \qquad \begin{array}{l}\text{Apply the distributive property} \\ (\text{recall } 30 = \frac{30}{1}).\end{array}$$

$$5x - 20 = 6x - 30 \qquad \text{Clear fractions.}$$

$$5x - 6x - 20 = 6x - 6x - 30 \qquad \text{Subtract } 6x \text{ from both sides.}$$

$$-x - 20 = -30$$

$$-x - 20 + 20 = -30 + 20 \qquad \text{Add 20 to both sides.}$$

$$-x = -10$$

$$\frac{-x}{-1} = \frac{-10}{-1} \qquad \text{Divide both sides by } -1.$$

$$x = 10$$

b.
$$\frac{1}{3}(x + 7) - \frac{1}{2}(x + 1) = 4 \qquad \text{The LCD of 3 and 2 is 6.}$$

$$6\left[\frac{1}{3}(x + 7) - \frac{1}{2}(x + 1)\right] = 6(4) \qquad \text{Multiply both sides by 6.}$$

$$\frac{\overset{2}{6}}{1}\left[\frac{1}{3}(x + 7)\right] - \frac{\overset{3}{6}}{1}\left[\frac{1}{2}(x + 1)\right] = 6(4) \qquad \text{Apply the distributive property.}$$

The product is 2. The product is 3.

$$2(x + 7) - 3(x + 1) = 24$$

$$2x + 14 - 3x - 3 = 24 \qquad \text{Clear parentheses.}$$

$$-x + 11 = 24 \qquad \text{Combine } like \text{ terms.}$$

Avoiding Mistakes

Through marginal notes labeled Avoiding Mistakes students are alerted to common errors and are shown methods to avoid them.

Avoiding Mistakes

Notice that on the left-hand side of this equation, the product of 6 and $\frac{1}{3}$ is taken first, and then the result of 2 is distributed through the parentheses. Similarly, the product of 6 and $-\frac{1}{2}$ is taken first. The result of -3 is then distributed through the parentheses.

References to Classroom Activities

Throughout each section of the Annotated Instructor's Edition, references are made to Classroom Activities. These references appear on the page relative to where the activities might be introduced in lecture. The activities are described in the Instructor's Resource Manual.

Calculator Connections

Optional Calculator Connections, appear throughout the text. These can be implemented at the instructor's discretion depending on the amount of emphasis placed on the calculator in the course. Many of the Calculator Connections include screen captures and include a set of exercises that provide an opportunity for students to apply the skills introduced. They are designed for use with a scientific and/or a graphing calculator.

Practice Exercises

The Practice Exercises comprise a variety of problem types.

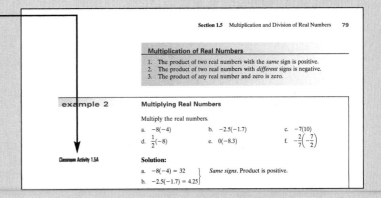

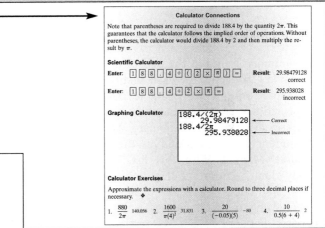

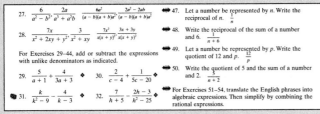

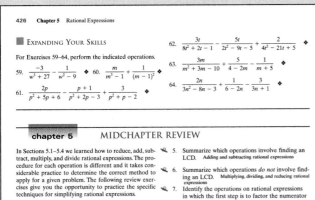

- **Writing Exercises** ✎ offer students an opportunity to conceptualize and communicate their understanding of algebra. These, along with the **translating expressions exercises** ⬌, enable students to strengthen their command of mathematical language and notation and improve their reading and writing skills.
- **Geometry Exercises** ▱ throughout the Practice Exercises encourage students to review and apply geometry concepts.
- **Review Problems** appear within the Practice Exercises to help students retain their knowledge of concepts previously learned. Each review problem is labeled with a section number, referencing the section where the problem type is introduced.
- **Applications** based on real world facts and figures motivate students and enable them to hone their problem-solving skills.
- **Exercises Keyed to Video** ▣ are labeled with an icon to help students and instructors identify those exercises that appear in the video series that accompanies *Beginning Algebra*.

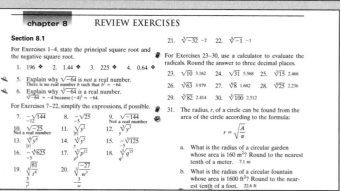

chapter 4   MIDCHAPTER REVIEW: "FACTORING STRATEGY"

1. What is meant by a prime factor? A prime factor cannot be factored further.
2. What is the first step in factoring any polynomial? Factor out the GCF.
3. When factoring a binomial, what pattern can you look for? Look for the difference of squares: $a^2 - b^2$.
4. When factoring a trinomial what pattern do you look for first before using the grouping method or trial-and-error method? Look for a perfect square trinomial: $a^2 + 2ab + b^2$ or $a^2 - 2ab + b^2$.
5. Are factorable polynomials factored completely in one step? Not all polynomials factor completely in one step. For example: $2x^4 - 32$.

Midchapter Reviews

Midchapter Reviews are provided to help solidify the foundation of concepts learned in the beginning of a chapter before expanding to new ideas presented later in the chapter.

End-of-Chapter Summary and Exercises

The **Summary**, found at the end of each chapter, outlines key concepts for each section and illustrates those concepts with examples. The Summary also provides a list of important terms that mirror those appearing in the Vocabulary Worksheets of the Student Portfolio found in the Instructor's Resource Manual. With this list, students can quickly identify important ideas and vocabulary to be reviewed before quizzes or exams.

    Following the Summary is a set of **Review Exercises** that are organized by section. A **Chapter Test** appears after each set of Review Exercises. Chapters 2–9 also include a **Cumulative Review** that follows the Chapter Test. These end-of-chapter materials provide students with ample opportunity to prepare for quizzes or exams.

- **Calculator Exercises** signify situations where a calculator would provide assistance for time-consuming calculations. These exercises were carefully designed to demonstrate the types of situations where a calculator is a handy tool rather than a "crutch." They are designed for use with either a scientific or a graphing calculator.
- **Expanding Your Skills**, found near the end of most Practice Exercise sets, challenge students' knowledge of the concepts presented.

chapter 8   SUMMARY

SECTION 8.1—INTRODUCTION TO ROOTS AND RADICALS

KEY CONCEPTS:	EXAMPLES:
b is a square root of a if $b^2 = a$.	The square roots of 16 are 4 and -4 because $(4)^2 = 16$ and $(-4)^2 = 16$.
The expression $\sqrt{a}$ represents the principal square root of a.	$\sqrt{16} = 4$ Because $4^2 = 16$
b is an nth-root of a if $b^n = a$.	$\sqrt[4]{16} = 2$ Because $2^4 = 16$
1. If n is a positive *even* integer and $a > 0$, then $\sqrt[n]{a}$ is the principal (positive) nth-root of a.	$\sqrt[3]{125} = 5$ Because $5^3 = 125$
2. If n is a positive *odd* integer, then $\sqrt[n]{a}$ is the nth-root of a.	
3. If n is any positive integer, then $\sqrt[n]{0} = 0$	

chapter 8   REVIEW EXERCISES

Section 8.1

For Exercises 1–4, state the principal square root and the negative square root.

1. 196 ✦
2. 1.44 ✦
3. 225 ✦
4. 0.64 ✦

5. Explain why $\sqrt{-64}$ is *not* a real number. There is no real number b such that $b^2 = -64$.
6. Explain why $\sqrt[3]{-64}$ *is* a real number. $\sqrt[3]{-64} = -4$ because $(-4)^3 = -64$.

For Exercises 7–22, simplify the expressions, if possible.

7. $-\sqrt{144}$ -12
8. $\frac{-\sqrt{25}}{-5}$ $\frac{5}{-5}$
9. $\sqrt{-144}$ Not a real number
10. $\frac{-\sqrt{25}}{|y|}$ Not a real number
11. $\sqrt[4]{y^2}$
12. $\sqrt[3]{y^3}$ y
13. $\sqrt[4]{y^4}$ $\frac{y}{|y|}$
14. $\sqrt[3]{y^3}$ y
15. $-\sqrt[3]{125}$ -5
16. $\frac{-\sqrt[4]{625}}{-5}$
17. $\sqrt[4]{p^{12}}$ p^3
18. $\sqrt[5]{q^{15}}$ q^3
19. $\sqrt{\frac{81}{t^8}}$ $\frac{3}{t^4}$
20. $\sqrt[3]{\frac{-27}{w^3}}$ $\frac{-3}{w}$
21. $\sqrt[5]{-32}$ -2
22. $\sqrt[4]{-1}$ -1

For Exercises 23–30, use a calculator to evaluate the radicals. Round the answer to three decimal places.

23. $\sqrt{10}$ 3.162
24. $\sqrt{31}$ 5.568
25. $\sqrt{15}$ 2.466
26. $\sqrt[3]{63}$ 3.979
27. $\sqrt[3]{8}$ 1.682
28. $\sqrt[3]{25}$ 2.236
29. $\sqrt[4]{82}$ 2.414
30. $\sqrt[4]{100}$ 2.512

31. The radius, r, of a circle can be found from the area of the circle according to the formula:

$$r = \sqrt{\frac{A}{\pi}}$$

a. What is the radius of a circular garden whose area is 160 m²? Round to the nearest tenth of a meter. 7.1 m
b. What is the radius of a circular fountain whose area is 1600 ft²? Round to the nearest tenth of a foot. 22.6 ft

chapter 8   TEST

1. State the conditions for a radical expression to be in simplified form. ✦

For Exercises 2–7, simplify the radicals, if possible. Assume all variables represent positive real numbers.

2. $\sqrt{242x^3}$ $11x\sqrt{2}$
3. $\sqrt{48y^4}$ $2y\sqrt{6y}$
4. $\sqrt{-64}$ Not a real number
5. $\sqrt{\frac{5a^6}{81}}$ $\frac{a\sqrt{5a^2}}{9}$
6. $\frac{9}{\sqrt{6}}$ $\frac{3\sqrt{6}}{2}$
7. $\frac{2}{\sqrt{5}+6}$ ✦

8. Translate the English phrases into algebraic expressions and simplify.

a. The sum of the square root of twenty-five and the cube of five. $\sqrt{25} + 5^3$; 130
b. The difference of the square of four and the fourth root of 16. $4^2 - \sqrt[4]{16}$; 14

9. Estimate the value of the following radicals. Then use your calculator to approximate the value to three decimal places.

a. $\sqrt{38}$ 6.164
b. $\sqrt{20}$ 2.115

10. A baseball player hits the ball at an angle of 30° with an initial velocity of 112 ft/s. The horizontal

CUMULATIVE REVIEW EXERCISES, CHAPTERS 1–8

For Exercises 1–2, simplify completely:

1. $\frac{|-3 - 12 \div 6 + 2|}{\sqrt{5^2 - 4^2}}$ 1

2. $\left(-\frac{4}{5} \div \frac{2}{15}\right)^2 + \frac{1}{6}$ $\frac{217}{6}$

3. Solve for y: $2 - 5[2y + 4] - (-3y - 1) = -(y + 5)$ $y = -2$

4. Solve for a: $2a + b + c = A$ $a = \frac{A - b - c}{2}$

5. Solve the inequality. Graph the solution set. Then write the solution in set-builder notation and in interval notation: $2x - 5(x + 1) < -x + 3$ ✦

6. The sum of two-thirds of a number and five equals the number. Find the number. The number is 15.

10. Perform the indicated operations:
$2(x - 3) - (3x + 4)(3x - 4)$ $-9x^2 + 2x + 10$

11. Perform the indicated operations:
$\left(\frac{1}{2}c + 4\right)^2$ $\frac{1}{4}c^2 + 4c + 16$

12. Divide:
$\frac{14x^3y - 7x^2y^2 + 28xy^2}{7x^2y^2}$ $\frac{2x}{y} - 1 + \frac{4}{x}$

In Exercises 13–15, factor completely:

13. $6ax + 2bx - 3ay - by$ $(3a + b)(2x - y)$
14. $m^4 - 81$ $(m - 3)(m + 3)(m^2 + 9)$
15. $50c^2 + 40c + 8$ $2(5c + 2)^2$
16. Solve for x: $10x^2 = x + 2$ $x = -\frac{2}{5}, x = \frac{1}{2}$

ANCILLARIES FOR THE INSTRUCTOR

Annotated Instructor's Edition

The Annotated Instructor's Edition (AIE) contains answers to all exercises and tests. The answers to most exercises are printed in green next to each problem (answers not appearing on the page are found in the back of the AIE in an appendix). This ancillary also provides valuable keys that serve as a useful guide to instructors as they assign homework problems and structure lessons. Icons and references are placed throughout the Practice Exercises so that instructors can easily identify problem types, such as writing exercises, geometry exercises, "translation" exercises, review exercises, challenge problems (Expanding Your Skills), and calculator exercises. Students *do not see* all of these icons in the Student Edition of the text. Students see only the *video icons*, the *calculator icons*, and the reference to *Expanding Your Skills*.

Instructor's Resource Manual

Written by the authors, the Instructor's Resource Manual (IRM) contains numerous classroom activities designed for group or individual work, a series of Internet activities, and materials for a student portfolio. Several **Classroom Activities** are available for each section in the text to be used as a complement to lecture. These short activities can be completed in 5 to 10 minutes. In addition, Internet activities called **Technology Connections** are provided for each chapter to give students working applications of topics covered in the chapter. For instructors who want to use the Classroom Activities or Technology Connections as cooperative learning lessons, the IRM also provides strategies for successful implementation of cooperative learning.

The **Student Portfolio** is another significant feature of the Instructor's Resource Manual. The materials in the portfolio provide guidelines for the student to maintain an organized notebook with notes, homework, tests, test corrections, a calendar, a record of grades, and vocabulary. The portfolio is a vehicle to reinforce students' responsibility for their own learning.

This supplement is available in print form, and it can also be accessed online at www.mhhe.com/miller_oneill.

Instructor's Solutions Manual

The *Instructor's Solutions Manual* contains comprehensive, worked-out solutions to all exercises in the Practice Exercise sets; the Midchapter Reviews; the end-of-chapter Review Exercises; the Chapter Tests; and the Cumulative Review Exercises. A *Student's Solutions Manual* is also available for sale to students. The *Student's Solutions Manual* contains worked-out solutions to the odd numbered Practice Exercises, Midchapter Reviews questions, Review Exercises, Chapter Test questions, and Cumulative Review exercises.

Student's Edition Prefatory "Help Yourself"

Encourage your students to read "Help Yourself," which is a prefatory section in the Student Edition of this book. "Help Yourself" offers students hints for success in mathematics courses. It begins with practical advice on preparing for class and for exams. Then, it encourages students to become familiar with their text via the Key Feature section. This walk-through outlines pedagogical features of the text along with sample pages. Following the Key Features, students are provided with a list of supplements and are given brief descriptions of each. Finally, in Putting It All Together, students are offered suggestions on how to use their text and accompanying supplements to achieve success in developmental mathematics. (*Note to instructors*: The Instructor's Preface is not included in the Student Edition.)

Instructor's Testing and Resource CD-ROM

This CD-ROM contains a computerized test bank that utilizes Brownstone Diploma® testing software. The computerized test bank allows you to create well-formatted quizzes or tests using a large bank of algorithmically generated and static questions. When creating a quiz or test, you can manually choose individual questions or have the software randomly select questions based on section, question type, difficulty level, and other criteria. Instructors also have the ability to add or edit test bank questions to create their own customized test bank. In addition to printed tests, the test generator can deliver tests over a local area network or the World Wide Web, with automatic grading.

Also available on the CD-ROM are preformatted tests that appear in two forms: Adobe Acrobat (pdf) and Microsoft Word files. These files are provided for convenient access to "ready to use" tests.

Online Learning Center

The Online Learning Center (OLC), located at www.mhhe.com/miller_oneill contains resources for students and instructors alike. The OLC consists of the Student Learning Site, the Instructor's Resource Site, and the Information Center.

Through the Instructor's Resource Site, instructors can access an electronic version of the Instructor's Resource Manual (the electronic version contains all the same elements as found in the print version), links to professional resources, a PowerPoint presentation (transparencies), printable tests, group projects, a link to PageOut, and more.

To access the Instructor's Resource Site, instructors must have a passcode that can be obtained by contacting a McGraw-Hill Higher Education representative.

The Student Learning Site is also passcode-protected. Passcodes for students can be found at the front of their texts, when newly purchased. *Passcodes are available free to students when they purchase a **new** text.* The Student Learning Site contains the student version of the Classroom Activities and Technology Connections that appear in the Instructor's Resource Manual, as well as the Student Portfolio. Students also have access to the group projects, tutorials, a formula card, NetTutor™, and more!

The Information Center can be accessed by students and instructors alike, without a passcode. Through the Information Center users can access general information about the text and its ancillaries.

NetTutor

NetTutor is a revolutionary system that enables students to interact with a live tutor over the World Wide Web by using NetTutor's Web-based, graphical chat capabilities. Students can also submit questions and receive answers, browse previously answered questions, and view previous live chat sessions.

NetTutor can be accessed on the Online Learning Center through the Student Learning Site.

PageOut

PageOut is McGraw-Hill's unique point-and-click course Website tool, enabling instructors to create a full-featured, professional quality course Website without knowing HTML coding. With PageOut instructors can post a syllabus online, assign McGraw-Hill Online Learning Center content, add links to important off-site resources, and maintain student results in the online grade book. Instructors can also send class announcements, copy a course site to share with colleagues, and upload original files. PageOut is free for every McGraw-Hill user. For those instructors who are short on time, a team is on hand, ready to help build a site!

Learn more about PageOut and other McGraw-Hill digital solutions at www.mhhe.com/solutions.

Miller/O'Neill Tutorial CD-ROM

This interactive CD-ROM is a self-paced tutorial specifically linked to the text that reinforces topics through unlimited opportunities to review concepts and practice problem solving. The CD-ROM provides section-specific animated lessons with accompanying audio, practice exercises that enable students to work through problems with step-by-step guidance available, concept-matching problems that test vocabulary skills as well as identification of properties and rules, and more. The results of students' work on homework or quizzes on the CD can be reported to a centralized grade book managed by the instructor on the World Wide Web. This browser-based CD requires virtually no computer training on the part of students and will run on both Windows and Macintosh computers. The tutorial CD-ROM is free to students who purchase a *new* text.

Miller/O'Neill Video Series (Videotapes or Video CDs)

The video series is based on problems taken directly from the Practice Exercises. The Practice Exercises contain icons (in both the student and the instructor edition) that show which problems from the text appear in the video series. A mathematics instructor presents selected problems and works through them, following the solution methodology employed in the text. The video series is also available on video CDs.

ALEKS®

ALEKS is ...

- A comprehensive course management system. It tells you exactly what your students know and don't know.

- Artificial intelligence. It totally individualized assessment and learning.

- Customizable. Click on or off each course topic.

- Web based. Use a standard browser for easy Internet access.

- Inexpensive. There are no set up fees or site license fees.

ALEKS (Assessment and LEarning in Knowledge Spaces) is an artificial intelligence-based system for individualized math learning available via the World Wide Web.

http://www.highedmath.aleks.com

ALEKS delivers precise, qualitative diagnostic assessments of students' math knowledge, guides them in the selection of appropriate new study material, and records their progress toward mastery of curricular goals in a robust course management system.

ALEKS interacts with the student much as a skilled human tutor would, moving between explanation and practice as needed, correcting and analyzing errors, defining terms and changing topics on request. By sophisticated modeling of a student's "knowledge state" for a given subject matter, ALEKS can focus clearly on what the student is most ready to learn next, thereby building a learning momentum that fuels success.

The ALEKS system was developed with a multi-million-dollar grant from the National Science Foundation. It has little to do with what is commonly thought of as educational software. The theory behind ALEKS is a specialized field of mathematical cognitive science called "Knowledge Spaces."

Knowledge Space theory, which concerns itself with the mathematical dynamics of knowledge acquisition, has been under development by researchers in cognitive science since the early 1980s. The Chairman and founder of ALEKS Corporation, Jean-Claude Falmagne, is an internationally recognized leader in scientific work in this field.

Please visit http://www.highedmath.aleks.com for a trial run.

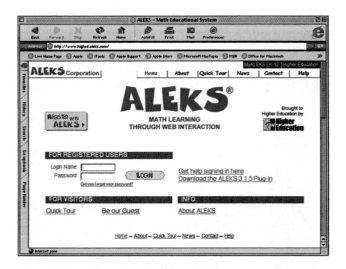

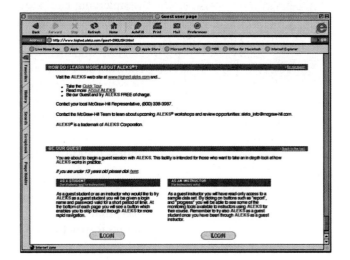

Reference: Fractions, Decimals, Percents, Geometry, and Study Skills

R

On January 3, 2000, the price per share of stock for Hershey Foods was $46\frac{3}{8}$, or, equivalently, $46.375. A year later, on January 2, 2001, the price per share rose to $61\frac{5}{8}$, or $61.625. The net increase is given by

$$\$61\tfrac{5}{8} - \$46\tfrac{3}{8} = \$15\tfrac{1}{4}$$

or equivalently, $15.25.

Therefore, if a stockholder bought 100 shares on January 3, 2000, and sold the stock on January 2, 2001, the increase in value is given by

$$100 \times \$15.25 = \$1525$$

The yield (or percent increase) during this 1-year period is found by dividing the difference in the selling price and original price by the original price.

$$\frac{\$15.25}{\$46.375} \approx 0.33, \text{ or } 33\%$$

Hence, the stockholder made 33% growth on the investment.

Operations on fractions, decimals, and percents as well as an introduction to geometry is the focus of this reference chapter.

For further information visit primefactors at

www.mhhe.com/miller_oneill

section

R.1   FRACTIONS

1. Basic Definitions

The study of algebra involves many of the operations and procedures used in arithmetic. Therefore, we begin this text by reviewing the basic operations of addition, subtraction, multiplication, and division on fractions and mixed numbers.

In day-to-day life, the numbers we use for counting are

the **natural numbers**: 1, 2, 3, 4, . . . and

the **whole numbers**: 0, 1, 2, 3, . . .

Whole numbers are used to count the number of whole units in a quantity. A fraction is used to express part of a whole unit. If a child gains $2\frac{1}{2}$ lb, the child has gained two whole pounds plus a portion of a pound. To express the additional half pound mathematically, we may use the fraction, $\frac{1}{2}$.

A Fraction and Its Parts

Fractions are numbers of the form $\frac{a}{b}$, where $\frac{a}{b} = a \div b$ and b does not equal zero. In the fraction $\frac{a}{b}$, the **numerator** is a, and the **denominator** is b.

The denominator of a fraction indicates how many equal parts divide the whole. The numerator indicates how many parts are being represented. For instance, suppose Jack wants to plant carrots in $\frac{2}{5}$ of a rectangular garden. He can divide the garden into five equal parts and use two of the parts for carrots (Figure R-1).

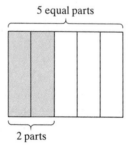

5 equal parts

2 parts   The shaded region represents $\frac{2}{5}$ of the garden.

Figure R-1

Definition of a Proper Fraction, an Improper Fraction, and a Mixed Number

1. If the numerator of a fraction is less than the denominator, the fraction is a **proper fraction**. A proper fraction represents a quantity that is less than a whole unit.
2. If the numerator of a fraction is greater than or equal to the denominator, then the fraction is an **improper fraction**. An improper fraction represents a quantity greater than or equal to a whole unit.
3. A **mixed number** is a whole number added to a fraction.

Proper Fractions: $\dfrac{3}{5}$ $\dfrac{1}{8}$

Improper Fractions: $\dfrac{7}{5}$ $\dfrac{8}{8}$

Mixed Numbers: $1\tfrac{1}{5}$ $2\tfrac{3}{8}$

2. Prime Factorization

To perform operations on fractions it is important to understand the concept of a factor. For example, when the numbers 2 and 6 are multiplied, the result (called the **product**) is 12.

$$2 \times 6 = 12$$

factors product

The numbers 2 and 6 are said to be **factors*** of 12. The number 12 is said to be factored when it is written as the product of two or more natural numbers. For example, 12 can be factored in several ways:

$$12 = 1 \times 12 \qquad 12 = 2 \times 6 \qquad 12 = 3 \times 4 \qquad 12 = 2 \times 2 \times 3$$

A natural number greater than 1 whose only factors are 1 and itself is called a **prime number**. The first several prime numbers are 2, 3, 5, 7, 11, and 13. A natural number greater than 1 that is not prime is called a **composite number**. That is, a composite number has factors other than itself and 1. The first several composite numbers are 4, 6, 8, 9, 10, 12, 14, 15, and 16.

The number 1 is neither prime nor composite.

---

example 1 **Writing a Natural Number as a Product of Prime Factors**

Write each number as a product of prime factors.

a. 12 b. 30

Solution:

a. $12 = 2 \times 2 \times 3$ Divide 12 by prime numbers until the number 1 is obtained. Or use a factor tree

$$\begin{array}{r|l} 2 & 12 \\ 2 & 6 \\ 3 & 3 \\ \hline & 1 \end{array}$$

---

*In this context, we refer only to natural number factors.

b. $30 = 2 \times 3 \times 5$

$$\begin{array}{r|l} 2 & 30 \\ 3 & 15 \\ 5 & 5 \\ & 1 \end{array}$$

3.   Reducing Fractions

The process of factoring numbers can be used to reduce fractions to lowest terms. A fractional portion of a whole can be represented by infinitely many fractions. For example, Figure R-2 shows that $\frac{1}{2}$ is equivalent to $\frac{2}{4}, \frac{3}{6}, \frac{4}{8}$, and so on.

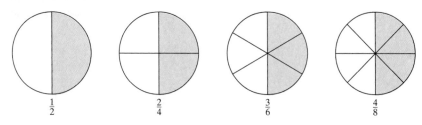

$$\frac{1}{2} \qquad \frac{2}{4} \qquad \frac{3}{6} \qquad \frac{4}{8}$$

Figure R-2

The fraction $\frac{1}{2}$ is said to be in lowest terms because the numerator and denominator share no common factor other than 1. The fraction $\frac{2}{4}$ is not in lowest terms because the numerator and denominator are both divisible by 2.

To reduce a fraction to lowest terms, the goal is to "divide out" common factors from both the numerator and denominator.

example 2

Classroom Activity R.1B

Reducing a Fraction to Lowest Terms

Reduce the fraction $\frac{12}{30}$ to lowest terms.

Solution:

From Example 1, we have $12 = 2 \times 2 \times 3$ and $30 = 2 \times 3 \times 5$. Hence,

$$\frac{12}{30} = \frac{2 \times 2 \times 3}{2 \times 3 \times 5}$$

The common factors of 2 and 3 may be "divided out" of the numerator and denominator.

$$= \frac{\overset{1}{\cancel{2}} \times 2 \times \overset{1}{\cancel{3}}}{\underset{1}{\cancel{2}} \times \underset{1}{\cancel{3}} \times 5} = \frac{2}{5}$$

Multiply $1 \times 2 \times 1 = 2$.

Multiply $1 \times 1 \times 5 = 5$.

example 3

Reducing a Fraction to Lowest Terms

Reduce $\frac{14}{42}$ to lowest terms.

Solution:

$$\frac{14}{42} = \frac{2 \times 7}{2 \times 3 \times 7}$$
Factor the numerator and denominator.

$$= \frac{\overset{1}{\cancel{2}} \times \overset{1}{7}}{\underset{1}{\cancel{2}} \times 3 \times \underset{1}{7}} = \frac{1}{3}$$

Multiply $1 \times 1 = 1$.

Multiply $1 \times 3 \times 1 = 3$.

4. Multiplying Fractions

Multiplying Fractions

If b is not zero and d is not zero, then

$$\frac{a}{b} \times \frac{c}{d} = \frac{a \times c}{b \times d}$$

To multiply fractions, multiply the numerators and multiply the denominators.

example 4

Multiplying Fractions

Multiply the fractions: $\frac{1}{4} \times \frac{1}{2}$

Solution:

$$\frac{1}{4} \times \frac{1}{2} = \frac{1 \times 1}{4 \times 2} = \frac{1}{8}$$
Multiply the numerators. Multiply the denominators.

Notice that the product $\frac{1}{4} \times \frac{1}{2}$ represents a quantity that is $\frac{1}{4}$ of $\frac{1}{2}$. Taking $\frac{1}{4}$ of a quantity is equivalent to dividing the quantity by 4. One-half of a pie divided into four pieces leaves pieces that each represent $\frac{1}{8}$ of the pie (Figure R-3).

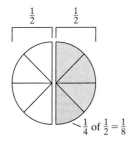

$\frac{1}{2}$ $\frac{1}{2}$

$\frac{1}{4}$ of $\frac{1}{2} = \frac{1}{8}$

Figure R-3

example 5

Multiplying Fractions

Multiply the fractions.

a. $\dfrac{7}{10} \times \dfrac{15}{14}$ b. $\dfrac{2}{13} \times \dfrac{13}{2}$ c. $5 \times \dfrac{1}{5}$

Solution:

a. $\dfrac{7}{10} \times \dfrac{15}{14} = \dfrac{7 \times 15}{10 \times 14}$

Multiply the numerators and multiply the denominators.

Divide out common factors in the numerator and denominator.

$= \dfrac{\overset{1}{\cancel{7}} \times 3 \times \overset{1}{\cancel{5}}}{2 \times \underset{1}{\cancel{5}} \times 2 \times \underset{1}{\cancel{7}}} = \dfrac{3}{4}$

— Multiply $1 \times 3 \times 1 = 3$.

— Multiply $2 \times 1 \times 2 \times 1 = 4$.

b. $\dfrac{2}{13} \times \dfrac{13}{2} = \dfrac{2 \times 13}{13 \times 2} = \dfrac{\overset{1}{\cancel{2}} \times \overset{1}{\cancel{13}}}{\underset{1}{\cancel{13}} \times \underset{1}{\cancel{2}}} = \dfrac{1}{1} = 1$

— Multiply $1 \times 1 = 1$.

— Multiply $1 \times 1 = 1$.

c. $5 \times \dfrac{1}{5} = \dfrac{5}{1} \times \dfrac{1}{5}$

The whole number 5 can be written as $\frac{5}{1}$.

$= \dfrac{\overset{1}{\cancel{5}} \times 1}{1 \times \underset{1}{\cancel{5}}} = \dfrac{1}{1} = 1$

Multiply and reduce.

5. Dividing Fractions

Before we divide fractions, we need to know how to find the reciprocal of a fraction. Notice from Example 4 that $\frac{2}{13} \times \frac{13}{2} = 1$ and $5 \times \frac{1}{5} = 1$. The numbers $\frac{2}{13}$ and $\frac{13}{2}$ are said to be reciprocals because their product is 1. Likewise the numbers 5 and $\frac{1}{5}$ are reciprocals.

The Reciprocal of a Number

Two numbers are **reciprocals** of each other if their product is 1. Therefore the reciprocal of the fraction

$$\dfrac{a}{b} \text{ is } \dfrac{b}{a} \qquad \text{because} \qquad \dfrac{a}{b} \times \dfrac{b}{a} = 1$$

Number	Reciprocal	Product
$\dfrac{2}{13}$	$\dfrac{13}{2}$	$\dfrac{2}{13} \times \dfrac{13}{2} = 1$
$\dfrac{1}{8}$	$\dfrac{8}{1}$ (or equivalently 8)	$\dfrac{1}{8} \times 8 = 1$
$6 \left(\text{or equivalently } \dfrac{6}{1} \right)$	$\dfrac{1}{6}$	$6 \times \dfrac{1}{6} = 1$

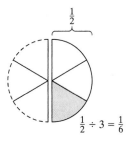

$$\frac{1}{2}$$

$$\frac{1}{2} \div 3 = \frac{1}{6}$$

Figure R-4

To understand the concept of dividing fractions, consider a pie that is half-eaten. Suppose the remaining half must be divided among three people, that is: $\frac{1}{2} \div 3$. However, dividing by 3 is equivalent to taking $\frac{1}{3}$ of the remaining $\frac{1}{2}$ of the pie (Figure R-4).

$$\frac{1}{2} \div 3 = \frac{1}{2} \cdot \frac{1}{3} = \frac{1}{6}$$

This example illustrates that *dividing two numbers is equivalent to multiplying the first number by the reciprocal of the second number.*

Division of Fractions

Let a, b, c, and d be numbers such that b, c, and d are not zero. Then,

multiply

$$\frac{a}{b} \div \frac{c}{d} = \frac{a}{b} \times \frac{d}{c}$$

reciprocal

To divide fractions, multiply the first fraction by the reciprocal of the second fraction.

example 6

Dividing Fractions

Divide the fractions.

a. $\dfrac{8}{5} \div \dfrac{3}{10}$ b. $\dfrac{12}{13} \div 6$

Classroom Activity R.1C

Solution:

a. $\dfrac{8}{5} \div \dfrac{3}{10} = \dfrac{8}{5} \times \dfrac{10}{3}$ Multiply by the reciprocal of $\frac{3}{10}$, which is $\frac{10}{3}$.

$$= \frac{8 \times \overset{2}{\cancel{10}}}{\underset{1}{\cancel{5}} \times 3} = \frac{16}{3}$$ Multiply and reduce to lowest terms.

b. $\dfrac{12}{13} \div 6 = \dfrac{12}{13} \div \dfrac{6}{1}$ Write the whole number 6 as $\frac{6}{1}$.

$$= \frac{12}{13} \times \frac{1}{6}$$ Multiply by the reciprocal of $\frac{6}{1}$, which is $\frac{1}{6}$.

$$\frac{\overset{2}{\cancel{12}} \times 1}{13 \times \underset{1}{\cancel{6}}} = \frac{2}{13}$$ Multiply and reduce to lowest terms.

6. Adding and Subtracting Fractions

Adding and Subtracting Fractions

Two fractions can be added or subtracted only if they have a common denominator. Let a, b, and c, be numbers such that b does not equal zero. Then,

$$\frac{a}{b} + \frac{c}{b} = \frac{a+c}{b} \qquad \text{and} \qquad \frac{a}{b} - \frac{c}{b} = \frac{a-c}{b}$$

To add or subtract fractions with the same denominator, add or subtract the numerators and write the result over the common denominator.

example 7   **Adding and Subtracting Fractions with the Same Denominator**

Add or subtract as indicated.

a. $\dfrac{1}{12} + \dfrac{7}{12}$ b. $\dfrac{13}{5} - \dfrac{3}{5}$

Solution:

a. $\dfrac{1}{12} + \dfrac{7}{12} = \dfrac{1+7}{12}$ Add the numerators.

$= \dfrac{8}{12}$ Simplify.

$= \dfrac{2}{3}$ Reduce.

b. $\dfrac{13}{5} - \dfrac{3}{5} = \dfrac{13-3}{5}$ Subtract the numerators.

$= \dfrac{10}{5}$ Simplify.

$= 2$ Reduce.

In Example 7, we added and subtracted fractions with the same denominators. To add or subtract fractions with different denominators we must first write them as equivalent fractions with a common denominator. A common denominator may be *any* common multiple of the denominators. We will use the least common multiple to form the **least common denominator (LCD)**.

Consider the fractions $\frac{1}{3}$ and $\frac{1}{2}$. A common denominator must be a multiple of 3 and a multiple of 2. Hence the product $3 \times 2 = 6$ can be used as a common denominator. Notice that the numbers 6, 12, 18, 24 and so on are all multiples of both 3 and 2. However, 6 is the *least* common multiple.

Consider the fractions $\frac{1}{4}$ and $\frac{1}{8}$. Notice that the product $4 \times 8 = 32$ is a common multiple of 4 and 8. However, the least common multiple of 4 and 8 is 8. To understand why, write each denominator as a product of prime factors.

$$4 = 2 \times 2 \qquad \text{and} \qquad 8 = 2 \times 2 \times 2$$

Both 4 and 8 are composed of repeated factors of 2. However, any number that is a multiple of $2 \times 2 \times 2 = 8$ is automatically a multiple of $2 \times 2 = 4$. Therefore, it is sufficient to use the factor 2 the maximum number of times it appears in either fraction. The LCD of $\frac{1}{4}$ and $\frac{1}{8}$ is $\underbrace{2 \times 2 \times 2}_{\text{3 factors of 2}} = 8$.

Steps to Finding the Least Common Denominator of Two Fractions

1. Write each denominator as a product of prime factors.
2. The LCD is the product of unique prime factors from both denominators. If a factor is repeated in either denominator, use that factor the maximum number of times it appears in either denominator.

example 8

Finding the LCD of Two Fractions

Find the LCD of $\frac{1}{9}$ and $\frac{1}{15}$.

Solution:

$9 = 3 \times 3$ and $15 = 3 \times 5$ Factor the denominators.

$\text{LCD} = 3 \times 3 \times 5 = 45$ The LCD is the product of the factors of 3 and 5, where 3 is repeated twice.

To add or subtract fractions, a common denominator must first be determined. Then it is necessary to convert each fraction to an equivalent fraction with the common denominator.

Writing Equivalent Fractions

To write a fraction as an equivalent fraction with a common denominator, multiply the numerator and denominator by the factors from the common denominator that are missing from the denominator of the original fraction.

Note: Multiplying the numerator and denominator by the *same* nonzero quantity will leave the value of the fraction unchanged.

example 9

Writing Equivalent Fractions

a. Write the fractions $\frac{1}{9}$ and $\frac{1}{15}$ as equivalent fractions with the LCD as the denominator.
b. Subtract $\frac{1}{9} - \frac{1}{15}$.

Solution:

From Example 8, we know that the LCD for $\frac{1}{9}$ and $\frac{1}{15}$ is 45.

a. $\dfrac{1}{9} = \dfrac{1 \times 5}{9 \times 5} = \dfrac{5}{45}$ Multiply numerator and denominator by 5. This creates a denominator of 45.

$\dfrac{1}{15} = \dfrac{1 \times 3}{15 \times 3} = \dfrac{3}{45}$ Multiply numerator and denominator by 3. This creates a denominator of 45.

b. $\dfrac{1}{9} - \dfrac{1}{15}$

$= \dfrac{5}{45} - \dfrac{3}{45}$ Write $\frac{1}{9}$ and $\frac{1}{15}$ as equivalent fractions with the same denominator.

$= \dfrac{2}{45}$ Subtract.

---

example 10   **Adding Fractions with Different Denominators**

Suppose Nakeysha ate $\frac{1}{2}$ of an ice-cream pie, and her friend Carla ate $\frac{1}{3}$ of the pie. How much of the ice-cream pie was eaten?

Classroom Activity R.1D   **Solution:**

$\dfrac{1}{2} + \dfrac{1}{3}$ The LCD is $3 \times 2 = 6$.

$= \dfrac{1 \times 3}{2 \times 3} + \dfrac{1 \times 2}{3 \times 2}$ Multiply numerator and denominator by the missing factors.

$= \dfrac{3}{6} + \dfrac{2}{6}$

$= \dfrac{5}{6}$ Add the fractions.

Together, Nakeysha and Carla ate $\frac{5}{6}$ of the ice-cream pie.

By converting the fractions $\frac{1}{2}$ and $\frac{1}{3}$ to the same denominator, we are able to add *like* size pieces of pie (Figure R-5).

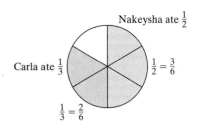

Figure R-5

example 11   **Adding and Subtracting Fractions**

Simplify: $\frac{1}{2} + \frac{5}{8} - \frac{11}{16}$.

Solution:

$$\frac{1}{2} + \frac{5}{8} - \frac{11}{16}$$

Factor the denominators: $2 = 2$, $8 = 2 \times 2 \times 2$, and $16 = 2 \times 2 \times 2 \times 2$.
The LCD is $2 \times 2 \times 2 \times 2 = 16$. Next write the fractions as equivalent fractions with the LCD.

$$= \frac{1 \times 2 \times 2 \times 2}{2 \times 2 \times 2 \times 2} + \frac{5 \times 2}{8 \times 2} - \frac{11}{16}$$

Multiply numerators and denominators by the factors missing from each denominator.

$$= \frac{8}{16} + \frac{10}{16} - \frac{11}{16}$$

$$= \frac{8 + 10 - 11}{16}$$

Add and subtract the numerators.

$$= \frac{7}{16}$$

Simplify.

7. Operations on Mixed Numbers

Recall that a mixed number is a whole number added to a fraction. The number $3\frac{1}{2}$ represents the sum of three wholes plus a half. That is, $3\frac{1}{2} = 3 + \frac{1}{2}$. For this reason, any mixed number can be converted to an improper fraction by using addition.

$$3\frac{1}{2} = 3 + \frac{1}{2} = \frac{6}{2} + \frac{1}{2} = \frac{7}{2}$$

Tip: A short-cut to writing a mixed number as an improper fraction is to multiply the whole number by the denominator of the fraction. Then add this value to the numerator of the fraction and write the result over the denominator.

$3\frac{1}{2} \longrightarrow$ Multiply the whole number by the denominator: $3 \times 2 = 6$.

Add the numerator: $6 + 1 = 7$.

Write the result over the denominator: $\frac{7}{2}$.

To add, subtract, multiply, or divide mixed numbers, we will first write the mixed number as an improper fraction.

example 12

Operations on Mixed Numbers

Perform the indicated operations.

a. $5\frac{1}{3} - 2\frac{1}{4}$ b. $7\frac{1}{2} \div 3$

Classroom Activity R.1E

Solution:

a. $5\frac{1}{3} - 2\frac{1}{4}$

$= \dfrac{16}{3} - \dfrac{9}{4}$ Write the mixed numbers as improper fractions.

$= \dfrac{16 \times 4}{3 \times 4} - \dfrac{9 \times 3}{4 \times 3}$ The LCD is 12. Multiply numerators and denominators by the missing factors from the denominators.

$= \dfrac{64}{12} - \dfrac{27}{12}$

$= \dfrac{37}{12}$ or $3\frac{1}{12}$ Subtract the fractions.

> **Tip:** An improper fraction can also be written as a mixed number. Both answers are acceptable. Note that
>
> $$\frac{37}{12} = \frac{36}{12} + \frac{1}{12} = 3 + \frac{1}{12}, \text{ or } 3\frac{1}{12}$$

b. $7\frac{1}{2} \div 3$

Avoiding Mistakes

Remember that when dividing (or multiplying) fractions, a common denominator is not necessary.

$= \dfrac{15}{2} \div \dfrac{3}{1}$ Write the mixed number and whole number as fractions.

$= \dfrac{\overset{5}{\cancel{15}}}{2} \times \dfrac{1}{\underset{1}{\cancel{3}}}$ Multiply by the reciprocal of $\frac{3}{1}$, which is $\frac{1}{3}$.

$= \dfrac{5}{2}$, or $2\frac{1}{2}$ The answer may be written as an improper fraction or as a mixed number.

section R.1 PRACTICE EXERCISES

For Exercises 1–8, identify the numerator and denominator of the fraction. Then determine if the fraction is a proper fraction or an improper fraction.

1. $\dfrac{7}{8}$ Numerator: 7; denominator: 8; proper

2. $\dfrac{2}{3}$ Numerator: 2; denominator: 3; proper

3. $\dfrac{9}{5}$ Numerator: 9; denominator: 5; improper

4. $\dfrac{5}{2}$ Numerator: 5; denominator: 2; improper

5. $\dfrac{6}{6}$ Numerator: 6; denominator: 6; improper

6. $\dfrac{4}{4}$ Numerator: 4; denominator: 4; improper

7. $\dfrac{12}{1}$ Numerator: 12; denominator: 1; improper

8. $\dfrac{5}{1}$ Numerator: 5; denominator: 1; improper

For Exercises 9–16, write a proper or improper fraction associated with the shaded region of each figure.

9. $\dfrac{3}{4}$

10.  $\dfrac{4}{5}$

11. 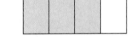 $\dfrac{4}{3}$

12.  $\dfrac{5}{4}$

13. $\dfrac{1}{6}$

14. $\dfrac{1}{8}$

15. $\dfrac{2}{2}$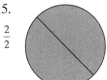

16. $\dfrac{4}{4}$

For Exercises 17–20, write both an improper fraction and a mixed number associated with the shaded region of each figure.

17. $\dfrac{5}{2}$ or $2\frac{1}{2}$

18. $\dfrac{5}{3}$ or $1\frac{2}{3}$

19. $\dfrac{6}{2}$ or 3

20. $\dfrac{6}{3}$ or 2

21. Explain the difference between the set of whole numbers and the set of natural numbers. The set of whole numbers includes the number 0 and the set of natural numbers does not.

22. Explain the difference between a proper fraction and an improper function. A proper fraction depicts a number less than one unit. An improper fraction depicts a number greater than or equal to a whole unit.

23. Write a fraction that reduces to $\frac{1}{2}$. (Answers may vary.) For example: $\dfrac{2}{4}$

24. Write a fraction that reduces to $\frac{1}{3}$. (Answers may vary.) For example: $\dfrac{2}{6}$

For Exercises 25–32, identify the number as either a prime number or a composite number.

25. 5 Prime

26. 9 Composite

27. 4 Composite

28. 2 Prime

29. 39 Composite

30. 23 Prime

31. 53 Prime

32. 51 Composite

For Exercises 33–40, write the number as a product of prime factors.

33. 36 $2 \times 2 \times 3 \times 3$

34. 70 $2 \times 5 \times 7$

35. 42 $2 \times 3 \times 7$

36. 35 5×7

37. 110 $2 \times 5 \times 11$

38. 136 $2 \times 2 \times 2 \times 17$

39. 135 $3 \times 3 \times 3 \times 5$

40. 105 $3 \times 5 \times 7$

For Exercises 41–52, reduce each fraction. Use prime factorization if necessary.

41. $\dfrac{3}{15}$ $\dfrac{1}{5}$

42. $\dfrac{8}{12}$ $\dfrac{2}{3}$

43. $\dfrac{6}{16}$ $\dfrac{3}{8}$

44. $\dfrac{12}{20}$ $\dfrac{3}{5}$

45. $\dfrac{42}{48}$ $\dfrac{7}{8}$

46. $\dfrac{35}{80}$ $\dfrac{7}{16}$

47. $\dfrac{48}{64}$ $\dfrac{3}{4}$

48. $\dfrac{32}{48}$ $\dfrac{2}{3}$

 49. $\dfrac{110}{176}$ $\dfrac{5}{8}$

50. $\dfrac{70}{120}$ $\dfrac{7}{12}$

51. $\dfrac{150}{200}$ $\dfrac{3}{4}$

52. $\dfrac{119}{210}$ $\dfrac{17}{30}$

For Exercises 53–54, identify if the statement is true or false. If it is false, rewrite as a true statement.

53. When multiplying or dividing fractions it is necessary to have a common denominator. False: When adding or subtracting fractions it is necessary to have a common denominator.

54. When dividing two fractions it is necessary to multiply the first fraction by the reciprocal of the second fraction. True

For Exercises 55–66, multiply or divide as indicated.

55. $\dfrac{10}{13} \times \dfrac{26}{15}$ $\dfrac{4}{3}$ or $1\frac{1}{3}$

56. $\dfrac{15}{28} \times \dfrac{7}{9}$ $\dfrac{5}{12}$

57. $\dfrac{3}{7} \div \dfrac{9}{14}$ $\dfrac{2}{3}$

58. $\dfrac{7}{25} \div \dfrac{1}{5}$ $\dfrac{7}{5}$ or $1\frac{2}{5}$

59. $\dfrac{9}{10} \times 5$ $\dfrac{9}{2}$ or $4\frac{1}{2}$

60. $\dfrac{3}{7} \times 14$ 6

61. $4\dfrac{3}{5} \div \dfrac{1}{10}$ 46

62. $2\dfrac{4}{5} \div \dfrac{7}{11}$ $\dfrac{22}{5}$ or $4\frac{2}{5}$

63. $3\dfrac{1}{5} \times \dfrac{7}{8}$ $\dfrac{14}{5}$ or $2\frac{4}{5}$

64. $2\dfrac{1}{2} \times \dfrac{4}{5}$ 2

65. $1\dfrac{2}{9} \div 6$ $\dfrac{11}{54}$

66. $2\dfrac{2}{5} \div \dfrac{2}{7}$ $\dfrac{42}{5}$ or $8\frac{2}{5}$

67. Stephen's take-home pay is $1200 a month. If his rent is $\frac{1}{4}$ of his pay, how much is his rent? $300

68. Gus decides to save $\frac{1}{3}$ of his pay each month. If his monthly pay is $2112, how much will he save each month? $704

69. A recipe for a casserole calls for $\frac{1}{3}$ of a dozen eggs. How many eggs are needed? Four eggs

70. Shontell had time to print out only $\frac{3}{5}$ of her book report before school. If the report is 10 pages long, how many pages did she print out? Six pages

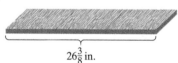 71. Gail buys 6 lb of mixed nuts to be divided into decorative jars that will each hold $\frac{3}{4}$ lb of nuts. How many jars will she be able to fill? Eight jars

Figure for Exercise 71

72. Natalie has 4 yd of material with which she can make holiday aprons. If it takes $\frac{1}{2}$ yd of material per apron, how many aprons can she make? Eight aprons

73. There are 4 cups of oatmeal in a box. If each serving is $\frac{1}{3}$ of a cup, how many servings are contained in the box? 12 servings

74. A board $26\frac{3}{8}$ in. long must be cut into three equal pieces. Find the length of each piece. $8\frac{19}{24}$ in.

$26\frac{3}{8}$ in.

Figure for Exercise 74

75. Richard bakes candy for the holidays. He requires $1\frac{3}{4}$ tsp of vanilla for each batch of candy. If a bottle of vanilla contains 21 tsp, how many batches can he make before he runs out of vanilla? 12 batches

76. A piece of rope $52\frac{1}{2}$ in. long must be cut into eight equal pieces. Find the length of each piece. $6\frac{9}{16}$ in.

❖ See Additional Answers Appendix   Writing   Translating Expression   Geometry   Scientific Calculator   Video

For Exercises 77–82, find the least common denominator for each pair of fractions.

77. $\frac{1}{6}, \frac{5}{24}$ 24

78. $\frac{1}{12}, \frac{11}{30}$ 60

79. $\frac{9}{20}, \frac{3}{8}$ 40

80. $\frac{13}{24}, \frac{7}{40}$ 120

81. $\frac{7}{10}, \frac{11}{45}$ 90

82. $\frac{1}{20}, \frac{1}{30}$ 60

For Exercises 83–106, add or subtract as indicated.

83. $\frac{5}{14} + \frac{1}{14}$ $\frac{3}{7}$

84. $\frac{9}{5} + \frac{1}{5}$ 2

85. $\frac{17}{24} - \frac{5}{24}$ $\frac{1}{2}$

86. $\frac{11}{18} - \frac{5}{18}$ $\frac{1}{3}$

87. $\frac{1}{8} + \frac{3}{4}$ $\frac{7}{8}$

88. $\frac{3}{16} + \frac{1}{2}$ $\frac{11}{16}$

89. $\frac{3}{8} - \frac{3}{10}$ $\frac{3}{40}$

90. $\frac{12}{35} - \frac{1}{10}$ $\frac{17}{70}$

91. $\frac{7}{26} - \frac{2}{13}$ $\frac{3}{26}$

92. $\frac{11}{24} - \frac{5}{16}$ $\frac{7}{48}$

93. $\frac{7}{18} + \frac{5}{12}$ $\frac{29}{36}$

94. $\frac{3}{16} + \frac{9}{20}$ $\frac{51}{80}$

95. $\frac{3}{4} - \frac{1}{20}$ $\frac{7}{10}$

96. $\frac{1}{6} - \frac{1}{24}$ $\frac{1}{8}$

97. $\frac{5}{12} + \frac{5}{16}$ $\frac{35}{48}$

98. $\frac{3}{25} + \frac{8}{35}$ $\frac{61}{175}$

99. $2\frac{1}{8} + 1\frac{3}{8}$ $\frac{7}{2}$ or $3\frac{1}{2}$

100. $1\frac{3}{14} + 1\frac{1}{14}$ $\frac{16}{7}$ or $2\frac{2}{7}$

 101. $1\frac{5}{6} - \frac{7}{8}$ $\frac{23}{24}$

102. $2\frac{1}{3} - \frac{5}{6}$ $\frac{3}{2}$ or $1\frac{1}{2}$

103. $1\frac{1}{6} + 3\frac{3}{4}$ $\frac{59}{12}$ or $4\frac{11}{12}$

104. $4\frac{1}{2} + 2\frac{2}{3}$ $\frac{43}{6}$ or $7\frac{1}{6}$

105. $1 - \frac{7}{8}$ $\frac{1}{8}$

106. $2 - \frac{3}{7}$ $\frac{11}{7}$ or $1\frac{4}{7}$

 107. A futon, when set up as a sofa, measures $3\frac{5}{6}$ ft wide. When it is opened to be used as a bed, the width is increased by $1\frac{3}{4}$ ft. What is the total width of this bed? $5\frac{7}{12}$ ft

Figure for Exercise 107

108. If Sally adds a lace that is $\frac{7}{8}$-in. wide to a skirt that is $20\frac{1}{2}$ in. long, what will be the final length of the skirt? $21\frac{3}{8}$ in.

109. Three children ate Cheerios for breakfast. Suzy had $\frac{1}{2}$ cup, Tracy had $\frac{1}{3}$ cup and Sheila had $\frac{3}{4}$ cup. How much cereal was eaten that morning? $1\frac{7}{12}$ cups

110. Jose ordered two seafood platters for a party. One platter has $1\frac{1}{2}$ lb of shrimp and the other has $\frac{3}{4}$ lb of shrimp. How many pounds does he have altogether? $2\frac{1}{4}$ lb

111. Ayako took a trip to the store $5\frac{1}{2}$ miles away. If she rode the bus for $4\frac{5}{6}$ miles and walked the rest of the way, how far did she have to walk? $\frac{2}{3}$ miles

112. Average rainfall in Tampa for the month of November is $2\frac{3}{4}$ in. One stormy weekend $3\frac{1}{8}$ in. of rain fell. How many inches of rain over the average is this? $\frac{3}{8}$ in.

113. Maria has 4 yd of material. If she sews a dress that requires $3\frac{1}{8}$ yd, how much material will she have left? $\frac{7}{8}$ yd

114. Pete started working out at the gym several months ago. His waist measured $38\frac{1}{2}$ in. when he began and is now $33\frac{3}{4}$ in. How many inches did he lose around his waist? $4\frac{3}{4}$ in.

115. Find the perimeter (distance around) the parking lot. $244\frac{1}{2}$ yd

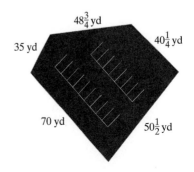

$48\frac{3}{4}$ yd

$40\frac{1}{4}$ yd

35 yd

70 yd

$50\frac{1}{2}$ yd

Figure for Exercise 115

116. A booth for display at a science fair is in the shape of a triangle. Find the perimeter of the triangle.   30 ft

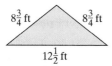

$8\frac{3}{4}$ ft $8\frac{3}{4}$ ft

$12\frac{1}{2}$ ft

Figure for Exercise 116

For Exercises 117–120, approximate the sum by rounding each fraction to the nearest whole number.

117. $\dfrac{7}{8} + \dfrac{16}{15}$
    Approximately 2

118. $\dfrac{6}{5} + \dfrac{21}{22}$
    Approximately 2

119. $\dfrac{21}{4} + \dfrac{98}{100} + \dfrac{80}{41}$
    Approximately 8

120. $\dfrac{29}{5} + \dfrac{51}{10} + \dfrac{7}{8}$
    Approximately 12

KEY TERMS:

composite number
denominator
factor
fraction
improper fraction
least common denominator (LCD)
mixed number
natural numbers
numerator
prime number
product
proper fraction
reciprocal
whole numbers

Concepts

1. **Introduction to a Place Value System**
2. **Converting Fractions to Decimals**
3. **Converting Decimals to Fractions**
4. **Converting Percents to Decimals and Fractions**
5. **Converting Decimals and Fractions to Percents**
6. **Applications of Percents**

section

R.2   DECIMALS AND PERCENTS

1. Introduction to a Place Value System

In a **place value** number system each digit in a numeral has a particular value determined by its location in the numeral (Figure R-6).

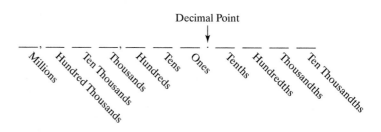

Figure R-6

For example, the number 197.215 represents

$$(1 \times 100) + (9 \times 10) + (7 \times 1) + \left(2 \times \frac{1}{10}\right) + \left(1 \times \frac{1}{100}\right) + \left(5 \times \frac{1}{1000}\right)$$

Each of the digits 1, 9, 7, 2, 1, and 5 is multiplied by 100, 10, 1, $\frac{1}{10}$, $\frac{1}{100}$, and $\frac{1}{1000}$, respectively, depending on its location in the numeral 197.215.

By obtaining a common denominator and adding fractions, we have

$$197.215 = 100 + 90 + 7 + \frac{200}{1000} + \frac{10}{1000} + \frac{5}{1000}$$

$$= 197 + \frac{215}{1000} \quad \text{or} \quad 197\frac{215}{1000}$$

Because 197.215 is equal to the mixed number $197\frac{215}{1000}$, we read 197.215 as one hundred ninety-seven *and* two hundred fifteen thousandths. The decimal point is read as the word *and*.

If there are no digits to the right of the decimal point, we usually omit the decimal point. The number 7125. is written simply as 7125, for example.

Classroom Activity R.2A

2. Converting Fractions to Decimals

In Section R.1, we learned that a fraction represents part of a whole unit. Likewise, the digits to the right of the decimal point represent a fraction of a whole unit. In this section we will learn how to convert a fraction to a decimal number and vice versa.

Converting a Fraction to a Decimal

To convert a fraction to a decimal, divide the numerator of the fraction by the denominator of the fraction.

example 1

Converting Fractions to Decimals

Convert each fraction to a decimal.

a. $\dfrac{7}{40}$ b. $\dfrac{2}{3}$ c. $\dfrac{13}{7}$

Solution:

a. $\dfrac{7}{40} = 0.175$

The number 0.175 is said to be a **terminating decimal** because there are no nonzero digits to the right of the last digit, 5.

$$
\begin{array}{r}
0.175 \\
40\overline{)7.000} \\
\underline{40} \\
300 \\
\underline{280} \\
200 \\
\underline{200} \\
0
\end{array}
$$

b. $\dfrac{2}{3} = 0.666\ldots$

$$\begin{array}{r} .666\ldots \\ 3\overline{)2.00000} \\ \underline{18} \\ 20 \\ \underline{18} \\ 20 \\ \underline{18} \\ 2\ldots \end{array}$$

The pattern $0.666\ldots$ continues indefinitely. Therefore, we say that this is a **repeating decimal**.

For a repeating decimal, a horizontal bar is often used to denote the repeating pattern after the decimal point.

Hence, $\frac{2}{3} = 0.\overline{6}$.

c. $\dfrac{13}{7} = 1.857142857142\ldots$

$$\begin{array}{r} 1.857142857142\ldots \\ 7\overline{)13.0000000000000} \end{array}$$

The number $1.857142857142\ldots$ is also a repeating pattern, and in this case the pattern includes several digits. Therefore, we write $\frac{13}{7} = 1.\overline{857142}$.

3. Converting Decimals to Fractions

To convert a terminating decimal to a fraction, we return to the basis of the place value system.

Converting a Terminating Decimal to a Fraction

1. Write the digits to the left of the decimal point as a whole number.
2. Write each digit to the right of the decimal point as a fraction over 10, 100, 1000 and so on, depending on its place position.
3. Convert to a single fraction by adding fractions.

example 2   **Converting Terminating Decimals to Fractions**

Convert each decimal to a fraction.

a. 0.0023   b. 50.06

Classroom Activity R.2B

Tip: We can also change a decimal to a fraction by writing it as read. The number 0.0023 is read as "23 ten-thousandths." Therefore,

$$0.0023 = \frac{23}{10,000}$$

Tip: The number $50 + \frac{6}{100}$ can be written as the mixed number $50\frac{6}{100}$ or in reduced form as $50\frac{3}{50}$.

Solution:

a. $0.0023 = 0 \times \frac{1}{10} + 0 \times \frac{1}{100} + 2 \times \frac{1}{1000} + 3 \times \frac{1}{10,000}$

$= \frac{2}{1000} + \frac{3}{10,000}$ Simplify.

$= \frac{20}{10,000} + \frac{3}{10,000}$ Find a common denominator.

$= \frac{23}{10,000}$ Add the fractions.

b. $50.06 = 50 + 0 \times \frac{1}{10} + 6 \times \frac{1}{100}$

$= 50 + \frac{6}{100}$ Simplify.

$= \frac{5000}{100} + \frac{6}{100}$ Find a common denominator.

$= \frac{5006}{100}$ Add the fractions.

$= \frac{2503}{50}$ Reduce to lowest terms.

Repeating decimals also can be written as fractions. However, the procedure to convert a repeating decimal to a fraction requires some knowledge of algebra. The list below shows some common repeating decimals and an equivalent fraction for each.

$0.\overline{1} = \frac{1}{9}$ $0.\overline{4} = \frac{4}{9}$ $0.\overline{7} = \frac{7}{9}$

$0.\overline{2} = \frac{2}{9}$ $0.\overline{5} = \frac{5}{9}$ $0.\overline{8} = \frac{8}{9}$

$0.\overline{3} = \frac{3}{9} = \frac{1}{3}$ $0.\overline{6} = \frac{6}{9} = \frac{2}{3}$ $0.\overline{9} = \frac{9}{9} = 1$

4.   Converting Percents to Decimals and Fractions

The concept of percent (%) is widely used in a variety of mathematical applications.

A stock decreased by 12% for the year.
The sales tax in a certain state is 6%.
Women represent 54% of graduating college seniors in the United States.

The word **percent** means "per hundred," and is denoted by the % symbol. It refers to the number of parts out of one hundred parts. A 6% sales tax for example means that 6¢ in tax is owed for every 100¢ spent.

The % symbol implies "division by 100," or equivalently, "multiplication by $\frac{1}{100} = 0.01$."

$$6\% = 6 \times (0.01) = 0.06 \qquad \text{or} \qquad 6\% = 6 \times \frac{1}{100} = \frac{6}{100} = \frac{3}{50}$$

Converting a Percent to a Decimal

To convert a percent to a decimal, remove the % symbol and multiply by 0.01.

Note: This is equivalent to removing the % symbol and moving the decimal point two places to the *left*, inserting zeros if necessary.

Converting a Percent to a Fraction

To convert a percent to a fraction, remove the % symbol and multiply by $\frac{1}{100}$. Reduce the resulting fraction to lowest terms if necessary.

---

example 3   **Converting Percents to Decimals**

Convert the percents to decimals.

a.   78%   b.   412%   c.   0.045%

Tip: Multiplying by 0.01 is equivalent to dividing by 100. This has the effect of moving the decimal point two places to the left.

Solution:

a. $78\% = 78 \times 0.01 = 0.78$
b. $412\% = 412 \times 0.01 = 4.12$
c. $0.045\% = 0.045 \times 0.01 = 0.00045$

---

example 4   **Converting Percents to Fractions**

Convert the percents to fractions.

a.   52%   b. $33\frac{1}{3}\%$ c.   6.5%

Solution:

a. $52\% = 52 \times \dfrac{1}{100}$ Drop the % symbol and multiply by $\frac{1}{100}$.

$= \dfrac{52}{100}$ Multiply.

$= \dfrac{13}{25}$ Reduce.

b. $33\frac{1}{3}\% = 33\frac{1}{3} \times \frac{1}{100}$ Drop the % symbol and multiply by $\frac{1}{100}$.

$= \frac{100}{3} \times \frac{1}{100}$ Write the mixed number as a fraction $33\frac{1}{3} = \frac{100}{3}$.

$= \frac{100}{300}$ Multiply the fractions.

$= \frac{1}{3}$ Reduce.

c. $6.5\% = 6.5 \times \frac{1}{100}$ Drop the % symbol and multiply by $\frac{1}{100}$.

$= \frac{65}{10} \times \frac{1}{100}$ Write 6.5 as an improper fraction.

$= \frac{65}{1000}$ Multiply the fractions.

$= \frac{13}{200}$ Reduce.

5.   Converting Decimals and Fractions to Percents

To convert a percent to a decimal or fraction, we drop the percent symbol and multiply by $0.01 = \frac{1}{100}$. To convert a decimal or fraction to a percent, we reverse the process.

Converting a Decimal to a Percent

To convert a decimal to a percent, multiply by 100 and attach the % symbol.

Note: This is equivalent to moving the decimal point two places to the *right* (inserting zeros if necessary) and attaching the % symbol.

Converting a Fraction to a Percent

To convert a fraction to a percent, first convert the fraction to a decimal (divide the numerator by the denominator). Then change the resulting decimal to a percent.

example 5   **Converting Decimals to Percents**

Convert the decimals to percents.

a.   0.92   b.   10.80   c.   0.005

Solution:

a. $0.92 = 0.92 \times 100\% = 92\%$ Multiply by 100 and attach the % symbol.

b. $10.80 = 10.80 \times 100\% = 1080\%$ Multiply by 100 and attach the % symbol.

c. $0.005 = 0.005 \times 100\% = 0.5\%$ Multiply by 100 and attach the % symbol.

---

example 6

Converting Fractions to Percents

Convert the fractions to percents.

a. $\dfrac{2}{5}$ b. $\dfrac{1}{16}$ c. $\dfrac{5}{3}$

Classroom Activity R.2C

Solution:

a. $\dfrac{2}{5} = 0.4$ Convert the fraction to a decimal.

  $= 0.4 \times 100\%$ Convert the decimal to a percent.

  $= 40\%$

b. $\dfrac{1}{16} = 0.0625$ Convert the fraction to a decimal.

  $= 0.0625 \times 100\%$ Convert the decimal to a percent.

  $= 6.25\%$

c. $\dfrac{5}{3} = 1.\overline{6}$ Convert the fraction to a decimal.

  $= 1.\overline{6} \times 100\%$ Convert the decimal to a percent.

  $= 166.\overline{6}\%$ or $166\frac{2}{3}\%$

> **Tip:** To multiply $1.\overline{6}$ by 100, write out several digits within the repeat cycle.
>
> $1.\overline{6} = 1.666\ldots \times 100\%$ Multiplying by 100 moves the
> $= 166.6\ldots\%$ decimal point two places to the right.
>
> $= 166.\overline{6}\%$, or $166\frac{2}{3}\%$

---

6.   Applications of Percents

Many applications involving percents involve finding a percent of some base number. For example, suppose a textbook is discounted 25%. If the book originally cost $60, find the amount of the discount.

In this example, we must find 25% of $60. In this context, the word *of* means multiply.

25% of $60
0.25 × 60 = 15   The amount of the discount is $15.

Note that the *decimal form* of a percentage is always used in calculations. Therefore, 25% was converted to 0.25 *before* multiplying by $60.

---

example 7 **Applying Percentages**

Shauna received a raise, so now her new salary is 105% of her old salary. Find Shauna's new salary if her old salary was $36,000 per year.

Solution:

The new salary is 105% of $36,000.

1.05 × 36,000 = 37,800   The new salary is $37,800 per year.

---

example 8 **Applying Percentages**

A couple leaves a 15% tip for a $32 dinner. Find the amount of the tip.

Solution:

The tip is 15% of $32.

0.15 × 32 = 4.80   The tip is $4.80.

In some applications it is necessary to convert a fractional part of a whole to a percentage of the whole.

---

example 9 **Finding a Percentage**

Union College in Schenectady, New York, accepts approximately 520 students each year from 3500 applicants. What percentage does 520 represent? Round to the nearest tenth of a percent.

Classroom Activity R.2D

Solution:

$$\frac{520}{3500} = 0.149$$ Convert the fractional part of the total number of applicants to decimal form.

$$= 0.149 \times 100\%$$ Convert the decimal to a percent.

$$= 14.9\%$$ Simplify.

Approximately 14.9% of the applicants to Union College are accepted.

Calculator Connections

Calculators only can display a limited number of digits on the calculator screen. Therefore, repeating decimals and terminating decimals with a large number of digits will be truncated or rounded to fit the calculator display. For example the fraction $\frac{2}{3} = 0.\overline{6}$ may be entered into the calculator as $\boxed{2}$ $\boxed{\div}$ $\boxed{3}$. The result may appear as 0.6666666667 or as 0.6666666666. The fraction $\frac{2}{11}$ equals the repeating decimal $0.\overline{18}$. However, the calculator converts $\frac{2}{11}$ to the terminating decimal 0.1818181818.

```
2/3
        .6666666667
2/11
        .1818181818
```

Calculator Exercises

Without using a calculator, find a repeating decimal to represent each of the following fractions. Then use a calculator to confirm your answer.

1. $\frac{4}{9}$ ❖   2. $\frac{7}{11}$ ❖   3. $\frac{3}{22}$ ❖   4. $\frac{5}{13}$ ❖

section R.2   PRACTICE EXERCISES

For Exercises 1–12, write the name of the place value for the digit in red.

1. 481.24
   Tens
2. 1345.42
   Tens
3. 2912.032
   Hundreds
4. 4208.03
   Hundreds
5. 2.381
   Tenths
6. 8.249
   Tenths
7. 21.413
   Hundredths
8. 82.794
   Hundredths
9. 32.43
   Ones
10. 78.04
    Ones
11. 0.38192
    Ten-thousandths
12. 0.89754
    Ten-thousandths

13. The first 10 Roman numerals are: I, II, III, IV, V, VI, VII, VIII, IX, X. Is this numbering system a place value system? Explain your answer. ❖

For Exercises 14–21, convert each fraction to a terminating decimal or a repeating decimal.

14. $\frac{7}{10}$ 0.7
15. $\frac{9}{10}$ 0.9
16. $\frac{9}{25}$ 0.36
17. $\frac{3}{25}$ 0.12
18. $\frac{11}{9}$ $1.\overline{2}$
19. $\frac{16}{9}$ $1.\overline{7}$
20. $\frac{7}{33}$ $0.\overline{21}$
21. $\frac{2}{11}$ $0.\overline{18}$

For Exercises 22–33, convert each decimal to a fraction or a mixed number.

22. 0.45 $\frac{9}{20}$
23. 0.65 $\frac{13}{20}$
24. 0.181 $\frac{181}{1000}$
25. 0.273 $\frac{273}{1000}$
26. 2.04 $\frac{51}{25}$ or $2\frac{1}{25}$
27. 6.02 $\frac{301}{50}$ or $6\frac{1}{50}$
28. 13.007 $\frac{13007}{1000}$ or $13\frac{7}{1000}$
29. 12.003 $\frac{12003}{1000}$ or $12\frac{3}{1000}$
30. $0.\overline{5}$ $\frac{5}{9}$
31. $0.\overline{8}$ $\frac{8}{9}$
32. $1.\overline{1}$ $\frac{10}{9}$ or $1\frac{1}{9}$
33. $2.\overline{3}$ $\frac{7}{3}$ or $2\frac{1}{3}$

For Exercises 34–41, convert each percent to a decimal and to a fraction.

34. The sale price is 30% off of the original price. $0.3, \frac{3}{10}$

35. An HMO (health maintenance organization) pays 80% of all doctors' bills. $0.8, \dfrac{4}{5}$

36. The building will be 75% complete by spring. $0.75, \dfrac{3}{4}$

37. Chan plants roses in 25% of his garden. $0.25, \dfrac{1}{4}$

38. The bank pays $3\frac{3}{4}$% interest on a checking account. $0.0375, \dfrac{3}{80}$

39. A credit union pays $4\frac{1}{2}$% interest on a savings account. $0.045, \dfrac{9}{200}$

40. Kansas received 15.7% of its annual rainfall in 1 week. $0.157, \dfrac{157}{1000}$

41. Social Security withholds 5.8% of an employee's gross pay. $0.058, \dfrac{29}{500}$

42. Explain how to convert a decimal to a percent.
Multiply by 100 and apply the % sign.

43. Explain how to convert a percent to a decimal.
Remove the % sign and divide by 100.

For Exercises 44–55, convert the decimal to a percent.

44. 0.05 5%
45. 0.06 6%

46. 0.90 90%
47. 0.70 70%

48. 1.2 120%
49. 4.8 480%

50. 7.5 750%
51. 9.3 930%

52. 0.135 13.5%
53. 0.536 53.6%

54. 0.003 0.3%
55. 0.002 0.2%

For Exercises 56–67, convert the fraction to a percent.

56. $\dfrac{3}{50}$ 6%
57. $\dfrac{23}{50}$ 46%

58. $\dfrac{9}{2}$ 450%
59. $\dfrac{7}{4}$ 175%

60. $\dfrac{5}{8}$ 62.5%
61. $\dfrac{1}{8}$ 12.5%

62. $\dfrac{5}{16}$ 31.25%
63. $\dfrac{7}{16}$ 43.75%

64. $\dfrac{5}{6}$ $83.\overline{3}\%$
65. $\dfrac{4}{15}$ $26.\overline{6}\%$

66. $\dfrac{14}{15}$ $93.\overline{3}\%$
67. $\dfrac{5}{18}$ $27.\overline{7}\%$

68. A suit that costs $140 is discounted by 30%. How much is the discount? $42

69. A community college has a 12% increase in enrollment over the previous year. If the enrollment last year was 10,800 students, how many students have enrolled this year? 12,096 students

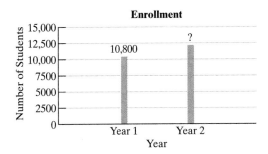

Figure for Exercise 69

70. A tip of 15% is left after a dinner that costs $54. How much is the tip? $8.10

71. A textbook costs $85. If the sales tax is 7%, how much is the tax on the textbook? $5.95

72. Jorge saves 6% of his pay for his summer vacation. If he makes $625 a week, how much will he save each week? $37.50

73. The following pie graph shows a family budget based on a net income of $2400 per month.

   a. Determine the amount spent on rent. $792

   b. Determine the amount spent on car payments. $408

c. Determine the amount spent on utilities.
   $192

d. How much more money is spent than saved?
   $1488

Monthly Budget

Figure for Exercise 73

 74. By the end of the year, Felipe will have 75% of his mortgage paid. If the mortgage was originally for $90,000, how much will have been paid at the end of the year? $67,500

 75. A certificate of deposit (CD) earns 8% interest in 1 year. If Mr. Patel has $12,000 invested in the CD, how much interest will he receive at the end of the year? $960

 76. On a state exit exam, a student can answer 30% of the problems incorrectly and still pass the test. If the exam has 40 problems, how many can a student answer incorrectly and still pass?
12 questions

KEY TERMS:

percent
place value
repeating decimal
terminating decimal

Concepts

1. Perimeter
2. Area
3. Volume
4. Angles
5. Triangles

R.3   INTRODUCTION TO GEOMETRY

1. Perimeter

In this section we present several facts and formulas that may be used throughout the text in applications of geometry. One of the most important uses of geometry involves the measurement of objects of various shapes. We begin with an introduction to perimeter, area, and volume for several common shapes and objects.

    Perimeter is defined as the distance around a figure. For a polygon (a figure with many sides) the perimeter is the sum of the lengths of the sides. For a circle the distance around the outside is called the **circumference**.

Rectangle   **Square**   **Triangle**   **Circle**

$P = 2\ell + 2w$ $P = 4s$ $P = a + b + c$ Circumference: $C = 2\pi r$

example 1

Finding Perimeter and Circumference

Find the perimeter or circumference as indicated. Use 3.14 for π.

a. Perimeter of the triangle

4 in. 2 in.

5 in.

b. Perimeter of the rectangle

3.1 ft

5.5 ft

c. Circumference of the circle

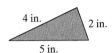

6 cm

Solution:

a. $P = a + b + c$

  $= 4 \text{ in.} + 2 \text{ in.} + 5 \text{ in.}$ Add the lengths of the sides.

  $= 11 \text{ in.}$ The perimeter is 11 in.

b. $P = 2\ell + 2w$

  $= 2(5.5 \text{ ft}) + 2(3.1 \text{ ft})$ Substitute $\ell = 5.5$ ft and $w = 3.1$ ft.

  $= 11 \text{ ft} + 6.2 \text{ ft}$

  $= 17.2 \text{ ft}$ The perimeter is 17.2 ft.

c. $C = 2\pi r$

  $= 2(3.14)(6 \text{ cm})$ Substitute 3.14 for π and $r = 6$ cm.

  $= 6.28(6 \text{ cm})$

  $= 37.68 \text{ cm}$ The circumference is 37.68 cm.

2. Area

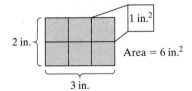

2 in.

3 in.

1 in.^{2}

Area = 6 in.^{2}

Figure R-7

The **area** of a geometric figure is the number of square units that can be enclosed within the figure. For example, the rectangle shown in Figure R-7 encloses 6 square inches (6 in.^{2}).

The formulas used to compute the area for several common geometric shapes are given here:

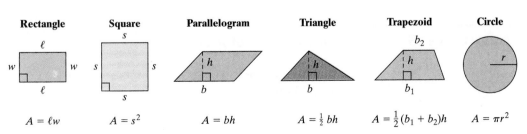

Rectangle	Square	Parallelogram	Triangle	Trapezoid	Circle
$A = \ell w$	$A = s^2$	$A = bh$	$A = \frac{1}{2} bh$	$A = \frac{1}{2}(b_1 + b_2)h$	$A = \pi r^2$

example 2

Finding Area

Find the area enclosed by each figure.

a.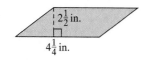
b.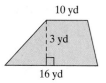

Solution:

> **Tip:** Notice that the units of area are always square units such as square inches (in.^{2}), square feet (ft^2), square yards (yd^2), square centimeters (cm^2), and so on.

a. $A = bh$ The figure is a parallelogram.

$= (4\frac{1}{4} \text{ in.})(2\frac{1}{2} \text{ in.})$ Substitute $b = 4\frac{1}{4}$ in. and $h = 2\frac{1}{2}$ in.

$= \left(\frac{17}{4} \text{ in.}\right)\left(\frac{5}{2} \text{ in.}\right)$

$= \frac{85}{8}$ in.^{2} or $10\frac{5}{8}$ in.^{2}

b. $A = \frac{1}{2}(b_1 + b_2)h$ The figure is a trapezoid.

$= \frac{1}{2}(16 \text{ yd} + 10 \text{ yd})(3 \text{ yd})$ Substitute $b_1 = 16$ yd, $b_2 = 10$ yd, and $h = 3$ yd.

$= \frac{1}{2}(26 \text{ yd})(3 \text{ yd})$

$= (13 \text{ yd})(3 \text{ yd})$

$= 39 \text{ yd}^2$ The area is 39 yd^2.

> **Tip:** Notice that several of the formulas presented thus far involve multiple operations. The order in which we perform the arithmetic is called the order of operations and is covered in detail in Section 1.2. For now, we will follow these guidelines in the order given below:
>
> 1. Perform operations within parentheses first.
> 2. Evaluate expressions with exponents.
> 3. Perform multiplication or division in order from left to right.
> 4. Perform addition or subtraction in order from left to right.

example 3

Finding Area of a Circle

Find the area of a circular fountain if the radius is 25 ft. Use 3.14 for π.

Solution:

$A = \pi r^2$

$= (3.14)(25 \text{ ft})^2$

$= (3.14)(625 \text{ ft}^2)$ Substitute 3.14 for π and $r = 25$ ft.

$= 1962.5 \text{ ft}^2$ The area of the fountain is 1962.5 ft^2.

3. Volume

The **volume** of a solid is the number of cubic units that can be enclosed within a solid. The solid shown in Figure R-8 contains 18 cubic inches (18 in.³).

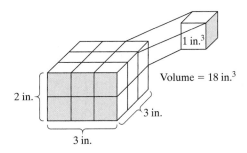

Figure R-8

The formulas used to compute the volume of several common rectangular solids are given here:

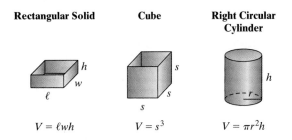

Rectangular Solid   **Cube**   **Right Circular Cylinder**

$V = \ell w h$ $V = s^3$ $V = \pi r^2 h$

Tip: Notice that the volume formulas for these three figures are given by the product of the area of the base and the height of the figure:

$V = \ell w h$ $V = s \cdot s \cdot s$ $V = \pi r^2 h$

Area of Rectangular Base   Area of Square Base   Area of Circular Base

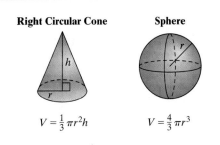

Right Circular Cone   **Sphere**

$V = \frac{1}{3}\pi r^2 h$ $V = \frac{4}{3}\pi r^3$

example 4   Finding Volume

Find the volume of each object.

a.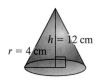

$1\frac{1}{2}$ ft

$1\frac{1}{2}$ ft

$1\frac{1}{2}$ ft

b.

$h = 12$ cm

$r = 4$ cm

Solution:

a. $V = s^3$ The object is a cube.

$= (1\frac{1}{2}\ \text{ft})^3$ Substitute $s = 1\frac{1}{2}$ ft.

$= \left(\frac{3}{2}\ \text{ft}\right)^3$

$= \left(\frac{3}{2}\ \text{ft}\right)\left(\frac{3}{2}\ \text{ft}\right)\left(\frac{3}{2}\ \text{ft}\right)$

$= \frac{27}{8}\ \text{ft}^3$, or $3\frac{3}{8}\ \text{ft}^3$

Tip: Notice that the units of volume are cubic units such as cubic inches (in.³), cubic feet (ft³), cubic yards (yd³), cubic centimeters (cm³), and so on.

b. $V = \frac{1}{3}\pi r^2 h$ The object is a right circular cone.

$= \frac{1}{3}(3.14)(4\ \text{cm})^2(12\ \text{cm})$ Substitute 3.14 for π, $r = 4$ cm, and $h = 12$ cm.

$= \frac{1}{3}(3.14)(16\ \text{cm}^2)(12\ \text{cm})$

$= 200.96\ \text{cm}^3$

example 5   Finding Volume in an Application

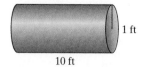

1 ft

10 ft

Classroom Activity R.3A

An underground gas tank is in the shape of a right circular cylinder.

a. Find the volume of the tank. Use 3.14 for π.

b. Find the cost to fill the tank with gasoline if gasoline costs $9/ft³.

Solution:

a. $V = \pi r^2 h$

$= (3.14)(1\ \text{ft})^2(10\ \text{ft})$ Substitute 3.14 for π, $r = 1$ ft, and $h = 10$ ft.

$= (3.14)(1\ \text{ft}^2)(10\ \text{ft})$

$= 31.4\ \text{ft}^3$ The tank holds 31.4 ft³ of gasoline.

b.   Cost = ($9/ft^3)(31.4 ft^3)

$\qquad$ = \$282.6 $\qquad\qquad$ It will cost \$282.6 to fill the tank.

4.   Angles

Applications involving angles and their measure come up often in the study of algebra, trigonometry, calculus, and applied sciences. The most common unit to measure an angle is the degree (°). Several angles and their corresponding degree measure are shown in Figure R-9.

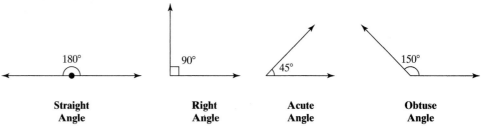

Figure R-9

- An angle that measures 90° is a **right angle** (right angles are often marked at the vertex with a square or corner symbol, □).
- An angle that measures 180° is called a **straight angle**.
- An angle that measures between 0° and 90° is called an **acute angle**.
- An angle that measures between 90° and 180° is called an **obtuse angle**.
- Two angles with the same measure are **equal angles** (or **congruent angles**).

The measure of an angle will be denoted by the symbol m written in front of the angle. Therefore, the measure of $\angle A$ is denoted $m(\angle A)$.

- Two angles are said to be **complementary** if their sum is 90°.
- Two angles are said to be **supplementary** if their sum is 180°.

$m(\angle x) + m(\angle y) = 90°$ $\qquad\qquad\qquad\qquad$ $m(\angle x) + m(\angle y) = 180°$

When two lines intersect, four angles are formed (Figure R-10). In Figure R-10, $\angle a$ and $\angle b$ are said to be a pair of **vertical angles**. Another set of vertical angles is the pair $\angle c$ and $\angle d$. An important property of vertical angles is that the measures of two vertical angles are *equal*. In the figure $m(\angle a) = m(\angle b)$ and $m(\angle c) = m(\angle d)$.

Parallel lines are lines that lie in the same plane and do not intersect. In Figure R-11, the lines L_1 and L_2 are parallel lines. If a line intersects two parallel lines, the

Figure R-10

line is called a **transversal**. In Figure R-11, the line m is a transversal and forms eight angles with the parallel lines L_1 and L_2.

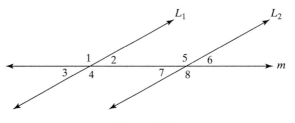

Figure R-11

The measures of angles 1–8 in Figure R-11 have the following special properties.

L_1 and L_2 are Parallel. Line m is a Transversal	Name of Angles	Property
	The following pairs of angles are called **alternate interior angles**:	Alternate interior angles are equal in measure.
	$\angle 2$ and $\angle 7$	$m(\angle 2) = m(\angle 7)$
	$\angle 4$ and $\angle 5$	$m(\angle 4) = m(\angle 5)$
	The following pairs of angles are called **alternate exterior angles**:	Alternate exterior angles are equal in measure.
	$\angle 1$ and $\angle 8$	$m(\angle 1) = m(\angle 8)$
	$\angle 3$ and $\angle 6$	$m(\angle 3) = m(\angle 6)$
	The following pairs of angles are called **corresponding angles**:	Corresponding angles are equal in measure.
	$\angle 1$ and $\angle 5$	$m(\angle 1) = m(\angle 5)$
	$\angle 2$ and $\angle 6$	$m(\angle 2) = m(\angle 6)$
	$\angle 3$ and $\angle 7$	$m(\angle 3) = m(\angle 7)$
	$\angle 4$ and $\angle 8$	$m(\angle 4) = m(\angle 8)$

---

example 6   Finding Unknown Angles in a Diagram

Find the measure of each angle and explain how the angle is related to the given angle of 70°.

a. $\angle a$
b. $\angle b$
c. $\angle c$
d. $\angle d$

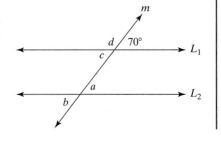

Classroom Activity R.3B

Solution:

a. $m(\angle a) = 70°$ $\angle a$ is a corresponding angle to the given angle of 70°.

b. $m(\angle b) = 70°$ $\angle b$ and the given angle of 70° are alternate exterior angles.

c. $m(\angle c) = 70°$ $\angle c$ and the given angle of 70° are vertical angles.

d. $m(\angle d) = 110°$ $\angle d$ is the supplement of the given angle of 70°.

5. Triangles

Triangles are categorized by the measures of the angles (Figure R-12) and by the number of equal sides or angles (Figure R-13).

- An **acute triangle** is a triangle in which all three angles are acute.
- A **right triangle** is a triangle in which one angle is a right angle.
- An **obtuse triangle** is a triangle in which one angle is obtuse.

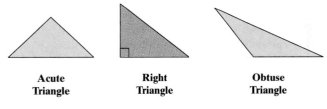

| Acute | Right | Obtuse |
| Triangle | Triangle | Triangle |

Figure R-12

- An **equilateral triangle** is a triangle in which all three angles (and all three sides) are equal.
- An **isosceles triangle** is a triangle in which two sides are equal (the angles opposite the equal sides are also equal).
- A **scalene triangle** is a triangle in which no sides (or angles) are equal.

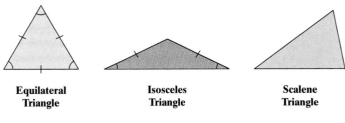

| Equilateral | Isosceles | Scalene |
| Triangle | Triangle | Triangle |

Figure R-13

The following important property is true for all triangles.

Sum of the Angles in a Triangle

The sum of the measures of the angles of a triangle is 180°.

example 7

Finding Unknown Angles in a Diagram

Find the measure of each
angle in the figure.

a. $\angle a$
b. $\angle b$
c. $\angle c$
d. $\angle d$
e. $\angle e$

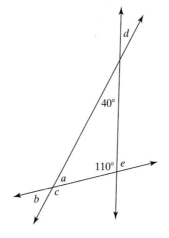

Classroom Activity R.3C

Solution:

a. $m(\angle a) = 30°$ The sum of the angles in a triangle is 180°.

b. $m(\angle b) = 30°$ $\angle a$ and $\angle b$ are vertical angles and are equal.

c. $m(\angle c) = 150°$ $\angle c$ and $\angle a$ are supplementary angles ($\angle c$ and $\angle b$ are also supplementary).

d. $m(\angle d) = 40°$ $\angle d$ and the given angle of 40° are vertical angles.

e. $m(\angle e) = 70°$ $\angle e$ and the given angle of 110° are supplementary angles.

---

section R.3   PRACTICE EXERCISES

 1. Identify which of the following units could be measures of perimeter.

a. Square inches (in.^{2})

b. Meters (m)

c. Cubic feet (ft^3)

d. Cubic meters (m^3)

e. Miles

f. Square centimeters (cm^2)

g. Square yards (yd^2)

h. Cubic inches (in.^{3})

i. Kilometers (km)

b, e, i

For Exercises 2–5, find the perimeter and area of each figure.

2.
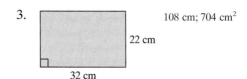
32 m; 60 m^2
6 m
10 m

3.
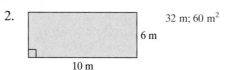
108 cm; 704 cm^2
22 cm
32 cm

4.
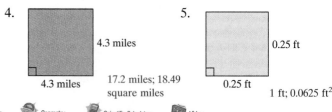
4.3 miles
4.3 miles
17.2 miles; 18.49 square miles

5.
0.25 ft
0.25 ft
1 ft; 0.0625 ft^2

6. Identify which of the following units could be measures of circumference.

    a. Square inches (in.^{2})

    b. Meters (m)

    c. Cubic feet (ft^3)

    d. Cubic meters (m^3)

    e. Miles

    f. Square centimeters (cm^2)

    g. Square yards (yd^2)

    h. Cubic inches (in.^{3})

    i. Kilometers (km)

    b, e, i

For Exercises 7–10, find the perimeter or circumference. Use 3.14 for π.

7.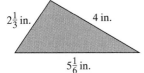

   $2\frac{1}{3}$ in.   4 in. $11\frac{1}{2}$ in. $5\frac{1}{6}$ in.

8.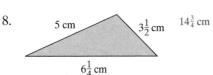

   5 cm $3\frac{1}{2}$ cm $14\frac{3}{4}$ cm $6\frac{1}{4}$ cm

9.

   10 ft

   31.4 ft

10.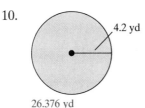

    4.2 yd

    26.376 yd

11. Identify which of the following units could be measures of area.

    a. Square inches (in.^{2})

    b. Meters (m)

    c. Cubic feet (ft^3)

    d. Cubic meters (m^3)

    e. Miles

    f. Square centimeters (cm^2)

    g. Square yards (yd^2)

    h. Cubic inches (in.^{3})

    i. Kilometers (km)

    a, f, g

For Exercises 12–21, find the area. Use 3.14 for π.

12.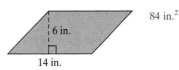

    6 in.   14 in.   84 in.^{2}

13.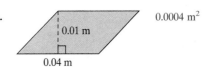

    0.01 m   0.04 m   0.0004 m^2

14.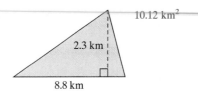

    2.3 km   8.8 km   10.12 km^2

15.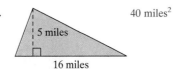

    5 miles   16 miles   40 miles^2

16.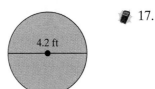

    4.2 ft

    13.8474 ft^2

17.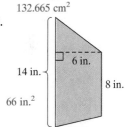

    6.5 cm

    132.665 cm^2

18.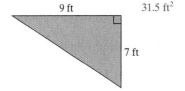

    8 in.   6 in.   14 in.

    66 in.^{2}

19.

    6 in.   14 in.   8 in.

    66 in.^{2}

20.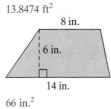

    9 ft   7 ft   31.5 ft^2

21.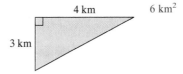

    4 km   3 km   6 km^2

22. Identify which of the following units could be measures of volume.

    a. Square inches (in.²)

    b. Meters (m)

    c. Cubic feet (ft³)

    d. Cubic meters (m³)

    e. Miles

    f. Square centimeters (cm²)

    g. Square yards (yd²)

    h. Cubic inches (in.³)

    i. Kilometers (km)
    c, d, h

For Exercises 23–26, find the volume of each figure. Use 3.14 for π.

23.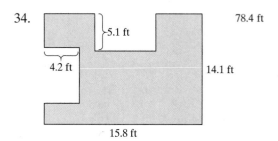
    2 ft
    6 ft
    75.36 ft³

24. 3.3 cm
    6.2 cm
    212.00652 cm³

25.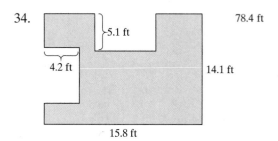
    4 in.
    39 in.³
    1½ in.
    6½ in.

26. 2⅛ cm
    24 7/16 cm³
    2 cm
    5¾ cm

27. Find the volume of a spherical balloon whose radius is 9 in. Use 3.14 for π.   3052.08 in.³

28. Find the volume of a spherical ball whose radius is 3 in. Use 3.14 for π.   113.04 in.³

29. Find the volume of a snow cone in the shape of a right circular cone whose radius is 3 cm and whose height is 12 cm. Use 3.14 for π.   113.04 cm³

30. Find the volume of a pile of gravel in the shape of a right circular cone whose radius is 10 yd and whose height is 18 yd. Use 3.14 for π.
    1884 yd³

31. Find the volume of a cube that is 3.2 ft on a side.   32.768 ft³

32. Find the volume of a cube that is 10.5 cm on a side.   1157.625 cm³

For Exercises 33–34, find the perimeter.

33.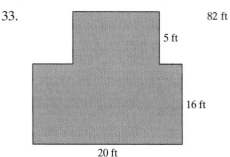
    82 ft
    5 ft
    16 ft
    20 ft

34.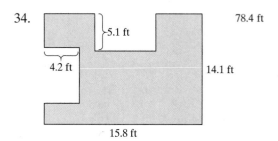
    78.4 ft
    5.1 ft
    4.2 ft
    14.1 ft
    15.8 ft

For Exercises 35–38, find the area of the shaded region. Use 3.14 for π.

35.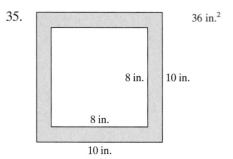
    36 in.²
    8 in.   10 in.
    8 in.
    10 in.

36.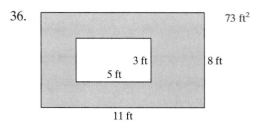
    73 ft²
    3 ft   8 ft
    5 ft
    11 ft

37.

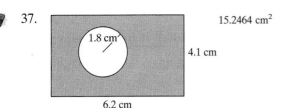

15.2464 cm²

38.

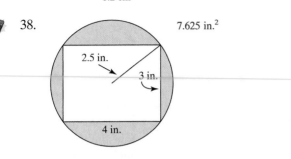

7.625 in.²

39. A wall measuring 20 ft by 8 ft can be painted for $50.

   a. What is the price per square foot? Round to the nearest cent. $0.31/ft²

   b. At this rate, how much would it cost to paint the remaining three walls that measure 20 ft by 8 ft, 16 ft by 8 ft, and 16 ft by 8 ft, respectively? Round to the nearest dollar. $129

40. Suppose it costs $320 to carpet a 16 ft by 12 ft room.

   a. What is the price per square foot? Round to the nearest cent. $1.67/ft²

   b. At this rate, how much would it cost to carpet a room that is 20 ft by 32 ft? $1068.80

41. If you were to purchase fencing for a garden, would you measure the perimeter or area of the garden? Perimeter

42. If you were to purchase sod (grass) for your front yard, would you measure the perimeter or area of the yard? Area

43. a. Find the area of a circular pizza that is 8 in. in diameter (the radius is 4 in.). Use 3.14 for π. 50.24 in.²

   b. Find the area of a circular pizza that is 12 in. in diameter (the radius is 6 in.). 113.04 in.²

   c. Assume that the 8-in. diameter and 12-in. diameter pizzas are both the same thickness. Which would provide more pizza, two 8-in. pizzas or one 12-in. pizza? One 12-in. pizza

44. Find the area of a circular stained glass window that is 16 in. in diameter. Use 3.14 for π. 200.96 in.²

45. Find the volume of a soup can in the shape of a right circular cylinder if its radius is 3.2 cm and its height is 9 cm. Use 3.14 for π. 289.3824 cm³

46. Find the volume of a coffee mug whose radius is 2.5 in. and whose height is 6 in. Use 3.14 for π. 117.75 in.³

For Exercises 47–54, answer True or False. If an answer is false, explain why.

47. The sum of the measures of two right angles equals the measure of a straight angle. True

48. Two right angles are complementary. False; they are supplementary.

49. Two right angles are supplementary. True

50. Two acute angles cannot be supplementary. True

51. Two obtuse angles cannot be supplementary. True

52. An obtuse angle and an acute angle can be supplementary. True

53. If a triangle is equilateral, then it is not scalene. True

54. If a triangle is isosceles, then it is also scalene. ❖

55. What angle is its own complement? 45°

56. What angle is its own supplement? 90°

57. If possible find two acute angles that are supplementary. Not possible

58. If possible find two acute angles that are complementary. Answers may vary. For example: 30°, 60°

59. If possible find an obtuse angle and an acute angle that are supplementary. Answers may vary. For example: 100°, 80°

60. If possible find two obtuse angles that are supplementary. Not possible

61. Refer to the figure.
    a. State all pairs of vertical angles.
        ∠1 and ∠3, ∠2 and ∠4
    b. State all pairs of supplementary angles.
        ∠1 and ∠2, ∠2 and ∠3, ∠3 and ∠4, ∠1 and ∠4
    c. If the measure of ∠4 is 80°, find the measures of ∠1, ∠2, and ∠3.
        $m(\angle 1) = 100°, m(\angle 2) = 80°, m(\angle 3) = 100°$

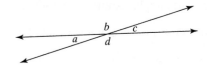

Figure for Exercise 61

62. Refer to the figure.
    a. State all pairs of vertical angles.
        ∠b and ∠d, ∠a and ∠c
    b. State all pairs of supplementary angles.
        ∠a and ∠b, ∠b and ∠c, ∠c and ∠d, ∠a and ∠d
    c. If the measure of ∠a is 25°, find the measures of ∠b, ∠c, and ∠d.
        $m(\angle b) = 155°, m(\angle c) = 25°, m(\angle d) = 155°$

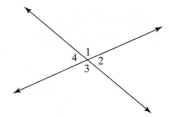

Figure for Exercise 62

For Exercises 63–70, find the complement of each angle.

63.	33°	64.	87°	65.	12°	66.	45°
	57°		3°		78°		45°
67.	30°	68.	20°	69.	70°	70.	60°
	60°		70°		20°		30°

For Exercises 71–78, find the supplement of each angle.

71.	33°	72.	87°	73.	122°	74.	90°
	147°		93°		58°		90°
75.	45°	76.	150°	77.	135°	78.	30°
	135°		30°		45°		150°

For Exercises 79–86, refer to the figure. Assume that L_1 and L_2 are parallel lines cut by the transversal, n.

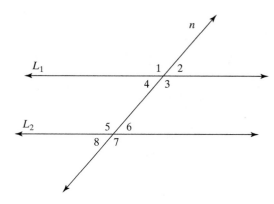

Figure for Exercises 79–86

79. $m(\angle 5) = m(\angle\underline{\ 7\ })$ Reason: Vertical angles are equal.

80. $m(\angle 5) = m(\angle\underline{\ 3\ })$ Reason: Alternate interior angles are equal.

81. $m(\angle 5) = m(\angle\underline{\ 1\ })$ Reason: Corresponding angles are equal.

82. $m(\angle 7) = m(\angle\underline{\ 3\ })$ Reason: Corresponding angles are equal.

83. $m(\angle 7) = m(\angle\underline{\ 1\ })$ Reason: Alternate exterior angles are equal.

84. $m(\angle 7) = m(\angle\underline{\ 5\ })$ Reason: Vertical angles are equal.

85. $m(\angle 3) = m(\angle\underline{\ 5\ })$ Reason: Alternate interior angles are equal.

86. $m(\angle 3) = m(\angle\underline{\ 1\ })$ Reason: Vertical angles are equal.

87. Find the measure of angles a–g in the figure. Assume that L_1 and L_2 are parallel and that n is a transversal. $m(\angle a) = 45°, m(\angle b) = 135°, m(\angle c) = 45°, m(\angle d) = 135°, m(\angle e) = 45°, m(\angle f) = 135°, m(\angle g) = 45°$

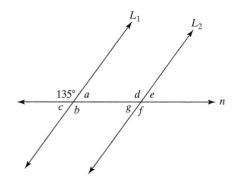

135° a   d e
c b   g f   n

Figure for Exercise 87

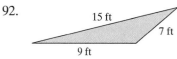

92.

15 ft
7 ft
9 ft

Scalene

93. Can a triangle be both a right triangle and an obtuse triangle? Explain. No, a 90° angle plus an angle greater than 90° would make the sum of the angles greater than 180°.

94. Can a triangle be both a right triangle and an isosceles triangle? Explain.
Yes, the legs of the right angle can be equal.

For Exercises 95–98, find the missing angles.

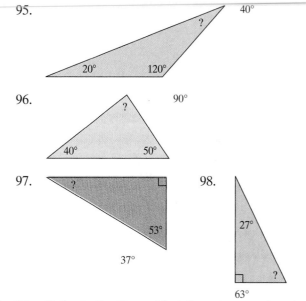

95.

40°
?
20°   120°

96.

90°
?
40°   50°

97.

?
53°
37°

98.

27°
?
63°

88. Find the measure of angles a–g in the figure. Assume that L_1 and L_2 are parallel and that n is a transversal. $m(\angle a) = 65°, m(\angle b) = 115°,$
$m(\angle c) = 115°, m(\angle d) = 65°, m(\angle e) = 115°, m(\angle f) = 115°, m(\angle g) = 65°$

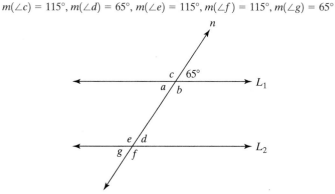

n

c 65°
a b L_1

e d
g f L_2

Figure for Exercise 88

For Exercises 89–92, identify the triangle as equilateral, isosceles, or scalene.

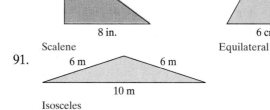

89.

6 in.   10 in.

8 in.
Scalene

90.

6 cm   6 cm

6 cm
Equilateral

91.

6 m   6 m

10 m
Isosceles

99. Refer to the figure. Find the measure of angles a–j. $m(\angle a) = 80°, m(\angle b) = 80°, m(\angle c) = 100°,$
$m(\angle d) = 100°, m(\angle e) = 65°, m(\angle f) = 115°, m(\angle g) = 115°, m(\angle h) = 35°,$
$m(\angle i) = 145°, m(\angle j) = 145°$

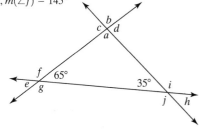

b
c d
a

f 65°   35°   i
e g   j h

Figure for Exercise 99

100. Refer to the figure. Find the measure of angles
a–j. $m(\angle a) = 120°, m(\angle b) = 60°, m(\angle c) = 60°,$
$m(\angle d) = 40°, m(\angle e) = 40°, m(\angle f) = 140°, m(\angle g) = 140°, m(\angle h) = 160°,$
$m(\angle i) = 20°, m(\angle j) = 160°$

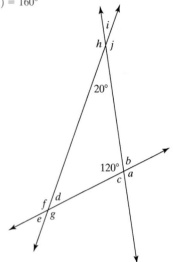

Figure for Exercise 100

101. Refer to the figure. Find the measure of angles
a–k. Assume that L_1 and L_2 are parallel.
$m(\angle a) = 70°, m(\angle b) = 65°, m(\angle c) = 65°, m(\angle d) = 110°, m(\angle e) = 70°,$
$m(\angle f) = 110°, m(\angle g) = 115°,$
$m(\angle h) = 115°, m(\angle i) = 65°,$
$m(\angle j) = 70°, m(\angle k) = 65°$

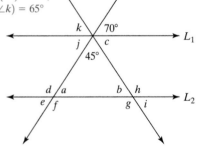

Figure for Exercise 101

102. Refer to the figure. Find the measure of angles
a–k. Assume that L_1 and L_2 are parallel.
$m(\angle a) = 75°, m(\angle b) = 65°, m(\angle c) = 40°, m(\angle d) = 65°, m(\angle e) = 65°,$
$m(\angle f) = 40°, m(\angle g) = 40°, m(\angle h) = 140°, m(\angle i) = 140°,$
$m(\angle j) = 115°, m(\angle k) = 115°$

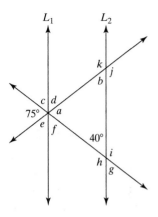

Figure for Exercise 102

Key Terms:

acute angle
acute triangle
alternate exterior angles
alternate interior angles
area of a circle
area of a parallelogram
area of a rectangle
area of a square
area of a trapezoid
area of a triangle
circumference
complementary angles
congruent angles
corresponding angles
equal angles
equilateral triangle
isosceles triangle
obtuse angle
obtuse triangle
parallel lines
perimeter of a polygon
perimeter of a rectangle
perimeter of a square
right angle
right triangle
scalene triangle
straight angle
supplementary angles
transversal
vertical angles
volume of a cube
volume of a rectangular solid
volume of a right circular cone
volume of a right circular cylinder
volume of a sphere

section

R.4 STUDY TIPS

In taking a course in algebra, you are making a commitment to yourself, your instructor, and your classmates. Following some or all of the study tips below can help you be successful in this endeavor. The features of this text that will assist you are printed in blue.

1. Before the Course

- Purchase the necessary materials for the course before the course begins or on the first day.
- Obtain a three-ring binder to keep and organize your notes, homework, tests, and any other materials acquired in the class. We call this type of notebook a portfolio.
- Arrange your schedule so that you have enough time to attend class and to do homework. A common rule of thumb is to set aside at least 2 hours for homework for every hour spent in class. That is, if you are taking a 4 credit-hour course, plan on at least 8 hours a week for homework. If you experience difficulty in mathematics, plan for more time. A 4 credit-hour course will then take *at least* 12 hours each week—about the same as a part-time job.
- Communicate with your employer and family members the importance of your success in this course so that they can support you.
- Be sure to find out the type of calculator (if any) that your instructor requires.

2. During the Course

- Read the section in the text *before* the lecture to familiarize yourself with the material and terminology.
- Attend every class and be on time.
- Take notes in class. Write down all of the examples that the instructor presents. Read the notes after class and add any comments to make your notes clearer to you. Use a tape recorder to record the lecture if needed.
- Ask questions in class.
- Read the section in the text *after* the lecture and pay special attention to the Tip boxes and Avoiding Mistakes boxes.
- Do homework every night. Even if your class does not meet everyday, you should still do some work every night to keep the material fresh in your mind.
- Check your homework with the answers that are supplied in the back of this text. Correct the exercises that do not match and circle or star the ones that you cannot correct yourself. This way you can easily find them and ask your instructor the next day.
- Write the definition and give an example of each Key Term usually found at the end of each Summary section at the end of the chapter.
- The Midchapter Reviews provide additional practice to master the first portion of the chapter. Sometimes the most difficult part of learning mathematics is retaining all that you learn. The Midchapter Reviews are excellent tools for retention of material.
- Form a study group with fellow students in your class, and exchange phone numbers. You will be surprised by how much you can learn by talking about mathematics with other students.
- If you use a calculator in your class, read the Calculator Connections boxes to learn how and when to use your calculator.

3. Preparation for Exams

- Look over your homework. Pay special attention to the exercises you have circled or starred to be sure that you have learned that concept.
- Read through the Summary at the end of the chapter. Be sure that you understand each concept and example. If not, go to the section in the text and reread that section.
- Give yourself enough time to take the Chapter Test uninterrupted. Then check the answers. For each problem you answered incorrectly, go to the Review Exercises and do all of the problems that are similar.
- To prepare for the final exam, complete the Cumulative Review Exercises at the end of each chapter, starting with Chapter 2. If you complete the cumulative reviews after finishing each chapter, then you will be preparing for the final throughout the course. The Cumulative Review Exercises are another excellent tool for helping you retain material.

4. Where to Go for Help

- At the first sign of trouble, see your instructor. Most instructors have specific office hours set aside to help students. Don't wait until after you have failed an exam to seek assistance.
- Get a tutor. Most colleges and universities have free tutoring available.
- When your instructor and tutor are unavailable, use the Student Solutions Manual for step-by-step solutions to the odd-numbered problems in the exercise sets.
- Work with another student from your class.
- Work on the computer. Many mathematics tutorial programs and websites are available on the Internet, including the one that accompanies this text: www.mhhe.com/miller_oneill

SET OF REAL NUMBERS

Many of the activities we perform every day are followed in a natural order. For example, we would not put on our shoes before pulling on our socks. Nor would a doctor begin surgery before giving an anesthetic.

In mathematics, it is also necessary to follow a prescribed order of operations to simplify an algebraic expression. For more information about the order of operations, visit calconline at

www.mhhe.com/miller_oneill

For example, the expression $24 - 6 \times 2$ is properly simplified by performing multiplication before subtraction:

$$24 - 6 \times 2 = 24 - 12 = 12$$

When parentheses are present, the expressions within parentheses must be performed first:

$$(24 - 6) \times 2 = 18 \times 2 = 36$$

In this chapter, the order of operations for simplifying algebraic expressions is discussed at length.

section

1.1   SETS OF NUMBERS AND THE REAL NUMBER LINE

1.   Real Number Line

The numbers we work with on a day-to-day basis are all part of the set of **real numbers**. The real numbers encompass zero, all positive, and all negative numbers, including those represented by fractions and decimal numbers. The set of real numbers can be represented graphically on a horizontal number line with a point labeled as 0. Positive real numbers are graphed to the right of 0, and negative real numbers are graphed to the left. Zero is neither positive nor negative. Each point on the number line corresponds to exactly one real number. For this reason, this number line is called the **real number line** (Figure 1-1).

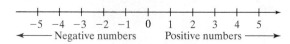

Figure 1-1

2.   Plotting Points on the Number Line

---

example 1

Plotting Points on the Real Number Line

Plot the points on the real number line that represent the following real numbers.

a. -3 b. $\dfrac{3}{2}$ c. -4.7 d. $\dfrac{16}{5}$

Solution:

a. Because -3 is negative, it lies three units to the left of zero.
b. The fraction $\frac{3}{2}$ can be expressed as the mixed number $1\frac{1}{2}$ which lies half-way between 1 and 2 on the number line.
c. The negative number -4.7 lies $\frac{7}{10}$ units to the left of -4 on the number line.
d. The fraction $\frac{16}{5}$ can be expressed as the mixed number $3\frac{1}{5}$ which lies $\frac{1}{5}$ unit to the right of 3 on the number line.

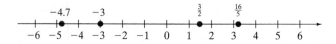

---

3.   Set of Real Numbers

In mathematics, a well-defined collection of elements is called a set. The symbols { } are used to enclose the elements of the set. For example, the set {A, B, C, D, E} represents the set of the first five letters of the alphabet.

Several sets of numbers are used extensively in algebra that are subsets (or part) of the set of real numbers. These are the

set of natural numbers
set of whole numbers
set of integers
set of rational numbers
set of irrational numbers

Definition of the Natural Numbers, Whole Numbers, and Integers

The set of **natural numbers** is $\{1, 2, 3, \ldots\}$
The set of **whole numbers** is $\{0, 1, 2, 3, \ldots\}$
The set of **integers** is $\{\ldots -3, -2, -1, 0, 1, 2, 3, \ldots\}$

Notice that the set of whole numbers includes the natural numbers. Therefore, every natural number is also a whole number. The set of integers includes the set of whole numbers. Therefore, every whole number is also an integer. The relationship among the elements of the natural numbers, whole numbers, and integers is shown in Figure 1-2.

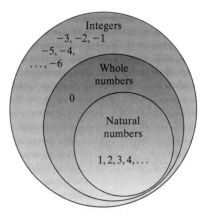

Figure 1-2

Fractions are also among the numbers we use frequently. A number that can be written as a fraction whose numerator is an integer and whose denominator is a nonzero integer is called a rational number.

Definition of the Rational Numbers

The set of **rational numbers** is the set of numbers that can be expressed in the form $\frac{p}{q}$, where both p and q are integers and q does not equal 0.

We also say that a rational number $\frac{p}{q}$ is a *ratio* of two integers, p and q, where q is not equal to zero.

---

example 2

Identifying Rational Numbers

Show that the following numbers are rational numbers by finding an equivalent ratio of two integers.

a. $-\dfrac{2}{3}$ b. -12 c. 0.5 d. $0.\overline{6}$

Classroom Activity 1.1A

Solution:

a. The fraction $-\frac{2}{3}$ is a rational number because it can be expressed as the ratio of -2 and 3.
b. The number -12 is a rational number because it can be expressed as the ratio of -12 and 1. That is, $-12 = \frac{-12}{1}$. In this example, we see that an integer is also a rational number.
c. The terminating decimal 0.5 is a rational number because it can be expressed as the ratio of 5 and 10. That is, $0.5 = \frac{5}{10}$. In this example we see that a terminating decimal is also a rational number.
d. The repeating decimal $0.\overline{6}$ is a rational number because it can be expressed as the ratio of 2 and 3. That is, $0.\overline{6} = \frac{2}{3}$. In this example we see that a repeating decimal is also a rational number.

Tip: Any rational number can be represented by a terminating decimal or by a repeating decimal.

Some real numbers, such as the number π, cannot be represented by the ratio of two integers. In decimal form, an irrational number is a nonterminating, nonrepeating decimal. The value of π, for example, can be approximated as $\pi \approx 3.1415926535897932$. However, the decimal digits continue indefinitely with no repeated pattern. Other examples of irrational numbers are the square roots of nonperfect squares, such as $\sqrt{3}$ (read as "the positive square root of 3"). The expression $\sqrt{3}$ is a number that when multiplied by itself equals 3.

Definition of the Irrational Numbers

The set of **irrational numbers** is the set of real numbers that are not rational.

Note: An irrational number cannot be written as a terminating decimal or as a repeating decimal.

The set of real numbers consists of both the rational and the irrational numbers. The relationship among these important sets of numbers is illustrated in Figure 1-3:

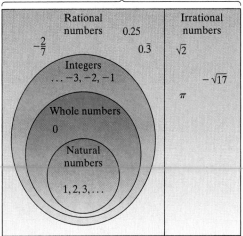

Figure 1-3

---

example 3   **Classifying Numbers by Set**

Check the set(s) to which each number belongs. The numbers may belong to more than one set.

	Natural Numbers	Whole Numbers	Integers	Rational Numbers	Irrational Numbers	Real Numbers
5						
$-\dfrac{47}{3}$						
1.48						
$\sqrt{7}$						
0						

Solution:

	Natural Numbers	Whole Numbers	Integers	Rational Numbers	Irrational Numbers	Real Numbers
5	✔	✔	✔	✔ (ratio of 5 and 1)		✔
$-\dfrac{47}{3}$				✔ (ratio of −47 and 3)		✔
1.48				✔ (ratio of 148 and 100)		✔
$\sqrt{7}$					✔	✔
0		✔	✔	✔ (ratio of 0 and 1)		✔

4.   Inequalities

The relative size of two real numbers can be compared using the real number line. Suppose a and b represent two real numbers. We say that a is less than b, denoted $a < b$, if a lies to the left of b on the number line.

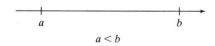

$$a < b$$

We say that a is greater than b, denoted $a > b$, if a lies to the right of b on the number line.

$$a > b$$

Table 1-1 summarizes the relational operators that compare two real numbers a and b.

Table 1-1	
Mathematical Expression	**Translation**
$a < b$	a is less than b.
$a > b$	a is greater than b.
$a \le b$	a is less than or equal to b.
$a \ge b$	a is greater than or equal to b.
$a = b$	a is equal to b.
$a \ne b$	a is not equal to b.
$a \approx b$	a is approximately equal to b.

The symbols $<$, $>$, $\le$, $\ge$, and $\ne$ are called inequality signs, and the expressions $a < b$, $a > b$, $a \le b$, $a \ge b$, and $a \ne b$ are called **inequalities**.

example 4   **Ordering Real Numbers**

The average temperatures (in degrees Celsius) for selected cities in the United States and Canada in January are shown in Table 1-2.

Table 1-2	
City	**Temp (°C)**
Prince George, British Columbia	−12.1
Corpus Christi, Texas	13.4
Parkersburg, West Virginia	−0.9
San Jose, California	9.7
Juneau, Alaska	−5.7
New Bedford, Massachusetts	−0.2
Durham, North Carolina	4.2

a. Plot a point on the real number line representing the temperature of each city.
b. Then compare the temperatures between the following cities and fill in the blank with the appropriate inequality sign: $<$ or $>$.

Solution:

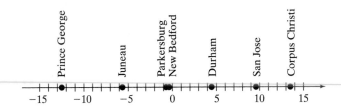

a. Temperature of San Jose $\boxed{<}$ temperature of Corpus Christi
b. Temperature of Juneau $\boxed{>}$ temperature of Prince George
c. Temperature of Parkersburg $\boxed{<}$ temperature of New Bedford
d. Temperature of Parkersburg $\boxed{>}$ temperature of Prince George

5. Opposite of a Real Number

To gain mastery of any algebraic skill, it is necessary to know the meaning of key definitions and key symbols. Two important definitions are the opposite of a real number and the absolute value of a real number.

Definition of the Opposite of a Real Number

Two numbers that are the same distance from 0 but on opposite sides of 0 on the number line are called **opposites** of each other. Symbolically, we denote the opposite of a real number a as $-a$.

example 5

Finding the Opposite of a Real Number

a. Find the opposite of 5.
b. Find the opposite of $-\frac{4}{7}$.
c. Evaluate $-(0.46)$.
d. Evaluate $-(-\frac{11}{3})$.

Solution:

a. The opposite of 5 is -5.

b. The opposite of $-\dfrac{4}{7}$ is $\dfrac{4}{7}$.

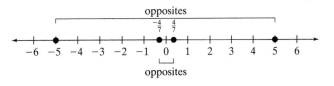

c. $-(0.46) = -0.46$ The expression $-(0.46)$ represents the opposite of 0.46.

d. $-\left(-\dfrac{11}{3}\right) = \dfrac{11}{3}$ The expression $-(-\frac{11}{3})$ represents the opposite of $-\frac{11}{3}$.

6.   Absolute Value of a Real Number

The concept of absolute value will be used to define the addition of real numbers in Section 1.3.

Informal Definition of the Absolute Value of a Real Number

The **absolute value** of a real number a, denoted $|a|$, is the distance between a and 0 on the number line.

Note: The absolute value of any real number is nonnegative.

For example, $|3| = 3$ and $|-3| = 3$.

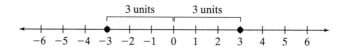

example 6   **Finding the Absolute Value of a Real Number**

Evaluate the absolute value expressions.

a. $|-4|$ b. $\left|\frac{1}{2}\right|$ c. $|-6.2|$ d. $|0|$

Classroom Activity 1.1B

Solution:

a. $|-4| = 4$ -4 is 4 units from 0 on the number line.

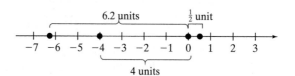

b. $\left|\frac{1}{2}\right| = \frac{1}{2}$ $\frac{1}{2}$ is $\frac{1}{2}$ unit from 0 on the number line.

c. $|-6.2| = 6.2$ -6.2 is 6.2 units from 0 on the number line.

d. $|0| = 0$ 0 is 0 units from 0 on the number line.

The absolute value of a number a is its distance from zero on the number line. The definition of $|a|$ may also be given symbolically depending on whether a is negative or nonnegative.

Definition of the Absolute Value of a Real Number

Let a be a real number. Then

1. If a is nonnegative (that is, $a \geq 0$), then $|a| = a$.
2. If a is negative (that is, $a < 0$), then $|a| = -a$.

This definition states that if a is a nonnegative number, then $|a|$ equals a itself. If a is a negative number, then $|a|$ equals the opposite of a. For example,

$|9| = 9$ Because 9 is positive, then $|9|$ equals the number 9 itself.

$|-7| = 7$ Because -7 is negative, then $|-7|$ equals the opposite of -7 which is 7.

Calculator Connections

Scientific and graphing calculators approximate irrational numbers by using rational numbers in the form of terminating decimals. For example, consider approximating π and $\sqrt{3}$:

Scientific Calculator:

Enter: [π] (or [2nd] [π])   **Result:** 3.141592654

Enter: [3] [$\sqrt{}$]   **Result:** 1.732050808

Graphing Calculator:

Enter: [2nd] [π] [ENTER]

Enter: [2nd] [$\sqrt{}$] [3] [ENTER]

```
π
              3.141592654
√(3
              1.732050808
```

Note that when writing approximations, we use the symbol, $\approx$.

$$\pi \approx 3.141592654 \quad \text{and}$$
$$\sqrt{3} \approx 1.73205808$$

---

section 1.1   PRACTICE EXERCISES

1. Plot the numbers on a real number line: $\{1, -2, -\pi, 0, -\frac{5}{2}, 5.1\}$ ❖

2. Plot the numbers on a real number line: $\{3, -4, \frac{1}{8}, -1.7, -\frac{4}{3}, 1.75\}$ ❖

For Exercises 3–22, describe each number as (a) a terminating decimal, (b) a repeating decimal, or (c) a nonterminating, nonrepeating decimal.

3. 0.29 a
4. 3.8 a
5. $\frac{1}{9}$ b
6. $\frac{1}{3}$ b
7. $\frac{1}{8}$ a
8. $\frac{1}{5}$ a
9. 5 a
10. 2 a
11. 2π c
12. 3π c
13. -0.125 a
14. -3.24 a
15. -3 a
16. -6 a
17. $\frac{7}{20}$ a
18. $\frac{5}{8}$ a

19. $0.\overline{2}$ b   20. $0.\overline{6}$ b   21. $\sqrt{6}$ c   22. $\sqrt{10}$
c

23. List all of the numbers from Exercises 3–22 that are rational numbers.
$0.29, 3.8, \frac{1}{9}, \frac{1}{3}, \frac{1}{8}, \frac{1}{5}, 5, 2, -0.125, -3.24, -3, -6, \frac{7}{20}, \frac{5}{8}, 0.\overline{2}, 0.\overline{6}$

24. List all of the numbers from Exercises 3–22 that are irrational numbers. $2\pi, 3\pi, \sqrt{6}, \sqrt{10}$

25. Describe the set of natural numbers. $\{1, 2, 3, 4, \ldots\}$

26. Describe the set of rational numbers.   The set of all numbers $\frac{p}{q}$ such that p and q are integers and q does not equal zero

27. Describe the set of irrational numbers.
The set of all real numbers that are not rational

28. Describe the set of whole numbers. $\{0, 1, 2, 3, 4, \ldots\}$

29. Describe the set of real numbers.   The set of all numbers that includes all rational and irrational numbers

30. Describe the set of integers.
$\{\ldots -3, -2, -1, 0, 1, 2, 3, \ldots\}$

31. List three numbers that are real numbers but not rational numbers.   For example: $\pi, -\sqrt{2}, \sqrt{3}$

32. List three numbers that are real numbers but not irrational numbers.   For example: $-\frac{1}{2}, 5, 0.\overline{3}$

33. List three numbers that are integers but not natural numbers.   For example: $-4, -1, 0$

34. List three numbers that are integers but not whole numbers.   For example: $-5, -2, -1$

35. List three numbers that are rational but not natural numbers.   For example: $-2, \frac{1}{2}, 0$

36. List three numbers that are rational but not integers.   For example: $-\frac{3}{4}, \frac{1}{2}, 0.206$

For Exercises 37–43, let $A = \{-\frac{3}{2}, \frac{3}{0}, \sqrt{11}, -4, 0.\overline{6}, \frac{0}{5}, \sqrt{7}, 1\}$

37. Are all of the numbers in set A real numbers?
No, $\frac{3}{0}$ is undefined.

38. List all of the rational numbers in set A. $-\frac{3}{2}, -4, 0.\overline{6}, \frac{0}{5}, 1$

39. List all of the whole numbers in set A. $\frac{0}{5}, 1$

40. List all of the natural numbers in set A.   1

41. List all of the irrational numbers in set A.
$\sqrt{11}, \sqrt{7}$

42. List all of the integers in set A. $-4, \frac{0}{5}, 1$

43. Plot the real numbers of set A on a number line.
❖

44. The elevations of selected cities in the United States are shown in the figure. Plot a point on

the real number line representing the elevation of each city. Then compare the elevations and fill in the blank with the appropriate inequality sign: $<$ or $>$. (A negative number indicates that the city is below sea level.)   ❖

Figure for Exercise 44

a. Elevation of Tucson ___$>$___ elevation of Cincinnati.

b. Elevation of New Orleans ___$<$___ elevation of Chicago.

c. Elevation of New Orleans ___$<$___ elevation of Houston.

d. Elevation of Chicago ___$>$___ elevation of Cincinnati.

45. The elevations of selected cities in the United States are given in the table. Plot a point on the real number line representing the elevation of each city. Then compare the elevations and fill in the blank with the appropriate inequality sign: $<$ or $>$.   ❖

City	Elevation (in feet)*
Dallas, TX	390
Kansas City, MO	720
Long Beach, CA	−7
Denver, CO	5130
Philadelphia, PA	0

Table for Exercise 45

*A negative number indicates that the city is below sea level.

a. Elevation of Kansas City ___$>$___ elevation of Dallas.

b. Elevation of Philadelphia ___$<$___ elevation of Kansas City.

c.   Elevation of Long Beach __<__ elevation of Philadelphia.

d.   Elevation of Dallas __<__ elevation of Denver.

46.   The LPGA Samsung World Championship of women's golf scores are given in the table. Plot a point on the real number line representing the score of each golfer, then compare the scores and fill in the blank with the appropriate inequality sign: < or >. ❖

LPGA Golfers	Final Score with Respect to Par
Annika Sorenstam	7
Laura Davies	−4
Lorie Kane	0
Cindy McCurdy	3
Se Ri Pak	−8

Table for Exercise 46

a.   Kane's score __>__ Pak's score.

b.   Sorenstam's score __>__ Davies' score.

c.   Pak's score __<__ McCurdy's score.

d.   Kane's score __>__ Davies' score.

47.   The LPGA Samsung World Championship of women's golf scores are given in the following table. Plot a point on the real number line representing the score of each golfer, then compare the scores and fill in the blank with the appropriate inequality sign: < or >. ❖

LPGA Golfers	Final Score with Respect to Par
Akiko Fukushima	−6
Juli Inkster	2
Karrie Webb	−7
Lorie Kane	0
Meg Mallon	−3

Table for Exercise 47

a.   Fukushima's score __<__ Mallon's score.

b.   Kane's score __>__ Webb's score.

c.   Inkster's score __>__ Mallon's score.

d.   Fukushima's __>__ Webb's score.

For Exercises 48–55, write the opposite and the absolute value for each number.

48.   18   49.   2   50.   −6.1   51.   −2.5
      −18, 18   −2, 2   6.1, 6.1   2.5, 2.5

52. $-\dfrac{5}{8}$ $\dfrac{5}{8}, \dfrac{5}{8}$ 53. $-\dfrac{1}{3}$ $\dfrac{1}{3}, \dfrac{1}{3}$ 54. $\dfrac{7}{3}$ $-\dfrac{7}{3}, \dfrac{7}{3}$ 55. $\dfrac{1}{9}$ $-\dfrac{1}{9}, \dfrac{1}{9}$

The opposite of a is denoted as $-a$. For Exercises 56–59, evaluate the opposites.

56. $-(-3)$ 3   57. $-(-5.1)$ 5.1

58. $-\left(-\dfrac{7}{3}\right)$ $\dfrac{7}{3}$ 59. $-(-7)$ 7

For Exercises 60–61, answer true or false. If a statement is false, explain why.

60.   If n is positive, then $|n|$ is negative.
      False, $|n|$ is never negative.
61.   If m is negative, then $|m|$ is negative.
      False, $|m|$ is never negative.

For Exercises 62–85, determine if the statements are true or false. Use the real number line to justify the answer.

62. $5 > 2$ True   63. $8 < 10$ True

64. $6 < 6$ False   65. $19 > 19$ False

66. $-7 \geq -7$ True   67. $-1 \leq -1$ True

68. $2 \leq 6$ True   69. $-5 \geq 0$ False

70. $-5 > -2$ False   71. $6 < -10$ False

72. $8 \neq 8$ False   73. $10 \neq 10$ False

74. $|-2| \geq |-1|$ True   75. $|3| \leq |-1|$ False

76. $|-5| = |5|$ True   77. $|-3| = |3|$ True

78. $|7| \neq |-7|$ False   79. $|-13| \neq |13|$ False

80. $-1 < |-1|$ True   81. $-6 < |-6|$ True

82. $|-8| \geq |8|$ True   83. $|-11| \geq |11|$ True

84. $|-2| \leq |2|$ True   85. $|-21| \leq |21|$ True

■ EXPANDING YOUR SKILLS

86.   For what numbers, a, is $-a$ positive?   For all $a < 0$

87.   For what numbers, a, is $|a| = a$?   For all $a \geq 0$

❖ See Additional Answers Appendix   ✎ Writing   ⇔ Translating Expression   Geometry   Scientific Calculator   Video

section

1.2   ORDER OF OPERATIONS

1.   Variables and Expressions

A **variable** is a symbol or letter such as x, y, and z, used to represent an unknown number. **Constants** are values that are not variable such as the numbers 3, -1.5, $\frac{2}{7}$, and π. An algebraic **expression** is a collection of variables and constants under algebraic operations. For example, $\frac{3}{x}$, $y + 7$, and $t - 1.4$ are algebraic expressions.

The symbols used to show the four basic operations of addition, subtraction, multiplication, and division are summarized in Table 1-3.

Table 1-3		
Operation	**Symbols**	**Translation**
Addition	$a + b$	**sum** of a and b a plus b b added to a b more than a a increased by b the total of a and b
Subtraction	$a - b$	**difference** of a and b a minus b b subtracted from a a decreased by b b less than a
Multiplication	$a \times b, a \cdot b, a(b), (a)b, (a)(b), ab$ (*Note*: We rarely use the notation $a \times b$ because the symbol, $\times$, might be confused with the variable, x.)	**product** of a and b a times b a multiplied by b
Division	$a \div b, \dfrac{a}{b}, a/b, b\overline{)a}$	**quotient** of a and b a divided by b b divided into a ratio of a and b a over b a per b

2.   Evaluating Algebraic Expressions

The value of an algebraic expression depends on the values of the variables within the expression.

example 1   Evaluating an Algebraic Expression

Evaluate the algebraic expression when $p = 4$ and $q = \frac{3}{4}$.

a. $100 - p$ b. $8q$

Classroom Activity 1.2A

Solution:

a. $100 - p$

   $100 - (\ \ )$ When substituting a number for a variable, use parentheses.

   $= 100 - (4)$ Substitute $p = 4$ in the parentheses.

   $= 96$ Subtract.

b. $8q$

   $= 8(\ \ )$ When substituting a number for a variable, use parentheses.

   $= 8\left(\dfrac{3}{4}\right)$ Substitute $q = \frac{3}{4}$ in the parentheses.

   $= \dfrac{8}{1} \cdot \dfrac{3}{4}$ Write the whole number as a fraction.

   $= \dfrac{24}{4}$ Multiply fractions.

   $= 6$ Reduce.

3.   Exponential Expressions

In algebra, repeated multiplication can be expressed using exponents. The expression, $4 \cdot 4 \cdot 4$ can be written as

In the expression 4^3, 4 is the base, and 3 is the exponent, or power. The exponent indicates how many factors of the base to multiply.

Definition of b^n

Let b represent any real number and n represent a positive integer. Then,

$$b^n = \underbrace{b \cdot b \cdot b \cdot b \ldots \cdot b}_{n\text{-factors of } b}$$

b^n is read as "b to the nth power."
b is called the **base** and n is called the **exponent**, or **power**.
b^2 is read as "b squared" and b^3 is read as "b cubed."

---

example 2   **Evaluating Exponential Expressions**

Translate the expression into words and then evaluate the expression.

a. 2^5 b. 5^2 c. $\left(\dfrac{3}{4}\right)^3$ d. 1^6

Solution:

a. The expression 2^5 is read as "two to the fifth power"
   $2^5 = (2)(2)(2)(2)(2) = 32.$

b. The expression 5^2 is read as "five to the second power" or "five, squared"
   $5^2 = (5)(5) = 25.$

c. The expression $\left(\frac{3}{4}\right)^3$ is read as "three-fourths to the third power" or "three-fourths, cubed"

$$\left(\frac{3}{4}\right)^3 = \left(\frac{3}{4}\right)\left(\frac{3}{4}\right)\left(\frac{3}{4}\right) = \frac{27}{64}$$

d. The expression 1^6 is read as "one to the sixth power"
   $1^6 = (1)(1)(1)(1)(1)(1) = 1.$

4. Square Roots

The reverse operation to squaring a number is to find its **square roots**. For example, finding a square root of 9 is equivalent to asking "what number(s) when squared equals 9?" The symbol, $\sqrt{\ }$, (called a radical sign) is used to find the *principal* square root of a number. By definition, the principal square root of a number is nonnegative. Therefore, $\sqrt{9}$, is the nonnegative number that when squared equals 9. Hence $\sqrt{9} = 3$ because 3 is nonnegative and $(3)^2 = 9$. Several more examples follow:

$$\sqrt{64} = 8 \qquad \text{Because } (8)^2 = 64$$
$$\sqrt{121} = 11 \qquad \text{Because } (11)^2 = 121$$
$$\sqrt{0} = 0 \qquad \text{Because } (0)^2 = 0$$

Tip: To simplify square roots, it is advisable to become familiar with the following squares and square roots.

$0^2 = 0 \rightarrow \sqrt{0} = 0$	$7^2 = 49 \rightarrow \sqrt{49} = 7$
$1^2 = 1 \rightarrow \sqrt{1} = 1$	$8^2 = 64 \rightarrow \sqrt{64} = 8$
$2^2 = 4 \rightarrow \sqrt{4} = 2$	$9^2 = 81 \rightarrow \sqrt{81} = 9$
$3^2 = 9 \rightarrow \sqrt{9} = 3$	$10^2 = 100 \rightarrow \sqrt{100} = 10$
$4^2 = 16 \rightarrow \sqrt{16} = 4$	$11^2 = 121 \rightarrow \sqrt{121} = 11$
$5^2 = 25 \rightarrow \sqrt{25} = 5$	$12^2 = 144 \rightarrow \sqrt{144} = 12$
$6^2 = 36 \rightarrow \sqrt{36} = 6$	$13^2 = 169 \rightarrow \sqrt{169} = 13$

Classroom Activity 1.2B

5. Order of Operations

When algebraic expressions contain numerous operations, it is important to evaluate the operations in the proper order. Parentheses ( ), brackets [ ], and braces { } are used for grouping numbers and algebraic expressions. It is important to recognize that operations must be done within parentheses and other grouping symbols first. Other grouping symbols include absolute value bars, radical signs, and fraction bars.

Order of Operations

1. Simplify expressions within parentheses and other grouping symbols first. These include absolute value bars, fraction bars, and radicals. If imbedded parentheses are present, start with the innermost parenthesis.
2. Evaluate expressions involving exponents and radicals.
3. Perform multiplication or division in the order that they occur from left to right.
4. Perform addition or subtraction in the order that they occur from left to right.

example 3

Applying the Order of Operations

Simplify the expressions.

a. $17 - 3 \cdot 2 + 2^2$

b. $\dfrac{1}{2}\left(\dfrac{5}{6} - \dfrac{3}{4}\right)$

c. $25 - 12 \div 3 \cdot 4$

d. $6.2 - |-2.1| + \sqrt{15 - 6}$

e. $28 - 2[(6 - 3)^2 + 4]$

f. $\dfrac{\sqrt{\frac{1}{4}} + \frac{3}{2}}{\frac{1}{2} - \frac{2}{5}}$

Solution:

a. $17 - 3 \cdot 2 + \underline{2^2}$

$= 17 - 3 \cdot 2 + 4$ Simplify exponents.

$= 17 - 6 + 4$ Multiply before adding or subtracting.

$= 11 + 4$ Add and subtract from left to right.

$= 15$

b. $\dfrac{1}{2}\left(\dfrac{5}{6} - \dfrac{3}{4}\right)$ Subtract fractions within the parentheses.

$= \dfrac{1}{2}\left(\dfrac{10}{12} - \dfrac{9}{12}\right)$ The least common denominator is 12.

$= \dfrac{1}{2}\left(\dfrac{1}{12}\right)$

$= \dfrac{1}{24}$ Multiply fractions.

c. $25 - 12 \div 3 \cdot 4$ Multiply or divide in order from left to right.

$= 25 - 4 \cdot 4$ Notice that the operation $12 \div 3$ is performed first (not $3 \cdot 4$).

$= 25 - 16$ Multiply $4 \cdot 4$ before subtracting.

$= 9$ Subtract.

d. $6.2 - |-2.1| + \sqrt{15 - 6}$

$= 6.2 - |-2.1| + \sqrt{9}$ Simplify within the square root.

$= 6.2 - (2.1) + 3$ Simplify the square root and absolute value.

$= 4.1 + 3$ Add or subtract from left to right.

$= 7.1$ Add.

e. $28 - 2[(6 - 3)^2 + 4]$

$= 28 - 2[(3)^2 + 4]$ Simplify within the inner parentheses first.

$= 28 - 2[(9) + 4]$ Simplify exponents.

$= 28 - 2[13]$ Add within the square brackets.

$= 28 - 26$ Multiply before subtracting.

$= 2$ Subtract.

f. $\dfrac{\sqrt{\frac{1}{4}} + \frac{3}{2}}{\frac{1}{2} - \frac{2}{5}}$ The fraction bar is an implied grouping symbol.

$$\frac{\left(\sqrt{\frac{1}{4}} + \frac{3}{2}\right)}{\left(\frac{1}{2} - \frac{2}{5}\right)}$$

Simplify numerator and denominator separately, then divide.

$= \dfrac{\frac{1}{2} + \frac{3}{2}}{\frac{1}{2} - \frac{2}{5}}$ Simplify the square root $\sqrt{\dfrac{1}{4}} = \dfrac{1}{2}$

$= \dfrac{\dfrac{4}{2}}{\dfrac{1 \cdot 5}{2 \cdot 5} - \dfrac{2 \cdot 2}{5 \cdot 2}}$ Add fractions in the numerator. The LCD for denominator is 10. Multiply each individual fraction by the factors missing from the denominator of the LCD.

$= \dfrac{2}{\dfrac{5}{10} - \dfrac{4}{10}}$

$= \dfrac{2}{\dfrac{1}{10}}$ Subtract fractions in the denominator.

$= 2 \div \dfrac{1}{10}$ Divide the fractions.

$= 2 \cdot \dfrac{10}{1}$ Multiply the first number by the reciprocal of the second number.

$= 20$

6. Translations

example 4

Translating from English Form to Algebraic Form

Translate each English phrase to an algebraic expression.

a. The quotient of x and 5
b. The difference of p and the square root of q
c. Seven less than n
d. Eight more than the absolute value of w

◆ **Avoiding Mistakes**

Recall that "a less than b" is translated as $b - a$. Therefore, the statement "seven less than n" must be translated as $n - 7$, not $7 - n$.

Solution:

a. $\dfrac{x}{5}$ or $x \div 5$
b. $p - \sqrt{q}$
c. $n - 7$
d. $|w| + 8$

example 5

Translating from English Form to Algebraic Form

Translate each English phrase into an algebraic expression. Then evaluate the expression for $a = 6$, $b = 4$, and $c = 20$.

a. The product of a and the square root of b
b. Twice the sum of b and c

Solution:

a. $a\sqrt{b}$ Translate.

    $= (\ )\sqrt{(\ )}$ Use parentheses to substitute a number for a variable.

    $= (6)\sqrt{(4)}$ Substitute $a = 6$ and $b = 4$.

    $= 6 \cdot 2$ Simplify the radical first.

    $= 12$ Multiply.

◆ **Avoiding Mistakes**

To compute "twice the sum of b and c" it is necessary to take the sum first and then multiply by 2. To ensure the proper order, the sum of b and c must be enclosed in parentheses. The proper translation is:

$2(b + c)$

b. $2(b + c)$ Translate.

    $= 2((\ ) + (\ ))$ Use parentheses to substitute a number for a variable.

    $= 2((4) + (20))$ Substitute $b = 4$ and $c = 20$.

    $= 2(24)$ Simplify within the parentheses first.

    $= 48$ Multiply.

Calculator Connections

On a calculator, we enter exponents higher than the second power by using the key labeled y^x or $\wedge$. For example, evaluate 2^4 and 10^6:

Scientific Calculator:

Enter: [2] $[y^x]$ [4] [=]   **Result:** 16

Enter: [10] $[y^x]$ [6] [=]   **Result:** 1000000

Graphing Calculator:

Most calculators also have the capability to enter several operations at once. However, it is important to note that fraction bars and radicals require user-defined parentheses to ensure that the proper order of operations is followed. For example, evaluate the following expressions on a calculator:

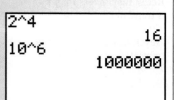

a. $130 - 2(5 - 1)^3$ b. $\dfrac{18 - 2}{11 - 9}$ c. $\sqrt{25 - 9}$

Scientific Calculator:

Enter: [130] [−] [2] [×] [(] [5] [−] [1] [)] $[y^x]$ [3] [=]   **Result:** 2

Enter: [(] [18] [−] [2] [)] [÷] [(] [11] [−] [9] [)] [=]   **Result:** 8

Enter: [(] [25] [−] [9] [)] $[\sqrt{\ }]$ **Result:** 4

Graphing Calculator:

```
130-2*(5-1)^3
                    2
(18-2)/(11-9)
                    8
√(25-9)
                    4
```

Calculator Exercises

Simplify the expression without the use of a calculator. Then enter the expression into the calculator to verify your answer. ❖

1. $\dfrac{4 + 6}{8 - 3}$ 2   2. $110 - 5(2 + 1) - 4$ 91   3. $100 - 2(5 - 3)^3$ 84

4. $3 + (4 - 1)^2$ 12   5. $(12 - 6 + 1)^2$ 49   6. $3 \cdot 8 - \sqrt{32 + 2^2}$ 18

7. $\sqrt{18 - 2}$ 4   8. $(4 \cdot 3 - 3 \cdot 3)^3$ 27   9. $\dfrac{20 - 3^2}{26 - 2^2}$ 0.5

section 1.2	PRACTICE EXERCISES

For Exercises 1–4, let
$C = \{7, \frac{1}{3}, 0, -2, -\sqrt{9}, \pi, 3.25, -0.\overline{1}, \sqrt{5}\}$ 1.1

1. Plot the integers from set C on a number line. ❖

2. Plot the whole numbers from set C on a number line. ❖

3. Plot the irrational numbers from set C on a number line. ❖

4. Plot the rational numbers from set C on a number line. ❖

5. The high temperatures (in degrees Celsius, °C) in Montreal, Quebec, Canada, for a week in January are given in the table.   1.1

Day	High (°C)
Sunday	−8
Monday	−9
Tuesday	−12
Wednesday	−8
Thursday	−9
Friday	−8
Saturday	−7

Table for Exercise 5

Fill in the blank with the appropriate symbol ($>$, $<$, $=$).

a. Temperature on Tuesday __<__ temperature on Friday.

b. Temperature on Wednesday __>__ temperature on Monday.

c. Temperature on Saturday __>__ temperature on Thursday.

d. Temperature on Tuesday __<__ temperature on Saturday.

6. The high temperatures for Edmonton, Alberta, Canada, for a week in January are given in the table.   1.1

Day	High (°C)
Sunday	1
Monday	2
Tuesday	2
Wednesday	5
Thursday	−2
Friday	−3
Saturday	−3

Table for Exercise 6

Fill in the blank with the appropriate symbol ($>$, $<$, $=$).

a. Temperature on Tuesday __>__ temperature on Friday.

b. Temperature on Wednesday __>__ temperature on Monday.

c. Temperature on Saturday __<__ temperature on Thursday.

d. Temperature on Friday __=__ temperature on Saturday.

For Exercises 7–18, evaluate the expressions for the given substitutions.

7. $y - 3$ when $y = 18$ 15

8. $3q$ when $q = 5$ 15

9. $\dfrac{15}{t}$ when $t = 5$ 3

10. $8 + w$ when $w = 12$ 20

11. $2c - 3$ when $c = 4$ 5

12. $8x + 2$ when $x = \dfrac{3}{4}$ 8

13. $5 + 6d$ when $d = \dfrac{2}{3}$ 9

14. $\dfrac{6}{5}h - 1$ when $h = 10$ 11

15. $p^2 + \dfrac{2}{9}$ when $p = \dfrac{2}{3}$ $\dfrac{2}{3}$

❖ See Additional Answers Appendix   Writing   ⬌ Translating Expression   Geometry   Scientific Calculator   Video

16. $z^3 - \dfrac{2}{27}$ when $z = \dfrac{2}{3}$ $\dfrac{2}{9}$

17. $5(x + 2.3)$ when $x = 1.1$ 17

18. $3(2.1 - y)$ when $y = 0.5$ 4.8

19. The area of a rectangle may be computed as $A = \ell w$, where ℓ is the length of the rectangle and w is the width. Find the area for the rectangular field.   57,600 ft^2

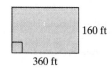

160 ft

360 ft

Figure for Exercise 19

20. The perimeter of the rectangular field from Exercise 19 may be computed as $P = 2\ell + 2w$. Find the perimeter.   1040 ft

21. The area of a trapezoid is given by $A = \frac{1}{2}(b_1 + b_2)h$, where b_1 and b_2 are the lengths of the two parallel sides and h is the height. Find the area of the window shown in the figure.   21 ft^2

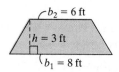

$b_2 = 6$ ft

$h = 3$ ft

$b_1 = 8$ ft

Figure for Exercise 21

22. The volume of a rectangular solid is given by $V = \ell wh$, where ℓ is the length of the box, w is the width, and h is the height. Find the volume of the box shown in the figure.   1000 yd^3

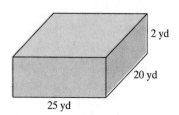

2 yd

20 yd

25 yd

Figure for Exercise 22

23. a. For the expression $5x^3$, what is the base for the exponent 3? x

    b. Does 5 have an exponent? If so, what is it?
    Yes, 1

24. a. For the expression $2y^4$, what is the base for the exponent 4? y

    b. Does 2 have an exponent? If so, what is it?
    Yes, 1

For Exercises 25–32, write each of the products using exponents.

25. $\dfrac{1}{6} \cdot \dfrac{1}{6} \cdot \dfrac{1}{6} \cdot \dfrac{1}{6}$ $\left(\dfrac{1}{6}\right)^4$

26. $10 \cdot 10 \cdot 10 \cdot 10 \cdot 10 \cdot 10$ 10^6

27. $a \cdot a \cdot a \cdot b \cdot b$ a^3b^2 28. $7 \cdot x \cdot x \cdot y \cdot y$ $7x^2y^2$

29. $5c \cdot 5c \cdot 5c \cdot 5c \cdot 5c$ $(5c)^5$

30. $3 \cdot w \cdot z \cdot z \cdot z \cdot z$ $3wz^4$

31. $8 \cdot y \cdot x \cdot x \cdot x \cdot x \cdot x \cdot x$ $8yx^6$ 32. $\dfrac{2}{3}t \cdot \dfrac{2}{3}t \cdot \dfrac{2}{3}t$ $\left(\dfrac{2}{3}t\right)^3$

For Exercises 33–40, write each expression in expanded form using the definition of an exponent.

33. x^3 $x \cdot x \cdot x$ 34. y^4 $y \cdot y \cdot y \cdot y$ 35. $(2b)^3$ $2b \cdot 2b \cdot 2b$ 36. $(8c)^2$ $8c \cdot 8c$

37. $10y^5$ $10 \cdot y \cdot y \cdot y \cdot y \cdot y$ 38. x^2y^3 $x \cdot x \cdot y \cdot y \cdot y$ 39. $2wz^2$ $2 \cdot w \cdot z \cdot z$ 40. $3a^3b$ $3a \cdot a \cdot a \cdot b$

For Exercises 41–48, simplify the expressions.

41. 5^2 25   42. 4^3 64   43. $\left(\dfrac{1}{7}\right)^2$ $\dfrac{1}{49}$ 44. $\left(\dfrac{1}{2}\right)^5$ $\dfrac{1}{32}$

45. $(0.25)^3$ 0.015625   46. $(0.8)^2$ 0.64   47. 2^6 64   48. 13^2 169

For Exercises 49–56, simplify the square roots.

49. $\sqrt{81}$ 9   50. $\sqrt{64}$ 8   51. $\sqrt{4}$ 2   52. $\sqrt{9}$ 3

53. $\sqrt{100}$ 10   54. $\sqrt{49}$ 7   55. $\sqrt{16}$ 4   56. $\sqrt{36}$ 6

For Exercises 57–84, use the order of operations to simplify the expressions.

57. $8 + 2 \cdot 6$ 20   58. $7 + 3 \cdot 4$ 19

59. $(8 + 2)6$ 60   60. $(7 + 3)4$ 40

61. $4 + 2 \div 2 \cdot 3 + 1$ 8

62. $5 + 6 \cdot 2 \div 4 - 1$ 7

63. $\dfrac{1}{4} \cdot \dfrac{2}{3} - \dfrac{1}{6}$ 0   64. $\dfrac{3}{4} \cdot \dfrac{2}{3} + \dfrac{2}{3}$ $\dfrac{7}{6}$

65. $\dfrac{9}{8} - \dfrac{1}{3} \cdot \dfrac{3}{4}$ $\dfrac{7}{8}$

66. $\dfrac{11}{6} - \dfrac{3}{8} \cdot \dfrac{4}{3}$ $\dfrac{4}{3}$

67. $3[5 + 2(8 - 3)]$ 45

68. $2[4 + 3(6 - 4)]$ 20

69. $10 + |-6|$ 16

70. $18 + |-3|$ 21

71. $21 - |8 - 2|$ 15

72. $12 - |6 - 1|$ 7

73. $2^2 + \sqrt{9} \cdot 5$ 19

74. $3^2 + \sqrt{16} \cdot 2$ 17

75. $\sqrt{9 + 16} - 2$ 3

76. $\sqrt{36 + 13} - 5$ 2

77. $\dfrac{7 + 3(8 - 2)}{(7 + 3)(8 - 2)}$ $\dfrac{5}{12}$

78. $\dfrac{16 - 8 \div 4}{4 + 8 \div 4 - 2}$ $\dfrac{7}{2}$

79. $\dfrac{15 - 5(3 \cdot 2 - 4)}{10 - 2(4 \cdot 5 - 16)}$ $\dfrac{5}{2}$

80. $\dfrac{5(7 - 3) + 8(6 - 4)}{4[7 + 3(2 \cdot 9 - 8)]}$ $\dfrac{9}{37}$

81. $[4^2 \cdot (6 - 4) \div 8] + [7 \cdot (8 - 3)]$ 39

82. $(18 \div \sqrt{4}) \cdot \{[(9^2 - 1) \div 2] - 15\}$ 225

83. $48 - 13 \cdot 3 + [(50 - 7 \cdot 5) + 2]$ 26

84. $80 \div 16 \cdot 2 + (6^2 - |-2|)$ 44

For Exercises 85–98, translate each English phrase into an algebraic expression.

85. The product of 3 and x $3x$

86. The sum of b and 6 $b + 6$

87. The quotient of x and 7 $\dfrac{x}{7}$ or $x \div 7$

88. Four divided by k $\dfrac{4}{k}$ or $4 \div k$

89. The difference of 2 and a $2 - a$

90. Three subtracted from t $t - 3$

91. x more than twice y $2y + x$

92. Nine decreased by the product of 3 and p $9 - 3p$

93. Four times the sum of x and 12 $4(x + 12)$

94. Twice the difference of x and 3 $2(x - 3)$

95. Twice x subtracted from 21 $21 - 2x$

96. The quotient of twice x and 11 $\dfrac{2x}{11}$ or $(2x) \div 11$

97. Fourteen less than t $t - 14$

98. Q less than 3 $3 - Q$

For Exercises 99–116, translate each algebraic expression into an English phrase. (Answers may vary.)

99. $5 + r$ The sum of 5 and r

100. $18 - x$ The difference of 18 and x

101. $s - 14$ The difference of s and 14

102. $y + 12$ The sum of y and 12

103. $\dfrac{5}{2p}$ The quotient of 5 and the product of 2 and p

104. xyz The product of x, y, and z

105. $7x + 1$ One more than the product of 7 and x

106. $c - 2d$ c decreased by the product of 2 and d

107. 5^2 5, squared

108. 6^3 6, cubed

109. $\sqrt{5}$ The square root of 5

110. $\sqrt{10}$ The square root of 10

111. 7^3 7, cubed

112. 10^2 10, squared

113. $2 + x^2$ The sum of 2 and the square of x

114. $z^2 + 16$ The sum of the square of z and 16

115. $3 + \sqrt{r}$ The sum of 3 and the square root of r

116. $21 - \sqrt{w}$ The difference of 21 and the square root of w

117. Some students use the following common memorization device (mnemonic) to help them remember the order of operations: the acronym PEMDAS or **P**lease **E**xcuse **M**y **D**ear **A**unt **S**ally to remember **P**arentheses, **E**xponents, **M**ultiplication, **D**ivision, **A**ddition, and **S**ubtraction. The problem with this mnemonic is that it suggests that multiplication is done before division and similarly, it suggests that addition is performed before subtraction. Explain why following this acronym may give the incorrect answer for the expressions:

    a. $36 \div 4 \cdot 3$ ❖   b. $36 - 4 + 3$ ❖

118. If you use the acronym **P**lease **E**xcuse **M**y **D**ear **A**unt **S**ally to remember the order of operations, what must you keep in mind about the last four steps?   ❖

119. Explain why the acronym **P**lease **E**xcuse **D**r. **M**ichael **S**mith's **A**unt could also be used as a memory device for the order of operations. ❖

EXPANDING YOUR SKILLS

For Exercises 120–123, use the order of operations to simplify the expressions.

120. $\dfrac{\sqrt{\frac{1}{9}} + \frac{2}{3}}{\sqrt{\frac{4}{25}} + \frac{3}{5}}$ 1

121. $\dfrac{5 - \sqrt{9}}{\sqrt{\frac{4}{9}} + \frac{1}{3}}$ 2

122. $\dfrac{|-2|}{|-10| - |2|}$ $\dfrac{1}{4}$

123. $\dfrac{|-4|^2}{2^2 + \sqrt{144}}$ 1

❖ See Additional Answers Appendix   Writing   Translating Expression   Geometry   Scientific Calculator   Video

section

1.3   ADDITION OF REAL NUMBERS

1.   Addition of Real Numbers and the Number Line

Adding real numbers can be visualized on the number line. To add a positive number, move to the right on the number line. To add a negative number, move to the left on the number line. The following example may help to illustrate the process.

On a winter day in Detroit, suppose the temperature starts out at 5 degrees Fahrenheit (5°F) at noon, and then drops 12° two hours (h) later when a cold front passes through. The resulting temperature can be represented by the expression $5° + (-12°)$. On the number line, start at 5 and count 12 units to the left (Figure 1-4). The resulting temperature at 2:00 P.M. is $-7°F$.

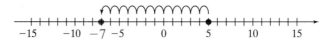

Figure 1-4

---

example 1   **Using the Number Line to Add Real Numbers**

Use the number line to add the numbers.

a. $-5 + 2$ b. $-1 + (-4)$ c. $4 + (-7)$

Solution:

a. $-5 + 2 = -3$

Start at -5 and count
2 units to the right.
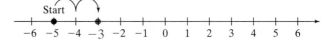

b. $-1 + (-4) = -5$

Start at -1 and count
4 units to the left.
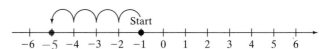

c. $4 + (-7) = -3$

Start at 4 and count 7
units to the left.
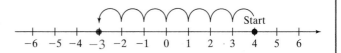

---

2.   Addition of Real Numbers

When adding large numbers or numbers that involve fractions or decimals, counting units on the number line can be cumbersome. Study the following example to determine a pattern for adding two numbers with the *same* sign.

$1 + 4 = 5$

$-1 + (-4) = -5$

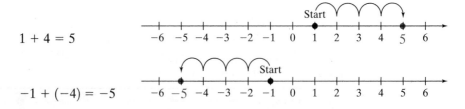

Adding Numbers with the *Same* Sign

To **add two numbers with the *same* sign**, add their absolute values and apply the common sign.

Study the following example to determine a pattern for adding two numbers with *different* signs.

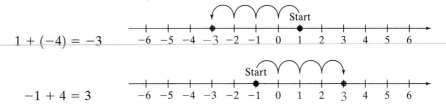

$1 + (-4) = -3$

$-1 + 4 = 3$

Adding Numbers with *Different* Signs

To **add two numbers with *different* signs**, subtract the smaller absolute value from the larger absolute value. Then apply the sign of the number having the larger absolute value.

example 2

Adding Real Numbers

Add the numbers.

a. $-12 + 15$ b. $7 + (-13)$ c. $-4.5 + (-2.1)$

Classroom Activity 1.3A

Solution:

a. $-12 + 15$ Different signs

Subtract the absolute values. Apply the sign of the larger absolute value.

$= +(|15| - |-12|) = +(15 - 12) = 3$

Sign of the number with the larger absolute value

Subtract the absolute values.

b. $7 + (-13)$ Different signs

Subtract the absolute values. Apply the sign of the larger absolute value.

$= -(|-13| - |7|) = -(13 - 7) = -6$

Sign of the number with the larger absolute value

Subtract the absolute values.

c. $-4.5 + (-2.1)$ Same signs. Add the absolute values and keep the common sign.

$$= -(|-4.5| + |-2.1|) = -(4.5 + 2.1) = -6.6$$

Keep the common sign.

Add the absolute values.

3. Translations

example 3

Translating Expressions Involving the Addition of Real Numbers

Translate each English phrase into an algebraic expression. Then simplify the result.

a. The sum of -12, -8, -9, and -1
b. Negative three-tenths added to $-\frac{7}{8}$
c. The sum of -12 and its opposite

Solution:

a. $-12 + (-8) + (-9) + (-1)$ Translate.

$= -20 + (-9) + (-1)$ Add from left to right.

$= -29 + (-1)$

$= -30$

b. $-\dfrac{7}{8} + \left(-\dfrac{3}{10}\right)$ Translate.

$= -\dfrac{35}{40} + \left(-\dfrac{12}{40}\right)$ Get a common denominator.

$= -\dfrac{47}{40}$ The numbers have the same signs. Add their absolute values and keep the common sign. $-\left(\frac{35}{40} + \frac{12}{40}\right)$.

c. $-12 + (12)$ Translate.

$= 0$ Add.

Tip: The sum of any number and its opposite is 0.

4.   Applications Involving Addition of Real Numbers

example 4   **Adding Real Numbers in Applications**

a.   A running back on a football team gains 4 yards (yd). On the next play, the quarterback is sacked and loses 13 yd. Write a mathematical expression to describe this situation and then simplify the result.

b.   A student has $120 in her checking account. After depositing her paycheck of $215, she writes a check for $255 to cover her portion of the rent and another check for $294 to cover her car payment. Write a mathematical expression to describe this situation and then simplify the result.

Classroom Activity 1.3B

Solution:

a. $4 + (-13)$ The loss of 13 yd can be interpreted as adding -13 yd.

$\quad = -9$ The football team has a net loss of 9 yd.

b. $\underline{120 + 215} + (-255) + (-294)$ Writing a check is equivalent to adding a negative amount to the bank account.

$\quad = \underline{335 + (-255)} + (-294)$ Use the order of operations. Add from left to right.

$\quad = \qquad 80 + (-294)$

$\quad = \qquad\qquad -214$ The student has overdrawn her account by $214.

---

section 1.3   PRACTICE EXERCISES

For Exercises 1–8, classify the numbers as rational or irrational.   1.1

1. $\dfrac{2}{9}$ 2. $-\dfrac{2}{3}$ 3. $-\dfrac{5}{8}$ 4. $\dfrac{1}{2}$

   Rational   Rational   Rational   Rational

5. π 6. $-\sqrt{11}$ 7. -4 8. 3

   Irrational   Irrational   Rational   Rational

Plot the points in set A on a number line. Then for Exercises 9–14 use the number line to place the appropriate inequality ($<$, $>$) between the expressions.   1.1

$$A = \left\{ -\dfrac{1}{3}, 0, \sqrt{3}, -4, \dfrac{1}{8}, -2\tfrac{3}{4}, \sqrt{25} \right\}$$

9. $0 \underline{\quad > \quad} -\dfrac{1}{3}$ 10. $\sqrt{3} \underline{\quad < \quad} \sqrt{25}$

11. $\sqrt{25} \underline{\quad > \quad} \sqrt{3}$ 12. $\dfrac{1}{8} \underline{\quad > \quad} -\dfrac{1}{3}$

13. $-2\tfrac{3}{4} \underline{\quad > \quad} -4$ 14. $0 \underline{\quad < \quad} \sqrt{3}$

For Exercises 15–46, add the integers.

15. $6 + (-3)$ 3   16. $8 + (-2)$ 6

17. $2 + (-5)$ -3 18. $7 + (-3)$ 4

19. $-19 + 2$ -17 20. $-25 + 18$ -7

21. $-4 + 11$ 7   22. $-3 + 9$ 6

23. $-16 + (-3)$ -19 24. $-12 + (-23)$ -35

25. $-2 + (-21)$ -23 26. $-13 + (-1)$ -14

27. $0 + (-5)$ -5 28. $0 + (-4)$ -4

29. $-3 + 0$ -3 30. $-8 + 0$ -8

31. $-16 + 16$ 0   32. $11 + (-11)$ 0

33. $41 + (-41)$ 0

34. $-15 + 15$ 0

35. $4 + (-9)$ -5

36. $6 + (-9)$ -3

37. $7 + (-2) + (-8)$ -3

38. $2 + (-3) + (-6)$ -7

39. $-17 + (-3) + 20$ 0

40. $-9 + (-6) + 15$ 0

41. $-3 + (-8) + (-12)$ -23

42. $-8 + (-2) + (-13)$ -23

 43. $-42 + (-3) + 45 + (-6)$ -6

44. $36 + (-3) + (-8) + (-25)$ 0

45. $-5 + (-3) + (-7) + 4 + 8$ -3

46. $-13 + (-1) + 5 + 2 + (-20)$ -27

47. The temperature in Minneapolis, Minnesota, began at $-5°F$ ($5°$ below zero) at 6:00 A.M. By noon the temperature had risen $13°$, and by the end of the day, the temperature had dropped $11°$ from its noon time high. Write an expression using addition that describes the change in temperatures during the day. Then evaluate the expression to give the temperature at the end of the day. $-5 + 13 + (-11), -3°$

48. The temperature in Toronto, Ontario, Canada, began at $4°F$. A cold front went through at noon, and the temperature dropped $9°$. By 4:00 P.M. the temperature had risen $2°$ from its noon time low. Write an expression using addition that describes the changes in temperature during the day. Then evaluate the expression to give the temperature at the end of the day. $4 + (-9) + 2, -3°$

49. During a football game, the University of Oklahoma's team gained 3 yd, lost 5 yd, and then gained 14 yd. Write an expression using addition that describes the team's total loss or gain and evaluate the expression. $3 + (-5) + 14$, 12-yd gain

50. During a football game, the Nebraska Cornhuskers lost 2 yd, gained 6 yd, and then lost 5 yd.

Write an expression using addition that describes the team's total loss or gain and evaluate the expression. $-2 + 6 + (-5)$, -1 yd or 1-yd loss

51. State the rule for adding two numbers with different signs.   ❖

52. State the rule for adding two numbers with the same signs.   To add two numbers with the same sign, add their absolute values and apply the sign.

For Exercises 53–72, add the rational numbers.

53. $23.81 + (-2.51)$ 21.3

54. $-9.23 + 10.53$ 1.3

55. $-\dfrac{2}{7} + \dfrac{1}{14}$ $-\dfrac{3}{14}$

56. $-\dfrac{1}{8} + \dfrac{5}{16}$ $\dfrac{3}{16}$

57. $-2.1 + \left(-\dfrac{3}{10}\right)$ -2.4 or $-\dfrac{12}{5}$

58. $-8.3 + \left(-\dfrac{9}{10}\right)$ -9.2 or $-\dfrac{46}{5}$

59. $\dfrac{3}{4} + (-0.5)$ $\dfrac{1}{4}$ or 0.25

60. $-\dfrac{3}{2} + 0.45$ $-\dfrac{21}{20}$ or -1.05

61. $8.23 + (-8.23)$ 0

62. $-7.5 + 7.5$ 0

63. $-\dfrac{7}{8} + 0$ $-\dfrac{7}{8}$

64. $0 + \left(-\dfrac{21}{22}\right)$ $-\dfrac{21}{22}$

65. $-\dfrac{2}{3} + \left(-\dfrac{1}{9}\right) + 2$ $\dfrac{11}{9}$

66. $-\dfrac{1}{4} + \left(-\dfrac{3}{2}\right) + 2$ $\dfrac{1}{4}$

67. $-47.36 + 24.28$ -23.08

68. $-0.015 + (0.0026)$ -0.0124

69. $516.816 + (-22.13)$ 494.686

70. $87.02 + (-93.19)$ -6.17

71. $-0.000617 + (-0.0015)$ -0.002117

72. $-5315.26 + (-314.89)$ -5630.15

73. Yoshima has $52.23 in her checking account. She writes a check for groceries for $52.95.

   a. Write an addition problem that expresses Yoshima's transaction. $52.23 + (-52.95)$

   b. Is Yoshima's account overdrawn?   Yes

74. Mohammad has $40.02 in his checking account. He writes a check for a pair of shoes for $40.96.

   a. Write an addition problem that expresses Mohammad's transaction. $40.02 + (-40.96)$

   b. Is Mohammad's account overdrawn?   Yes

75. In the game show Jeopardy a contestant responds to six questions with the following

outcomes: +\$100, +\$200, −\$500, +\$300, +\$100, −\$200

a. Write an expression using addition to describe the contestant's scoring activity.
$100 + 200 + (−500) + 300 + 100 + (−200)$

b. Evaluate the expression from part (a) to determine the contestant's final outcome.   \$0

76. A company that has been in business for 5 years has the following profit and loss record.
$−50,000 + (−32,000) + (−5000) + 13,000 + 26,000$

a. Write an expression using addition to describe the company's profit/loss activity.

b. Evaluate the expression from part (a) to determine the company's net profit or loss.
−\$48,000

Year	Profit/Loss ($)
1	−50,000
2	−32,000
3	−5000
4	13,000
5	26,000

Table for Exercise 76

For Exercises 77–82, evaluate the expression for $x = −3$, $y = −2$, and $z = 16$.

77. $x + y + \sqrt{z}$ −1

78. $2z + x + y$ 27

79. $y + 3\sqrt{z}$ 10

80. $−\sqrt{z} + y$ −6

81. $|x| + |y|$ 5

82. $z + x + |y|$ 15

For Exercises 83–92, translate the English phrase into an algebraic expression. Then evaluate the expression.

83. The sum of −6 and −10 $−6 + (−10) = −16$

84. The sum of −3 and 5 $−3 + 5 = 2$

85. Negative three increased by 8 $−3 + 8 = 5$

86. Twenty-one increased by 4 $21 + 4 = 25$

87. Seventeen more than −21 $−21 + 17 = −4$

88. Twenty-four more than −7 $−7 + 24 = 17$

89. Three times the sum of −14 and 20 $3(−14 + 20) = 18$

90. Two times the sum of 6 and −10 $2(6 + (−10)) = −8$

91. Five more than the sum of −7 and −2
$(−7 + (−2)) + 5 = −4$

92. Negative six more than the sum of 4 and −1
$(4 + (−1)) + (−6) = −3$

Concepts

1. Subtraction of Real Numbers

2. Translations

3. Applications Involving Subtraction

4. Applying the Order of Operations

section

1.4   SUBTRACTION OF REAL NUMBERS

1.   Subtraction of Real Numbers

In the previous section, we learned the rules for adding real numbers. Subtraction of real numbers is defined in terms of the addition process. For example, consider the following subtraction problem and the corresponding addition problem:

$$6 − 4 = 2 \quad \Leftrightarrow \quad 6 + (−4) = 2$$

In each case, we start at 6 on the number line and move to the left 4 units. That is, adding the opposite of 4 produces the same result as subtracting 4. This is true in general. To **subtract two real numbers**, add the opposite of the second number to the first number.

Subtraction of Real Numbers

If a and b are real numbers, then $a - b = a + (-b)$

$$10 - 4 = 10 + (-4) = 6$$
$$-10 - 4 = -10 + (-4) = -14$$
Subtracting 4 is the same as adding -4.

$$10 - (-4) = 10 + (4) = 14$$
$$-10 - (-4) = -10 + (4) = -6$$
Subtracting -4 is the same as adding 4.

example 1

Subtracting Real Numbers

Subtract the numbers.

a. $4 - (-9)$ b. $-6 - 9$ c. $-11 - (-5)$ d. $7 - 10$

Classroom Activity 1.4A

Solution:

a. $4 - (-9)$

$= 4 + (9) = 13$

Take the opposite of -9.
Change subtraction to addition.

b. $-6 - 9$

$= -6 + (-9) = -15$

Take the opposite of 9.
Change subtraction to addition.

c. $-11 - (-5)$

$= -11 + (5) = -6$

Take the opposite of -5.
Change subtraction to addition.

d. $7 - 10$

$= 7 + (-10) = -3$

Take the opposite of 10.
Change subtraction to addition.

2. Translations

example 2

Translating Expressions Involving Subtraction

Write an algebraic expression for each English phrase and then simplify the result.

a. The difference of -7 and -5
b. 12.4 subtracted from -4.7
c. -24 decreased by the sum of -10 and 13
d. Seven-fourths less than one-third

Classroom Activity 1.4B

Solution:

a. $-7 - (-5)$ Translate.

$= -7 + (5)$ Rewrite subtraction in terms of addition.

$= -2$ Simplify.

Tip: Recall that "*b* subtracted from *a*" is translated as $a - b$. Hence, -4.7 is written first and then 12.4.

Tip: Parentheses must be used around the sum of -10 and 13 so that -24 is decreased by the entire quantity $(-10 + 13)$.

b. $-4.7 - 12.4$ Translate.

   $= -4.7 + (-12.4)$ Rewrite subtraction in terms of addition.

   $= -17.1$ Simplify.

c. $-24 - (-10 + 13)$ Translate.

   $= -24 - (3)$ Simplify inside parentheses.

   $= -24 + (-3)$ Rewrite subtraction in terms of addition.

   $= -27$ Simplify.

d. $\dfrac{1}{3} - \dfrac{7}{4}$

   $= \dfrac{1}{3} + \left(-\dfrac{7}{4}\right)$ Rewrite subtraction in terms of addition.

   $= \dfrac{4}{12} + \left(-\dfrac{21}{12}\right)$ Get a common denominator.

   $= -\dfrac{17}{12}$ or $-1\frac{5}{12}$

3. Applications Involving Subtraction

example 3

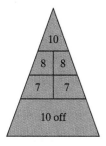

Figure 1-5

Using Subtraction of Real Numbers in an Application

The scoring area for a shuffleboard court is shown in Figure 1-5. A player slides a puck down the court and gains or loses points depending on the final position of the puck. Notice that if the puck lands at the bottom of the triangle, the player loses 10 points.

   Harold's first four pucks land as follows:

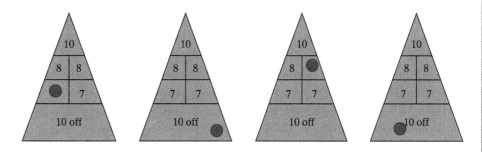

Write an expression to determine Harold's score. Then simplify the expression.

Solution:

$7 - 10 + 8 - 10$

$= 7 + (-10) + 8 + (-10)$ Rewrite subtraction in terms of addition.

$= -3 + 8 + (-10)$ Add from left to right.

$= 5 + (-10)$

$= -5$ Harold is down 5 points.

---

example 4

Using Subtraction of Real Numbers in an Application

The highest recorded temperature in North America was 134°F, recorded on July 10, 1913, in Death Valley, California. The lowest temperature of −81°F was recorded on February 3, 1947, in Snag, Yukon, Canada.

Find the difference between the highest and lowest recorded temperatures in North America.

Solution:

$134 - (-81)$

$= 134 + (81)$ Rewrite subtraction in terms of addition.

$= 215$ Add.

The difference between the highest and lowest temperatures is 215°F.

---

4. Applying the Order of Operations

example 5

Applying the Order of Operations

Simplify the expressions.

a. $-6 + \{10 - [7 - (-4)]\}$ b. $5 - \sqrt{35 - (-14)} - 2$

c. $\left(-\dfrac{5}{8} - \dfrac{2}{3}\right) - \left(\dfrac{1}{8} + 2\right)$ d. $-6 - |7 - 11| + (-3 + 7)^2$

Classroom Activity 1.4C

Solution:

a. $-6 + \{10 - [7 - (-4)]\}$ Work inside the inner brackets first.

$= -6 + \{10 - [7 + (4)]\}$ Rewrite subtraction in terms of addition.

$= -6 + [10 - (11)]$ Simplify the expression inside brackets.

$= -6 + [10 + (-11)]$ Rewrite subtraction in terms of addition.

$= -6 + (-1)$

$= -7$ Add.

b. $5 - \sqrt{35 - (-14)} - 2$ Work inside the radical first.

$= 5 - \sqrt{35 + (14)} - 2$ Rewrite subtraction in terms of addition.

$= 5 - \sqrt{49} - 2$

$= 5 - 7 - 2$ Simplify the radical.

$= 5 + (-7) + (-2)$ Rewrite subtraction in terms of addition.

$= -2 + (-2)$ Add from left to right.

$= -4$

c. $\left(-\dfrac{5}{8} - \dfrac{2}{3}\right) - \left(\dfrac{1}{8} + 2\right)$ Work inside the parentheses first.

$= \left[-\dfrac{5}{8} + \left(-\dfrac{2}{3}\right)\right] - \left(\dfrac{1}{8} + 2\right)$ Rewrite subtraction in terms of addition.

$= \left[-\dfrac{15}{24} + \left(-\dfrac{16}{24}\right)\right] - \left(\dfrac{1}{8} + \dfrac{16}{8}\right)$ Get a common denominator in each parentheses.

$= \left(-\dfrac{31}{24}\right) - \left(\dfrac{17}{8}\right)$ Add fractions in each parentheses.

$= \left(-\dfrac{31}{24}\right) + \left(-\dfrac{17}{8}\right)$ Rewrite subtraction in terms of addition.

$= -\dfrac{31}{24} + \left(-\dfrac{51}{24}\right)$ Get a common denominator.

$= -\dfrac{82}{24}$ Add.

$= -\dfrac{41}{12}$ Reduce.

d. $-6 - |7 - 11| + (-3 + 7)^2$ Simplify within absolute value and parentheses first.

$= -6 - |7 + (-11)| + (-3 + 7)^2$ Rewrite subtraction in terms of addition.

$= -6 - |-4| + (4)^2$

$= -6 - (4) + 16$ Simplify absolute value and exponent.

$= -6 + (-4) + 16$ Rewrite subtraction in terms of addition.

$= -10 + 16$ Add from left to right.

$= 6$

Calculator Connections

Most calculators can add, subtract, multiply, and divide signed numbers. It is important to note, however, that the key used for the negative sign is different from the key used for subtraction. On a scientific calculator, the $\boxed{+/-}$ key or $\boxed{+\circ-}$ key is used to enter a negative number or to change the sign of an existing number. On a graphing calculator, the $\boxed{(-)}$ key is used. These keys should not be confused with the $\boxed{-}$ key which is used for subtraction. For example, try simplifying the following expressions.

a. $-7 + (-4) - 6$ b. $-3.1 - (-0.5) + 1.1$

Scientific Calculator:

Enter: $\boxed{7}$ $\boxed{+/-}$ $\boxed{+}$ $\boxed{(}$ $\boxed{4}$ $\boxed{+/-}$ $\boxed{)}$ $\boxed{-}$ $\boxed{6}$ $\boxed{=}$ Result: -17

Enter: $\boxed{3}$ $\boxed{.}$ $\boxed{1}$ $\boxed{+/-}$ $\boxed{-}$ $\boxed{(}$ $\boxed{0}$ $\boxed{.}$ $\boxed{5}$ $\boxed{+/-}$ $\boxed{)}$ $\boxed{+}$ $\boxed{1}$ $\boxed{.}$ $\boxed{1}$ $\boxed{=}$

Result: -1.5

Graphing Calculator:

```
-7+(-4)-6
              -17
-3.1-(-0.5)+1.1
              -1.5
```

Calculator Exercises

Simplify the expression without the use of a calculator. Then use the calculator to verify your answer. ❖

1. $-8 + (-5)$ -13
2. $4 + (-5) + (-1)$ -2
3. $627 - (-84)$ 711
4. $-0.06 - 0.12$ -0.18
5. $-3.2 + (-14.5)$ -17.7
6. $-472 + (-518)$ -990
7. $-12 - 9 + 4$ -17
8. $209 - 108 + (-63)$ 38

---

section 1.4   PRACTICE EXERCISES

1. List five whole numbers.   1.1
    For example: $0, 1, 2, 3, 4$
2. List five rational numbers.   1.1
    For example: $-3, -\frac{1}{2}, 0, 1, 1.\overline{3}$
3. List five irrational numbers.   1.1
    For example: $-\sqrt{2}, \sqrt{3}, \sqrt{5}, -2\pi, \pi$
4. List five integers.   1.1
    For example: $-2, -1, 0, 1, 2$

5. List five natural numbers.   1.1
    For example: $1, 2, 3, 4, 5$
6. List five real numbers.   1.1
    For example: $-\sqrt{5}, -2, 0, \frac{3}{8}, 7.3$

⬅ For Exercises 7–10, translate each English phrase into an algebraic expression.   1.2

❖ See Additional Answers Appendix   ✎ Writing   ⬅ Translating Expression   ◆ Geometry   Scientific Calculator   Video

7. The square root of 6 $\sqrt{6}$

8. The square of x x^2

9. Negative seven increased by 10 $-7 + 10$

10. Two more than $-b$ $-b + 2$

For Exercises 11–18, fill in the blank to make each statement correct.

11. $5 - 3 = 5 +$ ___-3___

12. $8 - 7 = 8 +$ ___-7___

13. $-2 - 12 = -2 +$ ___-12___

14. $-4 - 9 = -4 +$ ___-9___

15. $7 - (-4) = 7 +$ ___4___

16. $13 - (-4) = 13 +$ ___4___

17. $-9 - (-3) = -9 +$ ___3___

18. $-15 - (-10) = -15 +$ ___10___

For Exercises 19–60, subtract the rational numbers.

19. $3 - 5$ -2
20. $9 - 12$ -3
21. $3 - (-5)$ 8
22. $9 - (-12)$ 21
23. $-3 - 5$ -8
24. $-9 - 12$ -21
25. $-3 - (-5)$ 2
26. $-9 - (-5)$ -4
27. $23 - 17$ 6
28. $14 - 2$ 12
29. $23 - (-17)$ 40
30. $14 - (-2)$ 16
31. $-23 - 17$ -40
32. $-14 - 2$ -16
33. $-23 - (-17)$ -6
34. $-14 - (-2)$ -12
35. $-6 - 14$ -20
36. $-9 - 12$ -21
37. $-7 - 17$ -24
38. $-8 - 21$ -29
39. $13 - (-12)$ 25
40. $20 - (-5)$ 25
41. $-14 - (-9)$ -5
42. $-21 - (-17)$ -4
43. $\frac{1}{2} - \frac{1}{10}$ $\frac{2}{5}$
44. $\frac{2}{7} - \frac{3}{14}$ $\frac{1}{14}$
45. $-\frac{11}{12} - \left(-\frac{1}{4}\right)$ $-\frac{2}{3}$
46. $-\frac{7}{8} - \left(-\frac{1}{6}\right)$ $-\frac{17}{24}$

47. $6.8 - (-2.4)$ 9.2
48. $7.2 - (-1.9)$ 9.1
49. $3.1 - 8.82$ -5.72
50. $1.8 - 9.59$ -7.79
51. $-4 - 3 - 2 - 1$ -10
52. $-10 - 9 - 8 - 7$ -34
53. $6 - 8 - 2 - 10$ -14
54. $20 - 50 - 10 - 5$ -45
55. $-36.75 - 14.25$ -51
56. $-84.21 - 112.16$ -196.37
57. $-112.846 + (-13.03) - 47.312$ -173.188
58. $-96.473 + (-36.02) - 16.617$ -149.11
59. $0.085 - (-3.14) + (0.018)$ 3.243
60. $0.00061 - (-0.00057) + (0.0014)$ 0.00258

For Exercises 61–70, translate each English phrase into an algebraic expression. Then evaluate the expression.

61. Six minus -7 $6 - (-7) = 13$
62. Eighteen minus -1 $18 - (-1) = 19$
63. Eighteen subtracted from 3 $3 - 18 = -15$
64. Twenty-one subtracted from 8 $8 - 21 = -13$
65. The difference of -5 and -11 $-5 - (-11) = 6$
66. The difference of -2 and -18 $-2 - (-18) = 16$
67. Negative thirteen subtracted from -1 $-1 - (-13) = 12$
68. Negative thirty-one subtracted from -19 $-19 - (-31) = 12$
69. Twenty less than -32 $-32 - 20 = -52$
70. Seven less than -3 $-3 - 7 = -10$

For Exercises 71–80, perform the indicated operations. Remember to perform addition or subtraction as they occur from left to right.

71. $6 + 8 - (-2) - 4 + 1$ 13
72. $-3 - (-4) + 1 - 2 - 5$ -5
73. $-1 - 7 + (-3) - 8 + 10$ -9
74. $13 - 7 + 4 - 3 - (-1)$ 8
75. $2 - (-8) + 7 + 3 - 15$ 5
76. $8 - (-13) + 1 - 9$ 13
77. $-6 + (-1) + (-8) + (-10)$ -25

78. $-8 + (-3) + (-5) + (-2)$ -18

79. $-6 - 1 - 8 - 10$ -25 80. $-8 - 3 - 5 - 2$ -18

 81. The highest mountain in the world is Mt. Everest, located in the Himalayas. Its height is 8848 meters (m) (29,028 ft). The lowest recorded depth in the ocean is located in the Marianas Trench in the Pacific Ocean. Its "height" relative to sea level is $-11{,}033$ m ($-36{,}198$ ft). Determine the difference in elevation, in meters, between the highest mountain in the world and the deepest ocean trench. (*Source: Information Please Almanac,* 1999)   19,881 m

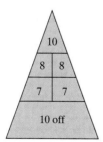

Figure for Exercise 81

82. The lowest point in North America is located in Death Valley, California, at an elevation of -282 ft (-86 m). The highest point in North America is Mt. McKinley, Alaska, at an elevation of 20,320 ft (6194 m). Find the difference in elevation, in feet, between the highest and lowest points in North America. (*Source: Information Please Almanac,* 1999)   20,602 ft

A shuffleboard court is shown here. In shuffleboard, 7, 8, or 10 points are scored if a player's puck lands on any of the five regions at the top of the triangle (see figure). If the puck lands at the base of the triangle, 10 points are taken off the score. (See Example 3)

Figure for Exercises 83–84

83. Jasper's first four pucks land as follows:

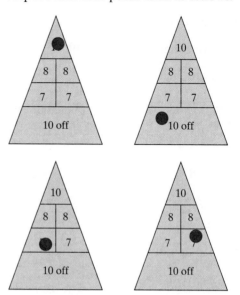

Write an expression that describes Jasper's score. Then evaluate the expression.
$10 - 10 + 7 + 7 = 14$

84. Ethyl's first four pucks land as follows:

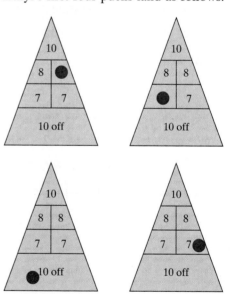

Write an expression that describes Ethel's score. Then evaluate the expression.
$8 + 7 - 10 + 7 = 12$

85. In Ohio, the highest temperature ever recorded was 113°F and the lowest was -39°F. Find the difference between the highest and lowest temperatures. (*Source: Information Please Almanac,* 1999)   152°

86. In Mississippi, the highest temperature ever recorded was 115°F and the lowest was −19°F. Find the difference between the highest and lowest temperatures. (*Source: Information Please Almanac*, 1999)   134°

For Exercises 87–94, evaluate the expressions for $a = -2, b = -6$, and $c = -1$.

87. $(a + b) - c$ −7
88. $(a - b) + c$ 3
89. $a - (b + c)$ 5
90. $a + (b - c)$ −7
91. $(a - b) - c$ 5
92. $(a + b) + c$ −9
93. $a - (b - c)$ 3
94. $a + (b + c)$ −9

For Exercises 95–100, evaluate the expression using the order of operations.

95. $\sqrt{29 + (-4)} - 7$ −2
96. $8 - \sqrt{98 + (-3) + 5}$ −2
97. $|10 + (-3)| - |-12 + (-6)|$ −11
98. $|6 - 8| + |12 - 5|$ 9
99. $\dfrac{3 - 4 + 5}{4 + (-2)}$ 2
100. $\dfrac{12 - 14 + 6}{6 + (-2)}$ 1

---

chapter 1   MIDCHAPTER REVIEW

1. State the rule for adding two negative numbers.
   Add their absolute values and apply a negative sign.
2. State the rule for adding a negative number to a positive number. Subtract the smaller absolute value from the larger absolute value. Apply the sign of the number with the larger absolute value.
3. State the rule for subtracting a negative number from a positive number.   Add their absolute values.
4. State the rule for subtracting a positive number from a negative number.
   Add their absolute values and apply a negative sign.

For Exercises 5–34, add or subtract as indicated.

5. $65 - 24$ 41
6. $42 - 29$ 13
7. $13 - (-18)$ 31
8. $22 - (-24)$ 46
9. $4.8 - 6.1$ −1.3
10. $3.5 - 7.1$ −3.6
11. $4 + (-20)$ −16
12. $5 + (-12)$ −7
13. $\dfrac{1}{3} - \dfrac{5}{12}$ $-\dfrac{1}{12}$
14. $\dfrac{3}{8} - \dfrac{1}{12}$ $\dfrac{7}{24}$

15. $-32 - 4$ −36
16. $-51 - 8$ −59
17. $-6 + (-6)$ −12
18. $-25 + (-25)$ −50
19. $-4 - \left(-\dfrac{5}{6}\right)$ $-\dfrac{19}{6}$
20. $-2 - \left(-\dfrac{2}{5}\right)$ $-\dfrac{8}{5}$
21. $-60 + 55$ −5
22. $-55 + 23$ −32
23. $-18 - (-18)$ 0
24. $-3 - (-3)$ 0
25. $-3.5 - 4.2$ −7.7
26. $-6.6 - 3.9$ −10.5
27. $-90 + (-24)$ −114
28. $-35 + (-21)$ −56
29. $-14 + (-2) - 16$ −32
30. $-25 + (-6) - 15$ −46
31. $-42 + 12 + (-30)$ −60
32. $-46 + 16 + (-40)$ −70
33. $-10 - 8 - 6 - 4 - 2$ −30
34. $-100 - 90 - 80 - 70 - 60$ −400

1.5   MULTIPLICATION AND DIVISION OF REAL NUMBERS

1. Multiplication of Real Numbers

Multiplication of real numbers can be interpreted as repeated addition.

example 1

Multiplying Real Numbers

Multiply the real numbers by writing the expressions as repeated addition.

a. $3(4)$ b. $3(-4)$

Solution:

a. $3(4) = 4 + 4 + 4 = 12$
b. $3(-4) = -4 + (-4) + (-4) = -12$

The results from Example 1 suggest that the product of two positive numbers is positive and the product of a positive number and a negative number is negative. Refer to Table 1-4 and observe the pattern across the bottom row.

Table 1-4

×	3	2	1	0	−1	−2	−3
2	6	4	2	0	−2	−4	−6

The product of two positive numbers is **positive**.   The product of a positive number and a negative number is **negative**.

The pattern along the bottom row shows that as 2 is multiplied by consecutively smaller integers, the product decreases by 2. As this pattern continues, notice that the product of a negative number and a positive number is negative.

To determine the sign of the product of two negative numbers, consider Table 1-5 and note the pattern across the bottom row.

Table 1-5

×	3	2	1	0	−1	−2	−3
−2	−6	−4	−2	0	2	4	6

The product of a positive number and a negative number is **negative**.   The product of two negative numbers is **positive**.

As −2 is multiplied by consecutively smaller integers, the product increases by 2. As this pattern continues, notice that the product of two negative numbers is positive.

Multiplication of Real Numbers

1. The product of two real numbers with the *same* sign is positive.
2. The product of two real numbers with *different* signs is negative.
3. The product of any real number and zero is zero.

---

example 2 **Multiplying Real Numbers**

Multiply the real numbers.

a. $-8(-4)$ b. $-2.5(-1.7)$ c. $-7(10)$

d. $\dfrac{1}{2}(-8)$ e. $0(-8.3)$ f. $-\dfrac{2}{7}\left(-\dfrac{7}{2}\right)$

Classroom Activity 1.5A

Solution:

a. $-8(-4) = 32$ *Same signs.* Product is positive.

b. $-2.5(-1.7) = 4.25$

c. $-7(10) = -70$ *Different signs.* Product is negative.

d. $\dfrac{1}{2}(-8) = -4$

e. $0(-8.3)$

   $= 0$ The product of any real number and zero is zero.

f. $-\dfrac{2}{7}\left(-\dfrac{7}{2}\right)$

   $= \dfrac{14}{14}$ Multiply. *Same signs.* Product is positive.

   $= 1$ Reduce.

The order of operation indicates that multiplication or division is performed in order from left to right.

$(-1)(-1)$	$(-1)(-1)(-1)$	$(-1)(-1)(-1)(-1)$	$(-1)(-1)(-1)(-1)(-1)$
$= 1$	$= (1)(-1)$	$= (1)(-1)(-1)$	$= (1)(-1)(-1)(-1)$
	$= -1$	$= (-1)(-1)$	$= (-1)(-1)(-1)$
		$= 1$	$= (1)(-1)$
			$= -1$

Tip: The pattern demonstrated in these examples indicates that

- The product of an even number of negative factors is positive.
- The product of an odd number of negative factors is negative.

2.   Exponential Expressions

Recall that for any real number b and any positive integer, n:

$$b^n = \underbrace{b \cdot b \cdot b \cdot b \dots \cdot b}_{n \text{ factors of } b}$$

Be particularly careful when evaluating exponential expressions involving negative numbers. An exponential expression with a negative base is written with parentheses around the base, such as $(-2)^4$.

To evaluate $(-2)^4$, the base -2 is multiplied four times:

$$(-2)^4 = (-2)(-2)(-2)(-2) = 16$$

If parentheses are *not* used, the expression -2^4 has a different meaning:

- The expression -2^4 has a base of 2 (not -2) and can be interpreted as $-1 \cdot 2^4$. Hence,

$$-2^4 = -1(2)(2)(2)(2) = -16$$

- The expression -2^4 can also be interpreted as the opposite of 2^4. Hence,

$$-2^4 = -(2 \cdot 2 \cdot 2 \cdot 2) = -16$$

example 3

Classroom Activity 1.5B

Evaluating Exponential Expressions

Simplify.

a. $(-5)^2$ b. -5^2 c. $\left(-\dfrac{1}{2}\right)^3$ d. -0.4^3

Solution:

a. $(-5)^2 = (-5)(-5) = 25$
b. $-5^2 = -1(5)(5) = -25$
c. $\left(-\dfrac{1}{2}\right)^3 = \left(-\dfrac{1}{2}\right)\left(-\dfrac{1}{2}\right)\left(-\dfrac{1}{2}\right) = -\dfrac{1}{8}$
d. $-0.4^3 = -1(0.4)(0.4)(0.4) = -0.064$

3.   Division of Real Numbers

Notice from Example 2 that $-\frac{2}{7}\left(-\frac{7}{2}\right) = 1$. Recall that two numbers are **reciprocals** if their product is 1. Symbolically, if a is a nonzero real number, then the reciprocal of a is $\frac{1}{a}$ because $a \cdot \frac{1}{a} = 1$. This definition also implies that a number and its reciprocal have the same sign.

The Reciprocal of a Real Number

Let a be a nonzero real number. Then, the **reciprocal** of a is $\frac{1}{a}$.

Recall that to subtract two real numbers, we add the opposite of the second number to the first number. In a similar way, division of real numbers is defined in terms of multiplication. To divide two real numbers, we multiply the first number by the reciprocal of the second number.

Division of Real Numbers

Let a and b be real numbers such that $b \neq 0$. Then, $a \div b = a \cdot \frac{1}{b}$.

Consider the quotient $10 \div 5$. The reciprocal of 5 is $\frac{1}{5}$, so we have

$$10 \div 5 = 2 \qquad \text{or equivalently,} \qquad 10 \cdot \overset{\text{multiply}}{\underset{\text{reciprocal}}{\frac{1}{5}}} = 2$$

Because division of real numbers can be expressed in terms of multiplication, then the sign rules that apply to multiplication also apply to division.

Division of Real Numbers

1. The quotient of two real numbers with the *same* sign is positive.
2. The quotient of two real numbers with *different* signs is negative.

---

example 4

Dividing Real Numbers

Divide the real numbers.

a. $200 \div (-10)$ b. $\dfrac{-48}{16}$ c. $\dfrac{-6.25}{-1.25}$ d. $\dfrac{-9}{-5}$

e. $15 \div -25$ f. $-\dfrac{3}{14} \div \dfrac{9}{7}$ g. $\dfrac{\frac{2}{5}}{-\frac{2}{5}}$

Solution:

a. $200 \div (-10) = -20$ *Different signs.* Quotient is negative.

b. $\dfrac{-48}{16} = -3$ *Different signs.* Quotient is negative.

c. $\dfrac{-6.25}{-1.25} = 5$ *Same signs.* Quotient is positive.

d. $\dfrac{-9}{-5} = \dfrac{9}{5}$ *Same signs.* Quotient is positive.

Tip: If the numerator and denominator of a fraction are both negative, then the quotient is positive. Therefore, $\frac{-9}{-5}$ can be simplified to $\frac{9}{5}$.

Tip: If the numerator and denominator of a fraction have opposite signs then the quotient will be negative. Therefore, a fraction has the same value whether the negative sign is written in the numerator, in the denominator, or in front of the fraction.

$$\frac{-3}{5} = \frac{3}{-5} = -\frac{3}{5}$$

e. $15 \div -25$ *Different signs.* Quotient is negative.

$$= \frac{15}{-25}$$

$$\longrightarrow = -\frac{3}{5}$$

f. $-\frac{3}{14} \div \frac{9}{7}$ *Different signs.* Quotient is negative.

$$= -\frac{3}{14} \cdot \frac{7}{9}$$ Multiply by the reciprocal of $\frac{9}{7}$ which is $\frac{7}{9}$.

$$= -\frac{21}{126}$$ Multiply fractions.

$$= -\frac{1}{6}$$ Reduce.

g. $\dfrac{\frac{2}{5}}{-\frac{2}{5}}$ This is equivalent to $\frac{2}{5} \div \left(-\frac{2}{5}\right)$.

$$= \frac{2}{5}\left(-\frac{5}{2}\right)$$ Multiply by the reciprocal of $-\frac{2}{5}$, which is $-\frac{5}{2}$.

$$= -1$$ Multiply fractions.

4.   Division Involving Zero

Multiplication can be used to check any division problem. If $\frac{a}{b} = c$, then $bc = a$ (provided that $b \neq 0$). For example,

$$\frac{8}{-4} = -2 \quad \rightarrow \quad \underline{\text{Check:}} \quad (-4)(-2) = 8 \; \checkmark$$

This relationship between multiplication and division can be used to investigate division problems involving the number zero.

1. The quotient of 0 and any nonzero number is 0. For example:

$$\frac{0}{6} = 0 \quad \text{because } 6 \cdot 0 = 0 \; \checkmark$$

2. The quotient of any nonzero number and 0 is undefined. For example,

$$\frac{6}{0} = ?$$

Finding the quotient $\frac{6}{0}$ is equivalent to asking, "what number times zero will equal 6?" That is, $(0)(?) = 6$. No real number satisfies this condition. Therefore, we say that division by zero is undefined.

3. The quotient of 0 and 0 cannot be determined. Evaluating an expression of the form $\frac{0}{0} = ?$ is equivalent to asking "what number times zero will equal 0?" That is, $(0)(?) = 0$. Any real number will satisfy this requirement; however, expressions involving $\frac{0}{0}$ are usually discussed in advanced mathematics courses.

Division Involving Zero

Let a represent a nonzero real number. Then,

1. $\dfrac{0}{a} = 0$ 2. $\dfrac{a}{0}$ is undefined

5. Applying the Order of Operations

example 5

Applying the Order of Operations

Simplify the expressions.

a. $-36 \div (-27) \div (-9)$ b. $-\dfrac{7}{8} \div (3\frac{1}{4}) \div (-2)$

Classroom Activity 1.5C

Solution:

a. $\underbrace{-36 \div (-27)} \div (-9)$

$= \dfrac{-36}{-27} \div -9$ Divide from left to right.

$= \dfrac{4}{3} \div -9$ Reduce.

$= \dfrac{4}{3}\left(-\dfrac{1}{9}\right)$ Multiply by the reciprocal of -9, which is $-\frac{1}{9}$.

$= -\dfrac{4}{27}$

b. $-\dfrac{7}{8} \div (3\frac{1}{4}) \div (-2)$ Divide from left to right.

$= -\dfrac{7}{8} \div \left(\dfrac{13}{4}\right) \div (-2)$ Change the mixed number to an improper fraction.

$= -\underbrace{\dfrac{7}{8}\left(\dfrac{4}{13}\right)} \div (-2)$ Multiply by the reciprocal of $\frac{13}{4}$, which is $\frac{4}{13}$.

$= -\dfrac{28}{104} \div (-2)$

$= -\dfrac{28}{104}\left(-\dfrac{1}{2}\right)$ Multiply by the reciprocal of -2, which is $-\frac{1}{2}$.

$= \dfrac{28}{208}$ *Same signs.* Product is positive.

$= \dfrac{7}{52}$ Reduce.

example 6   **Applying the Order of Operations**

Simplify the expressions.

a. $-8 + 8 \div (-2) \div (-6)$

b. $\dfrac{4 + \sqrt{30 - 5}}{-5 - 1}$

c. $\dfrac{24 - 2[-3 + (5 - 8)]^2}{2|-12 + 3|}$

Classroom Activity 1.5D

Solution:

a. $-8 + 8 \div (-2) \div (-6)$

$= -8 + (-4) \div (-6)$ Perform division before addition.

$= -8 + \dfrac{4}{6}$ The quotient of -4 and -6 is positive $\frac{4}{6}$ or $\frac{2}{3}$.

$= -\dfrac{8}{1} + \dfrac{2}{3}$ Write -8 as a fraction.

$= -\dfrac{24}{3} + \dfrac{2}{3}$ Get a common denominator.

$= -\dfrac{22}{3}$ Add.

b. $\dfrac{4 + \sqrt{30 - 5}}{-5 - 1}$ Simplify numerator and denominator separately.

$= \dfrac{4 + \sqrt{25}}{-6}$ Simplify within the radical and simplify the denominator.

$= \dfrac{4 + 5}{-6}$ Simplify the radical.

$= \dfrac{9}{-6}$

$= \dfrac{3}{-2}, \text{ or } -\dfrac{3}{2}$ Reduce.

c. $\dfrac{24 - 2[-3 + (5 - 8)]^2}{2|-12 + 3|}$ Simplify numerator and denominator separately.

$= \dfrac{24 - 2[-3 + (-3)]^2}{2|-9|}$ Simplify within the inner parentheses and absolute value.

$= \dfrac{24 - 2[-6]^2}{2(9)}$ Simplify within brackets, [   ]. Simplify the absolute value.

$$= \frac{24 - 2(36)}{2(9)}$$ Simplify exponents.

$$= \frac{24 - 72}{18}$$ Perform multiplication before subtraction.

$$= \frac{-48}{18}, \text{ or } -\frac{8}{3}$$ Reduce.

example 7   **Finding Average Temperature**

The low temperatures for Churchill Falls, Labrador, Canada, for a week in January are given in Table 1-6. Find the average low temperature for the week. Round to the nearest tenth of a degree.

Table 1-6

Day	Low Temp. (°C)
Mon.	−17
Tues.	−15
Wed.	−20
Thurs.	−21
Fri.	−24
Sat.	−24
Sun.	−24

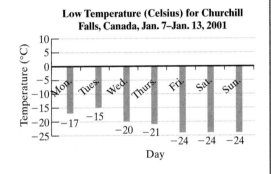

Low Temperature (Celsius) for Churchill Falls, Canada, Jan. 7–Jan. 13, 2001

Solution:

The average is found by adding the temperature values and dividing by the number of temperatures.

$$\frac{-17 + (-15) + (-20) + (-21) + (-24) + (-24) + (-24)}{7}$$ The average of seven temperatures.

$$= \frac{-145}{7}$$

$$\approx -20.7$$ The average low temperature for the week was −20.7°C.

Calculator Connections

Be particularly careful when raising a negative number to an even power on a calculator. For example, the expressions $(-4)^2$ and -4^2 have different values. That is, $(-4)^2 = 16$ and $-4^2 = -16$. Verify these expressions on a calculator.

Scientific Calculator:

To evaluate $(-4)^2$

Enter: ⟮ (⟯ ⟮ 4 ⟯ ⟮ +/− ⟯ ⟮) ⟯ ⟮ x^2 ⟯   **Result:**   16

To evaluate -4^2 on a scientific calculator, it is important to square 4 first and then take its opposite.

Enter: ⟮ 4 ⟯ ⟮ x^2 ⟯ ⟮ +/− ⟯   **Result:**   −16

Graphing Calculator:

```
(-4)²
                    16
-4²
                   -16



```

The graphing calculator allows for several methods of denoting the multiplication of two real numbers. For example, consider the product of −8 and 4.

```
-8*4
                   -32
-8(4)
                   -32
(-8)(4)
                   -32
```

Calculator Exercises

Simplify the expression without the use of a calculator. Then use the calculator to verify your answer.   ❖

1. $-6(5)$ −30

2. $\dfrac{-5.2}{2.6}$ −2

3. $(-5)(-5)(-5)(-5)$ 625

4. $(-5)^4$ 625

5. -5^4
   −625

6. -2.4^2
   −5.76

7. $(-2.4)^2$
   5.76

8. $(-1)(-1)(-1)$
   −1

9. $\dfrac{-8.4}{-2.1}$ 4

10. $90 \div (-5)(2)$ −36

For Exercises 1–8, determine if the expression is true or false. 1.1–1.4

1. $6 + (-2) > -5 + 6$ True

2. $\sqrt{100} - (-4) \leq 7 - (-13)$ True

3. $-21 - 4 \geq -\sqrt{4} - 12$ False

4. $-8 - \sqrt{16} \geq 9 - \sqrt{16}$ False

5. $|-6| + |-14| \leq |-3| + |-17|$ True

6. $16 - |-2| \leq |-3| - (-11)$ True

7. $\sqrt{36} - |-6| > 0$ False

8. $\sqrt{9} + |-3| \leq 0$ False

Multiplication can be thought of as repeated addition. For each product in Exercises 9–12, write an equivalent addition problem.

9. $5(4)$ $4 + 4 + 4 + 4 + 4$

10. $2(6)$ $6 + 6$

11. $3(-2)$ $(-2) + (-2) + (-2)$

12. $5(-6)$ $(-6) + (-6) + (-6) + (-6) + (-6)$

For Exercises 13–18, show how multiplication can be used to check the division problems.

13. $\frac{14}{-2} = -7$ $(-2)(-7) = 14$

14. $\frac{-18}{-6} = 3$ $(-6)3 = -18$

15. $\frac{0}{-5} = 0$ $-5 \cdot 0 = 0$

16. $\frac{0}{-4} = 0$ $-4 \cdot 0 = 0$

17. $\frac{6}{0}$ is undefined   No number multiplied by zero equals 6.

18. $\frac{-4}{0}$ is undefined   No number multiplied by zero equals -4.

For Exercises 19–86, multiply or divide as indicated.

19. $2 \cdot 3$ 6

20. $8 \cdot 6$ 48

21. $2(-3)$ -6

22. $8(-6)$ -48

23. $(-2)3$ -6

24. $(-8)6$ -48

25. $(-2)(-3)$ 6

26. $(-8)(-6)$ 48

27. $24 \div 3$ 8

28. $52 \div 2$ 26

29. $24 \div (-3)$ -8

30. $52 \div (-2)$ -26

31. $(-24) \div 3$ -8

32. $(-52) \div 2$ -26

33. $(-24) \div (-3)$ 8

34. $(-52) \div (-2)$ 26

35. $-6 \cdot 0$ 0

36. $-8 \cdot 0$ 0

37. $-18 \div 0$ Undefined

38. $-42 \div 0$ Undefined

39. $0\left(-\frac{2}{5}\right)$ 0

40. $0\left(-\frac{1}{8}\right)$ 0

41. $0 \div \left(-\frac{1}{10}\right)$ 0

42. $0 \div \left(\frac{4}{9}\right)$ 0

43. $\frac{-14}{-7}$ 2

44. $\frac{-21}{-3}$ 7

45. $\frac{13}{-65}$ $-\frac{1}{5}$

46. $\frac{7}{-77}$ $-\frac{1}{11}$

47. $\frac{-9}{6}$ $-\frac{3}{2}$

48. $\frac{-15}{10}$ $-\frac{3}{2}$

49. $\frac{-30}{-100}$ $\frac{3}{10}$

50. $\frac{-250}{-1000}$ $\frac{1}{4}$

51. $\frac{26}{-13}$ -2

52. $\frac{52}{-4}$ -13

53. $1.72(-4.6)$ -7.912

54. $361.3(-14.9)$ -5383.37

55. $-0.02(-4.6)$ 0.092

56. $-0.06(-2.15)$ 0.129

57. $\frac{14.4}{-2.4}$ -6

58. $\frac{50.4}{-6.3}$ -8

59. $\frac{-5.25}{-2.5}$ 2.1

60. $\frac{-8.5}{-27.2}$ 0.3125, or $\frac{5}{16}$

61. $(-3)^2$ 9

62. $(-7)^2$ 49

63. -3^2 -9

64. -7^2 -49

65. $\left(-\frac{2}{3}\right)^3$ $-\frac{8}{27}$

66. $\left(-\frac{1}{5}\right)^3$ $-\frac{1}{125}$

67. $(-0.2)^4$ 0.0016

68. $(-0.1)^4$ 0.0001

69. -0.2^4 -0.0016

70. -0.1^4 -0.0001

71. $-|-3|$ -3

72. $-|-5|$ -5

73. $-(-3)$ 3

74. $-(-5)$ 5

75. $-|7|$ -7

76. $-|8|$ -8

77. $|-7|$ 7

78. $|-8|$ 8

79. $(-2)(-5)(-3)$ -30

80. $(-6)(-1)(-10)$ -60

81. $(-8)(-4)(-1)(-3)$ 96

82. $(-6)(-3)(-1)(-5)$ 90

83. $100 \div (-10) \div (-5)$ 2

84. $150 \div (-15) \div (-2)$ 5

85. $-12 \div (-6) \div (-2)$ -1

86. $-36 \div (-2) \div 6$ 3

87. For 3 weeks Jim pays $2 a week for lottery tickets. Jim has one winning ticket for $3; write an expression that describes his net gain or loss. How much money has Jim won or lost?
$-2(3) + 3 = -3$, loss of $3

88. Stephanie pays $2 a week for 6 weeks for lottery tickets. Stephanie has one winning ticket for $5; write an expression that describes her net gain or loss. How much money has Stephanie won or lost?
$-2(6) + 5 = -7$, loss of $7

For Exercises 89–100, multiply or divide as indicated.

89. $87 \div (-3)$ -29

90. $96 \div (-6)$ -16

91. $-4(-12)$ 48

92. $(-5)(-11)$ 55

93. $2.8(-5.1)$
-14.28

94. $(7.21)(-0.3)$
-2.163

95. $(-6.8) \div (-0.02)$ 340

96. $(-12.3) \div (-0.03)$ 410

97. $\left(-\dfrac{2}{15}\right)\left(\dfrac{25}{3}\right)$ $-\dfrac{10}{9}$

98. $\left(-\dfrac{5}{16}\right)\left(\dfrac{4}{9}\right)$ $-\dfrac{5}{36}$

99. $\left(-\dfrac{7}{8}\right) \div \left(-\dfrac{9}{16}\right)$
$\dfrac{14}{9}$

100. $\left(-\dfrac{22}{23}\right) \div \left(-\dfrac{11}{3}\right)$
$\dfrac{6}{23}$

101. Is the expression $\frac{10}{5x}$ equal to 10/5x? Explain.
No, parentheses are required around the quantity $5x$; 10/(5x)

102. Is the expression 10/(5x) equal to $\frac{10}{5x}$? Explain.
Yes, the parentheses indicate that the divisor is the quantity $5x$.

For Exercises 103–110, translate the English phrase into an algebraic expression. Then evaluate the expression.

103. The product of -3.75 and 0.3 $-3.75(0.3) = -1.125$

104. The product of -0.4 and -1.258
$(-0.4)(-1.258) = 0.5032$

105. The quotient of $\frac{16}{5}$ and $\left(-\frac{8}{9}\right)$ $\dfrac{16}{5} \div \left(-\dfrac{8}{9}\right) = -\dfrac{18}{5}$

106. The quotient of $\left(-\frac{3}{14}\right)$ and $\frac{1}{7}$ $-\dfrac{3}{14} \div \dfrac{1}{7} = -\dfrac{3}{2}$

107. The number -0.4 plus the quantity 6 times -0.42 $-0.4 + 6(-0.42) = -2.92$

108. The number 0.5 plus the quantity -2 times 0.125 $0.5 + (-2)(0.125) = 0.25$

109. The number $-\frac{1}{4}$ minus the quantity 6 times $-\frac{1}{3}$ $-\dfrac{1}{4} - 6\left(-\dfrac{1}{3}\right) =$

110. Negative five minus the quantity $\left(-\frac{5}{6}\right)$ times $\frac{3}{8}$ $-5 - \left(-\dfrac{5}{6}\right)\dfrac{3}{8} = 0.75 \dfrac{75}{16}$

111. Evaluate the expressions in parts (a) and (b).
    a. $-4 - 3 - 2 - 1$ -10
    b. $-4(-3)(-2)(-1)$ 24
    c. Explain the difference between the operations in parts (a) and (b).
    In part (a) we subtract; in part (b) we multiply.

112. Evaluate the expressions in parts (a) and (b).
    a. $-10 - 9 - 8 - 7$ -34
    b. $-10(-9)(-8)(-7)$ 5040
    c. Explain the difference between the operations in parts (a) and (b).
    In part (a) we subtract; in part (b) we multiply.

For Exercises 113–126, perform the indicated operations.

113. $8 - 2^3 \cdot 5 + 3 - (-6)$ -23

114. $-14 \div (-7) - 8 \cdot 2 + 3^3$ 13

115. $-(2 - 8)^2 \div (-6) \cdot 2$ 12

116. $-(3 - 5)^2 \cdot 6 \div (-4)$ 6

117. $\dfrac{6(-4) - 2(5 - 8)}{-6 - 3 - 5}$ $\dfrac{9}{7}$

118. $\dfrac{3(-4) - 5(9 - 11)}{-9 - 2 - 3}$ $\dfrac{1}{7}$

119. $\dfrac{-4 + 5}{(-2) \cdot 5 + 10}$ Undefined

120. $\dfrac{-3 + 10}{2(-4) + 8}$ Undefined

121. $|-5| - |-7|$ -2

122. $|-8| - |-2|$ 6

123. $-|-1| - |5|$ -6

124. $-|-10| - |6|$ -16

125. $\dfrac{|2 - 9| - |5 - 7|}{10 - 15}$ -1

126. $\dfrac{|-2 + 6| - |3 - 5|}{13 - 11}$ 1

For Exercises 127–134, evaluate the expression for $x = -2$, $y = -4$, and $z = 6$.

127. $x^2 - 2y$ 12

128. $3y^2 - z$ 42

129. $4(2x - z)$ -40

130. $6(3x + y)$ -60

131. $\dfrac{3x + 2y}{y}$ $\dfrac{7}{2}$

132. $\dfrac{2z - y}{x}$ -8

133. $\dfrac{x + 2y}{x - 2y}$ $-\dfrac{5}{3}$

134. $\dfrac{x - z}{x^2 - z^2}$ $\dfrac{1}{4}$

135. Evaluate $x^2 + 6$ for $x = 2$, and for $x = -2$
For $x = 2$, 10; for $x = -2$, 10

136. Evaluate $y^3 + 6$ for $y = 2$ and for $y = -2$
For $y = 2$, 14; for $y = -2$, -2

section
1.6   PROPERTIES OF REAL NUMBERS AND SIMPLIFYING EXPRESSIONS

Concepts

1. Commutative Properties of Real Numbers
2. Associative Properties of Real Numbers
3. Identity and Inverse Properties of Real Numbers
4. Distributive Property of Multiplication over Addition
5. Simplifying Algebraic Expressions
6. Clearing Parentheses and Combining *Like* Terms

1.   Commutative Properties of Real Numbers

When getting dressed in the morning, it makes no difference whether you put on your left shoe first and then your right shoe, or vice versa. This example illustrates a process in which the order does not affect the outcome. Such a process or operation is said to be commutative.

In algebra, the operations of addition and multiplication are commutative because the order in which we add or multiply two real numbers does not affect the result. For example,

$$10 + 5 = 5 + 10 \quad \text{and} \quad 10 \cdot 5 = 5 \cdot 10$$

Commutative Properties of Real Numbers

If a and b are real numbers, then

1. $a + b = b + a$ **commutative property of addition**

2. $ab = ba$ **commutative property of multiplication**

It is important to note that although the operations of addition and multiplication are commutative, subtraction and division are *not* commutative. For example:

$$\underbrace{10 - 5}_{5} \neq \underbrace{5 - 10}_{-5} \quad \text{and} \quad \underbrace{10 \div 5}_{2} \neq \underbrace{5 \div 10}_{\frac{1}{2}}$$

example 1

Applying the Commutative Property of Addition

Use the commutative property of addition to rewrite each expression.

a. $-3 + (-7)$ b. $3x^3 + 5x^4$

Solution:

a. $-3 + (-7) = -7 + (-3)$
b. $3x^3 + 5x^4 = 5x^4 + 3x^3$

Recall that subtraction is not a commutative operation. However, if we rewrite the difference of two numbers, $a - b$, as $a + (-b)$, we can apply the commutative property of addition. This is demonstrated in the next example.

example 2

Applying the Commutative Property of Addition

Rewrite the expression in terms of addition. Then apply the commutative property of addition.

a. $5a - 3b$ b. $z^2 - \dfrac{1}{4}$

Solution:

a. $5a - 3b$

   $= 5a + (-3b)$ Rewrite subtraction as addition of $-3b$.

   $= -3b + 5a$ Apply the commutative property of addition.

b. $z^2 - \dfrac{1}{4}$

   $= z^2 + \left(-\dfrac{1}{4}\right)$ Rewrite subtraction as addition of $-\frac{1}{4}$.

   $= -\dfrac{1}{4} + z^2$ Apply the commutative property of addition.

example 3

Applying the Commutative Property of Multiplication

Use the commutative property of multiplication to rewrite each expression.

a. $12(-6)$ b. $x \cdot 4$

Solution:

a. $12(-6) = -6(12)$

b. $x \cdot 4 = 4 \cdot x$ (or simply $4x$)

2.   Associative Properties of Real Numbers

The associative property of real numbers states that the order in which three or more real numbers are grouped under addition or multiplication will not affect the outcome. For example,

$$(5 + 10) + 2 = 5 + (10 + 2) \qquad \text{and} \qquad (5 \cdot 10)2 = 5(10 \cdot 2)$$

$$15 + 2 = 5 + 12 \qquad\qquad\qquad (50)2 = 5(20)$$

$$17 = 17 \qquad\qquad\qquad\qquad 100 = 100$$

Associative Properties of Real Numbers

If a, b, and c represent real numbers, then

1. $(a + b) + c = a + (b + c)$ **associative property of addition**

2. $(ab)c = a(bc)$ **associative property of multiplication**

---

example 4

Applying the Associative Property of Multiplication

Use the associative property of multiplication to rewrite each expression. Then simplify the expression if possible.

a. $(5y)y$ b. $4(5z)$ c. $-\dfrac{3}{2}\left(-\dfrac{2}{3}w\right)$

Solution:

a. $(5y)y$

$= 5(y \cdot y)$ Apply the associative property of multiplication.

$= 5y^2$ Simplify.

b. $4(5z)$

$= (4 \cdot 5)z$ Apply the associative property of multiplication.

$= 20z$ Simplify.

c. $-\dfrac{3}{2}\left(-\dfrac{2}{3}w\right)$

$= \left[-\dfrac{3}{2}\left(-\dfrac{2}{3}\right)\right]w$ Apply the associative property of multiplication.

$= 1w$ Simplify.

$= w$

Note: In most cases, a detailed application of the associative property will not be shown when multiplying two expressions. Instead, the process will be written in one step, such as

$$(5y)y = 5y^2, \quad 4(5z) = 20z \qquad \text{and} \qquad -\frac{3}{2}\left(-\frac{2}{3}w\right) = w$$

3. Identity and Inverse Properties of Real Numbers

The number 0 has a special role under the operation of addition. Zero added to any real number does not change the number. Therefore, the number 0 is said to be the additive identity (also called the identity element of addition). For example,

$$-4 + 0 = -4 \qquad 0 + 5.7 = 5.7 \qquad 0 + \frac{3}{4} = \frac{3}{4}$$

The number 1 has a special role under the operation of multiplication. Any real number multiplied by 1 does not change the number. Therefore, the number 1 is said to be the multiplicative identity (also called the identity element of multiplication). For example,

$$(-8)1 = -8 \qquad 1(-2.85) = -2.85 \qquad 1\left(\frac{1}{5}\right) = \frac{1}{5}$$

Identity Properties of Real Numbers

If a is a real number, then

1. $a + 0 = 0 + a = a$ **identity property of addition**
2. $a \cdot 1 = 1 \cdot a = a$ **identity property of multiplication**

The sum of a number and its opposite equals 0. For example, $-12 + 12 = 0$. For any real number, a, the opposite of a (also called the **additive inverse** of a) is $-a$ and $a + (-a) = -a + a = 0$. The inverse property of addition states that the sum of any number and its additive inverse is the identity element, 0. For example,

Number	Additive Inverse (opposite)	Sum
9	-9	$9 + (-9) = 0$
-21.6	21.6	$-21.6 + 21.6 = 0$
$\dfrac{2}{7}$	$-\dfrac{2}{7}$	$\dfrac{2}{7} + \left(-\dfrac{2}{7}\right) = 0$

If b is a nonzero real number, then the reciprocal of b (also called the **multiplicative inverse** of b) is $\frac{1}{b}$. The inverse property of multiplication states that the product of b and its multiplicative inverse is the identity element, 1. Symbolically, we have $b \cdot \frac{1}{b} = \frac{1}{b} \cdot b = 1$. For example,

Number	Multiplicative Inverse (reciprocal)	Product
7	$\dfrac{1}{7}$	$7 \cdot \dfrac{1}{7} = 1$
3.14	$\dfrac{1}{3.14}$	$3.14\left(\dfrac{1}{3.14}\right) = 1$
$-\dfrac{3}{5}$	$-\dfrac{5}{3}$	$-\dfrac{3}{5}\left(-\dfrac{5}{3}\right) = 1$

<hr>

Inverse Properties of Real Numbers

If a is a real number and b is a nonzero real number, then

1. $a + (-a) = -a + a = 0$ **inverse property of addition**

2. $b \cdot \dfrac{1}{b} = \dfrac{1}{b} \cdot b = 1$ **inverse property of multiplication**

<hr>

Classroom Activity 1.6A

4. Distributive Property of Multiplication over Addition

The operations of addition and multiplication are related by an important property called the **distributive property of multiplication over addition**. Consider the expression $6(2 + 3)$. The order of operations indicates that the sum $2 + 3$ is evaluated first, and then the result is multiplied by 6:

$$6(2 + 3)$$
$$= 6(5)$$
$$= 30$$

Notice that the same result is obtained if the factor of 6 is multiplied by each of the numbers 2 and 3, and then their products are added:

$6(2 + 3)$ The factor of 6 is distributed to the numbers 2 and 3.

$$= 6(2) + 6(3)$$
$$= \quad 12 + 18$$
$$= \quad\quad 30$$

Tip: The mathematical definition of the distributive property is consistent with the everyday meaning of the word *distribute*. To distribute means to "spread out from one to many." In the mathematical context, the factor a is distributed to both b and c in the parentheses.

The distributive property of multiplication over addition states that this is true in general.

Distributive Property of Multiplication over Addition

If a, b, and c are real numbers, then

$$a(b + c) = ab + ac \quad\quad \text{and} \quad\quad (b + c)a = ab + ac$$

<hr>

example 5   Applying the Distributive Property

Apply the distributive property: $2(a + 6b + 7)$

Solution:

Tip: Notice that the parentheses are removed after the distributive property is applied. Sometimes this is referred to as clearing parentheses.

$2(a + 6b + 7)$

$= 2(a + 6b + 7)$

$= 2(a) + 2(6b) + 2(7)$ Apply the distributive property.

$= 2a + 12b + 14$ Simplify.

Because the difference of two expressions $a - b$ can be written in terms of addition as $a + (-b)$, the distributive property can be applied when the operation of subtraction is present within the parentheses. For example,

$5(y - 7)$

$= 5[y + (-7)]$ Rewrite subtraction as addition of -7.

$= 5[y + (-7)]$ Apply the distributive property.

$= 5(y) + 5(-7)$

$= 5y + (-35)$, or $5y - 35$ Simplify.

---

example 6

Applying the Distributive Property

Use the distributive property to rewrite each expression.

a. $-(-3a + 2b + 5c)$ b. $-6(2 - 4x)$

Classroom Activity 1.6B

Tip: Notice that a negative factor preceding the parentheses changes the signs of all terms to which it is multiplied.

$-1(-3a + 2b + 5c)$
$= +3a - 2b - 5c$

Solution:

a. $-(-3a + 2b + 5c)$

$= -1(-3a + 2b + 5c)$ The negative sign preceding the parentheses can be interpreted as taking the opposite of the quantity that follows or as
$-1(-3a + 2b + 5c)$

$= -1(-3a + 2b + 5c)$

$= -1(-3a) + (-1)(2b) + (-1)(5c)$ Apply the distributive property.

$= 3a + (-2b) + (-5c)$ Simplify.

$= 3a - 2b - 5c$

b. $-6(2 - 4x)$

$= -6[2 + (-4x)]$ Change subtraction to addition of $-4x$.

$= -6[2 + (-4x)]$

$= -6(2) + (-6)(-4x)$ Apply the distributive property. Notice that multiplying by -6 changes the signs of all terms to which it is applied.

$= -12 + 24x$ Simplify.

Note: In most cases, the distributive property will be applied without as much detail as shown in Examples 5 and 6. Instead, the distributive property will be applied in one step.

$2(a + 6b + 7)$ $-(3a + 2b + 5c)$ $-6(2 - 4x)$

1 step $= 2a + 12b + 14$ 1 step $= -3a - 2b - 5c$ 1 step $= -12 + 24x$

5. Simplifying Algebraic Expressions

An algebraic expression is the sum of one or more terms. A term is a constant or the product of a constant and one or more variables. For example, the expression

$$-7x^2 + xy - 100 \quad \text{or} \quad -7x^2 + xy + (-100)$$
$$\text{consists of the terms } -7x^2, xy, \text{ and } -100$$

The terms $-7x^2$ and xy are **variable terms** and the term -100 is called a **constant term**. It is important to distinguish between a term and the factors within a term. For example, the quantity xy is one term, and the values x and y are factors within the term. The constant factor in a term is called the numerical coefficient (or simply **coefficient**) of the term. In the terms $-7x^2, xy,$ and -100, the coefficients are $-7, 1,$ and -100 respectively.

Terms are said to be *like* terms if they each have the same variables, and the corresponding variables are raised to the same powers. For example,

Like Terms			*Unlike* Terms			
$-3b$	and	$5b$	$-5c$	and	$7d$	(different variables)
$17xy$	and	$-4xy$	$6xy$	and	$3x$	(different variables)
$9p^2q^3$	and	p^2q^3	$4p^2q^3$	and	$8p^3q^2$	(different powers)
$5w$	and	$2w$	$5w$	and	2	(different variables)
7	and	10	7	and	$10a$	(different variables)

---

example 7

Identifying Terms, Factors, Coefficients and *Like* Terms

a. List the terms of the expression $5x^2 - 3x + 2$
b. Identify the coefficient of the term $6yz^3$
c. Which of the pairs are *like* terms: $8b, 3b^2$ or $4c^2d, -6c^2d$

Classroom Activity 1.6C

Solution:

a. The terms of the expression $5x^2 - 3x + 2$ are $5x^2, -3x,$ and 2.
b. The coefficient of $6yz^2$ is 6.
c. $4c^2d$ and $-6c^2d$ are *like* terms.

---

6. Clearing Parentheses and Combining *Like* Terms

Two terms can be added or subtracted only if they are *like* terms. To add or subtract *like* terms, we use the distributive property as shown in the next example.

---

example 8

Using the Distributive Property to Add and Subtract *Like* Terms

Add or subtract as indicated.

a. $7x + 2x$
b. $-2p + 3p - p$

Solution:

a. $7x + 2x$

$= (7 + 2)x$ Apply the distributive property.

$= 9x$ Simplify.

b. $-2p + 3p - p$

$= -2p + 3p - 1p$ Note that $-p$ equals $-1p$.

$= (-2 + 3 - 1)p$ Apply the distributive property.

$= (0)p$ Simplify.

$= 0$

Although the distributive property is used to add and subtract *like* terms, it is tedious to write each step. Observe that adding or subtracting *like* terms is a matter of combining the coefficients and leaving the variable factors unchanged. This can be shown in one step, a shortcut that we will use throughout the text. For example,

$$7x + 2x = 9x \qquad -2p + 3p - 1p = 0p = 0 \qquad -3a - 6a = -9a$$

---

example 9   **Using the Distributive Property to Add and Subtract *Like* Terms**

a. $3yz + 5 - 2yz + 9$ b. $1.2w^3 + 5.7w^3$

Solution:

a. $3yz + 5 - 2yz + 9$

Tip: Notice that constants such as 5 and 9 are *like* terms.

$= 3yz - 2yz + 5 + 9$ Group *like* terms together.

$= 1yz + 14$ Combine *like* terms.

$= yz + 14$

b. $1.2w^3 + 5.7w^3$

$= 6.9w^3$ Combine *like* terms.

The next two examples illustrate how the distributive property is used to clear parentheses.

---

example 10   **Clearing Parentheses and Combining *Like* Terms**

Simplify by clearing parentheses and combining *like* terms: $5 - 2(3x + 7)$

Solution:

$5 - 2(3x + 7)$ The order of operations indicates that we must perform multiplication before subtraction.

It is important to understand that a factor of -2 (not 2) will be multiplied to all terms within the parentheses. To see why this is so, we may rewrite the subtraction in terms of addition.

$$= 5 + (-2)(3x + 7) \qquad \text{Change subtraction to addition.}$$

$$= 5 + (-2)(3x + 7) \qquad \text{A factor of } -2 \text{ is to be distributed to terms in the parentheses.}$$

$$= 5 + (-2)(3x) + (-2)(7) \qquad \text{Apply the distributive property.}$$

$$= 5 + (-6x) + (-14) \qquad \text{Simplify.}$$

$$= 5 + (-14) + (-6x) \qquad \text{Group } like \text{ terms together.}$$

$$= -9 + (-6x) \qquad \text{Combine } like \text{ terms.}$$

$$= -9 - 6x$$

---

example 11

Clearing Parentheses and Combining *Like* Terms

Simplify by clearing parentheses and combining *like* terms.

a. $10(5y + 2) - 6(y - 1)$ b. $\dfrac{1}{4}(4k + 2) - \dfrac{1}{2}(6k + 1)$

c. $-(4s - 6t) - (3t + 5s) - 2s$

Classroom Activity 1.6D

Solution:

a. $10(5y + 2) - 6(y - 1)$

$$= 50y + 20 - 6y + 6 \qquad \text{Apply the distributive property. Notice that a factor of } -6 \text{ is distributed through the second parentheses and changes the signs.}$$

$$= 50y - 6y + 20 + 6 \qquad \text{Group } like \text{ terms together.}$$

$$= 44y + 26 \qquad \text{Combine } like \text{ terms.}$$

b. $\dfrac{1}{4}(4k + 2) - \dfrac{1}{2}(6k + 1)$

$$= \dfrac{4}{4}k + \dfrac{2}{4} - \dfrac{6}{2}k - \dfrac{1}{2} \qquad \text{Apply the distributive property. Notice that a factor of } -\tfrac{1}{2} \text{ is distributed through the second parentheses and changes the signs.}$$

$$= k + \dfrac{1}{2} - 3k - \dfrac{1}{2} \qquad \text{Simplify fractions.}$$

$$= k - 3k + \dfrac{1}{2} - \dfrac{1}{2} \qquad \text{Group } like \text{ terms together.}$$

$$= -2k + 0 \qquad \text{Combine } like \text{ terms.}$$
$$= -2k$$

c. $-(4s - 6t) - (3t + 5s) - 2s$

$= -4s + 6t - 3t - 5s - 2s$ Apply the distributive property.

$= -4s - 5s - 2s + 6t - 3t$ Group *like* terms together.

$= -11s + 3t$ Combine *like* terms.

section 1.6   PRACTICE EXERCISES

For Exercises 1–16, perform the indicated operations.
1.3–1.5

1. $(-6) + 14$ 8

2. $(-2) + 9$ 7

3. $-13 - (-5)$ -8

4. $-1 - (-19)$ 18

5. $18 \div (-4)$ $-\frac{9}{2}$, or -4.5

6. $-27 \div 5$ $-\frac{27}{5}$, or -5.4

7. $-3 \cdot 0$ 0

8. $0(-15)$ 0

9. $\frac{1}{2} + \frac{3}{8}$ $\frac{7}{8}$

10. $\frac{7}{2} + \frac{5}{9}$ $\frac{73}{18}$

11. $\frac{25}{21} - \frac{6}{7}$ $\frac{1}{3}$

12. $\frac{8}{9} - \frac{1}{3}$ $\frac{5}{9}$

13. $\left(-\frac{3}{5}\right)\left(\frac{4}{27}\right)$ $-\frac{4}{45}$

14. $\left(\frac{1}{6}\right)\left(-\frac{8}{3}\right)$ $-\frac{4}{9}$

15. $\left(-\frac{11}{12}\right) \div \left(-\frac{5}{4}\right)$ $\frac{11}{15}$

16. $\left(-\frac{14}{15}\right) \div \left(-\frac{7}{5}\right)$ $\frac{2}{3}$

17. What is another name for multiplicative inverse?   Reciprocal

18. What is another name for additive inverse?   Opposite

19. What is the additive identity?   0

20. What is the multiplicative identity?   1

For Exercises 21–29, match the statements with the properties of multiplication and addition.

21. $6 \cdot \frac{1}{6} = 1$ b

22. $7(4 \cdot 9) = (7 \cdot 4)9$ f

23. $2(3 + k) = 6 + 2k$ i

24. $3 \cdot 7 = 7 \cdot 3$ c

25. $5 + (-5) = 0$ g

a. Commutative property of addition

b. Inverse property of multiplication

c. Commutative property of multiplication

d. Associative property of addition

e. Identity property of multiplication

26. $18 \cdot 1 = 18$ e

27. $(3 + 7) + 19 = 3 + (7 + 19)$ d

28. $23 + 6 = 6 + 23$ a

29. $3 + 0 = 3$ h

f. Associative property of multiplication

g. Inverse property of addition

h. Identity property of addition

i. Distributive property of multiplication over addition

For Exercises 30–45, name the property that justifies each statement.

30. $2 + (-3) = (-3) + 2$
Commutative property of addition

31. $-2 + 0 = -2$ Identity property of addition

32. $4(3a) = (4 \cdot 3)a$
Associative property of multiplication

 33. $\frac{7}{8} \cdot \frac{8}{7} = 1$
Inverse property of multiplication

34. $(-6 + 1) + 10 = -6 + (1 + 10)$
Associative property of addition

35. $8 + (-8) = 0$ Inverse property of addition

36. $-3(b + 7) = -3b - 21$
Distributive property of multiplication over addition

37. $5(12) = 12(5)$ Commutative property of multiplication

38. $0 + \left(-\frac{2}{3}\right) = \left(-\frac{2}{3}\right)$ Identity property of addition

 39. $2 + (4 + (-1)) = (2 + 4) + (-1)$
Associative property of addition

40. $-25(1) = -25$
Identity property of multiplication

41. $5(2y + 9) = 10y + 45$
Distributive property of multiplication over addition

42. $-3.82 + 3.82 = 0$ Inverse property of addition

43. $-8 + (-11) = -11 + (-8)$
Commutative property of addition

44. $(6 + 2) + (-5) = (-5) + (6 + 2)$
Commutative property of addition

45. $3 + \left(-\frac{1}{2} + 2\right) = \left(-\frac{1}{2} + 2\right) + 3$
Commutative property of addition

 See Additional Answers Appendix   Writing   Translating Expression   Geometry   Scientific Calculator   Video

For Exercises 46–53, determine if the expressions are equivalent. If two expressions are not equivalent, state why.

46. $3a + b, b + 3a$
Equivalent

47. $4y + 1, 1 + 4y$
Equivalent

48. $2c + 7, 9c$ ❖

49. $5z + 4, 9z$ ❖

50. $5x - 3, 3 - 5x$
quivalent; subtraction is not commutative.

51. $6d - 7, 7 - 6d$
Not equivalent; subtraction is not commutative.

52. $5x - 3, -3 + 5x$
Equivalent

53. $8 - 2x, -2x + 8$
Equivalent

54. Which grouping of terms is easier computationally,

$$(14\tfrac{2}{7} + 2\tfrac{1}{3}) + \tfrac{2}{3} \quad \text{or} \quad 14\tfrac{2}{7} + (2\tfrac{1}{3} + \tfrac{2}{3})?$$
$14\tfrac{2}{7} + (2\tfrac{1}{3} + \tfrac{2}{3})$ is easier.

55. Which grouping of terms is easier computationally,

$$(5\tfrac{1}{8} + 18\tfrac{2}{5}) + 1\tfrac{1}{5} \quad \text{or} \quad 5\tfrac{1}{8} + (18\tfrac{2}{5} + 1\tfrac{1}{5})?$$
$5\tfrac{1}{8} + (18\tfrac{2}{5} + 1\tfrac{1}{5})$ is easier.

For Exercises 56–69, use the distributive property to clear parentheses.

56. $6(5x + 1)$ $\quad 30x + 6$
57. $2(x + 7)$ $\quad 2x + 14$

58. $-2(a + 8)$ $\quad -2a - 16$
59. $-3(2z + 9)$ $\quad -6z - 27$

60. $3(5c - d)$ $\quad 15c - 3d$
61. $4(w - 13z)$ $\quad 4w - 52z$

62. $-7(y - 2)$ $\quad -7y + 14$
63. $-2(4x - 1)$ $\quad -8x + 2$

64. $-\dfrac{2}{3}(x - 6)$ $\quad -\tfrac{2}{3}x + 4$
65. $-\dfrac{1}{4}(2b - 8)$ $\quad -\tfrac{1}{2}b + 2$

66. $-(2p + 10)$
$\quad -2p - 10$
67. $-(7q + 1)$
$\quad -7q - 1$

68. $-(-3w - 5z)$
$\quad 3w + 5z$
69. $-(-7a - b)$
$\quad 7a + b$

The distributive property can be used to simplify a product of two numbers by writing one of the factors as a sum or difference. For example,

$$2(98) = 2(100 - 2) = 200 - 4 = 196$$

$$5(27) = 5(20 + 7) = 100 + 35 = 135$$

For Exercises 70–73, rewrite the expression in parentheses as a sum or difference. Then use the distributive property to evaluate the products without the use of a calculator.

70. $4(92)$ $\quad 368$
71. $3(81)$ $\quad 243$

72. $4(902)$ $\quad 3608$
73. $5(799)$ $\quad 3995$

For Exercises 74–77, for each polynomial, list the terms and their coefficients.

74. $3xy - 6x^2 + y - 17$ 75. $2x - y + 18xy + 5$

Term	Coefficient
$3xy$	3
$-6x^2$	-6
y	1
-17	-17

Term	Coefficient
$2x$	2
$-y$	-1
$18xy$	18
5	5

Table for Exercise 74 **Table for Exercise 75**

76. $x^4 - 10xy + 12 - y$ 77. $-x + 8y - 9x^2y - 3$

Term	Coefficient
x^4	1
$-10xy$	-10
12	12
$-y$	-1

Term	Coefficient
$-x$	-1
$8y$	8
$-9x^2y$	-9
-3	-3

Table for Exercise 76 **Table for Exercise 77**

78. Explain why $12x$ and $12x^2$ are not *like* terms.
The exponents on x are different.

79. Explain why $3x$ and $3xy$ are not *like* terms.
The variable factors are different.

80. Explain why $7z$ and $\sqrt{13}z$ are *like* terms.
The variables are the same and raised to the same power.

81. Explain why $2x$ and $8x$ are *like* terms.
The variables are the same and raised to the same power.

82. Write three different *like* terms.
For example: $5y, -2y, y$

83. Write three terms that are not *like*.
For example: $5y, -2x, 6$

For Exercises 84–91, simplify by combining *like* terms.

84. $5k - 10k - 12k + 16 + 7$ $\quad -17k + 23$

85. $-4p - 2p + 8p - 15 + 3$ $\quad 2p - 12$

86. $9x - 7y + 12x + 14y$ $\quad 21x + 7y$

87. $2y - 8z + y - 5z - 3y$ $\quad -13z$

88. $\dfrac{1}{4}a + b - \dfrac{3}{4}a - 5b$ $\quad -\tfrac{1}{2}a - 4b$

89. $\dfrac{2}{5} + 2t - \dfrac{3}{5} + t - \dfrac{6}{5}$ $\quad 3t - \tfrac{7}{5}$

90. $2.8z - 8.1z + 6 - 15.2$ $\quad -5.3z - 9.2$

91. $2.4 - 8.4w - 2w + 0.9$ $\quad -10.4w + 3.3$

For Exercises 92–115, simplify by clearing parentheses and combining *like* terms.

92. $-3(2x - 4) + 10$
   $-6x + 22$

93. $-2(4a + 3) - 14$
   $-8a - 20$

94. $4(w + 3) - 12$ $4w$

95. $5(2r + 6) - 30$ $10r$

96. $5 - 3(x - 4)$
   $-3x + 17$

97. $4 - 2(3x + 8)$
   $-6x - 12$

98. $-3(2t + 4) + 8(2t - 4)$ $10t - 44$

99. $-5(5y + 9) + 3(3y + 6)$ $-16y - 27$

100. $2(w - 5) - (2w + 8)$ -18

 101. $6(x + 3) - (6x - 5)$ 23

102. $-\dfrac{1}{3}(6t + 9) + 10$
   $-2t + 7$

103. $-\dfrac{3}{4}(8 + 4q) + 7$
   $-3q + 1$

104. $10(5.1a - 3.1) + 4$ $51a - 27$

105. $100(-3.14p - 1.05) + 212$ $-314p + 107$

106. $-4m + 2(m - 3) + 2m$ -6

107. $-3b + 4(b + 2) - 8b$ $-7b + 8$

108. $\dfrac{1}{2}(10q - 2) + \dfrac{1}{3}(2 - 3q)$ $4q - \dfrac{1}{3}$

109. $\dfrac{1}{5}(15 - 4p) - \dfrac{1}{10}(10p + 5)$ $-\dfrac{9}{5}p + \dfrac{5}{2}$

110. $7n - 2(n - 3) - 6 + n$ $6n$

111. $8k - 4(k - 1) + 7 - k$ $3k + 11$

112. $6(x + 3) - 12 - 4(x - 3)$ $2x + 18$

113. $5(y - 4) + 3 - 6(y - 7)$ $-y + 25$

114. $6.1(5.3z - 4.1) - 5.8$ $32.33z - 30.81$

115. $-3.6(1.7q - 4.2) + 14.6$ $-6.12q + 29.72$

▮ EXPANDING YOUR SKILLS

116. As a small child in school, the great mathematician Karl Friedrich Gauss (1777–1855) was said to have found the sum of the integers from 1 to 100 mentally:

$$1 + 2 + 3 + 4 + \cdots + 99 + 100$$

Rather than adding the numbers sequentially, he added the numbers in pairs:

$$(1 + 99) + (2 + 98) + (3 + 97) + \cdots$$

a. Use this technique to add the integers from 1 to 10. 55

$$1 + 2 + 3 + 4 + 5 + 6 + 7 + 8 + 9 + 10$$

b. Use this technique to add the integers from 1 to 20. 210

Concepts

1. **Introduction to Graphing**
2. **Rectangular Coordinate System**
3. **Plotting Points in a Rectangular Coordinate System**
4. **Applications of Plotting Points**

section

1.7 CONNECTIONS TO GRAPHING: RECTANGULAR COORDINATE SYSTEM

1. Introduction to Graphing

Mathematics is a powerful tool used by scientists and has directly contributed to the highly technical world we live in. Applications of mathematics have led to advances in the sciences, business, computer technology, and medicine.

One fundamental application of mathematics is the graphical representation of numerical information (or **data**). For example, Table 1-7 represents the number of clients admitted to a drug and alcohol rehabilitation program over a 24-month period from January 2000 through December 2001.

Table 1-7

	Month	Number of Clients		Month	Number of Clients
Jan. 2000	1	55	Jan. 2001	13	55
Feb.	2	62	Feb.	14	72
March	3	57	March	15	75
April	4	49	April	16	82
May	5	73	May	17	70
June	6	64	June	18	85
July	7	68	July	19	79
Aug.	8	60	Aug.	20	91
Sept.	9	70	Sept.	21	88
Oct.	10	72	Oct.	22	95
Nov.	11	69	Nov.	23	86
Dec.	12	68	Dec.	24	79

In table form, the information is difficult to picture and interpret. It appears that on a monthly basis, the number of clients fluctuates. However, when the data are represented in a graph, an upward trend is clear (Figure 1-6).

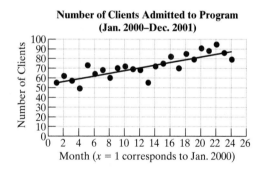

Number of Clients Admitted to Program (Jan. 2000–Dec. 2001)

Month ($x = 1$ corresponds to Jan. 2000)

Figure 1-6

From the increase in clients depicted in this graph, management for the reha- bilitation center might make plans for the future. If the trend continues, manage- ment might consider expanding its facilities and increasing its staff to accommodate the expected increase in clients.

2.   Rectangular Coordinate System

In the previous example, two variables are represented, time and the number of clients. To picture two variables, we use a graph with two number lines drawn at right angles to each other (Figure 1-7). This forms a **rectangular coordinate system**. The horizontal line is called the **x-axis**, and the vertical line is called the **y-axis**. The point where the lines intersect is called the **origin**. On the x-axis, the numbers to the right of the origin are positive and the numbers to the left are negative. On the y-axis, the numbers above the origin are positive and the numbers below are negative. The x- and y-axes divide the graphing area into four regions called **quadrants**.

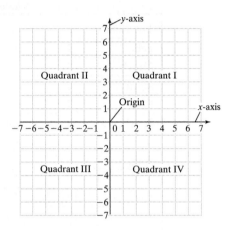

Figure 1-7

3.   Plotting Points in a Rectangular Coordinate System

Points graphed in a rectangular coordinate system are defined by two numbers as an **ordered pair**, (x, y). The first number (called the first coordinate, or the abscissa) is the horizontal position from the origin. The second number (called the second coordinate, or the ordinate) is the vertical position from the origin. The next example shows how points are plotted in a rectangular coordinate system.

example 1

Plotting Points in a Rectangular Coordinate System

Plot the points.

a. $(4, 5)$ b. $(-4, -5)$ c. $(-1, 3)$ d. $(3, -1)$

e. $\left(\dfrac{1}{2}, -\dfrac{7}{3}\right)$ f. $(-2, 0)$ g. $(0, 0)$ h. $(\pi, 1.1)$

Classroom Activity 1.7A

Solution:

a.   The ordered pair $(4, 5)$ indicates that $x = 4$ and $y = 5$. Move 4 units in the positive x-direction (4 units to the right), and from there move 5 units in the positive y-direction (5 units up). Then plot the point. The point is in Quadrant I.

b.   The ordered pair $(-4, -5)$ indicates that $x = -4$ and $y = -5$. Move 4 units in the negative x-direction (4 units to the left), and from there move 5 units in the negative y-direction (5 units down). Then plot the point. The point is in Quadrant III.

c.   The ordered pair $(-1, 3)$ indicates that $x = -1$ and $y = 3$. Move 1 unit to the left and 3 units up. The point is in Quadrant II.

d.   The ordered pair $(3, -1)$ indicates that $x = 3$ and $y = -1$. Move 3 units to the right and 1 unit down. The point is in Quadrant IV.

e.   The improper fraction $-\frac{7}{3}$ can be written as the mixed number $-2\frac{1}{3}$. Therefore, to plot the point $\left(\frac{1}{2}, -\frac{7}{3}\right)$ move to the right $\frac{1}{2}$ unit, and down $2\frac{1}{3}$ units. The point is in Quadrant IV.

f.   The point $(-2, 0)$ indicates $y = 0$. Therefore, the point is on the x-axis.

g.   The point $(0, 0)$ is at the origin.

h.   The irrational number, π, can be approximated as 3.14. Thus, the point $(\pi, 1.1)$ is located approximately 3.14 units to the right and 1.1 units up. The point is in Quadrant I.

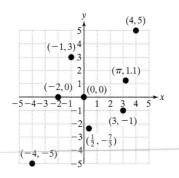

Figure 1-8

Tip: Notice that changing the order of the x- and y-coordinates changes the location of the point. The point $(-1, 3)$ for example is in Quadrant II, whereas $(3, -1)$ is in Quadrant IV (Figure 1-8). This is why points are represented by *ordered* pairs. The order is important.

4.   Applications of Plotting Points

example 2

Classroom Activity 1.7B

Plotting Points in an Application

The daily low temperatures (in degrees Fahrenheit) for Sudbury, Ontario, Canada, for the week of January 7–13, 2001, is given in Table 1-8.

Table 1-8

Date	Day Number, x	Temperature, (°F), y
Jan. 7, 2001	1	−3
Jan. 8, 2001	2	−5
Jan. 9, 2001	3	1
Jan. 10, 2001	4	6
Jan. 11, 2001	5	5
Jan. 12, 2001	6	0
Jan. 13, 2001	7	−4

a.   Write an ordered pair for each row in the table using the day number as the x-coordinate and the temperature as the y-coordinate.

b.   Plot the ordered pairs from part (a) on a rectangular coordinate system.

Solution:

a. $(1, -3)$ Each ordered pair represents the day number and the corre-
   $(2, -5)$ sponding low temperature for that day.
   $(3, 1)$
   $(4, 6)$
   $(5, 5)$
   $(6, 0)$
   $(7, -4)$

b.

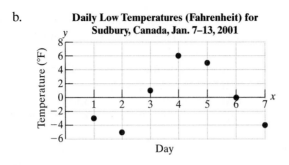

Daily Low Temperatures (Fahrenheit) for
Sudbury, Canada, Jan. 7–13, 2001

---

example 3

Plotting Points in an Application

The price per share of a stock (in dollars) over a period of eight days is shown in Figure 1-9.

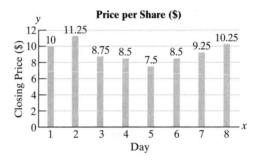

Figure 1-9

a. Use the information provided in the graph to write the ordered pairs corresponding to each point. Interpret the meaning of each ordered pair.

b. Compute the daily gain or loss for the stock as follows:

Gain or Loss = (current day's price) − (previous day's price)

Complete the following table, beginning on day 2. Then plot the corresponding points on a rectangular coordinate system to show daily gain or loss.

Day, x	Daily Gain or Loss, y	Ordered Pair
2		
3		
4		
5		
6		
7		
8		

Solution:

a.   (1, 10)   The price per share on day 1 was $10.

  (2, 11.25)   The price per share on day 2 was $11.25.

  (3, 8.75)   The price per share on day 3 was $8.75.

  (4, 8.50)   The price per share on day 4 was $8.50.

  (5, 7.50)   The price per share on day 5 was $7.50.

  (6, 8.50)   The price per share on day 6 was $8.50.

  (7, 9.25)   The price per share on day 7 was $9.25.

  (8, 10.25)   The price per share on day 8 was $10.25.

b.   To find the daily gain or loss, subtract the previous day's stock price from the current day's price.

  For Day 2:   ($11.25) − ($10.00) = $1.25

  For Day 3:   ($8.75) − ($11.25) = −$2.50

  For Day 4:   ($8.50) − ($8.75) = −$0.25

  For Day 5:   ($7.50) − ($8.50) = −$1.00

  For Day 6:   ($8.50) − ($7.50) = $1.00

  For Day 7:   ($9.25) − ($8.50) = $0.75

  For Day 8:   ($10.25) − ($9.25) = $1.00

Day, x	Daily Gain or Loss, y ($)	Ordered Pair
2	1.25	(2, 1.25)
3	−2.50	(3, −2.50)
4	−0.25	(4, −0.25)
5	−1.00	(5, −1.00)
6	1.00	(6, 1.00)
7	0.75	(7, 0.75)
8	1.00	(8, 1.00)

A graph of the daily gain or loss is shown in Figure 1-10.

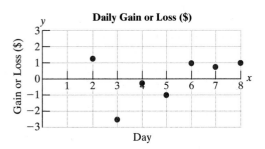

Figure 1-10

section 1.7   PRACTICE EXERCISES

For Exercises 1–10, simplify the expressions.   1.6

1. $-3(x - 2) + 5x$ $2x + 6$ 2. $8(2x - 3) + 9$ $16x - 15$

3. $5x + 2 - 7x - 3x$
   $-5x + 2$
4. $10 - 4x + 7x - 2 + x$
   $4x + 8$

5. $-(-7x + 9) + (3x - 1)$ $10x - 10$

6. $8x - (2x + 3) - 19$ $6x - 22$

7. $3(-4x) + 10 - 6x + 1$ $-18x + 11$

8. $x - 5(3 - x) + 12$ $6x - 3$

9. $\frac{1}{2}x + \frac{3}{4}x - 0.6 + 0.9$ $\frac{5}{4}x + 0.3$

10. $\frac{1}{3}x + 1.7 - \frac{4}{3}x + 3.2$ $-x + 4.9$

11. Plot the points on a rectangular coordinate system.   ❖

    a. $(2, 6)$ b. $(6, 2)$
    c. $(0, -3)$ d. $(-3, 0)$

12. Plot the points on a rectangular coordinate system.   ❖

    a. $(-1, 2)$ b. $(2, -1)$
    c. $(0, 7)$ d. $(7, 0)$

13. Plot the points on a rectangular coordinate system.   ❖

    a. $(4, 5)$ b. $(-4, 5)$
    c. $(4, -5)$ d. $(-4, -5)$

14. Plot the points on a rectangular coordinate system.   ❖

    a. $(2, 3)$ b. $(-2, 3)$
    c. $(2, -3)$ d. $(-2, -3)$

15. Plot the points on a rectangular coordinate system.   ❖

    a. $(-1, 5)$ b. $(0, 4)$ c. $\left(-2, -\frac{3}{2}\right)$
    d. $(2, -0.75)$ e. $(4, 2)$ f. $(-6.1, 0)$
    g. $(0, 0)$

16. Plot the points on a rectangular coordinate system.   ❖

    a. $(7, 0)$ b. $(-3, -1)$ c. $(0, 0)$
    d. $(0, 1.5)$ e. $(6, 1)$ f. $\left(-\frac{1}{4}, 4\right)$
    g. $\left(\frac{1}{4}, -4\right)$

17. Identify the quadrant in which the point $(13, -2)$ is found.   IV

18. Identify the quadrant in which the point $(25, 16)$ is found.   I

19. Identify the quadrant in which the point $(-8, 14)$ is found.   II

20. Identify the quadrant in which the point $(-82, -71)$ is found.   III

21. Explain why the point $(0, -5)$ is *not* located in Quadrant IV. $(0, -5)$ lies on the y-axis.

22. Explain why the point $(-1, 0)$ is *not* located in Quadrant I. $(-1, 0)$ lies on the x-axis.

23. Explain where the point $\left(\frac{7}{8}, 0\right)$ is located.
    $\left(\frac{7}{8}, 0\right)$ is located on the x-axis.

24. Explain where the point $\left(0, \frac{6}{5}\right)$ is located.
    $\left(0, \frac{6}{5}\right)$ is located on the y-axis.

25. A movie theater has kept records of popcorn sales versus movie attendance.

    a. Write the corresponding ordered pairs using the movie attendance as the x-variable and sales of popcorn as the y-variable. Interpret the meaning of each ordered pair.   ❖

    b. Plot the data points on a rectangular coordinate system.   ❖

Movie Attendance (number of people)	Sales of Popcorn ($)
250	225
175	193
315	330
220	209
450	570
400	480
190	185

Table for Exercise 25

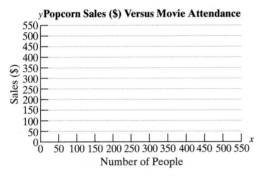

Figure for Exercise 25

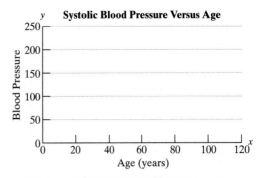

Figure for Exercise 26

26. The age and systolic blood pressure (in millimeters of mercury, mm Hg) for eight different women are given in the table.

   a. Write the corresponding ordered pairs using the woman's age as the *x*-variable and the systolic blood pressure as the *y*-variable. Interpret the meaning of each ordered pair. ❖

   b. Plot the data points on a rectangular coordinate system. ❖

Age (years)	Systolic Blood Pressure (mm Hg)
57	149
41	120
71	158
36	115
64	151
25	110
40	118
77	165

Table for Exercise 26

27. The following ordered pairs give the population of the U.S. colonies from 1700 to 1770. Let *x* represent the year, where *x* = 0 corresponds to 1700, *x* = 10 corresponds to 1710 and so on. Let *y* represent the population of the colonies.

   (0, 251000)   (10, 332000)   (20, 466000)

   (30, 629000)   (40, 906000)   (50, 1171000)

   (60, 1594000)   (70, 2148000)

   a. Interpret the meaning of each ordered pair. ❖

   b. Plot the points on a rectangular coordinate system. ❖

   (*Source: Information Please Almanac,* 1999)

28. The *poverty threshold* is defined by the federal government as an annual income at or below a certain value (defined yearly). The poverty threshold for selected years between 1960 and 1995 is given by the ordered pairs shown here. Let *x* represent the year where *x* = 0 corresponds to 1960, *x* = 5 corresponds to 1965 and so on. Let *y* represent the poverty threshold measured in dollars.

   (0, 1490)   (5, 1582)   (10, 1954)

   (20, 4190)   (30, 6652)   (35, 7763)

   a. Interpret the meaning of each ordered pair. ❖

   b. Plot the points on a rectangular coordinate system. ❖

   (*Source: Information Please Almanac,* 1999)

29. The following table shows the average temperature in degrees Celsius for Montreal Quebec, Canada, by month.

Month, (x)		Temperature °C, (y)
Jan.	1	−10.2
Feb.	2	−9.0
March	3	−2.5
April	4	5.7
May	5	13.0
June	6	18.3
July	7	20.9
Aug.	8	19.6
Sept.	9	14.8
Oct.	10	8.7
Nov.	11	2.0
Dec.	12	−6.9

Table for Exercise 29

a. Write the corresponding ordered pairs, letting $x = 1$ correspond to the month of January. ❖

b. Plot the ordered pairs on a rectangular coordinate system. ❖

30. The table shows the average temperature in degrees Fahrenheit for Fairbanks, Alaska, by month.

Month, (x)		Temperature °F, (y)
Jan.	1	−12.8
Feb.	2	−4.0
March	3	8.4
April	4	30.2
May	5	48.2
June	6	59.4
July	7	61.5
Aug.	8	56.7
Sept.	9	45.0
Oct.	10	25.0
Nov.	11	6.1
Dec.	12	−10.1

Table for Exercise 30

a. Write the corresponding ordered pairs, letting $x = 1$ correspond to the month of January. ❖

b. Plot the ordered pairs on a rectangular coordinate system. ❖

31. The price per share of a stock (in dollars) over a period of 8 days is shown in the graph.

Figure for Exercise 31

a. Use the information provided in the graph to write the ordered pairs corresponding to the price of the stock each day. Interpret the meaning of each ordered pair. ❖

b. Complete the table by providing the net gain or loss from the previous day. Begin on day 2. Then plot these points on a rectangular coordinate system to show daily gain or loss. (See Example 3)

Day, x	Daily Gain or Loss, y	Ordered Pair
2	3.25	(2, 3.25)
3	−1.25	(3, −1.25)
4	1.75	(4, 1.75)
5	−2.75	(5, −2.75)
6	1.25	(6, 1.25)
7	0.75	(7, 0.75)
8	2.25	(8, 2.25)

Table for Exercise 31

32. The price per share of a stock (in dollars) over a period of 8 days is shown in the graph.

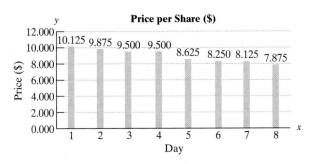

Figure for Exercise 32

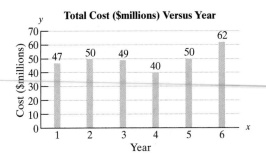

Total Cost ($millions) Versus Year

Figure for Exercise 33b

b. The company's total cost (in $millions) for each year is shown in the graph. Write the ordered pairs corresponding to the cost each year. Interpret the meaning of each ordered pair. ❖

a. Use the information provided in the graph to write the ordered pairs corresponding to the price of the stock each day. Interpret the meaning of each ordered pair. ❖

b. Complete the table by providing the net gain or loss from the previous day. Begin on day 2. Then plot these points on a rectangular coordinate system to show daily gain or loss. (See Example 3)

Day, x	Daily Gain or Loss, y	Ordered Pair
2	−0.250	(2, −0.250)
3	−0.375	(3, −0.375)
4	0	(4, 0)
5	−0.875	(5, −0.875)
6	−0.375	(6, −0.375)
7	−0.125	(7, −0.125)
8	−0.250	(8, −0.250)

Table for Exercise 32

c. The profit for the company is defined as: Profit = (income) − (cost). Complete the table to determine the company's profit each year. Then graph the points on a rectangular coordinate system.

Year, x	Profit ($millions), y
1	−5
2	7
3	−10
4	6
5	8
6	0

Table for Exercise 33c

33. a. The income (in $millions) for a small company over a period of 6 years is given in the graph. Write the ordered pairs corresponding to the income each year. Interpret the meaning of each ordered pair. ❖

34. a. The income (in $thousands) for an individual business over a period of 6 years is given in the graph. Write the ordered pairs corresponding to the income each year. Interpret the meaning of each ordered pair. ❖

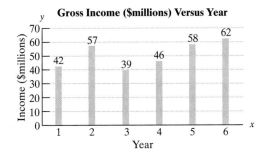

Figure for Exercise 33a

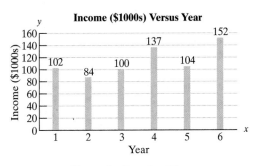

Figure for Exercise 34a

❖ See Additional Answers Appendix   Writing   Translating Expression   Geometry   Scientific Calculator   Video

b. The business's total cost (in $thousands) for each year is shown in the graph. Write the ordered pairs corresponding to the cost each year. Interpret the meaning of each ordered pair. ❖

Figure for Exercise 34b

c. The profit for the company is defined as: Profit = (income) − (cost). Complete the table to determine the company's profit each year. Then graph the points on a rectangular coordinate system.

Year, x	Profit ($thousands), y
1	17
2	−2
3	18
4	41
5	−6
6	44

Table for Exercise 34c

---

chapter 1   SUMMARY

SECTION 1.1—SETS OF NUMBERS AND THE REAL NUMBER LINE

KEY CONCEPTS:

Natural numbers: $\{1, 2, 3, \ldots\}$

Whole numbers: $\{0, 1, 2, 3, \ldots\}$

Integers: $\{\ldots -3, -2, -1, 0, 1, 2, 3, \ldots\}$

Rational numbers: The set of numbers that can be expressed in the form $\frac{p}{q}$, where p and q are integers and q does not equal 0.

Irrational numbers: The set of real numbers that are not rational.

Real numbers: The set of both the rational numbers and the irrational numbers.

EXAMPLES:

Some rational numbers: $\frac{1}{7}, -0.5, 0.\overline{3}$

Some irrational numbers: $\sqrt{7}, -\sqrt{2}, \pi$

The Real Number Line:

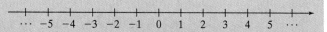

$a < b$	"a is less than b."
$a > b$	"a is greater than b."
$a \leq b$	"a is less than or equal to b."
$a \geq b$	"a is greater than or equal to b."

$5 < 7$	"5 is less than 7."
$-2 > -10$	"−2 is greater than −10."
$y \leq 3.4$	"y is less than or equal to 3.4."
$x \geq \frac{1}{2}$	"x is greater than or equal to $\frac{1}{2}$."

❖ See Additional Answers Appendix   Writing   Translating Expression   Geometry   Scientific Calculator   Video

Two numbers that are the same distance from zero but on opposite sides of zero on the number line are called opposites. The opposite of a is denoted $-a$.

5 and -5 are opposites.

The absolute value of a real number, a, denoted $|a|$, is the distance between a and 0 on the number line.

$$\text{If } a \geq 0, |a| = a$$
$$\text{If } a < 0, |a| = -a$$

$$|7| = 7$$
$$|-7| = 7$$

Key Terms:

absolute value	opposite
inequality	rational numbers
integers	real number line
irrational numbers	real numbers
natural numbers	whole numbers

Section 1.2—Order of Operations

Key Concepts:

A variable is a symbol or letter used to represent an unknown number.

A constant is a value that is not variable.

An algebraic expression is a collection of variables and constants under algebraic operations.

$$b^n = \underbrace{b \cdot b \cdot b \cdot b \dots b}_{n\text{-factors of } b}$$

b is the base,
n is the exponent

$\sqrt{x}$ is the positive square root of x.

The Order of Operations:

1. Simplify expressions within parentheses and other grouping symbols first.
2. Evaluate expressions involving exponents and radicals.
3. Do multiplication or division in the order that they occur from left to right.
4. Do addition or subtraction in the order that they occur from left to right.

Key Terms:

base	power
constant	product
difference	quotient
exponent	square root
expression	sum
order of operations	variable

Examples:

Variables: x, y, z, a, b

Constants: $2, -3, \pi$

Expressions: $2x + 5, 3a + b^2$

$$5^3 = 5 \cdot 5 \cdot 5 = 125$$

$$\sqrt{49} = 7$$

Simplify:

$$10 + 5(3 - 1)^2 - \sqrt{5 - 1}$$

$$= 10 + 5(2)^2 - \sqrt{4}$$
$$= 10 + 5(4) - 2$$
$$= 10 + 20 - 2$$
$$= 30 - 2$$
$$= 28$$

SECTION 1.3—ADDITION OF REAL NUMBERS

KEY CONCEPTS:

Addition of Real Numbers:

Same Signs:
Add the absolute values of the numbers and apply the common sign to the sum.

Different Signs:
Subtract the smaller absolute value from the larger absolute value. Then apply the sign of the number having the larger absolute value.

KEY TERMS:

adding two numbers with the same sign
adding two numbers with different signs

EXAMPLES:

$$-3 + (-4) = -7$$
$$-1.3 + (-9.1) = -10.4$$

$$-5 + 7 = 2$$
$$\frac{2}{3} + \left(-\frac{7}{3}\right) = -\frac{5}{3}$$

SECTION 1.4—SUBTRACTION OF REAL NUMBERS

KEY CONCEPTS:

Subtraction of Real Numbers:

Add the opposite of the second number to the first number. That is,

$$a - b = a + (-b)$$

KEY TERMS:

subtraction of real numbers

EXAMPLES:

$$7 - (-5) = 7 + (5) = 12$$

$$-3 - 5 = -3 + (-5) = -8$$

$$-11 - (-2) = -11 + (2) = -9$$

SECTION 1.5—MULTIPLICATION AND DIVISION OF REAL NUMBERS

KEY CONCEPTS:

Multiplication and Division of Real Numbers:

Same Signs:
Product is positive.
Quotient is positive.

Different Signs:
Product is negative.
Quotient is negative.

The reciprocal of a number a is $\frac{1}{a}$.

EXAMPLES:

$$(-5)(-2) = 10 \qquad \frac{-20}{-4} = 5$$

$$(-3)(7) = -21 \qquad \frac{-4}{8} = -\frac{1}{2}$$

The reciprocal of -6 is $-\frac{1}{6}$.

Multiplication and Division Involving Zero:

The product of any real number and 0 is 0.

The quotient of 0 and any nonzero real number is 0.

The quotient of any nonzero real number and 0 is undefined.

$$4 \cdot 0 = 0$$

$$0 \div 4 = 0$$

$$4 \div 0 \text{ is undefined}$$

KEY TERMS:

division by zero
multiplication or division of numbers with the different signs
multiplication or division of numbers with the same sign
reciprocal of a real number

SECTION 1.6—PROPERTIES OF REAL NUMBERS AND SIMPLIFYING EXPRESSIONS

KEY CONCEPTS:

The Properties of Real Numbers:

Commutative Property of Addition:
$$a + b = b + a$$

Associative Property of Addition:
$$(a + b) + c = a + (b + c)$$

Identity Property of Addition:
The number 0 is said to be the identity element for addition because:
$$0 + a = a \quad \text{and} \quad a + 0 = a$$

Inverse Property of Addition:
$$a + (-a) = 0 \quad \text{and} \quad -a + a = 0$$

Commutative Property of Multiplication:
$$ab = ba$$

Associative Property of Multiplication:
$$(ab)c = a(bc)$$

Identity Property of Multiplication:
The number 1 is said to be the identity element for multiplication because:
$$1 \cdot a = a \quad \text{and} \quad a \cdot 1 = a$$

EXAMPLES:

$$-5 + (-7) = (-7) + (-5)$$

$$(2 + 3) + 10 = 2 + (3 + 10)$$

$$0 + \frac{3}{4} = \frac{3}{4} \quad \text{and} \quad \frac{3}{4} + 0 = \frac{3}{4}$$

$$1.5 + (-1.5) = 0 \quad \text{and} \quad -1.5 + 1.5 = 0$$

$$(-3)(-4) = (-4)(-3)$$

$$(2 \cdot 3)6 = 2(3 \cdot 6)$$

$$1 \cdot 5 = 5 \quad \text{and} \quad 5 \cdot 1 = 5$$

Inverse Property of Multiplication:

$$a \cdot \frac{1}{a} = 1 \quad \text{and} \quad \frac{1}{a} \cdot a = 1$$

$$6 \cdot \frac{1}{6} = 1 \quad \text{and} \quad \frac{1}{6} \cdot 6 = 1$$

The Distributive Property of Multiplication over Addition:

$$a(b + c) = ab + ac$$

Simplify using the distributive property:

$$2(x + 4y) = 2x + 8y$$

$$-(a + 6b - 5c) = -a - 6b + 5c$$

A term is a constant or the product of a constant and one or more variables.

The coefficient of a term is the numerical factor of the term.

Like terms have the same variables, and the corresponding variables have the same powers.

Two terms can be added or subtracted if they are *like* terms. Sometimes it is necessary to clear parentheses before adding or subtracting *like* terms.

Examples of Terms:

$$-2x, \qquad\qquad 5yz^2$$

Coefficient is -2.   Coefficient is 5.

***Like* Terms:**

$$3x \text{ and } -5x, \, 4a^2b \text{ and } 2a^2b$$

Simplify:

$$-4d + 12d + d = 9d$$

Simplify:

$$-2w - 4(w - 2) + 3$$
$$= -2w - 4w + 8 + 3$$
$$= -6w + 11$$

KEY TERMS:

additive inverse
associative property of addition
associative property of multiplication
coefficient
commutative property of addition
commutative property of multiplication
constant term
distributive property of multiplication over addition
identity property of addition
identity property of multiplication
inverse property of addition
inverse property of multiplication
like terms
multiplicative inverse
variable term

SECTION 1.7—CONNECTIONS TO GRAPHING: RECTANGULAR COORDINATE SYSTEM

KEY CONCEPTS:

Graphical representation of numerical data is often helpful to study problems in real-world applications.

A rectangular coordinate system is made up of a horizontal line called the *x*-axis and a vertical line called the *y*-axis. The point where the lines meet is the origin. The four regions of the plane are called quadrants.

The point (x, y) is an ordered pair. The first element in the ordered pair is the point's horizontal position from the origin. The second element in the ordered pair is the point's vertical position from the origin.

KEY TERMS:

coordinate
data
ordered pair
origin
quadrant

rectangular coordinate
   system
x-axis
y-axis

EXAMPLES:

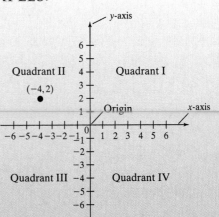

---

chapter 1   REVIEW EXERCISES

Section 1.1

1. Classify the following numbers by checking the boxes of the sets to which the numbers belong.

For Exercises 2–5, determine the absolute values.

2. $\left|\dfrac{1}{2}\right|$ $\dfrac{1}{2}$ 3. $|-6|$ 6   4. $|-\sqrt{7}|$ $\sqrt{7}$ 5. $|0|$ 0

	Real Number	Integer	Rational Number	Whole Number	Irrational Number	Natural Number
7	✓	✓	✓	✓		✓
$\dfrac{1}{3}$	✓		✓			
-4	✓	✓	✓			
0	✓	✓	✓	✓		
$-\sqrt{3}$	✓				✓	
$-0.\overline{2}$	✓		✓			
π	✓				✓	
1	✓	✓	✓	✓		✓

Table for Exercise 1

 See Additional Answers Appendix    Writing   Translating Expression   Geometry   Scientific Calculator   Video

For Exercises 6–13, identify whether the inequality is true or false.

6. $-6 > -1$
   False

7. $0 < -5$
   False

8. $-10 \leq 0$
   True

9. $5 \neq -5$
   True

10. $7 \geq 7$
    True

11. $7 \geq -7$
    True

12. $0 \leq -3$
    False

13. $-\dfrac{2}{3} \leq -\dfrac{2}{3}$ True

Section 1.2

 For Exercises 14–23, translate the English phrases into algebraic expressions.

14. The product of x and $\dfrac{2}{3}$ $x \cdot \dfrac{2}{3}$, or $\dfrac{2}{3}x$

15. The quotient of 7 and y $\dfrac{7}{y}$, or $7 \div y$

16. The sum of 2 and $3b$ $2 + 3b$

17. The difference of a and 5 $a - 5$

18. Two more than $5k$ $5k + 2$

19. Seven less than $13z$ $13z - 7$

20. The quotient of 6 and x, decreased by 18 $\dfrac{6}{x} - 18$

21. The product of y and 3, increased by 12 $3y + 12$

22. Three-eighths subtracted from z $z - \dfrac{3}{8}$

23. Five subtracted from two times p $2p - 5$

For Exercises 24–29, simplify the expressions.

24. 6^3 216

25. 15^2 225

26. $\sqrt{36}$ 6

27. $\left(\dfrac{1}{4}\right)^2$ $\dfrac{1}{16}$

28. $\dfrac{1}{\sqrt{100}}$ $\dfrac{1}{10}$

29. $\left(\dfrac{3}{2}\right)^3$ $\dfrac{27}{8}$

For Exercises 30–33, perform the indicated operations.

30. $15 - 7 \cdot 2 + 12$ 13

31. $|-11| + |5| - (7 - 2)$ 11

32. $4^2 - (5 - 2)^2$ 7

33. $22 - 3(8 \div 4)^2$ 10

For Exercises 34–37, evaluate the expressions with the given substitutions.

34. $3(x + 2) \div y$ for $x = 4$ and $y = 9$ 2

35. $a^2 - bc$ for $a = 6, b = 5$, and $c = 2$ 26

36. $w + xy - \sqrt{z}$ for $w = 12, x = 6$, $y = 5$, and $z = 25$ 37

37. $(u - v)^2 + (u^2 - v^2)$ for $u = 5$ and $v = 3$ 20

Section 1.3

For Exercises 38–48, add the rational numbers.

38. $-6 + 8$ 2

39. $14 + (-10)$ 4

40. $21 + (-6)$ 15

41. $-12 + (-5)$ -17

42. $\dfrac{2}{7} + \left(-\dfrac{1}{9}\right)$ $\dfrac{11}{63}$

43. $\left(-\dfrac{8}{11}\right) + \left(\dfrac{1}{2}\right)$ $-\dfrac{5}{22}$

44. $\left(-\dfrac{1}{10}\right) + \left(-\dfrac{5}{6}\right)$ $-\dfrac{14}{15}$

45. $\left(-\dfrac{5}{2}\right) + \left(-\dfrac{1}{5}\right)$ $-\dfrac{27}{10}$

46. $-8.17 + 6.02$ -2.15

47. $2.9 + (-7.18)$ -4.28

48. $2 + 5 + (-8) + (-7) + 0 + 13 + (-1)$ 4

49. Under what conditions will the expression $a + b$ be negative?   ❖

50. The high temperatures (in degrees Celsius) for the province of Alberta, Canada, during a week in January were $-8, -1, -4, -3, -4, 0$, and 7. What was the average high temperature for that week? Round to the nearest tenth of a degree. $-1.9°C$

Section 1.4

For Exercises 51–61, subtract the rational numbers.

51. $13 - 25$ -12

52. $31 - (-2)$ 33

53. $-8 - (-7)$ -1

54. $-2 - 15$ -17

55. $\left(-\dfrac{7}{9}\right) - \dfrac{5}{6}$ $-\dfrac{29}{18}$

56. $\dfrac{1}{3} - \dfrac{9}{8}$ $-\dfrac{19}{24}$

57. $7 - 8.2$ -1.2

58. $-1.05 - 3.2$ -4.25

59. $-16.1 - (-5.9)$ -10.2

60. $7.09 - (-5)$ 12.09

61. $6 - 14 - (-1) - 10 - (-21) - 5$ -1

62. Under what conditions will the expression $a - b$ be negative?   If $a < b$

 For Exercises 63–67, write an algebraic expression and simplify.

63. -18 subtracted from -7 $-7 - (-18) = 11$

64. The difference of -6 and 41 $-6 - 41 = -47$

65. Seven decreased by 13 $7 - 13 = -6$

66. Five subtracted from the difference of 20 and -7 $(20 - (-7)) - 5 = 22$

67. The sum of 6 and -12, decreased by 21 $(6 + (-12)) - 21 = -27$

68. In Nevada, the highest temperature ever recorded was 125°F and the lowest was -50°F. Find the difference between the highest and lowest temperatures. (*Source: Information Please Almanac* 1999) $125 - (-50) = 175, 175°F$

Section 1.5

For Exercises 69–82, multiply or divide as indicated.

69. $10(-17)$ -170

70. $(-7)13$ -91

71. $(-52) \div 26$ -2

72. $(-48) \div (-16)$ 3

73. $\dfrac{7}{4} \div \left(-\dfrac{21}{2}\right)$ $-\dfrac{1}{6}$

74. $\dfrac{2}{3}\left(-\dfrac{12}{11}\right)$ $-\dfrac{8}{11}$

75. $-\dfrac{21}{5} \cdot 0$ 0

76. $\dfrac{3}{4} \div 0$ Undefined

77. $0 \div (-14)$ 0

78. $\dfrac{0}{3} \cdot \dfrac{1}{8}$ 0

79. $(-0.45)(-5)$ 2.25

80. $(-2.1) \div (-0.07)$ 30

81. $\dfrac{-21}{14}$ $-\dfrac{3}{2}$

82. $\dfrac{-13}{-52}$ $\dfrac{1}{4}$

For Exercises 83–86, perform the indicated operations.

83. $9 - 4[-2(4 - 8) - 5(3 - 1)]$ 17

84. $\dfrac{8(-3) - 6}{-7 - (-2)}$ 6

85. $\dfrac{2}{3} - \left(\dfrac{3}{8} + \dfrac{5}{6}\right) \div \dfrac{5}{3}$ $-\dfrac{7}{120}$

86. $5.4 - (0.3)^2 \div 0.09$ 4.4

87. In statistics, the formula $x = \mu + z\sigma$ is used to find cut-off values for data that follow a bell-shaped curve. Find x if $\mu = 100$, $z = -1.96$ and $\sigma = 15$. 70.6

For Exercises 88–94, answer true or false. If a statement is false, explain why.

88. If n is positive, then $-n$ is negative.   True

89. If m is negative, then m^3 is negative.   True

90. If m is negative, then m^4 is negative.   False, any nonzero real number raised to an even power is positive.

91. If $m > 0$ and $n > 0$, then $mn > 0$.   True

92. If $p < 0$ and $q < 0$, then $pq < 0$.   False, the product of two negative numbers is positive.

93. A number and its reciprocal have the same signs.   True

94. A nonzero number and its opposite have different signs.   True

Section 1.6

For Exercises 95–102, answers may vary.

95. Give an example of the commutative property of addition.   For example: $2 + 3 = 3 + 2$

96. Give an example of the associative property of addition.   For example: $(2 + 3) + 4 = 2 + (3 + 4)$

97. Give an example of the inverse property of addition.   For example: $5 + (-5) = 0$

98. Give an example of the identity property of addition.   For example: $7 + 0 = 7$

99. Give an example of the commutative property of multiplication.   For example: $5 \cdot 2 = 2 \cdot 5$

100. Give an example of the associative property of multiplication.   For example: $(8 \cdot 2)10 = 8(2 \cdot 10)$

101. Give an example of the inverse property of multiplication.   For example: $3 \cdot \dfrac{1}{3} = 1$

102. Give an example of the identity property of multiplication.   For example: $8 \cdot 1 = 8$

103. Explain why $5x - 2y$ is the same as $-2y + 5x$. $5x - 2y = 5x + (-2y)$, then use the commutative property of addition.

104. Explain why $3a - 9y$ is the same as $-9y + 3a$. $3a - 9y = 3a + (-9y)$, then use the commutative property of addition.

105. List the terms of the expression: $3y + 10x - 12 + xy$ $3y, 10x, -12, xy$

106. Identify the coefficients for the terms listed in Exercise 105. $3, 10, -12, 1$

107. Simplify each expression by combining *like* terms.

   a. $3a + 3b - 4b + 5a - 10$ $8a - b - 10$

   b. $-6p + 2q + 9 - 13q - p + 7$ $-7p - 11q + 16$

108. Use the distributive property to clear the parentheses.

   a. $-2(4z + 9)$ $-8z - 18$
   b. $5(4w - 8y + 1)$ $20w - 40y + 5$

For Exercises 109–114, simplify the expression.

109. $2p - (p + 5) + 3$ $p - 2$

110. $6(h + 3) - 7h - 4$ $-h + 14$

111. $\frac{1}{2}(-6q) + q - 4\left(3q + \frac{1}{4}\right)$ $-14q - 1$

112. $0.3b + 12(0.2 - 0.5b)$ $-5.7b + 2.4$

113. $-4[2(x + 1) - (3x + 8)]$ $4x + 24$

114. $5[(7y - 3) + 3(y + 8)]$ $50y + 105$

Section 1.7

For Exercises 115–122, graph the points on a rectangular coordinate system.

115. $\left(\frac{1}{2}, 6\right)$ ❖   116. $(-1, 4)$ ❖   117. $(1, -1)$ ❖

118. $(0, 3)$ ❖   119. $(0, 0)$ ❖   120. $\left(-\frac{5}{8}, 0\right)$ ❖

121. $(-2, -2)$ ❖   122. $(3, 1)$ ❖

123. The price per share of a stock (in dollars) over a period of 8 days is shown in the graph.

Figure for Exercise 123

a. Use the information provided in the graph to write the ordered pairs corresponding to the price of the stock each day. Interpret the meaning of each ordered pair.   ❖

b. Complete the following table by providing the net gain or loss from the previous day. Begin on day 2. Then plot these points on a rectangular coordinate system to show daily gain or loss. (See Example 3 in Section 1.7)

Day, x	Daily Gain or Loss, y	Ordered Pair
2	2.25	(2, 2.25)
3	−0.50	(3, −0.50)
4	−1.00	(4, −1.00)
5	−2.25	(5, −2.25)
6	−0.25	(6, −0.25)
7	0	(7, 0)
8	1.75	(8, 1.75)

Table for Exercise 123

124. The number of people living below the poverty level (in millions) for selected years between 1965 and 1995 is given by the ordered pairs shown here. Let x represent the year, where $x = 0$ corresponds to 1965, $x = 5$ corresponds to 1970, and so on. Let y represent the number of people (in millions) living in below the poverty level. (*Source: U.S. Bureau of the Census*)

(0, 25.9)   (5, 29.3)   (10, 33.1)

(25, 33.6)   (30, 36.4)

a. Interpret the meaning of each ordered pair.   ❖

b. Plot the points on a rectangular coordinate system.   ❖

---

chapter 1   TEST

1. Is $0.\overline{315}$ a rational number or irrational number? Explain your reasoning. Rational, all repeating decimals are rational numbers.

2. Plot the points on a number line: $|3|, 0, -2, 0.5,$ $|-\frac{3}{2}|, \sqrt{16}.$ ❖

3. Use the number line in Exercise 2 to identify whether the statements are true or false.

    a. $|3| < -2$ False   b. $0 \le \left|-\dfrac{3}{2}\right|$ True

    c. $-2 < 0.5$ True   d. $|3| \ge \left|-\dfrac{3}{2}\right|$ True

4. Use the definition of exponents to expand the expressions:

    a. $(4x)^3$ $(4x)(4x)(4x)$ b. $4x^3$ $4 \cdot x \cdot x \cdot x$

5. a. Translate the expression into an English phrase: $2(a - b)$. (Answers may vary.) Twice the difference of a and b
   b. Translate the expression into an English phrase: $2a - b$. (Answers may vary.) The difference of twice a and b.

6. Translate the phrase into an algebraic expression: "The quotient of the square root of c and the square of d." $\dfrac{\sqrt{c}}{d^2}$ or $\sqrt{c} \div d^2$

For Exercises 7–18, perform the indicated operations.

7. $18 + (-12)$ 6   8. $21 - (-7)$ 28

9. $-\dfrac{1}{8} + \left(-\dfrac{3}{4}\right)$ $-\dfrac{7}{8}$ 10. $-10.06 - (-14.72)$ 4.66

11. $-84 \div 7$ -12 12. $38 \div 0$ Undefined

13. $7(-4)$ -28 14. $-22 \cdot 0$ 0

15. $(8 - 10)\dfrac{3}{2} + (-5)$ -8

16. $8 - [(2 - 4) - (8 - 9)]$ 9

17. $\dfrac{\sqrt{5^2 - 4^2}}{|-12 + 3|}$ $\dfrac{1}{3}$ 18. $\dfrac{|4 - 10|}{2 - 3(5 - 1)}$ $-\dfrac{3}{5}$

19. Identify the property that justifies each statement.

    a. $6(-8) = (-8)6$ Commutative property of multiplication

    b. $5 + 0 = 5$ Identity property of addition

    c. $(2 + 3) + 4 = 2 + (3 + 4)$ Associative property of addition

    d. $\dfrac{1}{7} \cdot 7 = 1$ Inverse property of multiplication

    e. $8[7(-3)] = (8 \cdot 7)(-3)$ Associative property of multiplication

For Exercises 20–21, simplify the expression.

20. $3k - 20 + (-9k) + 12$ $-6k - 8$

21. $4(p - 5) - (8p + 3)$ $-4p - 23$

22. The following table depicts a boy's height versus his age. Let x represent the boy's age and y represent his height.

Age (years), x	Height (inches), y
5	46
7	50
9	55
11	60

Table for Exercise 22

    a. Write the data as ordered pairs and interpret the meaning of each ordered pair. ❖

    b. Graph the ordered pairs on a rectangular co-ordinate system. ❖

    c. From the graph estimate the boy's height at age 10. ❖

    d. Predict the boy's height at age 12. ❖

    e. The data appear to follow an upward trend up to the boy's teenage years. Do you think this trend will continue? Would it be reasonable to use these data to predict the boy's height at age 25? ❖

❖ See Additional Answers Appendix   Writing   Translating Expression   Geometry   Scientific Calculator   Video

Linear Equations and Inequalities

The construction, from 1932 to 1937, of the Golden Gate Bridge in San Francisco is perhaps one of the greatest architectural achievements of the twentieth century. The total length of the bridge including the approaches spans 1.7 miles and has a total weight of 887,000 tons. The bridge has two main cables that pass over the tops of two large towers. Together, these cables contain 80,000 miles of steel wire—enough to circle the earth three times.

During the construction of the bridge, workers had to contend with foggy conditions, strong ocean currents, and powerful winds sweeping in from the Pacific Ocean. Eleven men lost their lives during construction.

The design and implementation of an engineering project of this magnitude requires heavy reliance on mathematics. In this chapter, we begin building a foundation of algebraic skills through the study of linear equations and related applications. For more information about applications of linear equations visit volbox and volcylinder at

www.mhhe.com/miller_oneill

section

2.1 ADDITION, SUBTRACTION, MULTIPLICATION, AND DIVISION PROPERTIES OF EQUALITY

1. Definition of a Linear Equation in One Variable

An **equation** is a mathematical statement that indicates that two algebraic expressions are equal. Therefore, an equation must contain an equal sign, =. Some examples of equations are:

$$x - 3 = 9 \qquad y^2 = 25 \qquad \frac{-3}{w} + \frac{1}{2} = 0$$

A **solution to an equation** is a value of the variable that makes the equation a true statement. That is, a solution to an equation makes the right-hand side of the equation equal to the left-hand side.

Equation:	Solution(s):	Check:
$x - 3 = 9$	$x = 12$	$(12) - 3 \stackrel{?}{=} 9$
		$9 = 9$ ✔

$$y^2 = 25 \qquad y = 5 \quad \text{and} \quad y = -5 \qquad (5)^2 \stackrel{?}{=} 25 \qquad\qquad (-5)^2 \stackrel{?}{=} 25$$
$$25 = 25 \text{ ✔} \qquad\qquad 25 = 25 \text{ ✔}$$

$$\frac{-3}{w} + \frac{1}{2} = 0 \qquad w = 6 \qquad \frac{-3}{(6)} + \frac{1}{2} = 0$$
$$-\frac{1}{2} + \frac{1}{2} \stackrel{?}{=} 0$$
$$0 = 0 \text{ ✔}$$

The set of all solutions to an equation is called the **solution set**. Hence the solution set for $x - 3 = 9$ is $\{12\}$, the solution set for $y^2 = 25$ is $\{5, -5\}$ and the solution set for $\frac{-3}{w} + \frac{1}{2} = 0$ is $\{6\}$.

In the study of algebra, you will encounter a variety of equations. In this chapter, we will focus on a specific type of equation called a linear equation in one variable.

Definition of a Linear Equation in One Variable

Let a and b be real numbers such that $a \neq 0$. A **linear equation in one variable** is an equation that can be written in the form

$$ax + b = 0$$

Because the variable has an implied exponent of 1, a linear equation is sometimes called a first-degree equation. The following equations are linear equations in one variable.

$$2x + 3 = 0 \qquad \frac{1}{5}a - \frac{2}{7} = 0 \qquad -3.5z + 14 = 0$$

2. Addition and Subtraction Properties of Equality

If two equations have the same solution set, then the equations are said to be equivalent. For example, the following equations are equivalent because the solution set for each equation is {6}.

Equivalent Equations: Check the Solution $x = 6$:

$2x - 5 = 7$ $2(6) - 5 = 7 \Rightarrow 12 - 5 = 7$ ✔

$2x = 12$ $2(6) = 12 \Rightarrow \quad 12 = 12$ ✔

$x = 6$ $6 = 6 \Rightarrow \quad 6 = 6$ ✔

To solve a linear equation, $ax + b = 0$, the goal is to find *all* values of x that make the equation true. One general strategy for solving an equation is to rewrite it as an equivalent but simpler equation. This process is repeated until the equation can be written in the form $x =$ number. The addition and subtraction properties of equality help us do this.

Addition and Subtraction Properties of Equality

Let a, b, and c represent algebraic expressions.

1. **Addition property of equality**:   If $a = b$,
   then, $a + c = b + c$

2. **Subtraction property of equality**:   If $a = b$,
   then, $a - c = b - c$

The addition and subtraction properties of equality indicate that adding or subtracting the same quantity to each side of an equation results in an equivalent equation. This is true because if two quantities are increased or decreased by the same amount, then the resulting quantities will also be equal (Figure 2-1).

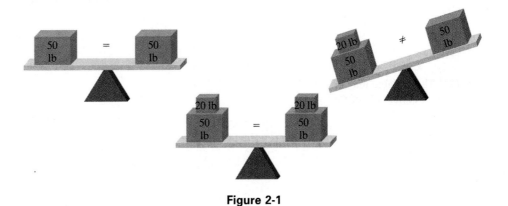

Figure 2-1

example 1   **Applying the Addition and Subtraction Properties of Equality**

Solve the equations.

a. $p - 4 = 11$ b. $w + 5 = -2$ c. $\dfrac{9}{4} = q - \dfrac{3}{4}$ d. $-1.2 + z = 4.6$

Classroom Activity 2.1B

Solution:

In each equation, the goal is to isolate the variable on one side of the equation. To accomplish this, we use the fact that the sum of a number and its opposite is zero and the difference of a number and itself is zero.

a. $\qquad p - 4 = 11$

$\qquad p - 4 + 4 = 11 + 4$ To isolate p, add 4 to both sides ($-4 + 4 = 0$).

$\qquad\qquad p + 0 = 15$ Simplify.

$\qquad\qquad\quad p = 15$

$\qquad$ Check: $p - 4 = 11$ Check the solution by substituting $p = 15$ back in

$\qquad\qquad 15 - 4 \overset{2}{=} 11$ the original equation.

$\qquad\qquad\qquad 11 \overset{2}{=} 11 \ ✔$

b. $\qquad w + 5 = -2$

$\qquad w + 5 - 5 = -2 - 5$ To isolate w, subtract 5 from both sides. ($5 - 5 = 0$).

$\qquad\qquad w + 0 = -7$ Simplify.

$\qquad\qquad\quad w = -7$

$\qquad$ Check: $w + 5 = -2$ Check the solution by substituting $w = -7$ back

$\qquad\qquad -7 + 5 \overset{2}{=} -2$ in the original equation.

$\qquad\qquad\quad -2 \overset{2}{=} -2 \ ✔$

Tip: The variable may be isolated on either side of the equation.

c. $\qquad \dfrac{9}{4} = q - \dfrac{3}{4}$

$\qquad \dfrac{9}{4} + \dfrac{3}{4} = q - \dfrac{3}{4} + \dfrac{3}{4}$ To isolate q, add $\frac{3}{4}$ to both sides ($-\frac{3}{4} + \frac{3}{4} = 0$).

$\qquad\qquad \dfrac{12}{4} = q + 0$ Simplify.

$\qquad\qquad\quad 3 = q$ or equivalently, $q = 3$

$\qquad$ Check: $\dfrac{9}{4} = q - \dfrac{3}{4}$ Check the solution.

$\qquad\qquad \dfrac{9}{4} \overset{2}{=} 3 - \dfrac{3}{4}$ Substitute $q = 3$ in the original equation.

$$\frac{9}{4} \overset{?}{=} \frac{12}{4} - \frac{3}{4} \qquad \text{Get a common denominator.}$$

$$\frac{9}{4} \overset{?}{=} \frac{9}{4} \ \checkmark$$

d. $\qquad -1.2 + z = 4.6$

$-1.2 + 1.2 + z = 4.6 + 1.2 \qquad$ To isolate z, add 1.2 to both sides ($-1.2 +$

$0 + z = 5.8 \qquad\qquad\qquad$ $1.2 = 0$).

$z = 5.8$

Check: $-1.2 + z = 4.6 \qquad$ Check the equation.

$-1.2 + 5.8 \overset{?}{=} 4.6 \qquad$ Substitute $z = 5.8$ in the original equation.

$4.6 \overset{?}{=} 4.6 \ \checkmark$

3.   Multiplication and Division Properties of Equality

Adding or subtracting the same quantity to both sides of an equation results in an equivalent equation. In a similar way, multiplying or dividing both sides of an equation by the same nonzero quantity also results in an equivalent equation. This is stated formally as the multiplication and division properties of equality.

Multiplication and Division Properties of Equality

Let a, b, and c represent algebraic expressions.

1. **Multiplication property of equality:**   If $\qquad a = b$,

then, $ac = bc$

2. **Division property of equality:**   If $\qquad a = b$

then, $\dfrac{a}{c} = \dfrac{b}{c}$ (provided $c \neq 0$)

To understand the multiplication property of equality, suppose we start with a true equation such as $10 = 10$. If both sides of the equation are multiplied by a constant such as 3, the result is also a true statement (Figure 2-2).

$10 = 10$

$3 \cdot 10 = 3 \cdot 10$

$30 = 30$

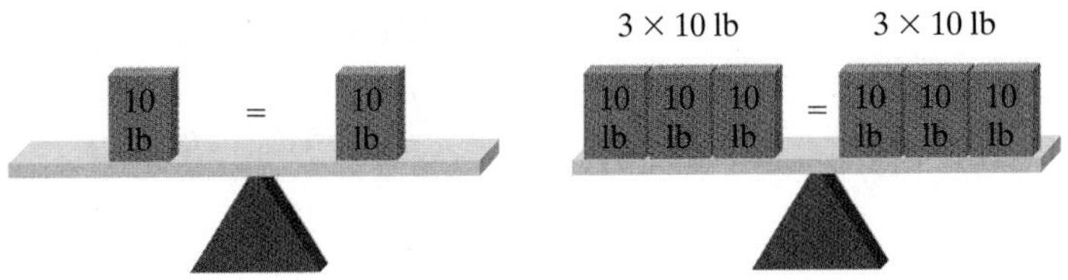

Figure 2-2

Similarly, if the equation is divided by a nonzero real number such as 2, the result is also a true statement (Figure 2-3).

$$10 = 10$$

$$\frac{10}{2} = \frac{10}{2}$$

$$5 = 5$$

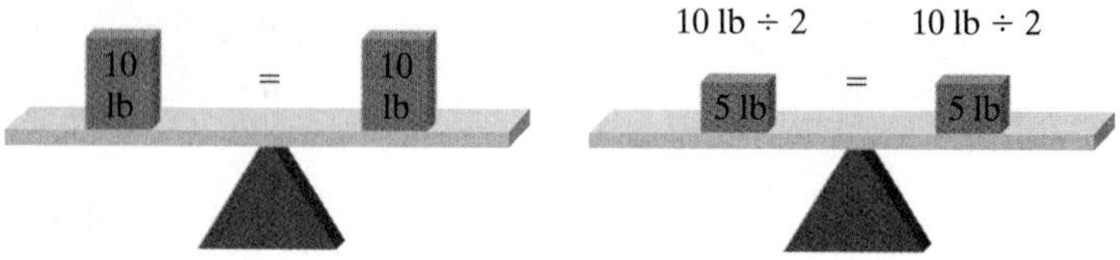

10 lb ÷ 2   10 lb ÷ 2

Figure 2-3

Tip: Recall that the product of a number and its reciprocal is 1. For example:

$$\frac{1}{5}(5) = 1$$

$$\frac{3}{2} \cdot \frac{2}{3} = 1,$$

$$-\frac{7}{2}\left(-\frac{2}{7}\right) = 1$$

To solve an equation in the variable x, the goal is to write the equation in the form $x =$ number. In particular, notice that we desire the coefficient of x to be 1. That is, we want to write the equation as $1x =$ number. Therefore, to solve an equation such as $5x = 15$, we can multiply both sides of the equation by the reciprocal of the x-term coefficient. In this case, multiply both sides by the reciprocal of 5, which is $\frac{1}{5}$.

$$5x = 15$$

$$\frac{1}{5}(5x) = \frac{1}{5}(15) \qquad \text{Multiply by } \tfrac{1}{5}.$$

$$1x = 3 \qquad \text{The coefficient of the } x\text{-term is now 1.}$$

$$x = 3$$

Tip: Recall that the quotient of a nonzero real number and itself is 1. For example:

$$\frac{5}{5} = 1$$

$$\frac{\frac{3}{2}}{\frac{3}{2}} = 1$$

$$\frac{-3.5}{-3.5} = 1$$

The division property of equality can also be used to solve the equation $5x = 15$ by dividing both sides by the coefficient of the x-term. In this case, divide both sides by 5 to make the coefficient of x equal to 1.

$$5x = 15$$

$$\frac{5x}{5} = \frac{15}{5} \qquad \text{Divide by 5.}$$

$$1x = 3 \qquad \text{The coefficient on the } x\text{-term is now 1.}$$

$$x = 3$$

example 2   Applying the Multiplication and Division Properties of Equality

Solve the equations using the multiplication or division properties of equality.

a. $12x = 60$ b. $-\frac{2}{9}q = \frac{1}{3}$ c. $-24.752 = -2.72z$

d. $\frac{d}{6} = -4$ e. $-x = 8$

Classroom Activity 2.1C

Solution:

a. $12x = 60$

$$\frac{12x}{12} = \frac{60}{12}$$ To obtain a coefficient of 1 for the x-term, divide both sides by 12.

$$1x = 5$$ Simplify.

$$x = 5$$ Check: $12x = 60$

$$12(5) \overset{?}{=} 60$$

$$60 \overset{?}{=} 60 \checkmark$$

b. $$-\frac{2}{9}q = \frac{1}{3}$$

$$\left(-\frac{9}{2}\right)\left(-\frac{2}{9}q\right) = \frac{1}{3}\left(-\frac{9}{2}\right)$$ To obtain a coefficient of 1 for the q-term, multiply by the reciprocal of $-\frac{2}{9}$, which is $-\frac{9}{2}$.

$$1q = -\frac{3}{2}$$ Simplify. The product of a number and its reciprocal is 1.

$$q = -\frac{3}{2}$$ Check: $-\frac{2}{9}q = \frac{1}{3}$

$$-\frac{2}{9}\left(-\frac{3}{2}\right) \overset{?}{=} \frac{1}{3}$$

$$\frac{1}{3} = \frac{1}{3} \checkmark$$

Tip: When applying the multiplication or division properties of equality to obtain a coefficient of 1 for the variable term, we will generally use the following convention:

- If the coefficient of the variable term is expressed as a fraction, we will usually multiply both sides by its reciprocal.
- Otherwise, we will divide both sides by the coefficient itself.

c. $-24.752 = -2.72z$

$$\frac{-24.752}{-2.72} = \frac{-2.72z}{-2.72}$$ To obtain a coefficient of 1 for the z-term, divide by -2.72.

$$9.1 = 1z$$ Simplify.

$$9.1 = z \qquad \text{or equivalently,} \qquad z = 9.1$$

$$\underline{\text{Check:}} \; -24.752 = -2.72z$$

$$-24.752 \overset{?}{=} -2.72(9.1)$$

$$-24.752 = -24.752 \; \checkmark$$

d. $\dfrac{d}{6} = -4$

$\dfrac{1}{6}d = -4$ $\frac{d}{6}$ is equivalent to $\frac{1}{6}d$.

$\dfrac{6}{1} \cdot \dfrac{1}{6}d = -4 \cdot \dfrac{6}{1}$ To obtain a coefficient of 1 for the d-term, multiply by the reciprocal of $\frac{1}{6}$, which is $\frac{6}{1}$.

$1d = -24$ Simplify.

$d = -24$ $\underline{\text{Check:}} \; \dfrac{d}{6} = -4$

$\dfrac{-24}{6} \overset{?}{=} -4$

$-4 = -4 \; \checkmark$

e. $-x = 8$ Note that $-x$ is equivalent to $-1 \cdot x$.

$-1x = 8$

$\dfrac{-1x}{-1} = \dfrac{8}{-1}$ To obtain a coefficient of 1 for the x-term, divide by -1.

$x = -8$ $\underline{\text{Check:}} \; -x = 8$

$-(-8) \overset{?}{=} 8$

$8 = 8 \; \checkmark$

It is important to distinguish between cases where the addition or subtraction properties of equality should be used to isolate a variable versus those in which the multiplication or division properties of equality should be used. Remember the goal is to obtain a coefficient of 1 for the variable. Compare the equations:

$$5 + x = 20 \qquad \text{and} \qquad 5x = 20$$

In the first equation, the relationship between 5 and x is addition. Therefore, we want to reverse the process by subtracting 5 from both sides. In the second equation, the relationship between 5 and x is multiplication. To isolate x, we reverse the process by dividing by 5 or equivalently, multiplying by the reciprocal, $\frac{1}{5}$.

$$5 + x = 20 \qquad\qquad \text{and} \qquad 5x = 20$$

$$5 - 5 + x = 20 - 5 \qquad\qquad \dfrac{5x}{5} = \dfrac{20}{5}$$

$$x = 15 \qquad\qquad\qquad x = 4$$

4. Translations

example 3

Translating Linear Equations

Write an algebraic equation to represent each English sentence. Then solve the equation.

a. The quotient of a number and 4 is 6.
b. The product of a number and 4 is 6.
c. Negative twelve is equal to the sum of -5 and a number.
d. The value 1.4 subtracted from a number is 5.7.

Classroom Activity 2.1D

Solution:

For each case we will let x represent the unknown number.

a. The quotient of a number and 4 is 6.

$$\frac{x}{4} = 6$$

$$4 \cdot \frac{x}{4} = 4 \cdot 6 \qquad \text{Multiply both sides by 4.}$$

$$\frac{4}{1} \cdot \frac{x}{4} = 4 \cdot 6$$

$$x = 24 \qquad \underline{\text{Check:}} \; \frac{24}{4} = 6 \; \checkmark$$

b. The product of a number and 4 is 6.

$$4x = 6$$

$$\frac{4x}{4} = \frac{6}{4} \qquad \text{Divide both sides by 4.}$$

$$x = \frac{3}{2} \qquad \underline{\text{Check:}} \; 4\left(\tfrac{3}{2}\right) = 6 \; \checkmark$$

c. Negative twelve is equal to the sum of -5 and a number.

$$-12 = -5 + x$$

$$-12 + 5 = -5 + 5 + x \qquad \text{Add 5 to both sides.}$$

$$-7 = x \qquad \underline{\text{Check:}} \; -12 = -5 + (-7) \; \checkmark$$

d. The value 1.4 subtracted from a number is 5.7.

$$x - 1.4 = 5.7$$

$$x - 1.4 + 1.4 = 5.7 + 1.4 \qquad \text{Add 1.4 to both sides.}$$

$$x = 7.1 \qquad \underline{\text{Check}}: 7.1 - 1.4 = 5.7 \checkmark$$

5. Applications of Linear Equations

example 4   **Using Linear Equations in an Application**

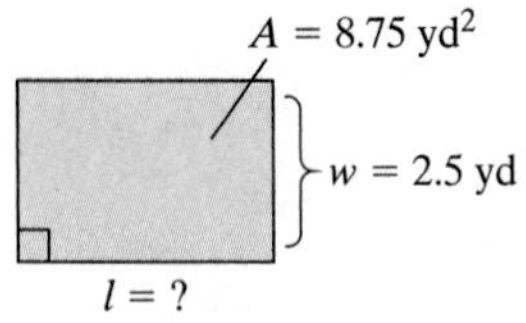

$A = 8.75$ yd^2

$w = 2.5$ yd

$l = ?$

Figure 2-4

The area, A, of a rectangle is given by the formula $A = lw$, where l represents the length of the rectangle and w represents the width. Find the length of a rectangular tablecloth whose width is 2.5 yd and whose area is 8.75 yd^2 (Figure 2-4).

Solution:

$$A = lw$$

$$8.75 = l \cdot (2.5) \qquad \text{Substitute } A = 8.75 \text{ and } w = 2.5, \text{ and solve for } l.$$

$$\frac{8.75}{2.5} = \frac{l \cdot (2.5)}{2.5} \qquad \text{To obtain a coefficient of 1 for } l, \text{ divide both sides by 2.5.}$$

$$3.5 = l$$

The length of the tablecloth is 3.5 yd.

---

section 2.1   PRACTICE EXERCISES

For Exercises 1–8, identify the following as either an expression or an equation.

1. $5x + 3x$ Expression

2. $x - 4 + 5x$ Expression

3. $8x + 2 = 7$ Equation

4. $9 = 2x - 4$ Equation

5. $3x^2 + x = -3$ Equation

6. $7x^3 + x = 10$ Equation

7. $8 - x^3 + 12x$ Expression

8. $7x - 5 + x^2$ Expression

9. Explain how to determine if a number is a solution to an equation.   Substitute the value into the equation and determine if the right-hand side is equal to the left-hand side.

10. Is $x = 3$ a solution to the equation $x^2 - 6 = 5$?   No

11. Is $x = 3$ a solution to the equation $x^2 - 2 = 0$?   No

12. Is $x = -2$ a solution to the equation $3x + 9 = 3$?   Yes

13. Is $x = -4$ a solution to the equation $x + 5 = 1$?   Yes

14. Determine if $x = 2$ is a solution to any of the following equations:

a. $2x + 1 = 5$ Yes

b. $x^2 - 2 = 3$ No

c. $x - 5 = 0$ No

d. $\dfrac{2}{x} + 3 = 4$ Yes

15. Determine if $x = 0$ is a solution to any of the following equations:

a. $5x + 2 = 6$ No

b. $4 = x + 4$ Yes

c. $x^2 - 4 = 4$ No

d. $6x^2 + 1 = 1$ Yes

16. Explain how to check an equation.   Substitute the answer into the equation and determine if the right-hand side is equal to the left-hand side.

❖ See Additional Answers Appendix   🖊 Writing   ⬅ Translating Expression   📖 Geometry   Scientific Calculator   Video

For Exercises 17–36, solve the equation using the addition or subtraction property of equality. Be sure to check your answers.

17. $x + 6 = 5$
$x = -1$

18. $x - 2 = 10$
$x = 12$

19. $q - 14 = 6$
$q = 20$

20. $w + 3 = -5$
$w = -8$

21. $2 + m = -15$
$m = -17$

22. $-6 + n = 10$
$n = 16$

23. $-23 = y - 7$
$y = -16$

24. $-9 = -21 + b$
$b = 12$

25. $5 = z - \dfrac{1}{2}$ $z = \dfrac{11}{2}$ or $5\dfrac{1}{2}$

26. $-7 = p + \dfrac{2}{3}$
$p = -\dfrac{23}{3}$ or $-7\dfrac{2}{3}$

27. $x + \dfrac{5}{2} = \dfrac{1}{2}$ $x = -2$

28. $x - \dfrac{2}{3} = \dfrac{7}{3}$ $x = 3$

29. $4.1 = 2.8 + a$
$a = 1.3$

30. $5.1 = -2.5 + y$
$y = 7.6$

31. $4 + c = 4$
$c = 0$

32. $-13 + b = -13$
$b = 0$

33. $-6.02 + c = -8.15$ $c = -2.13$

34. $p + 0.035 = -1.12$ $p = -1.155$

35. $3.245 + t = -0.0225$ $t = -3.2675$

36. $-1.004 + k = 3.0589$ $k = 4.0629$

For Exercises 37–42, write an algebraic equation to represent each English sentence. (Let x represent the unknown number.) Then solve the equation.

37. The sum of negative eight and a number is forty-two. $-8 + x = 42, x = 50$

38. The sum of thirty-one and a number is thirteen. $31 + x = 13, x = -18$

39. The difference of a number and negative six is eighteen. $x - (-6) = 18, x = 12$

40. The sum of negative twelve and a number is negative fifteen. $-12 + x = -15, x = -3$

41. The sum of a number and $\dfrac{5}{8}$ is $\dfrac{13}{8}$. $x + \dfrac{5}{8} = \dfrac{13}{8}, x = 1$

42. The difference of a number and $\dfrac{2}{3}$ is $\dfrac{1}{3}$. $x - \dfrac{2}{3} = \dfrac{1}{3}, x = 1$

For Exercises 43–62, solve the equations using the multiplication or division property of equality. Be sure to check your answers.

43. $6x = 54$ $x = 9$

44. $2w = 8$ $w = 4$

45. $12 = -3p$ $p = -4$

46. $6 = -2q$ $q = -3$

47. $-5y = 0$ $y = 0$

48. $-3k = 0$ $k = 0$

49. $-\dfrac{y}{5} = 3$ $y = -15$

50. $-\dfrac{z}{7} = 1$ $z = -7$

51. $\dfrac{4}{5} = -t$ $t = -\dfrac{4}{5}$

52. $-\dfrac{3}{7} = -h$ $h = \dfrac{3}{7}$

53. $\dfrac{2}{5}a = -4$ $a = -10$

54. $\dfrac{3}{8}b = -9$ $b = -24$

55. $-\dfrac{1}{5}b = -\dfrac{4}{5}$ $b = 4$

56. $-\dfrac{3}{10}w = \dfrac{2}{5}$ $w = -\dfrac{4}{3}$

57. $-41 = -x$ $x = 41$

58. $32 = -y$ $y = -32$

59. $3.81 = -0.03p$ $p = -127$

60. $2.75 = -0.5q$ $q = -5.5$

61. $5.82y = -15.132$
$y = -2.6$

62. $-32.3x = -0.4522$
$x = 0.014$

For Exercises 63–66, write an algebraic equation to represent each English sentence. (Let x represent the unknown number.) Then solve the equation.

63. The product of a number and seven is negative sixty-three. $x \cdot 7 = -63$ or $7x = -63, x = -9$

64. The product of negative three and a number is 24. $-3x = 24, x = -8$

65. The quotient of a number and 12 is one-third. $\dfrac{x}{12} = \dfrac{1}{3}, x = 4$

66. Eighteen is equal to the quotient of a number and two. $18 = \dfrac{x}{2}, x = 36$

A relationship among the variables distance (d), rate (r), and time (t) is given by the formula, $d = rt$ (distance equals rate times time). Use this formula to solve for the indicated variables in Exercises 67–70.

67. If Jonas runs at a rate of 4.5 mph, how long will it take him to run 6 miles? $1\dfrac{1}{3}$ h

68. If Tatyana walks at a rate of 3 mph, how long will it take her to walk 8 miles? $2\dfrac{2}{3}$ h

69. An in-line skater skated 6 miles in 45 min ($\dfrac{3}{4}$ h). What was her average speed? 8 mph

70. A skate-boarder skated 3 miles in 30 min ($\dfrac{1}{2}$ h). What was his average speed? 6 mph

❖ See Additional Answers Appendix   ✎ Writing   ⬌ Translating Expression   Geometry   Scientific Calculator   Video

The formula to find the area of a rectangle is given by $A = lw$ (area equals length times width). Use this formula to solve for the indicated variables in Exercises 71–74.

71. Find the length of a rectangle with width 23 cm and area 322 cm². *14 cm*

72. Find the length of a rectangle with width 10 in. and area 150 in². *15 in.*

73. Find the width of the rectangular table shown in the figure.   *4.04 ft*

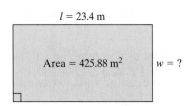

$l = 4.5$ ft

Area = 18.18 ft² $w = ?$

Figure for Exercise 73

74. Find the width of the rectangular garden shown in the figure.   *18.2 m*

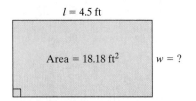

$l = 23.4$ m

Area = 425.88 m² $w = ?$

Figure for Exercise 74

For Exercises 75–102, solve the equation using the appropriate property of equality.

75. $a - 9 = 1$ $a = 10$

76. $b - 2 = -4$ $b = -2$

77. $1 = -9x$ $x = -\dfrac{1}{9}$

78. $-4 = -2k$ $k = 2$

79. $-\dfrac{2}{3}h = 8$ $h = -12$

80. $\dfrac{3}{4}p = 15$ $p = 20$

81. $\dfrac{2}{3} + t = 8$ $t = \dfrac{22}{3}$ or $7\frac{1}{3}$

82. $\dfrac{3}{4} + y = 15$ $y = \dfrac{57}{4}$ or $14\frac{1}{4}$

83. $\dfrac{r}{3} = -12$ $r = -36$

84. $\dfrac{d}{-4} = 5$ $d = -20$

85. $k + 16 = 32$ $k = 16$

86. $-18 = -9 + t$ $t = -9$

87. $16k = 32$ $k = 2$

88. $-18 = -9t$ $t = 2$

89. $7 = -4q$ $q = -\dfrac{7}{4}$ or $-1\frac{3}{4}$

90. $-3s = 10$ $s = -\dfrac{10}{3}$ or $-3\frac{1}{3}$

91. $-4 + q = 7$ $q = 11$

92. $s - 3 = 10$ $s = 13$

93. $-\dfrac{1}{3}d = 12$ $d = -36$

94. $-\dfrac{2}{5}m = 10$ $m = -25$

95. $4 = \dfrac{1}{2} + z$ $z = \dfrac{7}{2}$ or $3\frac{1}{2}$

96. $3 = \dfrac{1}{4} + p$ $p = \dfrac{11}{4}$ or $2\frac{3}{4}$

97. $1.2y = 4.8$ $y = 4$

98. $4.3w = 8.6$ $w = 2$

99. $4.8 = 1.2 + y$ $y = 3.6$

100. $8.6 = w - 4.3$ $w = 12.9$

101. $0.0034 = y - 0.405$ $y = 0.4084$

102. $-0.98 = m + 1.0034$ $m = -1.9834$

SOLVING LINEAR EQUATIONS

Concepts

1. **Solving Linear Equations Involving Multiple Steps**

2. **Steps to Solve a Linear Equation in One Variable**

3. **Conditional Equations, Identities, and Contradictions**

1. Solving Linear Equations Involving Multiple Steps

In Section 2.1 we studied a one-step process to solve linear equations by using the addition, subtraction, multiplication, and division properties of equality. In this section, we will learn how to solve linear equations that require multiple steps. When solving an equation always keep in mind that the ultimate goal is to isolate the variable.

example 1

Solving a Linear Equation

Solve the equation: $-2w - 7 = 11$

Classroom Activity 2.2A

Solution:

$$-2w - 7 = 11$$

$-2w - 7 + 7 = 11 + 7$ Add 7 to both sides of the equation. This isolates the w-term.

$$-2w = 18$$

$$\frac{-2w}{-2} = \frac{18}{-2}$$

$$w = -9$$

Next, apply the division property of equality to obtain a coefficient of 1 for w. Divide by -2 on both sides.

<u>Check</u>:

$$-2w - 7 = 11$$

$-2(-9) - 7 \overset{?}{=} 11$ Substitute $w = -9$ in the original equation.

$$18 - 7 \overset{?}{=} 11$$

$$11 = 11 \; \checkmark$$

In the next example, the variable x appears on both sides of the equation. In this case, apply the addition or subtraction properties of equality to collect the variable terms on one side of the equation and the constant terms on the other side.

example 2

Solving a Linear Equation

Solve the equation: $6x - 4 = 2x - 8$

Solution:

To isolate x, we must first "move" all x-terms to one side of the equation by using the addition or subtraction property of equality.

$$6x - 4 = 2x - 8$$

$6x - 2x - 4 = 2x - 2x - 8$ Subtract $2x$ from both sides leaving $0x$ on the right-hand side.

$4x - 4 = 0x - 8$ Simplify.

$4x - 4 = -8$ The x-terms have now been combined on one side of the equation.

$$4x - 4 + 4 = -8 + 4$$

$$4x = -4$$

$$\frac{4x}{4} = \frac{-4}{4}$$

$$x = -1$$

Add 4 to both sides of the equation. This combines the constant terms on the other side of the equation.

To obtain a coefficient of 1 for x, divide both sides of the equation by 4.

Check:

$$6x - 4 = 2x - 8$$

$$6(-1) - 4 \stackrel{?}{=} 2(-1) - 8$$

$$-6 - 4 \stackrel{?}{=} -2 - 8$$

$$-10 = -10$$

Tip: It is important to note that the variable may be isolated on either side of the equation. We will solve the equation from Example 2 again, this time isolating the variable on the right-hand side.

$$6x - 4 = 2x - 8$$

$$6x - 6x - 4 = 2x - 6x - 8 \qquad \text{Subtract } 6x \text{ on both sides.}$$

$$0x - 4 = -4x - 8$$

$$-4 = -4x - 8$$

$$-4 + 8 = -4x - 8 + 8 \qquad \text{Add 8 to both sides.}$$

$$4 = -4x$$

$$\frac{4}{-4} = \frac{-4x}{-4} \qquad \text{Divide both sides by } -4.$$

$$-1 = x, \quad \text{or equivalently, } x = -1$$

2. Steps to Solve a Linear Equation in One Variable

In some cases it is necessary to simplify both sides of a linear equation before applying the properties of equality. Therefore, we offer the following steps to solve a linear equation in one variable.

> ### Steps to Solve a Linear Equation in One Variable
>
> 1. Simplify both sides of the equation by clearing parentheses and combining *like* terms.
> 2. Use the addition and subtraction properties of equality to collect the variable terms on one side of the equation.
> 3. Use the addition and subtraction properties of equality to collect the constant terms on the other side of the equation.
> 4. Use the multiplication and division properties of equality to make the coefficient of the variable term equal to 1.
> 5. Check your answer.

example 3 **Solving Linear Equations**

Solve the equations:

a. $\frac{1}{5}x + 3 = 2$ b. $2.2y - 8.3 = 6.2y + 12.1$ c. $7 + 3 = 2(p - 3)$

Classroom Activity 2.2B

Solution:

a. $\frac{1}{5}x + 3 = 2$ **Step 1:** The right- and left-hand sides are already simplified.

Step 2: All variable terms are already on one side of the equation.

$\frac{1}{5}x + 3 - 3 = 2 - 3$ **Step 3:** Use the subtraction property of equality to collect the constant terms on the other side of the equation.

$\frac{1}{5}x = -1$

$5\left(\frac{1}{5}x\right) = 5(-1)$ **Step 4:** Use the multiplication property of equality to make the coefficient of the x-term equal to 1.

$1x = -5$

$x = -5$ **Step 5:** <u>Check:</u>

$$\frac{1}{5}x + 3 = 2$$

$$\frac{1}{5}(-5) + 3 \overset{?}{=} 2$$

$$-1 + 3 \overset{?}{=} 2$$

$$2 = 2 ✔$$

b. $\quad 2.2y - 8.3 = 6.2y + 12.1$

Step 1: The right- and left-hand sides are already simplified.

$2.2y - 2.2y - 8.3 = 6.2y - 2.2y + 12.1$

$-8.3 = 4.0y + 12.1$

Step 2: Subtract $2.2y$ from both sides to collect the variable terms on one side of the equation.

$-8.3 - 12.1 = 4.0y + 12.1 - 12.1$

$-20.4 = 4.0y$

Step 3: Subtract 12.1 from both sides to collect the constant terms on the other side.

$$\frac{-20.4}{4.0} = \frac{4.0y}{4.0}$$

$-5.1 = y$

Step 4: To obtain a coefficient of 1 for the y-term, divide both sides of the equation by 4.0.

$y = -5.1$

Step 5: <u>Check:</u>

$2.2y - 8.3 = 6.2y + 12.1$

$2.2(-5.1) - 8.3 \stackrel{?}{=} 6.2(-5.1) + 12.1$

$-11.22 - 8.3 \stackrel{?}{=} -31.62 + 12.1$

$-19.52 = -19.52 ✔$

c. $\quad 7 + 3 = 2(p - 3)$

$10 = 2p - 6$

Step 1: Simplify both sides of the equation by clearing parentheses and combining *like* terms.

Step 2: The variable terms are already on one side.

$10 + 6 = 2p - 6 + 6$

Step 3: Add 6 to both sides to collect the constant terms on the other side.

$16 = 2p$

$$\frac{16}{2} = \frac{2p}{2}$$

Step 4: Divide both sides by 2 to obtain a coefficient of 1 for p.

$8 = p$ or $p = 8$

Step 5: The check is left to the reader.

The next example illustrates the procedure for solving linear equations when multiple steps of simplification are necessary.

example 4   **Solving Linear Equations**

Solve the equations:

a. $2 + 7x - 5 = 6(x + 3) + 2x$
b. $2[9 - (z - 3) + 4z] = 4z - 5(z + 2) - 8$

Solution:

a. $2 + 7x - 5 = 6(x + 3) + 2x$

$-3 + 7x = 6x + 18 + 2x$ **Step 1:** Add *like* terms on the left. Clear parentheses on the right.

$-3 + 7x = 8x + 18$ Combine *like* terms.

$-3 + 7x - 7x = 8x - 7x + 18$ **Step 2:** Subtract $7x$ from both sides.

$-3 = x + 18$ Simplify.

$-3 - 18 = x + 18 - 18$ **Step 3:** Subtract 18 from both sides.

$-21 = x$ **Step 4:** Because the coefficient of the x-term is already 1, there is no need to apply the multiplication or division property of equality.

$x = -21$

Step 5: Check:

$$2 + 7x - 5 = 6(x + 3) + 2x$$
$$2 + 7(-21) - 5 \stackrel{?}{=} 6(-21 + 3) + 2(-21)$$
$$2 - 147 - 5 \stackrel{?}{=} 6(-18) - 42$$
$$-145 - 5 \stackrel{?}{=} -108 - 42$$
$$-150 = -150 \; ✔$$

b. $2[9 - (z - 3) + 4z] = 4z - 5(z + 2) - 8$

$2(9 - z + 3 + 4z) = 4z - 5z - 10 - 8$ **Step 1:** Clear parentheses.

$2(12 + 3z) = -z - 18$ Combine *like* terms.

$24 + 6z = -z - 18$ Clear parentheses.

$24 + 6z + z = -z + z - 18$ **Step 2:** Add z to both sides.

$24 + 7z = -18$

$24 - 24 + 7z = -18 - 24$ **Step 3:** Subtract 24 from both sides.

$7z = -42$

$\dfrac{7z}{7} = \dfrac{-42}{7}$ **Step 4:** Divide both sides by 7.

$z = -6$ **Step 5:** The check is left for the reader.

3.   Conditional Equations, Identities, and Contradictions

The solutions to a linear equation are the values of x that make the equation a true statement. A linear equation may have one unique solution, no solution, or an infinite number of solutions.

I. Conditional Equations

An equation that is true for some values of the variable but false for other values is called a **conditional equation**. The equation $x + 4 = 6$, for example, is true on the condition that $x = 2$. For other values of x, the statement $x + 4 = 6$ is false.

II. Contradictions

Some equations have no solution, such as $x + 1 = x + 2$. There is no value of x, that when increased by 1 will equal the same value increased by 2. If we tried to solve the equation by subtracting x from both sides, we get the contradiction $1 = 2$. This indicates that the equation has no solution. An equation that has no solution is called a **contradiction**.

$$x + 1 = x + 2$$
$$x - x + 1 = x - x + 2$$
$$1 = 2 \quad \text{(contradiction)} \qquad \text{No solution.}$$

III. Identities

An equation that has all real numbers as its solution set is called an **identity**. For example, consider the equation, $x + 4 = x + 4$. Because the left- and right-hand sides are identically equal, any real number substituted for x will result in equal quantities on both sides. If we subtract x from both sides of the equation, we get the identity $4 = 4$. In such a case, the solution is the set of all real numbers.

$$x + 4 = x + 4$$
$$x - x + 4 = x - x + 4$$
$$4 = 4 \quad \text{(identity)} \qquad \text{The solution is all real numbers.}$$

---

example 5   **Identifying Conditional Equations, Contradictions, and Identities**

Identify each equation as a conditional equation, a contradiction, or an identity. Then describe the solution.

a. $4k - 5 = 2(2k - 3) + 1$ b. $2(b - 4) = 2b - 7$ c. $3x + 7 = 2x - 5$

Classroom Activity 2.2C

Solution:

a.
$$4k - 5 = 2(2k - 3) + 1$$
$$4k - 5 = 4k - 6 + 1 \qquad \text{Clear parentheses.}$$
$$4k - 5 = 4k - 5 \qquad \text{Combine } like \text{ terms.}$$
$$4k - 4k - 5 = 4k - 4k - 5 \qquad \text{Subtract } 4k \text{ from both sides.}$$
$$-5 = -5$$

This is an identity. The solution is all real numbers.

b. $2(b - 4) = 2b - 7$

$2b - 8 = 2b - 7$ Clear parentheses.

$2b - 2b - 8 = 2b - 2b - 7$ Subtract $2b$ from both sides.

$-8 = -7$

This is a contradiction. There is no solution.

c. $3x + 7 = 2x - 5$

$3x - 2x + 7 = 2x - 2x - 5$ Subtract $2x$ from both sides.

$x + 7 = -5$ Simplify.

$x + 7 - 7 = -5 - 7$ Add 7 to both sides.

$x = -12$

This is a conditional equation. The solution is $x = -12$. (The equation is true only on the condition that $x = -12$.)

---

section 2.2   PRACTICE EXERCISES

For Exercises 1–8, simplify the expressions by clearing parentheses and combining *like* terms.   2.1

1. $-3(4t) + 5t - 6$ 2. $8(2x) - 3x + 9$
   $-7t - 6$ $13x + 9$

3. $5z + 2 - 7z - 3z$ $-5z + 2$

4. $10 - 4w + 7w - 2 + w$ $4w + 8$

5. $-(-7p + 9) + (3p - 1)$ $10p - 10$

6. $8y - (2y + 3) - 19$ $6y - 22$

7. $5(3a) + 5(3 + a)$ 8. $-2(6 + b) - 2(6b)$
   $20a + 15$ $-14b - 12$

9. Explain the difference between simplifying an expression and solving an equation. ❖

For Exercises 10–17, solve the equations using the addition, subtraction, multiplication, or division property of equality.   2.1

10. $5w = -30$ $w = -6$ 11. $-7y = 21$ $y = -3$

12. $x + 8 = -15$ $x = -23$ 13. $z - 23 = -28$ $z = -5$

14. $6 = a - \dfrac{7}{8}$ $a = \dfrac{55}{8}$ or $6\frac{7}{8}$ 15. $12 = b + \dfrac{1}{5}$ $b = \dfrac{59}{5}$ or $11\frac{4}{5}$

16. $-\dfrac{9}{8} = -\dfrac{3}{4}k$ $k = \dfrac{3}{2}$ or $1\frac{1}{2}$ 17. $-\dfrac{3}{10} = -6h$ $h = \dfrac{1}{20}$

18. Which properties of equality would you apply to solve the equation $2x + 6 = 0$?
First use the subtraction property, then the division property.

19. Which properties of equality would you apply to solve the equation $-2x - 6 = 0$?
First use the addition property, then the division property.

For Exercises 20–45, solve the equations using the steps outlined in the text.

20. $6z + 1 = 13$ $z = 2$ 21. $5x + 2 = -13$ $x = -3$

22. $3y - 4 = 14$ $y = 6$ 23. $-7w - 5 = -19$ $w = 2$

24. $-2p + 8 = 3$ $y = \dfrac{5}{2}$ or $2\frac{1}{2}$ 25. $4q + 5 = 2$ $q = -\dfrac{3}{4}$

26. $6 = 7m - 1$ $m = 1$ 27. $-9 = 4n - 1$ $n = -2$

28. $-\dfrac{1}{2} - 4x = 8$
   $x = -\dfrac{17}{8}$ or $-2\frac{1}{8}$
                        29. $2b - \dfrac{1}{4} = 5$ $b = \dfrac{21}{8}$ or $2\frac{5}{8}$

30. $0.2x + 3.1 = -5.3$ 31. $-1.8 + 2.4a = -6.6$
   $x = -42$ $a = -2$

32. $\dfrac{5}{8} = \dfrac{1}{4} - \dfrac{1}{2}p$ $p = -\dfrac{3}{4}$ 33. $\dfrac{6}{7} = \dfrac{1}{7} + \dfrac{5}{3}r$ $r = \dfrac{3}{7}$

34. $7w - 6w + 1 = 10 - 4$ $w = 5$

35. $5v - 3 - 4v = 13$ $v = 16$

36. $11h - 8 - 9h = -16$ $h = -4$

37. $6u - 5 - 8u = -7$ $u = 1$

❖ See Additional Answers Appendix   ✎ Writing   ⬌ Translating Expression   Geometry   Scientific Calculator   Video

38. $3a + 7 = 2a - 19$ 39. $6b - 20 = 14 + 5b$
     $a = -26$ $b = 34$
40. $-4r - 28 = -78 - r$ $r = \dfrac{50}{3}$ or $16\frac{2}{3}$

41. $-6x - 7 = -3 - 8x$ $x = 2$

42. $-2z - 8 = -z$ $z = -8$ 43. $-7t + 4 = -6t$ $t = 4$

44. $\dfrac{5}{6}x + \dfrac{2}{3} = -\dfrac{1}{6}x - \dfrac{5}{3}$ 45. $\dfrac{3}{7}x - \dfrac{1}{4} = -\dfrac{4}{7}x - \dfrac{5}{4}$
     $x = -\dfrac{7}{3}$ or $-2\frac{1}{3}$ $x = -1$

For Exercises 46–61, solve the equations using the steps outlined in the text.

46. $3(2p - 4) = 15$ $p = \dfrac{9}{2}$ or $4\frac{1}{2}$ 47. $4(t + 15) = 20$ $t = -10$

48. $6(3x + 2) - 10 = -4$ $x = -\dfrac{1}{3}$

49. $4(2k + 1) - 1 = 5$ $k = \dfrac{1}{4}$

50. $2(y - 3) - y = 6$ $y = 12$ 51. $4(w - 5) - 3w = 2$ $w = 22$

52. $17(s + 3) = 4(s - 10) + 13$ $s = -6$

53. $5(4 + p) = 3(3p - 1) - 9$ $p = 8$

54. $6(3t - 4) + 10 = 5(t - 2) - (3t + 4)$ $t = 0$

55. $-5y + 2(2y + 1) = 2(5y - 1) - 7$ $y = 1$

56. $-2[(4p + 1) - (3p - 1)] = 5(3 - p) - 9$ $p = \dfrac{10}{3}$ or $3\frac{1}{3}$

57. $5 - (6k + 1) = 2[(5k - 3) - (k - 2)]$ $k = \dfrac{3}{7}$

58. $0.2w - 0.47 = 0.53 - 0.2(2w - 13)$ $w = 6$

59. $0.4z - 0.15 = 0.65 - 0.3(6 - 2z)$ $z = 5$

60. $3(-0.9n + 0.5) = -3.5n + 1.3$ $n = -0.25$

61. $7(0.4m - 0.1) = 5.2m + 0.86$ $m = -0.65$

62. A conditional linear equation has (choose one): One solution, no solution, or infinitely many solutions.   One solution

63. An equation that is a contradiction has (choose one): One solution, no solution, or infinitely many solutions.   No solution

64. An equation that is an identity has (choose one): One solution, no solution, or infinitely many solutions.   Infinitely many solutions

For Exercises 65–74, identify as a conditional equation, a contradiction, or an identity. Then describe the solution.

65. $2(k - 7) = 2k - 13$ Contradiction; no solution

66. $5h + 4 = 5(h + 1) - 1$ Identity; all real numbers

67. $7x + 3 = 6(x - 2)$ Conditional equation; $x = -15$ 68. $3y - 1 = 1 + 3y$ Contradiction; no solution

69. $3 - 5.2p = -5.2p + 3$ Identity; all real numbers

70. $2(q + 3) = 4q + q - 9$ Conditional equation; $q = 5$

71. $5(h - 1) - 1 = 2(h - 3) + 3h$ Identity; all real numbers

72. $2(5w - 3) - 5w = 5w - 6$ Identity; all real numbers

73. $2(5x + 7) - 2x = 2(7 - x)$ Conditional equation; $x = 0$

74. $6(2y - 1) - y = -(6 + y)$ Conditional equation; $y = 0$

▧ EXPANDING YOUR SKILLS

75. Suppose $x = -5$ is a solution to the equation $x + a = 10$. Find the value of a. $a = 15$

76. Suppose $x = 6$ is a solution to the equation $x + a = -12$. Find the value of a. $a = -18$

77. Suppose $x = 3$ is a solution to the equation $ax = 12$. Find the value of a. $a = 4$

78. Suppose $x = 11$ is a solution to the equation $ax = 49.5$. Find the value of a. $a = 4.5$

For Exercises 79–80, identify as a conditional equation, a contradiction, or an identity.

79. $2(q + 1) - 1.5 = 5(0.4q + 0.16)$ Contradiction

80. $1.1(0.2k - 0.6) - 0.06 = 0.22(k + 1) + 0.05$ Contradiction

Concepts

1. Problem-Solving Strategies
2. Translations Involving Linear Equations
3. Consecutive Integer Problems
4. Applications of Linear Equations

section

2.3   APPLICATIONS OF LINEAR EQUATIONS: INTRODUCTION TO PROBLEM SOLVING

1.   Problem-Solving Strategies

Linear equations can be used to solve many real-world applications. However, with "word problems" students often do not know where to start. To help organize the problem-solving process, we offer the following guidelines:

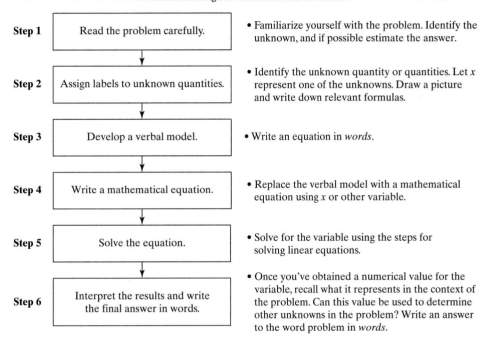

Problem-Solving Flowchart for Word Problems

Step 1   Read the problem carefully.
• Familiarize yourself with the problem. Identify the unknown, and if possible estimate the answer.

Step 2   Assign labels to unknown quantities.
• Identify the unknown quantity or quantities. Let x represent one of the unknowns. Draw a picture and write down relevant formulas.

Step 3   Develop a verbal model.
• Write an equation in *words*.

Step 4   Write a mathematical equation.
• Replace the verbal model with a mathematical equation using x or other variable.

Step 5   Solve the equation.
• Solve for the variable using the steps for solving linear equations.

Step 6   Interpret the results and write the final answer in words.
• Once you've obtained a numerical value for the variable, recall what it represents in the context of the problem. Can this value be used to determine other unknowns in the problem? Write an answer to the word problem in *words*.

2.   Translations Involving Linear Equations

We begin our work with word problems with practice translating between an English sentence and an algebraic equation. Recall from Section 1.2, that several key words translate to the algebraic operations of addition, subtraction, multiplication, and division.

Addition: $a + b$	**Subtraction: $a - b$**
The sum of a and b	The difference of a and b
a plus b	a minus b
b added to a	b subtracted from a
b more than a	a decreased by b
a increased by b	b less than a
the total of a and b	

Multiplication: $a \cdot b$	Division: $a \div b$
The product of a and b	The quotient of a and b
a times b	a divided by b
a multiplied by b	b divided into a
	The ratio of a and b
	a over b
	a per b

example 1

Translating Linear Equations

The sum of a number and negative eleven is negative fifteen. Find the number.

Solution:

	Step 1: Read the problem.
Let x represent the unknown number.	**Step 2:** Label the unknown.

$$\overset{\text{the sum of}}{\text{(a number)}} + \overset{\text{is}}{(-11)} = (-15)$$

Step 3: Develop a verbal model.

$$x + (-11) = -15$$

Step 4: Write an equation.

$$x + (-11) + 11 = -15 + 11$$

Step 5: Solve the equation.

$$x = -4$$

The number is -4.

Step 6: Write the final answer in words.

example 2

Translating Linear Equations

Forty less than five times a number is fifty-two less than the number. Find the number.

Solution:

Step 1: Read the problem.

Let x represent the unknown number.

Step 2: Label the unknown.

$$\begin{pmatrix} 5 \text{ times} \\ \text{a number} \end{pmatrix} - (40) = \begin{pmatrix} \text{the} \\ \text{number} \end{pmatrix} - (52)$$

Step 3: Develop a verbal model.

$$5x \quad - 40 = \quad x \quad - 52$$

Step 4: Write an equation.

$$5x - 40 = x - 52$$

Step 5: Solve the equation.

$$5x - x - 40 = x - x - 52$$

$$4x - 40 = -52$$

Avoiding Mistakes

It is important to remember that subtraction is not a commutative operation. Therefore, the order in which two real numbers are subtracted affects the outcome. The expression "forty less than five times a number" must be translated as: $5x - 40$ (not $40 - 5x$). Similarly, "fifty-two less than the number" must be translated as: $x - 52$ (not $52 - x$).

$$4x - 40 + 40 = -52 + 40$$

$$4x = -12$$

$$\frac{4x}{4} = \frac{-12}{4}$$

$$x = -3$$

The number is -3.   **Step 6:** Write the final answer in words.

---

example 3   **Translating Linear Equations**

Twice the sum of a number and six is two more than three times the number. Find the number.

Classroom Activity 2.3A   **Solution:**

	Step 1: Read the problem.
Let x represent the unknown number.	**Step 2:** Label the unknown.

◆ **Avoiding Mistakes**

It is important to enclose "the sum of a number and six" within parentheses so that the entire quantity is multiplied by 2. Forgetting the parentheses would imply that the x-term only is multiplied by 2.

Correct: $2(x + 6)$

twice   the sum   is   2 more than
 ↓   ↓   ↓   ↓
2 $(x + 6)$ $=$ $3x + 2$
                    ↑
              three times
              a number

Step 3: Develop a verbal model.

Step 4: Write an equation.

$$2(x + 6) = 3x + 2$$

$$2x + 12 = 3x + 2$$

$$2x - 2x + 12 = 3x - 2x + 2$$

$$12 = x + 2$$

$$12 - 2 = x + 2 - 2$$

$$10 = x$$

Step 5: Solve the equation.

The number is 10.   **Step 6:** Write the final answer in words.

---

3. Consecutive Integer Problems

The word *consecutive* means "following one after the other in order without gaps." The numbers 6, 7, 8 are examples of three **consecutive integers**. The numbers $-4, -2, 0, 2$, are examples of **consecutive even integers**. The numbers 23, 25, 27 are examples of **consecutive odd integers**.

Notice that any two consecutive integers differ by 1. Therefore, if x represents an integer, then $x + 1$ represents the next consecutive integer (Figure 2-5).

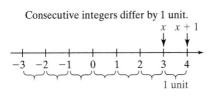

Consecutive integers differ by 1 unit.

Figure 2-5

Any two consecutive even integers differ by 2. Therefore, if x represents an even integer, then $x + 2$ represents the next consecutive even integer (Figure 2-6).

Consecutive even integers differ by 2 units.

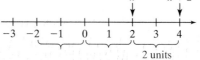

Figure 2-6

Likewise, any two consecutive odd integers differ by 2. If x represents an odd integer, then $x + 2$ is the next odd integer (Figure 2-7).

Consecutive odd integers differ by 2 units.

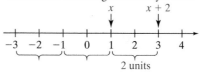

Figure 2-7

---

example 4   **Solving an Application Involving Consecutive Integers**

The sum of two consecutive odd integers is -188. Find the integers.

Solution:

	Step 1: Read the problem.
Let x represent the first odd integer.	**Step 2:** Label the variables.
$x + 2$ represents the second odd integer.	

$$\left(\begin{array}{c}\text{First}\\\text{integer}\end{array}\right) + \left(\begin{array}{c}\text{second}\\\text{integer}\end{array}\right) = (\text{total})$$

Step 3: Write an equation in words.

$$x \quad + \quad (x + 2) \quad = -188$$

Step 4: Write a mathematical equation.

$$x + (x + 2) = -188$$
$$2x + 2 = -188$$

Step 5: Solve for x.

$$2x + 2 - 2 = -188 - 2$$
$$2x = -190$$
$$\frac{2x}{2} = \frac{-190}{2}$$
$$x = -95$$

The first integer is $x = -95$.

Step 6: Interpret the results and write the answer in words.

The second integer is $x + 2 = -95 + 2 = -93$.

The two integers are -95 and -93.

> **Tip:** With word problems it is advisable to check that the answer is reasonable.
> The numbers -95 and -93 are consecutive odd integers. Furthermore, their sum is -188 as desired.

---

example 5

Solving an Application Involving Consecutive Integers

Ten times the smallest of three consecutive integers is twenty-two more than three times the sum of the integers. Find the integers.

Classroom Activity 2.3B

Solution:

Step 1: Read the problem.

Let x represent the first integer.

$x + 1$ represents the second consecutive integer.

$x + 2$ represents the third consecutive integer.

Step 2: Label the variables.

$$\begin{pmatrix} 10 \text{ times} \\ \text{the first} \\ \text{integer} \end{pmatrix} = \begin{pmatrix} 3 \text{ times} \\ \text{the sum of} \\ \text{the integers} \end{pmatrix} + 22$$

Step 3: Write an equation in words.

$$10x = 3\underbrace{[(x) + (x + 1) + (x + 2)]}_{\text{the sum of the integers}} + 22$$

(10 times the first integer   is   3 times ... 22 more than)

Step 4: Write a mathematical equation.

$$10x = 3(x + x + 1 + x + 2) + 22$$

Step 5: Solve the equation.

$$10x = 3(3x + 3) + 22$$

Clear parentheses.

$$10x = 9x + 9 + 22$$

Combine *like* terms.

$$10x = 9x + 31$$

$$10x - 9x = 9x - 9x + 31$$

$$x = 31$$

Isolate the x-terms on one side.

The first integer is $x = 31$.

The second integer is $x + 1 = 31 + 1 = 32$.

The third integer is $x + 2 = 31 + 2 = 33$.

Step 6: Interpret the results and write the answer in words.

The three integers are 31, 32, and 33.

4.   Applications of Linear Equations

example 6

Using a Linear Equation in an Application

As of June 6, 1998, the two films with the largest total box office revenues were *Titanic* and *Star Wars*. Together the two films grossed $1043 million. If *Titanic* made $121 million more than *Star Wars*, find the total box office revenue for each film. (*Source: Information Please Almanac, 1999*)

Classroom Activity 2.3C

Solution:

Step 1: Read the problem.

Let x represent the revenue from *Star Wars* (in $millions).

Step 2: Label the unknowns.

$x + 121$ represents the revenue from *Titanic* (in $millions).

$$\left(\begin{array}{c}\text{Revenue from}\\ \textit{Star Wars}\end{array}\right) + \left(\begin{array}{c}\text{revenue from}\\ \textit{Titanic}\end{array}\right) = \left(\begin{array}{c}\text{total}\\ \text{revenue}\end{array}\right)$$

Step 3: Develop a verbal equation.

$$x \qquad + \quad (x + 121) \quad = \quad 1043$$

Step 4: Write a mathematical equation.

$$x + (x + 121) = 1043$$

Step 5: Solve the equation.

$$2x + 121 = 1043$$
$$2x + 121 - 121 = 1043 - 121$$
$$2x = 922$$
$$\frac{2x}{2} = \frac{922}{2}$$
$$x = 461$$

Revenue from *Star Wars*: $x = 461.$

Revenue from *Titanic*: $x + 121 = 461 + 121 = 582.$

Step 6: Interpret the results and write the final answer in words.

As of June 6, 1998, the revenue made from *Star Wars* was $461 million and the revenue made from *Titanic* was $582 million.

---

section 2.3   PRACTICE EXERCISES

For Exercises 1–6, write an algebraic expression to represent the English sentence. Then solve the equation.
2.1

1. The sum of a number and sixteen is negative thirty-one. Find the number. $x + 16 = -31, x = -47$

2. The sum of a number and negative twenty-one is fourteen. Find the number. $x + (-21) = 14, x = 35$

3. The difference of a number and six is negative three. Find the number. $x - 6 = -3, x = 3$

4. The difference of a number and negative four is negative twelve. Find the number.
$x - (-4) = -12, x = -16$

5. Sixteen less than a number is negative one. Find the number. $x - 16 = -1, x = 15$

6. Ten less than a number is negative thirteen. Find the number. $x - 10 = -13, x = -3$

For Exercises 7–12, identify as a conditional equation, a contradiction, or an identity. Then describe the solution.   2.2

7. $4(t - 2) = 1 + 4t$
Contradiction; no solution

8. $2x - 3 = 2(5 + x)$
Contradiction; no solution

9. $-5(y + 4) = 15$
Conditional equation; $y = -7$

10. $4(p - 1) + 8 = -10$
Conditional equation; $p = -\dfrac{7}{2}$

11. $7(2m - 2) - 5m = 9m - 14$ Identity; all real numbers

12. $-5n - 2(n + 4) = -7n - 8$ Identity; all real numbers

For Exercises 13–24, use the problem-solving flowchart (page 141) to solve the problems.

13. Twice the sum of a number and seven is eight. Find the number.   The number is -3.

14. Twice the sum of a number and negative two is sixteen. Find the number.   The number is 10.

15. Five times the difference of a number and three is four less than four times the number. Find the number.   The number is 11.

16. Three times the difference of a number and seven is one less than twice the number. Find the number.   The number is 20.

17. A number added to five is the same as twice the number. Find the number.   The number is 5.

18. Three times a number is the same as the difference of twice the number and seven. Find the number.   The number is -7.

19. The sum of six times a number and ten is equal to the difference of the number and fifteen. Find the number. The number is −5.

20. The difference of fourteen and three times a number is the same as the sum of the number and negative ten. Find the number. The number is 6.

21. If three is added to five times a number, the result is forty-three more than the number. Find the number. The number is 10.

22. If seven is added to three times a number, the result is thirty-one more than the number. The number is 12.

23. If the difference of a number and four is tripled, the result is six more than the number. Find the number. The number is 9.

24. Twice the sum of a number and eleven is twenty-two less than three times the number. Find the number. The number is 44.

For Exercises 25–30, use the problem-solving flowchart (page 141) to solve the problems.

25. In 1983, 104 more Democrats than Republicans were in the U.S. House of Representatives. If the total number of representatives in the House was 434, find the number of representatives from each party. There were 165 Republicans and 269 Democrats.

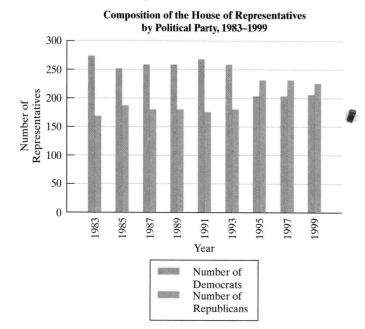

Composition of the House of Representatives by Political Party, 1983–1999

Figure for Exercises 25, 26

26. In 1995, 12 more Republicans than Democrats were in the U.S. House of Representatives. If the House had a total of 434 representatives from these two parties, find the number of Democrats and the number of Republicans. There were 211 Democrats and 223 Republicans.

27. A board is 86 cm in length and must be cut so that one piece is 20 cm longer than the other piece. Find the length of each piece. The lengths of the pieces are 33 cm and 53 cm.

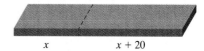

x $x + 20$

Figure for Exercise 27

28. A rope is 54 in. in length and must be cut into two pieces. If one piece must be twice as long as the other, find the length of each piece. The lengths of the pieces are 18 in. and 36 in.

29. The longest river in Africa is the Nile. It is 2455 km longer than the Congo River, also in Africa. The sum of the lengths of these rivers is 11,195 km. What is the length of each river? The Congo River is 4370 km long, and the Nile River is 6825 km.

Nile

Congo

Figure for Exercise 29

30. The average depth of the Gulf of Mexico is three times the depth of the Red Sea. The difference between the average depths is 1078 m. What is the average depth of the Gulf of Mexico and the average depth of the Red Sea? The average depth for the Red Sea is 539 m and that of the Gulf of Mexico is 1617 m.

31. a. If x represents the smallest of three consecutive integers, write an expression to represent each of the next two consecutive integers. $x + 1, x + 2$

    b. If x represents the largest of three consecutive integers, write an expression to represent each of the previous two consecutive integers. $x − 1, x − 2$

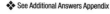

32. a. If x represents the smallest of three consecutive even integers, write an expression to represent each of the next two consecutive even integers. $x + 2, x + 4$

    b. If x represents the largest of three consecutive even integers, write an expression to represent each of the previous two consecutive even integers. $x - 2, x - 4$

33. a. If x represents the smallest of three consecutive odd integers, write an expression to represent each of the next two consecutive odd integers. $x + 2, x + 4$

    b. If x represents the largest of three consecutive odd integers, write an expression to represent each of the previous two consecutive odd integers. $x - 2, x - 4$

34. Is it possible to find two consecutive integers that differ by 5? Explain.
    No, two consecutive integers always differ by 1.

35. The sum of two consecutive integers is -67. Find the integers. The integers are -34 and -33.

36. The sum of two consecutive odd integers is 28. Find the integers. The integers are 13 and 15.

37. The sum of the page numbers on two facing pages in a book is 941. What are the page numbers? The page numbers are 470 and 471.

Figure for Exercise 37

38. Three raffle tickets are represented by three consecutive integers. If the sum of the three integers is 2,666,031, find the numbers.
    The ticket numbers are 888,676; 888,677; and 888,678.

Figure for Exercise 38

39. Three consecutive odd integers are such that 3 times the smallest is 9 more than twice the largest. Find the three numbers.
    The numbers are 17, 19, and 21.

40. Three consecutive even integers are such that the sum of the two larger integers is 232 more than three times the smallest integer. Find the three integers. The integers are -226, -224, and -222.

41. The perimeter of a triangle is 42 in. The lengths of the sides are represented by three consecutive integers. Find the lengths of the sides of the triangle. The sides are 13 in., 14 in., and 15 in.

42. The perimeter of a triangle is 54 m. The lengths of the sides are represented by three consecutive even integers. Find the lengths of the three sides.
    The sides are 16 m, 18 m, and 20 m.

43. The perimeter of a pentagon (a five-sided polygon) is 80 in. The five sides are represented by consecutive integers. Find the measures of the sides. The sides are 14 in., 15 in., 16 in., 17 in., and 18 in.

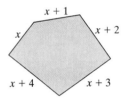

Figure for Exercise 43

44. The perimeter of a pentagon (a five-sided polygon) is 95 in. The five sides are represented by consecutive integers. Find the measures of the sides. The sides are 17 in., 18 in., 19 in., 20 in., and 21 in.

45. The area of Greenland is 201,900 km^2 less than three times the area of New Guinea. What is the area of New Guinea if the area of Greenland is 2,175,600 km^2? The area of New Guinea is 792,500 km^2.

46. The deepest point in the Pacific Ocean is 676 m more than twice the deepest point in the Arctic Ocean. If the deepest point in the Pacific is 10,920 m, how many meters is the deepest point in the Arctic Ocean?
    The deepest point in the Arctic Ocean is 5122 m.

47. Asia and Africa are the two largest continents in the world. The land area of Asia is approximately 14,514,000 km^2 larger than the land area of Africa. Together their total area is 74,644,000 km^2. Find the land area of Asia and the land area of Africa.   The area of Africa is 30,065,000 km^2. The area of Asia is 44,579,000 km^2.

48. Mt. Everest, the highest mountain in the world is 2654 m higher than Mt. McKinley, the highest mountain in the United States. If the sum of their heights is 15,042 m, find the height of each mountain.   Mt. McKinley is 6194 m high. Mt. Everest is 8848 m high.

■ EXPANDING YOUR SKILLS

49. In the 1998–1999 hockey season, the three highest paid NHL hockey players were Sergei Fedorov (Detroit Red Wings), Paul Kariya (Anaheim Mighty Ducks), and Eric Lindros (Philadelphia Flyers). Fedorov received $5.5 million more than Kariya, and Lindros received the same salary as Kariya. If the total of all three salaries was $31 million, how much did each player earn?   Kariya and Lindros earned $8.5 million, and Fedorov earned $14 million.

50. In the 1998–1999 hockey season, Mats Sundin (Toronto Maple Leafs), Curtis Joseph (Toronto Maple Leafs), and Dominik Hasek (Buffalo Sabres) earned a total of $19.8 million. If Hasek made $3 million less than twice what Joseph made, and Sundin made $0.8 million more than Joseph, how much did each player earn?   Joseph earned $5.5 million, Hasek earned $8 million, and Sundin earned $6.3 million.

Concepts

1. **Clearing Fractions and Decimals**

2. **Steps to Solving a Linear Equation**

3. **Solving Basic Percent Equations**

4. **Applications Involving Commission**

5. **Applications Involving Sales Tax**

6. **Applications Involving Simple Interest**

section
2.4   LINEAR EQUATIONS: CLEARING FRACTIONS AND DECIMALS

1. Clearing Fractions and Decimals

Linear equations that contain fractions can be solved in different ways. The first procedure, illustrated here, uses the method outlined in Section 2.2.

$$\frac{5}{6}x - \frac{3}{4} = \frac{1}{3}$$

$$\frac{5}{6}x - \frac{3}{4} + \frac{3}{4} = \frac{1}{3} + \frac{3}{4}$$ To isolate the variable term, add $\frac{3}{4}$ to both sides.

$$\frac{5}{6}x = \frac{4}{12} + \frac{9}{12}$$ Find the common denominator on the right-hand side.

$$\frac{5}{6}x = \frac{13}{12}$$ Simplify.

$$\frac{6}{5}\left(\frac{5}{6}x\right) = \frac{6}{5}\left(\frac{13}{12}\right)$$ Multiply by the reciprocal of $\frac{5}{6}$, which is $\frac{6}{5}$.

$$x = \frac{13}{10}$$

Sometimes it is simpler to solve an equation with fractions by eliminating the fractions first using a process called **clearing fractions**. To clear fractions in the equation $\frac{5}{6}x - \frac{3}{4} = \frac{1}{3}$, we can multiply both sides of the equation by the least common denominator (LCD) of all terms in the equation. In this case, the LCD of 6, 4, and 3 is 12. Because each denominator in the equation is a factor of 12, we can reduce common factors to leave integer coefficients for each term.

---

example 1

Solving a Linear Equation by Clearing Fractions

Solve the equation $\frac{5}{6}x - \frac{3}{4} = \frac{1}{3}$ by clearing fractions first.

Solution:

$$\frac{5}{6}x - \frac{3}{4} = \frac{1}{3}$$

$$12\left(\frac{5}{6}x - \frac{3}{4}\right) = 12\left(\frac{1}{3}\right)$$ Multiply both sides of the equation by the LCD, 12.

$$\frac{\overset{2}{\cancel{12}}}{1}\left(\frac{5}{6}x\right) - \frac{\overset{3}{\cancel{12}}}{1}\left(\frac{3}{4}\right) = \frac{\overset{4}{\cancel{12}}}{1}\left(\frac{1}{3}\right)$$ Apply the distributive property (recall that $12 = \frac{12}{1}$).

$$2(5x) - 3(3) = 4(1)$$ Reduce common factors to clear the fractions.

$$10x - 9 = 4$$

$$10x - 9 + 9 = 4 + 9$$ Add 9 to both sides.

$$10x = 13$$

$$\frac{10x}{10} = \frac{13}{10}$$ Divide both sides by 10.

$$x = \frac{13}{10}$$ Simplify.

Tip: The fractions in this equation can be eliminated by multiplying both sides of the equation by *any* common multiple of the denominators. For example, try multiplying both sides of the equation by 24:

$$24\left(\frac{5}{6}x - \frac{3}{4}\right) = 24\left(\frac{1}{3}\right)$$

$$\frac{\overset{4}{\cancel{24}}}{1}\left(\frac{5}{6}x\right) - \frac{\overset{6}{\cancel{24}}}{1}\left(\frac{3}{4}\right) = \frac{\overset{8}{\cancel{24}}}{1}\left(\frac{1}{3}\right)$$

$$20x - 18 = 8$$

$$20x = 26$$

$$\frac{20x}{20} = \frac{26}{20}$$

$$x = \frac{13}{10}$$

The same procedure used to clear fractions in an equation can be used to clear decimals. For example, consider the equation $0.05x + 0.25 = 0.2$. Because any terminating decimal can be written as a fraction, the equation can be interpreted as $\frac{5}{100}x + \frac{25}{100} = \frac{2}{10}$. A convenient common denominator for all terms in this equation is 100. Therefore, we can multiply the original equation by 100 to clear decimals.

---

example 2

Solving a Linear Equation by Clearing Decimals

Solve the equation $0.05x + 0.25 = 0.2$ by clearing decimals first.

Solution:

$$0.05x + 0.25 = 0.2$$

$$100(0.05x + 0.25) = 100(0.2) \qquad \text{Multiply both sides of the equation by 100.}$$

$$100(0.05x) + 100(0.25) = 100(0.2) \qquad \text{Apply the distributive property.}$$

$$5x + 25 = 20 \qquad \text{Simplify (decimals have been cleared).}$$

$$5x + 25 - 25 = 20 - 25 \qquad \text{Subtract 25 from both sides.}$$

$$5x = -5$$

$$\frac{5x}{5} = \frac{-5}{5} \qquad \text{Divide both sides by 5.}$$

$$x = -1 \qquad \text{Simplify.}$$

Tip: Notice that multiplying a decimal number by 100 has the effect of moving the decimal point two places to the right. Similarly, multiplying by 10 moves the decimal point one place to the right, multiplying by 1000 moves the decimal point three places to the right, and so on.

This equation can be checked by hand or by using a calculator.

$$0.05x + 0.25 = 0.2$$

$$0.05(-1) + 0.25 \stackrel{?}{=} 0.2$$

$$-0.05 + 0.25 \stackrel{?}{=} 0.2$$

$$0.2 = 0.2 \checkmark$$

---

2. Steps to Solving a Linear Equation

In this section, we combine the process for clearing fractions and decimals with the general strategies for solving linear equations. To solve a linear equation, it is important to follow the steps listed below.

Steps for Solving a Linear Equation in One Variable

1. Consider clearing fractions or decimals (if any are present) by multiplying both sides of the equation by a common denominator of all terms.
2. Simplify both sides of the equation by clearing parentheses and combining *like* terms.
3. Use the addition and subtraction properties of equality to collect the variable terms on one side of the equation.

4. Use the addition and subtraction properties of equality to collect the constant terms on the other side of the equation.
5. Use the multiplication and division properties of equality to make the coefficient of the variable term equal to 1.
6. Check your answer.

example 3

Solving Linear Equations

a. $\frac{1}{6}x - \frac{2}{3} = \frac{1}{5}x - 1$

b. $\frac{1}{3}(x + 7) - \frac{1}{2}(x + 1) = 4$

Classroom Activity 2.4A

Solution:

a.

$$\frac{1}{6}x - \frac{2}{3} = \frac{1}{5}x - 1$$ The LCD of 6, 3, and 5 is 30.

$$30\left(\frac{1}{6}x - \frac{2}{3}\right) = 30\left(\frac{1}{5}x - 1\right)$$ Multiply by the LCD, 30.

$$\frac{\overset{5}{\cancel{30}}}{1} \cdot \frac{1}{6}x - \frac{\overset{10}{\cancel{30}}}{1} \cdot \frac{2}{3} = \frac{\overset{6}{\cancel{30}}}{1} \cdot \frac{1}{5}x - 30(1)$$ Apply the distributive property (recall $30 = \frac{30}{1}$).

$$5x - 20 = 6x - 30$$ Clear fractions.

$$5x - 6x - 20 = 6x - 6x - 30$$ Subtract $6x$ from both sides.

$$-x - 20 = -30$$

$$-x - 20 + 20 = -30 + 20$$ Add 20 to both sides.

$$-x = -10$$

$$\frac{-x}{-1} = \frac{-10}{-1}$$ Divide both sides by -1.

$$x = 10$$

b.

$$\frac{1}{3}(x + 7) - \frac{1}{2}(x + 1) = 4$$ The LCD of 3 and 2 is 6.

$$6\left[\frac{1}{3}(x + 7) - \frac{1}{2}(x + 1)\right] = 6(4)$$ Multiply both sides by 6.

$$\frac{6}{1}\left[\frac{1}{3}(x + 7)\right] - \frac{6}{1}\left[\frac{1}{2}(x + 1)\right] = 6(4)$$ Apply the distributive property.

The product is 2. The product is -3.

$$2(x + 7) - 3(x + 1) = 24$$

$$2x + 14 - 3x - 3 = 24$$ Clear parentheses.

$$-x + 11 = 24$$ Combine *like* terms.

◆ **Avoiding Mistakes**

Notice that on the left-hand side of this equation, the product of 6 and $\frac{1}{3}$ is taken first, and then the result of 2 is distributed through the parentheses. Similarly, the product of 6 and $-\frac{1}{2}$ is taken first. The result of -3 is then distributed through the parentheses.

$$-x + 11 - 11 = 24 - 11 \qquad \text{Subtract 11.}$$

$$-x = 13$$

$$\frac{-x}{-1} = \frac{13}{-1} \qquad \text{Divide by } -1.$$

$$x = -13 \qquad \text{The check is left to the reader.}$$

3. Solving Basic Percent Equations

Recall from Section R.2 that the word **percent** means "per hundred." Therefore, 60% means $\frac{60}{100}$, or equivalently in decimal form, $60\% = 0.60$.

example 4

Solving Basic Percent Equations

a. What percent of 60 is 25.2?
b. 8.2 is 125% of what number?

Classroom Activity 2.4B

Solution:

a. Let x represent the unknown percentage.

What percent of 60 is 25.2?

$$x \quad \cdot \quad 60 = 25.2$$

$$60x = 25.2$$

$$\frac{60x}{60} = \frac{25.2}{60}$$

$$x = 0.42, \text{ or } 42\%$$

25.2 is 42% of 60.

Step 1: Read the problem.
Step 2: Label the variables.
Step 3: Create a verbal model.
Step 4: Write a mathematical equation.
Step 5: Solve the equation.
Step 6: Interpret the results and write the answer in words.

b. Let x represent the unknown number.

8.2 is 125% of what number?

$$8.2 = 1.25 \cdot x$$

⬢ Avoiding Mistakes

Be sure to use the decimal form of a percentage within an equation.

$$125\% = 1.25$$

$$8.2 = 1.25x$$

$$\frac{8.2}{1.25} = \frac{1.25x}{1.25}$$

$$6.56 = x$$

8.2 is 125% of 6.56.

Step 1: Read the problem.
Step 2: Label the variables.
Step 3: Create verbal model.
Step 4: Write a mathematical equation.
Step 5: Solve the equation.
Step 6: Interpret the results and write the answer in words.

4. Applications Involving Commission

example 5

Using Percentages to Find Commission

Terrance works for a prestigious clothing shop and earns $10,000 per year plus 12.5% commission on sales. What is Terrance's total commission if his total in sales is $260,000?

Classroom Activity 2.4C

Solution: **Step 1:** Read the problem.

Let x represent the total commission.   **Step 2:** Label the variable.

$$\begin{pmatrix} \text{Total} \\ \text{commission} \end{pmatrix} = \begin{pmatrix} \text{commission} \\ \text{rate} \end{pmatrix}\begin{pmatrix} \text{total} \\ \text{sales} \end{pmatrix}$$

Step 3: Write an equation in words.

$$x \quad = \quad (0.125)(\$260,000)$$

Step 4: Write a mathematical equation.

$$= \$32,500$$

Step 5: Solve the equation.

Terrance made $32,500 in commission.

Step 6: Interpret the results and write the answer in words.

5. Applications Involving Sales Tax

example 6

Applying Percentages

A video game is purchased for a total of $48.15 including sales tax. If the tax rate is 7%, find the original price of the video game before sales tax.

Classroom Activity 2.4D

Solution: **Step 1:** Read the problem.

Let x represent the price of the video game.   **Step 2:** Label variables.

Tip: The equation in Example 6 could have been solved easily without clearing decimals.

$0.07x$ represents the amount of sales tax.

$$\begin{pmatrix} \text{Original} \\ \text{price} \end{pmatrix} + \begin{pmatrix} \text{sales} \\ \text{tax} \end{pmatrix} = \begin{pmatrix} \text{total} \\ \text{cost} \end{pmatrix}$$

Step 3: Write a verbal equation.

$$x + 0.07x = 48.15$$

$$1.07x = 48.15$$

$$\frac{1.07x}{1.07} = \frac{48.15}{1.07}$$

$$x = 45$$

This illustrates the point that the decision to clear decimals (or fractions) is a matter of preference.

$$x \quad + \quad 0.07x \quad = \quad \$48.15$$

Step 4: Write a mathematical equation.

$$1.07x = 48.15$$

Step 5: Solve for x.

$$100(1.07x) = 100(48.15)$$

Multiply by 100 to clear decimals.

$$107x = 4815$$

$$\frac{107x}{107} = \frac{4815}{107}$$

Divide both sides by 107.

$$x = 45$$

Step 6: Interpret the results and write the answer in words.

The original price was $45.

6. Applications Involving Simple Interest

One important application of percentages is in computing simple interest on a loan or on an investment.

Banks hold large quantities of money for its customers. However, because all bank customers are unlikely to withdraw all their money on a single day, a bank does not keep all the money in cash. Instead, it keeps some cash for day-to-day transactions but invests the remaining portion of the money. Because a bank uses its customer's money to make investments, and because it wants to attract more customers the bank pays interest on the money. **Simple interest** is interest that is earned on principal (the original amount of money invested in an account). The following formula is used to compute simple interest:

$$\left(\begin{array}{c}\text{Simple}\\\text{interest}\end{array}\right) = \left(\begin{array}{c}\text{principal}\\\text{invested}\end{array}\right)\left(\begin{array}{c}\text{annual}\\\text{interest rate}\end{array}\right)\left(\begin{array}{c}\text{time}\\\text{in years}\end{array}\right)$$

This formula is often written symbolically as $I = Prt$.

For example, to find the simple interest earned on $2000 invested at 7.5% interest for 3 years, we have

$$I = P \cdot r \cdot t$$
$$\text{Interest} = (\$2000)(0.075)(3)$$
$$= \$450$$

---

example 7

Applying Simple Interest

Jorge wants to save money for his daughter's college education. If Jorge needs to have $4340 at the end of 4 years, how much money would he need to invest at a 6% simple interest rate?

Classroom Activity 2.4E

Solution:

Let x represent the original amount invested.

$$\left(\begin{array}{c}\text{Original}\\\text{principal}\end{array}\right) + (\text{interest}) = (\text{total})$$

$$(P) \quad + \quad (Prt) \quad = (\text{total})$$
$$x \quad + \quad x(0.06)(4) = 4340$$

Step 1: Read the problem.

Step 2: Label the variables.

Step 3: Write an equation in words.

Recall that interest is computed by the formula $I = Prt$.

Step 4: Write a mathematical equation.

$$x + 0.24x = 4340$$

$$1.24x = 4340$$ **Step 5:** Solve the equation.

$$\frac{1.24x}{1.24} = \frac{4340}{1.24}$$

$$x = 3500$$

The original investment should be \$3500. **Step 6:** Interpret the results and write the answer in words.

section 2.4 PRACTICE EXERCISES

For Exercises 1–10, solve the equation.

1. $25x = -15$ 1.1 $x = -\frac{3}{5}$

2. $42 = 6y$ 1.1 $y = 7$

3. $34 = m - 12$ 1.1 $m = 46$

4. $-19 + n = 14$ 1.1 $n = 33$

5. $5(x + 2) - 3 = 4x + 5$ 1.2 $x = -2$

6. $-2(2x - 4x) = 6 + 18$ 1.2 $x = 6$

7. $2 + 3(b - 6) - 2b = 6$ 1.2 $b = 22$

8. $6(z + 2) - 12 = 14$ 1.2 $z = \frac{7}{3}$

9. $3(2y + 3) - 4(-y + 1) = 7y - 10$ 1.2 $y = -5$

10. $-(3w + 4) + 5(w - 2) - 3(6w - 8) = 10$ 1.2 $w = 0$

11. Solve the equation and describe the solution set:
    $7x + 2 = 7(x - 12)$ 1.2   No solution

12. Solve the equation and describe the solution set:
    $2(3x - 6) = 3(2x - 4)$ 1.2   All real numbers

For Exercises 13–16, determine which of the values could be used to clear fractions or decimals in the given equation.

13. $\frac{2}{3}x - \frac{1}{6} = \frac{x}{9}$ Values: 6, 9, 12, 18, 24, 36   18, 36

14. $\frac{1}{4}x - \frac{2}{7} = \frac{1}{2}x + 2$ Values: 4, 7, 14, 21, 28, 42   28

15. $0.02x + 0.5 = 0.35x + 1.2$ Values: 10; 100; 1000; 10,000   100; 1000; 10,000

16. $0.003 - 0.002x = 0.1x$ Values: 10; 100; 1000; 10,000   1000; 10,000

For Exercises 17–28, solve the equation.

17. $\frac{1}{2}x + 3 = 5$ $x = 4$

18. $\frac{1}{3}y - 4 = 9$ $y = 39$

19. $\frac{1}{6}y + 2 = \frac{5}{12}$ $y = -\frac{19}{2}$

20. $\frac{2}{15}z + 3 = \frac{7}{5}$ $z = -12$

21. $\frac{1}{3}q + \frac{3}{5} = \frac{1}{15}q - \frac{2}{5}$ $q = -\frac{15}{4}$

22. $\frac{3}{7}x - 5 = \frac{24}{7}x + 7$ $x = -4$

23. $\frac{12}{5}w + 7 = 31 - \frac{3}{5}w$ $w = 8$

24. $-\frac{1}{9}p - \frac{5}{18} = -\frac{1}{6}p + \frac{1}{3}$ $p = 11$

25. $\frac{2}{3}(x + 4) = \frac{2}{3}x - \frac{1}{6}$ No solution

26. $\frac{1}{9}(3 - a) = \frac{1}{6} - \frac{1}{9}a$ No solution

27. $\frac{1}{6}(2c - 1) = \frac{1}{3}c - \frac{1}{6}$ All real numbers

28. $\frac{3}{4}b - \frac{1}{2} = -\frac{1}{4}(2 - 3b)$ All real numbers

29. Negative eight is half of a number. Find the number.   The number is -16.

30. One third of a number equals five. Find the number.   The number is 15.

❖ See Additional Answers Appendix   Writing   Translating Expression   Geometry   Scientific Calculator   Video

31. The sum of $\frac{2}{5}$ and twice a number is the same as the sum of $\frac{11}{5}$ and the number. Find the number. The number is $\frac{9}{5}$.

32. The difference of three times a number and $\frac{5}{9}$ is the same as the sum of twice the number and $\frac{1}{9}$. Find the number. The number is $\frac{2}{3}$.

 33. The sum of twice a number and $\frac{3}{4}$ is the same as the difference of four times the number and $\frac{1}{8}$. Find the number. The number is $\frac{7}{16}$.

34. The difference of a number and $-\frac{11}{12}$ is the same as the difference of three times the number and $\frac{1}{6}$. Find the number. The number is $\frac{13}{24}$.

For Exercises 35–40, solve the equation.

35. $9.2y - 4.3 = 50.9$ $y = 6$

36. $-6.3x + 1.5 = -4.8$ $x = 1$

 37. $21.1w + 4.6 = 10.9w + 35.2$ $w = 3$

38. $0.05z + 0.2 = 0.15z - 10.5$ $z = 107$

39. $0.2p - 1.4 = 0.2(p - 7)$ All real numbers

40. $0.5(3q + 87) = 1.5q + 43.5$ All real numbers

 For Exercises 41–52, find the missing values.

41. 45 is what percent of 360?   12.5%

42. 338 is what percent of 520?   65%

43. 544 is what percent of 640?   85%

44. 576 is what percent of 800?   72%

45. What is 0.5% of 150?   0.75

46. What is 9.5% of 616?   58.52

47. What is 42% of 740?   310.8

48. What is 56% of 280?   156.8

49. 177 is 20% of what number?   885

50. 126 is 15% of what number?   840

51. 275 is 12.5% of what number?   2200

52. 594 is 45% of what number?   1320

 The number of AIDS cases reported in the United States in 1997 by race or ethnic group is shown in the figure. Use this figure to answer Exercises 53–56. Round the answers to the nearest tenth of a percent.

Number of AIDS Cases 1997–United States

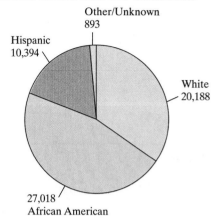

Source: Centers for Disease Control.

Figure for Exercises 53–56

53. What percentage of reported AIDS cases in the United States are African American?   46.2%

54. What percentage of reported AIDS cases in the United States are Hispanic?   17.8%

55. What percentage of reported U.S. AIDS cases are white?   34.5%

56. What percentage of reported U.S. AIDS cases are described as being in the other or unknown category?   1.5%

 For Exercises 57–70, solve for the unknown quantity.

57. Molly buys a golf outfit that costs $74.95. If the sales tax rate is 7%, for how much should she write her check? $80.20

58. Patrick purchased golf shoes for $85.98. If the sales tax rate is 6%, how much was charged to his Visa credit card? $91.14

59. The sales tax for a screwdriver set came to $1.04. If the sales tax rate is 6.5%, what was the price of the screwdriver? $16.00

60. The sales tax for a picture frame came to $1.32. If the sales tax rate is 5.5%, what was the price of the picture frame? $24.00

61. Sun Lei bought a laptop computer over the Internet for $1800. The total cost, including tax, came to $1890. What is the sales tax rate?   5%

62. Jamie purchased a compact disc and paid $18.26. If the disc price is $16.99, what is the sales tax rate (round to the nearest tenth of a percent)?   7.5%

63. A dress is marked at 30% off. If the sale price is $20.97, what was the original price of the dress? $29.96

64. A jacket is on sale for $53.60. If this represents a 20% discount, what was the original price of the jacket? $67.00

65. The local car dealership pays its sales personnel a commission of 25% of the dealer profit on each car. The dealer made a profit of $18,250 on the cars Joëlle sold last month. What was her commission last month? $4562.50

66. Dan sold a beachfront home for $650,000. If his commission rate is 4%, what did he earn on the sale of that home? $26,000

67. A salesperson at *You Bought It* discount store earns 3% commission on all appliances that he sells. If Geoff's commission for the month was $116.37, how much did he sell? $3879

68. Anna makes a commission at an appliance store. In addition to her base salary, she earns a 2.5% commission on her sales. If Anna's commission for a month was $260, how much did she sell? $10,400

69. For selling software, Tom received a bonus commission based on sales over $500. If he received $180 in commission for selling a total of $2300 worth of software, what is his commission rate?   10%

70. In addition to an hourly salary, Jessica earns a commission for selling ice cream bars at the beach. If she sells $708 worth of ice cream and receives a commission of $56.64, what is her commission rate?   8%

For Exercises 71–76, solve these equations involving simple interest.

71. How much interest will Pam earn in 4 years if she invests $3000 in an account that pays 3.5% simple interest? $420

72. How much interest will Roxanne have to pay if she borrows $2000 for 2 years at a simple interest rate of 4%? $160

73. Bob borrowed some money for 1 year at 5% simple interest. If he had to pay back a total of $1260, how much did he originally borrow? $1200

74. Mike borrowed some money for 2 years at 6% simple interest. If he had to pay back a total of $3640, how much did he originally borrow? $3250

75. If $1500 grows to $1950 after 5 years, find the simple interest rate.   6%

76. If $9000 grows to $10,440 in 2 years, find the simple interest rate.   8%

EXPANDING YOUR SKILLS

For Exercises 77–80, solve the equation.

77. $\dfrac{1}{2}a + 0.4 = -0.7 - \dfrac{3}{5}a$ $a = -1$

78. $\dfrac{3}{4}c - 0.11 = 0.23(c - 5)$ $c = -2$

79. $0.8 + \dfrac{7}{10}b = \dfrac{3}{2}b - 0.8$ $b = 2$

80. $0.78 - \dfrac{1}{25}h = \dfrac{3}{5}h - 0.5$ $h = 2$

81. Diane sells women's sportswear at a department store. She earns a regular salary and, as a bonus, she receives a commission of 4% on all merchandise sold over $200. If Diane earned an extra $25.80 last week in commission, how much merchandise did she sell over $200? $645

82. Bob's position in men's formal wear pays by commission. He earns 5% for the first $1000 he sells and 6% on sales thereafter. If Bob's commission last week was $170, how much merchandise did he sell? $3000 total

83. A sweater is on sale for 15% off. With a sales tax rate of 5%, the sales tax amounts to $2.55. What was the original price of the sweater before the sale? $60

84. A set of golf clubs is on sale for 25% off. With a sales tax rate of 7%, the sales tax amounts to $21.00. What was the original price of the clubs? $400

---

chapter 2   MIDCHAPTER REVIEW

For Exercises 1–24, solve the equation.

1. $2b + 23 = 6b - 5$
   $b = 7$

2. $-x = 7$
   $x = -7$

3. $\dfrac{y}{4} = -2$ $y = -8$

4. $10p - 9 + 2p - 3 = 8p - 18$ $p = -\dfrac{3}{2}$, or $-1\frac{1}{2}$

5. $0.5(2a - 3) - 0.1 = 0.4(6 + 2a)$ $a = 20$

6. $-\dfrac{5}{9}w + \dfrac{11}{12} = \dfrac{23}{36}$ $w = \dfrac{1}{2}$

7. $-6x = 0$ $x = 0$

8. $15.2q = -2.4q - 176$ $q = -10$

9. $9.8h + 2 = 3.8h + 20$ $h = 3$

10. $-k - 41 = 3 - k$ No solution

11. $\dfrac{1}{4}(x + 4) = \dfrac{1}{5}(2x + 3)$ $x = \dfrac{8}{3}$, or $2\frac{2}{3}$

12. $7y + 3(2y + 5) = 10y + 17$ $y = \dfrac{2}{3}$

13. $2z - 7 = 2(z - 13)$
    No solution

14. $x - 17.8 = -21.3$
    $x = -3.5$

15. $\dfrac{4}{5}w = 10$ $w = \dfrac{25}{2}$, or $12\frac{1}{2}$

16. $5c + 25 = 20$ $c = -1$

17. $4b - 8 - b = -3b + 2(3b - 4)$ All real numbers

18. $36 = 6z + 9$ $z = \dfrac{9}{2}$, or $4\frac{1}{2}$

19. $-3a + 1 = 19$ $a = -6$

20. $-5(1 - x) + x = -(6 - 2x) + 6$ $x = \dfrac{5}{4}$, or $1\frac{1}{4}$

21. $3(4h - 2) - (5h - 8) = 8 - (2h + 3)$ $h = \dfrac{1}{3}$

22. $1.72w - 0.04w = 0.42$ $w = 0.25$

23. $\dfrac{3}{8}t - \dfrac{5}{8} = \dfrac{1}{2}t + \dfrac{1}{8}$ $t = -6$

24. $3(8x - 1) + 10 = 6(5 + 4x) - 23$ All real numbers

 For Exercises 25–28, write a problem in words that translates to the given equation. (Answers will vary)

25. $x + 18 = 3(x + 2)$ ❖

26. $x - 5 = 2(x + 1)$
    ❖

27. $2(x - 5) = \dfrac{1}{3}x$

28. $\dfrac{1}{2}(x + 4) = 2x$

Twice the difference of a number and five is the same as one third of the number.

One half the sum of a number and four is the same as twice the number.

section
2.5   APPLICATIONS OF GEOMETRY

1.   Literal Equations

Formulas (or **literal equations**) are equations that contain several variables. For example, the perimeter of a triangle (distance around the triangle) can be found by the formula $P = a + b + c$, where a, b, and c are the lengths of the sides (Figure 2-8).

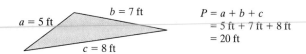

$$P = a + b + c$$
$$= 5\text{ ft} + 7\text{ ft} + 8\text{ ft}$$
$$= 20\text{ ft}$$

Figure 2-8

In this section, we will learn how to rewrite formulas to solve for a different variable within the formula. Suppose, for example, that the perimeter of a triangle is known and two of the sides are known (say sides a and b). Then the third side, c, can be found by subtracting the lengths of the known sides from the perimeter (Figure 2-9).

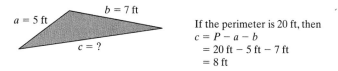

If the perimeter is 20 ft, then
$$c = P - a - b$$
$$= 20\text{ ft} - 5\text{ ft} - 7\text{ ft}$$
$$= 8\text{ ft}$$

Figure 2-9

To solve a formula for a different variable, we use the same properties of equality outlined in the earlier sections of this chapter. For example, consider the two equations $2x + 3 = 11$ and $wx + y = z$. Suppose we want to solve for x in each case:

$2x + 3 = 11$		$wx + y = z$	
$2x + 3 - 3 = 11 - 3$	Subtract 3.	$wx + y - y = z - y$	Subtract y.
$2x = 8$		$wx = z - y$	
$\dfrac{2x}{2} = \dfrac{8}{2}$	Divide by 2.	$\dfrac{wx}{w} = \dfrac{z - y}{w}$	Divide by w.
$x = 4$		$x = \dfrac{z - y}{w}$	

The equation on the left has only one variable and we are able to simplify the equation to find a numerical value for x. The equation on the right has multiple variables. Because we do not know the values of w, y, and z, we are not able to simplify further. The value of x is left as a formula in terms of w, y, and z.

example 1

Solving Formulas for an Indicated Variable

Solve the formulas for the indicated variables.

a. $d = rt$ for t

b. $5x + 2y = 12$ for y

Classroom Activity 2.5A

Solution:

a. $d = rt$ for t The goal is to isolate the variable t.

Tip: The original equation $d = rt$ represents the distance traveled, d, in terms of the rate of speed, r, and the time of travel, t.

$$\frac{d}{r} = \frac{rt}{r}$$ Because the relationship between r and t is multiplication, we reverse the process by dividing both sides by r.

The equation $t = \frac{d}{t}$ repre-sents the same relationship among the variables, however, the time of travel is expressed in terms of the distance and rate.

$$\frac{d}{r} = t, \text{ or equivalently } t = \frac{d}{r}$$

b. $5x + 2y = 12$ for y The goal is to solve for y.

$$5x - 5x + 2y = 12 - 5x$$ Subtract $5x$ from both sides to isolate the y-term.

$$2y = -5x + 12$$

Tip: On the right-hand side we chose to write the variable term first $-5x + 12$ as is customary. However, it is also correct to write the ex-pression as $12 - 5x$.

$$\frac{2y}{2} = \frac{-5x + 12}{2}$$ Divide both sides by 2 to isolate y.

$$y = \frac{-5x + 12}{2}$$

Tip: The expression $\dfrac{-5x + 12}{2}$

can also be written with the divisor 2 applied individually to each term in the numera-tor. Hence the answer may appear in several different forms. Each is correct:

$$y = \frac{-5x + 12}{2} \quad \text{or} \quad y = \frac{-5x}{2} + \frac{12}{2} \quad \text{or} \quad y = -\frac{5}{2}x + 6$$

example 2

Solving Formulas for an Indicated Variable

The formula $C = \frac{5}{9}(F - 32)$ is used to find the temperature, C, in degrees Celsius for a given temperature expressed in degrees Fahrenheit, F.

a.   Use the formula $C = \frac{5}{9}(F - 32)$ to find the equivalent Celsius temperature for 86°F.

b.   Solve the formula $C = \frac{5}{9}(F - 32)$ for F.

c.   Use the equation found in part (b) to find the equivalent Fahrenheit tem-perature for 30°C.

Solution:

a. $C = \dfrac{5}{9}(F - 32)$

$ = \dfrac{5}{9}(86 - 32)$ Substitute $F = 86$.

$ = \dfrac{5}{9}(54)$ Simplify.

$ = 30$

30°C is equivalent to 86°F.

b. $C = \dfrac{5}{9}(F - 32)$

$9C = 9 \cdot \dfrac{5}{9}(F - 32)$ Multiply by 9 to clear fractions.

$9C = 5(F - 32)$ Simplify.

$9C = 5F - 160$ Clear parentheses.

$9C + 160 = 5F - 160 + 160$ Add 160 to both sides.

$9C + 160 = 5F$

$\dfrac{9C + 160}{5} = \dfrac{5F}{5}$ Divide both sides by 5 to isolate F.

$\dfrac{9C + 160}{5} = F$

The answer may be written in several forms:

$$F = \dfrac{9C + 160}{5} \quad \text{or} \quad F = \dfrac{9C}{5} + \dfrac{160}{5} \quad \text{or} \quad F = \dfrac{9}{5}C + 32$$

c.   Using $F = \dfrac{9}{5}C + 32$, substitute 30° for C.

$F = \dfrac{9}{5}(30) + 32$

$ = 54 + 32$

$ = 86$

86°F is equivalent to 30°C.   This is consistent with the result from part (a).

■────────────────

2.   Geometry Applications: Perimeter

In Section R.3, we presented numerous facts and formulas related to geometry. Sometimes these are needed to solve applications in geometry.

example 3

Solving a Geometry Application Involving Perimeter

The length of a rectangular lot is 1 m less than twice the width. If the perimeter is 190 m, find the length and width.

Classroom Activity 2.5B

Solution:

	Step 1: Read the problem.
Let x represent the width of the rectangle.	**Step 2:** Label the variables.

Then $2x - 1$ represents the length.

x

$2x - 1$

$$P = 2l + 2w$$ **Step 3:** Perimeter formula

$$190 = 2(2x - 1) + 2(x)$$ **Step 4:** Write an equation in terms of x.

$$190 = 4x - 2 + 2x$$ **Step 5:** Solve for x.

$$190 = 6x - 2$$

$$192 = 6x$$

$$\frac{192}{6} = \frac{6x}{6}$$

$$32 = x$$

The width is $x = 32$.

The length is $2x - 1 = 2(32) - 1 = 63$.   **Step 6:** Interpret the results and write the answer in words.

The width of the rectangular lot is 32 m and the length is 63 m.

3. Geometry Applications: Complementary Angles

example 4

Solving a Geometry Application Involving Complementary Angles

Two complementary angles are drawn such that one angle is 4° more than seven times the other angle. Find the measure of each angle.

Classroom Activity 2.5C

Solution:

	Step 1: Read the problem.
Let x represent the measure of one angle.	**Step 2:** Label the variables.

Then $7x + 4$ represents the measure of the other angle.

The angles are complementary, so their sum must be 90°.

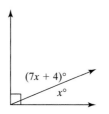

$$\left(\begin{array}{c}\text{Measure of}\\ \text{first angle}\end{array}\right) + \left(\begin{array}{c}\text{measure of}\\ \text{second angle}\end{array}\right) = 90°$$ **Step 3:** Create a verbal equation.

$$x \quad + \quad 7x + 4 \quad = 90$$ **Step 4:** Write a mathematical equation.

$$8x + 4 = 90$$ **Step 5:** Solve for x.

$$8x = 86$$

$$\frac{8x}{8} = \frac{86}{8}$$

$$x = 10.75$$

Step 6: Interpret the results and write the answer in words.

One angle is $x = 10.75$.

The other angle is $7x + 4 = 7(10.75) + 4 = 79.25$.

The angles are 10.75° and 79.25°.

■

4. Geometry Applications: Angles Inscribed within a Triangle

example 5

Solving a Geometry Application

One angle in a triangle is twice as large as the smallest angle. The third angle is 10° more than seven times the smallest angle. Find the measure of each angle.

Solution:

Step 1: Read the problem.

Let x represent the measure of the smallest angle.   **Step 2:** Label the variables.

Then $2x$ and $7x + 10$ represent the measures of the other two angles.

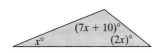

The sum of the angles must be 180°.

$$x + 2x + (7x + 10) = 180$$

$$x + 2x + 7x + 10 = 180$$

$$10x + 10 = 180$$

$$10x = 170$$

$$x = 17$$

The smallest angle is $x = 17$.

The other angles are $2x = 2(17) = 34$

$$7x + 10 = 7(17) + 10 = 129$$

The angles are 17°, 34°, and 129°.

Step 3: Create a verbal equation.

Step 4: Write a mathematical equation.

Step 5: Solve for x.

Step 6: Interpret the results and write the answer in words.

5. Geometry Applications: Circumference

example 6

Solving a Geometry Application Involving Circumference

The distance around a circular garden is 188.4 ft. Find the radius to the nearest tenth of a foot (Figure 2-10).

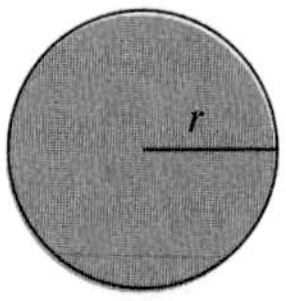

$c = 188.4$ ft

Figure 2-10

Solution:

$$C = 2\pi r \qquad \text{Use the formula for the circumference of a circle.}$$

$$188.4 = 2\pi r \qquad \text{Substitute 188.4 for } C.$$

$$\frac{188.4}{2\pi} = \frac{2\pi r}{2\pi} \qquad \text{Divide both sides by } 2\pi.$$

$$r = \frac{188.4}{2\pi} \approx 30.0$$

The radius is approximately 30.0 ft.

Tip: In Example 6, we could have solved the equation $C = 2\pi r$ for the variable r first before substituting the value of C.

$$C = 2\pi r \qquad\qquad r = \frac{C}{2\pi}$$

$$\frac{C}{2\pi} = \frac{2\pi r}{2\pi} \qquad\qquad r = \frac{188.4}{2\pi} \approx 30.0$$

$$\frac{C}{2\pi} = r$$

Calculator Connections

Note that parentheses are required to divide 188.4 by the quantity 2π. This guarantees that the calculator follows the implied order of operations. Without parentheses, the calculator would divide 188.4 by 2 and then multiply the result by π.

Scientific Calculator

Enter: $\boxed{1}\,\boxed{8}\,\boxed{8}\,\boxed{.}\,\boxed{4}\,\boxed{\div}\,\boxed{(}\,\boxed{2}\,\boxed{\times}\,\boxed{\pi}\,\boxed{)}\,\boxed{=}$ **Result:** 29.98479128 correct

Enter: $\boxed{1}\,\boxed{8}\,\boxed{8}\,\boxed{.}\,\boxed{4}\,\boxed{\div}\,\boxed{2}\,\boxed{\times}\,\boxed{\pi}\,\boxed{=}$ **Result:** 295.938028 incorrect

Graphing Calculator

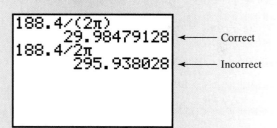

```
188.4/(2π)
        29.98479128    ⟵ Correct
188.4/2π
       295.938028      ⟵ Incorrect
```

Calculator Exercises

Approximate the expressions with a calculator. Round to three decimal places if necessary. ❖

1. $\dfrac{880}{2\pi}$ 140.056
2. $\dfrac{1600}{\pi(4)^2}$ 31.831
3. $\dfrac{20}{(-0.05)(5)}$ −80
4. $\dfrac{10}{0.5(6+4)}$ 2

section 2.5   PRACTICE EXERCISES

For Exercises 1–16, solve the equation.

1. $3 + z = 10$ <u>2.1</u> $z = 7$ 2. $4 = 7 + w$ <u>2.1</u> $w = -3$

3. $\dfrac{k}{5} = 7$ <u>2.1</u> $k = 35$ 4. $\dfrac{m}{6} = -2$ <u>2.1</u> $m = -12$

5. $-2a + 1 = 7$
   <u>2.2</u> $a = -3$

6. $-3b - 5 = 19$
   <u>2.2</u> $b = -8$

7. $2 + 3(b - 6) - 3b = 6$ <u>2.2</u>   No solution

8. $6(z + 2) - 12 = 14 + 6z$ <u>2.2</u>   No solution

9. $3(2y + 3) - 4(-y + 1) = 7y - 10$ <u>2.2</u> $y = -5$

10. $-(3w + 4) + 5(w - 2) - 3(6w - 8) = 10$
    <u>2.2</u> $w = 0$

11. $\dfrac{1}{2}(x - 3) + \dfrac{3}{4} = 3x - \dfrac{3}{4}$ <u>2.4</u> $x = 0$

12. $\dfrac{5}{6}x + \dfrac{1}{2} = \dfrac{1}{4}(x - 4)$ <u>2.4</u> $x = -\dfrac{18}{7}$, or $-2\dfrac{4}{7}$

13. $2p - 5 + 4p = 6(p + 2) - 17$ <u>2.3</u>   All real numbers

14. $3h - (h - 1) = 2(h + 3) - 5$ <u>2.3</u>   All real numbers

15. $0.5(y + 2) - 0.3 = 0.4y + 0.5$ <u>2.4</u> $y = -2$

16. $-0.02(1 - 4m) = 0.6 + 0.18m$ <u>2.4</u> $m = -6.2$

For Exercises 17–26, solve for the indicated variable.

17. $P = a + b + c$ for a $a = P - b - c$

18. $P = a + b + c$ for b $b = P - a - c$

19. $x = y - z$ for y
    $y = x + z$

20. $c + d = e$ for d
    $d = e - c$

21. $p = 250 + q$ for q
    $q = p - 250$

22. $y = 35 + x$ for x
    $x = y - 35$

23. $d = rt$ for t $t = \dfrac{d}{r}$

24. $d = rt$ for r $r = \dfrac{d}{t}$

25. $PV = nrt$ for t $t = \dfrac{PV}{nr}$

26. $P_1V_1 = P_2V_2$ for V_1 $V_1 = \dfrac{P_2V_2}{P_1}$

For Exercises 27–38, solve for the indicated variable.

27. $x - y = 5$ for x
    $x = 5 + y$

28. $x + y = -2$ for y
    $y = -2 - x$

29. $3x + y = -19$ for y $y = -19 - 3x$

30. $x - 6y = -10$ for x $x = 6y - 10$

31. $2x + 3y = 6$ for x ❖

32. $5x + 2y = 10$ for y ❖

33. $-2x - y = 9$ for x $x = \dfrac{9 + y}{-2}$, or $-\dfrac{9 + y}{2}$

34. $3x - y = -13$ for x $x = \dfrac{-13 + y}{3}$

35. $4x - 3y = 12$ for y $y = \dfrac{12 - 4x}{-3}$, or $-\dfrac{12 - 4x}{3}$

36. $6x - 3y = 4$ for x $x = \dfrac{3y + 4}{6}$

37. $ax + by = c$ for x $x = \dfrac{c - by}{a}$

38. $ax + by = c$ for y $y = \dfrac{c - ax}{b}$

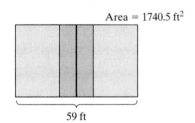 For Exercises 39–50, solve the geometry formulas for the indicated variables.

39. a. A rectangle has length l and width w. Write a formula for the area. $A = lw$

    b. Solve the formula for the width, w. $w = \dfrac{A}{l}$

    c. The area of a rectangular volleyball court is 1740.5 ft^2 and the length is 59 ft. Find the width.   The width is 29.5 ft.

Area = 1740.5 ft^2

59 ft

Figure for Exercise 39

40. a. A parallelogram has height h and base b. Write a formula for the area. $A = bh$

    b. Solve the formula for the base, b. $b = \dfrac{A}{h}$

    c. Find the base of the parallelogram pictured if the area is 40 m^2.   The base is 8 m.

5 m

$b = ?$

Figure for Exercise 40

41. a. A rectangle has length l and width w. Write a formula for the perimeter. $P = 2l + 2w$

    b. Solve the formula for the length, l. $l = \dfrac{P - 2w}{2}$

    c. The perimeter of the soccer field at Giants Stadium is 338 m. If the width is 66 m, find the length.   The length is 103 m.

Perimeter = 338 m

66 m

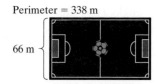

Figure for Exercise 41

42. a. The length of each side of a square is *s*. Write a formula for the perimeter of the square. $P = 4s$

    b. Solve the formula for the length of a side, *s*. $s = \dfrac{P}{4}$

    c. The Pyramid of Khufu (known as the Great Pyramid) at Giza has a square base. If the distance around the bottom is 921.6 m, find the length of the sides at the bottom of the pyramid. The length of each side at the bottom of the pyramid is 230.4 m.

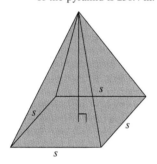

Figure for Exercise 42

43. a. A triangle has height *h* and base *b*. Write a formula for the area. $A = \dfrac{1}{2}bh$

    b. Solve the formula for the height, *h*. $h = \dfrac{2A}{b}$

    c. Find the height of the triangle pictured if the area is 12 km². The height is 4 km.

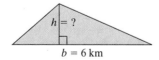

$h = ?$

$b = 6$ km

Figure for Exercise 43

44. a. A circle has a radius of *r*. Write a formula for the area. $A = \pi r^2$

    b. Solve the formula for π. $\pi = \dfrac{A}{r^2}$

    c. The area of a circle is 113 in.² and the radius is 6 in. Use these values to approximate the value of π. Round to two decimal places. $\pi \approx 3.14$

45. a. A circle has a radius of *r*. Write a formula for the circumference. $C = 2\pi r$

    b. Solve the formula for the radius, *r*. $r = \dfrac{C}{2\pi}$

    c. The circumference of the circular Buckingham Fountain in Chicago is approximately 880 ft. Find the radius. Round to the nearest foot. The radius is approximately 140 ft.

Figure for Exercise 45

46. a. A rectangular solid has height *h*, width *w*, and length *l*. Write a formula for the volume of the solid. $V = lwh$

    b. Solve the formula for the height, *h*. $h = \dfrac{V}{lw}$

    c. A rectangular box mounted on the back of a pick-up truck holds 45 ft³ of cargo space. If the length and width of the box are 4.5 ft and 5.0 ft, respectively, find the height of the box. The height of the box is 2 ft.

47. The volume of a cylinder is given by $V = \pi r^2 h$. Solve for the height, *h*. $h = \dfrac{V}{\pi r^2}$

48. The volume of a sphere is given by $V = \frac{4}{3}\pi r^3$. Solve for r^3. $r^3 = \dfrac{3V}{4\pi}$

49. The area of a trapezoid is given by $A = \frac{1}{2}(B + b)h$. Solve for *B*. $B = \dfrac{2A}{h} - b$ or $\dfrac{2A - bh}{h}$

50. The volume of a pyramid with a rectangular base is given by $V = \frac{1}{3}lwh$. Solve for the height, *h*. $h = \dfrac{3V}{lw}$

For Exercises 51–69, use the problem-solving flowchart (page 141) from Section 2.3.

51. The perimeter of a rectangular garden is 24 ft. The length is 2 ft more than the width. Find the length and the width of the garden. The length is 7 ft and the width is 5 ft.

52. In a small rectangular wallet photo, the width is 7 cm less than the length. If the border (perimeter) of the photo is 34 cm, find the length and width.   The length is 12 cm and the width is 5 cm.

53. A builder buys a rectangular lot of land such that the length is 5 m less than two times the width. If the perimeter is 590 m, find the length and the width.   The length is 195 m and the width is 100 m.

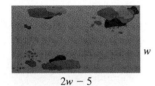

2w − 5

w

Figure for Exercise 53

54. The perimeter of a rectangular pool is 140 yd. If the length is 10 yd more than the width, find the length and the width.
The length is 40 yd and the width is 30 yd.

w + 10

w

Figure for Exercise 54

55. The largest angle in a triangle is three times the smallest angle. The middle angle is two times the smallest angle. Given that the sum of the angles in a triangle is 180°, find the measure of each angle.
The measures of the angles are 30°, 60°, and 90°.

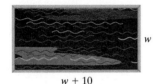

(3x)°
(2x)°
x°

Figure for Exercise 55

56. The smallest angle in a triangle is 90° less than the largest angle. The middle angle is 60° less than the largest angle. Find the measure of each angle.   The measures of the angles are 20°, 50°, and 110°.

57. The smallest angle in a triangle is half the largest angle. The middle angle is 30° less than the largest angle. Find the measure of each angle.
The measures of the angles are 42°, 54°, and 84°.

58. The largest angle of a triangle is three times the middle angle. The smallest angle is 10° less than

the middle angle. Find the measure of each angle.   The measures of the angles are 38°, 28°, and 114°.

59. Find the value of x and the measure of each angle labeled in the figure.
x = 17; the measures of the angles are 34° and 56°.

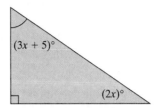

(3x + 5)°
(2x)°

Figure for Exercise 59

60. Find the value of y and the measure of each angle labeled in the figure.
y = 26.5; the measures of the angles are 28.5° and 61.5°.

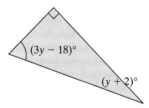

(3y − 18)°
(y + 2)°

Figure for Exercise 60

61. Sometimes memory devices are helpful for remembering mathematical facts. Recall that the sum of two complementary angles is 90°. That is, two complementary angles when added together form a right triangle or "corner." The words **C**omplementary and **C**orner both start with the letter **C**. Derive your own memory device for remembering that the sum of two supplementary angles is 180°.   Adjacent **S**upplementary angles form a **S**traight angle. The words *supplementary* and *straight* both begin with the same letter.

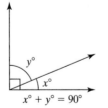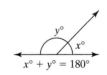

y°
x°
x° + y° = 90°

y°
x°
x° + y° = 180°

Complementary angles form a "corner"   Supplementary angles . . .

62. Two angles are complementary. One angle is 20° less than the other angle. Find the measures of the angles.   The angles are 55° and 35°.

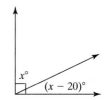

Figure for Exercise 62

 63. Two angles are complementary. One angle is twice as large as the other angle. Find the measures of the angles. The angles are 30° and 60°.

64. Two angles are complementary. One angle is 4° less than three times the other angle. Find the measures of the angles. The angles are 23.5° and 66.5°.

65. Two angles are supplementary. One angle is three times as large as the other angle. Find the measures of the angles. The angles are 45° and 135°.

66. Two angles are supplementary. One angle is twice as large as the other angle. Find the measures of the angles. The angles are 60° and 120°.

67. Two angles are supplementary. One angle is 6° more than four times the other. Find the measures of the two angles. The angles are 34.8° and 145.2°.

68. Find the measures of the vertical angles labeled in the figure by first solving for *x*.
x = 20; the vertical angles measure 37°.

Figure for Exercise 68

69. Find the measures of the vertical angles labeled in the figure by first solving for *y*.
y = 40; the vertical angles measure 146°.

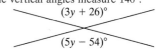

Figure for Exercise 69

■ EXPANDING YOUR SKILLS

For Exercises 70–74, find the indicated perimeters, areas, or volumes. Be sure to include the proper units and round each answer to two decimal places.

70. a. Find the area of a circle with radius 11.5 m.
415.48 m²
   b. Find the volume of a right circular cylinder with radius 11.5 m and height 25 m. 10386.89 m³

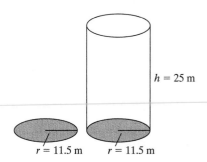

h = 25 m

r = 11.5 m   *r* = 11.5 m

Figure for Exercise 70

71. a. Find the area of a circle with radius 3.25 ft.
33.18 ft²
   b. Find the volume of a right circular cylinder with radius 3.25 ft and height 8 cm. 265.46 ft³

72. a. Find the area of a parallelogram with base 30 in. and height 12 in. 360 in.²

   b. Find the area of a triangle with base 30 in. and height 12 in. 180 in.²

   c. Compare the areas found in parts (a) and (b). The area of the triangle is one half the area of the parallelogram.

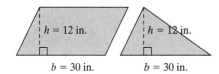

h = 12 in.   *h* = 12 in.

b = 30 in.   *b* = 30 in.

Figure for Exercise 72

73. a. Find the area of a parallelogram with base 7 m and height 4 m. 28 m²

   b. Find the area of a triangle with base 7 m and height 4 m. 14 m²

   c. Compare the areas found in parts (a) and (b). The area of the triangle is one half the area of the parallelogram.

74. a. Find the volume of a right circular cylinder with radius 6 ft and height 5 ft. Round to two decimals. 565.49 ft³

   b. Find the volume of a right circular cone with radius 6 ft and height 5 ft. 188.50 ft³

   c. Compare the volumes found in parts (a) and (b). The volume of the cone is one third the volume of the cylinder.

section

2.6   MORE APPLICATIONS OF LINEAR EQUATIONS

1. Applications Involving Ticket Sales

Application problems that involve finding a combination of two or more quantities to meet specified constraints is called a mixture problem. In the next example, we are asked to find the number of adults and the number of children that attended a play. The constraints are that exactly 282 people were present and that the total revenue was $1046. Algebra is used to find the right combination of the number of adults and number of children that attended the play.

example 1

Solving a Mixture Problem Involving Ticket Sales

Two hundred eighty-two people attended a recent performance of *Cinderella*. Adult tickets sold for $5 and children's tickets sold for $3. Find the number of adults and the number of children that attended the play if the total revenue was $1046.

Solution: **Step 1:**   Read the problem.

This example has two unknowns: the number of adults and the number of children.

We can let x represent either quantity. However, regardless of the choice of x, the other unknown will be represented as $282 - x$. For example,

If x represents the number of children, then

$$\begin{pmatrix} \text{Number} \\ \text{of adults} \end{pmatrix} = \begin{pmatrix} \text{total number} \\ \text{of people} \end{pmatrix} - \begin{pmatrix} \text{number of} \\ \text{children} \end{pmatrix}$$

$$\text{Number of adults} = \qquad 282 \qquad - \qquad x$$

Let x represent the number of children.   **Step 2:**   Label the variables.

Then, $282 - x$ represents the number of adults.

The children's tickets are $3 each, so the value of all children's tickets is $3x$.

The adult tickets are $5 each, so the value of all adult tickets is $5(282 - x)$.

Sometimes it is helpful to organize the information in a chart to label the variables.

	Children's Tickets	**Adult Tickets**	**Total**
Number of tickets	x	$282 - x$	282
Value of tickets	$3x$	$5(282 - x)$	1046

The total revenue is equal to the sum of the revenue from the children's tickets and from the adult tickets. Using the bottom row of the chart, we have

$$\left(\begin{array}{c}\text{Value of}\\\text{children's tickets}\end{array}\right) + \left(\begin{array}{c}\text{value of}\\\text{adult tickets}\end{array}\right) = \left(\begin{array}{c}\text{total value}\\\text{of tickets}\end{array}\right)$$

Step 3: Create the verbal equation.

$$3x \quad + \quad 5(282 - x) \quad = \quad 1046$$

Step 4: Write a mathematical equation.

$$3x + 5(282 - x) = 1046$$
$$3x + 1410 - 5x = 1046$$

Step 5: Solve for x.

$$-2x + 1410 = 1046$$
$$-2x = -364$$
$$\frac{-2x}{-2} = \frac{-364}{-2}$$
$$x = 182$$

Tip: Check the answer. 182 children's tickets and 100 adult tickets make 282 total tickets sold.
   Furthermore, 182 children's tickets at \$3 each amounts to \$546. One hundred adult tickets at \$5 each amounts to \$500. Therefore, the total revenue is \$1046 as desired.

The number of children is 182.

Step 6: Interpret the results and write the answer in words.

The number of adults is $282 - x = 282 - 182 = 100$.

There were 182 children and 100 adults that attended the performance.

2.   Applications Involving Principal and Interest

example 2

Solving an Application Involving Principal and Interest

Shana invests some money in an account that earns 5% simple interest and three times that amount in an account that earns 7% simple interest. If the total interest is \$390 at the end of 1 year, find the amount invested in each account.

Classroom Activity 2.6A

Solution:

In this application, we are "mixing" money between two accounts to find the correct combination that yields exactly \$390 in interest. We can set up a chart to label the variables. Recall that simple interest is computed by the formula: $I = Prt$. The time of the investment is 1 year, so we have: $I = Pr$

Step 1: Read the problem.

Let x represent the amount invested at 5%.

Step 2: Label the variables.

Then $3x$ represents the amount invested at 7%.

	5% Account	7% Account	Total
Principal	x	$3x$	
Interest	$0.05x$	$0.07(3x)$	390

The total interest is equal to the sum of the interest from the 5% account and from the 7% account. Using the bottom row of the chart, we have

$$\left(\begin{array}{c}\text{Interest earned}\\ \text{in 5\% account}\end{array}\right) + \left(\begin{array}{c}\text{interest earned}\\ \text{in 7\% account}\end{array}\right) = \left(\begin{array}{c}\text{total}\\ \text{interest}\end{array}\right)$$

Step 3: Create a verbal equation.

$$0.05x \quad + \quad 0.07(3x) \quad = \quad 390$$

Step 4: Write a mathematical equation.

$$0.05x + 0.07(3x) = 390$$

$$0.05x + 0.21x = 390$$

Step 5: Solve for x.

$$0.26x = 390$$

$$\frac{0.26x}{0.26} = \frac{390}{0.26}$$

$$x = 1500$$

The amount invested at 5% is $1500.

Step 6: Interpret the results and write the answer in words.

Tip: Check your answer. First notice that $4500 is three times $1500. Furthermore, if Shana invested $1500 at 5%, the interest from that account is 0.05($1500) = $75. Likewise $4500 invested at 7% produces 0.07($4500) = $315 in interest. Therefore, the total interest is $75 + $315 = $390, as expected.

The amount invested at 7% is $3x = 3(\$1500) = \$4500$.

Shana invested $1500 in the 5% account and $4500 in the 7% account.

3. Applications Involving Distance, Rate, and Time

The formula: (distance) = (rate)(time) or simply, $d = rt$, relates the distance traveled to the rate of travel and the time of travel.

For example, if a car travels at 60 mph for 3 hours, then

$$d = (60 \text{ mph})(3 \text{ hours})$$
$$= 180 \text{ miles}$$

If a car travels at 60 mph for x hours, then

$$d = (60 \text{ mph})(x \text{ hours})$$
$$= 60x \text{ miles}$$

example 3   Solving an Application Involving Distance, Rate, and Time

Two families that live 270 miles apart plan to meet for an afternoon picnic. To share the driving, they want to meet somewhere between their two homes. Both families leave at 9:00 A.M., but one family averages 12 mph faster than the other family. If the families meet at the designated spot $2\frac{1}{2}$ hours later, determine

a.   The average rate of speed for each family
b.   How far each family traveled to the picnic

Classroom Activity 2.6B

Solution: Step 1:   Read the problem and draw a sketch.

For simplicity, we will call the two families, Family A and Family B. Let Family A be the family that travels at the slower rate (Figure 2-11).

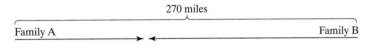

Figure 2-11

The following chart is helpful for organizing the information given in this problem. Let x represent the rate of Family A.

Step 2:   Label the variables.

	Distance	Rate	Time
Family A		x	2.5
Family B		$x + 12$	2.5

To complete the first column, we can use the relationship $d = rt$. The distance traveled by Family A is: $d_A = (x)(2.5)$. The distance traveled by Family B is $d_B = (x + 12)(2.5)$.

	Distance	Rate	Time
Family A	2.5x	x	2.5
Family B	2.5(x + 12)	$x + 12$	2.5

To set up an equation, recall that the total distance between the two families is given as 270 miles.

$$\begin{pmatrix} \text{Distance} \\ \text{traveled by} \\ \text{Family A} \end{pmatrix} + \begin{pmatrix} \text{distance} \\ \text{traveled by} \\ \text{Family B} \end{pmatrix} = \begin{pmatrix} \text{total} \\ \text{distance} \end{pmatrix}$$

Step 3:   Create a verbal equation.

$$2.5x + 2.5(x + 12) = 270$$

Step 4:   Write a mathematical equation.

$$2.5x + 2.5(x + 12) = 270$$

$$2.5x + 2.5x + 30 = 270 \qquad \text{Step 5: Solve for } x.$$

$$5.0x + 30 = 270$$

$$5x = 240$$

$$\frac{5x}{5} = \frac{240}{5}$$

$$x = 48$$

a.   Family A traveled 48 (mph).   **Step 6:** Interpret the

Family B traveled $x + 12 = 48 + 12 = 60$ (mph).   results and write
                                                          the answer in
                                                          words.

b.   To compute the distance each family traveled, use $d = rt$:

Family A traveled: (48 mph)(2.5 h) = 120 miles

Family B traveled: (60 mph)(2.5 h) = 150 miles

4.   Ratio and Proportion

Sometimes linear equations appear when solving applications that involve proportions.

Definition of Ratio and Proportion

1.   The **ratio** of a to b is $\frac{a}{b}$ $(b \neq 0)$ and can also be expressed as $a{:}b$ or $a \div b$.
2.   An equation that equates two ratios is called a **proportion**. Therefore, if $b \neq 0$ and $d \neq 0$, then $\frac{a}{b} = \frac{c}{d}$ is a proportion.

One method of solving a proportion is to use the technique of clearing fractions.

example 4   **Solving a Proportion**

Solve the proportion. $\dfrac{3}{10} = \dfrac{x}{57}$

Solution:

$$\frac{3}{10} = \frac{x}{57} \qquad \text{The LCD is 570.}$$

$$570\left(\frac{3}{10}\right) = 570\left(\frac{x}{57}\right) \qquad \text{Multiply both sides by the LCD.}$$

$$\overset{57}{\cancel{570}}\left(\frac{3}{\underset{1}{\cancel{10}}}\right) = \overset{10}{\cancel{570}}\left(\frac{x}{\underset{1}{\cancel{57}}}\right) \qquad \text{Reduce common factors to clear fractions.}$$

$$57(3) = 10(x)$$

$$171 = 10x \qquad \text{Solve the resulting equation.}$$

$$\frac{171}{10} = \frac{10x}{10}$$

$$x = 17.1 \qquad \text{The solution checks in the original equation.}$$

Tip: For any proportion

$$\frac{a}{b} = \frac{c}{d} \quad (b \neq 0, d \neq 0)$$

the cross products, ad and bc, are equal. Hence $ad = bc$. Equating the cross products is a quick way to clear fractions in a proportion.* Consider the proportion in Example 4:

$$\frac{3}{10} = \frac{x}{57}$$

$$3 \cdot 57 = 10 \cdot x$$

$$171 = 10x$$

$$\frac{171}{10} = \frac{10x}{10}$$

$$17.1 = x \quad \text{or} \quad x = 17.1$$

*It is important to realize that this technique is valid only for proportions.

5. Applications Involving Ratio and Proportion

example 5

Solving an Application Involving Proportions

A 4-oz serving of orange juice contains 106 g of water. How many grams of water are in a 30-oz container of orange juice?

Classroom Activity 2.6C

Solution:

Step 1: Read the problem.

Let x represent the number of grams of water in 30 oz of orange juice.

Step 2: Label the variables.

Write two equivalent ratios depicting the number of grams of water to the number of ounces of orange juice.

Step 3: Create a verbal model.

$$\underbrace{\dfrac{\text{Grams of water}}{\text{ounces of o. j.}}}_{} \longrightarrow \dfrac{106 \text{ g}}{4 \text{ oz}} = \dfrac{x}{30 \text{ oz}} \longleftarrow \underbrace{\dfrac{\text{Grams of water}}{\text{ounces of o. j.}}}_{}$$

$$\dfrac{106}{4} = \dfrac{x}{30}$$

Step 4: Write a mathematical equation.

$$106(30) = 4x$$

Step 5: Equate the cross products.

$$3180 = 4x$$

Solve the equation.

$$\dfrac{3180}{4} = \dfrac{4x}{4}$$

$$795 = x \text{ or } x = 795$$

There is 795 g of water in 30 oz of orange juice.

Step 6: Interpret the results and write the answer in words.

6. Applications Involving Similar Triangles

Proportions are used in geometry with **similar triangles**. Two triangles are said to be similar if they have equal angles. In such a case, the lengths of the corresponding sides are proportional. The triangles in Figure 2-12 are similar. Therefore, the following ratios are equivalent.

$$\dfrac{a}{x} = \dfrac{b}{y} = \dfrac{c}{z}$$

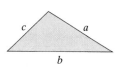
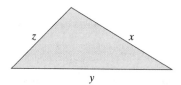

Figure 2-12

---

example 6

Using Similar Triangles in an Application

The shadow cast by a yardstick is 2 ft long. The shadow cast by a tree is 11 ft long. Find the height of the tree.

Solution:

Step 1: Read the problem.

Let x represent the height of the tree.

Step 2: Label the variables.

We will assume that the measurements were taken at the same time of day. Therefore, the angle of the sun is the same on both objects, and we can set up similar triangles (Figure 2-13).

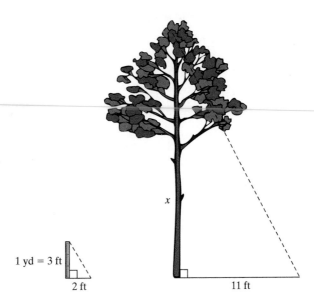

1 yd = 3 ft

2 ft

x

11 ft

Figure 2-13

Step 3: Create a verbal model.

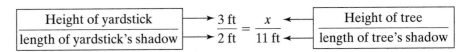

$$\frac{\text{Height of yardstick}}{\text{length of yardstick's shadow}} \longrightarrow \frac{3 \text{ ft}}{2 \text{ ft}} = \frac{x}{11 \text{ ft}} \longleftarrow \frac{\text{Height of tree}}{\text{length of tree's shadow}}$$

$$\frac{3}{2} = \frac{x}{11}$$ **Step 4:** Write a mathematical equation.

$3(11) = 2x$ **Step 5:** Equate the cross products.

$33 = 2x$ Solve the equation.

$$\frac{33}{2} = \frac{2x}{2}$$

$16.5 = x$ **Step 6:** Interpret the results and write the answer in words.

The tree is 16.5 ft high.

section 2.6   PRACTICE EXERCISES

For Exercises 1–10, solve for the indicated variable. 2.5

1. $3x + 7 = 13$ for x $x = 2$

2. $V = \pi r^2 h$ for r^2 $r^2 = \dfrac{V}{\pi h}$

3. $4x + 5y = 20$ for y $y = \dfrac{20 - 4x}{5}$

4. $4b - 2b + 8 = 3(b - 6) - 9$ for b $b = 35$

5. $0.3(a + 14.2) = 7.1a - 2.2$ for a $a = 0.95$

6. $A = 4\pi r^2$ for r^2 $r^2 = \dfrac{A}{4\pi}$

7. $\dfrac{2}{3}(p + 9) = \dfrac{5}{6}p$ for p $p = 36$

8. $-7 = -4r - 6$ for r $r = \dfrac{1}{4}$

9. $V = \dfrac{1}{3}lwh$ for h $h = \dfrac{3V}{lw}$

10. $-x + 9y = 16$ for y $y = \dfrac{x + 16}{9}$

11. Adult tickets to a school play cost $6 per ticket.

    a. If 200 adult tickets are sold, how much revenue is produced? $1200

    b. If x adult tickets are sold, write an expression that represents the revenue. $6x$

    c. If $750 - x$ adult tickets are sold, write an expression that represents the revenue.
       $6(750 - x)$ or $4500 - 6x$

12. Children's tickets to a movie cost $3.50 per ticket.

    a. If a theater sells 600 children's tickets, how much revenue is produced? $2100

    b. If a theater sells x children's tickets, write an expression that represents the revenue. $3.5x$

    c. If a theater sells $2x$ children's tickets, write an expression that represents the revenue.
       $3.5(2x)$ or $7x$

13. The local church had an ice cream social and sold tickets for $3 and $2. When the social was over, 81 tickets had been sold totaling $215. How many of each type of ticket did the church sell?
    53 tickets were sold at $3 and 28 tickets were sold at $2.

	$3 Tickets	$2 Tickets	Total
Number of tickets			
Value of tickets			

Table for Exercise 13

14. A high school had two raffles to raise funds to purchase books for the library. One raffle offered tickets at $0.50 to win a new portable radio and the other offered tickets for $2 to win a new bike. There were 224 tickets sold bringing in $251.50 for the library. How many of each type of ticket were sold?
    131 tickets were sold at $0.50 and 93 tickets were sold at $2.

15. A savings account earns 6% simple interest per year.

    a. If a woman invests $5000 in the account, how much interest will she earn after 1 year? $300

    b. If a woman has x dollars invested in the account, write an expression that represents the total interest earned after 1 year. $0.06x$

    c. If a woman has $20,000 - x$ invested in the account, write an expression that represents the total interest earned after 1 year.
       $0.06(20,000 - x)$ or $1200 - 0.06x$

16. A money market account earns 3.5% simple interest per year.

    a. If a man invests $4000 in the account, how much interest will he earn after 1 year? $140

    b. If a man has x dollars invested in the account, write an expression that represents the total interest earned after 1 year. $0.035x$

    c. If a man has $10,000 - x$ invested in the account, write an expression that represents the total interest earned after 1 year.
       $0.035(10,000 - x)$ or $350 - 0.035x$

17. Nora has an account with her bank that pays 6% annual interest. She also has a savings account with her credit union that pays 8% annual interest. She received a total of $104 in interest in 1 year. If there is $800 more in the bank account than in the credit union, how much is in each account?   There is $400 in the 8% account and $1200 in the 6% account.

	6% Account	8% Account	Total
Principal			
Interest			

Table for Exercise 17

18. Bob has $8000 to invest. He puts part of the money in an account that pays 8% annual interest and invests the rest at 12% annual interest. If his total interest after 1 year was $840, how much did he invest at 12%?   Bob invested $5000 at 12%.

19. How can $8750 be divided between a 6% account and an 8% account so that each account earns the same amount of simple interest after 1 year? $5000 should be invested in the 6% account and $3750 should be invested in the 8% account.

20. Iacco has a total of $20,000 to invest. She invests part in an account earning 5% simple interest and part in an account earning 3% simple interest. How much money is invested in each account if the interest from each account is the same at the end of 1 year? $7500 is invested in the 5% account, and $12,500 is invested in the 3% account.

21. A certain granola mixture is 10% peanuts.

    a. If a container has 20 lb of granola, how many pounds of peanuts are in it?   2 lb

    b. If a container has x pounds of granola, write an expression that represents the number of pounds of peanuts in the granola. $0.10x$

    c. If a container has $x + 3$ pounds of granola, write an expression that represents the number of pounds of peanuts.
    $0.10(x + 3) = 0.10x + 0.30$

22. A certain blend of coffee sells for $9.00 per pound.

    a. If a container has 20 lb of coffee, how much will it cost? $180

    b. If a container has x pounds of coffee, write an expression that represents the cost. $9.00x$

    c. If a container has $40 - x$ pounds of this coffee, write an expression that represents the cost. $9.00(40 - x) = 360 - 9x$

23. The Coffee Company wishes to mix coffee worth $12 per pound with coffee worth $8 per pound to produce 50 lb of coffee worth $8.80 per pound. How many pounds of the $12 coffee and how many pounds of the $8 coffee must be used?
Ten pounds of coffee sold at $12 per pound and 40 lb of coffee sold at $8 per pound.

	$12 Coffee	$8 Coffee	Total
Number of pounds			
Value of coffee			

Table for Exercise 23

24. The Nut House sells pecans worth $4 per pound and cashews worth $6 per pound. How many pounds of pecans and how many pounds of cashews must be mixed to form 16 lb of a nut mixture worth $4.50 per pound?
12 lb of pecans and 4 lb of cashews

25. Sally wishes to mix raisins with granola to sell in decorated containers at a flea market. She can get raisins for $1.69 per pound and granola for $2.59 per pound. How many pounds of raisins and how many pounds of granola should she use to make 6 lb of a mixture worth $2.29?
2 lb of raisins and 4 lb of granola

26. Gina is going to a party and was asked to bring 3 lb of mixed nuts that includes cashews and peanuts. If the cashews cost $6.00 per pound and the peanuts cost $1.50 per pound, how many pounds of each did she buy if her total cost was $9.00?   Gina bought 1 lb of cashews and 2 lb of peanuts.

27. a. If a car travels 60 mph for 5 h, find the distance traveled.   300 miles

    b. If a car travels at x miles per hour for 5 h, write an expression that represents the distance traveled. $5x$

    c. If a car travels at $x + 12$ mph for 5 h, write an expression that represents the distance traveled. $5(x + 12)$ or $5x + 60$

28. a. If a plane travels at 550 mph for 2.5 h, find the distance traveled.   1375 miles

    b. If a plane travels at x miles per hour for 2.5 h, write an expression that represents the distance traveled. $2.5x$

    c. If a plane travels at $x - 100$ mph for 2.5 h, write an expression that represents the distance traveled. $2.5(x - 100)$ or $2.5x - 250$

29. A car travels 55 mph for 4 h, and another car travels the same distance at 40 mph. For how many hours does the second car travel?
The car travels 5.5 h.

 ❖ See Additional Answers Appendix   Writing   Translating Expression   Geometry   Scientific Calculator   Video

30. A car leaves a bus station and travels 55 mph. A bus leaves the same station, traveling 65 mph in the same direction $\frac{1}{2}$ h later. How long will it take the bus to overtake the car?
    It would take the bus 2.75 h to overtake the car.

31. Two cars are 144 miles apart and traveling toward each other on the same road. They meet in $1\frac{1}{2}$ h. One car is traveling 4 mph faster than the other. What is the average speed of each car?
    The slower car travels at 46 mph and the faster car travels at 50 mph.

	Distance	Rate	Time
Faster car			
Slower car			

Table for Exercise 31

32. Two cars are 190 miles apart and traveling toward each other along the same road. They meet in 2 h. One car is traveling 5 mph slower than the other. What is the average speed of each car?
    The cars travel at 50 mph and 45 mph.

33. A Piper Cub airplane has an average air speed that is 10 mph faster than a Cessna 150 airplane. If the combined distance traveled by these two small airplanes is 690 miles after 3 h, what is the average speed of each plane?
    The Cessna's speed is 110 mph and the Piper Cub's speed is 120 mph.

34. A Cessna 182 airplane has an average air speed that is 50 mph slower than a Mooney airplane. If the combined distance traveled by these two small airplanes is 825 miles after $2\frac{1}{2}$ h, what is the average speed of each plane?
    The Mooney travels 190 mph, and the Cessna travels 140 mph.

35. Two boats travelling the same direction leave a harbor at noon. After 3 h they are 60 miles apart. If one boat travels twice as fast as the other, find the average rate of each boat.
    The rates of the boats are 20 mph and 40 mph.

36. Two canoes travel down a river, starting at 9:00. One canoe travels twice as fast as the other. After 3.5 h, the canoes are 5.25 miles apart. Find the average rate of each canoe.
    The rates of the canoes are 1.5 mph and 3 mph.

For Exercises 37–46, solve the proportions.

37. $\frac{3}{4} = \frac{x}{20}$ $x = 15$

38. $\frac{2}{7} = \frac{y}{28}$ $y = 8$

39. $\frac{z}{26} = \frac{1}{2}$ $z = 13$

40. $\frac{w}{40} = \frac{1}{4}$ $w = 10$

41. $\frac{3}{7} = \frac{a}{133}$ $a = 57$

42. $\frac{4}{11} = \frac{b}{132}$ $b = 48$

43. $\frac{9}{2} = \frac{k}{84}$ $k = 378$

44. $\frac{15}{4} = \frac{m}{68}$ $m = 255$

45. $\frac{2.4}{17.5} = \frac{p}{1505}$ $p = 206.4$

46. $\frac{1.7}{8.5} = \frac{q}{467.5}$ $q = 93.5$

47. One tablespoon of fertilizer is used for $\frac{1}{2}$ gal of water. How much fertilizer is needed for 15 gal of water?   30 tbsp of fertilizer is needed.

48. To make an insecticide for plants, it is necessary to mix 3 tbsp of chemical for every 2 gal of water. How many tablespoons of chemical are needed for 12 gal of water?   18 tbsp of chemical is needed.

49. If 12 out of 80 M&Ms are red, then how many red M&Ms would you expect to find in a bag containing 200 M&Ms?   30 red M&Ms

50. In a can of mixed nuts there are approximately three peanuts out of every five nuts. If the can holds 400 nuts, how many are expected to be peanuts?   240 peanuts

51. In a sample of ballots, it is found that 8 ballots out of 5000 are incorrectly marked and must be thrown out. How many incorrectly marked ballots would be expected in a state where 2,600,000 ballots are cast?
    4160 incorrectly marked ballots are expected.

52. To conduct an election poll, a sample of 2000 voters is randomly selected. It is found that 100 of the people surveyed would not vote for either the Republican or Democratic presidential nominee. How many voters would not be expected to vote for the Democratic or Republican presidential nominee in a county where 800,000 votes are cast?   40,000 voters would not be expected to vote for either the Republican or Democratic presidential nominee.

Triangles that are the same shape but different sizes are said to be similar. The corresponding sides of similar triangles are proportional. Use this fact to find the values of the missing sides, x and y, in Exercises 53–54.

53.
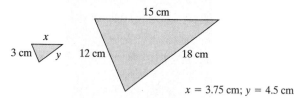
$x = 3.75$ cm; $y = 4.5$ cm

See Additional Answers Appendix   Writing   Translating Expression   Geometry   Scientific Calculator   Video

54.

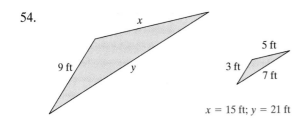

$x = 15$ ft; $y = 21$ ft

55. To estimate the height of a light pole, a mathematics student measures the length of a shadow cast by a meterstick and the length of the shadow cast by the light pole. Find the height of the light pole (see figure).   The height of the pole is 7 m.

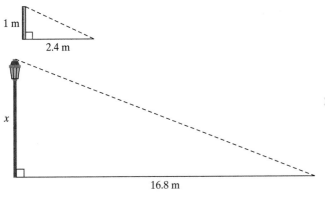

Figures for Exercise 55

56. To estimate the height of a building, a student measures the length of a shadow cast by a yardstick and the length of the shadow cast by the building (see figure). Find the height of the building.   The height is 36 ft.

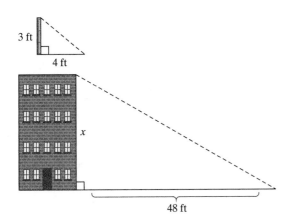

Figures for Exercise 56

▓ EXPANDING YOUR SKILLS

57. Nickels are worth $0.05 each.

  a. If a piggy bank contains 70 nickels, how much money is this? $3.50

  b. If a bank contains x nickels, write an expression that represents the total value. $0.05x$

  c. If a bank contains $30 + x$ nickels, write an expression that represents the total value.
  $0.05(30 + x)$ or $1.5 + 0.05x$

58. Quarters are worth $0.25 each.

  a. If a child has 12 quarters, how much money is this? $3.00

  b. If a child has x quarters, write an expression that represents the total value. $0.25x$

  c. If a child has $20 - x$ quarters, write an expression that represents the total value.
  $0.25(20 - x) = 5 - 0.25x$

59. Jean-Paul has 12 coins in his pocket consisting of quarters and dimes. The total value of these coins is $1.65. How many quarters and dimes does Jean-Paul have?
Jean-Paul has three quarters and nine dimes.

	Quarters	Dimes	Total
Number of coins			
Value of coins			

Table for Exercise 59

60. In Anna's purse there are 11 coins consisting of nickels and dimes. If the total value of these coins is $0.95, how many of each type of coin are there?   Anna has three nickels and eight dimes.

61. A bank customer withdraws $200 from her account in $10 and $20 bills. She requests twice as many $20 bills as $10 bills. How many of each kind of bill does she receive?
She receives four $10 bills and eight $20 bills.

62. A woman withdraws $250 from her account in $5, $10, and $20 bills. If she receives four $10 bills and twice as many $5 bills as $20 bills, how many of each type of bill does she receive?
She receives seven $20 bills, fourteen $5 bills, and four $10 bills.

section

2.7   LINEAR INEQUALITIES

1. Solutions to Inequalities

Recall that $a < b$ (equivalently $b > a$) means that a lies to the left of b on the number line. The statement $a > b$ (equivalently $b < a$) means that a lies to the right of b on the number line. If $a = b$, then a and b are represented by the same point on the number line.

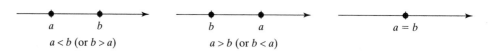

A Linear Inequality in One Variable

A linear inequality in one variable, x, is defined as any relationship of the form:

$$ax + b < 0, ax + b \le 0, ax + b > 0, \text{ or } ax + b \ge 0, \text{ where } a \ne 0.$$

The following inequalities are linear equalities in one variable.

$$2x - 3 < 0 \qquad -4z - 3 > 0 \qquad a \le 4 \qquad 5.2y \ge 10.4$$

The number line is a useful tool to visualize the solution set of an equation or inequality. For example, the solution set to the equation $x = 2$ is $\{2\}$ and may be graphed as a single point on the number line.

$$x = 2 \qquad \begin{array}{c} \\ \hline -6\;-5\;-4\;-3\;-2\;-1\;\;0\;\;1\;\;2\;\;3\;\;4\;\;5\;\;6 \end{array}$$

The solution set to an inequality is the set of real numbers that make the inequality a true statement. For example, the solution set to the inequality $x \ge 2$ is all real numbers 2 or greater. Because the solution set has an infinite number of values, we cannot list all of the individual solutions. However, we can graph the solution set on the number line.

$$x \ge 2 \qquad \begin{array}{c} \\ \hline -6\;-5\;-4\;-3\;-2\;-1\;\;0\;\;1\;\;2\;\;3\;\;4\;\;5\;\;6 \end{array}$$

The square bracket symbol, [, is used on the graph to indicate that the point $x = 2$ is included in the solution set. By convention, square brackets, either [or], are used to *include* a point on a graph. Parentheses, (or), are used to *exclude* a point on a graph.

The solution set of the inequality $x > 2$ includes the real numbers greater than 2 but not including 2. Therefore, a (symbol is used on the graph to indicate that $x = 2$ is not included.

$$x > 2 \qquad \begin{array}{c} \\ \hline -6\;-5\;-4\;-3\;-2\;-1\;\;0\;\;1\;\;2\;\;3\;\;4\;\;5\;\;6 \end{array}$$

2. Graphing Linear Inequalities

example 1

Graphing Linear Inequalities

Graph the solution sets.

a. $x > -1$ b. $c \leq \dfrac{7}{3}$

Solution:

a. $x > -1$

The solution set is the set of all real numbers strictly greater than -1. Therefore, we graph the region on the number line to the right of -1. Because $x = -1$ is not included in the solution set, we use the (symbol at $x = -1$.

b. $c \leq \frac{7}{3}$ is equivalent to $c \leq 2\frac{1}{3}$.

The solution set is the set of all real numbers less than or equal to $2\frac{1}{3}$. Therefore, graph the region on the number line to the left of and including $2\frac{1}{3}$. Use the symbol] to indicate that $c = 2\frac{1}{3}$ is included in the solution set.

Tip: Some textbooks use a closed circle or an open circle (● or ○) rather than a bracket or parenthesis to denote inclusion or exclusion of a value on the real number line. For example the solution sets for the inequalities $x > -1$ and $c \leq \frac{7}{3}$ are graphed here.

$x > -1$

$c \leq \frac{7}{3}$

     A statement that involves more than one inequality is called a compound inequality. One type of compound inequality is used to indicate that one number is between two others. For example, the inequality $-2 < x < 5$ means that $-2 < x$ and $x < 5$. In words, this is easiest to understand if we read the variable first: x is greater than -2 and x is less than 5. The numbers satisfied by these two conditions are those between -2 and 5.

example 2

Graphing a Compound Inequality

Graph the solution set of the inequality: $-4.1 < y \le -1.7$

Solution:

$-4.1 < y \le -1.7$ means that

$-4.1 < y$ and $y \le -1.7$

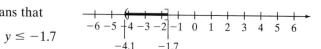

-4.1 -1.7

Shade the region of the number line greater than -4.1 and less than or equal to -1.7. These are the values of y between -4.1 and -1.7 (including $y = -1.7$).

3. Set-Builder Notation and Interval Notation

Graphing the solution set to an inequality is one way to define the set. Two other methods are to use **set-builder notation** or **interval notation**.

Set-Builder Notation

The solution to the inequality $x \ge 2$ can be expressed in set-builder notation as follows:

$$\{x \mid x \ge 2\}$$

the set of all x such that x is greater than or equal to 2

Interval Notation

To understand interval notation, first think of a number line extending infinitely far to the right and infinitely far to the left. Sometimes we use the infinity symbol, ∞, or negative infinity symbol, $-\infty$, to label the far right and far left ends of the number line (Figure 2-14).

$-\infty$ ————————————————→ ∞
0

Figure 2-14

To express the solution set of an inequality in interval notation, sketch the graph first. Then use the endpoints to define the interval.

Inequality	Graph	Interval Notation
$x \ge 2$	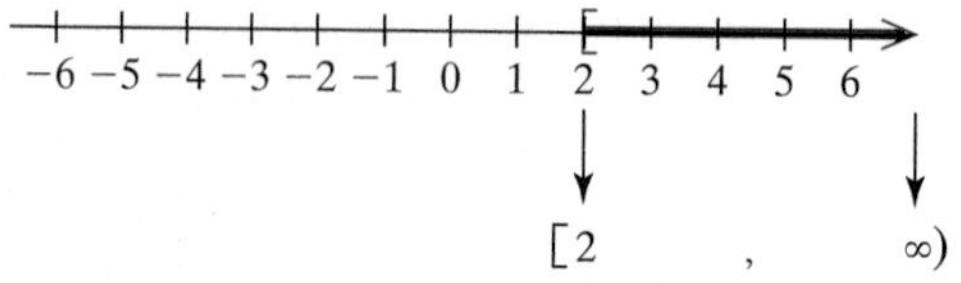 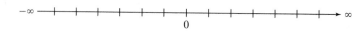	$[2, \infty)$

$[2$, $\infty)$

The graph of the solution set $x \ge 2$ begins at 2 and extends infinitely far to the right. The corresponding interval notation begins at 2 and extends to ∞. Notice that a square bracket [is used at 2 for both the graph and the interval notation. A parenthesis is always used at ∞ (and for $-\infty$), because there is no endpoint.

Using Interval Notation

- The endpoints used in interval notation are always written from left to right. That is, the smaller number is written first, followed by a comma, followed by the larger number.
- A parenthesis, (or), indicates that an endpoint is excluded from the set.
- A square bracket, [or], indicates that an endpoint is included in the set.
- Parentheses, (and), are always used with $-\infty$ and ∞.

example 3

Using Set-Builder Notation and Interval Notation

Complete the chart.

Set-Builder Notation	Graph	Interval Notation
	−6 −5 −4 −3 −2 −1 0 1 2 3 4 5 6	
		$\left[-\frac{1}{2}, \infty\right)$
$\{y \mid -2 \le y < 4\}$		

Classroom Activity 2.7A

Solution:

Set-Builder Notation	Graph	Interval Notation
$\{x \mid x < -3\}$	−6 −5 −4 −3 −2 −1 0 1 2 3 4 5 6	$(-\infty, -3)$
$\{x \mid x \ge -\frac{1}{2}\}$	−6 −5 −4 −3 −2 −1 0 1 2 3 4 5 6 $-\frac{1}{2}$	$\left[-\frac{1}{2}, \infty\right)$
$\{y \mid -2 \le y < 4\}$	−6 −5 −4 −3 −2 −1 0 1 2 3 4 5 6	$[-2, 4)$

4. Addition and Subtraction Properties of Inequality

The process to solve a linear inequality is very similar to the method used to solve linear equations. Recall that adding or subtracting the same quantity to both sides of an equation results in an equivalent equation. The addition and subtraction properties of inequality state that the same is true for an inequality.

Addition and Subtraction Properties of Inequality

Let a, b, and c represent real numbers.

1. ***Addition Property of Inequality:**   If $a < b$,
   then $a + c < b + c$

2. ***Subtraction Property of Inequality:**   If $a < b$,
   then $a - c < b - c$

*These properties may also be stated for $a \leq b$, $a > b$, and $a \geq b$.

To illustrate the addition and subtraction properties of inequality, consider the inequality $10 < 20$. If we subtract or add a real number such as 5 to both sides, the left-hand side will still be less than the right-hand side (Figure 2-15).

$$10 < 20 \qquad\qquad 10 < 20$$

$$10 + 5 < 20 + 5 \qquad 10 - 5 < 20 - 5$$

$$15 < 25 \qquad\qquad 5 < 15$$

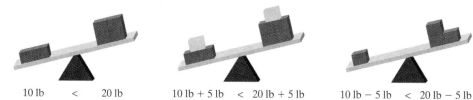

10 lb   <   20 lb   10 lb + 5 lb   <   20 lb + 5 lb   10 lb − 5 lb   <   20 lb − 5 lb

Figure 2-15

example 4   Solving a Linear Inequality

Solve the inequality and graph the solution set. Express the solution set in set-builder notation and in interval notation.

$$-2p + 5 < -3p + 6$$

Solution:

$$-2p + 5 < -3p + 6$$

$$-2p + 3p + 5 < -3p + 3p + 6 \qquad \text{Addition property of inequality (add } 3p \text{ to both sides).}$$

$$p + 5 < 6 \qquad \text{Simplify.}$$

$$p + 5 - 5 < 6 - 5 \qquad \text{Subtraction property of inequality.}$$

$$p < 1$$

Graph:

$$\xleftarrow{\hspace{1cm}} \begin{array}{ccccccccccccc} -6 & -5 & -4 & -3 & -2 & -1 & 0 & 1 & 2 & 3 & 4 & 5 & 6 \end{array} \xrightarrow{\hspace{1cm}}$$

Set-builder notation: $\{p \mid p < 1\}$

Interval notation: $(-\infty, 1)$

Tip: The solution to an inequality gives a set of values that make the original inequality true. Therefore, you can test your final answer by using **test points**. That is, pick a value in the proposed solution set and verify that it makes the original inequality true. Furthermore, any test point picked outside the solution set should make the original inequality false. For example,

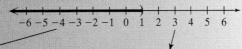

Pick $x = -4$ as an arbitrary test point within the proposed solution set.

$$-2p + 5 < -3p + 6$$
$$-2(-4) + 5 \overset{?}{<} -3(-4) + 6$$
$$8 + 5 \overset{?}{<} 12 + 6$$
$$13 < 18 \;✔\quad \text{True}$$

Pick $x = 3$ as an arbitrary test point outside the proposed solution set.

$$-2p + 5 < -3p + 6$$
$$-2(3) + 5 \overset{?}{<} -3(3) + 6$$
$$-6 + 5 \overset{?}{<} -9 + 6$$
$$-1 \overset{?}{<} -3 \quad \text{False}$$

5. Multiplication and Division Properties of Inequality

Multiplying both sides of an equation by the same quantity results in an equivalent equation. However, the same is not always true for an inequality. If you multiply or divide an inequality by a negative quantity, the direction of the inequality symbol must be reversed.

For example, consider multiplying or dividing the inequality, $4 < 5$ by -1.

Multiply/Divide by -1

$$4 < 5$$
$$-4 > -5$$

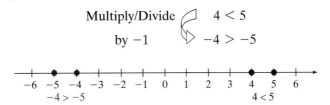

Figure 2-16

The number 4 lies to the left of 5 on the number line. However, -4 lies to the right of -5 (Figure 2-16). Changing the sign of two numbers changes their relative position on the number line. This is stated formally in the multiplication and division properties of inequality.

Multiplication and Division Properties of Inequality

Let a, b, and c represent real numbers. Then

*If c is positive and $a < b$, then

$$ac < bc \text{ and } \frac{a}{c} < \frac{b}{c}$$

*If c is negative and $a < b$, then

$$ac > bc \text{ and } \frac{a}{c} > \frac{b}{c}$$

The second statement indicates that if both sides of an inequality are multiplied or divided by a negative quantity, the inequality sign must be reversed.

*These properties may also be stated for $a \leq b$, $a > b$, and $a \geq b$.

---

example 5   **Solving a Linear Inequality**

Solve the inequality $-5x - 3 \leq 12$. Graph the solution set and write the answer in interval notation.

Classroom Activity 2.7B

Solution:

$$-5x - 3 \leq 12$$

$$-5 - 3 + 3 \leq 12 + 3 \qquad \text{Add 3 to both sides.}$$

$$-5x \leq 15$$

$$\frac{-5x}{-5} \geq \frac{15}{-5} \qquad \text{Divide by } -5. \text{ Reverse the direction of the inequality sign.}$$

$$x \geq -3$$

Interval notation: $[-3, \infty)$

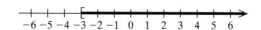

Tip: The inequality $-5x - 3 \leq 12$, could have been solved by isolating x on the right-hand side of the inequality. This would create a positive coefficient on the variable term and eliminate the need to divide by a negative number.

$$-5x - 3 \leq 12$$

$$-3 \leq 5x + 12$$

$$-15 \leq 5x \qquad \text{Notice that the coefficient of } x \text{ is positive.}$$

$$\frac{-15}{5} \leq \frac{5x}{5} \qquad \text{Do not reverse the inequality sign, because we are dividing by a positive number.}$$

$$-3 \leq x, \text{ or equivalently, } x \geq -3$$

example 6

Solving Linear Inequalities

Solve the inequality $-\frac{1}{4}k + \frac{1}{6} \leq 2 + \frac{2}{3}k$. Graph the solution set and write the answer in interval notation.

Classroom Activity 2.7C

Solution:

$$-\frac{1}{4}k + \frac{1}{6} \leq 2 + \frac{2}{3}k$$

$$12\left(-\frac{1}{4}k + \frac{1}{6}\right) \leq 12\left(2 + \frac{2}{3}k\right)$$ Multiply both sides by 12 to clear fractions. (Because we multiplied by a positive number, the inequality sign is not reversed.)

$$\frac{12}{1}\left(-\frac{1}{4}k\right) + \frac{12}{1}\left(\frac{1}{6}\right) \leq 12(2) + \frac{12}{1}\left(\frac{2}{3}k\right)$$ Apply the distributive property.

$$-3k + 2 \leq 24 + 8k$$ Simplify.

$$-3k - 8k + 2 \leq 24 + 8k - 8k$$ Subtract $8k$ from both sides.

$$-11k + 2 \leq 24$$

$$-11k + 2 - 2 \leq 24 - 2$$ Subtract 2 from both sides.

$$-11k \leq 22$$

$$\frac{-11k}{-11} \geq \frac{22}{-11}$$ Divide both sides by -11. Reverse the inequality sign.

$$k \geq -2$$

Graph: $-4\ -3\ -2\ -1\ \ 0\ \ 1\ \ 2\ \ 3\ \ 4$

Interval notation: $[-2, \infty)$

6.   Solving Inequalities of the Form $a < x < b$

To solve a compound inequality of the form $a < x < b$ we can work with the inequality as a three-part inequality and isolate the variable, x, as demonstrated in the following example.

example 7

Solving a Compound Inequality of the Form $a < x < b$

Solve the inequality: $-3 \leq 2x + 1 < 7$. Graph the solution and write the answer in interval notation.

Solution:

To solve the compound inequality $-3 \leq 2x + 1 < 7$ isolate the variable x in the middle. The operations performed on the middle portion of the inequality must also be performed on the left-hand side and right-hand side.

$$-3 \leq 2x + 1 < 7$$

$-3 - 1 \leq 2x + 1 - 1 < 7 - 1$ Subtract 1 from all three parts of the inequality.

$-4 \leq 2x < 6$ Simplify.

$\dfrac{-4}{2} \leq \dfrac{2x}{2} < \dfrac{6}{2}$ Divide by 2 in all three parts of the inequality.

$-2 \leq x < 3$

Graph:

Interval notation: $[-2, 3)$

7. Applications of Linear Inequalities

Table 2-1 provides several commonly used translations to express inequalities.

Table 2-1	
English Phrase	**Mathematical Inequality**
a is less than b	$a < b$
a is greater than b a exceeds b	$a > b$
a is less than or equal to b a is at most b a is no more than b	$a \leq b$
a is greater than or equal to b a is at least b a is no less than b	$a \geq b$

example 8   Translating Expressions Involving Inequalities

Translate the English phrases into mathematical inequalities:

a. Claude's annual salary, s, is no more than $40,000.
b. A citizen must be at least 18 years old to vote. (Let x represent a citizens' age.)

Solution:

a. $s \leq 40,000$
b. $x \geq 18$

Linear inequalities are found in a variety of applications. See how the next example can help you determine the minimum grade you need on an exam to get an A in your math course.

---

example 9

Solving an Application with Linear Inequalities

To earn an A in a math class, Alsha must average at least 90 on all of her tests. Suppose Alsha has scored 79, 86, 93, 90, and 95 on her first five math tests. Determine the minimum score she needs on her sixth test to get an A in the class.

Classroom Activity 2.7D

Solution:

Let x represent the score on the final exam.   Label the variable.

$$\left(\begin{array}{c}\text{Average of}\\ \text{all tests}\end{array}\right) \geq 90$$ Create a verbal model.

$$\frac{79 + 86 + 93 + 90 + 95 + x}{6} \geq 90$$ The average score is found by taking the sum of the test scores and dividing by the number of scores.

$$\frac{443 + x}{6} \geq 90$$ Simplify.

$$6\left(\frac{443 + x}{6}\right) \geq (90)6$$ Multiply both sides by 6 to clear fractions.

$$443 + x \geq 540$$ Solve the inequality.

$$x \geq 540 - 443$$ Subtract 443 from both sides.

$$\geq 97$$ Interpret the results.

Alsha must score at least 97 on her sixth exam to receive an A in the course.

---

section 2.7   PRACTICE EXERCISES

1. a. Simplify the expression: $3(x + 2) - (2x - 7)$
        x + 13
    b. Simplify the expression:
        $-(5x - 1) - 2(x + 6)$ *−7x − 11*
    c. Solve the equation: $3(x + 2) - (2x - 7) =$
        $-(5x - 1) - 2(x + 6)$ 2.2 *x = −3*

2. a. Simplify the expression: $6 - 8(x + 3) + 5x$
        −3x − 18
    b. Simplify the expression: $5x - (2x - 5) + 13$
        3x + 18
    c. Solve the equation: $6 - 8(x + 3) + 5x =$
        $5x - (2x - 5) + 13$ 2.2 *x = −6*

3. The sum of a number and twenty-eight is the same as the difference of two times the number and negative three. Find the number.
    2.3 The number is 25.

4. A bus and a car leave a bus stop at the same time, one traveling north and the other traveling south. If they travel at an average rate of 35 mph and 50 mph, respectively, how long will it take for them to be 204 miles apart assuming no stops?   2.6 It will take 2.4 h.

 See Additional Answers Appendix   Writing   Translating Expression   Geometry   Scientific Calculator   Video

5. The Ryder Cup is an annual golf tournament between the United States and Europe. In 1985 Europe won with five points more than the United States. If the total number of points scored was 28, how many points did each team score?   <u>2.3</u>   The United States scored $11\frac{1}{2}$ points, and Europe scored $16\frac{1}{2}$ points.

6. One angle of a triangle measures twice the smallest angle. The third angle is six times the measure of the smallest angle. Find the measures of each angle in the triangle.   <u>2.5</u>   The angles are $20°$, $40°$, and $120°$.

For Exercises 7–12, graph each inequality and write the set in interval notation.

Set-Builder Notation	Graph	Interval Notation
7. $\{x\|x \geq 6\}$		$[6, \infty)$
8. $\left\{x\|\frac{1}{2} < x \leq 4\right\}$		$\left(\frac{1}{2}, 4\right]$
9. $\{x\|x \leq 2.1\}$		$(-\infty, 2.1]$
10. $\left\{x\|x > \frac{7}{3}\right\}$		$\left(\frac{7}{3}, \infty\right)$
11. $\{x\|-2 < x \leq 7\}$		$(-2, 7]$
12. $\{x\|x < -5\}$		$(-\infty, -5)$

For Exercises 13–18, write each set in set-builder notation and in interval notation.

Set-Builder Notation	Graph	Interval Notation
13. $\left\{x\|x > \frac{3}{4}\right\}$		$\left(\frac{3}{4}, \infty\right)$
14. $\{x\|x \leq -0.3\}$		$(-\infty, -0.3]$
15. $\{x\|-1 < x < 8\}$		$(-1, 8)$
16. $\{x\|x \geq 0\}$		$[0, \infty)$
17. $\{x\|x < -14\}$		$(-\infty, -14)$
18. $\{x\|0 < x \leq 9\}$		$(0, 9]$

For Exercises 19–24, graph each set and write the set in set-builder notation.

Set-Builder Notation	Graph	Interval Notation
19. $\{x\|x \geq 18\}$		$[18, \infty)$
20. $\{x\|-10 \leq x \leq -2\}$		$[-10, -2]$
21. $\{x\|x < -0.6\}$		$(-\infty, -0.6)$
22. $\left\{x\|x < \frac{5}{3}\right\}$		$\left(-\infty, \frac{5}{3}\right)$
23. $\{x\|-3.5 \leq x < 7.1\}$		$[-3.5, 7.1)$
24. $\{x\|x \geq -10\}$		$[-10, \infty)$

25. Let x represent a student's average in a math class. The grading scale is given here.

A	$93 \leq x \leq 100$
B+	$89 \leq x < 93$
B	$84 \leq x < 89$
C+	$80 \leq x < 84$
C	$75 \leq x < 80$
F	$0 \leq x < 75$

a. Write the range of scores corresponding to each letter grade in interval notation.   ❖

b. If Stephan's average is 84.01, what grade will he receive?   ❖

c. If Estella's average is 79.89, what grade will she receive?   ❖

26. Let x represent a student's average in a science class. The grading scale is given here.

A	$90 \leq x \leq 100$
B+	$86 \leq x < 90$
B	$80 \leq x < 86$
C+	$76 \leq x < 80$
C	$70 \leq x < 76$
D+	$66 \leq x < 70$
D	$60 \leq x < 66$
F	$0 \leq x < 60$

a. Write the range of scores corresponding to each letter grade in interval notation.   ❖

b. If Jacque's average is 89.99, what is her grade?   ❖

c. If Marc's average is 66.01, what is his grade?   ❖

For Exercises 27–36, translate the English phrase into a mathematical inequality.

27. The speed of a car, s, was at least 110 km/h.
$s \geq 110$

28. The length of a fish, L, was at least 10 in. $L \geq 10$

29. Tasha's average test score, t, exceeded 90. $t > 90$

30. The wind speed, w, exceeded 75 mph. $w > 75$

31. The height of a cave, h, was no more than 2 ft.
    $h \leq 2$

32. The temperature of the water in Blue Spring, t, is no more than 72°F. $t \leq 72$

33. The temperature on the tennis court, t, was no less than 100°F. $t \geq 100$

34. The length of the hike, L, was no less than 8 km.
    $L \geq 8$

35. The depth, d, of a certain pool was at most 10 ft.
    $d \leq 10$

36. The amount of rain, a, in a recent storm was at most 2 in. $a \leq 2$

For Exercises 37–44, solve the equation in part (a). For part (b), solve the inequality and graph the solution set.

37. a. $x + 3 = 6$ ❖   38. a. $y - 6 = 12$ ❖
    b. $x + 3 > 6$ ❖   b. $y - 6 \geq 12$ ❖

39. a. $p - 4 = 9$ ❖   40. a. $k + 8 = 10$ ❖
    b. $p - 4 \leq 9$ ❖   b. $k + 8 < 10$ ❖

41. a. $4c = -12$ ❖   42. a. $5d = -35$ ❖
    b. $4c < -12$ ❖   b. $5d > -35$ ❖

43. a. $-10z = 15$ ❖   44. a. $-2w = 14$ ❖
    b. $-10z \leq 15$ ❖   b. $-2w < 14$ ❖

For Exercises 45–48, determine whether the given number is a solution to the inequality.

45. $-2x + 5 < 4$ $x = -2$ No

46. $-3y - 7 > 5$ $y = 6$ No

47. $4(p + 7) - 1 > 2 + p$ $p = 1$ Yes

48. $3 - k < 2(-1 + k)$ $k = 4$ Yes

For Exercises 49–84, solve the inequality. Graph the solution set and write the set in interval notation.

49. $x + 5 \leq 6$ ❖   50. $y - 7 < 6$ ❖

51. $q - 7 > 3$ ❖   52. $r + 4 \geq -1$ ❖

53. $4 < 1 + z$ ❖   54. $3 > z - 6$ ❖

55. $2 \geq a - 6$ ❖   56. $7 \leq b + 12$ ❖

57. $3c > 6$ ❖   58. $4d \leq 12$ ❖

59. $-3c > 6$ ❖   60. $-4d \leq 12$ ❖

61. $-h \leq -14$ ❖   62. $-q > -7$ ❖

63. $12 \geq -\dfrac{x}{2}$ ❖   64. $6 < -\dfrac{m}{3}$ ❖

65. $-2 \leq p + 1 < 4$ ❖   66. $0 < k + 7 < 6$ ❖

67. $-3 < 6h - 3 < 12$ ❖   68. $-6 \leq 4a - 2 \leq 12$ ❖

69. $0.6z \geq 54$ ❖   70. $-0.7w > 28$ ❖

71. $-\dfrac{2}{3}y < 6$ ❖   72. $\dfrac{3}{4}x \leq -12$ ❖

73. $-2x - 4 \leq 11$ ❖   74. $-3x + 1 > 0$ ❖

75. $-7b - 3 \leq 2b$ ❖   76. $3t \geq 7t - 35$ ❖

77. $4n + 2 < 6n + 8$ ❖   78. $2w - 1 \leq 5w + 8$ ❖

79. $8 - 6(x - 3) > -4x + 12$ ❖

80. $3 - 4(h - 2) > -5h + 6$ ❖

81. $\dfrac{7}{6}p + \dfrac{4}{3} \geq \dfrac{11}{6}p - \dfrac{7}{6}$ ❖

82. $\dfrac{1}{3}w - \dfrac{1}{2} \leq \dfrac{5}{6}w + \dfrac{1}{2}$ ❖

83. $-1.2a - 0.4 < -0.4a + 2$ ❖

84. $-0.4c + 1.2 > -2c - 0.4$ ❖

85. The average summer rainfall for Miami, Florida, for June, July, and August is 7.4 in. per month. If Miami receives 5.9 in. of rain in June and 6.1 in. in July, how much rain is required in August to exceed the 3-month summer average?
    More than 10.2 in. of rain is needed.

❖ See Additional Answers Appendix   ✎ Writing   ⬌ Translating Expression   Geometry   Scientific Calculator   Video

86. The average winter snowfall for Burlington, Vermont, for December, January, and February is 18.7 in. per month. If Burlington received 22 in. of snow in December and 24 in. in January, how much snow is required in February to exceed the 3-month winter average?
More than 10.1 in. of snow is needed.

87. An artist paints wooden birdhouses. She buys the birdhouses for $9 each. However, for large orders, the price per birdhouse is discounted by a percentage off the original price. Let x represent the number of birdhouses ordered. The corresponding discount is given in the table.

Size of Order	Discount
$x \le 49$	0%
$50 \le x \le 99$	5%
$100 \le x \le 199$	10%
$x \ge 200$	20%

Table for Exercise 87

a. If the artist makes an order for 190 birdhouses, compute the total cost. ❖

b. Which costs more 190 bird houses or 200 birdhouses? Explain your answer. ❖

88. A wholesaler sells T-shirts to a surf shop at $8 per shirt. However, for large orders, the price per shirt is discounted by a percentage off the original price. Let x represent the number of shirts ordered. The corresponding discount is given in the table.

Number of Shirts Ordered	Discount
$x \le 24$	0%
$25 \le x \le 49$	2%
$50 \le x \le 99$	4%
$100 \le x \le 149$	6%
$x \ge 150$	8%

Table for Exercise 88

a. If the surf shop orders 50 shirts, compute the total cost. $384

b. Which costs more, 148 shirts or 150 shirts? Explain your answer. It costs $1112.96 for 148 shirts and $1104.00 for 150 shirts. 150 shirts cost less than 148 shirts because the discount is higher.

89. Maggie sells lemonade at an art show. She has a fixed cost of $75 to cover the registration fee for the art show. In addition, her cost to produce each lemonade is $0.17. If x represents the number of lemonades, then the total cost to produce x lemonades is given by:

$$\text{Cost} = 75 + 0.17x$$

If Maggie sells each lemonade for $2, then her revenue (the amount she brings in) for selling x lemonades is given by:

$$\text{Revenue} = 2.00x$$

a. Write an inequality that expresses the number of lemonades, x, that Maggie must sell to make a profit. Profit is realized when the revenue is greater than the cost (Revenue > cost). $2.00x > 75 + 0.17x$

b. Solve the inequality in part (a). $x > 41$; profit occurs when more than 41 lemonades are sold.

90. Two rental car companies rent subcompact cars at a discount. Company A rents for $14.95 per day plus 22 cents per mile. Company B rents for $18.95 a day plus 18 cents per mile. Let x represent the number of miles driven in one day.

The cost to rent a subcompact car for one day from Company A is:

$$\text{Cost}_A = 14.95 + 0.22x$$

The cost to rent a subcompact car for one day from Company B is:

$$\text{Cost}_B = 18.95 + 0.18x$$

a. Write an inequality that expresses the number of miles, x, for which the daily cost to rent from Company A is less than the daily cost to rent from Company B

$$(\text{Cost}_A < \text{Cost}_B)$$
$14.95 + 0.22x < 18.95 + 0.18x$

b. Solve the inequality in part (a).
$x < 100$; company A costs less than company B if the mileage is less than 100 miles.

▪ EXPANDING YOUR SKILLS

For Exercises 91–96, solve the inequality. Graph the solution set and write the set in interval notation.

91. $3(x + 2) - (2x - 7) \le (5x - 1) - 2(x + 6)$ ❖

92. $6 - 8(y + 3) + 5y > 5y - (2y - 5) + 13$ ❖

93. $-2 - \dfrac{w}{4} \le \dfrac{1 + w}{3}$ ❖

94. $\dfrac{z - 3}{4} - 1 > \dfrac{z}{2}$ ❖

95. $-0.703 < 0.122p - 2.472$ ❖

96. $3.88 - 1.335t \ge 5.66$ ❖

Concepts

1. **Definition of a Linear Equation in Two Variables**
2. **Solutions to Linear Equations in Two Variables**
3. **Graphing Linear Equations in Two Variables by Plotting Points**
4. **Applications of Linear Equations in Two Variables**

section

2.8 CONNECTIONS TO GRAPHING: LINEAR EQUATIONS IN TWO VARIABLES

1. Definition of a Linear Equation in Two Variables

Recall that an equation in the form $ax + b = 0$, where $a \ne 0$, is called a linear equation in one variable. A solution to such an equation is a value of x that makes the equation a true statement. For example, $3x + 6 = 0$ has a solution of $x = -2$.

In this section we will look at linear equations in two variables.

Definition of a Linear Equation in Two Variables

Let a, b, and c be real numbers such that a and b are not both zero. Then, an equation that can be written in the form:

$$ax + by = c$$

is called a **linear equation in two variables**.

---

2. Solutions to Linear Equations in Two Variables

The equation $x + y = 5$ is a linear equation in two variables. A solution to such an equation is an ordered pair (x, y) that makes the equation a true statement. Several solutions to the equation $x + y = 5$ are listed here:

Solution:	Check:
(x, y)	$x + y = 5$
$(2, 3)$	$(2) + (3) = 5$ ✔
$(1, 4)$	$(1) + (4) = 5$ ✔
$(0, 5)$	$(0) + (5) = 5$ ✔
$(-2, 7)$	$(-2) + (7) = 5$ ✔

By graphing these ordered pairs, we see that the solution points line-up (Figure 2-17).

 ❖ See Additional Answers Appendix   Writing   Translating Expression   Geometry   Scientific Calculator   Video

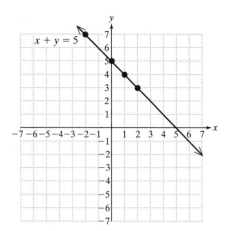

Figure 2-17

Notice that there are infinitely many solutions to the equation $x + y = 5$ so they cannot all be listed. Therefore, to visualize all solutions to the equation $x + y = 5$, we draw the line through the points in the graph. Every point on the line represents an ordered pair solution to the equation $x + y = 5$, and the line represents the set of *all* solutions to the equation.

3. Graphing Linear Equations in Two Variables by Plotting Points

The word *linear* means "relating to or resembling a line." It is not surprising then that the solution set for any linear equation in two variables forms a line in a rectangular coordinate system. Because two points determine a line, to graph a linear equation it is sufficient to find two solution points and draw the line between them. This process is demonstrated in the next example.

---

example 1

Graphing a Linear Equation

Given the linear equation $4x + 2y = 8$,

a. Complete the table.

x	y
4	
	2

b. Graph the ordered pairs found in part (a). Draw the line through the two points to represent all solutions to the equation.

Solution:

a. The first row of the table provides a value for x. Substitute $x = 4$ into the equation and solve for the corresponding value of y.

x	y
4	
	2

The second row of the table provides a value for y. Substitute $y = 2$ into the equation and solve for the corresponding value of x.

$4x + 2y = 8$

$4(4) + 2y = 8$

$16 + 2y = 8$

$2y = -8$

$\dfrac{2y}{2} = \dfrac{-8}{2}$

$y = -4$

$4x + 2y = 8$

$4x + 2(2) = 8$

$4x + 4 = 8$

$4x = 4$

$\dfrac{4x}{4} = \dfrac{4}{4}$

$x = 1$

x	y
4	-4
1	2

b. The ordered pairs corresponding to the table values are $(4, -4)$ and $(1, 2)$. Graph the points and the line defined by the points (Figure 2-18).

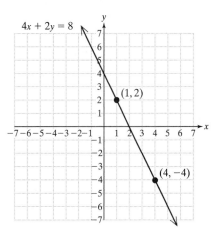

Figure 2-18

The values for x and y given in the table were picked arbitrarily by the authors. Any two values could have been used, and the corresponding points would form the same line. It is important to note that once a value for x has been chosen, the corresponding y-value is determined by the equation. Similarly, once a value for y has been chosen, the corresponding x-value is determined by the equation.

Tip: To check that a line is graphed correctly, determine a third ordered pair solution to the equation and verify that the points line up.

To find a third point on the line $4x + 2y = 8$, choose a different value of x or y. If we choose $x = 2$, we have:

$$4(2) + 2y = 8$$

$$8 + 2y = 8$$

$$2y = 0$$

$$\frac{2y}{2} = \frac{0}{2}$$

$$y = 0$$

The ordered pair $(2, 0)$ is a solution.

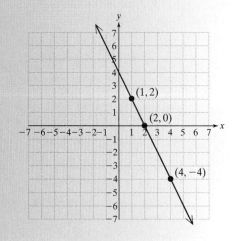

---

example 2

Graphing a Linear Equation

Given the equation $y = -\dfrac{1}{3}x + 1$,

a.   Complete the table.

x	y
6	
	4
−2	

b.   Graph the ordered pairs found in part (a). Draw the line through the points to represent all solutions of the equation.

Classroom Activity 2.8A

Solution:

a.

x	y
6	
	4
−2	

Substitute: $x = 6$

$$y = -\frac{1}{3}x + 1$$

$$= -\frac{1}{3}(6) + 1$$

$$= -2 + 1$$

$$= -1$$

Substitute: $y = 4$

$$y = -\frac{1}{3}x + 1$$

$$4 = -\frac{1}{3}x + 1$$

$$3 = -\frac{1}{3}x$$

$$(-3)3 = (-3)\left(-\frac{1}{3}x\right)$$

$$-9 = x$$

Substitute: $x = -2$

$$y = -\frac{1}{3}x + 1$$

$$= -\frac{1}{3}(-2) + 1$$

$$= \frac{2}{3} + 1$$

$$= \frac{5}{3}$$

The completed table is

x	y
6	−1
−9	4
−2	$\dfrac{5}{3}$

b. The ordered pairs corresponding to the table values are $(6, -1), (-9, 4),$ and $(-2, \frac{5}{3})$. Note that $\frac{5}{3} = 1\frac{2}{3}$. Therefore, the point $(-2, \frac{5}{3})$ is equivalent to $(-2, 1\frac{2}{3})$. See Figure 2-19.

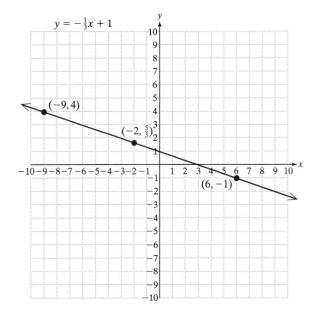

Figure 2-19

4.   Applications of Linear Equations in Two Variables

example 3

Applying a Linear Equation in Two Variables

Exercise physiologists can estimate the percentage of body fat for an individual by taking skin fold measurements from the abdomen, triceps, and thigh. Skin fold measurements are measured in millimeters with a pinching device called calipers. Based on experimental data, scientists have found that the following equation relates skin fold measurements, x, (in millimeters) to $y\%$ body fat.

$$y = 0.33x + 0.25 \qquad \text{for } 5 \le x \le 45$$

Tip: The condition $5 \le x \le 45$ indicates that the formula is valid for x values (skin fold measurements) between 5 mm and 45 mm, inclusive.

For parts (a)–(c), use the linear equation to find the percentage of body fat for the indicated skin fold measurements. Write the answers as ordered pairs and interpret the meaning of each ordered pair.

a. 15 mm ($x = 15$)
b. 25 mm ($x = 25$)
c. 40 mm ($x = 40$)
d. Graph the line using the ordered pairs from parts (a)–(c).
e. If a person has 10.15% body fat ($y = 10.15$), what is the corresponding skin fold measurement?

Classroom Activity 2.8B

Solution:

a. The variable x represents skin fold, so substitute $x = 15$ into the equation and solve for y.

$$y = 0.33x + 0.25$$
$$= 0.33(15) + 0.25$$
$$= 4.95 + 0.25$$
$$= 5.20$$

The ordered pair $(15, 5.20)$ indicates that a person with a 15 mm skin fold will have approximately 5.2% body fat.

b. $y = 0.33x + 0.25$

$= 0.33(25) + 0.25$ Substitute $x = 25$ into the equation and solve for y.

$= 8.25 + 0.25$

$= 8.50$

The ordered pair $(25, 8.50)$ indicates that a person with a 25-mm skin fold will have approximately 8.5% body fat.

c. $y = 0.33x + 0.25$

$= 0.33(40) + 0.25$ Substitute $x = 40$ into the equation and solve for y.

$= 13.2 + 0.25$

$= 13.45$

The ordered pair $(40, 13.45)$ indicates that a person with a 40-mm skin fold will have approximately 13.45% body fat.

d. In application problems, a linear equation may have restrictions on the x-variable. It would not make sense, for instance, to have a negative skin fold measurement. In this case skin fold measurements are restricted to $5 \le x \le 45$. For this reason the line is drawn only for x-values between 5 and 45 mm (Figure 2-20).

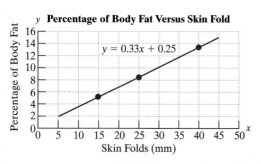

Percentage of Body Fat Versus Skin Fold

$y = 0.33x + 0.25$

Skin Folds (mm)

Percentage of Body Fat

Figure 2-20

e. Because the variable y represents $y\%$ body fat, substitute $y = 10.15$ into the equation and solve for x.

$$y = 0.33x + 0.25$$

$$10.15 = 0.33x + 0.25 \qquad \text{Substitute } y = 10.15.$$

$$9.9 = 0.33x \qquad \text{Solve for } x.$$

$$\frac{9.9}{0.33} = \frac{0.33x}{0.33}$$

$$30 = x \qquad \begin{array}{l}\text{A person with } 10.15\% \text{ body fat will have a 30-mm} \\ \text{skin fold.}\end{array}$$

Calculator Connections

A viewing window of a graphing calculator shows a portion of a rectangular coordinate system. The standard viewing window for many calculators shows the x-axis between -10 and 10 and the y-axis between -10 and 10 (Figure 2-21). Furthermore, the scale defined by the "tic" marks on both the x- and y-axes is usually set to 1.

The "Standard Viewing Window"

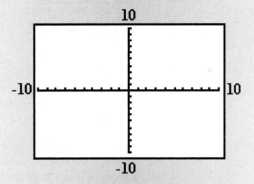

Figure 2-21

To graph an equation in x and y on a graphing calculator, the equation must be written with the y-variable isolated. Therefore, the equation $x + y = 5$ must first be written as $y = -x + 5$ (or $y = 5 - x$) before it can be entered into a graphing calculator. The *Graph* option displays the graph of the line.

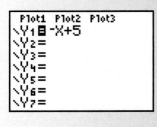

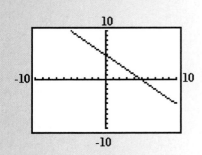

To graph the equation $4x + 2y = 8$, first solve for y. Enter the equation into the calculator and select the *Graph* option.

$$4x + 2y = 8$$

$$2y = -4x + 8$$

$$\frac{2y}{2} = \frac{-4x}{2} + \frac{8}{2}$$

$$y = -2x + 4$$

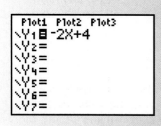
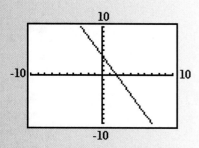

To graph the equation $y = -\frac{1}{3}x + 1$, use parentheses around the fraction $\frac{1}{3}$.

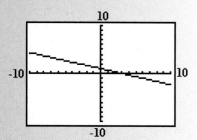

Sometimes the standard viewing window does not provide an adequate display for the graph of an equation. For example, the graph of $y = -x + 15$ is visible only in a small portion of the upper right corner of the standard viewing window.

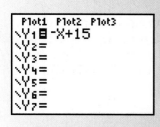
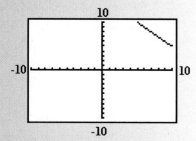

To see where this line crosses the x- and y-axes, we can change the viewing window to accommodate larger values of x and y. Most calculators have a *Range* feature or *Window* feature that allows the user to change the minimum and maximum x- and y-values.

To get a better picture of the equation $y = -x + 15$, change the minimum x-value to -10 and the maximum x-value to 20. Similarly, use a minimum y-value of -10 and a maximum y-value of 20.

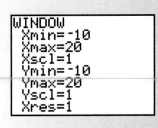

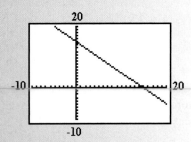

Calculator Exercises

Use a graphing calculator to graph the equations on the standard viewing window. Then graph the equation again on the suggested window.

1. $y = -3x - 2$
   a. Standard viewing window ❖
   b. Window defined by: $-5 \le x \le 5$ and $-5 \le y \le 5$ ❖

2. $y = 0.33x - 0.25$
   a. Standard viewing window ❖
   b. Window defined by: $-2 \le x \le 2$ and $-2 \le y \le 2$ ❖

3. $2x + y = 5$
   a. Standard viewing window ❖
   b. Window defined by: $-1 \le x \le 8$ and $-4 \le y \le 8$ ❖

4. $3x - 2y = 32$
   a. Standard viewing window ❖
   b. Window defined by: $-10 \le x \le 20$ and $-20 \le y \le 10$ ❖

5. $x + 2y = -50$
   a. Standard viewing window ❖
   b. Window defined by $-60 \le x \le 20$ and $-30 \le y \le 20$ ❖

section 2.8 PRACTICE EXERCISES

For Exercises 1–4, substitute the given number to determine if it is a solution to the equation. _2.1_

1. $3x + 4 = 13$; $x = 3$
   3 is a solution.
2. $x - 5 = 2$; $x = 7$
   7 is a solution.
3. $-2 - 3y = 10$; $y = -4$ −4 is a solution.
4. $6 + 2y = 0$; $y = 3$ 3 is not a solution.

For Exercises 5–12, solve the linear equation in one variable. _2.2_

5. $2 + 5y = 10$ $y = \dfrac{8}{5}$
6. $3 - 4y = -6$ $y = \dfrac{9}{4}$
7. $2x + 5 = 10$ $x = \dfrac{5}{2}$
8. $3x - 4 = -6$ $x = -\dfrac{2}{3}$
9. $y = 2(-1) + 12$ $y = 10$
10. $y = 3(2) - 8$ $y = -2$
11. $8 = 2x + 12$ $x = -2$
12. $10 = 3x - 8$ $x = 6$

For Exercises 13–18, substitute the given ordered pair to determine if it is a solution to the equation.

13. $x + y = -5$; $(-2, -3)$ Yes
14. $x - y = 6$; $(8, 2)$ Yes
15. $y = 3x - 2$; $(1, 1)$ Yes
16. $y = -\dfrac{1}{3}x + 3$; $(-3, 4)$ Yes
17. $y = -\dfrac{5}{2}x + 5$; $(-2, 0)$ No
18. $4x + 5y = 20$; $(-5, -4)$ No

For Exercises 19–34, complete the table, and graph the corresponding ordered pairs. Draw the line defined by the points to represent all solutions to the equation.

19. $x + y = 12$ ❖

x	y
5	7
8	4
6	6
3	9

20. $x + y = 14$ ❖

x	y
1	13
12	2
2	12
6	8

21. $x - y = 6$ ❖

x	y
3	−3
0	−6
6	0
7	1

22. $x - y = -1$ ❖

x	y
3	4
−1	0
1	2
2	3

23. $x + 4y = 8$ ❖

x	y
8	0
0	2
12	−1

24. $x - 3y = 6$ ❖

x	y
0	−2
6	0
3	−1

25. $2x - 3y = 6$ ❖

x	y
0	−2
3	0
2	$-\dfrac{2}{3}$

26. $4x + 2y = 8$ ❖

x	y
0	4
2	0
3	−2

27. $y = 5x + 1$ ❖

x	y
1	6
2	11
−1	−4

28. $y = -3x - 3$ ❖

x	y
−2	3
−1	0
−4	9

29. $y = \dfrac{2}{7}x - 5$ ❖

x	y
7	−3
−14	−9
0	−5

30. $y = -\dfrac{3}{5}x - 2$ ❖

x	y
0	−2
5	−5
10	−8

31. $5x + 3y = 12$ ❖

x	y
1	$\frac{7}{3}$
0	4
−1	$\frac{17}{3}$

32. $4x − 3y = 6$ ❖

x	y
−2	$-\frac{14}{3}$
$\frac{9}{2}$	4
0	−2

a.   Find y when $x = 13$.   ❖

b.   Find x when $y = 279.80$.   ❖

c.   Write the ordered pairs from parts (a) and (b) and interpret their meaning in the context of the problem.   ❖

d.   Graph the ordered pairs and the line defined by the points.   ❖

33. $y = -3.47x - 5.81$
❖

x	y
3.1	−16.567
4	−19.69
−1.6	−0.258

34. $y = -0.24x + 4.62$
❖

x	y
4.5	3.54
3	3.9
−8.8	6.732

37. The value of a car depreciates the minute that it is driven off of the lot. For a Hyundai Accent, the value of the car is given by the equation $y = -1531x + 11,599$ $(x \geq 0)$ where y is the value of the car in dollars, x years after its purchase.

a.   Find y when $x = 1$.   ❖

b.   Find x when $y = 7006$.   ❖

c.   Write the ordered pairs from parts (a) and (b) and interpret their meaning in the context of the problem.   ❖

d.   Graph the ordered pairs and the line defined by the points.   ❖

35. The students in the ninth grade at Atlantic High School pick up aluminum cans to be recycled. The current value of aluminum is $0.69 per pound. If the students pay $20 to rent a truck to haul the cans, then the following equation expresses the amount of money that they earn, y, given the number of pounds of aluminum, x.

$$y = 0.69x - 20 \quad (x \geq 0)$$

a.   Let $x = 55$ and solve for y.   ❖

b.   Let $y = 80.05$ and solve for x.   ❖

c.   Write the ordered pairs from parts (a) and (b) and interpret their meaning in the context of the problem.   ❖

d.   Graph the ordered pairs and the line defined by the points.   ❖

36. The store "CDs R US" sells all compact discs for $13.99. The following equation represents the revenue, y, (in dollars) generated by selling x CDs.

$$y = 13.99x \quad (x \geq 0)$$

38. The enrollment in Catholic schools in the United States declined after 1970. This decline can be approximated by the linear equation

$$y = -94,378x + 4,363,000$$

where y is the total enrollment in Catholic schools and x is the number of years after 1970 (that is, $x = 0$ corresponds to 1970, $x = 1$ corresponds to 1971, and so on) (*Source: Information Please Almanac*, 1999)

a.   Find y when x is 5.   ❖

b.   Find x when y is 3,136,086. Round to the nearest year.   ❖

c.   Write the ordered pairs from parts (a) and (b) and interpret their meaning in the context of the problem.   ❖

d.   Graph the ordered pairs and the line defined by the points.   ❖

39. The total sales for the Emerson Electric Co. from 1994 to 1999 is given by the equation $y = 1136x + 8790$. The value of y is in millions of dollars and x is the time in years after 1994. That is, $x = 0$ corresponds to 1994, $x = 1$ corresponds to 1995, and so on.

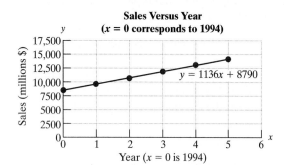

Sales Versus Year
($x = 0$ corresponds to 1994)

$y = 1136x + 8790$

(*Source:* Emerson Electric Co., 1999 Annual Report.)

Figure for Exercise 39

a. Find the amount in sales reported for the year 1995. $9926 million

b. Find the amount in sales reported for the year 1998. $13,334 million

c. Use the equation to find the year in which the sales equaled $12,198 million.   1997

d. Use the equation to find the year in which the sales equaled $14,470 million.   1999

40. The net earning for the Emerson Electric Co. from 1994 to 1999 is given by the equation $y = 105x + 800$. The value of y is in millions of dollars and x is the time in years after 1994. That is, $x = 0$ corresponds to 1994, $x = 1$ corresponds to 1995, and so on.

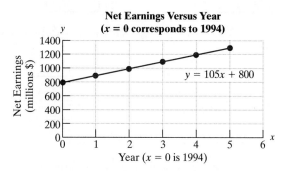

Net Earnings Versus Year
($x = 0$ corresponds to 1994)

$y = 105x + 800$

(*Source:* Emerson Electric Co., 1999 Annual Report.)

Figure for Exercise 40

a. Find the net earnings for the year 1994. $800 million

b. Find the net earnings for the year 1997. $1115 million

c. Use the equation to find the year in which the net earnings equaled 1010 million dollars.   1996

d. Use the equation to find the year in which the net earnings equaled 1325 million dollars.   1999

chapter 2   SUMMARY

SECTION 2.1—ADDITION, SUBTRACTION, MULTIPLICATION, AND DIVISION PROPERTIES OF EQUALITY

KEY CONCEPTS:

An equation is an algebraic statement that indicates two expressions are equal. A solution to an equation is a value of the variable that makes the equation a true statement. The set of all solutions to an equation is the solution set of the equation.

A linear equation in one variable can be written in the form $ax + b = 0$, where $a \neq 0$.

Addition Property of Equality:

If $a = b$, then $a + c = b + c$

Subtraction Property of Equality:

If $a = b$, then $a - c = b - c$

Multiplication Property of Equality:

If $a = b$, then $ac = bc$

Division Property of Equality:

If $a = b$, then $\dfrac{a}{c} = \dfrac{b}{c}$ $(c \neq 0)$

KEY TERMS:

addition property of equality
division property of equality
equation
linear equation in one variable
multiplication property of equality
solution set
solution to an equation
subtraction property of equality

EXAMPLES:

An example of an equation: $2x + 1 = 9$
The solution is $x = 4$.

The equation $5x + 20 = 0$ is a linear equation in one variable.

Solve:

$$x - 5 = 12$$
$$x - 5 + 5 = 12 + 5$$
$$x = 17$$

Solve:

$$z + 1.44 = 2.33$$
$$z + 1.44 - 1.44 = 2.33 - 1.44$$
$$z = 0.89$$

Solve:

$$\frac{3}{4}x = 12$$
$$\frac{4}{3} \cdot \frac{3}{4}x = 12 \cdot \frac{4}{3}$$
$$x = 16$$

Solve:

$$16 = 8y$$
$$\frac{16}{8} = \frac{8y}{8}$$
$$2 = y$$

SECTION 2.2—SOLVING LINEAR EQUATIONS

KEY CONCEPTS:

Steps for Solving a Linear Equation in One Variable:

1. Simplify both sides of the equation by clearing parentheses and collecting *like* terms.
2. Use the addition and subtraction properties of equality to collect the variable terms on one side of the equation.
3. Use the addition and subtraction properties of equality to collect the constant terms on the other side of the equation.
4. Use the multiplication and division properties of equality to make the coefficient of the variable term equal to 1.
5. Check your answer.

A conditional equation is true for some values of the variable but is false for other values.

An equation that has all real numbers as its solution set is an identity.

An equation that has no solution is a contradiction.

KEY TERMS:

conditional equation
contradiction
identity

EXAMPLES:

Solve the equation:

$$5y + 6 = 3(y - 1) + 2$$

$5y + 6 = 3y - 3 + 2$	Clear parentheses.
$5y + 6 = 3y - 1$	Combine *like* terms.
$2y + 6 = -1$	Isolate variable term.
$2y = -7$	Isolate constant term.
$y = -\dfrac{7}{2}$	Divide by the coefficient, 2.

Check:

$$5\left(-\frac{7}{2}\right) + 6 \stackrel{?}{=} 3\left(-\frac{7}{2} - 1\right) + 2$$

$$-\frac{35}{2} + 6 \stackrel{?}{=} -\frac{21}{2} - 3 + 2$$

$$-\frac{35}{2} + \frac{12}{2} \stackrel{?}{=} -\frac{21}{2} - \frac{6}{2} + \frac{4}{2}$$

$$-\frac{23}{2} = -\frac{23}{2} \checkmark$$

$x + 5 = 7$ is a conditional equation because it is true only on the condition that $x = 2$.

Solve:

$$x + 4 = 2(x + 2) - x$$
$$x + 4 = 2x + 4 - x$$
$$x + 4 = x + 4$$
$$4 = 4 \quad \text{is an identity}$$

The solution is all real numbers.

Solve:

$$y - 5 = 2(y + 3) - y$$
$$y - 5 = 2y + 6 - y$$
$$y - 5 = y + 6$$
$$-5 = 6 \quad \text{is a contradiction}$$

There is no solution.

SECTION 2.3—APPLICATIONS OF LINEAR EQUATIONS: INTRODUCTION TO PROBLEM SOLVING

KEY CONCEPTS:

Problem-Solving Steps for Word Problems

1. Read the problem carefully.
2. Assign labels to unknown quantities.
3. Develop a verbal model.
4. Write a mathematical equation.
5. Solve the equation.
6. Interpret the results and write the answer in words.

KEY TERMS:

consecutive even integers
consecutive integers
consecutive odd integers
problem-solving flow chart

EXAMPLES:

Consecutive integer problem:

The perimeter of a triangle is 54 m. The lengths of the sides are represented by three consecutive even integers. Find the lengths of the three sides.

1. Read the problem.
2. Let x represent one side, $x + 2$ represent the second side, and $x + 4$ represent the third side.

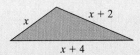

3. (First side) + (second side) + (third side) = perimeter
4. $x + (x + 2) + (x + 4) = 54$
5. $3x + 6 = 54$

$$3x = 48$$

$$x = 16$$

$$x + 2 = 18$$

$$x + 4 = 20$$

6. The lengths of the three sides are 16 m, 18 m, and 20 m.

SECTION 2.4—LINEAR EQUATIONS: CLEARING FRACTIONS AND DECIMALS

KEY CONCEPTS:

To Clear Fractions (or Decimals) in an Equation

Multiply both sides of the equation by a common denominator of all the fractions.

EXAMPLES:

Solve the equation:

$$-1.2x - 5.1 = 16.5 \qquad \text{Multiply both sides by 10.}$$

$$10(-1.2x - 5.1) = 10(16.5)$$

$$-12x - 51 = 165$$

$$-12x = 216$$

$$\frac{-12x}{-12} = \frac{216}{-12}$$

$$x = -18$$

Steps for Solving a Linear Equation in One Variable

1. Consider clearing fractions or decimals if any exist in the equation.
2. Simplify both sides of the equation by clearing parentheses and combining *like* terms.
3. Use the addition and subtraction properties of equality to collect the variable terms on one side of the equation.
4. Use the addition and subtraction properties of equality to collect the constant terms on the other side of the equation.
5. Use the multiplication and division properties of equality to make the coefficient of the variable term equal to 1.
6. Check your answer.

Solve the equation:

$$\frac{1}{2}(x - 4) - \frac{3}{4}(x + 2) = \frac{1}{4}$$

$$4\left[\frac{1}{2}(x - 4) - \frac{3}{4}(x + 2)\right] = 4\left(\frac{1}{4}\right)$$

$$2(x - 4) - 3(x + 2) = 1$$

$$2x - 8 - 3x - 6 = 1$$

$$-x - 14 = 1$$

$$-x = 15$$

$$x = -15$$

Check:

$$\frac{1}{2}(-15 - 4) - \frac{3}{4}(-15 + 2) = \frac{1}{4}$$

$$\frac{1}{2}(-19) - \frac{3}{4}(-13) = \frac{1}{4}$$

$$\frac{-19}{2} + \frac{39}{4} = \frac{1}{4}$$

$$\frac{-38}{4} + \frac{39}{4} = \frac{1}{4} ✔$$

Applications Involving Percents:

Simple interest: $I = Prt$

KEY TERMS:

clearing fractions
percent
simple interest

John Li invests $5400 at 8.5% simple interest. How much interest does he make after 5 years?

Solution:

$$I = Prt$$
$$= (\$5400)(0.085)(5)$$
$$= \$2295$$

John makes $2295 in interest after 5 years.

SECTION 2.5—APPLICATIONS OF GEOMETRY

KEY CONCEPTS:

A literal equation is an equation that has more than one variable. Often such an equation can be manipulated to solve for different variables.

EXAMPLES:

Given:

$$P = 2a + b \qquad \text{Solve for } b.$$

$$P - 2a = 2a - 2a + b$$

$$P - 2a = b$$

$$\text{or} \qquad b = P - 2a$$

Formulas from Section R.3 can be used in applications involving geometry.

KEY TERM:

literal equations

Solve:

Find the length of a side of a square whose perimeter is 28 ft.

Use the formula $P = 4s$. Substitute 28 for P and solve:

$$P = 4s$$

$$28 = 4s$$

$$7 = s$$

The length of a side of the square is 7 ft.

Solve:

Two complementary angles are drawn such that one angle is 5° more than three times the other angle. Find the measures of the two angles.

Let x represent one angle.

Let $3x + 5$ represent the other angle.

$$x + (3x + 5) = 90$$

$$4x + 5 = 90$$

$$4x = 85$$

$$x = 21.25$$

$$3x + 5 = 68.75$$

The two angles are 21.25° and 68.75°.

SECTION 2.6—MORE APPLICATIONS OF LINEAR EQUATIONS

KEY CONCEPTS:

The following types of word problems are introduced in Section 2.6.

1. Mixture problems
2. Applications involving distance, rate, and time
3. Ratio and proportions

EXAMPLES:

Mixture problem:

The amount Estella invested in a 6% savings account is $3500 less than the amount she invested in a 10% savings account. At the end of one year, she received $750 in simple interest. Find the amount Estella invested in each account.

Let x represent the amount at 10%.

Then, $x - 3500$ is the amount at 6%.

The ratio of a to b is $\frac{a}{b}$ ($b \neq 0$). A proportion is an equation that equates two ratios.

$$\frac{a}{b} = \frac{c}{d} \quad (b \neq 0, d \neq 0)$$

In a proportion, the cross products are equal, $ad = bc$. For example:

$$\frac{x}{4} = \frac{15}{12}$$

$$12x = 4(15)$$

$$\frac{12x}{12} = \frac{60}{12}$$

$$x = 5$$

KEY TERMS:

proportion
ratio
similar triangles

$$\left(\begin{array}{c}\text{Interest from} \\ 10\% \text{ account}\end{array}\right) + \left(\begin{array}{c}\text{interest from} \\ 6\% \text{ account}\end{array}\right) = \left(\begin{array}{c}\text{total} \\ \text{interest}\end{array}\right)$$

$0.10x + 0.06(x - 3500) = 750$ Multiply both sides of the equation by 100.

$$10x + 6(x - 3500) = 75,000$$

$$10x + 6x - 21,000 = 75,000$$

$$16x = 96,000$$

$$\frac{16x}{16} = \frac{96,000}{16}$$

$$x = 6000$$

$$x - 3500 = 2500$$

$6000 was invested at 10% and $2500 was invested at 6%.

Distance, rate, and time problem:

A driver makes a trip in a rain storm in 6 h. In good weather she increases her speed by 10 mph and the trip only takes 5 h. Find the driver's speed during the rain.

	d	r	t
Rain	$6x$	x	6
No rain	$5(x + 10)$	$x + 10$	5

The distance traveled is the same.

$$6x = 5(x + 10)$$

$$6x = 5x + 50$$

$$x = 50$$

The driver drove 50 mph in the rain.

SECTION 2.7—LINEAR INEQUALITIES

KEY CONCEPTS:

A linear inequality in one variable, x, is any relationship in the form: $ax + b < 0$, $ax + b > 0$, $ax + b \leq 0$, or $ax + b \geq 0$, where $a \neq 0$.

The solution set to an inequality can be expressed as a graph or in set-builder notation or in interval notation.

When graphing an inequality or when writing interval notation a parenthesis, (or), is used to denote that an endpoint is *not included* in a solution set. A square bracket, [or], is used to show that an endpoint *is included* in a solution set. Parenthesis (or) are always used with $-\infty$ and ∞.

The inequality $a < x < b$ is used to show that x is greater than a and less than b. That is, x is *between* a and b.

Multiplying or dividing an inequality by a negative quantity requires the direction of the inequality sign to be reversed.

Test points can be used to verify that a solution to an inequality is correct. Substitute a test point from the proposed solution set into the original inequality. The test point should make the inequality true.

KEY TERMS:

addition property of inequality
division property of inequality
interval notation
linear inequality in one variable
multiplication property of inequality
set-builder notation
subtraction property of inequality
test point

EXAMPLES:

$3x - 1 < 11$ is a linear inequality.

Set-Builder Notation	Interval Notation	Graph
$\{x \mid x > a\}$	(a, ∞)	
$\{x \mid x \geq a\}$	$[a, \infty)$	
$\{x \mid x < a\}$	$(-\infty, a)$	
$\{x \mid x \leq a\}$	$(-\infty, a]$	
$\{x \mid a < x < b\}$	(a, b)	

Solve for x:

$$-2x + 6 \geq 14$$

$-2x + 6 - 6 \geq 14 - 6$ Subtract 6.

$-2x \geq 8$ Simplify.

$\dfrac{-2x}{-2} \leq \dfrac{8}{-2}$ Divide by -2. Reverse the inequality sign.

$$x \leq -4$$

Set-builder notation: $\{x \mid x \leq -4\}$

Graph:

Interval notation: $(-\infty, -4]$

Test point: $x = -5$

Original inequality: $-2x + 6 \geq 14$

Substitute:

$$-2(-5) + 6 \overset{?}{\geq} 14$$
$$10 + 6 \overset{?}{\geq} 14$$
$$16 \geq 14 \quad \text{True}$$

SECTION 2.8—CONNECTIONS TO GRAPHING: LINEAR EQUATIONS IN TWO VARIABLES

KEY CONCEPTS:

An equation written in the form $ax + by = c$ (where a and b are not both zero) is a linear equation in two variables.

A solution to a linear equation in x and y is an ordered pair (x, y) that makes the equation a true statement. The graph of the set of all solutions of a linear equation in two variables is a line in a rectangular coordinate system.

A linear equation can be graphed by finding at least two solutions and graphing the line through the points.

KEY TERM:

linear equation in two variables

EXAMPLES:

The equation $2x + y = 2$ is a linear equation in two variables.

Complete the table for the equation $2x + y = 2$:

x	y
0	
	4
3	

$$2x + y = 2 \qquad 2x + y = 2 \qquad 2x + y = 2$$
$$2(0) + y = 2 \qquad 2x + (4) = 2 \qquad 2(3) + y = 2$$
$$0 + y = 2 \qquad 2x = -2 \qquad 6 + y = 2$$
$$y = 2 \qquad x = -1 \qquad y = -4$$

x	y
0	2
−1	4
3	−4

Graph the line defined by the points in the table.

The ordered pairs are $(0, 2)$, $(-1, 4)$ and $(3, -4)$.

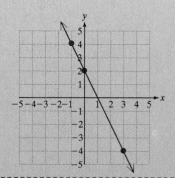

| chapter 2 | REVIEW EXERCISES |

Section 2.1

1. Label the following as either an expression or an equation:

   a. $3x + y = 10$
      Equation
   b. $9x + 10y - 2xy$
      Expression
   c. $4(x + 3) = 12$
      Equation
   d. $-5x = 7$
      Equation

2. Explain how to determine whether an equation is linear.   A linear equation can be written in the form $ax + b = 0$, $a \neq 0$.

3. Identify which equations are linear.

   a. $4x^2 + 8 = -10$
      Nonlinear
   b. $x + 18 = 72$
      Linear
   c. $-3 + 2y^2 = 0$
      Nonlinear
   d. $-4p - 5 = 6p$
      Linear

4. For the equation, $4y + 9 = -3$, determine if the given numbers are solutions.

   a. $y = 3$ No
   b. $y = 0$ No
   c. $y = -3$ Yes
   d. $y = -2$ No

For Exercises 5–12, solve the equation using the addition property, subtraction property, multiplication property, or division property of equality.

5. $a + 6 = -2$ $a = -8$

6. $6 = z - 9$ $z = 15$

7. $-\dfrac{3}{4} + k = \dfrac{9}{2}$ $k = \dfrac{21}{4}$ or $5\frac{1}{4}$

8. $0.1r = 7$ $r = 70$

9. $-5x = 21$ $x = -\dfrac{21}{5}$ or $-4\frac{1}{5}$

10. $\dfrac{t}{3} = -20$ $t = -60$

11. $-\dfrac{2}{5}k = \dfrac{4}{7}$ $k = -\dfrac{10}{7}$ or $-1\frac{3}{7}$

12. $-m = -27$ $m = 27$

13. The quotient of a number and negative six is equal to negative ten. Find the number.
    The number is 60.

14. The difference of a number and $-\frac{1}{8}$ is $\frac{5}{12}$. Find the number.   The number is $\dfrac{7}{24}$.

15. Four subtracted from a number is negative twelve. Find the number.   The number is -8.

16. Six subtracted from a number is negative eight. Find the number.   The number is -2.

Section 2.2

For Exercises 17–28, solve the equation.

17. $4d + 2 = 6$ $d = 1$

18. $5c - 6 = -9$ $c = -\dfrac{3}{5}$

19. $-7c = -3c - 9$ $c = \dfrac{9}{4}$ or $2\frac{1}{4}$

20. $-28 = 5w + 2$ $w = -6$

21. $\dfrac{b}{3} + 1 = 0$ $b = -3$

22. $\dfrac{2}{3}h - 5 = 7$ $h = 18$

23. $-3p + 7 = 5p + 1$ $p = \dfrac{3}{4}$

24. $4t - 6 = -12t + 16$ $t = \dfrac{11}{8}$

25. $4a - 9 = 3(a - 3)$ $a = 0$

26. $3(2c + 5) = -2(c - 8)$ $c = \dfrac{1}{8}$

27. $7b + 3(b - 1) + 16 = 2(b + 8)$ $b = \dfrac{3}{8}$

28. $2 + (17 - x) + 2(x - 1) = 4(x + 2) - 8$ $n = \dfrac{17}{3}$ or $5\frac{2}{3}$

29. Explain the difference between an equation that is a contradiction and an equation that is an identity.   A contradiction has no solution and an identity is true for all real numbers.

30. Label each equation as a conditional equation, a contradiction, or an identity.

    a. $x + 3 = 3 + x$
       Identity
    b. $3x - 19 = 2x + 1$
       Conditional equation
    c. $5x + 6 = 5x - 28$
       Contradiction
    d. $2x - 8 = 2(x - 4)$
       Identity
    e. $-8x - 9 = -8(x - 9)$
       Contradiction

Section 2.3

31. The sum of twice a number and four is forty. Find the number.   The number is 18.

32. The difference of a number and -5 is -11. Find the number.   The number is -16.

33. Twelve added to the sum of a number and two is forty-four. Find the number.   The number is 30.

34. Twenty added to the sum of a number and six is thirty-seven. Find the number.   The number is 11.

35. Three times a number is the same as the difference of twice the number and seven. Find the number.   The number is -7.

36. Eight less than five times a number is forty-eight less than the number. Find the number.   The number is -10.

37. The minimum salary for a Major League baseball player in 1985 was $60,000. This was twice the minimum salary in 1980. What was the minimum salary in 1980?   The minimum salary was $30,000 in 1980.

❖ See Additional Answers Appendix    Writing   Translating Expression   Geometry   Scientific Calculator   Video

38. The perimeter of a triangle is 78 in. The lengths of the sides are represented by three consecutive integers. Find the lengths of the sides of the triangle.   The sides are 25 in., 26 in., and 27 in.

39. The perimeter of a pentagon (a five-sided polygon) is 190 in. The five sides are represented by consecutive integers. Find the measures of the sides.   The sides are 36 in., 37 in., 38 in., 39 in., and 40 in.

Section 2.4

For Exercises 40–49, solve the equation by first clearing the fractions or decimals, if necessary.

40. $-\frac{1}{4}(2 - 3p) = \frac{3}{4}$ $p = \frac{5}{3}$ or $1\frac{2}{3}$

41. $\frac{1}{5}(-x + 3) = -\frac{1}{10}(2x - 1)$ No solution

42. $\frac{1}{4}y - \frac{3}{4} = \frac{1}{2}y + 1$ $y = -7$ 43. $17.3 - 2.7q = 10.55$ $q = 2.5$

44. $4.9z + 4.6 = 3.2z - 2.2$ $z = -4$

45. $5.74a + 9.28 = 2.24a - 5.42$ $a = -4.2$

46. $100 - (t - 6) = -(t - 1)$ No solution

47. $5t - (2t + 14) = 3t - 14$ All real numbers

48. $9 - 6(2z + 1) = -3(4z - 1)$ All real numbers

49. $3r + 2[3(r - 1) + 2] = 2(3r + 4)$ $r = \frac{10}{3}$, or $3\frac{1}{3}$

For Exercises 50–55, solve the problems involving percents.

50. What is 35% of 68?   23.8

51. What is 4% of 720?   28.8

52. 53.5 is what percent of 428?   12.5%

53. 68.4 is what percent of 72?   95%

54. 24 is 15% of what number?   160

55. 8.75 is 0.5% of what number?   1750

56. A novel originally selling at $29.99 is on sale for 12% off. What is the sale price of the book? Round to the nearest cent.   The sale price is $26.39.

57. What would be the total price (including tax) of a novel that sells for $26.39, if the sales tax rate is 7%?   The total price is $28.24.

58. In one week a salesperson receives $238 in commission. If this represents 14% of sales, how much did she sell?   She sold $1700.

59. Anna Tsao invested $3000 in an account paying 8% simple interest.
    a. How much interest will she earn in $3\frac{1}{2}$ years? $840
    b. What will her balance be at that time? $3840

60. The Super Bowl XXXI in 1997 was between the Green Bay Packers and New England Patriots. Green Bay ended the game with 35 points. This score was $1\frac{2}{3}$ times the number of points scored by New England. What was New England's score and who won?   New England scored 21 points. Green Bay won.

Section 2.5

For Exercises 61–68, solve for the indicated variable.

61. $C = K - 273$ for K $K = C + 273$

62. $K = C + 273$ for C $C = K - 273$

63. $P = 4s$ for s $s = \frac{P}{4}$ 64. $P = 3s$ for s $s = \frac{P}{3}$

65. $y = mx + b$ for x $x = \frac{y - b}{m}$ 66. $a + bx = c$ for x $x = \frac{c - a}{b}$

67. $2x + 5y = -2$ for y $y = \frac{-2 - 2x}{5}$

68. $-3x + y = 8$ for x $x = \frac{8 - y}{-3}$ or $-\frac{8 - y}{3}$

For Exercises 69–74, use the appropriate geometry formula to solve the problem.

69. Find the height of a right circular cone whose volume is 47.8 in.^{3} and whose radius is 3 in. Round to the nearest tenth of an inch.   The height is 5.1 in.

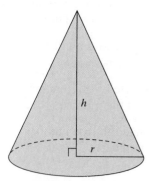

Figure for Exercise 69

70. Find the height of a parallelogram whose area is 42 m² and whose base is 6 m. *The height is 7 m.*

71. Pat is planning to extend the patio on the back of his house. He has drawn the shape shown in the figure, and he needs to know the width of the rectangular portion so that the total area will be approximately 85 ft². (Round the answer to the nearest foot.) *The width is 5 ft.*

Radius 4 ft

x

Length 12 ft

Figure for Exercise 71

72. The smallest angle of a triangle is 2° more than $\frac{1}{4}$ of the largest angle. The middle angle is 2° less than the largest angle. Find the measures of each angle. *The angles are 22°, 78°, 80°.*

73. One angle is 6° less than twice a second angle. If the two angles are complementary, what are their measures? *The angles are 32° and 58°.*

74. A rectangular window has width 1 ft less than its length. The perimeter is 18 ft. Find the length and the width of the window. *The length is 5 ft and the width is 4 ft.*

Section 2.6

75. Peggy invests some money in a savings account that pays 3.5% simple interest and buys some stock that yields 12%. Because of the volatility of the stock market, she invests half as much money in the stock as she invests in her savings account. If her total interest for the first year is $285.00, how much did Peggy invest in the savings account and how much did she invest in the stock? *Peggy invested $3000 in savings and $1500 in stocks.*

76. At a day care center, Suzanne wants to buy a treat for each of the 24 children. Ice cream on a stick costs $1.50 each and Popsicles cost $1.00 each. If Suzanne spends $29.00, how many of each type of treat does she buy? *She buys 10 ice cream on a stick and 14 Popsicles.*

77. A biker and jogger leave home at the same time, one traveling east and the other traveling west. If the biker maintains an average rate of 9 mph and the jogger maintains an average rate of 5 mph, how long will it take for them to be $10\frac{1}{2}$ miles apart? *It will take $\frac{3}{4}$ h.*

78. The O'Neill family is going to have a picnic with the Miller family. The two families live 35 miles apart. The O'Neill's travel in a small, four-cylinder car and drive an average of 5 mph slower than the Miller family. If it takes the families 20 min ($\frac{1}{3}$ h) to meet at a point between them, how fast does each family travel? *The Millers travel 55 mph, and the O'Neills travel 50 mph.*

For Exercises 79–80, solve the proportion.

79. $\frac{5}{3} = \frac{m}{2}$ $m = \frac{10}{3}$ or $3\frac{1}{3}$ 80. $\frac{7}{10} = \frac{n}{25}$ $n = 17.5$

81. A bag of mulch will cover about 16 ft² of flowerbeds. How many bags would be needed to cover 80 ft²? *Five bags will be needed.*

82. If a prescription calls for two pills every 6 h, how many pills will be needed for 5 days (120 h)? *Forty pills will be needed.*

Section 2.7

83. Graph the inequalities and write the set in interval notation.

a. $\{x \mid x > -2\}$ ❖   b. $\left\{x \mid x \le \frac{1}{2}\right\}$ ❖

c. $\{x \mid -1 < x \le 4\}$ ❖

84. A landscaper buys potted geraniums from a nursery at a price of $5 per plant. However, for large orders, the price per plant is discounted by a percentage off the original price. Let x represent the number of potted plants ordered. The corresponding discount is given in the following table.

Number of Plants	Discount (%)
$x \le 99$	0
$100 \le x \le 199$	2
$200 \le x \le 299$	4
$x \ge 300$	6

Table for Exercise 84

a. Find the cost to purchase 130 plants. *$637*

b. Which costs more, 300 plants or 295 plants? Explain your answer. *Three hundred plants cost $1410 and 295 plants cost $1416. 300 plants give a higher discount.*

For Exercises 85–92, solve the inequality. Graph the solution set and express the answer in interval notation.

85. $c + 6 < 23$ ❖

86. $3w - 4 > -5$ ❖

87. $-2x - 7 \geq 5$ ❖

88. $5(y + 2) \leq -4$ ❖

89. $-\dfrac{3}{7}a \leq -21$ ❖

90. $1.3 > 0.4t - 12.5$ ❖

91. $4k + 23 < 7k - 31$ ❖

92. $\dfrac{6}{5}h - \dfrac{1}{5} \leq \dfrac{3}{10} + h$ ❖

93. The summer average rainfall for Bermuda for June, July, and August is 5.3 in. per month. If Bermuda receives 6.3 in. of rain in June and 7.1 in. in July, how much rain is required in August to exceed the 3-month summer average? More than 2.5 in. is required.

94. Reggie sells hot dogs at a ballpark. He has a fixed cost of $33 to use the concession stand at the park. In addition, the cost for each hot dog is $0.40. If x represents the number of hot dogs sold, then the total cost is given by

$$\text{Cost} = 33 + 0.40x$$

If Reggie sells each hot dog for $1.50, then his revenue (the amount he brings in) for selling x hot dogs is given by

$$\text{Revenue} = 1.50x$$

a. Write an inequality that expresses the number of hot dogs, x, that Reggie must sell to make a profit. Profit is realized when the revenue is greater than the cost (revenue > cost). $1.50x > 33 + 0.4x$

b. Solve the inequality in part (a). $x > 30$; a profit is realized if more than 30 hot dogs are sold.

Section 2.8

For Exercises 95–96, complete the table. Sketch the line through the points to represent all solutions to the equation.

95. $x + y = 9$ ❖

96. $x - 2y = 6$ ❖

x	y
5	4
10	−1
0	9

x	y
6	0
8	1
−2	−4

97. In 1999, Hurricane Irene dumped rain on West Palm Beach, Florida at a rate of 1.5 in. per hour. The total amount of rain, y (in inches), after x hours is given by the equation

$$y = 1.5x \quad (x \geq 0)$$

If the total amount of rain that fell was 16.2 in., use the equation to determine how long it rained. It rained 10.8 h.

98. When traveling to a foreign country, it is necessary to convert your money to the currency of the country you are visiting. The exchange rates for various currencies around the world change on a daily basis. In November 1999, $1 was equal to £0.6098 (English pounds).

If x represents the amount of money in U.S. dollars, then the corresponding value in English pounds, y, is given by the equation

$$y = 0.6098x \quad (x \geq 0)$$

a. If a tourist has $135 U.S., use the equation to determine the corresponding value in English pounds. ❖

b. Use the equation to complete the table. Round to the nearest hundredth.

$US Dollars x	£ (English pounds) y
10	6.10
25	15.25
100	60.98
500	304.90

Table for Exercise 98

c. Write the table values from part (b) as ordered pairs. Interpret each ordered pair. ❖

d. A business traveler stays in London for a week. The hotel charge is £436. How many $US does this represent? ❖

99. At a certain gas station, gasoline costs $1.54 per gallon. If x gallons of gasoline are purchased, the total cost, y, (in dollars) is given by the equation

$$y = 1.54x \quad (x \geq 0)$$

a. Use the equation to find the cost of purchasing 12.4 gallons of gas. ❖

b. Use the equation to complete the table.

Number of Gallons x	Cost ($) y
5.0	7.70
7.5	11.55
10.0	15.40
12.5	19.25
15.0	23.10

Table for Exercise 99

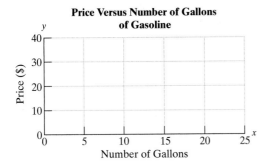

Price Versus Number of Gallons of Gasoline

Figure for Exercise 99

c. Write the table values from part (b) as ordered pairs and graph the ordered pairs on a rectangular coordinate system. ❖

d. How many gallons did a person purchase if the cost of the gasoline is $10.01? ❖

e. How many gallons did a person purchase if the cost of the gasoline is $26.18? ❖

100. A phone company charges 6 cents per minute ($0.06/min) of long distance time. If x represents the length of a call in minutes, then the cost, y, (in dollars) is given by the equation

$$y = 0.06x \quad (x \geq 0)$$

a. Use the equation to find the cost of 22 min of long distance. ❖

b. Use the equation to complete the table.

Number of Minutes of Long Distance x	Cost ($) y
5	0.30
15	0.90
25	1.50
35	2.10
45	2.70

Table for Exercise 100

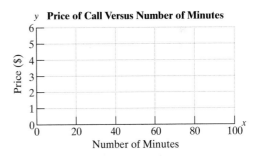

Price of Call Versus Number of Minutes

Figure for Exercise 100

c. Write the table values from part (b) as ordered pairs and graph the ordered pairs on a rectangular coordinate system. ❖

d. How many minutes did a person talk if the long distance charge is $2.22? ❖

e. How many minutes did a person talk if the long distance charge is $4.68? ❖

chapter 2   TEST

For Exercises 1–10, solve the equation.

1. $t + 3 = -13$ $t = -16$

2. $8 = p - 4$ $p = 12$

3. $\dfrac{t}{8} = -\dfrac{2}{9}$ $t = -\dfrac{16}{9}$

4. $-3x + 5 = -2$ $x = \dfrac{7}{3}$

5. $2(p - 4) = p + 7$ $p = 15$

6. $2 + d = 2 - 3(d - 5) - 2$ $d = \dfrac{13}{4}$

7. $\dfrac{1}{2} - q = -q + \dfrac{1}{2}$ All real numbers

8. $3h + 1 = 3(h + 1)$ No solution

9. $-\dfrac{3}{8}p + 2 = \dfrac{3}{16}(4 - 2p)$ No solution

10. $0.5c - 1.9 = 2.8 + 0.6c$ $c = -47$

11. Solve the equation for y: $3x + y = -4$
    $y = -3x - 4$

12. Solve the equation for r: $C = 2\pi r$ $r = \dfrac{C}{2\pi}$

13. The perimeter of a pentagon (a five-sided polygon) is 315 in. The five sides are represented by consecutive integers. Find the measures of the sides.   The sides are 61 in., 62 in., 63 in., 64 in., and 65 in.

14. In the 1997–1998 season, a couple purchased two NHL hockey tickets and two NBA basketball tickets for $153.92. A hockey ticket cost $4.32 more than a basketball ticket. What were the prices of the individual tickets?   The basketball tickets were $36.32, and the hockey tickets were $40.64.

15. The total bill for a pair of basketball shoes (including sales tax) is $87.74. If the tax rate is 7%, find the cost of the shoes before tax.
    The cost was $82.00.

16. A couple has $7200 to invest between two accounts, one that earns 8% simple interest and the other that earns 10% simple interest. If they earn $656 in interest after 1 year, how much was invested in each account?   They invested $3200 in the 8% account and $4000 in the 10% account.

17. Two angles are complementary. One angle is 26° more than the other angle. What are the measures of the angles?   The angles are 32° and 58°.

18. Two bikers leave the Harley-Davidson shop, traveling on the same road and in the same direction. One biker drives 5 mph faster than the other. The faster biker stops after $1\frac{1}{2}$ h but the slower biker stops after 1 h. If the two bikers are $32\frac{1}{2}$ miles apart, what was the speed of each biker?   Their speeds were 50 mph and 55 mph.

19. There are 3 grams of fat in one serving of oatmeal (0.5 cup). How many grams of fats are in 0.75 cup of oatmeal?   There are 4.5 g of fat.

20. Graph the inequalities and write the set in interval notation.

    a. $\{x \mid x < 0\}$ ❖

    b. $\{x \mid -2 \le x < 5\}$ ❖

For Exercises 21–22, solve the inequality. Graph the solution and write the solution set in interval notation.

21. $5x + 14 > -2x$ ❖

22. $2(3 - x) \ge 14$ ❖

23. The average winter snowfall for Syracuse, New York, for December, January, and February is 27.5 in. per month. If Syracuse receives 24 in. of snow in December and 32 in. in January, how much snow is required in February to exceed the 3-month average?   More than 26.5 in. is required.

24. If x represents an adult's age, then the person's maximum recommended heart rate, y, during exercise is approximated by the equation

    $$y = 220 - x \quad (x \ge 18)$$

    a. Use the equation to find the maximum recommended heart rate for a person who is 18 years old.   ❖

    b. Use the equation to complete the table.

Age x (years)	Maximum Heart Rate, y (beats per minute)
20	200
30	190
40	180
50	170
60	160

Table for Exercise 24

❖ See Additional Answers Appendix   Writing   Translating Expression   Geometry   Scientific Calculator   Video

Maximum Heart Rate Versus Age

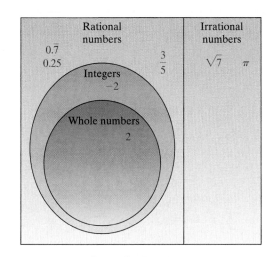

Figure for Exercise 24

c.   Write the table values from part (b) as ordered pairs, and graph the ordered pairs on a rectangular coordinate system. Interpret the meaning of each ordered pair in the context of the problem.   ❖

d.   How old is a person whose maximum recommended heart rate is 192 beats per minute? ❖

CUMULATIVE REVIEW EXERCISES, CHAPTERS 1–2

For Exercises 1–8, perform the indicated operations.

1. $-12 + 5 - (-3)$ -4

2. $\left| -\dfrac{1}{5} + \dfrac{7}{10} \right|$ $\dfrac{1}{2}$

3. $19.8 \div (-7.2)$ -2.75

4. $5 - 2[3 - (4 - 7)]$ -7

5. $-\dfrac{2}{3} + \left(\dfrac{1}{2} \right)^2$ $-\dfrac{5}{12}$

6. $-3^2 + (-5)^2$ 16

7. $\sqrt{5 - (-20)} - 3^2$ 4

8. $\dfrac{-3.5 + 2.8 \div 0.4}{[-1 + 2(-8)] + 4(3 - 5)^2}$ -3.5

9.   Place the numbers in the appropriate location in the diagram:

$$\left\{ \sqrt{7}, -2, 2, \dfrac{3}{5}, \pi, 0.\overline{7}, 0.25 \right\}$$

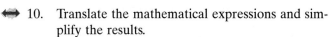 10.   Translate the mathematical expressions and simplify the results.

a.   One-third of three-fourths. $\dfrac{1}{3} \cdot \dfrac{3}{4} = \dfrac{1}{4}$

b.   The square root of the difference of five squared and nine. $\sqrt{5^2 - 9} = 4$

11.   Plot the points on a rectangular coordinate system. State the quadrant in which the point lies or whether the point lies on a coordinate axis.   ❖

a. $(-2, 5)$

b. $\left(\dfrac{7}{3}, -\dfrac{11}{4} \right)$

c. $(-\sqrt{9}, -\sqrt{25})$

d. $(4.2, 0)$

e. $(0, 0)$

12.   List the terms of the expression: $-7x^2y + 4xy - 6$ $-7x^2y, 4xy, -6$

13.   Identify the coefficient of the term: ab 1

14.   Simplify: $-4[2x - 3(x + 4)] + 5(x - 7)$ $9x + 13$

15.   Solve for x: $-2.5x - 5.2 = 12.8$ $x = -7.2$

16.   Solve for p: $-5(p - 3) + 2p = 3(5 - p)$
All real numbers

17.   The sum of two consecutive odd integers is 156. Find the integers.   The numbers are 77 and 79.

18.   The total bill for a man's three-piece suit (including sales tax) is $374.50. If the tax rate is 7%, find the cost of the suit before tax.
The cost before tax was $350.00.

Figure for Exercise 9

19. The area of a triangle is 41 cm². Find the height of the triangle if the base is 12 cm.
The height is $6\frac{5}{6}$ cm.

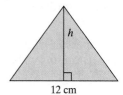

Figure for Exercise 19

20. Rachael invests twice as much money in an account earning 9% simple interest as she does in an account earning 4% simple interest. If the total interest is $792 after 1 year, find the amount originally invested in each account. Rachael invests $3600 in the 4% account and $7200 in the 9% account.

21. An athlete bicycles an average of 12 mph faster than she runs. She runs 0.8 h and bicycles for 1.2 h for a total distance of 30.4 miles. Find her average rate running and her average rate bicycling. Her average running rate is 8 mph and her average biking rate is 20 mph.

22. Solve the inequality. Graph the solution set on a number line and express the solution in interval notation: $-2x - 3(x + 1) < 7$. ❖

23. If there are 21 candies in a $1\frac{1}{2}$-lb box of chocolates, how many candies are in a 2-lb box?
There are 28 candies in the box.

24. The monthly number of admissions for a hospice program is given in the following table.

    a. Plot the ordered pairs on a rectangular coordinate system. ❖

b. Between which two consecutive months did the largest increase in admissions occur? ❖

c. What is the overall percentage increase between the first month and the last month? Round to the nearest percent. ❖

Month x	Number of Admissions y
1	15
2	18
3	19
4	25
5	22
6	27
7	30
8	32
9	33
10	33
11	35
12	37

Table for Exercise 24

25. During a tropical storm, Galveston, Texas received $\frac{3}{4}$ in. of rain per hour. The amount of rain, y, (in inches) is given by the equation $y = \frac{3}{4}x$, where x is the time in hours. If the total amount of rain that fell was 6 in., use the equation to determine how long it rained. It rained 8 h.

POLYNOMIALS AND PROPERTIES OF EXPONENTS

The Pythagorean theorem states that the sum of the squares of the legs of a right triangle equals the square of the hypotenuse:

$$a^2 + b^2 = c^2$$

b (leg) c (hypotenuse) a (leg)

The applications of the Pythagorean theorem extend to many different fields such as construction, navigation, engineering, and physics.

For example, suppose two cables must be used to support an antenna tower at points 30 ft and 75 ft up the tower. Each cable is fastened at a point 40 ft from the base of the tower. The Pythagorean theorem can be used to show that the lengths of the cables are 50 ft and 85 ft.

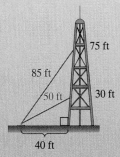

75 ft
85 ft
50 ft 30 ft
40 ft

$$(40)^2 + (30)^2 = (50)^2 \qquad (40)^2 + (75)^2 = (85)^2$$

$$1600 + 900 = 2500 \qquad 1600 + 5625 = 7225$$

$$2500 = 2500 ✔ \qquad 7225 = 7225 ✔$$

A proof of the Pythagorean theorem is given at pythagproof. For more information visit

www.mhhe.com/miller_oneill

section

3.1 EXPONENTS: MULTIPLYING AND DIVIDING COMMON BASES

1. Review of Exponential Notation

Recall that an **exponent** is used to show repeated multiplication of the **base**.

Definition of b^n

Let b represent any real number and n represent a positive integer. Then,

$$b^n = \underbrace{b \cdot b \cdot b \cdot b \ldots \cdot b}_{n \text{ factors of } b}$$

---

example 1 **Evaluating Expressions with Exponents**

For each expression, identify the exponent and base. Then evaluate the expression.

a. 6^2 b. $\left(-\dfrac{1}{2}\right)^3$ c. 0.8^4

Solution:

Expression	Base	Exponent	Result
a. 6^2	6	2	$(6)(6) = 36$
b. $\left(-\dfrac{1}{2}\right)^3$	$-\dfrac{1}{2}$	3	$\left(-\dfrac{1}{2}\right)\left(-\dfrac{1}{2}\right)\left(-\dfrac{1}{2}\right) = -\dfrac{1}{8}$
c. 0.8^4	0.8	4	$(0.8)(0.8)(0.8)(0.8) = 0.4096$

Note that if no exponent is explicitly written for an expression, then the expression has an implied exponent of 1. For example,

$$x = x^1$$
$$y = y^1$$
$$5 = 5^1$$

2. Evaluating Expressions with Exponents

Recall from Section 1.2 that particular care must be taken when evaluating exponential expressions involving negative numbers. An exponential expression with a negative base is written with parentheses around the base, such as $(-3)^2$.

To evaluate $(-3)^2$, we have: $(-3)^2 = (-3)(-3) = 9$

If no parentheses are present, the expression -3^2, is the *opposite* of 3^2, or equivalently, $-1 \cdot 3^2$.

Hence: $-3^2 = -1(3^2) = -1(3)(3) = -9$

example 2

Evaluating Expressions with Exponents

Evaluate each expression.

a. -5^4 b. $(-5)^4$ c. $(-2)^3$ d. -2^3

Solution:

a. -5^4
    $= -1 \cdot 5^4$
    $= -1 \cdot 5 \cdot 5 \cdot 5 \cdot 5$
    $= -625$

b. $(-5)^4$
    $= (-5)(-5)(-5)(-5)$
    $= 625$

c. $(-2)^3$
    $= (-2)(-2)(-2)$
    $= -8$

d. -2^3
    $= -1 \cdot 2^3$
    $= -1 \cdot 2 \cdot 2 \cdot 2$
    $= -8$

---

example 3

Evaluating Expressions with Exponents

Evaluate each expression for $a = 2$ and $b = -3$

a. $5a^2$ b. $(5a)^2$ c. $5ab^2$ d. $(b + a)^2$

Classroom Activity 3.1A

Solution:

a. $5a^2$

    $= 5(\ )^2$ Use parentheses to substitute a number for a variable.

    $= 5(2)^2$ Substitute $a = 2$.

    $= 5(4)$ Simplify.

    $= 20$

b. $(5a)^2$

    $= [5(\ )]^2$ Use parentheses to substitute a number for a variable.

    $= [5(2)]^2$ Substitute $a = 2$.

    $= (10)^2$ Simplify inside the parentheses first.

    $= 100$

c. $5ab^2$

    $= 5(2)(-3)^2$ Substitute $a = 2, b = -3$.

    $= 5(2)(9)$ Simplify.

    $= 90$ Multiply.

Tip: In the expression $5ab^2$, the exponent, 2, applies only to the variable b. The constant, 5, and the variable, a, both have an implied exponent of 1.

Avoiding Mistakes

Be sure to follow the order of operations. In Example 3(d), it would not be correct to square the terms within the parentheses before adding.

d. $(b + a)^2$

$= [(-3) + (2)]^2$ Substitute $b = -3$ and $a = 2$.

$= (-1)^2$ Simplify within the parentheses first.

$= 1$

3. Multiplying and Dividing Common Bases

In this section, we investigate the effect of multiplying or dividing two quantities with the same base. For example, consider the expressions: x^5x^2 and $\frac{x^5}{x^2}$. Simplifying each expression, we have:

$$x^5x^2 = (x \cdot x \cdot x \cdot x \cdot x)(x \cdot x) = \overbrace{x \cdot x \cdot x \cdot x \cdot x \cdot x \cdot x}^{7 \text{ factors of } x} = x^7$$

$$\frac{x^5}{x^2} = \frac{x \cdot x \cdot x \cdot \cancel{x} \cdot \cancel{x}}{\cancel{x} \cdot \cancel{x}} = \frac{x \cdot x \cdot x}{1} = x^3$$

These examples suggest that to multiply two quantities with the same base, we add the exponents. To divide two quantities with the same base, we subtract the exponent in the denominator from the exponent in the numerator. These rules are stated formally as Properties 1 and 2 of exponents.

Multiplication of Like Bases

Assume that $a \neq 0$ is a real number and that m and n represent positive integers. Then,

$$\text{Property 1:} \quad a^m a^n = a^{m+n}$$

Division of Like Bases

Assume that $a \neq 0$ is a real number and that m and n represent positive integers such that $m > n$. Then,

$$\text{Property 2:} \quad \frac{a^m}{a^n} = a^{m-n}$$

4. Simplifying Expressions with Exponents

example 4

Simplifying Expressions with Exponents

Simplify the expressions.

a. w^3w^4 b. $2^3 2^4$ c. $\dfrac{t^{14}}{t^{12}}$ d. $\dfrac{5^{14}}{5^{12}}$ e. $\dfrac{z^4 z^5}{z^3}$ f. $\dfrac{10^7}{10^2 \cdot 10}$

Classroom Activity 3.1B

Solution:

a. $w^3 w^4$

$= w^{3+4}$ Add the exponents (the base is unchanged).

$= w^7$

 **Avoiding Mistakes**

When we use Property 1 to add exponents, the base does not change. In Example 4(b), we have $2^3 2^4 = 2^7$.

b. $2^3 2^4$

$= 2^{3+4}$ Add the exponents (the base is unchanged).

$= 2^7$ or 128

c. $\dfrac{t^{14}}{t^{12}}$

$= t^{14-12}$ Subtract the exponents (the base is unchanged).

$= t^2$

d. $\dfrac{5^{14}}{5^{12}}$

$= 5^{14-12}$ Subtract the exponents (the base is unchanged).

$= 5^2$ or 25

e. $\dfrac{z^4 z^5}{z^3}$

$= \dfrac{z^{4+5}}{z^3}$ Add the exponents in the numerator (the base is unchanged).

$= \dfrac{z^9}{z^3}$

$= z^{9-3}$ Subtract the exponents (the base is unchanged).

$= z^6$

f. $\dfrac{10^7}{10^2 \cdot 10}$

$= \dfrac{10^7}{10^2 \cdot 10^1}$ Note that 10 is equivalent to 10^1.

$= \dfrac{10^7}{10^{2+1}}$ Add the exponents in the denominator (the base is unchanged).

$= \dfrac{10^7}{10^3}$

$= 10^{7-3}$ Subtract the exponents (the base is unchanged).

$= 10^4$ or 10,000   Simplify.

example 5

Simplifying Expressions with Exponents

Use the commutative and associative properties of real numbers and the properties of exponents to simplify the expressions.

a. $(3p^2q^4)(2pq^5)$

b. $\dfrac{16w^9z^3}{3w^8z}$

Classroom Activity 3.1C

Solution:

a. $(3p^2q^4)(2pq^5)$

$= (3 \cdot 2)(p^2p)(q^4q^5)$ Apply the associative and commutative properties of multiplication to group coefficients and like bases.

$= (3 \cdot 2)p^{2+1}q^{4+5}$ Add the exponents when multiplying like bases.

$= 6p^3q^9$ Simplify.

b. $\dfrac{16w^9z^3}{3w^8z}$

$= \left(\dfrac{16}{3}\right)\left(\dfrac{w^9}{w^8}\right)\left(\dfrac{z^3}{z}\right)$ Group like coefficients and factors.

$= \left(\dfrac{16}{3}\right)w^{9-8}z^{3-1}$ Subtract the exponents when dividing like bases.

$= \left(\dfrac{16}{3}\right)wz^2$ or $\dfrac{16wz^2}{3}$ Simplify.

5. Applications of Exponents

Recall that simple interest on an investment or loan is computed by the formula $I = Prt$, where P is the amount of principal, r is the interest rate, and t is the time in years. Simple interest is based only on the original principal. However, in most day-to-day applications, the interest computed on money invested or borrowed is compound interest. **Compound interest** is computed on the original principal and on the interest already accrued.

Suppose $1000 is invested at 8% interest for 3 years. Compare the total amount in the account if the money earns simple interest versus if the interest is compounded annually.

Simple Interest

The simple interest earned is given by $I = Prt$

$= (1000)(0.08)(3)$

$= \$240$

Thus, the total amount in the account after 3 years is $1240.

Compound Interest (annual)

To compute annual compound interest over a period of three years, compute the interest earned in the first, second, and third years, separately. Then add the results.

Table 3-1 shows that the interest earned the first year is based on the original principal only. The interest earned the second year is based on the original principal and on the first year interest. The interest earned the third year is based on the original principal and on the interest earned the first and second years.

Table 3-1		
Year	**Interest Earned $I = Prt$**	**Total Amount in the Account**
First year	$I = (\$1000)(0.08)(1) = \$80$	$\$1000 + \$80 = \$1080$
Second year	$I = (\$1080)(0.08)(1) = \$86.40$	$\$1080 + \$86.40 = \$1166.40$
Third year	$I = (\$1166.40)(0.08)(1) = \$93.31$	$\$1166.40 + \$93.31 = \$1259.71$

The difference in the account balance for interest compounded annually versus for simple interest is $\$1259.71 - \$1240 = \$19.71$.

The total amount, A, in an account earning compound interest may be computed quickly using the following formula:

$$A = P\left(1 + \frac{r}{n}\right)^{nt}$$

where P is the amount of principal, r is the annual interest rate (expressed in decimal form), n is the number of times interest is compounded per year, and, t is the number of years.

For example, for $1000 invested at 8% interest compounded annually for 3 years, we have

$P = 1000$

$r = 0.08$

$n = 1$ Annual interest is compounded once a year.

$t = 3$

$$A = P\left(1 + \frac{r}{n}\right)^{nt}$$

$$A = 1000\left(1 + \frac{0.08}{1}\right)^{(1)(3)}$$
$$= 1000(1.08)^3$$
$$= 1000(1.259712)$$
$$= 1259.712$$

Rounding to the nearest cent, we have $A = \$1259.71$, as expected.

Because it is tedious to compute the interest after each compound period, the formula for compounding interest streamlines the process. Using this formula, note that interest may be compounded

Annually	(1 time per year, $n = 1$)
Semiannually	(2 times per year, $n = 2$)
Quarterly	(4 times per year, $n = 4$)
Monthly	(12 times per year, $n = 12$)
Daily	(365 times per year, $n = 365$)

example 6   **Using Exponents in an Application**

Find the amount in an account after 2 years if the initial investment is $7000, invested at 9% interest compounded quarterly (four times per year).

Classroom Activity 3.1D

Solution:

Identify the values for each variable.

$$P = 7000$$

$$r = 0.09$$ Note that the decimal form of a percent is used for calculations.

$$n = 4$$

$$t = 2$$

$$A = P\left(1 + \frac{r}{n}\right)^{nt}$$

$$= 7000\left(1 + \frac{0.09}{4}\right)^{(4)(2)}$$ Substitute.

$$= 7000(1 + 0.0225)^8$$ Simplify inside the parentheses.

$$= 7000(1.0225)^8$$

$$= 7000(1.194831142)$$ Approximate $(1.0225)^8$.

$$= 8363.82$$ Multiply (round to the nearest cent).

The amount in the account after 2 years is $8363.82.

Calculator Connections

In Example 6, it was necessary to evaluate the expression $(1.0225)^8$. Recall that the $\boxed{\wedge}$ or $\boxed{y^x}$ key may be used to enter expressions with exponents.

Scientific Calculator

Enter: $\boxed{1}\ \boxed{.}\ \boxed{0}\ \boxed{2}\ \boxed{2}\ \boxed{5}\ \boxed{y^x}\ \boxed{8}\ \boxed{=}$ **Result:** 1.194831142

Graphing Calculator

```
1.0225^8
        1.194831142
```

Calculator Exercises

Use a calculator to evaluate the expressions.

1. $(1.06)^5$ ❖

2. $(1.02)^{40}$ ❖

3. $5000(1.06)^5$ ❖

4. $2000(1.02)^{40}$ ❖

5. $3000\left(1 + \dfrac{0.06}{12}\right)^{(12)(2)}$ ❖

6. $1000\left(1 + \dfrac{0.05}{4}\right)^{(4)(3)}$ ❖

---

section 3.1 PRACTICE EXERCISES

For this exercise set, assume all variables represent non-zero real numbers.

For Exercises 1–8, identify the base and the exponent.

1. r^4 Base: r;
exponent: 4

2. c^3 Base: c;
exponent: 3

3. 5^2 Base: 5;
exponent: 2

4. 3^5 Base: 3;
exponent: 5

5. $(-4)^8$ Base:
-4; exponent: 8

6. $(-1)^4$ Base:
-1; exponent: 4

7. x Base: x;
exponent: 1

8. q Base: q;
exponent: 1

9. What base corresponds to the exponent 5 in the expression $x^3y^5z^2$? y

10. What base corresponds to the exponent 2 in the expression w^3v^2? v

11. What base corresponds to the exponent 6 in the expression $4x^6$? x

12. What base corresponds to the exponent 3 in the expression $2y^3$? y

13. Evaluate the two expressions and compare the answers. Do the expressions have the same value? Why or why not? -5^2 and $(-5)^2$
No; $-5^2 = -25$ and $(-5)^2 = 25$

14. Evaluate the two expressions and compare the answers. Do the expressions have the same value? Why or why not? -3^4 and $(-3)^4$
No; $-3^4 = -81$ and $(-3)^4 = 81$

15. Evaluate the two expressions and compare the answers. Do the expressions have the same value? Why or why not? -2^5 and $(-2)^5$
Yes; $-2^5 = -32$ and $(-2)^5 = -32$

16. Evaluate the two expressions and compare the answers. Do the expressions have the same value? Why or why not? -5^3 and $(-5)^3$
Yes; $-5^3 = -125$ and $(-5)^3 = -125$

17. Evaluate the two expressions and compare the answers: $(\frac{1}{2})^3$ and $\frac{1}{2^3}$. $\left(\dfrac{1}{2}\right)^3 = \dfrac{1}{8}$ and $\dfrac{1}{2^3} = \dfrac{1}{8}$

18. Evaluate the two expressions and compare the answers: $(\frac{1}{5})^2$ and $\frac{1}{5^2}$. $\left(\dfrac{1}{5}\right)^2 = \dfrac{1}{25}$ and $\dfrac{1}{5^2} = \dfrac{1}{25}$

19. Evaluate the two expressions and compare the answers: $(\frac{3}{10})^2$ and $(0.3)^2$. $\left(\dfrac{3}{10}\right)^2 = \dfrac{9}{100}$ and $(0.3)^2 = 0.09$

20. Evaluate the two expressions and compare the answers: $(\frac{7}{10})^3$ and $(0.7)^3$ $\left(\dfrac{7}{10}\right)^3 = \dfrac{343}{1000}$ and $(0.7)^3 = 0.343$

For Exercises 21–26, use the geometry formulas found in Section R.3.

21. Find the area of the pizza shown in the figure. Round to the nearest square inch.   201 in.²

Figure for Exercise 21

22. Find the area of a circular pool 50 ft in diameter. Round to the nearest square foot.   1963 ft²

23. Find the volume of the sphere shown in the figure. Round to the nearest cubic centimeter.   113 cm³

$r = 3$ cm

Figure for Exercise 23

24. Find the volume of a spherical balloon that is 10 in. in diameter. Round to the nearest cubic inch.   524 in.³

25. The employees at a craft shop make square napkins out of decorative holiday material. To make one napkin, a square piece of material 24 in. on a side is required. How many square inches of material are required to make 60 napkins? How many square feet is this if 1 ft^2 = 144 in.^{2}?
    34,560 in.^{2}, or 240 ft^2

26. A construction company must pave a square parking lot. Find the area if the lot is 120 ft on a side. What is the area of the lot in square yards if 1 yd^2 = 9 ft^2?   14,400 ft^2, or 1600 yd^2

The amount, A, in an account earning compound interest is dependent on the following variables: the initial principal, P; the annual interest rate, r; the number of times interest is compounded per year, n; and the length of time, t, in years. The amount in the account is given by:

$$A = P\left(1 + \frac{r}{n}\right)^{nt}$$

Use this formula for Exercises 27–30.

27. Find the amount in an account after 2 years if the initial investment is $5000, invested at 7% interest compounded quarterly (four times per year). $5744.41

28. Find the amount in an account after 5 years if the initial investment is $2000, invested at 4% interest compounded semiannually (two times per year). $2437.99

29. Find the amount in an account after 3 years if the initial investment is $4000, invested at 6% interest compounded semiannually (two times per year). $4776.21

30. Find the amount in an account after 4 years if the initial investment is $10,000, invested at 5% interest compounded quarterly (four times per year). $12,198.90

31. Expand the following expressions first. Then simplify using exponents.
    a. $x^4 \cdot x^3$ b. $5^4 \cdot 5^3$
    $(x \cdot x \cdot x \cdot x)(x \cdot x \cdot x) = x^7$ $(5 \cdot 5 \cdot 5 \cdot 5)(5 \cdot 5 \cdot 5) = 5^7$

32. Expand the following expressions first. Then simplify using exponents.
    a. $y^2 \cdot y^4$ b. $3^2 \cdot 3^4$
    $(y \cdot y)(y \cdot y \cdot y \cdot y) = y^6$ $(3 \cdot 3)(3 \cdot 3 \cdot 3 \cdot 3) = 3^6$

For Exercises 33–42, simplify the expressions. Write the answers in exponent form.

33. $z^5 z^3$ z^8 34. $w^4 w^7$ w^{11}

35. $a \cdot a^8$ a^9 36. $p^4 p$ p^5

37. $4^5 \cdot 4^9$ 4^{14} 38. $6^7 \cdot 6^5$ 6^{12}

39. $9^4 \cdot 9$ 9^5 40. $12 \cdot 12^6$ 12^7

41. $c^5 c^2 c^7$ c^{14} 42. $b^7 b^2 b^8$ b^{17}

43. Expand the following expressions. Then reduce.
    a. $\dfrac{p^8}{p^3}$ $\dfrac{p \cdot p \cdot p \cdot p \cdot p \cdot p \cdot p \cdot p}{p \cdot p \cdot p}$ $= p^5$ b. $\dfrac{8^8}{8^3}$ $\dfrac{8 \cdot 8 \cdot 8 \cdot 8 \cdot 8 \cdot 8 \cdot 8 \cdot 8}{8 \cdot 8 \cdot 8}$ $= 8^5$

44. Expand the following expressions. Then reduce.
    a. $\dfrac{w^5}{w^2}$ $\dfrac{w \cdot w \cdot w \cdot w \cdot w}{w \cdot w}$ $= w^3$ b. $\dfrac{4^5}{4^2}$ $\dfrac{4 \cdot 4 \cdot 4 \cdot 4 \cdot 4}{4 \cdot 4}$ $= 4^3$

For Exercises 45–54, simplify the expressions. Write the answers in exponent form.

45. $\dfrac{x^8}{x^6}$ x^2 46. $\dfrac{z^5}{z^4}$ z

47. $\dfrac{a^{10}}{a}$ a^9 48. $\dfrac{b^{12}}{b}$ b^{11}

49. $\dfrac{7^{13}}{7^6}$ 7^7 50. $\dfrac{2^6}{2^4}$ 2^2

51. $\dfrac{5^8}{5}$ 5^7 52. $\dfrac{3^5}{3}$ 3^4

53. $\dfrac{y^{13}}{y^{12}}$ y 54. $\dfrac{w^7}{w^6}$ w

For Exercises 55–68, simplify the expressions. Write the answers in exponent form.

55. $\dfrac{h^3 h^8}{h^7}$ h^4 56. $\dfrac{n^5 n^4}{n^2}$ n^7

57. $\dfrac{x^9 x}{x^5}$ x^5 58. $\dfrac{k k^3}{k^2}$ k^2

59. $\dfrac{7^2 \cdot 7^6}{7}$ 7^7 60. $\dfrac{5^3 \cdot 5^8}{5}$ 5^{10}

61. $\dfrac{x^{13}}{x^3 x^4}$ x^6 62. $\dfrac{t^{10}}{t^5 t^3}$ t^2

63. $\dfrac{10^{20}}{10^3 \cdot 10^8}$ $\quad 10^9$

64. $\dfrac{3^{15}}{3^2 \cdot 3^{10}}$ $\quad 3^3$

65. $\dfrac{6^8 \cdot 6^5}{6^2 \cdot 6}$ $\quad 6^{10}$

66. $\dfrac{2^{14} \cdot 2}{2^3 \cdot 2^6}$ $\quad 2^6$

67. $\dfrac{z^3 z^{11}}{z^4 z^6}$ $\quad z^4$

68. $\dfrac{w^{12} w^2}{w^4 w^5}$ $\quad w^5$

For Exercises 69–80, use the commutative and associative properties of real numbers and the properties of exponents to simplify the expressions.

69. $(5a^2 b)(8a^3 b^4)$ $\quad 40a^5 b^5$

70. $(10xy^3)(3x^4 y)$ $\quad 30x^5 y^4$

71. $(r^6 s^4)(13r^2 s)$ $\quad 13r^8 s^5$

72. $(6p^2 q^8)(7p^5 q^3)$ $\quad 42p^7 q^{11}$

73. $\left(\dfrac{2}{3}m^{13}n^8\right)(24m^7 n^2)$ $\quad 16m^{20}n^{10}$

74. $\left(\dfrac{1}{4}c^6 d^6\right)(28c^2 d^7)$ $\quad 7c^8 d^{13}$

75. $\dfrac{14c^4 d^5}{7c^3 d}$ $\quad 2cd^4$

76. $\dfrac{36h^5 k^2}{9h^3 k}$ $\quad 4h^2 k$

77. $\dfrac{2x^3 y^5}{8xy^3}$ $\quad \dfrac{x^2 y^2}{4}$

78. $\dfrac{13w^8 z^3}{26w^2 z}$ $\quad \dfrac{w^6 z^2}{2}$

79. $\dfrac{25h^3 jk^5}{12h^2 k}$ $\quad \dfrac{25hjk^4}{12}$

80. $\dfrac{15m^5 np^{12}}{4mp^9}$ $\quad \dfrac{15m^4 np^3}{4}$

■ EXPANDING YOUR SKILLS

For Exercises 81–88, simplify the expressions using the addition or subtraction rules of exponents. Assume that a, b, m, and n represent positive integers.

81. $x^n x^{n+1}$ $\quad x^{2n+1}$

82. $y^a y^{2a}$ $\quad y^{3a}$

83. $p^{3m+5} p^{-m-2}$ $\quad p^{2m+3}$

84. $q^{4b-3} q^{-4b+4}$ $\quad q$

85. $\dfrac{z^{b+1}}{z^b}$ $\quad z$

86. $\dfrac{w^{5n+3}}{w^{2n}}$ $\quad w^{3n+3}$

87. $\dfrac{r^{3a+3}}{r^{3a}}$ $\quad r^3$

88. $\dfrac{t^{3+2m}}{t^{2m}}$ $\quad t^3$

---

section

3.2 MORE PROPERTIES OF EXPONENTS

Concepts

1. Power Rule for Exponents

2. The Properties

$(ab)^m = a^m b^m$ and

$\left|\dfrac{a}{b}\right|^m = \dfrac{a^m}{b^m}$

3. Simplifying Expressions with Exponents

1. Power Rule for Exponents

The expression $(x^2)^3$ indicates that the quantity x^2 is cubed.

$$(x^2)^3 = (x^2)(x^2)(x^2) = x^6$$

From this example, it appears that to raise a base to successive powers, we multiply the exponents. This is stated formally as the **power rule for exponents**.

Power Rule for Exponents

Assume that $a \neq 0$ is a real number and that m and n represent positive integers. Then,

$$\text{Property 3:} \quad (a^m)^n = a^{m \cdot n}$$

example 1 **Simplifying Expressions with Exponents**

Simplify the expressions.

a. $(s^4)^2$ b. $(3^4)^2$ c. $(x^2 x^5)^4$

 See Additional Answers Appendix   Writing   Translating Expression   Geometry   Scientific Calculator   Video

Solution:

a. $(s^4)^2$

$\quad = s^{4 \cdot 2}$ Multiply exponents (the base is unchanged).

$\quad = s^8$

b. $(3^4)^2$

$\quad = 3^{4 \cdot 2}$ Multiply exponents (the base is unchanged).

$\quad = 3^8$ or 6561

c. $(x^2 x^5)^4$

$\quad = (x^7)^4$ Simplify inside the parentheses by adding exponents.

$\quad = x^{7 \cdot 4}$ Multiply exponents (the base is unchanged).

$\quad = x^{28}$

2. The Properties $(ab)^m = a^m b^m$ and $\left(\dfrac{a}{b}\right)^m = \dfrac{a^m}{b^m}$

Consider the following expressions and their simplified forms:

$$(xy)^3 = (xy)(xy)(xy) = (x \cdot x \cdot x)(y \cdot y \cdot y) = x^3 y^3$$

$$\left(\frac{x}{y}\right)^3 = \left(\frac{x}{y}\right)\left(\frac{x}{y}\right)\left(\frac{x}{y}\right) = \left(\frac{x \cdot x \cdot x}{y \cdot y \cdot y}\right) = \frac{x^3}{y^3}$$

The expressions were simplified using the commutative and associative properties of multiplication. The simplified forms for each expression could have been reached in one step by applying the exponent to each factor inside the parentheses.

Power of a Product and Power of a Quotient

Assume that a and b are real numbers such that $b \neq 0$. Let m represent a positive integer. Then,

$$\text{Property 4:} \quad (ab)^m = a^m b^m$$

$$\text{Property 5:} \quad \left(\frac{a}{b}\right)^m = \frac{a^m}{b^m}$$

◆ **Avoiding Mistakes**

The power rule of exponents can be applied to a product of bases but in general cannot be applied to a sum or difference of bases.

$(ab)^n = a^n b^n$
Power rule can be applied.

$(a + b)^n \neq a^n + b^n$
Power rule *cannot* be applied.

Applying these properties of exponents, we have

$$(xy)^3 = x^3 y^3 \qquad \text{and} \qquad \left(\frac{x}{y}\right)^3 = \frac{x^3}{y^3}$$

3. Simplifying Expressions with Exponents

example 2

Simplifying Expressions with Exponents

Simplify the expressions.

a. $(-2xyz)^4$ b. $(5x^2y^7)^3$ c. $\left(\dfrac{1}{3xy^4}\right)^2$

Classroom Activity 3.2A

Solution:

a. $(-2xyz)^4$

$= (-2)^4x^4y^4z^4$ or $16x^4y^4z^4$ Raise each factor within parentheses to the fourth power.

b. $(5x^2y^7)^3$

$= 5^3(x^2)^3(y^7)^3$ Raise each factor within parentheses to the third power.

$= 125x^6y^{21}$ Multiply exponents and simplify.

c. $\left(\dfrac{1}{3xy^4}\right)^2$

$= \dfrac{1^2}{3^2x^2(y^4)^2}$ Square each factor within parentheses.

$= \dfrac{1}{9x^2y^8}$ Multiply exponents and simplify.

The properties of exponents can be used along with the properties of real numbers to simplify complicated expressions.

example 3

Simplifying Expressions with Exponents

Simplify the expressions.

a. $\dfrac{(x^2)^6(x^3)}{(x^7)^2}$ b. $(3cd^2)(2cd^3)^3$ c. $\left(\dfrac{x^7yz^4}{8xz^3}\right)^2(2xz^5)^3$

Solution:

a. $\dfrac{(x^2)^6(x^3)}{(x^7)^2}$ Clear parentheses by applying the power rule.

$$= \frac{x^{2 \cdot 6} x^3}{x^{7 \cdot 2}}$$ Multiply exponents.

$$= \frac{x^{12} x^3}{x^{14}}$$

$$= \frac{x^{12+3}}{x^{14}}$$ Add exponents in the numerator.

$$= \frac{x^{15}}{x^{14}}$$

$$= x^{15-14}$$ Subtract exponents.

$$= x$$ Simplify.

b. $(3cd^2)(2cd^3)^3$ Clear parentheses by applying the power rule.

$$= 3cd^2 \cdot 2^3 c^3 d^9$$ Raise each factor in the second parentheses to the third power.

$$= 3 \cdot 2^3 cc^3 d^2 d^9$$ Group like factors.

$$= 24c^{1+3} d^{2+9}$$ Add exponents from like factors.

$$= 24c^4 d^{11}$$ Simplify.

c. $\left(\dfrac{x^7 y z^4}{8xz^3}\right)^2 (2xz^5)^3$

$$= \left(\frac{x^{7-1} y z^{4-3}}{8}\right)^2 (2xz^5)^3$$ Simplify inside the first parentheses by subtracting exponents from like factors.

$$= \left(\frac{x^6 yz}{8}\right)^2 (2xz^5)^3$$

$$= \frac{(x^6)^2 y^2 z^2}{8^2} \cdot 2^3 x^3 (z^5)^3$$ Apply the power rule of exponents.

$$= \frac{x^{12} y^2 z^2}{64} \cdot 8x^3 z^{15}$$

$$= \frac{x^{12} y^2 z^2}{64} \cdot \frac{8x^3 z^{15}}{1}$$ Write the second expression as a fraction.

$$= \frac{8}{64} \cdot x^{12} x^3 y^2 z^2 z^{15}$$ Group like factors.

$$= \frac{1}{8} \cdot x^{12+3} y^2 z^{2+15}$$ Add the exponents from like factors. Reduce $\frac{8}{64} = \frac{1}{8}$.

$$= \frac{1}{8} x^{15} y^2 z^{17} \text{ or } \frac{x^{15} y^2 z^{17}}{8}$$ Simplify.

section 3.2 PRACTICE EXERCISES

For this exercise set assume all variables represent nonzero real numbers.

For Exercises 1–8, simplify. <u>3.1</u>

1. $4^2 \cdot 4^7$ 4^9

2. $5^8 \cdot 5^3 \cdot 5$ 5^{12}

3. $a^{13} \cdot a \cdot a^6$ a^{20}

4. $y^{14}y^3$ y^{17}

5. $\dfrac{d^{13}d}{d^5}$ d^9

6. $\dfrac{3^8 \cdot 3}{3^2}$ 3^7

7. $\dfrac{7^{11}}{7^5}$ 7^6

8. $\dfrac{z^4}{z^3}$ z

9. Explain when to add exponents versus when to multiply exponents. ❖

10. Explain when to add exponents versus when to subtract exponents. ❖

For Exercises 11–22, simplify and write answers in exponent form.

11. $(5^3)^4$ 5^{12}

12. $(2^8)^7$ 2^{56}

13. $(12^3)^2$ 12^6

14. $(6^4)^4$ 6^{16}

15. $(y^7)^2$ y^{14}

16. $(z^6)^4$ z^{24}

17. $(w^5)^5$ w^{25}

18. $(t^3)^6$ t^{18}

19. $(a^2a^4)^6$ a^{36}

20. $(z \cdot z^3)^2$ z^8

21. $(y^3y^4)^2$ y^{14}

22. $(w^5w)^4$ w^{24}

23. Evaluate the two expressions and compare the answers: $(2^2)^3$ and $(2^3)^2$. They are both equal to 2^6.

24. Evaluate the two expressions and compare the answers: $(4^4)^2$ and $(4^2)^4$. They are both equal to 4^8.

25. Evaluate the two expressions and compare the answers. Which expression is greater? Why?

$2^{(2^4)}$ and $(2^2)^4$

$2^{(2^4)} = 2^{16}$ $(2^2)^4 = 2^8; 2^{(2^4)}$ is greater than $(2^2)^4$.

26. Evaluate the two expressions and compare the answers. Which expression is greater? Why?

$3^{(2^4)}$ and $(3^2)^4$

$3^{(2^4)} = 3^{16}$ $(3^2)^4 = 3^8; 3^{(2^4)}$ is greater than $(3^2)^4$

For Exercises 27–38, use the appropriate property to clear the parentheses.

27. $\left(\dfrac{2}{3}\right)^3$ $\dfrac{8}{27}$

28. $\left(\dfrac{1}{6}\right)^2$ $\dfrac{1}{36}$

29. $\left(\dfrac{1}{4}\right)^2$ $\dfrac{1}{16}$

30. $\left(\dfrac{3}{4}\right)^3$ $\dfrac{27}{64}$

31. $\left(\dfrac{x}{y}\right)^5$ $\dfrac{x^5}{y^5}$

32. $\left(\dfrac{w}{z}\right)^7$ $\dfrac{w^7}{z^7}$

33. $\left(\dfrac{1}{t}\right)^4$ $\dfrac{1}{t^4}$

34. $\left(\dfrac{2}{r}\right)^4$ $\dfrac{16}{r^4}$

35. $(-3a)^4$ $81a^4$

36. $(-2x)^5$ $-32x^5$

37. $(-3abc)^3$ $-27a^3b^3c^3$

38. $(-5xyz)^2$ $25x^2y^2z^2$

For Exercises 39–56, simplify the expressions.

39. $(6u^2v^4)^3$ $216\,u^6v^{12}$

40. $(3a^5b^2)^6$ $729a^{30}b^{12}$

41. $5(x^2y)^4$ $5x^8y^4$

42. $18(u^3v^4)^2$ $18u^6v^8$

43. $\left(\dfrac{4}{rs^4}\right)^5$ $\dfrac{1024}{r^5s^{20}}$

44. $\left(\dfrac{2}{h^7k}\right)^3$ $\dfrac{8}{h^{21}k^3}$

45. $\left(\dfrac{3p}{q^3}\right)^5$ $\dfrac{343p^5}{q^{15}}$

46. $\left(\dfrac{5x^2}{y^3}\right)^4$ $\dfrac{625x^8}{y^{12}}$

47. $\dfrac{y^8(y^3)^4}{(y^2)^3}$ y^{14}

48. $\dfrac{(w^3)^2(w^4)^5}{(w^4)^2}$ w^{18}

49. $\dfrac{(5a^3b)^4(a^2b)^4}{(5ab)^2}$ $25a^{18}b^6$

50. $\dfrac{(6s^3)^2(s^4t^5)^2}{(3s^4t^2)^2}$ $4s^6t^6$

51. $\dfrac{(21x^5y)(2x^8y^4)}{14xy}$ $3x^{12}y^4$

52. $\dfrac{(4u^3v^3)(9u^4v)}{12u^5v^2}$ $3u^2v^2$

53. $\left(\dfrac{2c^3d^4}{3c^2d}\right)^2(3c^4d^2)^3$ $12c^{14}d^{12}$

54. $(10x^4yz^2)^3\left(\dfrac{x^3y^5z}{5xy^2}\right)^2$ $40x^{16}y^9z^8$

55. $(2c^3d^2)^5\left(\dfrac{c^6d^8}{4c^2d}\right)^3$ $\dfrac{c^{27}d^{31}}{2}$

56. $\left(\dfrac{s^5t^6}{2s^2t}\right)^2(10s^3t^3)^2$ $25s^{12}t^{16}$

EXPANDING YOUR SKILLS

For Exercises 57–64, simplify the expressions using the addition or subtraction properties of exponents. Assume that a, b, m, and n represent positive integers.

57. $(x^m)^2$ x^{2m}

58. $(y^3)^n$ y^{3n}

59. $(5a^{2n})^3$ $125a^{6n}$

60. $(3b^4)^m$ 3^mb^{4m}

61. $\left(\dfrac{m^2}{n^3}\right)^b$ $\dfrac{m^{2b}}{n^{3b}}$

62. $\left(\dfrac{x^5}{y^3}\right)^m$ $\dfrac{x^{5m}}{y^{3m}}$

63. $\left(\dfrac{3a^3}{5b^4}\right)^n$ $\dfrac{3^na^{3n}}{5^nb^{4n}}$

64. $\left(\dfrac{4m^6}{3n^2}\right)^b$ $\dfrac{4^bm^{6b}}{3^bn^{2b}}$

❖ See Additional Answers Appendix   Writing   Translating Expression   Geometry   Scientific Calculator   Video

section

3.3   DEFINITIONS OF b^0 AND b^{-n}

1. Definitions of b^0 and b^{-n}

In the previous two sections, we learned several rules that allow us to manipulate expressions containing *positive* integer exponents. In this section, we present two definitions that can be used to simplify expressions with negative exponents or with an exponent of zero.

Let m and n be positive integers. Recall that Property 2 of exponents states that

$$\frac{b^m}{b^n} = b^{m-n}$$

provided that the exponent in the numerator is greater than the exponent in the denominator, $m > n$. If the condition $m > n$ is not imposed, then the difference of exponents $m - n$ may be zero or negative. Therefore, we want to define the expressions b^0 and b^{-n} so that the properties of exponents can be extended to include zero and negative exponents. For example, we know that

$$1 = \frac{5}{5} \qquad 1 = \frac{5^2}{5^2} \qquad 1 = \frac{5^3}{5^3}$$

and so on. If we subtract exponents in any of these expressions, the result is 5^0.

Subtract exponents.

$$1 = \frac{5^3}{5^3} = 5^{3-3} = 5^0. \qquad \text{Therefore, we will define } 5^0 \text{ as 1.}$$

The same logic may be applied to any nonzero base, b, so we define $b^0 = 1$.

Definition of b^0

Let b be a real number such that $b \neq 0$. Then, $b^0 = 1$.

Tip: The expression $b^0 = 1$ provided b is *not* zero. Therefore, the expression 0^0 cannot be simplified by this rule.

For $x \neq 0$ and $y \neq 0$, the following expressions all equal 1:

$$4^0 = 1 \qquad \left(-\frac{1}{2}\right)^0 = 1 \qquad x^0 = 1 \qquad (xy)^0 = 1$$

Next, for a positive integer n, we want to define the expression b^{-n} so that the properties of exponents can be extended to include negative exponents.

Consider the expression $\dfrac{x^4}{x^7} = \dfrac{x \cdot x \cdot x \cdot x}{x \cdot x \cdot x \cdot x \cdot x \cdot x \cdot x} = \dfrac{1}{x^3}$

Subtract exponents.

Hence, $x^{-3} = \dfrac{1}{x^3}$

By subtracting exponents, we have $\dfrac{x^4}{x^7} = x^{4-7} = x^{-3}$

This example illustrates the following rule.

Definition of b^{-n}

Let n be an integer and b be a real number such that $b \neq 0$. Then,

$$b^{-n} = \left(\frac{1}{b}\right)^n = \frac{1}{b^n}$$

This definition indicates that to evaluate an expression with a negative exponent, we must take the *reciprocal of the base* and make the exponent *positive*. For example,

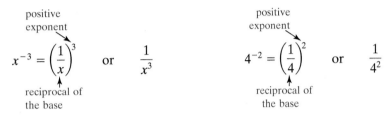

positive exponent

$$x^{-3} = \left(\frac{1}{x}\right)^3 \qquad \text{or} \qquad \frac{1}{x^3}$$

reciprocal of the base

positive exponent

$$4^{-2} = \left(\frac{1}{4}\right)^2 \qquad \text{or} \qquad \frac{1}{4^2}$$

reciprocal of the base

---

example 1

Simplifying Expressions with Negative and Zero Exponents

Simplify the expressions. Write the answer with positive exponents only. Assume all variables represent nonzero real numbers.

a. c^{-3} b. $\left(\dfrac{3}{4}\right)^{-2}$ c. $(50)^0 - 2^{-2}$ d. $\dfrac{p^4}{p^6}$

Classroom Activity 3.3A

Solution:

a. c^{-3}

$$= \left(\frac{1}{c}\right)^3 \qquad \text{Take the reciprocal of the base and make the exponent positive.}$$

$$= \frac{1}{c^3} \qquad \text{Simplify.}$$

b. $\left(\dfrac{3}{4}\right)^{-2}$

$= \left(\dfrac{4}{3}\right)^{2}$ Take the reciprocal of the base and make the exponent positive.

$= \dfrac{4^2}{3^2}$ Simplify.

$= \dfrac{16}{9}$

c. $(50)^0 - 2^{-2}$ Note that $(50)^0 = 1$.

$= 1 - \left(\dfrac{1}{2}\right)^{2}$ Take the reciprocal of the base, and make the exponent positive.

$= 1 - \dfrac{1}{4}$ Simplify.

$= \dfrac{4}{4} - \dfrac{1}{4}$ Get a common denominator.

$= \dfrac{3}{4}$ Simplify.

d. $\dfrac{p^4}{p^6}$

$= p^{4-6}$ Subtract exponents.

$= p^{-2}$ Simplify.

$= \left(\dfrac{1}{p}\right)^{2}$ Take the reciprocal of the base and make the exponent positive.

$= \dfrac{1}{p^2}$

2.   Properties of Integer Exponents: A Summary

Table 3-2 summarizes the properties and definitions of integer exponents.

Table 3-2

Properties of Integer Exponents Assume that a and b are real numbers ($b \neq 0$) and that m and n represent integers.		
Property	**Example**	**Details/Notes**
Multiplication of Like Bases 1. $b^m b^n = b^{m+n}$	$b^2 b^4 = b^{2+4} = b^6$	$b^2 b^4 = (b \cdot b)(b \cdot b \cdot b \cdot b) = b^6$
Division of Like Bases 2. $\dfrac{b^m}{b^n} = b^{m-n}$	$\dfrac{b^5}{b^2} = b^{5-2} = b^3$	$\dfrac{b^5}{b^2} = \dfrac{\cancel{b} \cdot \cancel{b} \cdot b \cdot b \cdot b}{\cancel{b} \cdot \cancel{b}} = b^3$
The Power Rule 3. $(b^m)^n = b^{m \cdot n}$	$(b^4)^2 = b^{4 \cdot 2} = b^8$	$(b^4)^2 = (b \cdot b \cdot b \cdot b)(b \cdot b \cdot b \cdot b) = b^8$
Power of a Product 4. $(ab)^m = a^m b^m$	$(ab)^3 = a^3 b^3$	$(ab)^3 = (ab)(ab)(ab)$ $= (a \cdot a \cdot a)(b \cdot b \cdot b) = a^3 b^3$
Power of a Quotient 5. $\left(\dfrac{a}{b}\right)^m = \dfrac{a^m}{b^m}$	$\left(\dfrac{a}{b}\right)^3 = \dfrac{a^3}{b^3}$	$\left(\dfrac{a}{b}\right)^3 = \left(\dfrac{a}{b}\right)\left(\dfrac{a}{b}\right)\left(\dfrac{a}{b}\right) = \dfrac{a \cdot a \cdot a}{b \cdot b \cdot b} = \dfrac{a^3}{b^3}$
Definitions Assume that b is a real number ($b \neq 0$) and that n represents an integer.		
Definition	**Example**	**Details/Notes**
$b^0 = 1$	$(4)^0 = 1$	Any nonzero quantity raised to the zero power equals 1.
$b^{-n} = \left(\dfrac{1}{b}\right)^n = \dfrac{1}{b^n}$	$b^{-5} = \left(\dfrac{1}{b}\right)^5 = \dfrac{1}{b^5}$	To simplify a negative exponent, take the reciprocal of the base and make the exponent positive.

3. Simplifying Expressions with Exponents

example 2

Simplifying Expressions with Exponents

Simplify the following expressions. Write the answer with positive exponents only. Assume all variables are nonzero.

a. $\dfrac{x^2 x^{-7}}{x^3}$ b. $\dfrac{z^2}{w^{-4} w^4 z^{-8}}$ c. $(-4ab^{-2})^{-3}$ d. $\left(\dfrac{2p^{-4} q^3}{5p^2 q}\right)^{-1}$

Solution:

a. $\dfrac{x^2 x^{-7}}{x^3}$

$= \dfrac{x^{2+(-7)}}{x^3}$ Add the exponents in the numerator.

$= \dfrac{x^{-5}}{x^3}$ Simplify.

$= x^{-5-3}$ Subtract the exponents.

$= x^{-8}$

$= \left(\dfrac{1}{x}\right)^8$ Simplify the negative exponent.

$= \dfrac{1}{x^8}$

b. $\dfrac{z^2}{w^{-4}w^4 z^{-8}}$

$= \dfrac{z^2}{w^{-4+4}z^{-8}}$ Add the exponents in the denominator.

$= \dfrac{z^2}{w^0 z^{-8}}$

$= \dfrac{z^2}{(1)z^{-8}}$ Recall that $w^0 = 1$.

$= z^{2-(-8)}$ Subtract the exponents.

$= z^{10}$ Simplify.

c. $(-4ab^{-2})^{-3}$

$= (-4)^{-3}a^{-3}(b^{-2})^{-3}$ Apply the power rule of exponents.

$= (-4)^{-3}a^{-3}b^6$

$= \left(\dfrac{1}{-4}\right)^3\left(\dfrac{1}{a}\right)^3 b^6$ Simplify the negative exponents.

$= \dfrac{1}{-64} \cdot \dfrac{1}{a^3} \cdot b^6$ Simplify.

$= -\dfrac{b^6}{64a^3}$ Multiply fractions.

Classroom Activity 3.3B

d. $\left(\dfrac{2p^{-4}q^3}{5p^2q}\right)^{-1}$

The negative exponent outside the parentheses can be eliminated by taking the reciprocal of the quantity within the parentheses.

$= \left(\dfrac{5p^2q}{2p^{-4}q^3}\right)^1$

Take the reciprocal of the base and make the exponent positive.

$= \dfrac{5p^2q}{2p^{-4}q^3}$

$= \dfrac{5p^{2-(-4)}q^{1-3}}{2}$

Subtract the exponents.

$= \dfrac{5p^6q^{-2}}{2}$

Simplify.

$= \dfrac{5p^6}{2} \cdot \dfrac{1}{q^2}$

Simplify the negative exponent.

$= \dfrac{5p^6}{2q^2}$

Simplify.

example 3

Simplifying an Expression with Exponents

Simplify the expression $2^{-1} + 3^{-1} + 5^0$. Write the answer with positive exponents only.

Solution:

$2^{-1} + 3^{-1} + 5^0$

$= \dfrac{1}{2} + \dfrac{1}{3} + 1$ Simplify negative exponents. Simplify $5^0 = 1$.

$= \dfrac{3}{6} + \dfrac{2}{6} + \dfrac{6}{6}$ Get a common denominator.

$= \dfrac{11}{6}$ Simplify.

4. Translations Involving Exponents

example 4

Translating Expressions Involving Exponents

Translate each phrase into a mathematical expression and simplify the expression. Write the answer with positive exponents only.

a. The quotient of y to the negative fourth power and the square of x
b. The cube of the square of y

Solution:

a. $\dfrac{y^{-4}}{x^2}$

$= y^{-4} \cdot \dfrac{1}{x^2}$ Write the quotient as a product.

$= \dfrac{1}{y^4} \cdot \dfrac{1}{x^2}$ Simplify the negative exponent.

$= \dfrac{1}{x^2 y^4}$

b. $(y^2)^3$

$= y^6$ Apply the power rule of exponents.

---

section 3.3   PRACTICE EXERCISES

For this set of exercises, assume all variables represent nonzero real numbers.

For Exercises 1–10, simplify the expressions.
3.1, 3.2

1. $b^3 b^8$ b^{11}

2. $c^7 c^2$ c^9

3. $\dfrac{x^6}{x^2}$ x^4

4. $\dfrac{y^9}{y^8}$ y

5. $\dfrac{9^4 \cdot 9^8}{9}$ 9^{11}

6. $\dfrac{3^{14}}{3^3 \cdot 3^5}$ 3^6

7. $(6ab^3c^2)^5$ $7776a^5b^{15}c^{10}$

8. $(7w^7z^2)^4$ $2401x^{28}z^8$

9. $\left(\dfrac{s^2 t^5}{4}\right)^3$ $\dfrac{s^6 t^{15}}{64}$

10. $\left(\dfrac{5k^3}{h^7}\right)^2$ $\dfrac{25k^6}{h^{14}}$

11. Simplify.

  a. 8^0 1

  b. $\dfrac{8^4}{8^4}$ 1

12. Simplify.

  a. d^0 1

  b. $\dfrac{d^3}{d^3}$ 1

For Exercises 13–26, simplify the expression.

13. p^0 1

14. k^0 1

15. 5^0 1

16. 2^0 1

17. -4^0 -1

18. -1^0 -1

19. $(-6)^0$ 1

20. $(-2)^0$ 1

21. $(8x)^0$ 1

22. $(-3y^3)^0$ 1

 23. $-7x^0$ -7

24. $6y^0$ 6

25. ab^0 a

26. pq^0 p

27. Simplify and write the answers with positive exponents.

  a. t^{-5} $\dfrac{1}{t^5}$

  b. $\dfrac{t^3}{t^8}$ $\dfrac{1}{t^5}$

28. Simplify and write the answers with positive exponents.

  a. 4^{-3} $\dfrac{1}{4^3}$

  b. $\dfrac{4^2}{4^5}$ $\dfrac{1}{4^3}$

29. Explain what is wrong with the following logic.

$\dfrac{x^4}{x^{-6}} = x^{4-6} = x^{-2}$ $\dfrac{x^4}{x^{-6}} = x^{4-(-6)} = x^{10}$

30. Explain what is wrong with the following logic.

$\dfrac{y^5}{y^{-3}} = y^{5-3} = y^{-2}$ $\dfrac{y^5}{y^{-3}} = y^{5-(-3)} = y^8$

31. Explain what is wrong with the following logic.

$2a^{-3} = \dfrac{1}{2a^3}$ $2a^{-3} = 2 \cdot \dfrac{1}{a^3} = \dfrac{2}{a^3}$

32. Explain what is wrong with the following logic.

$$5b^{-2} = \frac{1}{5b^2} \qquad 5b^{-2} = 5 \cdot \frac{1}{b^2} = \frac{5}{b^2}$$

For Exercises 33–76, simplify the expression. Write the answer with positive exponents only.

33. $\left(\dfrac{2}{7}\right)^{-3}$ $\dfrac{343}{8}$

34. $\left(\dfrac{5}{4}\right)^{-1}$ $\dfrac{4}{5}$

35. $\left(-\dfrac{1}{5}\right)^{-2}$ 25

36. $\left(-\dfrac{1}{3}\right)^{-3}$ -27

37. a^{-3} $\dfrac{1}{a^3}$

38. c^{-5} $\dfrac{1}{c^5}$

39. 12^{-1} $\dfrac{1}{12}$

40. 4^{-2} $\dfrac{1}{16}$

41. $(4b)^{-2}$ $\dfrac{1}{16b^2}$

42. $(3z)^{-1}$ $\dfrac{1}{3z}$

43. $6x^{-2}$ $\dfrac{6}{x^2}$

44. $7y^{-1}$ $\dfrac{7}{y}$

45. $w^{-4}w^{-2}$ $\dfrac{1}{w^6}$

46. $z^{-3}z^{-1}$ $\dfrac{1}{z^4}$

47. $x^{-8}x^4$ $\dfrac{1}{x^4}$

48. s^5s^{-6} $\dfrac{1}{s}$

49. $a^{-8}a^8$ 1

50. q^3q^{-3} 1

51. $y^{17}y^{-13}$ y^4

52. $b^{20}b^{-14}$ b^6

53. $(m^{-6}n^9)^3$ $\dfrac{n^{27}}{m^{18}}$

54. $(c^4d^{-5})^{-2}$ $\dfrac{d^{10}}{c^8}$

55. $(-3j^{-5}k^6)^4$ $\dfrac{81k^{24}}{j^{20}}$

56. $(6xy^{-11})^{-3}$ $\dfrac{y^{33}}{216x^3}$

57. $\dfrac{p^3}{p^9}$ $\dfrac{1}{p^6}$

58. $\dfrac{q^2}{q^{10}}$ $\dfrac{1}{q^8}$

59. $\dfrac{r^{-5}}{r^{-2}}$ $\dfrac{1}{r^3}$

60. $\dfrac{s^{-4}}{s^3}$ $\dfrac{1}{s^7}$

61. $\dfrac{7^3}{7^2 \cdot 7^8}$ $\dfrac{1}{7^7}$

62. $\dfrac{3^4 \cdot 3}{3^7}$ $\dfrac{1}{9}$

63. $\dfrac{a^{-1}b^2}{a^3b^8}$ $\dfrac{1}{a^4b^6}$

64. $\dfrac{k^{-4}h^{-1}}{k^6h}$ $\dfrac{1}{k^{10}h^2}$

65. $\dfrac{w^{-8}(w^2)^{-5}}{w^3}$ $\dfrac{1}{w^{21}}$

66. $\dfrac{p^2p^{-7}}{(p^2)^3}$ $\dfrac{1}{p^{11}}$

67. $(-8y^{-12})(2y^{16}z^{-2})$ $\dfrac{-16y^4}{z^2}$

68. $(5p^{-2}q^5)(-2p^{-4}q^{-1})$ $\dfrac{-10q^4}{p^6}$

69. $\dfrac{-18a^{10}b^6}{108a^{-2}b^6}$ $-\dfrac{a^{12}}{6}$

70. $\dfrac{-35x^{-4}y^{-3}}{-21x^2y^{-3}}$ $\dfrac{5}{3x^6}$

71. $\dfrac{(-4c^{12}d^7)^2}{(5c^{-3}d^{10})^{-1}}$ $80c^{21}d^{24}$

72. $\dfrac{(s^3t^{-2})^4}{(3s^{-4}t^6)^{-2}}$ $9s^4t^4$

73. $\left(\dfrac{2}{p^6p^3}\right)^{-3}$ $\dfrac{p^{27}}{8}$

74. $\left(\dfrac{5x}{x^7}\right)^{-2}$ $\dfrac{x^{12}}{25}$

75. $\left(\dfrac{5cd^{-3}}{10d^5}\right)^{-1}$ $\dfrac{2d^8}{c}$

76. $\left(\dfrac{4m^{10}n^4}{2m^{12}n^{-2}}\right)^{-1}$ $\dfrac{m^2}{2n^6}$

▪ EXPANDING YOUR SKILLS

For Exercises 77–82, simplify the expression.

77. $5^{-1} + 2^{-2}$ $\dfrac{9}{20}$

78. $4^{-2} + 8^{-1}$ $\dfrac{3}{16}$

79. $10^0 - 10^{-1}$ $\dfrac{9}{10}$

80. $3^0 - 3^{-2}$ $\dfrac{8}{9}$

81. $\dfrac{4^{-1} + 3^{-2}}{1 + 2^{-3}}$ $\dfrac{26}{81}$

82. $\dfrac{2^{-3} + 4^{-1}}{5^{-1} + 1}$ $\dfrac{5}{16}$

Concepts

1. **Introduction to Scientific Notation**
2. **Writing Numbers in Scientific Notation**
3. **Writing Numbers without Scientific Notation**
4. **Multiplying and Dividing Numbers in Scientific Notation**
5. **Applications of Scientific Notation**

section

3.4 SCIENTIFIC NOTATION

1. Introduction to Scientific Notation

In many applications in mathematics, it is necessary to work with very large or very small numbers. For example, the number of movie tickets sold in the United States and Canada in 1999 is estimated to be 1,680,000,000. The weight of a flea is approximately 0.00066 lb. To avoid writing numerous zeros in very large or small numbers, scientific notation was devised as a shortcut. Scientific notation is useful when performing calculations and when comparing the relative sizes of very large or very small numbers.

The principle behind scientific notation is to use a power of 10 to express the magnitude of the number. Consider the following powers of 10:

$$10^0 = 1$$

$$10^1 = 10 \qquad 10^{-1} = \frac{1}{10^1} = \frac{1}{10} = 0.1$$

$$10^2 = 100 \qquad 10^{-2} = \frac{1}{10^2} = \frac{1}{100} = 0.01$$

$$10^3 = 1000 \qquad 10^{-3} = \frac{1}{10^3} = \frac{1}{1000} = 0.001$$

$$10^4 = 10,000 \qquad 10^{-4} = \frac{1}{10^4} = \frac{1}{10,000} = 0.0001$$

In the base-10 numbering system, each place value to the left and right of the decimal point represents a different power of 10 (Figure 3-1).

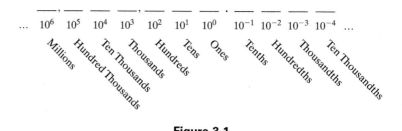

$$\ldots \; 10^6 \quad 10^5 \quad 10^4 \quad 10^3 \quad 10^2 \quad 10^1 \quad 10^0 \quad 10^{-1} \; 10^{-2} \; 10^{-3} \; 10^{-4} \; \ldots$$

Millions   Hundred Thousands   Ten Thousands   Thousands   Hundreds   Tens   Ones   Tenths   Hundredths   Thousandths   Ten Thousandths

Figure 3-1

Therefore, a number such as 4000 can be written as 4.0×1000, or equivalently, 4.0×10^3. Similarly, the number 0.07 can be written as $7.0 \times \frac{1}{100}$, or equivalently, 7.0×10^{-2}.

Definition of a Number Written in Scientific Notation

A number expressed in the form: $a \times 10^n$, where $1 \le |a| < 10$ and n is an integer is said to be written in **scientific notation**.

The numbers 4.0×10^3 and 7.0×10^{-2} are both expressed in scientific notation. To write a positive number in scientific notation, we apply the following guidelines:

1. Move the decimal point so that its new location is to the right of the first nonzero digit. The number should now be greater than or equal to 1 but less than 10. Count the number of places that the decimal point is moved.
2. If the original number is *large* (greater than or equal to 10), use the number of places the decimal point was moved as a *positive* power of 10.

$$450,000$$
$$\text{5 places} \qquad = 4.5 \times 100,000 = 4.5 \times 10^5$$

3. If the original number is *small* (between 0 and 1), use the number of places the decimal point was moved as a *negative* power of 10.

$$0.0002 \quad = 2.0 \times 0.0001 = 2.0 \times 10^{-4}$$

4 places

4. If the original number is greater than or equal to 1 but less than 10, use 0 as the power of 10.

$$7.592 = 7.592 \times 10^{0}$$ *Note*: A number between 1 and 10 is seldom written in scientific notation.

2. Writing Numbers in Scientific Notation

---

example 1

Writing Numbers in Scientific Notation

Write the numbers in scientific notation.

a. 53,000

b. 0.00053

Solution:

a. $53,000. = 5.3 \times 10^{4}$ To write 53,000 in scientific notation, the decimal point must be moved four places to the left. Because 53,000 is larger than 10, a *positive* power of 10 is used.

b. $0.00053 = 5.3 \times 10^{-4}$ To write 0.00053 in scientific notation, the decimal point must be moved four places to the right. Because 0.00053 is less than 1, a *negative* power of 10 is used.

---

example 2

Writing Numbers in Scientific Notation

Write the numerical values in scientific notation.

a. The number of movie tickets sold in the United States and Canada in 1999 is estimated to be 1,680,000,000.
b. The weight of a flea is approximately 0.00066 lb.
c. The temperature on a January day in Fargo dropped to $-43°F$.
d. A bench is 8.2 ft long.

Solution:

a. $1,680,000,000 = 1.68 \times 10^{9}$
b. $0.00066 \text{ lb} = 6.6 \times 10^{-4} \text{ lb}$
c. $-43°F = -4.3 \times 10^{1} °F$
d. $8.2 \text{ ft} = 8.2 \times 10^{0} \text{ ft}$

Tip: For a number written in scientific notation, the power of 10 is sometimes called the **order of magnitude** (or simply the magnitude) of the number.

- The order of magnitude of the number of movie tickets sold in 1999 in the United States and Canada is $\$10^9$ (billions of dollars).
- The mass of a flea is on the order of 10^{-4} lb (ten-thousandths of a pound).

3. Writing Numbers without Scientific Notation

example 3

Writing Numbers without Scientific Notation

Write the numerical values without scientific notation.

a. The mass of a proton is approximately 1.67×10^{-24} g.
b. The "nearby" star Vega is approximately 1.552×10^{14} miles from earth.

Classroom Activity 3.4A

Solution:

a. 1.67×10^{-24} g $= (1.67 \times 0.000\ 000\ 000\ 000\ 000\ 000\ 000\ 001)$ g
$= 0.000\ 000\ 000\ 000\ 000\ 000\ 000\ 001\ 67$ g

Because the power of 10 is negative, the value of 1.67×10^{-24} is a decimal number between 0 and 1. Move the decimal point 24 places to the *left*.

b. 1.552×10^{14} miles $= (1.552 \times 100,000,000,000,000)$ miles
$= 155,200,000,000,000$ miles

Because the power of 10 is a positive integer, the value of 1.552×10^{14} is a large number greater than 10. Move the decimal point 14 places to the *right*.

4. Multiplying and Dividing Numbers in Scientific Notation

To multiply or divide two numbers in scientific notation, use the commutative and associative properties of multiplication to group the powers of 10. For example:

$$400 \times 2000 = (4 \times 10^2)(2 \times 10^3) = (4 \cdot 2) \times (10^2 \cdot 10^3) = 8 \times 10^5$$

$$\frac{0.00054}{150} = \frac{5.4 \times 10^{-4}}{1.5 \times 10^2} = \left(\frac{5.4}{1.5}\right) \times \left(\frac{10^{-4}}{10^2}\right) = 3.6 \times 10^{-6}$$

example 4

Multiplying and Dividing Numbers in Scientific Notation

a. $(8.7 \times 10^4)(2.5 \times 10^{-12})$ 　　b. $\dfrac{4.25 \times 10^{13}}{8.5 \times 10^{-2}}$

Solution:

a. $(8.7 \times 10^4)(2.5 \times 10^{-12})$

$\quad = (8.7 \cdot 2.5) \times (10^4 \cdot 10^{-12})$ Commutative and associative properties of multiplication

$\quad = 21.75 \times 10^{-8}$ The number 21.75 is not in proper scientific notation because 21.75 is not between 1 and 10.

$\quad = (2.175 \times 10^1) \times 10^{-8}$ Rewrite 21.75 as 2.175×10^1.

$\quad = 2.175 \times (10^1 \times 10^{-8})$ Associative property of multiplication

$\quad = 2.175 \times 10^{-7}$ Simplify.

b. $\dfrac{4.25 \times 10^{13}}{8.5 \times 10^{-2}}$

$\quad = \left(\dfrac{4.25}{8.5}\right) \times \left(\dfrac{10^{13}}{10^{-2}}\right)$ Commutative and associative properties

$\quad = 0.5 \times 10^{15}$ The number 0.5×10^{15} is not in proper scientific notation because 0.5 is not between 1 and 10.

$\quad = (5.0 \times 10^{-1}) \times 10^{15}$ Rewrite 0.5 as 5.0×10^{-1}.

$\quad = 5.0 \times (10^{-1} \times 10^{15})$ Associative property of multiplication

$\quad = 5.0 \times 10^{14}$ Simplify.

5.   Applications of Scientific Notation

example 5

Applying Scientific Notation

If a spacecraft travels at 1.6×10^4 mph, how long will it take the craft to travel to Mars if the distance is approximately 8.0×10^7 miles?

Classroom Activity 3.4B

Solution:

Since $d = rt$, then $t = \dfrac{d}{r}$

$$t = \frac{8.0 \times 10^7 \text{ miles}}{1.6 \times 10^4 \text{ miles/hour}}$$

$$= \left(\frac{8.0}{1.6}\right) \times \left(\frac{10^7}{10^4}\right) \text{h}$$

$$= 5.0 \times 10^3 \text{ h}$$

The time required to travel to Mars is approximately $5.0 \times 10^3 = 5000$ h, or 208 days.

example 6

Applying Scientific Notation

Figure 3-2 depicts the total annual revenue by the Microsoft Corporation from June 1995 to June 1999.

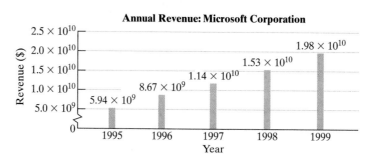

Figure 3-2

What is the difference in annual revenue between 1998 and 1999?

Solution:

To add or subtract two numbers in scientific notation, the numbers must have the *same magnitude*.

$$(\$1.98 \times 10^{10}) - (\$1.53 \times 10^{10}) = \$0.45 \times 10^{10} = \$4.5 \times 10^{9}$$

Microsoft had a \$4.5 billion increase in revenue between 1998 and 1999.

Calculator Connections

Both scientific and graphing calculators can perform calculations involving numbers written in scientific notation. Most calculators use an $\boxed{\text{EE}}$ key or an $\boxed{\text{EXP}}$ key to enter the power of 10.

Scientific Calculator

Enter: $\boxed{2}$ $\boxed{.}$ $\boxed{7}$ $\boxed{\text{EE}}$ (or $\boxed{\text{EXP}}$) $\boxed{5}$ $\boxed{=}$ **Result:**   270000

Enter: $\boxed{7}$ $\boxed{.}$ $\boxed{1}$ $\boxed{\text{EE}}$ (or $\boxed{\text{EXP}}$) $\boxed{3}$ $\boxed{+\bigcirc-}$ $\boxed{=}$ **Result:**   0.0071

Graphing Calculator

```
2.7E5
             270000
7.1E-3
              .0071
```

We recommend that you use parentheses to enclose each number written in scientific notation when performing calculations. Try using your calculator to perform the calculations from Example 4.

a. $(8.7 \times 10^4)(2.5 \times 10^{-12})$ b. $\dfrac{4.25 \times 10^{13}}{8.5 \times 10^{-2}}$

Scientific Calculator

Enter: (8 . 7 EE 4) × (2 . 5 EE 1 2 +○−)
= **Result**: 0.000000218

Enter: (4 . 2 5 EE 1 3) ÷ (8 . 5 EE 2 +○−)
) = **Result**: 5E14

Notice that the answer to part (b) is shown on the calculator in scientific notation. The calculator does not have enough room to display 14 zeros.

Graphing Calculator

```
(8.7E4)*(2.5E-12
)
             2.175E-7
(4.25E13)/(8.5E-
2)
                5E14
```

Calculator Exercises

Use a calculator to perform the indicated operations:

1. $(5.2 \times 10^6)(4.6 \times 10^{-3})$ ❖   2. $(2.19 \times 10^{-8})(7.84 \times 10^{-4})$ ❖

3. $\dfrac{4.76 \times 10^{-5}}{2.38 \times 10^9}$ ❖   4. $\dfrac{8.5 \times 10^4}{4.0 \times 10^{-1}}$ ❖

5. $\dfrac{(9.6 \times 10^7)(4.0 \times 10^{-3})}{2.0 \times 10^{-2}}$ ❖   6. $\dfrac{(5.0 \times 10^{-12})(6.4 \times 10^{-5})}{(1.6 \times 10^{-8})(4.0 \times 10^2)}$ ❖

7. $7.8 \times 10^{-3} + 9.4 \times 10^{-3}$ ❖   8. $4.3 \times 10^8 + 8.2 \times 10^8$ ❖

9. $6.43 \times 10^5 - 1.91 \times 10^5$ ❖   10. $2.01 \times 10^{-6} - 3.16 \times 10^{-6}$ ❖

❖ See Additional Answers Appendix   Writing   Translating Expression   Geometry   Scientific Calculator   Video

section 3.4   PRACTICE EXERCISES

For Exercises 1–12, simplify the expression. Assume all variables represent nonzero real numbers.   3.1–3.3

1. a^3a^{-4} $\dfrac{1}{a}$

2. b^5b^8 b^{13}

3. $10^3 \cdot 10^{-4}$ $\dfrac{1}{10}$

4. $10^5 \cdot 10^8$ 10^{13}

5. $\dfrac{x^3}{x^6}$ $\dfrac{1}{x^3}$

6. $\dfrac{y^2}{y^7}$ $\dfrac{1}{y^5}$

7. $\dfrac{10^3}{10^6}$ $\dfrac{1}{10^3}$

8. $\dfrac{10^2}{10^7}$ $\dfrac{1}{10^5}$

9. $\dfrac{z^9z^4}{z^3}$ z^{10}

10. $\dfrac{w^{-2}w^5}{w^{-1}}$ w^4

11. $\dfrac{10^9 \cdot 10^4}{10^3}$ 10^{10}

12. $\dfrac{10^{-2} \cdot 10^5}{10^{-1}}$ 10^4

13. Explain how you would write the number 0.000 000 000 23 in scientific notation.   Move the decimal point between 2 and 3 and multiply by 10^{-10}; 2.3×10^{-10}.

14. Explain how you would write the number 23,000,000,000,000 in scientific notation.   Move the decimal point between 2 and 3 and multiply by 10^{13}; 2.3×10^{13}.

15. Write the numerical values in scientific notation: In the world's largest tanker disaster, Amoco Cadiz spilled 68,000,000 gal of oil off Portsall, France, causing widespread environmental damage over 100 miles of Brittany coast. 6.8×10^7 gal; 1.0×10^2 miles

16. Write the numerical values in scientific notation: The human heart pumps about 1400 L of blood per day. That would mean that it pumps approximately 10,000,000 L per year. 1.4×10^3 L; 1.0×10^7 L

For Exercises 17–20, write the numbers in scientific notation.

17. The mass of a proton is approximately 0.000 000 000 000 000 000 000 0017 g. 1.7×10^{-24} g

18. The estimated wealth of Bill Gates in 1999 was $130,150,000,000. $\$1.3015 \times 10^{11}$

19. The number of shares of Microsoft Corporation owned by Bill Gates in 1999 was 141,159,990. 1.4115999×10^8 shares

20. One gram is equivalent to 0.0035 oz. 3.5×10^{-3} oz

21. Explain how you would write the number 3.1×10^{-9} without scientific notation.   Move the decimal nine places to the left; 0.000 000 0031

22. Explain how you would write the number 3.1×10^9 without scientific notation.   Move the decimal nine places to the right; 3,100,000,000

For Exercises 23–26, write the numbers without scientific notation.

23. One picogram (pg) is equal to 1×10^{-12} g.   0. 000 000 000 001 g

24. A nanometer (nm) is approximately 3.94×10^{-8} in.   0. 000 000 0394 in.

25. A normal diet contains between 1.6×10^3 Cal and 2.8×10^3 Cal per day.   1600 calories and 2800 calories

26. The total land area of Texas is approximately 2.62×10^5 square miles.   262,000 square miles

For Exercises 27–42, multiply or divide as indicated. Write the answers in scientific notation.

27. $(2.5 \times 10^6)(2.0 \times 10^{-2})$ 5.0×10^4

28. $(2.0 \times 10^{-7})(3.0 \times 10^{13})$ 6.0×10^6

29. $\dfrac{9.0 \times 10^{-6}}{4.0 \times 10^7}$ 2.25×10^{-13}

30. $\dfrac{7.0 \times 10^{-2}}{5.0 \times 10^9}$ 1.4×10^{-11}

31. $(8.0 \times 10^{10})(4.0 \times 10^3)$ 3.2×10^{14}

32. $(6.0 \times 10^{-4})(3.0 \times 10^{-2})$ 1.8×10^{-5}

33. $(3.2 \times 10^{-4})(7.6 \times 10^{-7})$ 2.432×10^{-10}

34. $(5.9 \times 10^{12})(3.6 \times 10^9)$ 2.124×10^{22}

35. $\dfrac{2.1 \times 10^{11}}{7.0 \times 10^{-3}}$ 3.0×10^{13}

36. $\dfrac{1.6 \times 10^{14}}{8.0 \times 10^{-5}}$ 2.0×10^{18}

37. $\dfrac{5.7 \times 10^{-2}}{9.5 \times 10^{-8}}$ 6.0×10^{5}

38. $\dfrac{2.72 \times 10^{-6}}{6.8 \times 10^{-4}}$ 4.0×10^{-3}

39. $6{,}000{,}000{,}000 \times 0.0000000023$ 1.38×10^{1}

40. $0.000055 \times 40{,}000$ 2.2×10^{0}

41. $\dfrac{0.0000000003}{6000}$ 5.0×10^{-14}

42. $\dfrac{420{,}000}{0.0000021}$ 2.0×10^{11}

43. If a piece of paper is 3.0×10^{-3} in. thick, how thick is a stack of 1.25×10^{3} pieces of paper?
    3.75 in.

44. A box of staples contains 5.0×10^{3} staples and weighs 15 oz. How much does one staple weigh? Write your answer in scientific notation.
    3.0×10^{-3} oz

45. In the year 2000, $\$6.0 \times 10^{8}$ was spent on 350,000 30-second television commercials for campaign ads for political candidates in the United States. Based on these figures, determine the average cost of a 30-second television commercial. (*Source:* Television Bureau of Advertising)
    Approximately $1714 per commercial

46. A state lottery had a jackpot of $\$5.2 \times 10^{7}$. This week the winner was a group of office employees that included 13 people. How much would each person receive? $\$4.0 \times 10^{6}$ or $4,000,000

47. Dinosaurs became extinct about 65 million years ago.

    a. Write the number 65 million in scientific notation. 6.5×10^{7}

    b. How many days is 65 million years?
       2.3725×10^{10} days

    c. How many hours is 65 million years?
       5.694×10^{11} h

    d. How many seconds is 65 million years?
       2.04984×10^{15} s

48. The earth is 111,600,000 km from the sun.

    a. Write the number 111,600,000 in scientific notation. 1.116×10^{8}

    b. If there are 1000 m in a kilometer, how many meters is the earth from the sun? 1.116×10^{11} m

    c. If there are 100 cm in a meter, how many centimeters is the earth from the sun?
       1.116×10^{13} cm

For Exercises 49–56, find the sum or difference. Write the answer in scientific notation.

49. $(2.0 \times 10^{7}) + (4.0 \times 10^{7})$ 6.0×10^{7}

50. $(3.0 \times 10^{-5}) + (2.0 \times 10^{-5})$ 5.0×10^{-5}

51. $(6.4 \times 10^{-12}) - (2.9 \times 10^{-12})$ 3.5×10^{-12}

52. $(4.5 \times 10^{4}) - (2.9 \times 10^{4})$ 1.6×10^{4}

53. $(5.0 \times 10^{6}) + (9.0 \times 10^{6})$ 1.4×10^{7}

54. $(4.0 \times 10^{-9}) + (8.0 \times 10^{-9})$ 1.2×10^{-8}

55. $(7.1 \times 10^{-3}) - (6.8 \times 10^{-3})$ 3.0×10^{-4}

56. $(3.3 \times 10^{8}) - (2.5 \times 10^{8})$ 8.0×10^{7}

57. In 1999, Wal-Mart Stores, Inc. had 7.28×10^{5} employees and the United States Postal Service had 7.0×10^{5} employees. What is the difference in the number of employees? Write the answer in scientific notation. (*Source:* Dun and Bradstreet)
    2.8×10^{4} employees

58. In 1997 General Motors' net income was $\$6.698 \times 10^{9}$, and the net income for the Ford Motor Company was $\$6.92 \times 10^{9}$. What is the difference in their net incomes? Write the answer in scientific notation. (*Source:* Dow Jones Global Indexes) $\$2.22 \times 10^{8}$

59. Refer to Exercise 58. If Chrysler had a net income of $\$2.805 \times 10^{9}$, what was the total income of the three major automobile manufacturers in 1997? Write the answer in scientific notation.
    $\$1.6423 \times 10^{10}$

60. If the Government of the United States had 1.94×10^{6} employees in 1999 and the United States Postal Service had 7.0×10^{5} employees, what was the total number of employees from these two employers in 1999? Write the answer in scientific notation. (*Hint*: $7.0 \times 10^{5} = 0.7 \times 10^{6}$) (*Source:* U.S. Department of Labor)
    2.64×10^{6} employees

EXPANDING YOUR SKILLS

61. The top world exporters are shown in the graph (as of June 1999). The values are in billions of dollars. (*Source:* International Labour Office)

Top Exporters (billions per year)

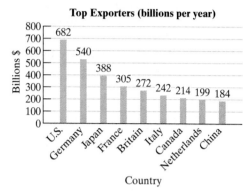

Country

Figure for Exercise 61

a. Use scientific notation to express the amount of money (in dollars) that each country exports per year. ❖

b. What is the total dollar amount from exports for the nine countries combined? Use scientific notation to express your final answer. ❖

c. How many more dollars in exports does Japan produce than China? ❖

d. What percentage of the total exports does Canada represent? (Round to the nearest percent.) ❖

62. Of all the workers in the United States who earn minimum wage, the pie chart shows a breakdown by educational level. (*Source:* U.S. Department of Labor)

Minimum-Wage Work Force by Educational Level

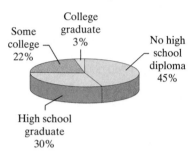

Figure for Exercise 62

a. In a group of 1.5×10^6 minimum-wage workers, how many would you expect to have no high school diploma? 6.75×10^5 workers

b. In a group of 1.5×10^6 minimum-wage workers, how many would you expect to be college graduates? 4.5×10^4 workers

63. As of October, 1999, the Hershey Foods Corporation reported $\$2.87 \times 10^9$ in annual revenue. This was a 10% loss from the figure reported 9 months earlier. What was the revenue 9 months earlier?   Approximately $\$3.19 \times 10^9$

64. As of September, 1999, the IBM Corporation reported $\$6.337 \times 10^{10}$ in annual revenue. This was a 12% increase from the figure reported 9 months earlier. What was the revenue 9 months earlier?   Approximately $\$5.66 \times 10^{10}$

---

chapter 3   MIDCHAPTER REVIEW

For Exercises 1–22 use all of the properties and definitions presented in the previous four sections. Simplify each expression and write the answers with positive exponents only. Assume that all variables represent nonzero real numbers.

1. $\left(\dfrac{1}{2}\right)^{-1} + \left(\dfrac{1}{3}\right)^0$ 3

2. $\left(\dfrac{1}{4}\right)^0 - \left(\dfrac{1}{5}\right)^{-1}$ -4

3. $(2^5 b^{-3})^{-3}$ $\dfrac{b^9}{2^{15}}$

4. $(3^{-2} y^3)^{-2}$ $\dfrac{81}{y^6}$

5. $\left(\dfrac{3x}{2y}\right)^{-4}$ $\dfrac{16y^4}{81x^4}$

6. $\left(\dfrac{6c}{5d^3}\right)^{-2}$ $\dfrac{25d^6}{36c^2}$

 ❖ See Additional Answers Appendix   Writing   Translating Expression   Geometry   Scientific Calculator   Video

7. $(3ab^2)(a^2b)^3$ $3a^7b^5$

8. $(4x^2y^3)^3(xy^2)$ $64x^7y^{11}$

9. $\left(\dfrac{xy^2}{x^3y}\right)^4$ $\dfrac{y^4}{x^8}$

10. $\left(\dfrac{a^3b}{a^5b^3}\right)^5$ $\dfrac{1}{a^{10}b^{10}}$

11. $\dfrac{(t^{-2})^3}{t^{-4}}$ $\dfrac{1}{t^2}$

12. $\dfrac{(p^3)^{-4}}{p^{-5}}$ $\dfrac{1}{p^7}$

13. $\left(\dfrac{2w^2x^3}{3y^0}\right)^3$ $\dfrac{8w^6x^9}{27}$

14. $\left(\dfrac{5a^0b^4}{4c^3}\right)^2$ $\dfrac{25b^8}{16c^6}$

15. $\dfrac{q^3r^{-2}}{s^{-1}t^5}$ $\dfrac{q^3s}{r^2t^5}$

16. $\dfrac{n^{-3}m^2}{p^{-3}q^{-1}}$ $\dfrac{m^2p^3q}{n^3}$

17. $\dfrac{(y^{-3})^2(y^5)}{(y^{-3})^{-4}}$ $\dfrac{1}{y^{13}}$

18. $\dfrac{(w^2)^{-4}(w^{-2})}{(w^5)^{-4}}$ w^{10}

19. $\left(\dfrac{-2a^2b^{-3}}{a^{-4}b^{-5}}\right)^{-3}$ $-\dfrac{1}{8a^{18}b^6}$

20. $\left(\dfrac{-3x^{-4}y^3}{2x^5y^{-2}}\right)^{-2}$ $\dfrac{4x^{18}}{9y^{10}}$

$\dfrac{k^8}{5h^6}$ 21. $(5h^{-2}k^0)^3(5k^{-2})^{-4}$

22. $(6m^3n^{-5})^{-4}(6m^0n^{-2})^5$ $\dfrac{6n^{10}}{m^{12}}$

23. Evaluate the following expressions. Do you see a pattern? $10^0, 10^1, 10^2, 10^3, 10^4$ 1, 10, 100, 1000, 10,000

24. Evaluate the following expressions. Do you see a pattern? $10^0, 10^{-1}, 10^{-2}, 10^{-3}, 10^{-4}$
1, 0.1, 0.01, 0.001, 0.0001

For Exercises 25–26, express the numbers in scientific notation.

25. In July 1999 the population of the United States was 272,330,000. 2.7233×10^8

26. In the year 2003, there will be about 28 million adults in America using the Internet for business from home. 2.8×10^7

For Exercises 27–28, express the numbers without scientific notation.

27. A micrometer (μm) is 1×10^{-6} m.   0.000001

28. A micrometer (μm) is 1×10^{-3} mm.   0.001

29. The probability of cell mutation of *Escherichia coli* (*E. coli*) is 1×10^{-7} per cell division. If 2×10^{10} cells are reproduced in the human body in a day, what is the probable number of mutations per day? (Multiply the probability of a cell mutation times the numbers of cells.) Write your answer in expanded form. $2 \times 10^3 = 2000$ mutations

---

Concepts

1. Introduction to Polynomials

2. Translations and Applications of Polynomials

3. Addition of Polynomials

4. Subtraction of Polynomials

5. Polynomials and Applications to Geometry

3.5   ADDITION AND SUBTRACTION OF POLYNOMIALS

1.   Introduction to Polynomials

One commonly used algebraic expression is called a polynomial. A **polynomial** in one variable, x, is defined as a sum of terms of the form ax^n, where a is a real number and the exponent, n, is a nonnegative integer. For each term, a is called the **coefficient**, and n is called the **degree of the term**. For example:

Term (expressed in the form ax^n)	Coefficient	Degree
$-12z^7$	-12	7
$x^3 \rightarrow$ rewrite as $1x^3$	1	3
$10w \rightarrow$ rewrite as $10w^1$	10	1
$7 \rightarrow$ rewrite as $7x^0$	7	0

If a polynomial has exactly one term, it is categorized as a **monomial**. A two-term polynomial is called a **binomial**, and a three-term polynomial is called a

trinomial. Usually the terms of a polynomial are written in descending order according to degree. The term with highest degree is called the **leading term**, and its coefficient is called the **leading coefficient**. The **degree of a polynomial** is the largest degree of all of its terms. Thus, the leading term determines the degree of the polynomial.

	Expression	Descending Order	Leading Coefficient	Degree of Polynomial
Monomials	$-3x^4$	$-3x^4$	-3	4
	17	17	17	0
Binomials	$4y^3 - 6y^5$	$-6y^5 + 4y^3$	-6	5
	$\dfrac{1}{2} - \dfrac{1}{4}c$	$-\dfrac{1}{4}c + \dfrac{1}{2}$	$-\dfrac{1}{4}$	1
Trinomials	$4p - 3p^3 + 8p^6$	$8p^6 - 3p^3 + 4p$	8	6
	$7a^4 - 1.2a^8 + 3a^3$	$-1.2a^8 + 7a^4 + 3a^3$	-1.2	8

example 1

Identifying the Parts of a Polynomial

Given: $4.5a - 2.7a^{10} + 1.6 - 3.7a^5$

a. List the terms of the polynomial, and state the coefficient and degree of each term.
b. Write the polynomial in descending order.
c. State the degree of the polynomial and the leading coefficient.

Classroom Activity 3.5A

Solution:

a. term: $4.5a$ coefficient: 4.5   degree: 1

term: $-2.7a^{10}$ coefficient: -2.7 degree: 10

term: 1.6   coefficient: 1.6   degree: 0

term: $-3.7a^5$ coefficient: -3.7 degree: 5

b. $-2.7a^{10} - 3.7a^5 + 4.5a + 1.6$

c. The degree of the polynomial is 10 and the leading coefficient is -2.7.

Polynomials may have more than one variable. In such a case, the degree of a term is the sum of the exponents of the variables contained in the term. For example, the term, $32x^2y^5z$, has degree 8 because the exponents applied to x, y, and z are 2, 5, and 1, respectively. The following polynomial has a degree of 11 because the highest degree of its terms is 11.

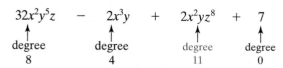

$$32x^2y^5z \quad - \quad 2x^3y \quad + \quad 2x^2yz^8 \quad + \quad 7$$

degree 8   degree 4   degree 11   degree 0

2.   Translations and Applications of Polynomials

---

example 2   **Translating from an English Phrase to an Algebraic Expression**

a.   Write a mathematical expression for: The difference of the square of y and the square of x.

b.   Write a monomial of degree 3, coefficient of -24, and variable w.

Solution:

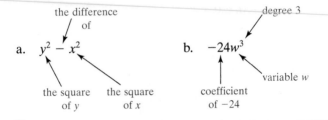

a. $y^2 - x^2$ b. $-24w^3$

---

example 3   **Using Polynomials in an Application**

A child throws a ball upward and the height of the ball, h, (in feet) can be computed by the following equation:

$$h = -32t^2 + 64t + 2$$ where t is the time (in seconds) after the ball is released.

a.   Find the height of the ball after 0.5 s, 1 s, and 1.5 s.

b.   Find the height of the ball at the time of release.

Solution:

a. $h = -32t^2 + 64t + 2$

$= -32(0.5)^2 + 64(0.5) + 2$ Substitute $t = 0.5$.

$= -32(0.25) + 32 + 2$

$= -8 + 32 + 2$

$= 26$ The height of the ball after 0.5 s is 26 ft.

$h = -32t^2 + 64t + 2$

$= -32(1)^2 + 64(1) + 2$ Substitute $t = 1$.

$= -32(1) + 64 + 2$

$= -32 + 64 + 2$

$= 34$ The height of the ball after 1 s is 34 ft.

$$h = -32t^2 + 64t + 2$$

$$= -32(1.5)^2 + 64(1.5) + 2 \qquad \text{Substitute } t = 1.5.$$

$$= -32(2.25) + 96 + 2$$

$$= -72 + 96 + 2$$

$$= 26 \qquad\qquad\qquad \text{The height of the ball after 1.5 s is 26 ft.}$$

b. $h = -32t^2 + 64t + 2$

$$= -32(0)^2 + 64(0) + 2 \qquad \text{At the time of release, } t = 0.$$

$$= 0 + 0 + 2$$

$$= 2 \qquad\qquad\qquad \text{The height of the ball at the time of release is 2 ft.}$$

3. Addition of Polynomials

Recall that two terms are said to be *like* terms if they each have the same variables, and the corresponding variables are raised to the same powers.

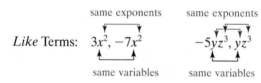

Like Terms: $3x^2, -7x^2$ $-5yz^3, yz^3$

Un*like* Terms: $9z^2, 12z^6$ $\dfrac{1}{3}w^6, \dfrac{2}{5}p^6$ $4y, 7$

Recall that the distributive property is used to add or subtract *like* terms. For example,

$$3x^2 + 9x^2 - 2x^2$$

$$= (3 + 9 - 2)x^2 \qquad \text{Apply the distributive property.}$$

$$= (10)x^2 \qquad\qquad \text{Simplify.}$$

$$= 10x^2$$

---

example 4 **Adding Polynomials**

Add the polynomials.

a. $3x^2y + 5x^2y$ b. $(-3c^3 + 5c^2 - 7c) + (11c^3 + 6c^2 + 3)$

c. $\left(\dfrac{1}{4}w^2 - \dfrac{2}{3}w\right) + \left(\dfrac{3}{4}w^2 + \dfrac{1}{6}w - \dfrac{1}{2}\right)$

Solution:

a. $3x^2y + 5x^2y$

$\quad = (3+5)x^2y$ Apply the distributive property.

$\quad = (8)x^2y$

$\quad = 8x^2y$ Simplify.

Tip: Although the distributive property is used to combine *like* terms, the process is simplified by combining the coefficients of *like* terms.

b. $(-3c^3 + 5c^2 - 7c) + (11c^3 + 6c^2 + 3)$

$\quad = -3c^3 + 11c^3 + 5c^2 + 6c^2 - 7c + 3$ Clear parentheses and group *like* terms.

$\quad = 8c^3 + 11c^2 - 7c + 3$ Combine *like* terms.

Tip: Polynomials can also be added by combining *like* terms in columns. The sum of the polynomials from Example 4b is shown here.

$\quad\quad -3c^3 + 5c^2 - 7c + 0$ Place holders such as 0 and $0c$ may be used to help
$\quad + \ 11c^3 + 6c^2 + 0c + 3$ line-up *like* terms.
$\quad\quad\overline{\ \ 8c^3 + 11c^2 - 7c + 3}$

c. $\left(\dfrac{1}{4}w^2 - \dfrac{2}{3}w\right) + \left(\dfrac{3}{4}w^2 + \dfrac{1}{6}w - \dfrac{1}{2}\right)$

$\quad = \dfrac{1}{4}w^2 + \dfrac{3}{4}w^2 - \dfrac{2}{3}w + \dfrac{1}{6}w - \dfrac{1}{2}$ Clear parentheses and group *like* terms.

$\quad = \dfrac{1}{4}w^2 + \dfrac{3}{4}w^2 - \dfrac{4}{6}w + \dfrac{1}{6}w - \dfrac{1}{2}$ Get common denominators for *like* terms.

$\quad = \dfrac{4}{4}w^2 - \dfrac{3}{6}w - \dfrac{1}{2}$ Add *like* terms.

$\quad = w^2 - \dfrac{1}{2}w - \dfrac{1}{2}$ Simplify.

4. Subtraction of Polynomials

The opposite (or additive inverse) of a real number a is $-a$. Similarly, if A is a polynomial, then $-A$ is its opposite.

example 5   Finding the Opposite of a Polynomial

Find the opposite of the polynomials.

a. $5x$ b. $3a - 4b - c$ c. $5.5y^4 - 2.4y^3 + 1.1y - 3$

Solution:

Tip: Notice that the sign of each term is changed when finding the opposite of a polynomial.

a. The opposite of $5x$ is $-(5x)$, or $-5x$.

b. The opposite of $3a - 4b - c$ is $-(3a - 4b - c)$ or equivalently, $-3a + 4b + c$.

c. The opposite of $5.5y^4 - 2.4y^3 + 1.1y - 3$ is $-(5.5y^4 - 2.4y^3 + 1.1y - 3)$, or equivalently, $-5.5y^4 + 2.4y^3 - 1.1y + 3$.

Subtraction of two polynomials is similar to subtracting real numbers. Add the opposite of the second polynomial to the first polynomial.

Definition of Subtraction of Polynomials

If A and B are polynomials, then $A - B = A + (-B)$.

example 6

Subtracting Polynomials

Subtract the polynomials.

a. $(-4p^4 + 5p^2 - 3) - (11p^2 + 4p - 6)$
b. $(a^2 - 2ab + 7b^2) - (-8a^2 - 6ab + 2b^2)$

Solution:

a. $(-4p^4 + 5p^2 - 3) - (11p^2 + 4p - 6)$

$= (-4p^4 + 5p^2 - 3) + (-11p^2 - 4p + 6)$ Add the opposite of the second polynomial.

$= -4p^4 + 5p^2 - 11p^2 - 4p - 3 + 6$ Group *like* terms.

$= -4p^4 - 6p^2 - 4p + 3$ Combine *like* terms.

Tip: Two polynomials can also be subtracted in columns by adding the opposite of the second polynomial to the first polynomial. Place holders (shown in red) may be used to help line up *like* terms.

$$
\begin{array}{r}
-4p^4 + 0p^3 + \ 5p^2 + 0p - 3 \\
-\underline{(0p^4 + 0p^3 + 11p^2 + 4p - 6)}
\end{array}
\quad \text{add the opposite} \quad
\begin{array}{r}
-4p^4 + 0p^3 + \ 5p^2 + 0p - 3 \\
+\ \underline{-0p^4 - 0p^3 - 11p^2 - 4p + 6} \\
-4p^4 \qquad - \ 6p^2 - 4p + 3
\end{array}
$$

Hence the difference of the polynomials is $-4p^4 - 6p^2 - 4p + 3$.

b. $(a^2 - 2ab + 7b^2) - (-8a^2 - 6ab + 2b^2)$

$= (a^2 - 2ab + 7b^2) + (8a^2 + 6ab - 2b^2)$ Add the opposite of the second polynomial.

$= a^2 + 8a^2 - 2ab + 6ab + 7b^2 - 2b^2$ Group *like* terms.

$= 9a^2 + 4ab + 5b^2$ Combine *like* terms.

Tip: Recall that $a - b = a + (-b)$, or equivalently, $a + -1b$. Therefore, subtraction of polynomials can be simplified by applying the distributive property to clear parentheses.

$$(a^2 - 2ab + 7b^2) - (-8a^2 - 6ab + 2b^2)$$

$$= a^2 - 2ab + 7b^2 - 1(-8a^2 - 6ab + 2b^2)$$

$$= a^2 - 2ab + 7b^2 + 8a^2 + 6ab - 2b^2 \qquad \text{Apply the distributive property.}$$

$$= a^2 + 8a^2 - 2ab + 6ab + 7b^2 - 2b^2 \qquad \text{Group } like \text{ terms.}$$

$$= 9a^2 + 4ab + 5b^2 \qquad \text{Combine } like \text{ terms.}$$

example 7

Subtracting Polynomials

Subtract: $\frac{1}{3}t^4 + \frac{1}{2}t^2$ from $t^2 - 4$ and simplify the result.

Classroom Activity 3.5B

Solution:

To subtract a from b, we write $b - a$. Thus, to subtract $\overbrace{\frac{1}{3}t^4 + \frac{1}{2}t^2}^{a}$ from $\overbrace{t^2 - 4}^{b}$, we have

$$\overset{b}{(t^2 - 4)} \overset{-}{-} \overset{a}{\left(\frac{1}{3}t^4 + \frac{1}{2}t^2\right)}$$

$$= t^2 - 4 - \frac{1}{3}t^4 - \frac{1}{2}t^2 \qquad \text{Apply the distributive property.}$$

$$= -\frac{1}{3}t^4 + t^2 - \frac{1}{2}t^2 - 4 \qquad \text{Group } like \text{ terms in descending order.}$$

$$= -\frac{1}{3}t^4 + \frac{2}{2}t^2 - \frac{1}{2}t^2 - 4 \qquad \text{The } t^2\text{-terms are the only } like \text{ terms.}$$

$$\qquad\qquad\qquad\qquad\qquad \text{Get a common denominator for the } t^2\text{-terms.}$$

$$= -\frac{1}{3}t^4 + \frac{1}{2}t^2 - 4 \qquad \text{Add } like \text{ terms.}$$

5. Polynomials and Applications to Geometry

example 8

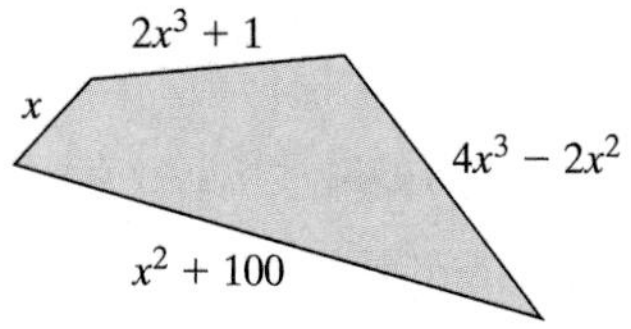

Figure 3-3

Adding Polynomials in Geometry

Find a polynomial that represents the perimeter of the polygon in Figure 3-3.

Solution:

The perimeter of a polygon is the sum of the lengths of the sides.

$$P = (x) + (2x^3 + 1) + (4x^3 - 2x^2) + (x^2 + 100)$$

$$= x + 2x^3 + 1 + 4x^3 - 2x^2 + x^2 + 100 \qquad \text{Clear parentheses.}$$

$$= 2x^3 + 4x^3 - 2x^2 + x^2 + x + 1 + 100 \qquad \text{Group } like \text{ terms.}$$

$$= 6x^3 - x^2 + x + 101 \qquad \text{Combine } like \text{ terms.}$$

The polynomial $6x^3 - x^2 + x + 101$ represents the perimeter of the figure.

example 9

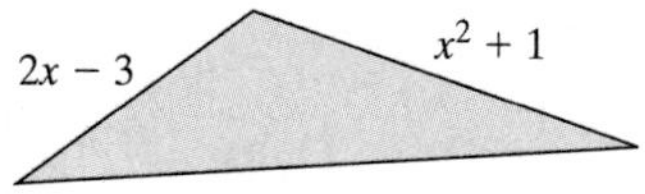

Figure 3-4

Classroom Activity 3.5C

Subtracting Polynomials in Geometry

If the perimeter of the triangle in Figure 3-4 can be represented by the polynomial $2x^2 + 5x + 6$, find a polynomial that represents the length of the missing side.

Solution:

The missing side of the triangle can be found by subtracting the sum of the two known sides from the perimeter.

$$\begin{pmatrix} \text{Length} \\ \text{of missing} \\ \text{side} \end{pmatrix} = (\text{perimeter}) - \begin{bmatrix} \text{sum of the} \\ \text{two known sides} \end{bmatrix}$$

$$\begin{pmatrix} \text{Length} \\ \text{of missing} \\ \text{side} \end{pmatrix} = (2x^2 + 5x + 6) - [(2x - 3) + (x^2 + 1)]$$

$$= 2x^2 + 5x + 6 - [2x - 3 + x^2 + 1] \qquad \text{Clear inner parentheses.}$$

$$= 2x^2 + 5x + 6 - (x^2 + 2x - 2) \qquad \text{Combine } like \text{ terms within [ ].}$$

$$= 2x^2 + 5x + 6 - x^2 - 2x + 2 \qquad \text{Apply the distributive property.}$$

$$= 2x^2 - x^2 + 5x - 2x + 6 + 2 \qquad \text{Group } like \text{ terms.}$$

$$= x^2 + 3x + 8 \qquad \text{Combine } like \text{ terms.}$$

section 3.5   PRACTICE EXERCISES

For Exercises 1–6, simplify the expression. Assume all variables represent nonzero real numbers.   3.1–3.3

1. $\dfrac{p^3 \cdot 4p}{p^2}$ $4p^2$

2. $(3x)(5x^{-4})$ $\dfrac{15}{x^3}$

3. $(6y^{-3})(2y^9)$ $12y^6$

4. $\dfrac{8t^{-6}}{4t^{-2}}$ $\dfrac{2}{t^4}$

5. $\dfrac{8^3 \cdot 8^{-4}}{8^{-2} \cdot 8^6}$ $\dfrac{1}{8^5}$

6. $\dfrac{3^4 \cdot 3^{-8}}{3^{12} \cdot 3^{-4}}$ $\dfrac{1}{3^{12}}$

7. Explain the difference between 3.0×10^7 and 3^7.   3.4 ❖

8. Explain the difference between 4.0×10^{-2} and 4^{-2}.   3.4 ❖

For Exercises 9–10, simplify and write the answers in scientific notation.   3.4

9. $(8,200,000)(0.000\ 016)$ 1.312×10^2

10. $\dfrac{9000}{0.000003}$ 3.0×10^9

11. Write a binomial of degree 3. (Answers may vary.)   For example: $x^3 + 6$

12. Write a trinomial of degree 6. (Answers may vary.)   For example: $5x^6 + x - 4$

13. Write a monomial of degree 5. (Answers may vary.)   For example: $8x^5$

14. Write a monomial of degree 1. (Answers may vary.)   For example: $3x$

15. Write a trinomial with the leading coefficient -6. (Answers may vary.)   For example: $-6x^2 + 2x + 5$

16. Write a binomial with the leading coefficient 13. (Answers may vary.)   For example: $13x + 7$

17. Write the polynomial in descending order:
$$6 + 7x^2 - 7x^4 + 9x$$
$-7x^4 + 7x^2 + 9x + 6$

18. Write the polynomial in descending order:
$$\tfrac{1}{2}y + y^2 - 12y^4 + y^3 - 6$$
$-12y^4 + y^3 + y^2 + \tfrac{1}{2}y - 6$

For Exercises 19–30, categorize the expression as a monomial, a binomial, or a trinomial.

19. $10a^2 + 5a$ Binomial

20. $7z + 13z^2 - 15$ Trinomial

21. $6x^2$ Monomial

22. 9 Monomial

23. $2t - t^4 - 5t$ Trinomial

24. $7x + 2$ Binomial

25. $12y^4 - 3y + 1$ Trinomial

26. $5bc^2$ Monomial

27. 23 Monomial

28. $4 - 2c$ Binomial

29. $w^4 - w^2$ Binomial

30. $-32xyz$ Monomial

31. Identify the degree for each monomial in Exercises 19–30.   Exercise 21: degree 2   Exercise 22: degree 0   Exercise 26: degree 3   Exercise 27: degree 0   Exercise 30: degree 3

32. Identify the coefficient for each monomial in Exercises 19–30.   Exercise 21: 6   Exercise 22: 9   Exercise 26: 5   Exercise 27: 23   Exercise 30: -32

33. Identify the degree for each binomial in Exercises 19–30.   Exercise 19: degree 2   Exercise 24: degree 1   Exercise 28: degree 1   Exercise 29: degree 4

34. Identify the degree for each trinomial in Exercises 19–30.   Exercise 20: degree 2   Exercise 23: degree 4   Exercise 25: degree 4

35. Identify the leading coefficient for each trinomial in Exercises 19–30.   Exercise 20: 13   Exercise 23: -1 Exercise 25: 12

36. Identify the leading coefficient for each binomial in Exercises 19–30.   Exercise 19: 10   Exercise 24: 7   Exercise 28: -2 Exercise 29: 1

37. A ball is dropped off a building and the height of the ball, h, (in feet) can be computed by the equation:
$$h = -16t^2 + 150$$
where t is the time (in seconds) after the ball is released.

   a. Find the height of the ball after 1 s, 1.5 s, and 2 s.   134 ft, 114 ft, 86 ft

   b. Find the height of the building by determining the height at the time of release.   150 ft

38. An object is dropped off a building and the height of the object, h, (in meters) can be computed by the equation:
$$h = -4.9t^2 + 45$$
where t is the time (in seconds) after the object is released.

❖ See Additional Answers Appendix   Writing   Translating Expression   Geometry   Scientific Calculator   Video

a. Find the height of the object after 1 s, 1.5 s, and 2 s. 40.1 m, 33.975 m, 25.4 m

b. Find the height of the building by determining the height at the time of release. 45 m

39. Explain why the terms $3x$ and $3x^2$ are not *like* terms. The exponents on the x-factors are different.

40. Explain why the terms $4w^3$ and $4z^3$ are not *like* terms. The variables are different.

For Exercises 41–52, add the polynomials.

41. $23x^2y + 12x^2y$ $35x^2y$

42. $-5ab^3 + 17ab^3$ $12ab^3$

43. $(6y + 3x) + (4y - 3x)$ $10y$

44. $(2z - 5h) + (-3z + h)$ $-z - 4h$

45. $3b^2 + (5b^2 - 9)$ $8b^2 - 9$

46. $4c + (3 - 10c)$ $-6c + 3$

47. $(7y^2 + 2y - 9) + (-3y^2 - y)$ $4y^2 + y - 9$

48. $(-3w^2 + 4w - 6) + (5w^2 + 2)$ $2w^2 + 4w - 4$

49. $\left(z - \dfrac{8}{3}\right) + \left(\dfrac{4}{3}z^2 - z + 1\right)$ $\frac{4}{3}z^2 - \frac{5}{3}$

50. $\left(-\dfrac{7}{5}r + 1\right) + \left(-\dfrac{3}{5}r^2 + \dfrac{7}{5}r + 1\right)$ $-\frac{3}{5}r^2 + 2$

51. $(7.9t^3 + 2.6t - 1.1) + (-3.4t^2 + 3.4t - 3.1)$
$7.9t^3 - 3.4t^2 + 6t - 4.2$

52. $(0.34y^2 + 1.23) + (3.42y - 7.56)$ $0.34y^2 + 3.42y - 6.33$

53. Find a polynomial that represents the perimeter of the figure. $4y^3 + 2y^2 + 2$

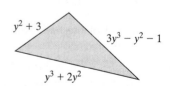

$y^2 + 3$

$3y^3 - y^2 - 1$

$y^3 + 2y^2$

Figure for Exercise 53

54. Find a polynomial that represents the perimeter of the figure. $9t^3 + t^2$

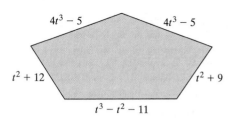

$4t^3 - 5$ $4t^3 - 5$

$t^2 + 12$ $t^2 + 9$

$t^3 - t^2 - 11$

Figure for Exercise 54

For Exercises 55–62, find the opposite of each polynomial.

55. $4h - 5$ $-4h + 5$

56. $5k - 12$ $-5k + 12$

57. $-2m^2 + 3m - 15$
$2m^2 - 3m + 15$

58. $-n^2 - 6n + 9$
$n^2 + 6n - 9$

59. $3v^3 + 5v^2 + 10v + 22$
$-3v^3 - 5v^2 - 10v - 22$

60. $7u^4 + 3v^2 + 17$
$-7u^4 - 3v^2 - 17$

61. $-9t^4 - 8t - 39$ $9t^4 + 8t + 39$

62. $-5r^5 - 3r^3 - r - 23$ $5r^5 + 3r^3 + r + 23$

For Exercises 63–78, subtract the polynomials.

63. $4a^3b^2 - 12a^3b^2$
$-8a^3b^2$

64. $5yz^4 - 14yz^4$
$-9yz^4$

65. $-32x^3 - 21x^3$
$-53x^3$

66. $-23c^5 - 12c^5$
$-35c^5$

67. $(7a - 7) - (12a - 4)$ $-5a - 3$

68. $(4x + 3v) - (-3x + v)$ $7x + 2v$

69. $(4k + 3) - (-12k - 6)$ $16k + 9$

70. $(3h - 15) - (8h - 13)$ $-5h - 2$

71. $25s - (23s - 14)$ $2s + 14$

72. $3x^2 - (-x^2 - 12)$ $4x^2 + 12$

73. $(5t^2 - 3t - 2) - (2t^2 + t + 1)$ $3t^2 - 4t - 3$

74. $(k^2 + 2k + 1) - (3k^2 - 6k + 2)$ $-2k^2 + 8k - 1$

75. $\left(\dfrac{2}{3}h^2 - \dfrac{1}{5}h - \dfrac{3}{4}\right) - \left(\dfrac{4}{3}h^2 - \dfrac{4}{5}h + \dfrac{7}{4}\right)$ $-\frac{2}{3}h^2 + \frac{3}{5}h - \frac{5}{2}$

76. $\left(\dfrac{3}{8}p^3 - \dfrac{5}{7}p^2 - \dfrac{2}{5}\right) - \left(\dfrac{5}{8}p^3 - \dfrac{2}{7}p^2 + \dfrac{7}{5}\right)$ $-\frac{1}{4}p^3 - \frac{3}{7}p^2 - \frac{9}{5}$

77. $(4.5x^4 - 3.1x^2 - 6.7) - (2.1x^4 + 4.4x)$
$2.4x^4 - 3.1x^2 - 4.4x - 6.7$

78. $(1.3c^3 + 4.8) - (4.3c^2 - 2c - 2.2)$
$1.3c^3 - 4.3c^2 + 2c + 7$

79. If the perimeter of the figure can be represented by the polynomial $5a^2 - 2a + 1$, find a polynomial that represents the length of the missing side. $3a^2 - 3a + 5$

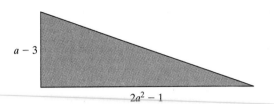

$a - 3$

$2a^2 - 1$

Figure for Exercise 79

80. If the perimeter of the figure can be represented by the polynomial $6w^3 - 2w - 3$, find a polynomial that represents the length of the missing side. $4w^3 - 6w^2 - 2w$

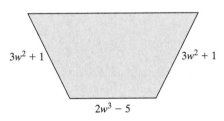

$3w^2 + 1$ $3w^2 + 1$

$2w^3 - 5$

Figure for Exercise 80

81. Subtract $(3x^3 - 5x + 10)$ from $(-2x^2 + 6x - 21)$.
$-3x^3 - 2x^2 + 11x - 31$

82. Subtract $(7a^5 - 2a^3 - 5a)$ from $(3a^5 - 9a^2 + 3a - 8)$. $-4a^5 + 2a^3 - 9a^2 + 8a - 8$

83. Find the difference of $(4b^3 + 6b - 7)$ and $(-12b^2 + 11b + 5)$. $4b^3 + 12b^2 - 5b - 12$

84. Find the difference of $(-5y^2 + 3y - 21)$ and $(-4y^2 - 5y + 23)$. $-y^2 + 8y - 44$

For Exercises 85–90, perform the indicated operation.

85. $(2ab^2 + 9a^2b) + (7ab^2 - 3ab + 7a^2b)$
$9ab^2 - 3ab + 16a^2b$

86. $(8x^2y - 3xy - 6xy^2) + (3x^2y - 12xy)$
$11x^2y - 15xy - 6xy^2$

87. $(4z^5 + z^3 - 3z + 13) - (-z^4 - 8z^3 + 15)$
$4z^5 + z^4 + 9z^3 - 3z - 2$

88. $(-15t^4 - 23t^2 + 16t) - (21t^3 + 18t^2 + t)$
$-15t^4 - 21t^3 - 41t^2 + 15t$

89. $(9x^4 + 2x^3 - x + 5) + (9x^3 - 3x^2 + 8x + 3) - (7x^4 - x + 12)$ $2x^4 + 11x^3 - 3x^2 + 8x - 4$

90. $(-6y^3 - 9y^2 + 23) - (7y^2 + 2y - 11) + (3y^3 - 25)$ $-3y^3 - 16y^2 - 2y + 9$

---

section

3.6   MULTIPLICATION OF POLYNOMIALS

Concepts

1. Multiplication of Monomials

2. Multiplication of Polynomials

3. Special Case Products: Difference of Squares and Perfect Square Trinomials

4. Multiplication of Polynomials and Applications to Geometry

1. Multiplication of Monomials

The properties of exponents covered in Sections 3.1–3.4 can be used to simplify many algebraic expressions including the multiplication of monomials. To multiply monomials, first use the associative and commutative properties of multiplication to group coefficients and like bases. Then simplify the result by using the properties of exponents.

example 1   **Multiplying Monomials**

Multiply the monomials.

a. $(3x^4)(4x^2)$ b. $(-4c^5d)(2c^2d^3e)$ c. $\left(\dfrac{1}{3}a^4b^3\right)\left(\dfrac{3}{4}b^7\right)$

Solution:

a. $(3x^4)(4x^2)$

$\qquad = (3 \cdot 4)(x^4x^2)$ Group coefficients and like bases.

$\qquad = 12x^6$ Add the exponents and simplify.

b. $(-4c^5d)(2c^2d^3e)$

$\qquad = (-4 \cdot 2)(c^5c^2)(dd^3)(e)$ Group coefficients and like bases.

$\qquad = -8c^7d^4e$ Simplify.

c. $\left(\dfrac{1}{3}a^4b^3\right)\left(\dfrac{3}{4}b^7\right)$

$\qquad = \left(\dfrac{1}{3} \cdot \dfrac{3}{4}\right)(a^4)(b^3b^7)$ Group coefficients and like bases.

$\qquad = \dfrac{1}{4}a^4b^{10}$ Simplify.

2. Multiplication of Polynomials

The distributive property is used to multiply polynomials: $a(b + c) = ab + ac$.

example 2   **Multiplying a Polynomial by a Monomial**

Multiply the polynomials

a. $2t(4t - 3)$ b. $-3a^2\left(-4a^2 + 2a - \dfrac{1}{3}\right)$

Solution:

a. $2t(4t - 3)$ Multiply each term of the polynomial by $2t$.

$\qquad = (2t)(4t) + 2t(-3)$ Apply the distributive property.

$\qquad = 8t^2 - 6t$ Simplify each term.

b. $-3a^2\left(-4a^2 + 2a - \dfrac{1}{3}\right)$ Multiply each term of the polynomial by $-3a^2$.

$= (-3a^2)(-4a^2) + (-3a^2)(2a) + (-3a^2)\left(-\dfrac{1}{3}\right)$ Apply the distributive property.

$= 12a^4 - 6a^3 + a^2$ Simplify each term.

Thus far, we have illustrated polynomial multiplication involving monomials. Next, the distributive property will be used to multiply polynomials with more than one term.

$(x + 3)(x + 5) = (x + 3)x + (x + 3)5$ Apply the distributive property.

$= (x + 3)x + (x + 3)5$ Apply the distributive property again.

$= x \cdot x + 3 \cdot x + x \cdot 5 + 3 \cdot 5$

$= x^2 + 3x + 5x + 15$

$= x^2 + 8x + 15$ Combine *like* terms.

Note: Using the distributive property results in multiplying each term of the first polynomial by each term of the second polynomial.

$(x + 3)(x + 5) = x \cdot x + x \cdot 5 + 3 \cdot x + 3 \cdot 5$

$= x^2 + 5x + 3x + 15$

$= x^2 + 8x + 15$

---

example 3   **Multiplying a Polynomial by a Polynomial**

Multiply the polynomials.

a. $(c - 7)(c + 2)$ b. $(10x + 3y)(2x - 4y)$ c. $(y - 2)(3y^2 + y - 5)$

Solution:

a. $(c - 7)(c + 2)$ Multiply each term in the first polynomial by each term in the second.

$= (c)(c) + (c)(2) + (-7)(c) + (-7)(2)$ Apply the distributive property.

$= c^2 + 2c - 7c - 14$ Simplify.

$= c^2 - 5c - 14$ Combine *like* terms.

Tip: Notice that the product of two *binomials* equals the sum of the products of the **F**irst terms, the **O**uter terms, the **I**nner terms and the **L**ast terms. The acronym, **FOIL** (First Outer Inner Last) can be used as a memory device to multiply two binomials.

$$
(c - 7)(c + 2) = (c)(c) + (c)(2) + (-7)(c) + (-7)(2)
$$
$$
= c^2 + 2c - 7c - 14
$$
$$
= c^2 - 5c - 14
$$

Classroom Activity 3.6A

b. $(10x + 3y)(2x - 4y)$ — Multiply each term in the first polynomial by each term in the second.

$$= (10x)(2x) + (10x)(-4y) + (3y)(2x) + (3y)(-4y)$$ — Apply the distributive property.

$$= 20x^2 - 40xy + 6xy - 12y^2$$ — Simplify each term.

$$= 20x^2 - 34xy - 12y^2$$ — Combine *like* terms.

◆ Avoiding Mistakes

It is important to note that the acronym **FOIL** does not apply to Example 3c because the product does not involve two binomials.

c. $(y - 2)(3y^2 + y - 5)$ — Multiply each term in the first polynomial by each term in the second.

$$= (y)(3y^2) + (y)(y) + (y)(-5) + (-2)(3y^2) + (-2)(y) + (-2)(-5)$$

$$= 3y^3 + y^2 - 5y - 6y^2 - 2y + 10$$ — Simplify each term.

$$= 3y^3 - 5y^2 - 7y + 10$$ — Combine *like* terms.

Tip: Multiplication of polynomials can be performed vertically by a process similar to column multiplication of real numbers. For example,

```
    235              3y^2 +  y - 5
  × 21            ×          y - 2
    235             -6y^2 - 2y + 10
   4700           3y^3 +  y^2 - 5y +  0
   4935           3y^3 - 5y^2 - 7y + 10
```

Note: When multiplying by the column method, it is important to *align like* terms vertically before adding terms.

3. Special Case Products: Difference of Squares and Perfect Square Trinomials

In some cases the product of two binomials take on a special pattern.

I. The first special case occurs when multiplying the sum and difference of the same two terms. For example:

$$(2x + 3)(2x - 3)$$
$$= 2x^2 - 6x + 6x - 9$$
$$= 4x^2 - 9$$

Notice that the middle terms are opposites. This leaves only the difference between the square of the first term and the square of the second term. For this reason, the product is called a difference of squares.

Note: The sum and difference of the same two terms are called **conjugates**. Thus, the expressions $2x + 3$ and $2x - 3$ are conjugates of each other.

II. The second special case involves the square of a binomial. For example:

$$(3x + 7)^2$$
$$= (3x + 7)(3x + 7)$$
$$= 9x^2 + 21x + 21x + 49$$
$$= 9x^2 + 42x + 49$$
$$= (3x)^2 + 2(3x)(7) + (7)^2$$

When squaring a binomial, the product will be a trinomial called a perfect square trinomial. The first and third terms are formed by squaring the terms of the binomial. The middle term equals twice the product of the terms in the binomial.

Note: The expression $(3x - 7)^2$ also expands to a perfect square trinomial, but the middle term will be negative:

$$(3x - 7)(3x - 7) = 9x^2 - 21x - 21x + 49 = 9x^2 - 42x + 49$$

Special Case Product Formulas

1. $(a + b)(a - b) = a^2 - b^2$ The product is called a **difference of squares**.

2. $(a + b)^2 = a^2 + 2ab + b^2$
$(a - b)^2 = a^2 - 2ab + b^2$ The product is called a **perfect square trinomial**.

You should become familiar with these special case products because they will be used again in the next chapter to factor polynomials.

example 4   **Finding Special Products**

Use the special product formulas to multiply the polynomials.

a. $(x - 9)(x + 9)$ b. $\left(\frac{1}{2}p - 6\right)\left(\frac{1}{2}p + 6\right)$

Solution:

a. $(x - 9)(x + 9)$ Apply the formula: $(a + b)(a - b) = a^2 - b^2$.

$$a^2 - b^2$$

$$= (x)^2 - (9)^2$$ Substitute $a = x$ and $b = 9$.

$$= x^2 - 81$$

Tip: The product of two conjugates can be checked by applying the distributive property:

$$(x - 9)(x + 9) = x^2 + 9x - 9x - 81$$
$$= x^2 - 81$$

b. $\left(\dfrac{1}{2}p - 6\right)\left(\dfrac{1}{2}p + 6\right)$ Apply the formula: $(a + b)(a - b) = a^2 - b^2$.

$$a^2 - b^2$$

$$= \left(\dfrac{1}{2}p\right)^2 - (6)^2$$ Substitute $a = \dfrac{1}{2}p, b = 6$.

$$= \dfrac{1}{4}p^2 - 36$$ Simplify each term.

---

example 5   **Finding Special Products**

Use the special product formulas to multiply the polynomials.

a. $(3w - 4)^2$ b. $(5x^2 + 2)^2$

Solution:

a. $(3w - 4)^2$ Apply the formula:
    $(a - b)^2 = a^2 - 2ab + b^2$.

$$a^2 - 2ab + b^2$$

$$= (3w)^2 - 2(3w)(4) + (4)^2$$ Substitute $a = 3w, b = 4$.

$$= 9w^2 - 24w + 16$$ Simplify each term.

Tip: The square of a binomial can be checked by explicitly writing the product of the two binomials and applying the distributive property:

$$(3w - 4)^2 = (3w - 4)(3w - 4) = 9w^2 - 12w - 12w + 16$$
$$= 9w^2 - 24w + 16$$

b. $(5x^2 + 2)^2$
                                            Apply the formula:
                                            $(a + b)^2 = a^2 + 2ab + b^2$.

$$\underbrace{a^2 + 2ab + b^2}$$

$$= (5x^2)^2 + 2(5x^2)(2) + (2)^2 \qquad \text{Substitute } a = 5x^2, b = 2.$$

$$= 25x^4 + 20x^2 + 4 \qquad \text{Simplify each term.}$$

4. Multiplication of Polynomials and Applications to Geometry

example 6

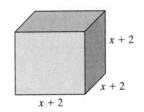

Figure 3-5

Classroom Activity 3.6B

Using Special Case Products in an Application of Geometry

Find a polynomial that represents the volume of the cube (Figure 3-5).

Solution:

$$\text{Volume} = (\text{length})(\text{width})(\text{height})$$

$$V = (x + 2)(x + 2)(x + 2) \qquad \text{or} \qquad V = (x + 2)^3$$

To expand $(x + 2)(x + 2)(x + 2)$, multiply the first two factors. Then multiply the result by the last factor.

$$V = \underline{(x + 2)(x + 2)}(x + 2)$$
$$= (x^2 + 4x + 4)(x + 2) \longleftarrow$$

Tip: $(x + 2)(x + 2) = (x + 2)^2$ and results in a perfect square trinomial.

$$(x + 2)^2 = (x)^2 + 2(x)(2) + (2)^2$$
$$= x^2 + 4x + 4$$

$$= (x^2)(x) + (x^2)(2) + (4x)(x) + (4x)(2) + (4)(x) + (4)(2) \qquad \text{Apply the distributive property.}$$

$$= x^3 + 2x^2 + 4x^2 + 8x + 4x + 8 \qquad \text{Group } like \text{ terms.}$$

$$= x^3 + 6x^2 + 12x + 8 \qquad \text{Combine } like \text{ terms.}$$

The volume of the cube can be represented by

$$V = (x + 2)^3 = x^3 + 6x^2 + 12x + 8.$$

example 7   Using the Product of Polynomials in Geometry

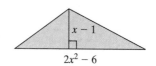

Find a polynomial that represents the area of the triangle (Figure 3-6).

Figure 3-6

Solution:

The area of a triangle is found by $A = \frac{1}{2}bh$. In this case, $b = 2x^2 - 6$ and $h = x - 1$; therefore, $A = \frac{1}{2}(2x^2 - 6)(x - 1)$.

$A = \dfrac{1}{2}(2x^2 - 6)(x - 1)$ Apply the distributive property.

$= (x^2 - 3)(x - 1)$

$= (x^2 - 3)(x - 1)$ Multiply each term in the first polynomial by each term in the second.

$= (x^2)(x) + (x^2)(-1) + (-3)(x) + (-3)(-1)$ Apply the distributive property.

$= x^3 - x^2 - 3x + 3$

The area of the triangle can be represented by the polynomial $x^3 - x^2 - 3x + 3$.

---

section 3.6   PRACTICE EXERCISES

For Exercises 1–12, simplify the expressions (if possible).
3.2, 3.5

1. $4x + 5x$ $9x$
2. $2y^2 - 4y^2$ $-2y^2$
3. $(4x)(5x)$ $20x^2$
4. $(2y^2)(-4y^2)$ $-8y^4$
5. $-5a^3b - 2a^3b$ $-7a^3b$
6. $7uvw^2 + uvw^2$ $8uvw^2$
7. $(-5a^3b)(-2a^3b)$ $10a^6b^2$
8. $(7uvw^2)(uvw^2)$ $7u^2v^2w^4$
9. $-c + 4c^2$ $4c^2 - c$
10. $3t + 3t^3$ $3t^3 + 3t$
11. $(-c)(4c^2)$ $-4c^3$
12. $(3t)(3t^3)$ $9t^4$

For Exercises 13–20, perform the indicated operations. Write the answers in scientific notation.   3.4

13. $(4.3 \times 10^6) + (2.3 \times 10^6)$ 6.6×10^6
14. $(3.9 \times 10^{-5}) + (1.5 \times 10^{-5})$ 5.4×10^{-5}
15. $(4.3 \times 10^6)(2.3 \times 10^6)$ 9.89×10^{12}
16. $(3.9 \times 10^{-5})(1.5 \times 10^{-5})$ 5.85×10^{-10}

17. $(-2.1 \times 10^{-12}) - (9.3 \times 10^{-12})$ -1.14×10^{-11}
18. $(4.4 \times 10^9) - (5.4 \times 10^9)$ -1.0×10^9
19. $(-2.1 \times 10^{-12})(9.3 \times 10^{-12})$ -1.953×10^{-23}
20. $(4.4 \times 10^9)(5.4 \times 10^9)$ 2.376×10^{19}

For Exercises 21–44, multiply the expressions.

21. $8(4x)$ $32x$
22. $-2(6y)$ $-12y$
23. $-10(5z)$ $-50z$
24. $7(3p)$ $21p$
25. $(x^{10})(4x^3)$ $4x^{13}$
26. $(a^{13}b^4)(12ab^4)$ $12a^{14}b^8$
27. $(4m^3n^7)(-3m^6n)$ $-12m^9n^8$
28. $(2c^7d)(-c^3d^{11})$ $-2c^{10}d^{12}$
29. $8pq(2pq - 3p + 5q)$ $16p^2q^2 - 24p^2q + 40pq^2$
30. $5ab(2ab + 6a - 3b)$ $10a^2b^2 + 30a^2b - 15ab^2$
31. $(k^2 - 13k - 6)(-4k)$ $-4k^3 + 52k^2 + 24k$
32. $(h^2 + 5h - 12)(-2h)$ $-2h^3 - 10h^2 + 24h$
33. $-15pq(3p^2 + p^3q^2 - 2q)$ $-45p^3q - 15p^4q^3 + 30pq^2$

34. $-4u^2v(2u - 5uv^3 + v)$ $-8u^3v + 20u^3v^4 - 4u^2v^2$

35. $(y - 10)(y + 9)$ 36. $(x + 5)(x - 6)$
$y^2 - y - 90$ $x^2 - x - 30$

37. $(m - 12)(m - 2)$ 38. $(n - 7)(n - 2)$
$m^2 - 14m + 24$ $n^2 - 9n + 14$

39. $(3x + 4)(x + 8)$ 40. $(7y + 1)(3y + 5)$
$3x^2 + 28x + 32$ $21y^2 + 38y + 5$

41. $(5s + 3)(s^2 + s - 2)$ 42. $(t - 4)(2t^2 - t + 6)$
$5s^3 + 8s^2 - 7s - 6$ $2t^3 - 9t^2 + 10t - 24$

43. $(3w - 2)(9w^2 + 6w + 4)$ $27w^3 - 8$

44. $(z + 5)(z^2 - 5z + 25)$ $z^3 + 125$

For Exercises 45–52, multiply the expressions.

45. $(3a - 4b)(3a + 4b)$ 46. $(5y + 7x)(5y - 7x)$
$9a^2 - 16b^2$ $25y^2 - 49x^2$

47. $(9k + 6)(9k - 6)$ 48. $(2h - 5)(2h + 5)$
$81k^2 - 36$ $4h^2 - 25$

49. $\left(\dfrac{1}{2} - t\right)\left(\dfrac{1}{2} + t\right)$ $\dfrac{1}{4} - t^2$ 50. $\left(r + \dfrac{1}{4}\right)\left(r - \dfrac{1}{4}\right)$ $r^2 - \dfrac{1}{16}$

51. $(u^3 + 5v)(u^3 - 5v)$ 52. $(8w^2 - x)(8w^2 + x)$
$u^6 - 25v^2$ $64w^4 - x^2$

For Exercises 53–60, square the binomials.

53. $(a + b)^2$ $a^2 + 2ab + b^2$ 54. $(a - b)^2$ $a^2 - 2ab + b^2$

55. $(x - y)^2$ $x^2 - 2xy + y^2$ 56. $(x + y)^2$ $x^2 + 2xy + y^2$

57. $(2c + 5)^2$ $4c^2 + 20c + 25$ 58. $(5d - 9)^2$ $25d^2 - 90d + 81$

59. $(3t^2 - 4s)^2$ 60. $(u^2 + 4v)^2$
$9t^4 - 24st^2 + 16s^2$ $u^4 + 8u^2v + 16v^2$

61. a. Evaluate $(2 + 4)^2$ by working within the parentheses first.   36
    b. Evaluate $2^2 + 4^2$.   20
    c. Compare the answers to parts (a) and (b) and make a conjecture about $(a + b)^2$ and $a^2 + b^2$. $(a + b)^2 \neq a^2 + b^2$ in general

62. a. Evaluate $(6 - 5)^2$ by working within the parentheses first.   1
    b. Evaluate $6^2 - 5^2$.   11
    c. Compare the answers to parts (a) and (b) and make a conjecture about $(a - b)^2$ and $a^2 - b^2$. $(a - b)^2 \neq a^2 - b^2$ in general

63. Find a polynomial expression that represents the area of the rectangle shown in the figure. $4x^2 - 25$

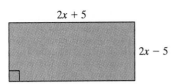

$2x + 5$

$2x - 5$

Figure for Exercise 63

64. Find a polynomial expression that represents the area of the rectangle shown in the figure. $36 - y^2$

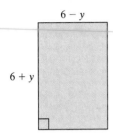

$6 - y$

$6 + y$

Figure for Exercise 64

65. Find a polynomial expression that represents the area of the square shown in the figure.
$16p^2 + 40p + 25$

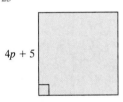

$4p + 5$

Figure for Exercise 65

66. Find a polynomial expression that represents the area of the square shown in the figure.
$49q^2 - 42q + 9$

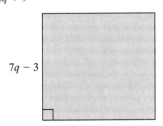

$7q - 3$

Figure for Exercise 66

For Exercises 67–88, multiply the expressions.

67. $(7x + y)(7x - y)$ 68. $(9w - 4z)(9w + 4z)$
$49x^2 - y^2$ $81w^2 - 16z^2$

69. $(5s + 3t)^2$ 70. $(5s - 3t)^2$
$25s^2 + 30st + 9t^2$ $25s^2 - 30st + 9t^2$

71. $(7x - 3y)(3x - 8y)$ 72. $(5a - 4b)(2a - b)$
    $21x^2 - 65xy + 24y^2$ $10a^2 - 13ab + 4b^2$

73. $\left(\dfrac{2}{3}t + 2\right)(3t + 4)$ 74. $\left(\dfrac{1}{5}s + 6\right)(5s - 3)$
    $2t^2 + \dfrac{26}{3}t + 8$ $s^2 + \dfrac{147}{5}s - 18$

75. $(5z + 3)(z^2 + 4z - 1)$ $5z^3 + 23z^2 + 7z - 3$

76. $(2k - 5)(2k^2 + 3k + 5)$ $4k^3 - 4k^2 - 5k - 25$

77. $\left(\dfrac{1}{3}m - n\right)^2$ 78. $\left(\dfrac{2}{5}p - q\right)^2$
    $\dfrac{1}{9}m^2 - \dfrac{2}{3}mn + n^2$ $\dfrac{4}{25}p^2 - \dfrac{4}{5}pq + q^2$

79. $6w^2(7w - 14)$ 80. $4v^3(v + 12)$
    $42w^3 - 84w^2$ $4v^4 + 48v^3$

81. $(4y - 8.1)(4y + 8.1)$ 82. $(2h + 2.7)(2h - 2.7)$
    $16y^2 - 65.61$ $4h^2 - 7.29$

83. $(3c^2 + 4)(7c^2 - 8)$ 84. $(5k^3 - 9)(k^3 - 2)$
    $21c^4 + 4c^2 - 32$ $5k^6 - 19k^3 + 18$

85. $(3.1x + 4.5)^2$ 86. $(2.5y + 1.1)^2$
    $9.61x^2 + 27.9x + 20.25$ $6.25y^2 + 5.5y + 1.21$

87. $(k - 4)^3$ 88. $(h + 3)^3$
    $k^3 - 12k^2 + 48k - 64$ $h^3 + 9h^2 + 27h + 27$

89. Find a polynomial that represents the area of the triangle shown in the figure. $15a^5 - 6a^2$

    (Recall: $A = \frac{1}{2}bh$)

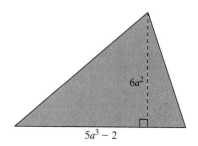

Figure for Exercise 89

90. Find a polynomial that represents the area of the triangle shown in the figure. $3t^3 - 12t$

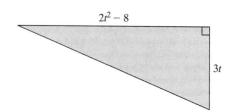

Figure for Exercise 90

91. Find a polynomial that represents the volume of the cube shown in the figure.

    (Recall: $V = s^3$) $27p^3 - 135p^2 + 225p - 125$

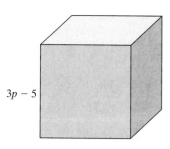

Figure for Exercise 91

92. Find a polynomial that represents the volume of the rectangular solid shown in the figure.

    (Recall: $V = lwh$) $r^3 - 15r^2 + 63r - 49$

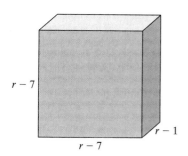

Figure for Exercise 92

EXPANDING YOUR SKILLS

93. What binomial when multiplied by $(3x + 5)$ will produce a product of $6x^2 - 11x - 35$? *Hint*: Let the quantity $(a + b)$ represent the unknown binomial. Then find a and b such that $(3x + 5)(a + b) = 6x^2 - 11x - 35$. $2x - 7$

94. What binomial when multiplied by $(2x - 4)$ will produce a product of $2x^2 + 8x - 24$? $x + 6$

section

3.7   DIVISION OF POLYNOMIALS

1.   Division by a Monomial

Division of polynomials will be presented in this section as two separate cases: The first case illustrates **division by a monomial** divisor. The second case illustrates division by a polynomial with two or more terms.

To divide a polynomial by a monomial, divide each individual term in the polynomial by the divisor and simplify the result.

Dividing a Polynomial by a Monomial

If a, b, and c are polynomials such that $c \neq 0$, then

$$\frac{a+b}{c} = \frac{a}{c} + \frac{b}{c} \qquad \text{Similarly,} \qquad \frac{a-b}{c} = \frac{a}{c} - \frac{b}{c}$$

example 1

Dividing a Polynomial by a Monomial

Divide the polynomials.

a. $\dfrac{5a^3 - 10a^2 + 20a}{5a}$

b. $(12y^2z^3 - 15yz^2 + 6y^2z) \div (-6y^2z)$

Classroom Activity 3.7A

Solution:

a. $\dfrac{5a^3 - 10a^2 + 20a}{5a}$

$= \dfrac{5a^3}{5a} - \dfrac{10a^2}{5a} + \dfrac{20a}{5a}$ Divide each term in the numerator by $5a$.

$= a^2 - 2a + 4$ Simplify each term using the properties of exponents.

b. $(12y^2z^3 - 15yz^2 + 6y^2z) \div (-6y^2z)$

$= \dfrac{12y^2z^3 - 15yz^2 + 6y^2z}{-6y^2z}$

$= \dfrac{12y^2z^3}{-6y^2z} - \dfrac{15yz^2}{-6y^2z} + \dfrac{6y^2z}{-6y^2z}$ Divide each term by $-6y^2z$.

$= -2z^2 + \dfrac{5z}{2y} - 1$ Simplify each term.

2. Long Division

If the divisor has two or more terms, a **long division** process similar to the division of real numbers is used. Take a minute to review the long division process for real numbers by dividing 5074 by 31.

$$
31\overline{)5074}
\qquad
\begin{array}{r}
1 \\
31\overline{)5074} \\
-31 \\
\hline
19
\end{array}
\quad\text{Subtract.}
$$

$$
\begin{array}{r}
16 \\
31\overline{)5074} \\
-31\downarrow \\
\hline
197 \\
-186 \\
\hline
11
\end{array}
$$

Bring down next column and repeat the process.

Subtract.

$$
\begin{array}{r}
163 \longleftarrow \text{Quotient} \\
31\overline{)5074} \\
-31 \\
\hline
197 \\
-186 \\
\hline
114 \\
-93 \\
\hline
21 \longleftarrow \text{Remainder}
\end{array}
$$

Therefore $5074 \div 31 = 163\frac{21}{31}$

A similar procedure is used for long division of polynomials as shown in the next example.

---

example 2

Using Long Division to Divide Polynomials

Divide the polynomials using long division: $(2x^2 - x + 3) \div (x - 3)$

Solution:

$$x - 3\overline{)2x^2 - x + 3}$$

Divide the leading term in the dividend by the leading term in the divisor.

$$\frac{2x^2}{x} = 2x$$

This is the first term in the quotient.

$$
\begin{array}{r}
2x \\
x - 3\overline{)2x^2 - x + 3} \\
-(2x^2 - 6x)
\end{array}
$$

Multiply $2x$ by the divisor.

$$2x(x - 3) = 2x^2 - 6x$$

and subtract the result.

$$\begin{array}{r} 2x \phantom{{}-x+3} \\ x-3\overline{)2x^2-x+3} \\ \underline{-2x^2+6x} \\ 5x \phantom{{}+3} \end{array}$$

Subtract the quantity $2x^2-6x$. To do this, add the opposite.

$$\begin{array}{r} 2x+5 \\ x-3\overline{)2x^2-x+3} \\ \underline{-2x^2+6x} \downarrow \\ 5x+3 \end{array}$$

Bring down the next column and repeat the process.

Divide the leading term by x: $(5x)/x = 5$.
Place 5 in the quotient.

$$\begin{array}{r} 2x+5 \\ x-3\overline{)2x^2-x+3} \\ \underline{-2x^2+6x} \\ 5x+3 \\ -(5x-15) \end{array}$$

Multiply the divisor by 5: $5(x-3) = 5x-15$ and subtract the result.

$$\begin{array}{r} 2x+5 \\ x-3\overline{)2x^2-x+3} \\ \underline{-2x^2+6x} \\ 5x+3 \\ \underline{-5x+15} \\ 18 \end{array}$$

Subtract the quantity $5x-15$ by adding the opposite.

The remainder is 18.

Summary:

The quotient is $2x+5$
The remainder is 18
The divisor is $x-3$
The dividend is $2x^2-x+3$

The solution to a long division problem is usually written in the form:

$$\text{quotient} + \frac{\text{remainder}}{\text{divisor}}$$

Hence

$$(2x^2-x+3) \div (x-3) = 2x+5+\frac{18}{x-3}$$

The division of polynomials can be checked in the same fashion as the division of real numbers. To check, we have

$$\text{Dividend} = (\text{divisor})(\text{quotient}) + \text{remainder}$$

$$2x^2-x+3 \overset{?}{=} (x-3)(2x+5)+(18)$$
$$\overset{?}{=} 2x^2-6x+5x-15+(18)$$
$$\overset{?}{=} 2x^2-x+3 \ \checkmark$$

example 3

Using Long Division to Divide Polynomials

Divide the polynomials using long division: $(2w^3 + 8w^2 - 16) \div (2w + 4)$

Classroom Activity 3.7B

Solution:

First note that the dividend has a missing power of w and can be written as $2w^3 + 8w^2 + 0w - 16$. The term $0w$ is a place holder for the missing term. It is helpful to use the place holder to keep the powers of w lined up.

$$
\begin{array}{r}
w^2 \\
2w + 4 \overline{)\, 2w^3 + 8w^2 + 0w - 16} \\
-(2w^3 + 4w^2)
\end{array}
$$

Divide $2w^3 \div 2w = w^2$. This is the first term of the quotient.
Then multiply $w^2(2w + 4) = 2w^3 + 4w^2$.

$$
\begin{array}{r}
w^2 \\
2w + 4 \overline{)\, 2w^3 + 8w^2 + 0w - 16} \\
-2w^3 - 4w^2 \\
\hline
4w^2 + 0w
\end{array}
$$

Subtract by adding the opposite.

Bring down the next column and repeat the process.

$$
\begin{array}{r}
w^2 + 2w \\
2w + 4 \overline{)\, 2w^3 + 8w^2 + 0w - 16} \\
-2w^3 - 4w^2 \\
\hline
4w^2 + 0w \\
-(4w^2 + 8w)
\end{array}
$$

Divide $4w^2$ by the leading term in the divisor. $4w^2 \div 2w = 2w$. Place $2w$ in the quotient.
Multiply $2w(2w + 4) = 4w^2 + 8w$.

$$
\begin{array}{r}
w^2 + 2w \\
2w + 4 \overline{)\, 2w^3 + 8w^2 + 0w - 16} \\
-2w^3 - 4w^2 \\
\hline
4w^2 + 0w \\
-4w^2 - 8w \\
\hline
-8w - 16
\end{array}
$$

Subtract by adding the opposite.

Bring down the next column and repeat.

$$
\begin{array}{r}
w^2 + 2w - 4 \\
2w + 4 \overline{)\, 2w^3 + 8w^2 + 0w - 16} \\
-2w^3 - 4w^2 \\
\hline
4w^2 + 0w \\
-4w^2 - 8w \\
\hline
-8w - 16 \\
-(-8w - 16)
\end{array}
$$

Divide $-8w$ by the leading term in the divisor. $-8w \div 2w = -4$. Place -4 in the quotient.
Multiply $-4(2w + 4) = -8w - 16$.

$$
\begin{array}{r}
w^2 + 2w - 4 \\
2w + 4 \overline{)\, 2w^3 + 8w^2 + 0w - 16} \\
-2w^3 - 4w^2 \\
\hline
4w^2 + 0w \\
-4w^2 - 8w \\
\hline
-8w - 16 \\
8w + 16 \\
\hline
0
\end{array}
$$

Subtract by adding the opposite.

The remainder is 0.

The quotient is $w^2 + 2w - 4$ and the remainder is 0.

In Example 3 the remainder is zero. Therefore, we say that $2w + 4$ divides *evenly* into $2w^3 + 8w^2 - 16$. For this reason, the divisor and quotient are factors of $2w^3 + 8w^2 - 16$. To check, we have

$$\text{Dividend} = (\text{divisor})(\text{quotient}) + \text{remainder}$$

$$2w^3 + 8w^2 - 16 \overset{?}{=} (2w + 4)(w^2 + 2w - 4) + 0$$
$$\overset{?}{=} 2w^3 + 4w^2 - 8w + 4w^2 + 8w - 16$$
$$\overset{?}{=} 2w^3 + 8w^2 - 16 ✔$$

---

example 4   **Using Long Division to Divide Polynomials**

Divide the polynomials using long division.

$$\frac{y^4 + 2y - 5}{y^2 + 1}$$

Solution:

Both the dividend and the divisor have missing powers of y. Leave place holders.

$$y^2 + 0y + 1 \overline{)y^4 + 0y^3 + 0y^2 + 2y - 5}$$

$$\begin{array}{r} y^2 \\ y^2 + 0y + 1 \overline{)y^4 + 0y^3 + 0y^2 + 2y - 5} \\ -(y^4 + 0y^3 + y^2) \end{array}$$

Divide $y^4 \div y^2 = y^2$. This is the first term of the quotient.

Multiply $y^2(y^2 + 0y + 1) = y^4 + 0y^3 + y^2$

$$\begin{array}{r} y^2 \\ y^2 + 0y + 1 \overline{)y^4 + 0y^3 + 0y^2 + 2y - 5} \\ -y^4 - 0y^3 - y^2 \\ \hline -y^2 + 2y - 5 \end{array}$$

Subtract by adding the opposite.

Bring down the next columns.

$$\begin{array}{r} y^2 -1 \\ y^2 + 0y + 1 \overline{)y^4 + 0y^3 + 0y^2 + 2y - 5} \\ -y^4 - 0y^3 - y^2 \\ \hline -y^2 + 2y - 5 \\ -(-y^2 - 0y - 1) \end{array}$$

Divide $-y^2 \div y^2 = -1$.

Multiply $-1(y^2 + 0y + 1) = -y^2 - 0y - 1.$

$$\begin{array}{r} y^2 -1 \\ y^2 + 0y + 1 \overline{)y^4 + 0y^3 + 0y^2 + 2y - 5} \\ -y^4 - 0y^3 - y^2 \\ \hline -y^2 + 2y - 5 \\ y^2 + 0y + 1 \\ \hline 2y - 4 \end{array}$$

Subtract by adding the opposite.

Remainder

Therefore,

$$\frac{y^4 + 2y - 5}{y^2 + 1} = y^2 - 1 + \frac{2y - 4}{y^2 + 1}$$

Check:

$$\text{Dividend} = (\text{divisor})(\text{quotient}) + \text{remainder}$$

$$y^4 + 2y - 5 \stackrel{?}{=} (y^2 + 1)(y^2 - 1) + (2y - 4)$$
$$\stackrel{?}{=} y^4 + y^2 - y^2 - 1 + 2y - 4$$
$$\stackrel{?}{=} y^4 + 2y - 5 \checkmark$$

Tip: Recall that

- Long division is used when the divisor has *two or more terms*.
- If the divisor has *one term* (when the divisor is a monomial) then divide each term in the dividend by the monomial divisor.

---

example 5

Determining Whether Long Division Is Necessary

Determine whether long division is necessary for each division of polynomials.

a. $\dfrac{2p^5 - 8p^4 + 4p - 16}{p^2 - 2p + 1}$

b. $\dfrac{2p^5 - 8p^4 + 4p - 16}{2p^2}$

c. $(3z^3 - 5z^2 + 10) \div (15z^3)$

d. $(3z^3 - 5z^2 + 10) \div (3z + 1)$

Solution:

a. $\dfrac{2p^5 - 8p^4 + 4p - 16}{p^2 - 2p + 1}$ The divisor has three terms. Use long division.

b. $\dfrac{2p^5 - 8p^4 + 4p - 16}{2p^2}$ The divisor has one term. No long division.

c. $(3z^3 - 5z^2 + 10) \div (15z^3)$ The divisor has one term. No long division.

d. $(3z^3 - 5z^2 + 10) \div (3z + 1)$ The divisor has two terms. Use long division.

---

section 3.7   PRACTICE EXERCISES

For Exercises 1–10, perform the indicated operations. 3.5, 3.6

1. $(6z^5 - 2z^3 + z - 6) - (10z^4 + 2z^3 + z^2 + z)$
   $6z^5 - 10z^4 - 4z^3 - z^2 - 6$

2. $(7a^2 + a - 6) + (2a^2 + 5a + 11)$ $9a^2 + 6a + 5$

3. $(10x + y)(x - 3y)$
   $10x^2 - 29xy - 3y^2$

4. $8b^2(2b^2 - 5b + 12)$
   $16b^4 - 40b^3 + 96b^2$

5. $(2w^3 + 5)^2$ $4w^6 + 20w^3 + 25$

6. $\left(\dfrac{4}{3}y^2 - \dfrac{1}{2}y + \dfrac{3}{8}\right) - \left(\dfrac{1}{3}y^2 + \dfrac{1}{4}y - \dfrac{1}{8}\right)$ $y^2 - \dfrac{3}{4}y + \dfrac{1}{2}$

7. $\left(\dfrac{7}{8}w - 1\right)\left(\dfrac{7}{8}w + 1\right)$ $\dfrac{49}{64}w^2 - 1$

8. $(0.3a^4 - 1.9a^3 + 2.1a - 8.1) + (6.4a^4 + 2.7a^2 + 9.2a + 7.1)$ $6.7a^4 - 1.9a^3 + 2.7a^2 + 11.3a - 1$

9. $(a + 3)(a^2 - 3a + 9)$ $a^3 + 27$

10. $(2x + 1)(5x - 3)$ $10x^2 - x - 3$

11. There are two methods for dividing polynomials. Explain when long division is used.   Use long division when the divisor is a polynomial with two or more terms.

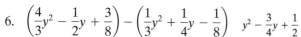 ❖ See Additional Answers Appendix    Writing   Translating Expression   Geometry   Scientific Calculator   Video

12. Explain how to check a polynomial division problem. Multiply the quotient by the divisor and add the remainder.

13. a. Divide $5t^2 + 6t$

$$\frac{15t^3 + 18t^2}{3t}$$

   b. Check by multiplying the quotient by the divisor.

14. a. Divide: $(-9y^4 + 6y^2 - y) \div (3y)$ $3y^3 + 2y - \dfrac{1}{3}$

   b. Check by multiplying the quotient by the divisor.

For Exercises 15–28, divide the polynomials.

15. $(6a^2 + 4a - 14) \div (2)$ $3a^2 + 2a - 7$

16. $\dfrac{4b^2 + 16b - 12}{4}$ $b^2 + 4b - 3$

17. $\dfrac{3p^3 - p^2}{p}$ $3p^2 - p$

18. $(7q^4 + 5q^2) \div q$ $7q^3 + 5q$

19. $(4m^2 + 8m) \div 4m^2$ $1 + \dfrac{2}{m}$

20. $\dfrac{n^2 - 8}{n}$ $n - \dfrac{8}{n}$

21. $\dfrac{14y^4 - 7y^3 + 21y^2}{-7y^2}$ $-2y^2 + y - 3$

22. $(25a^5 - 5a^4 + 15a^3 - 5a) \div (-5a)$ $-5a^4 + a^3 - 3a^2 + 1$

23. $(4x^3 - 24x^2 - x + 8) \div (4x)$ $x^2 - 6x - \dfrac{1}{4} + \dfrac{2}{x}$

24. $\dfrac{20w^3 + 15w^2 - w + 5}{10w}$ $2w^2 + \dfrac{3}{2}w - \dfrac{1}{10} + \dfrac{1}{2w}$

25. $\dfrac{-a^3b^2 + a^2b^2 - ab^3}{-a^2b^2}$ $a - 1 + \dfrac{b}{a}$

26. $(3x^4y^3 - x^2y^2 - xy^3) \div (-x^2y^2)$ $-3x^2y + 1 + \dfrac{y}{x}$

27. $(6t^4 - 2t^3 + 3t^2 - t + 4) \div (2t^3)$ $3t - 1 + \dfrac{3}{2t} - \dfrac{1}{2t^2} + \dfrac{2}{t^3}$

28. $\dfrac{2y^3 - 2y^2 + 3y - 9}{2y^2}$ $y - 1 + \dfrac{3}{2y} - \dfrac{9}{2y^2}$

29. a. Divide: $(z^2 + 7z + 11) \div (z + 5)$ $z + 2 + \dfrac{1}{z + 5}$

   b. Check by multiplying the quotient by the divisor and adding the remainder.

30. a. Divide $2w + 1 + \dfrac{7}{w - 4}$

$$\frac{2w^2 - 7w + 3}{w - 4}$$

   b. Check by multiplying the quotient by the divisor and adding the remainder.

For Exercises 31–50, divide the polynomials.

31. $\dfrac{t^2 + 4t + 3}{t + 1}$ $t + 3$

32. $(3x^2 + 8x + 4) \div (x + 2)$ $3x + 2$

33. $(7b^2 - 3b - 4) \div (b - 1)$ $7b + 4$

34. $\dfrac{w^2 - w - 2}{w - 2}$ $w + 1$

35. $\dfrac{5k^2 - 29k - 6}{5k + 1}$ $k - 6$

36. $(4y^2 + 25y - 21) \div (4y - 3)$ $y + 7$

37. $(4p^3 + 12p^2 + p - 12) \div (2p + 3)$ $2p^2 + 3p - 4$

38. $\dfrac{12a^3 - 2a^2 - 17a - 5}{3a + 1}$ $4a^2 - 2a - 5$

39. $\dfrac{k^2 - k - 6}{k + 1}$ $k - 2 + \dfrac{-4}{k + 1}$

40. $(h^2 + 3h + 1) \div (h + 2)$ $h + 1 + \dfrac{-1}{h + 2}$

41. $(4x^3 - 8x^2 + 15x - 16) \div (2x - 3)$ $2x^2 - x + 6 + \dfrac{2}{2x - 3}$

42. $\dfrac{3b^3 + b^2 + 17b - 49}{3b - 5}$ $b^2 + 2b + 9 + \dfrac{-4}{3b - 5}$

43. $\dfrac{a^2 + 9}{a + 3}$ $a - 3 + \dfrac{18}{a + 3}$

44. $(m^2 + 3) \div (m + 3)$ $m - 3 + \dfrac{12}{m + 3}$

45. $(w^4 + 5w^3 - 5w^2 - 15w + 7) \div (w^2 - 3)$ $w^2 + 5w - 2 + \dfrac{1}{w^2 - 3}$

46. $\dfrac{p^4 - p^3 - 4p^2 - 2p - 15}{p^2 + 2}$ $p^2 - p - 6 + \dfrac{-3}{p^2 + 2}$

47. $\dfrac{2n^4 + 5n^3 - 11n^2 - 20n + 12}{2n^2 + 3n - 2}$ $n^2 + n - 6$

48. $(6y^4 - 5y^3 - 8y^2 + 16y - 8) \div (2y^2 - 3y + 2)$ $3y^2 + 2y - 4$

49. $(5x^3 - 4x - 9) \div (5x^2 + 5x + 1)$ $x - 1 + \dfrac{-8}{5x^2 + 5x + 1}$

50. $\dfrac{3a^3 - 5a + 16}{3a^2 - 6a + 7}$ $a + 2 + \dfrac{2}{3a^2 - 6a + 7}$

51. Explain why $(x^3 - 8) \div (x - 2)$ is *not* $(x^2 + 4)$.
    To check, multiply $(x - 2)(x^2 + 4) = x^3 - 2x^2 + 4x - 8$, which does not equal $x^3 - 8$.

❖ See Additional Answers Appendix   ✎ Writing   ⬌ Translating Expression   Geometry   Scientific Calculator   ▶ Video

52. Explain why $(y^3 + 27) \div (y + 3)$ is *not* $(y^2 + 9)$. ❖

For Exercises 53–64, determine which method to use to divide the polynomials: monomial division or long division. Then use that method to divide the polynomials.

53. $\dfrac{9a^3 + 12a^2}{3a}$ Monomial division; $3a^2 + 4a$

54. $\dfrac{3y^2 + 17y - 12}{y + 6}$ Long division; $3y - 1 + \dfrac{-6}{y + 6}$

55. $(p^3 + p^2 - 4p - 4) \div (p^2 - p - 2)$ Long division; $p + 2$

56. $(q^3 + 1) \div (q + 1)$ Long division; $q^2 - q + 1$

57. $\dfrac{t^4 + t^2 - 16}{t + 2}$ Long division; $t^3 - 2t^2 + 5t - 10 + \dfrac{4}{t + 2}$

58. $\dfrac{-8m^5 - 4m^3 + 4m^2}{-2m^2}$ Monomial division; $4m^3 + 2m - 2$

59. $(w^4 + w^2 - 5) \div (w^2 - 2)$ Long division; $w^2 + 3 + \dfrac{1}{w^2 - 2}$

60. $(2k^2 + 9k + 7) \div (k + 1)$ Long division; $2k + 7$

61. $\dfrac{n^3 - 64}{n - 4}$ ❖

62. $\dfrac{15s^2 + 34s + 28}{5s + 3}$ ❖

63. $(9r^3 - 12r^2 + 9) \div (-3r^2)$ Monomial division; $-3r + 4 - \dfrac{3}{r^2}$

64. $(6x^4 - 16x^3 + 15x^2 - 5x + 10) \div (3x + 1)$ Long division; $2x^3 - 6x^2 + 7x - 4 + \dfrac{14}{3x + 1}$

EXPANDING YOUR SKILLS

For Exercises 65–72, divide the polynomials and note any patterns.

65. $(x^2 - 1) \div (x - 1)$ $x + 1$

66. $(x^3 - 1) \div (x - 1)$ $x^2 + x + 1$

67. $(x^4 - 1) \div (x - 1)$ $x^3 + x^2 + x + 1$

68. $(x^5 - 1) \div (x - 1)$ $x^4 + x^3 + x^2 + x + 1$

69. $x^2 \div (x - 1)$ ❖

70. $x^3 \div (x - 1)$ ❖

71. $x^4 \div (x - 1)$ $x^3 + x^2 + x + 1 + \dfrac{1}{x - 1}$

72. $x^5 \div (x - 1)$ $x^4 + x^3 + x^2 + x + 1 + \dfrac{1}{x - 1}$

Concepts

1. **Introduction to Nonlinear Graphs**

2. **Graphing Nonlinear Equations in Two Variables**

3. **Applications of Nonlinear Equations**

section

3.8 CONNECTIONS TO GRAPHING: INTRODUCTION TO NONLINEAR GRAPHS

1. Introduction to Nonlinear Graphs

In Section 2.8, we learned that an equation that can be written in the form $ax + by = c$ (where a and b are not both zero) is a linear equation in two variables. The graph of a linear equation is a line in a rectangular coordinate system (Figure 3-7).

$$2x - 4y = 8 \qquad\qquad 2x + y = 1$$

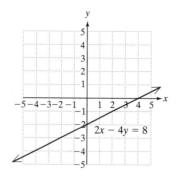

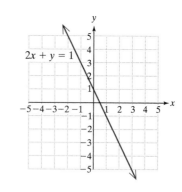

Figure 3-7

Linear equations in two variables are important relationships that come up often in mathematical applications. However, there are many situations where two variables are related nonlinearly. A nonlinear relationship between two variables is not depicted by a straight line in a rectangular coordinate system. For example, the temperature during a 24-h period for a given day in Los Angeles is given in Table 3-3. The corresponding graph of the data is not linear (not a straight line). See Figure 3-8.

Table 3-3		
Time	**Number of Hours after 12:00 A.M.** x	**Temperature (°F)** y
12:00 A.M.	0	64
2:00 A.M.	2	62
4:00 A.M.	4	61
6:00 A.M.	6	58
8:00 A.M.	8	60
10:00 A.M.	10	66
12:00 P.M.	12	76
2:00 P.M.	14	84
4:00 P.M.	16	79
6:00 P.M.	18	75
8:00 P.M.	20	72
10:00 P.M.	22	70
12:00 A.M.	24	65

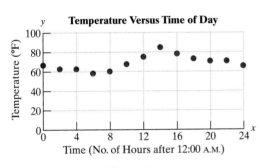

Temperature Versus Time of Day

Time (No. of Hours after 12:00 A.M.)

Figure 3-8

The temperature, y, for a given time, x, may be approximated by the equation

$$y = -0.026x^3 + 0.81x^2 - 4.6x + 62$$

where x is the number of hours after 12:00 A.M. and $0 \le x \le 24$.

Notice that this equation is nonlinear because it cannot be written in the form $ax + by = c$. Each variable term in a linear equation must have a degree of 1, whereas the equation $y = -0.026x^3 + 0.81x^2 - 4.6x + 62$ has terms of higher degree.

Classroom Activity 3.8A

---

example 1   **Evaluating a Nonlinear Equation**

Use the equation $y = -0.026x^3 + 0.81x^2 - 4.6x + 62$ to approximate the temperature at the following times. Round to the nearest tenth of a degree. (In this equation, y is the temperature in degrees Fahrenheit and x is the number of hours after 12:00 A.M.)

a. 12:00 A.M.
b. 7:00 A.M.
c. 11:00 A.M.
d. 3:00 P.M.
e. 11:00 P.M.

Solution:

a. At 12:00 A.M., $x = 0$

$y = -0.026x^3 + 0.81x^2 - 4.6x + 62$

$= -0.026(0)^3 + 0.81(0)^2 - 4.6(0) + 62$ Substitute $x = 0$.

$= 62$ At 12:00 A.M., the temperature is approximately 62°F.

b. Because 7:00 A.M. is 7 h after 12:00 A.M., then $x = 7$.

$y = -0.026x^3 + 0.81x^2 - 4.6x + 62$

$= -0.026(7)^3 + 0.81(7)^2 - 4.6(7) + 62$ Substitute $x = 7$.

$= 60.6$ At 7:00 A.M., the temperature is approximately 60.6°F.

c. Because 11:00 A.M. is 11 h after 12:00 A.M., then $x = 11$.

$y = -0.026x^3 + 0.81x^2 - 4.6x + 62$

$= -0.026(11)^3 + 0.81(11)^2 - 4.6(11) + 62$ Substitute $x = 11$.

$= 74.8$ At 11:00 A.M., the temperature is approximately 74.8°F.

d. Because 3:00 P.M. is 15 h after 12:00 A.M., then $x = 15$.

$y = -0.026x^3 + 0.81x^2 - 4.6x + 62$

$= -0.026(15)^3 + 0.81(15)^2 - 4.6(15) + 62$ Substitute $x = 15$.

$= 87.5$ At 3:00 P.M., the temperature is approximately 87.5°F.

e. Because 11:00 P.M. is 23 h after 12:00 A.M., then $x = 23$.

$y = -0.026x^3 + 0.81x^2 - 4.6x + 62$

$= -0.026(23)^3 + 0.81(23)^2 - 4.6(23) + 62$ Substitute $x = 23$.

$= 68.3$ At 11:00 P.M., the temperature is approximately 68.3°F.

2. Graphing Nonlinear Equations in Two Variables

Recall that the graph of a linear equation in two variables is a line. Therefore, to graph a linear equation, it is sufficient to find two ordered pair solutions to the equation and draw the line between the points. However, a nonlinear equation is not a straight line. Consequently, a sufficient number of points must be graphed so that a pattern or shape can be established.

example 2

Graphing a Nonlinear Equation

Given the equation: $y = x^2 - 1$

a. Complete the table to find several solutions to the equation.
b. Graph the ordered pairs corresponding to the table values. Then sketch a curve defined by the ordered pairs.

x	y
0	
1	
2	
3	
−1	
−2	
−3	

Classroom Activity 3.8B

Solution:

a. Substitute each value of x and solve for y.

x	y
0	
1	
2	
3	
−1	
−2	
−3	

$y = x^2 - 1$

$y = (0)^2 - 1 \qquad y = -1$
$y = (1)^2 - 1 \qquad y = 0$
$y = (2)^2 - 1 \qquad y = 3$
$y = (3)^2 - 1 \qquad y = 8$
$y = (-1)^2 - 1 \qquad y = 0$
$y = (-2)^2 - 1 \qquad y = 3$
$y = (-3)^2 - 1 \qquad y = 8$

x	y
0	−1
1	0
2	3
3	8
−1	0
−2	3
−3	8

b. The ordered pairs corresponding to the values in the table are

$(0, -1)$

$(1, 0)$

$(2, 3)$

$(3, 8)$

$(-1, 0)$

$(-2, 3)$

$(-3, 8)$

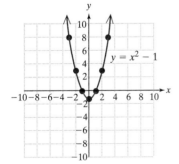

Figure 3-9

A graph through the points is shown in Figure 3-9.

3. Applications of Nonlinear Equations

example 3

Using a Nonlinear Equation in Two Variables in an Application

A ball is thrown straight up from a point 4 ft off the ground. At first, the height of the ball increases. However, the force of gravity slows the ball down until it eventually stops its upward motion and then drops back to the ground. The height of the ball in feet, y, can be represented by the following equation:

$$y = -16t^2 + 64t + 4 \qquad \text{where } t \text{ is the time in seconds after the ball is released } (t \geq 0).$$

a. Complete the table of values. Notice that the variable t is used in place of x.

Time (s) t	Height (ft) y
0	
1	
2	
3	
4	

b. Write the ordered pairs corresponding to the values in the table. Interpret the meaning of each ordered pair in the context of the problem.

c. Graph the ordered pairs and sketch a curve defined by the points.

Solution:

a. Substitute each value for t and solve for y.

$$y = -16t^2 + 64t + 4$$

Substitute $t = 0$. $y = -16(0)^2 + 64(0) + 4 \rightarrow y = 4$

Substitute $t = 1$. $y = -16(1)^2 + 64(1) + 4 \rightarrow y = 52$

Substitute $t = 2$. $y = -16(2)^2 + 64(2) + 4 \rightarrow y = 68$

Substitute $t = 3$. $y = -16(3)^2 + 64(3) + 4 \rightarrow y = 52$

Substitute $t = 4$. $y = -16(4)^2 + 64(4) + 4 \rightarrow y = 4$

The completed table is

Time (s) t	Height (ft) y
0	4
1	52
2	68
3	52
4	4

b. The ordered pairs corresponding to the table values are

$(0, 4)$ At 0 s (the time of launch), the ball is 4 ft above the ground.

$(1, 52)$ After 1 s, the ball is 52 ft above the ground.

$(2, 68)$ After 2 s, the ball is 68 ft above the ground.

$(3, 52)$ After 3 s, the ball is 52 ft above the ground.

$(4, 4)$ After 4 s, the ball is 4 ft above the ground.

c.

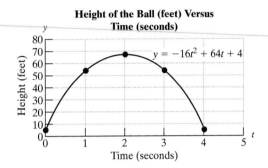

Height of the Ball (feet) Versus Time (seconds)

Figure 3-10

Notice that the equation $y = -16t^2 + 64t + 4$ is nonlinear (Figure 3-10). If the ball had followed a linear path, it would have continued to rise forever. The nonlinear equation provides a better tool for describing the path of the ball in the presence of gravity.

---

example 4

Using a Nonlinear Equation in Two Variables in an Application

The number of people (in thousands) in the United States expected to be over 100 years old has been increasing each year since 1994 (*Source:* U.S. Census Bureau). The equation

$$y = 0.40x^2 + 2.1x + 50$$

can be used to approximate the number of thousands of people, y, expected to be over 100 years old, x years after 1994. That is, $x = 0$ corresponds to 1994, $x = 1$ corresponds to 1995 and so on (Figure 3-11).

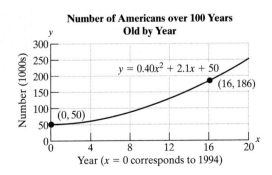

Number of Americans over 100 Years Old by Year

Figure 3-11

a. Use the equation to approximate the number of people that were over 100 years old in the year 1994.

b. Use the equation to predict the number of people expected to be over 100 years old in the year 2010.

c. Confirm the solutions to parts (a) and (b) with the graph of $y = 0.40x^2 + 2.1x + 50$.

Classroom Activity 3.8C

Solution:

a. $y = 0.40x^2 + 2.1x + 50$

$= 0.40(0)^2 + 2.1(0) + 50$ The year 1994 corresponds to $x = 0$.

$= 50$

In the year 1994, the number of Americans over the age of 100 was approximately 50,000.

b. $y = 0.40x^2 + 2.1x + 50$ The year 2010 is 16 years after 1994.

$= 0.40(16)^2 + 2.1(16) + 50$ Substitute $x = 16$.

$= 186$

In the year 2010, we expect approximately 186,000 Americans to be over the age of 100.

c. From part (a), when $x = 0$, $y = 50$. This corresponds to the ordered pair $(0, 50)$ on the graph.
From part (b), when $x = 16$, $y = 186$. This corresponds to the ordered pair $(16, 186)$ on the graph.

Calculator Connections

In Example 2 of this section, we graphed the equation $y = x^2 - 1$ by completing a table of points and sketching the curve defined by the points. Once an equation is entered into a calculator, many graphing calculators have a *Table* feature in which the y-values are computed for user-defined values of x (consult your user's manual). In most cases, the user must define the parameters for the table using a *TBLSET* menu. A starting value for x, *TblStart*, is required as well as an increment, ΔTbl, by which to increase (or decrease) the value of x.

The *Graph* option can be used to view the graph of the equation $y = x^2 - 1$.

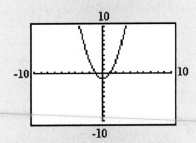

Calculator Exercises

Graph the nonlinear equations on the standard viewing window.

1. $y = x^2 - 4$ *Hint*: Enter the equation as $Y = x^2 - 4$. ❖

2. $y = x^3$ *Hint*: Enter the equation as $Y = x^3$. ❖

3. $y = x^2 - 4x + 5$ *Hint*: Enter the equation as $Y = x^2 - 4x + 5$. ❖

4. $y = -x^2 - 2x + 3$ *Hint*: Enter the equation as $Y = -x^2 - 2x + 3$. ❖

5. $y = (x - 2)^3$ *Hint*: Enter the equation as $Y = (x - 2)^3$. ❖

Graph the equation on the suggested viewing window.

6. $y = \dfrac{1}{5}x^2 - 20$ ❖

    Window: $-25 \le x \le 25$ and $-25 \le y \le 25$

7. $y = -x^3 - 5x^2 + 6x + 1$ ❖

    Window: $-10 \le x \le 10$ and $-40 \le y \le 40$

8. $y = -0.026x^3 + 0.81x^2 - 4.6x + 62$ (from Example 1) ❖

    Window: $0 \le x \le 24$ and $0 \le y \le 100$

9. $y = 0.40x^2 + 2.1x + 50$ (from Example 4) ❖

    Window: $0 \le x \le 20$ and $0 \le y \le 300$

section 3.8 PRACTICE EXERCISES

For Exercises 1–4, graph the linear equation by completing the table and plotting the corresponding points. 2.8

1. $x - y = 2$ ❖

x	y
3	1
4	2
0	-2

Table for Exercise 1

2. $x + y = -5$ ❖

x	y
2	-7
0	-5
-3	-2

Table for Exercise 2

 3. $2x + 5y = 11$ ❖

x	y
$\frac{11}{2}$	0
3	1
8	-1

Table for Exercise 3

4. $4x - 3y = 13$ ❖

x	y
3	$-\frac{1}{3}$
4	1
0	$-\frac{13}{3}$

Table for Exercise 4

 5. Which is a graph of a linear equation? b

a.

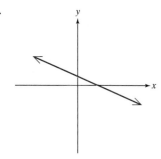

b.

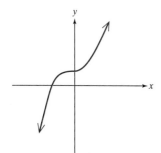

c.

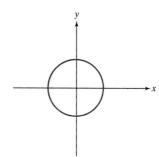

6. Which is a graph of a linear equation? c

a.

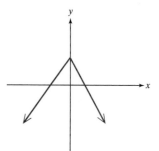

b.

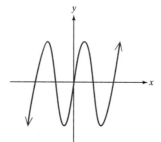

c.

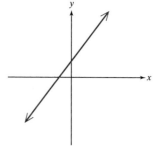

For Exercises 7–16, determine if the equation is linear or nonlinear.

7. $4x + 7y = 0$
   Linear

8. $y = 3x - 9$
   Linear

9. $-x + y^2 = 12$
   Nonlinear

10. $x^2 + 3y = -17$
    Nonlinear

11. $y^2 = 4x - 2$
    Nonlinear

12. $y = x^2 + 5$
    Nonlinear

13. $\frac{1}{4}x + \frac{1}{2}y = 6$
    Linear

14. $\frac{2}{3}y - \frac{3}{4}x = 1$
    Linear

15. $x = 2y - 5$
    Linear

16. $x = -5 + y$
    Linear

17. Given the equation $y = x^2 + 2$:

    a. Complete the table.

    b. Graph the ordered pairs corresponding to the values in the table. ❖

    c. Sketch a curve defined by the points in the graph. ❖

x	y
0	2
1	3
2	6
3	11
-1	3
-2	6
-3	11

    Table for Exercise 17

18. Given the equation $y = x^3$:

    a. Complete the table.

    b. Graph the ordered pairs corresponding to the values in the table. ❖

    c. Sketch a curve defined by the points in the graph. ❖

x	y
0	0
1	1
2	8
-1	-1
-2	-8

    Table for Exercise 18

19. Given the equation $y = x^3 + 1$:

    a. Complete the table.

    b. Graph the ordered pairs corresponding to the values in the table. ❖

    c. Sketch a curve defined by the points in the graph. ❖

x	y
0	1
1	2
2	9
-1	0
-2	-7

    Table for Exercise 19

20. Given the equation $y = x^2$:

    a. Complete the table.

    b. Graph the ordered pairs corresponding to the values in the table. ❖

    c. Sketch a curve defined by the points in the graph. ❖

x	y
0	0
1	1
2	4
-1	1
-2	4

    Table for Exercise 20

21. An airplane that is 1 mile (in ground distance) from the airport begins its descent from an altitude of 1000 ft (see the figure). Let x represent the ground distance (in miles) from the plane to the airport. Let y represent the altitude of the plane (in feet). The path of the plane can be represented by the equation

    $$y = -2000x^3 + 3000x^2 \qquad \text{where } 0 \le x \le 1.$$

Airplane Altitude Versus Horizontal Ground Distance

$$y = -2000x^3 + 3000x^2$$

Altitude (feet) vs Ground Distance (miles)

Figure for Exercise 21

Time (s)	Height (ft)
t	y
0	0
2	192
4	256
6	192
8	0

Table for Exercise 22

a. Complete the table.

Ground Distance (miles)	Altitude (ft)
x	y
0.00	0
0.25	156.25
0.50	500
0.75	843.75
1.00	1000

Table for Exercise 21

b. Write the ordered pairs corresponding to the values in the table. Interpret each ordered pair in the context of the problem. Do the ordered pairs fall on the curve drawn in the figure?   ❖

c. Why would a linear path not be suitable for the airplane to follow for its landing?   ❖

22. A rocket is launched from ground level with an initial velocity of 128 ft/s. The height of the rocket, y, changes with time, t, according to the equation

$$y = -16t^2 + 128t$$ where the height, y, is measured in feet and the time, t, is measured in seconds and $0 \le t \le 8$.

a. Complete the table, and graph the corresponding ordered pairs. Note that the variable t is used in place of x.   ❖

b. Interpret the meaning of each ordered pair found in part (a) in the context of the problem.   ❖

c. Why are there two different times at which the rocket is 192 ft above the ground?   ❖

23. The distance, d, that an object travels in free fall is given by the equation

$$d = 16t^2$$ where d is measured in feet and t is the time in seconds after release.

a. Is this equation linear or nonlinear?   ❖

b. Complete the table.

t	d
0	0
1	16
2	64
3	144
4	256
5	400

Table for Exercise 23

c. Write the ordered pairs corresponding to the entries in the table. (In this problem the variable t is used in place of x and d is used in place of y.) Interpret the meaning of each ordered pair in the context of this problem.   ❖

d. Verify your ordered pairs from part (c) in the graph of the equation.

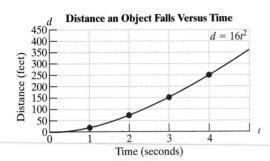

Figure for Exercise 23

24. The distance, d, that an object travels in free fall is given by the equation

$d = 4.9t^2$ where d is measured in meters and t is the time in seconds after release.

a. Is this equation linear or nonlinear? ❖

b. Complete the table.

t	d
0	0
1	4.9
2	19.6
3	44.1
4	78.4

Table for Exercise 24

c. Write the ordered pairs corresponding to the entries in the table. (In this problem the variable t is used in place of x, and d is used in place of y.) Interpret the meaning of each ordered pair in the context of this problem. ❖

d. Verify your ordered pairs from part (c) in the graph of the equation.

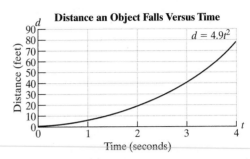

Figure for Exercise 24

25. A certain sports arena holds 20,000 seats. If a sports franchise charges x dollars per ticket for a game, then it can be expected to sell $20{,}000 - 400x$ tickets. The total revenue, R, is equal to the number of tickets sold times the price or equivalently,

$R = (20{,}000 - 400x)x$ $0 \le x \le 50$

a. Is the equation $R = (20{,}000 - 400x)x$ linear or nonlinear? ❖

b. Complete the table.

Cost per Ticket ($) x	Revenue ($) R
0	0
10	160,000
20	240,000
30	240,000
40	160,000
50	0

Table for Exercise 25

c. Write the ordered pairs corresponding to the entries in the table. (In this problem the variable R is used in place of y.) Interpret the meaning of each ordered pair in the context of this problem. ❖

d. Verify your ordered pairs from part (c) in the graph of the equation.

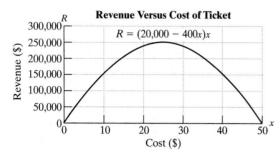

Figure for Exercise 25

d. Verify your ordered pairs from part (c) in the graph of the equation.

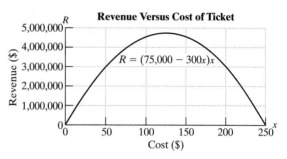

Figure for Exercise 26

26. A stadium holds 75,000 seats. If a concert promoter charges x dollars per ticket for a concert, then she can be expected to sell $75,000 - 300x$ tickets. The total revenue, R, is equal to the number of tickets sold times the price or equivalently,

$$R = (75,000 - 300x)x \quad 0 \le x \le 250$$

a. Is the equation $R = (75,000 - 300x)x$ linear or nonlinear? ❖

b. Complete the table.

Cost per Ticket ($) x	Revenue ($) R
0	0
25	1,687,500
50	3,000,000
100	4,500,000
250	0

Table for Exercise 26

c. Write the ordered pairs corresponding to the entries in the table. (In this problem the variable R is used in place of y.) Interpret the meaning of each ordered pair in the context of this problem. ❖

27. A rectangular box is to be constructed such that the width is $\frac{1}{2}$ as long as the length and the height is $\frac{1}{4}$ of the length. The volume of the box can be expressed as

$$V = lwh \quad \text{or}$$

$$= x\left(\frac{1}{2}x\right)\left(\frac{1}{4}x\right) \quad \text{or} \quad V = \frac{1}{8}x^3 \quad (x \ge 0)$$

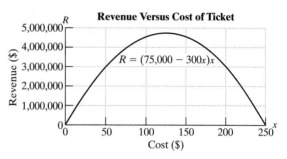

Figure for Exercise 27

a. Is the equation $V = \frac{1}{8}x^3$ linear or nonlinear? ❖

b. Complete the table.

x (in.)	V (in.^{3})
0	0
1	$\frac{1}{8}$
2	1
3	$\frac{27}{8}$
4	8
5	$\frac{125}{8}$

Table for Exercise 27

c. Write the ordered pairs corresponding to the entries in the table. (In this problem the variable V is used in place of y.) Graph the ordered pairs and sketch the curve defined by the points. ❖

28. The area of a square, A, is given by the equation $A = x^2 (x \geq 0)$, where x is the length of the sides.

    a. Is the equation $A = x^2$ linear or nonlinear? ❖

    [square figure with sides labeled x]

    Figure for Exercise 28

    b. Complete the table.

x (in.)	A (in.²)
0	0
1	1
2	4
3	9
4	16
5	25

    Table for Exercise 28

    c. Write the ordered pairs corresponding to the entries in the table. (In this problem the variable A is used in place of y.) Graph the ordered pairs and sketch the curve defined by the points. ❖

    d. What is the area of a square when the length of its sides is 6 ft? If the length of the sides is doubled, by how many times is the area increased? ❖

29. A ball is thrown at a 30° angle from the horizontal. The height of the ball, y, (in feet) is given by the equation

    $y = -16t^2 + 30t + 4$ where t is the time in seconds after release.

    a. Is the equation $y = -16t^2 + 30t + 4$ linear or nonlinear? ❖

    b. Complete the table.

t	y
0	4
0.5	15
1.0	18
1.5	13
2.0	0

    Table for Exercise 29

    c. Write the ordered pairs corresponding to the entries in the table. (In this problem the variable t is used in place of x.) Interpret the meaning of each ordered pair in the context of this problem. ❖

    d. Verify your ordered pairs from part (b) in the graph of the equation.

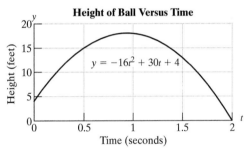

    Figure for Exercise 29

30. A baseball is thrown at a 30° angle from the horizontal. The height of the ball, y, (in feet) is given by the equation

    $y = -16t^2 + 25t + 5$ where t is the time in seconds after release.

    a. Is the equation $y = -16t^2 + 25t + 5$ linear or nonlinear? ❖

    b. Complete the table.

Time t	Height y
0	5
0.5	13.5
1.0	14
1.5	6.5
1.7	1.26

    Table for Exercise 30

c. Write the ordered pairs corresponding to the entries in the table. (In this problem the variable t is used in place of x.) Interpret the meaning of each ordered pair in the context of this problem. ❖

d. Verify your ordered pairs from part (b) in the graph of the equation.

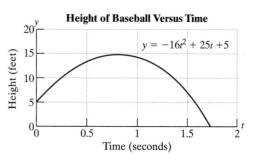

Figure for Exercise 30

---

chapter 3   SUMMARY

SECTION 3.1—EXPONENTS: MULTIPLYING AND DIVIDING COMMON BASES

KEY CONCEPTS:

Definition

$$b^n = \underbrace{b \cdot b \cdot b \cdot b \cdots b}_{n \text{ factors of } b}$$ b is the base,
n is the exponent

Multiplying Common Bases

$$a^m a^n = a^{m+n} \quad (m, n \text{ positive integers})$$

Dividing Common Bases

$$\frac{a^m}{a^n} = a^{m-n} \quad (a \neq 0, m, n, \text{ positive integers})$$

KEY TERMS:

base
compound interest
division of like bases
exponent
multiplication of like bases

EXAMPLES:

$3^4 = 3 \cdot 3 \cdot 3 \cdot 3 = 81$ 3 is the base

4 is the exponent

Compare:

$$(-5)^2 \quad \text{versus} \quad -5^2$$

versus
$(-5)^2 = (-5)(-5) = 25$

$-5^2 = -1(5^2) = -1(5)(5) = -25$

Simplify:

$$x^3 \cdot x^4 \cdot x^2 \cdot x = x^{10}$$

Simplify:

$$\frac{c^4 d^{10}}{cd^5} = c^{4-1}d^{10-5} = c^3 d^5$$

SECTION 3.2—MORE PROPERTIES OF EXPONENTS

KEY CONCEPTS:

Power Rule for Exponents

$$(a^m)^n = a^{mn} \quad (a \neq 0, m, n \text{ positive integers})$$

Power of a Product and Power of a Quotient

Assume m and n are positive integers and a and b are real numbers where $b \neq 0$. Then,

$$(ab)^m = a^m b^m \quad \text{and} \quad \left(\frac{a}{b}\right)^m = \frac{a^m}{b^m}$$

KEY TERMS:

$(ab)^m = a^m b^m$

$\left(\frac{a}{b}\right)^m = \frac{a^m}{b^m}$

power rule for exponents

EXAMPLES:

Simplify:

$$(x^4)^5 = x^{20}$$

Simplify:

$$(4uv^2)^3 = 4^3 u^3 (v^2)^3 = 64u^3 v^6$$

Simplify:

$$\left(\frac{p^5 q^3}{5pq^2}\right)^2 = \left(\frac{p^{5-1} q^{3-2}}{5}\right)^2 = \left(\frac{p^4 q}{5}\right)^2$$

$$= \frac{p^8 q^2}{25}$$

SECTION 3.3—DEFINITIONS OF b^0 AND b^{-n}

KEY CONCEPTS:

Definitions

If b is a real number such that $b \neq 0$, then:

1. $b^0 = 1$

2. $b^{-n} = \left(\frac{1}{b}\right)^n = \frac{1}{b^n}$

KEY TERMS:

b^0

b^{-n}

EXAMPLES:

Simplify:

$$4^0 = 1$$

Simplify:

$$y^{-7} = \frac{1}{y^7}$$

Simplify:

$$\left(\frac{2a^3 b}{a^{-2} c^{-4}}\right)^{-2} (a^0 b)^3$$

$$= \left(\frac{2a^{3-(-2)} b}{c^{-4}}\right)^{-2} (1 \cdot b)^3$$

$$= \left(\frac{2a^5 b}{c^{-4}}\right)^{-2} (b)^3 = \frac{2^{-2} a^{-10} b^{-2}}{c^8} \cdot b^3$$

$$= \frac{1}{2^2 a^{10} b^2 c^8} \cdot b^3 = \frac{b^{3-2}}{4a^{10} c^8}$$

$$= \frac{b}{4a^{10} c^8}$$

SECTION 3.4—SCIENTIFIC NOTATION

KEY CONCEPTS:

A number written in scientific notation is expressed in the form:

$a \times 10^n$ where $1 \leq |a| < 10$ and n is an integer. The value 10^n is sometimes called the order of magnitude or simply the magnitude of the number.

KEY TERMS:

order of magnitude
scientific notation

EXAMPLES:

Write the numbers in scientific notation:

$$35,000 = 3.5 \times 10^4$$

$$0.000\,000\,548 = 5.48 \times 10^{-7}$$

Multiply:

$$(3.5 \times 10^4)(2.0 \times 10^{-6})$$
$$= 7.0 \times 10^{-2}$$

Divide:

$$\frac{8.4 \times 10^{-9}}{2.1 \times 10^3} = 4.0 \times 10^{-9-3} = 4.0 \times 10^{-12}$$

SECTION 3.5—ADDITION AND SUBTRACTION OF POLYNOMIALS

KEY CONCEPTS:

A polynomial in one variable is a finite sum of terms of the form ax^n, where a is a real number and the exponent, n is a nonnegative integer. For each term, a is called the coefficient of the term and n is the degree of the term. The term with highest degree is the leading term and its coefficient is called the leading coefficient. The degree of the polynomial is the largest degree of all its terms.

    To add or subtract polynomials, add or subtract *like* terms.

KEY TERMS:

addition of polynomials
binomial
coefficient
degree of a polynomial
degree of a term
leading coefficient
leading term
like terms
monomial
polynomial
subtraction of polynomials
trinomial

EXAMPLES:

Given:
$$4x^5 - 8x^3 + 9x - 5$$

Coefficients of each term: $4, -8, 9, -5$

Degree of each term: $5, 3, 1, 0$

Leading term: $4x^5$

Leading coefficient: 4

Degree of polynomial: 5

Perform the indicated operations:

$$(2x^4 - 5x^3 + 1) - (x^4 + 3) + (x^3 - 4x - 7)$$
$$= 2x^4 - 5x^3 + 1 - x^4 - 3 + x^3 - 4x - 7$$
$$= 2x^4 - x^4 - 5x^3 + x^3 - 4x + 1 - 3 - 7$$
$$= x^4 - 4x^3 - 4x - 9$$

SECTION 3.6—MULTIPLICATION OF POLYNOMIALS

KEY CONCEPTS:

Multiplying Monomials

Use the commutative and associative properties of multiplication to group coefficients and like bases.

EXAMPLES:

Multiply:

$$(5a^2b)(-2ab^3)$$
$$= (5 \cdot -2)(a^2a)(bb^3)$$
$$= -10a^3b^4$$

Multiplying Polynomials

Multiply each term in the first polynomial by each term in the second.

Multiply:

$$-3ab^2(2a^2 - 5ab)$$
$$= (-3ab^2)(2a^2) + (-3ab^2)(-5ab)$$
$$= -6a^3b^2 + 15a^2b^3$$

Multiply:

$$(y + 3)(2y - 5)$$
$$= 2y^2 - 5y + 6y - 15$$
$$= 2y^2 + y - 15$$

Multiply:

$$(x - 2)(3x^2 - 4x + 11)$$
$$= 3x^3 - 4x^2 + 11x - 6x^2 + 8x - 22$$
$$= 3x^3 - 10x^2 + 19x - 22$$

Product of Conjugates

Results in a difference of squares

$$(a + b)(a - b) = a^2 - b^2$$

Multiply:

$$(3w - 4v)(3w + 4v)$$
$$= (3w)^2 - (4v)^2$$
$$= 9w^2 - 16v^2$$

Square of a Binomial

Results in a perfect square trinomial

$$(a + b)^2 = a^2 + 2ab + b^2$$
$$(a - b)^2 = a^2 - 2ab + b^2$$

Multiply:

$$(5c - 8d)^2$$
$$= (5c)^2 - 2(5c)(8d) + (8d)^2$$
$$= 25c^2 - 80cd + 64d^2$$

KEY TERMS:

conjugates
difference of squares
perfect square trinomial

SECTION 3.7—DIVISION OF POLYNOMIALS

KEY CONCEPTS:

Division of Polynomials

1. Division by a monomial, use the properties:

$$\frac{a + b}{c} = \frac{a}{c} + \frac{b}{c} \quad \text{and} \quad \frac{a - b}{c} = \frac{a}{c} - \frac{b}{c}$$

2. If the divisor has more than one term, use long division.

KEY TERMS:

division by a monomial
long division

EXAMPLES:

Divide:

$$\frac{-3x^2 - 6x + 9}{-3x}$$

$$= \frac{-3x^2}{-3x} - \frac{6x}{-3x} + \frac{9}{-3x}$$

$$= x + 2 - \frac{3}{x}$$

Divide:

$$(3x^2 - 5x + 1) \div (x + 2)$$

$$
\begin{array}{r}
3x - 11 \\
x + 2 \overline{)3x^2 - 5x + 1} \\
-(3x^2 + 6x) \\
\hline
-11x + 1 \\
-(-11x - 22) \\
\hline
23
\end{array}
$$

$$3x - 11 + \frac{23}{x + 2}$$

SECTION 3.8—CONNECTIONS TO GRAPHING: INTRODUCTION TO NONLINEAR GRAPHS

KEY CONCEPTS:

An equation written in the form $ax + by = c$, where a and b are not both zero, is a linear equation in two variables. An equation that cannot be written in this form is a nonlinear equation.

EXAMPLES:

The equation $5x - 4y = 8$ is linear.
The equation $5x^2 - 4y = 8$ is nonlinear.
This graph is nonlinear.

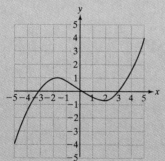

This graph is linear.

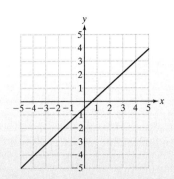

A solution to a nonlinear equation is an ordered pair that makes the equation true.

A nonlinear equation can be graphed by making a table of points and sketching the curve defined by the points.

KEY TERM:

nonlinear equations in two variables

Complete the table for the equation:

$y = x^2 - 1$. Then sketch the curve defined by the table of points.

x	y
0	
1	
2	
3	
-1	
-2	
-3	

$y = 0^2 - 1 = -1$

$y = 1^2 - 1 = 0$

$y = 2^2 - 1 = 3$

$y = 3^2 - 1 = 8$

$y = (-1)^2 - 1 = 0$

$y = (-2)^2 - 1 = 3$

$y = (-3)^2 - 1 = 8$

x	y
0	-1
1	0
2	3
3	8
-1	0
-2	3
-3	8

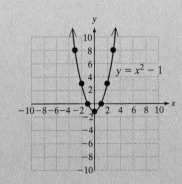

chapter 3	REVIEW EXERCISES

Section 3.1

For Exercises 1–4, identify the base and the exponent.

1. 5^3 ❖   2. x^4 ❖   3. $(-2)^0$ ❖   4. y ❖

5. Evaluate the expressions.

   a. 6^2 b. $(-6)^2$ c. -6^2

      36   36 -36

6. Evaluate the expressions.

   a. 4^3 b. $(-4)^3$ c. -4^3

      64   -64 -64

For Exercises 7–18, simplify and write the answers in exponent form. Assume that all variables represent nonzero real numbers.

7. $5^3 \cdot 5^{10}$ 5^{13} 8. $a^7 a^4$ a^{11}

9. $x \cdot x^6 \cdot x^2$ x^9 10. $6^3 \cdot 6 \cdot 6^5$ 6^9

11. $\dfrac{10^7}{10^4}$ 10^3 12. $\dfrac{y^{14}}{y^8}$ y^6

13. $\dfrac{b^9}{b}$ b^8 14. $\dfrac{7^8}{7}$ 7^7

15. $\dfrac{k^2 k^3}{k^4}$ k 16. $\dfrac{8^4 \cdot 8^7}{8^{11}}$ 1

17. $\dfrac{2^8 \cdot 2^{10}}{2^3 \cdot 2^7}$ 2^8 18. $\dfrac{q^3 q^{12}}{q q^8}$ q^6

19. Explain why $2^2 \cdot 4^4$ does *not* equal 8^6. ❖

20. Explain why $\dfrac{10^5}{5^2}$ does *not* equal 2^3. ❖

For Exercises 21–22, use the formula

$$A = P\left(1 + \dfrac{r}{n}\right)^{nt}$$

21. Find the amount in an account after 3 years if the initial investment is $6000, invested at 6% interest compounded semiannually (twice a year). $7164.31

22. Find the amount in an account after 2 years if the initial investment is $20,000, invested at 5% interest compounded quarterly (four times a year). $22,089.72

Section 3.2

For Exercises 23–40, simplify the expressions. Write the answers in exponent form. Assume all variables represent nonzero real numbers.

23. $(7^3)^4$ 7^{12} 24. $(c^2)^6$ c^{12}

25. $(p^4 p^2)^3$ p^{18} 26. $(9^5 \cdot 9^2)^4$ 9^{28}

27. $\left(\dfrac{a}{b}\right)^2$ $\dfrac{a^2}{b^2}$ 28. $\left(\dfrac{1}{3}\right)^4$ $\dfrac{1}{3^4}$

29. $\left(\dfrac{5}{c^2 d^5}\right)^2$ $\dfrac{5^2}{c^4 d^{10}}$ 30. $\left(-\dfrac{m^2}{4n^6}\right)^5$ $-\dfrac{m^{10}}{4^5 n^{30}}$

31. $(2ab^2)^4$ $2^4 a^4 b^8$ 32. $(-x^7 y)^2$ $x^{14} y^2$

33. $\left(\dfrac{-3x^3}{5y^2 z}\right)^3$ $\dfrac{-3^3 x^9}{5^3 y^6 z^3}$ 34. $\left(\dfrac{r^3}{s^2 t^6}\right)^5$ $\dfrac{r^{15}}{s^{10} t^{30}}$

35. $\dfrac{a^4 (a^2)^8}{(a^3)^3}$ a^{11} 36. $\dfrac{(8^3)^4 \cdot 8^{10}}{(8^4)^5}$ 8^2

37. $\dfrac{(4h^2 k)^2 (h^3 k)^4}{(2hk^3)^2}$ $4h^{14}$ 38. $\dfrac{(p^3 q)^3 (2p^2 q^4)^4}{(8p)(pq^3)^2}$ $2p^{14} q^{13}$

39. $(6x^2 y^4)^3 \left(\dfrac{2x^4 y^3}{4xy^2}\right)^2$

    $54x^{12} y^{14}$

40. $\left(\dfrac{a^4 b^6}{ab^4}\right)^3 (4a^2 b)^5$ $4^5 a^{19} b^{11}$

Section 3.3

For Exercises 41–62, simplify the expressions. Assume all variables represent nonzero real numbers.

41. 8^0 1   42. $(-b)^0$ 1

43. 1^0 1   44. $-x^0$ -1

45. $2y^0$ 2   46. $(2y)^0$ 1

47. z^{-5} $\dfrac{1}{z^5}$ 48. 10^{-4} $\dfrac{1}{10^4}$

49. $(6a)^{-2}$ $\dfrac{1}{36a^2}$ 50. $6a^{-2}$ $\dfrac{6}{a^2}$

51. $4^0 + 4^{-2}$ $\dfrac{17}{16}$ 52. $9^{-1} + 9^0$ $\dfrac{10}{9}$

53. $t^{-6} t^{-2}$ $\dfrac{1}{t^8}$ 54. $r^8 r^{-9}$ $\dfrac{1}{r}$

55. $\dfrac{12x^{-2} y^3}{6x^4 y^{-4}}$ $\dfrac{2y^7}{x^6}$ 56. $\dfrac{8ab^{-3} c^0}{10a^{-5} b^{-4} c^{-1}}$ $\dfrac{4a^6 bc}{5}$

57. $(-2m^2 n^{-4})^{-4}$ $\dfrac{n^{16}}{16m^8}$ 58. $(3u^{-5} v^2)^{-3}$ $\dfrac{u^{15}}{27v^6}$

❖ See Additional Answers Appendix   Writing   ⟷ Translating Expression   Geometry   Scientific Calculator   Video

59. $\dfrac{(k^{-6})^{-2}(k^3)}{5k^{-6}k^0}$ $\quad \dfrac{k^{21}}{5}$

60. $\dfrac{(3h)^{-2}(h^{-5})^{-3}}{h^{-4}h^8}$ $\quad \dfrac{h^9}{9}$

61. $\dfrac{7^0}{3^{-1}-6^{-1}}$ $\quad 6$

62. $\dfrac{2^0-2^{-1}}{2^{-1}-2^{-2}}$ $\quad 2$

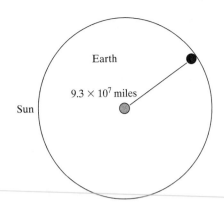

Figure for Exercise 71

Section 3.4

63. Write the numbers in scientific notation.

   a. The federal debt was estimated to be $5,914,800,000,000 in the year 2001. 5.9148×10^{12}

   b. The width of a piece of paper is 0.0042 in. 4.2×10^{-3} in.

   c. The area of the Pacific Ocean is 166,241,000 km². 1.66241×10^8 km²

64. Write the numbers without scientific notation.

   a. The pH of 10 means the hydrogen ion concentration is 1×10^{-10}. 0.000 000 0001

   b. The Social Security income is estimated to be 5.811×10^{11} in the year 2002. $581,100,000,000

   c. A fund-raising event for neurospinal research raised 2.56×10^5. $256,000

For Exercises 65–68, perform the indicated operations. Write the answers in scientific notation.

65. $(4.1 \times 10^{-6})(2.3 \times 10^{11})$
   9.43×10^5

66. $\dfrac{9.3 \times 10^3}{6.0 \times 10^{-7}}$
   1.55×10^{10}

67. $\dfrac{2000}{0.000008}$
   2.5×10^8

68. $(0.000078)(21,000,000)$
   1.638×10^3

69. Use your calculator to evaluate 5^{20}. Why is scientific notation necessary on your calculator to express the answer? 9.5367×10^{13}. This number is too big to fit on most calculator displays.

70. Use your calculator to evaluate $(0.4)^{30}$. Why is scientific notation necessary on your calculator to express the answer? 1.1529×10^{-12}. This number is too small to fit on most calculator displays.

71. The average distance between the earth and sun is 9.3×10^7 miles.

   a. If the earth's orbit is approximated by a circle, find the total distance the earth travels around the sun in one orbit. (*Hint*: The circumference of a circle is given by $C = 2\pi r$.) Express the answer in scientific notation. 5.84×10^8 miles

   b. If the earth makes one complete trip around the sun in 1 year (365 days = 8.76×10^3 h), find the average speed that the earth travels around the sun in miles per hour. Express the answer in scientific notation. 6.67×10^4 mph

72. The average distance between the planet Mercury and the sun is 3.6×10^7 miles.

   a. If Mercury's orbit is approximated by a circle, find the total distance Mercury travels around the sun in one orbit. (*Hint*: The circumference of a circle is given by $C = 2\pi r$.) Express the answer in scientific notation. 2.26×10^8 miles

   b. If Mercury makes one complete trip around the sun in 88 days (2.112×10^3 h), find the average speed that Mercury travels around the sun in miles per hour. Express the answer in scientific notation. 1.07×10^5 mph

Section 3.5

73. For the polynomial $7x^4 - x + 6$

   a. Classify as a monomial, a binomial, or a trinomial. Trinomial

   b. Identify the degree of the polynomial. 4

   c. Identify the leading coefficient. 7

❖ See Additional Answers Appendix　　🖉 Writing　　⬅ Translating Expression　　Geometry　　Scientific Calculator　　Video

74. For the polynomial $2y^3 - 5y^7$
    a. Classify as a monomial, a binomial, or a trinomial.   Binomial
    b. Identify the degree of the polynomial.   7
    c. Identify the leading coefficient. -5

For Exercises 75–80, add or subtract as indicated.

75. $(4x + 2) + (3x - 5)$ $7x - 3$

76. $(7y^2 - 11y - 6) - (8y^2 + 3y - 4)$ $-y^2 - 14y - 2$

77. $(9a^2 - 6) - (-5a^2 + 2a)$ $14a^2 - 2a - 6$

78. $(8w^4 - 6w + 3) + (2w^4 + 2w^3 - w + 1)$
    $10w^4 + 2w^3 - 7w + 4$

79. $\left(5x^3 - \dfrac{1}{4}x^2 + \dfrac{5}{8}x + 2\right) + \left(\dfrac{5}{2}x^3 + \dfrac{1}{2}x^2 - \dfrac{1}{8}x\right)$ ❖

80. $(-0.02b^5 + b^4 - 0.7b + 0.3) + (0.03b^5 - 0.1b^3 + b + 0.03)$ $0.01b^5 + b^4 - 0.1b^3 + 0.3b + 0.33$

81. Subtract $(9x^2 + 4x + 6)$ from $(7x^2 - 5x)$.
    $-2x^2 - 9x - 6$

82. Find the difference of $(x^2 - 5x - 3)$ and $(6x^2 + 4x + 9)$. $-5x^2 - 9x - 12$

83. Write a trinomial of degree 2 with a leading coefficient of -5. (Answers may vary.)
    For example: $-5x^2 + 2x - 4$

84. Write a binomial of degree 6 with leading coefficient 6. (Answers may vary.)
    For example: $6x^6 + 8$

Section 3.6

For Exercises 85–100, multiply the expressions.

85. $(25x^4y^3)(-3x^2y)$ $-75x^6y^4$ 86. $(9a^6)(2a^2b^4)$ $18a^8b^4$

87. $5c(3c^3 - 7c + 5)$
    $15c^4 - 35c^2 + 25c$

88. $(x^2 + 5x - 3)(-2x)$
    $-2x^3 - 10x^2 + 6x$

89. $(5k - 4)(k + 1)$
    $5k^2 + k - 4$

90. $(4t - 1)(5t + 2)$
    $20t^2 + 3t - 2$

91. $(q + 8)(6q - 1)$
    $6q^2 + 47q - 8$

92. $(2a - 6)(a + 5)$
    $2a^2 + 4a - 30$

93. $\left(7a + \dfrac{1}{2}\right)\left(7a + \dfrac{1}{2}\right)$ $49a^2 + 7a + \dfrac{1}{4}$

94. $(b - 4)^2$ $b^2 - 8b + 16$

95. $(4p^2 + 6p + 9)(2p - 3)$ $8p^3 - 27$

96. $(2w - 1)(-w^2 - 3w - 4)$ $-2w^3 - 5w^2 - 5w + 4$

97. $(b - 4)(b + 4)$
    $b^2 - 16$

98. $\left(\dfrac{1}{3}r^4 - s^2\right)\left(\dfrac{1}{3}r^4 + s^2\right)$
    $\dfrac{1}{9}r^8 - s^4$

99. $(-7z^2 + 6)^2$ $49z^4 - 84z^2 + 36$

100. $(2h + 3)(h^4 - h^3 + h^2 - h + 1)$
    $2h^5 + h^4 - h^3 + h^2 - h + 3$

Section 3.7

For Exercises 101–116, divide the polynomials.

101. $\dfrac{20y^3 - 10y^2}{5y}$ $4y^2 - 2y$

102. $(18a^3b^2 - 9a^2b - 27ab^2) \div 9ab$ $2a^2b - a - 3b$

103. $(12x^4 - 8x^3 + 4x^2) \div (-4x^2)$ $-3x^2 + 2x - 1$

104. $\dfrac{10z^7w^4 - 15z^3w^2 - 20zw}{-20z^2w}$ $-\dfrac{z^5w^3}{2} + \dfrac{3zw}{4} + \dfrac{1}{z}$

105. $\dfrac{x^2 + 7x + 10}{x + 5}$ $x + 2$

106. $(2t^2 + t - 10) \div (t - 2)$ $2t + 5$

107. $(2p^2 + p - 16) \div (2p + 7)$ $p - 3 + \dfrac{5}{2p + 7}$

108. $\dfrac{5a^2 + 27a - 22}{5a - 3}$ $a + 6 + \dfrac{-4}{5a - 3}$

109. $\dfrac{b^3 - 125}{b - 5}$ $b^2 + 5b + 25$

110. $(z^3 + 4z^2 + 5z + 20) \div (z^2 + 5)$ $z + 4$

111. $(y^4 - 4y^3 + 5y^2 - 3y + 2) \div (y^2 + 3)$ $y^2 - 4y + 2 + \dfrac{9y - 4}{y^2 + 3}$

112. $\dfrac{8x^3 + 27}{2x + 3}$ $4x^2 - 6x + 9$

113. $\dfrac{6x^3 - 5x^2 + 5}{3x + 2}$ $2x^2 - 3x + 2 + \dfrac{1}{3x + 2}$

114. $(4y^2 + 6y) \div (2y - 1)$ $2y + 4 + \dfrac{4}{2y - 1}$

115. $(3t^4 - 8t^3 + t^2 - 4t - 5) \div (3t^2 + t + 1)$ $t^2 - 3t + 1 + \dfrac{-2t - 6}{3t^2 + t + 1}$

116. $\dfrac{2w^4 + w^3 + 4w - 3}{2w^2 - w + 3}$ $w^2 + w - 1$

Section 3.8

117. Which is a graph of a linear equation?   a, c

    a.

    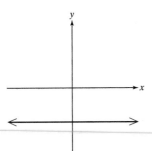

    b.

    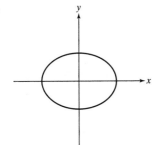

    c.

    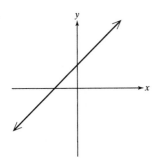

118. Given the equation $y = x^2 - 2$:

    a. Complete the table.

x	y
0	−2
1	−1
2	2
3	7
−1	−1
−2	2
−3	7

    Table for Exercise 118

    b. Graph the ordered pairs corresponding to the values in the table, and sketch a curve defined by the points.   ❖

119. Given the equation $y = x^3 - 1$:

    a. Complete the table.

x	y
0	−1
1	0
2	7
3	26
−1	−2
−2	−9
−3	−28

    Table for Exercise 119

    b. Graph the ordered pairs corresponding to the values in the table, and sketch a curve defined by the point.   ❖

120. A rock is dropped off a 400-ft cliff. The height of the rock, y, (in feet) is given by the equation

    $$y = -16t^2 + 400 \qquad \text{where } t \text{ is the time in seconds after release.}$$

    a. Is the equation $y = -16t^2 + 400$ linear or nonlinear? ❖

    b. Complete the table.

t	y
0	400
1.0	384
2.0	336
2.5	300
3.0	256
4.0	144
5.0	0

    Table for Exercise 120

    c. Write the ordered pairs corresponding to the entries in the table. (In this problem the variable t is used in place of x.) Interpret the meaning of each ordered pair in the context of this problem. ❖

    d. Verify your ordered pairs from part (c) in the graph of the equation.

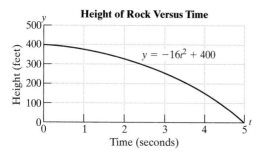

Figure for Exercise 120

---

chapter 3   TEST

Assume all variables represent nonzero real numbers.

1. Expand the expression using the definition of exponents, then simplify: $\dfrac{3^4 \cdot 3^3}{3^6}$

    $\dfrac{(3\cdot3\cdot3\cdot3)\cdot(3\cdot3\cdot3)}{3\cdot3\cdot3\cdot3\cdot3\cdot3} = 3$

For Exercises 2–11, simplify the expression. Write the answer with positive exponents only.

2. $9^5 \cdot 9$ 9^6

3. $\dfrac{q^{10}}{q^2}$ q^8

4. $(3a^2b)^3$ $27a^6b^3$

5. $\left(\dfrac{2x}{y^3}\right)^4$ $\dfrac{16x^4}{y^{12}}$

6. $(-7)^0$ 1

7. c^{-3} $\dfrac{1}{c^3}$

8. $\dfrac{14^3 \cdot 14^9}{14^{10} \cdot 14}$ 14

9. $\dfrac{(s^2t)^3(7s^4t)^4}{(7s^2t^3)^2}$ $49s^{18}t$

10. $(2a^0b^{-6})^2(2a^{-2}b^4)^{-3}$ $\dfrac{a^6}{2b^{24}}$

11. $\left(\dfrac{6a^{-5}b}{8ab^{-2}}\right)^{-2}\left(\dfrac{1}{2a^{-3}b}\right)$ $\dfrac{8a^{15}}{9b^7}$

12. a. Write the number in scientific notation: 43,000,000,000 4.3×10^{10}

    b. Write the number without scientific notation: 5.6×10^{-6} 0.000 0056

13. The average amount of water flowing over Niagara Falls is 1.68×10^5 m³/min.

    a. How many cubic meters of water flow over the falls in one day? 2.4192×10^8 m³

    b. How many cubic meters of water flow over the falls in one year? 8.83008×10^{10} m³

14. Write the polynomial in descending order: $4x + 5x^3 - 7x^2 + 11$. $5x^3 - 7x^2 + 4x + 11$

    a. Identify the degree of the polynomial.   3

❖ See Additional Answers Appendix   Writing   Translating Expression   Geometry   Scientific Calculator   Video

b. Identify the leading coefficient of the polynomial.   5

15. Perform the indicated operations.

$(7w^2 - 11w - 6) + (8w^2 + 3w + 4) - (-9w^2 - 5w + 2)$ $24w^2 - 3w - 4$

For Exercises 16–20, multiply the polynomials.

16. $-2x^3(5x^2 + x - 15)$
$-10x^5 - 2x^4 + 30x^3$

17. $(4a - 3)(2a - 1)$
$8a^2 - 10a + 3$

18. $(4y - 5)(y^2 - 5y + 3)$ $4y^3 - 25y^2 + 37y - 15$

19. $(2 + 3b)(2 - 3b)$
$4 - 9b^2$

20. $(5z - 6)^2$
$25z^2 - 60z + 36$

21. Find the perimeter and the area of the rectangle shown in the figure.   Perimeter: $12x - 2$; area: $5x^2 - 13x - 6$

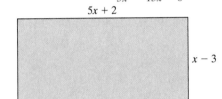

Figure for Exercise 21

22. Divide:

a. $(-12x^8 + x^6 - 8x^3) \div (4x^2)$ $-3x^6 + \dfrac{x^4}{4} - 2x$

b. $\dfrac{2y^2 - 13y + 21}{y - 3}$ $2y - 7$

23. The sum, S, of the first n natural numbers is given by the equation

$$S = \frac{1}{2}n^2 + \frac{1}{2}n$$

a. Is the equation linear or nonlinear?   Nonlinear

b. Find the sum:
$1 + 2 + 3 + 4 + 5 + 6 + 7 + 8 + 9 + 10$ 55

c. Use the equation $S = \frac{1}{2}n^2 + \frac{1}{2}n$ to find the sum of the first 10 natural numbers ($n = 10$).   55

d. Use the equation to find the sum of the first 20 natural numbers and the sum of the first 100 natural numbers.   First 20: 210; first 100: 5050

24. Given the equation $y = 2x^2$:

a. Complete the table.

x	y
0	0
−1	2
−2	8
−3	18
1	2
2	8
3	18

Table for Exercise 24

b. Graph the ordered pairs found in part (a), and sketch a graph through the points.   ❖

CUMULATIVE REVIEW EXERCISES, CHAPTERS 1–3

For Exercises 1–3, simplify completely.

1. $-5 - \dfrac{1}{2}[4 - 3(-7)]$ $-\dfrac{35}{2}$

2. $|-3^2 + 5|$ 4

3. $\dfrac{-3 - \sqrt{14 - (-2)} + 3^2}{-3.44 + 1.2^2}$ 4

4. Translate the phrase into a mathematical expression and simplify:

The difference of the square of five and the square root of four. $5^2 - \sqrt{4}, 23$

5. Which of the following are rational numbers?
$\left\{-7, \dfrac{0}{4}, 2, 0.8, \sqrt{100}, \sqrt{101}\right\}$ $-7, \dfrac{0}{4}, 2, 0.8, \sqrt{100}$

6. Solve for x: $\dfrac{1}{2}(x - 6) + \dfrac{2}{3} = \dfrac{1}{4}x$ $x = \dfrac{28}{3}$

7. Solve for y: $-2y - 3 = -5(y - 1) + 3y$ No solution

8. For a point in a rectangular coordinate system, in which quadrant are both the x- and y-coordinates negative?   Quadrant III

❖ See Additional Answers Appendix   ✎ Writing   ⬌ Translating Expression   Geometry   Scientific Calculator   Video

9. For a point in a rectangular coordinate system, on which axis is the x-coordinate zero and the y-coordinate nonzero?   *y*-axis

10. In a triangle, one angle is 23° more than the smallest angle. The third angle is 10° more than the sum of the other two angles. Find the measure of each angle.   The measures are 31°, 54°, 95°.

11. A salesperson makes 3% commission on the sale of merchandise. If his total commission at the end of a week is $360, what was the value of the merchandise he sold?   He sold $12,000 worth of merchandise.

12. A farmer wants to enclose a rectangular lot that is five times as long as it is wide. One of the shorter sides of the lot is adjacent to a barn and does not require fencing. The farmer plans to fence the other three sides. If the farmer has 264 ft of fencing available, what should the dimensions of the lot be?   The dimensions are 24 ft × 120 ft.

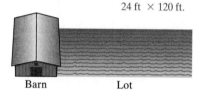

Barn   Lot

Figure for Exercise 12

13. A snow storm lasts for 9 h and dumps snow at a rate of $1\frac{1}{2}$ in./h. If there was already 6 in. of snow on the ground before the storm, the snow depth is given by the equation:

$$y = \frac{3}{2}x + 6$$ where y is the snow depth in inches and x is the time in hours.

   a. Find the snow depth after 4 h.   ❖

   b. Find the snow depth at the end of the storm.   ❖

   c. How long had it snowed when the total depth of snow was $14\frac{1}{4}$ in.?   ❖

   d. Complete the table and graph the corresponding ordered pairs.   ❖

Time (h) x	Snow Depth (in.) y
0	6
2	9
4	12
6	15
8	18
9	19.5

Table for Exercise 13

14. Solve the inequality. Graph the solution set on the real number line and express the solution in interval notation. $2 - 3(2x + 4) \le -2x - (x - 5)$
   $[-5, \infty)$

For Exercises 15–19, perform the indicated operations.

15. $(2x^2 + 3x - 7) - (-3x^2 + 12x + 8)$ $5x^2 - 9x - 15$

16. $(2y + 3z)(-y - 5z)$ $-2y^2 - 13yz - 15z^2$

17. $(4t - 3)^2$ $16t^2 - 24t + 9$

18. $\left(\frac{2}{5}a + \frac{1}{3}\right)\left(\frac{2}{5}a - \frac{1}{3}\right)$ $\frac{4}{25}a^2 - \frac{1}{9}$

19. $(7x - 8) - (x + 3)^2$ $-x^2 + x - 17$

For Exercises 20–21, divide the polynomials.

20. $(12a^4b^3 - 6a^2b^2 + 3ab) \div (-3ab)$ $-4a^3b^2 + 2ab - 1$

21. $\dfrac{4m^3 - 5m + 2}{m - 2}$ $4m^2 + 8m + 11 + \dfrac{24}{m - 2}$

For Exercises 22–24, use the properties of exponents to simplify the expressions. Write the answers with positive exponents only. Assume all variables represent nonzero real numbers.

22. $\dfrac{(x^2)^3(x^4x^{-6})}{x^5}$ $\dfrac{1}{x}$

23. $\left(\dfrac{2c^2d^4}{8cd^6}\right)^2$ $\dfrac{c^2}{16d^4}$

24. $\dfrac{10a^{-2}b^{-3}}{5a^0b^{-6}}$ $\dfrac{2b^3}{a^2}$

25. Write the following numbers in scientific notation.

    a. 407,100,000
        4.071×10^8

    b. 0.000 004 071
        4.071×10^{-6}

26. Write the following numbers without scientific notation.

    a. 3.89×10^{-4}
        0.000 389

    b. 4.5×10^{12}
        4,500,000,000,000

27. Perform the indicated operations and write the final answer in scientific notation. 2.788×10^{-2}

$$\frac{(8.2 \times 10^{-2})(6.8 \times 10^{-6})}{2.0 \times 10^{-5}}$$

FACTORING POLYNOMIALS

Ignoring air resistance, the distance, d (in feet), that a skydiver falls in d seconds is approximated by the formula:

$$d = 16t^2$$

After 1 second, the skydiver will have fallen 16 ft. After 2 seconds the skydiver will have fallen 64 ft, and after 3 seconds, the skydiver will have fallen 144 ft. Notice that each consecutive 1-second interval results in a larger increase in the distance fallen as shown in the graph.

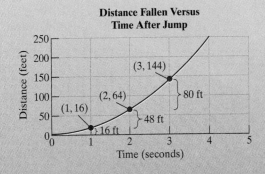

Distance Fallen Versus Time After Jump

To graph this and other quadratic relationships, visit xyplot at

www.mhhe.com/miller_oneill

section

4.1 GREATEST COMMON FACTOR AND FACTORING BY GROUPING

1. Introduction to Factoring

Chapter 4 is devoted to a mathematical operation called **factoring**. To factor an integer means to write the integer as a product of two or more integers. To factor a polynomial means to express the polynomial as a product of two or more polynomials.

In the product $2 \cdot 5 = 10$, for example, 2 and 5 are factors of 10.

In the product $(3x + 4)(2x - 1) = 6x^2 + 5x - 4$, the quantities $(3x + 4)$ and $(2x - 1)$ are factors of $6x^2 + 5x - 4$.

2. Greatest Common Factor of Two Integers

We begin our study of factoring by factoring integers. The number 20 for example can be factored as $1 \cdot 20, 2 \cdot 10, 4 \cdot 5$, or $2 \cdot 2 \cdot 5$. The product $2 \cdot 2 \cdot 5$ (or equivalently $2^2 \cdot 5$) consists only of prime numbers and is called the **prime factorization**.

The **greatest common factor** (denoted **GCF**) of two or more integers is the greatest factor common to each integer. To find the greatest common factor of two integers, it is often helpful to express the numbers as a product of prime factors as shown in the next example.

---

example 1

Identifying the GCF of Two Integers

Find the greatest common factor of each pair of integers.

a.   24 and 36   b.   105 and 40

Solution:

First find the prime factorization of each number. Then find the product of common factors.

a.
$$
\begin{array}{ll}
2\,\underline{|24} & 2\,\underline{|36} \\
2\,\underline{|12} & 2\,\underline{|18} \\
2\,\underline{|6} & 3\,\underline{|9} \\
3\,\underline{|3} & 3\,\underline{|3} \\
1 & 1
\end{array}
$$

Factors of $24 = (2 \cdot 2) \cdot 2 \cdot (3)$

Factors of $36 = (2 \cdot 2) \cdot 3 \cdot (3)$

The numbers 24 and 36 share two factors of 2 and one factor of 3. Therefore, the greatest common factor is $2 \cdot 2 \cdot 3 = 12$.

b.
$$
\begin{array}{ll}
5\,\underline{|105} & 5\,\underline{|40} \\
3\,\underline{|21} & 2\,\underline{|8} \\
7\,\underline{|7} & 2\,\underline{|4} \\
1 & 2\,\underline{|2} \\
 & 1
\end{array}
$$

Factors of $105 = 3 \cdot 7 \cdot (5)$

Factors of $40 = 2 \cdot 2 \cdot 2 \cdot (5)$

The greatest common factor is 5.

3. GCF of Two or More Monomials

example 2

Identifying the Greatest Common Factor of Two or More Polynomials

Find the GCF among each group of terms.

a. $7x^3, 14x^2, 21x^4$ b. $15a^4b, 25a^3b^2$ c. $8c^2d^7e, 6c^3d^4$

Solution:

List the factors of each term.

a. $7x^3 = 7 \cdot x \cdot x \cdot x$

    $14x^2 = 2 \cdot 7 \cdot x \cdot x$ The GCF is $7x^2$.

    $21x^4 = 3 \cdot 7 \cdot x \cdot x \cdot x \cdot x$

b. $15a^4b = 3 \cdot 5 \cdot a \cdot a \cdot a \cdot a \cdot b$

    $25a^3b^2 = 5 \cdot 5 \cdot a \cdot a \cdot a \cdot b \cdot b$ The GCF is $5a^3b$.

Tip: Notice that the expressions $15a^4b$ and $25a^3b^2$ share factors of 5, a, and b. The GCF is the product of the common factors, where each factor is raised to the lowest power to which it occurs in the original expressions.

$$\left.\begin{array}{l} 15a^4b = 3 \cdot 5a^4b \\ 25a^3b^2 = 5^2a^3b^2 \end{array}\right\} \quad \left.\begin{array}{l} \text{Lowest power of 5 is 1:} \quad 5^1 \\ \text{Lowest power of } a \text{ is 3:} \quad a^3 \\ \text{Lowest power of } b \text{ is 1:} \quad b^1 \end{array}\right\} \quad \text{The GCF is } 5a^3b.$$

c. $\left.\begin{array}{l} 8c^2d^7e = 2^3c^2d^7e \\ 6c^3d^4 = 2 \cdot 3c^3d^4 \end{array}\right\}$ The common factors are 2, c, and d.

    $\left.\begin{array}{l} \text{The lowest power of 2 is 1:} \quad 2^1 \\ \text{The lowest power of } c \text{ is 2:} \quad c^2 \\ \text{The lowest power of } d \text{ is 4:} \quad d^4 \end{array}\right\}$ The GCF is $2c^2d^4$.

Sometimes two polynomials share a common binomial factor as shown in the next example.

example 3

Finding the Greatest Common Binomial Factor

Find the greatest common factor between the terms: $3x(a + b)$ and $2y(a + b)$

Solution:

$3x(a + b)$
$2y(a + b)$ The only common factor is the binomial $(a + b)$. The GCF is $(a + b)$.

4.   Factoring out the Greatest Common Factor

The process of factoring a polynomial is the reverse process of multiplying polynomials. Both operations use the distributive property: $ab + ac = a(b + c)$.

Multiply

$$5y(y^2 + 3y + 1) = 5y(y^2) + 5y(3y) + 5y(1)$$
$$= 5y^3 + 15y^2 + 5y$$

Factor

$$5y^3 + 15y^2 + 15y = 5y(y^2) + 5y(3y) + 5y(1)$$
$$= 5y(y^2 + 3y + 1)$$

Steps to Removing the Greatest Common Factor

1.   Identify the GCF of all terms of the polynomial.
2.   Write each term as the product of the GCF and another factor.
3.   Use the distributive property to remove the GCF.

Note: To check the factorization, multiply the polynomials to remove parentheses.

example 4

Factoring out the Greatest Common Factor

Factor out the GCF.

a. $4x - 20$ b. $6w^2 + 3w$ c. $15y^3 + 12y^4$ d. $9a^4b - 18a^5b + 27a^6b$

Solution:

Tip: Any factoring problem can be checked by multiplying the factors:

Check: $4(x - 5) = 4x - 20$ ✔

a. $4x - 20$ The GCF is 4.

$= 4(x) - 4(5)$ Write each term as the product of the GCF and another factor.

$= 4(x - 5)$ Use the distributive property to factor out the GCF.

b. $6w^2 + 3w$ The GCF is $3w$.

$\quad = 3w(2w) + 3w(1)$ Write each term as the product of $3w$ and another factor.

$\quad = 3w(2w + 1)$ Use the distributive property to factor out the GCF.

Check: $3w(2w + 1) = 6w^2 + 3w$ ✔

c. $15y^3 + 12y^4$ The GCF is $3y^3$.

$\quad = 3y^3(5) + 3y^3(4y)$ Write each term as the product of $3y^3$ and another factor.

$\quad = 3y^3(5 + 4y)$ Use the distributive property to factor out the GCF.

Check: $3y^3(5 + 4y) = 15y^3 + 12y^4$ ✔

d. $9a^4b - 18a^5b + 27a^6b$ The GCF is $9a^4b$.

$\quad = 9a^4b(1) - 9a^4b(2a) + 9a^4b(3a^2)$ Write each term as the product of $9a^4b$ and another factor.

$\quad = 9a^4b(1 - 2a + 3a^2)$ Use the distributive property to factor out the GCF.

Check: $9a^4b(1 - 2a + 3a^2) = 9a^4b - 18a^5b + 27a^6b$ ✔

5.   Factoring out a Negative Factor

Sometimes it is advantageous to factor out the *opposite* of the GCF when the leading coefficient of the polynomial is negative. This is demonstrated in the next example. Notice that this *changes the signs* of the remaining terms inside the parentheses.

example 5

Factoring out a Negative Factor

Factor out the quantity $-4pq$ from the polynomial $-12p^3q - 8p^2q^2 + 4pq^3$

Solution:

$\quad -12p^3q - 8p^2q^2 + 4pq^3$ The GCF is $4pq$. However, in this case, we will factor out the *opposite* of the GCF, $-4pq$.

$\quad = -4pq(3p^2) + (-4pq)(2pq) + (-4pq)(-q^2)$ Write each term as the product of $-4pq$ and another factor.

$$= -4pq[3p^2 + 2pq + (-q^2)]$$

Factor out $-4pq$. Notice that each sign within the trinomial has changed.

$$= -4pq(3p^2 + 2pq - q^2)$$

To verify that this is the correct factorization and that the signs are correct, multiply factors.

Check: $-4pq(3p^2 + 2pq - q^2) = -12p^3q - 8p^2q^2 + 4pq^3$ ✔

6. Factoring out a Binomial Factor

The distributive property may also be used to factor out a common factor that consists of more than one term as shown in the next example.

example 6

Classroom Activity 4.1A

Factoring out a Binomial Factor

Factor out the greatest common factor: $2w(x + 3) - 5(x + 3)$

Solution:

$2w(x + 3) - 5(x + 3)$

The greatest common factor is the quantity $(x + 3)$.

$= (x + 3)(2w) - (x + 3)(5)$

Write each term as the product of $(x + 3)$ and another factor.

$= (x + 3)(2w - 5)$

Use the distributive property to factor out the GCF.

Check: $(x + 3)(2w - 5) = (x + 3)(2w) + (x + 3)(-5)$
$= 2w(x + 3) - 5(x + 3)$ ✔

7.   Factoring by Grouping

When two binomials are multiplied, the product before simplifying contains four terms. For example:

$$(x + 4)(3a + 2b) = (x + 4)(3a) + (x + 4)(2b)$$

$$= (x + 4)(3a) + (x + 4)(2b)$$

$$= 3ax + 12a + 2bx + 8b$$

In the next example, we learn how to reverse this process. That is, given a four-term polynomial, we will factor it as a product of two binomials. The process is called **factoring by grouping**.

example 7

Factoring by Grouping

Factor by grouping: $3ax + 12a + 2bx + 8b$

Solution:

$3ax + 12a + 2bx + 8b$ | **Step 1:** Identify and factor out the GCF from all four terms. In this case, the GCF is 1.

$= 3ax + 12a \mid + 2bx + 8b$ | Group the first pair of terms and the second pair of terms.

$= 3a(x + 4) + 2b(x + 4)$ | **Step 2:** Factor out the GCF from each pair of terms. *Note*: The two terms now share a common binomial factor of $(x + 4)$.

$= (x + 4)(3a + 2b)$ | **Step 3:** Factor out the common binomial factor.

Check: $(x + 4)(3a + 2b) = 3ax + 2bx + 12a + 8b$ ✔

Note: Step 2 results in two terms with a common binomial factor. If the two binomials are different, Step 3 cannot be performed. In such a case, the original polynomial may not be factorable by grouping.

Tip: One frequently asked question when factoring is whether the order can be switched between the factors. The answer is yes. Because multiplication is commutative, the order in which the factors are written does not matter.

$$(x + 4)(3a + 2b) = (3a + 2b)(x + 4)$$

Steps to Factoring by Grouping

To factor a four-term polynomial by grouping:

1. Identify and factor out the GCF from all four terms.
2. Factor out the GCF from the first pair of terms. Factor out the GCF from the second pair of terms. (Sometimes it is necessary to factor out the opposite of the GCF.)
3. If the two terms share a common binomial factor, factor out the binomial factor.

example 8

Factoring by Grouping

Factor the polynomials by grouping.

a. $ax + ay - bx - by$ b. $16w^4 - 40w^3 - 12w^2 + 30w$

Classroom Activity 4.1B

Solution:

a. $ax + ay - bx - by$

Step 1: Identify and factor out the GCF from all four terms. In this case, the GCF is 1.

$= ax + ay \mid - bx - by$

Group the first pair of terms and the second pair of terms.

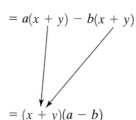

$= a(x + y) - b(x + y)$

Step 2: Factor out a from the first pair of terms.

Factor out $-b$ from the second pair of terms. (This causes sign changes within the second parentheses.)

$= (x + y)(a - b)$

Step 3: Factor out the common binomial factor.

Check: $(x + y)(a - b) = x(a) + x(-b) + y(a) + y(-b)$
$= ax - bx + ay - by$ ✔

Avoiding Mistakes

In Step 2, the expression $a(x + y) - b(x + y)$ is not yet factored because it is a *difference*, not a product. To factor the expression, you must carry it one step further.

$a(x + y) - b(x + y)$
$= (x + y)(a - b)$

The factored form must be represented as a product.

b. $16w^4 - 40w^3 - 12w^2 + 30w$

Step 1: Identify and factor out the GCF from all four terms. In this case, the GCF is $2w$.

$= 2w(8w^3 - 20w^2 - 6w + 15)$

$= 2w[8w^3 - 20w^2 \mid - 6w + 15]$

Group the first pair of terms and the second pair of terms.

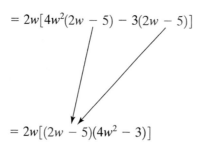

$= 2w[4w^2(2w - 5) - 3(2w - 5)]$

Step 2: Factor out $4w^2$ from the first pair of terms.

Factor out -3 from the second pair of terms. (This causes sign changes within the second parentheses.)

$= 2w[(2w - 5)(4w^2 - 3)]$

Step 3: Factor out the common binomial factor.

Check: $2w[(2w - 5)(4w^2 - 3)] = 2w(8w^3 - 20w^2 - 6w + 15)$
$= 16w^4 - 40w^3 - 12w^2 + 30w$ ✔

section 4.1   ## PRACTICE EXERCISES

For Exercises 1–12, identify the greatest common factor between each pair of terms.

1. $28, 63$ 7
2. $24, 40$ 8
3. $42, 30$ 6
4. $18, 52$ 2
5. $2a^2b, 3ab^2$ ab
6. $3x^3y^2, 5xy^4$ xy^2
7. $12w^3z, 16w^2z$ $4w^2z$
8. $20cd, 15c^3d$ $5cd$
9. $7(x - y), 9(x - y)$ $(x - y)$
10. $(2a - b), 3(2a - b)$ $(2a - b)$
11. $14(3x + 1)^2, 7(3x + 1)$ $7(3x + 1)$
12. $a^2(w + z), a^3(w + z)^2$ $a^2(w + z)$

13. a. Use the distributive property to multiply $3(x - 2y)$. $3x - 6y$
    b. Use the distributive property to factor $3x - 6y$. $3(x - 2y)$

14. a. Use the distributive property to multiply $a^2(5a + b)$. $5a^3 + a^2b$
    b. Use the distributive property to factor $5a^3 + a^2b$. $a^2(5a + b)$

For Exercises 15–38, factor out the GCF.

15. $4p + 12$ $4(p + 3)$
16. $3q - 15$ $3(q - 5)$
17. $5c^2 - 10c$ $5c(c - 2)$
18. $24d + 16d^3$ $8d(3 + 2d^2)$
19. $x^5 + x^3$ $x^3(x^2 + 1)$
20. $y^2 - y^4$ $y^2(1 - y^2)$
21. $t^4 - 4t$ $t(t^3 - 4)$
22. $7r^3 - r^5$ $r^3(7 - r^2)$
23. $2ab + 4a^3b$ $2ab(1 + 2a^2)$
24. $5u^3v^2 - 5uv$ $5uv(u^2v - 1)$
25. $38x^2y - 19x^2y^4$ $19x^2y(2 - y^3)$
26. $100a^5b^3 + 16a^2b$ $4a^2b(25a^3b^2 + 4)$
27. $42p^3q^2 + 14pq^2 - 7p^4q^4$ $7pq^2(6p^2 + 2 - p^3q^2)$
28. $8m^2n^3 - 24m^2n^2 + 4m^3n$ $4m^2n(2n^2 - 6n + m)$
29. $t^5 + 2rt^3 - 3t^4 + 4r^2t^2$ $t^2(t^3 + 2rt - 3t^2 + 4r^2)$
30. $u^2v + 5u^3v^2 - 2u^2 + 8uv$ $u(uv + 5u^2v^2 - 2u + 8v)$
31. $13(a + 6) - 4b(a + 6)$ $(a + 6)(13 - 4b)$
32. $7(x^2 + 1) - y(x^2 + 1)$ $(x^2 + 1)(7 - y)$
33. $8v(w^2 - 2) + (w^2 - 2)$ $(w^2 - 2)(8v + 1)$

34. $t(r + 2) + (r + 2)$ $(r + 2)(t + 1)$
35. $21x(x + 3) + 7x^2(x + 3)$ $7x(x + 3)^2$
36. $5y^3(y - 2) - 20y(y - 2)$ $5y(y - 2)(y^2 - 4)$
37. $6(z - 1)^3 + 7z(z - 1)^2 - (z - 1)$
    $(z - 1)(13z^2 - 19z + 5)$
38. $4(q + 5)^2 + 5q(q + 5) - (q + 5)$ $(q + 5)(9q + 19)$

39. The formula $P = 2l + 2w$ represents the perimeter, P, of a rectangle given the length, l, and the width, w. Factor out the GCF and write an equivalent formula in factored form. $P = 2(l + w)$

40. The formula $P = 2a + 2b$ represents the perimeter, P, of a parallelogram given the base, b, and an adjacent side, a. Factor out the GCF and write an equivalent formula in factored form.
    $P = 2(a + b)$

41. The formula $S = 2\pi r^2 + 2\pi rh$ represents the surface area, S, of a cylinder with radius, r, and height, h. Factor out the GCF and write an equivalent formula in factored form. $S = 2\pi r(r + h)$

42. The formula $A = P + Prt$ represents the total amount of money, A, in an account that earns simple interest at a rate, r, for t years. Factor out the GCF and write an equivalent formula in factored form. $A = P(1 + rt)$

43. For the polynomial $-2x^3 - 4x^2 + 8x$
    a. Factor out $-2x$. $-2x(x^2 + 2x - 4)$
    b. Factor out $2x$. $2x(-x^2 - 2x + 4)$

44. For the polynomial $-9y^5 + 3y^3 - 12y$
    a. Factor out $-3y$. $-3y(3y^4 - y^2 + 4)$
    b. Factor out $3y$. $3y(-3y^4 + y^2 - 4)$

45. Factor out -1 from the polynomial $-8t^2 - 9t - 2$. $-1(8t^2 + 9t + 2)$

46. Factor out -1 from the polynomial $-6x^3 - 2x - 5$. $-1(6x^3 + 2x + 5)$

47. Factor out -1 from the polynomial $-4y^3 + 5y - 7$. $-1(4y^3 - 5y + 7)$

48. Factor out -1 from the polynomial $-w^2 + w - 5$. $-1(w^2 - w + 5)$

For Exercises 49–56, factor out the opposite of the greatest common factor.

49. $-15p^3 - 30p^2$
$-15p^2(p + 2)$

50. $-24m^3 - 12m^4$
$-12m^3(2 + m)$

51. $-q^4 + 2q^2 - 9q$
$-q(q^3 - 2q + 9)$

52. $-r^3 + 9r^2 - 5r$
$-r(r^2 - 9r + 5)$

53. $-7x - 6y - 2z$
$-1(7x + 6y + 2z)$

54. $-4a + 5b - c$
$-1(4a - 5b + c)$

55. $-3(2c + 5) - 4c(2c + 5)$ $-1(2c + 5)(3 + 4c)$

56. $-6n(4n - 1) - 7(4n - 1)$ $-1(4n - 1)(6n + 7)$

For Exercises 57–72, factor by grouping.

57. $8a^2 - 4ab + 6ac - 3bc$ $(2a - b)(4a + 3c)$

58. $4x^3 + 3x^2y + 4xy^2 + 3y^3$ $(4x + 3y)(x^2 + y^2)$

59. $3q + 3p + qr + pr$ $(q + p)(3 + r)$

60. $xy - xz + 7y - 7z$ $(y - z)(x + 7)$

61. $6x^2 + 3x + 4x + 2$ $(2x + 1)(3x + 2)$

62. $4y^2 + 8y + 7y + 14$ $(y + 2)(4y + 7)$

63. $2t^2 + 6t - 5t - 15$ $(t + 3)(2t - 5)$

64. $2p^2 - p - 6p + 3$ $(2p - 1)(p - 3)$

65. $6y^2 - 2y - 9y + 3$ $(3y - 1)(2y - 3)$

66. $5a^2 + 30a - 2a - 12$ $(a + 6)(5a - 2)$

67. $b^4 + b^3 - 4b - 4$ $(b + 1)(b^3 - 4)$

68. $8w^5 + 12w^2 - 10w^3 - 15$ $(2w^3 + 3)(4w^2 - 5)$

69. $3j^2k + 15k + j^2 + 5$ $(j^2 + 5)(3k + 1)$

70. $2ab^2 - 6ac + b^2 - 3c$ $(b^2 - 3c)(2a + 1)$

71. $14w^6x^6 + 7w^6 - 2x^6 - 1$ $(2x^6 + 1)(7w^6 - 1)$

72. $18p^4q - 9p^5 - 2q + p$ $(2q - p)(9p^4 - 1)$

For Exercises 73–78, factor out the GCF first. Then factor by grouping.

73. $15x^4 + 15x^2y^2 + 10x^3y + 10xy^3$ $5x(x^2 + y^2)(3x + 2y)$

74. $2a^3b - 4a^2b + 32ab - 64b$ $2b(a - 2)(a^2 + 16)$

75. $4abx - 4b^2x - 4ab + 4b^2$ $4b(a - b)(x - 1)$

76. $p^2q - pq^2 - rp^2q + rpq^2$ $pq(p - q)(1 - r)$

77. $6st^2 - 18st - 6t^4 + 18t^3$ $6t(t - 3)(s - t^2)$

78. $15j^3 - 10j^2k - 15j^2k^2 + 10jk^3$ $5j(3j - 2k)(j - k^2)$

EXPANDING YOUR SKILLS

79. Write a polynomial that has a GCF of $3x$. (Answers may vary.)   For example: $6x^2 + 9x$

80. Write a polynomial that has a GCF of $7y$. (Answers may vary.)   For example: $14y - 21y^3 + 7y^2$

81. Write a polynomial that has a GCF of $4p^2q$. (Answers may vary.)   For example: $16p^4q^2 + 8p^3q - 4p^2q$

82. Write a polynomial that has a GCF of $2ab^2$. (Answers may vary.)   For example: $18a^2b^3 - 2ab^2$

Concepts

1. Grouping Method to Factor Trinomials

2. Factoring Trinomials with a Leading Coefficient of 1

3. Prime Polynomials

section

4.2   FACTORING TRINOMIALS: GROUPING METHOD

We have already learned how to factor out the GCF from a polynomial and how to factor a four-term polynomial by grouping. As we work through this chapter, we will expand our knowledge of factoring by learning how to factor trinomials and binomials.

There are two commonly used methods for factoring trinomials. The grouping method (or "ac" method) is presented here and the trial-and-error method is presented in the next section.

1. Grouping Method to Factor Trinomials

The product of two binomials results in a four-term expression that can sometimes be simplified to a trinomial. To factor the trinomial, we want to reverse the process.

Multiply

Multiply the binomials. Add the middle terms.

$$(2x + 3)(x + 2) = \longrightarrow 2x^2 + 4x + 3x + 6 = \longrightarrow 2x^2 + 7x + 6$$

Factor

$$2x^2 + 7x + 6 = \longrightarrow 2x^2 + 4x + 3x + 6 = \longrightarrow (2x + 3)(x + 2)$$

Rewrite the middle term as Factor by grouping.
a sum or difference of terms.

To factor a trinomial, $ax^2 + bx + c$, by the grouping method, we rewrite the middle term, bx, as a sum or difference of terms. The goal is to produce a four-term polynomial that can be factored by grouping. The process is outlined in the following box.

Grouping Method Factor $ax^2 + bx + c$ $(a \neq 0)$

1. Multiply the coefficients of the first and last terms (ac).
2. Find two integers whose product is ac and whose sum is b. (If no pair of integers can be found, then the trinomial cannot be factored further and is called a **prime polynomial**.)
3. Rewrite the middle term bx as the sum of two terms whose coefficients are the integers found in Step 2.
4. Factor by grouping.

The grouping method for factoring trinomials is illustrated in the next example. However, before we begin, keep these two important guidelines in mind:

- For any factoring problem you encounter, always factor out the GCF from all terms first.
- To factor a trinomial, write the trinomial in the form $ax^2 + bx + c$.

---

example 1

Factoring a Trinomial by the Grouping Method

Factor the trinomial by the grouping method: $2x^2 + 7x + 6$

Solution:

$2x^2 + 7x + 6$ Factor out the GCF from all terms. In this case, the GCF is 1.

$2x^2 + 7x + 6$ **Step 1:** The trinomial is written in the form $ax^2 + bx + c$.

$a = 2, b = 7, c = 6$ Find the product $ac = (2)(6) = 12$.

$$\underline{12} \qquad \underline{12}$$

$$1 \cdot 12 \qquad (-1)(-12)$$

$$2 \cdot 6 \qquad (-2)(-6)$$

$$3 \cdot 4 \qquad (-3)(-4)$$

$2x^2 + 7x + 6$

$$= 2x^2 + 3x + 4x + 6$$

$$= 2x^2 + 3x \mid + 4x + 6$$

$$= x(2x + 3) + 2(2x + 3)$$

$$= (2x + 3)(x + 2)$$

Check: $(2x + 3)(x + 2) = 2x^2 + 4x + 3x + 6$
$$= 2x^2 + 7x + 6 ✔$$

Step 2: List all the factors of ac and search for the pair whose sum equals the value of b.

That is, list the factors of 12 and find the pair whose sum equals 7.

The numbers 3 and 4 satisfy both conditions: $3 \cdot 4 = 12$ and $3 + 4 = 7$.

Step 3: Write the middle term of the trinomial as the sum of two terms whose coefficients are the selected pair of numbers: 3 and 4.

Step 4: Factor by grouping.

Tip: One frequently asked question is whether the order matters when we rewrite the middle term of the trinomial as two terms (Step 3). The answer is no. From the previous example, the two middle terms in Step 3 could have been reversed to obtain the same result:

$$2x^2 + 7x + 6$$
$$= 2x^2 + 4x + 3x + 6$$
$$= 2x(x + 2) + 3(x + 2)$$
$$= (x + 2)(2x + 3)$$

This example also points out that the order in which two factors are written does not matter. The expression $(x + 2)(2x + 3)$ is equivalent to $(2x + 3)(x + 2)$ because multiplication is a commutative operation.

example 2

Factoring Trinomials by the Grouping Method

Factor the trinomial by the grouping method: $-2x + 8x^2 - 3$

Solution:

$-2x + 8x^2 - 3$

First rewrite the polynomial in the form $ax^2 + bx + c$.

$= 8x^2 - 2x - 3$

The GCF is 1.

$a = 8, b = -2, c = -3$ **Step 1:** Find the product $ac = (8)(-3) = -24$.

$\underline{-24}$	$\underline{-24}$
$-1 \cdot 24$	$-6 \cdot 4$
$-2 \cdot 12$	$-8 \cdot 3$
$-3 \cdot 8$	$-12 \cdot 2$
$-4 \cdot 6$	$-24 \cdot 1$

Step 2: List all the factors of -24 and find the pair of factors whose sum equals -2.

The numbers -6 and 4 satisfy both conditions: $(-6)(4) = -24$ and $-6 + 4 = -2$.

$= 8x^2 - 2x - 3$

Step 3: Write the middle term of the trinomial as two terms whose coefficients are the selected pair of numbers, -6 and 4.

$= 8x^2 - 6x + 4x - 3$

$= 8x^2 - 6x + 4x - 3$ **Step 4:** Factor by grouping.

$= 2x(4x - 3) + 1(4x - 3)$

$= (4x - 3)(2x + 1)$

$\underline{\text{Check:}} \ (4x - 3)(2x + 1) = 8x^2 + 4x - 6x - 3$
$$= 8x^2 - 2x - 3 \ ✔$$

---

example 3   Factoring a Trinomial by the Grouping Method

Factor the trinomial by the grouping method: $10x^3 - 85x^2 + 105x$

Solution:

$10x^3 - 85x^2 + 105x$ The GCF is $5x$.

$= 5x(2x^2 - 17x + 21)$ The trinomial is in the form $ax^2 + bx + c$.

$a = 2, b = -17, c = 21$ **Step 1:** Find the product $ac = (2)(21) = 42$.

$\underline{42}$	$\underline{42}$
$1 \cdot 42$	$(-1)(-42)$
$2 \cdot 21$	$(-2)(-21)$
$3 \cdot 14$	$(-3)(-14)$
$6 \cdot 7$	$(-6)(-7)$

Step 2: List all the factors of 42 and find the pair whose sum equals -17.

The numbers -3 and -14 satisfy both conditions: $(-3)(-14) = 42$ and $-3 + (-14) = -17$.

$= 5x(2x^2 - 17x + 21)$

Step 3: Write the middle term of the trinomial as two terms whose coefficients are the selected pair of numbers, -3 and -14.

$= 5x(2x^2 - 3x - 14x + 21)$

$$= 5x(2x^2 - 3x \mid - 14x + 21) \qquad \textbf{Step 4:} \quad \text{Factor by grouping.}$$

$$= 5x[x(2x - 3) - 7(2x - 3)]$$

$$= 5x[(2x - 3)(x - 7)]$$

$$= 5x(2x - 3)(x - 7)$$

Check: $5x(2x - 3)(x - 7) = 5x(2x^2 - 14x - 3x + 21)$
$$= 5x(2x^2 - 17x + 21)$$
$$= 10x^3 - 85x^2 + 105x \ ✔$$

Tip: Notice when the GCF is removed from the original trinomial, the new trinomial has smaller coefficients. This makes the factoring process simpler because the product ac is smaller. It is much easier to list the factors of 42 than the factors of 1050.

Original trinomial	**With the GCF factored out**
$10x^3 - 85x^2 + 105x$	$5x(2x^2 - 17x + 21)$
$ac = (10)(105) = 1050$	$ac = (2)(21) = 42$

In most cases it is easier to factor a trinomial with a positive leading coefficient. If the leading coefficient is negative, a factor of -1 can be removed to change the sign of the leading coefficient as well as the coefficients of the remaining terms.

example 4

Factoring a Trinomial by the Grouping Method

Factor the trinomial by the grouping method: $-a^2 - 7ab + 18b^2$

Classroom Activity 4.2A

Solution:

$-a^2 - 7ab + 18b^2$

$= -1(a^2 + 7ab - 18b^2)$
            Factor out -1.

                Step 1:   Find the product $ac = -18$.

                Step 2:   The numbers 9 and -2 have a product of -18 and a sum of 7.

$= -1(a^2 + 9ab - 2ab - 18b^2)$ **Step 3:**   Rewrite the middle term $7ab$ as $9ab - 2ab$.

$= -1(a^2 + 9ab \mid - 2ab - 18b^2)$ **Step 4:**   Factor by grouping.

$= -1[a(a + 9b) - 2b(a + 9b)]$

$= -1[(a + 9b)(a - 2b)]$

Check: $-1[(a + 9b)(a - 2b)] = -1[a^2 - 2ab + 9ab - 18b^2]$
$$= -1[a^2 + 7ab - 18b^2]$$
$$= -a^2 - 7ab + 18b^2 \qquad ✔$$

2. Factoring Trinomials with a Leading Coefficient of 1

The grouping method is a general method to factor trinomials of the form $ax^2 + bx + c$. If the leading coefficient, a, is equal to 1, then the process can be simplified. First note that if $a = 1$, then the product $ac = 1c = c$. Therefore, in Step 2, we find two integers whose product is c and whose sum is b. Then, after rewriting the middle term and factoring by grouping, we have two binomial factors whose leading terms are x and whose constant terms are the integers found in Step 2.

     In the next example, we will factor trinomials of the form $x^2 + bx + c$ (leading coefficient of 1) to illustrate this process.

---

example 5   **Factoring Trinomials with a Leading Coefficient of 1**

Factor the trinomials.

a. $w^2 + 10w + 9$ b. $x^2 - 2x - 15$

Solution:

a. $w^2 + 10w + 9$ The GCF is 1.

The trinomial is in the form $ax^2 + bx + c$, where $a = 1$.

Step 1:   Because $a = 1$, the product $ac = c = 9$.

Step 2:   Find two numbers whose product is 9 and whose sum is 10. The numbers are 9 and 1.

$= w^2 + 9w + 1w + 9$ **Step 3:**   Rewrite the middle term as a sum of terms.

$= w(w + 9) + 1(w + 9)$ **Step 4:**   Factor by grouping.

$= (w + 9)(w + 1)$

Tip: The constants in the binomial factors are the two integers found in Step 2.

Check: $(w + 9)(w + 1) = w^2 + w + 9w + 9$
$= w^2 + 10w + 9$ ✔

b. $x^2 - 2x - 15$ The GCF is 1.

The trinomial is in the form $ax^2 + bx + c$, where $a = 1$.

Step 1:   Because $a = 1$, the product $ac = c = -15$.

Step 2:   Find two numbers whose product is -15 and whose sum is -2. The numbers are -5 and 3.

$= x^2 - 5x + 3x - 15$ **Step 3:**   Rewrite the middle term as a sum of terms.

Tip: The constants in the binomial factors are the two integers, −5 and 3, found in Step 2.

$= x(x - 5) + 3(x - 5)$ **Step 4:**   Factor by grouping.

$= (x - 5)(x + 3)$

Check: $(x - 5)(x + 3) = x^2 + 3x - 5x - 15$

$= x^2 - 2x - 15$

3.   Prime Polynomials

It should be noted that not every trinomial is factorable by the methods presented in this text.

---

example 6

Factoring a Trinomial by the Grouping Method

Factor the trinomial by the grouping method: $2p^2 - 8p + 3$

Classroom Activity 4.2B

Solution:

$2p^2 - 8p + 3$ **Step 1:**   The GCF is 1.

Step 2:   The product $ac = 6$.

$\underline{6}$ $\underline{6}$ **Step 3:**   List the factors of 6. Notice that no pair of factors has a sum of −8. Therefore, the trinomial cannot be factored.

$1 \cdot 6$ $(-1)(-6)$

$2 \cdot 3$ $(-2)(-3)$

The trinomial $2p^2 - 8p + 3$ is prime.

---

section 4.2   P<small>RACTICE</small> E<small>XERCISES</small>

For Exercises 1–10, factor out the GCF.   4.1

1. $8p^9 + 24p^3$ $8p^3(p^6 + 3)$ 2. $5q^4 - 10q^5$ $5q^4(1 - 2q)$

3. $12u^2 - 6u$ $6u(2u - 1)$ 4. $15t^2 + 5t$ $5t(3t + 1)$

5. $9x^2y + 12xy^2 - 15x^2y^2$ $3xy(3x + 4y - 5xy)$

6. $15ab^3 - 10a^2bc + 25b^2c^3$ $5b(3ab^2 - 2a^2c + 5bc^3)$

7. $5x(x - 2) - 2(x - 2)$ $(x - 2)(5x - 2)$

8. $8(y + 5) + 9y(y + 5)$ $(y + 5)(8 + 9y)$

9. $4p(3p - q) + 5(3p - q)$ $(3p - q)(4p + 5)$

10. $y(5y + 1) - 8(5y + 1)$ $(5y + 1)(y - 8)$

For Exercises 11–16, factor by grouping.   4.1

11. $2c^2 + 2cd + 3cd + 3d^2$ $(c + d)(2c + 3d)$

12. $p^2 - 2pq - pq + 2q^2$ $(p - 2q)(p - q)$

13. $2u - 10 - uv + 5v$ 14. $3r - 6 + rs - 2s$

$(u - 5)(2 - v)$ $(r - 2)(3 + s)$

15. $6a^2 + 24a - 12a - 48$ $6(a + 4)(a - 2)$

16. $5b^2 + 30b - 10b - 60$ $5(b + 6)(b - 2)$

17. What is a prime polynomial?   A polynomial that cannot be factored is prime.

18. How do you determine if a trinomial, $ax^2 + bx + c$, is prime?   A trinomial, $ax^2 + bx + c$, is prime if there are no two integers whose product is ac and whose sum is b.

 See Additional Answers Appendix   Writing    Translating Expression   Geometry   Scientific Calculator   Video

For Exercises 19–26, find the pair of integers whose product and sum are given.

19. Product: 12   Sum: 13   12, 1

20. Product: 12   Sum: 7   3, 4

21. Product: 8   Sum: −9   −8, −1

22. Product: −4   Sum: −3   −4, 1

23. Product: −6   Sum: 5   −1, 6

24. Product: −18   Sum: 7   9, −2

25. Product: −72   Sum: −6   −12, 6

26. Product: 36   Sum: −12   −6, −6

For Exercises 27–64, factor the trinomials using the grouping method.

27. $3x^2 + 13x + 4$
$(x + 4)(3x + 1)$

28. $2y^2 + 7y + 6$
$(2y + 3)(y + 2)$

29. $4w^2 - 9w + 2$
$(w - 2)(4w - 1)$

30. $2p^2 - 3p - 2$
$(p - 2)(2p + 1)$

31. $2m^2 + 5m - 3$
$(m + 3)(2m - 1)$

32. $6n^2 + 7n - 3$
$(2n + 3)(3n - 1)$

33. $8k^2 - 6k - 9$
$(4k + 3)(2k - 3)$

34. $9h^2 - 12h + 4$
$(3h - 2)^2$

35. $4k^2 - 20k + 25$
$(2k - 5)^2$

36. $16h^2 + 24h + 9$
$(4h + 3)^2$

37. $5x^2 + x + 7$ Prime

38. $4y^2 - y + 2$ Prime

39. $4p^2 + 5pq - 6q^2$
$(p + 2q)(4p - 3q)$

40. $6u^2 - 19uv + 10v^2$
$(3u - 2v)(2u - 5v)$

41. $15m^2 + mn - 2n^2$
$(5m + 2n)(3m - n)$

42. $12a^2 + 11ab - 5b^2$
$(4a + 5b)(3a - b)$

43. $3r^2 - rs - 14s^2$
$(r + 2s)(3r - 7s)$

44. $3h^2 + 19hk - 14k^2$
$(h + 7k)(3h - 2k)$

45. $2x^2 - 13xy + y^2$
Prime

46. $3p^2 + 20pq - q^2$
Prime

47. $q^2 - 11q + 10$
$(q - 10)(q - 1)$

48. $a^2 + 7a - 18$
$(a + 9)(a - 2)$

49. $r^2 - 6r - 40$
$(r + 4)(r - 10)$

50. $s^2 - 10s - 24$
$(s - 12)(s + 2)$

51. $x^2 + 6x - 7$
$(x + 7)(x - 1)$

52. $y^2 + 5y - 24$
$(y + 8)(y - 3)$

53. $m^2 - 13m + 42$
$(m - 6)(m - 7)$

54. $n^2 - 9n + 20$
$(n - 5)(n - 4)$

55. $a^2 + 9a + 20$
$(a + 5)(a + 4)$

56. $b^2 + 13b + 42$
$(b + 6)(b + 7)$

57. $t^2 + 5t + 5$ Prime

58. $s^2 - 6s + 3$ Prime

59. $p^2 + 20pq + 100q^2$
$(p + 10q)^2$

60. $c^2 - 14cd + 49d^2$
$(c - 7d)^2$

61. $x^2 - xy - 42y^2$
$(x - 7y)(x + 6y)$

62. $a^2 - 13ab + 40b^2$
$(a - 5b)(a - 8b)$

63. $r^2 + 8rs + 15s^2$
$(r + 5s)(r + 3s)$

64. $u^2 + 2uv - 15v^2$
$(u + 5v)(u - 3v)$

65. Is the expression $(2x + 4)(x - 7)$ factored completely? Explain why or why not.   No, $(2x + 4)$ contains a common factor of 2.

66. Is the expression $(3x + 1)(5x - 10)$ factored completely? Explain why or why not.   No, $(5x - 10)$ contains a common factor of 5.

For Exercises 67–76, first factor out a common factor. Then factor the trinomials by using the grouping method.

67. $6x^2 + 20x + 14$
$2(3x + 7)(x + 1)$

68. $4r^3 + 3r^2 - 10r$
$r(4r - 5)(r + 2)$

69. $2p^3 - 38p^2 + 120p$
$2p(p - 15)(p - 4)$

70. $4q^3 - 4q^2 - 80q$
$4q(q - 5)(q + 4)$

71. $x^2y^2 + 14x^2y + 33x^2$ $x^2(y + 3)(y + 11)$

72. $a^2b^2 + 13ab^2 + 30b^2$ $b^2(a + 10)(a + 3)$

73. $-k^2 - 7k - 10$
$-1(k + 2)(k + 5)$

74. $-m^2 - 15m + 34$
$-1(m - 2)(m + 17)$

75. $-3n^2 - 3n + 90$
$-3(n + 6)(n - 5)$

76. $-2h^2 + 28h - 90$
$-2(h - 9)(h - 5)$

For Exercises 77–84, write the polynomials in descending order. Then factor the polynomials.

77. $3 - 14z + 16z^2$
$(2z - 1)(8z - 3)$

78. $10w + 1 + 16w^2$
$(8w + 1)(2w + 1)$

79. $b^2 + 16 - 8b$
$(b - 4)^2$

80. $1 + q^2 - 2q$
$(q - 1)^2$

81. $25x - 5x^2 - 30$
$-5(x - 2)(x - 3)$

82. $20a - 18 - 2a^2$
$-2(a - 1)(a - 9)$

83. $-6 - t + t^2$
$(t - 3)(t + 2)$

84. $-6 + m + m^2$
$(m + 3)(m - 2)$

section

4.3   FACTORING TRINOMIALS: TRIAL-AND-ERROR METHOD

In Section 4.2, the grouping method was presented for factoring trinomials. In this section, we offer another method to factor trinomials, called the **trial-and-error method**. You and your instructor must determine which method is best for you.

1.   Trial-and-Error Method for Factoring Trinomials

To understand the basis of factoring trinomials of the form $ax^2 + bx + c$, first consider the multiplication of two binomials:

$$\overset{\text{Product of } 2\,\cdot\,1}{} \qquad \overset{\text{Product of } 3\,\cdot\,2}{}$$
$$(2x + 3)(1x + 2) = 2x^2 + \underline{\textbf{4x + 3x}} + 6 = 2x^2 + 7x + 6$$

Sum of products of inner
terms and outer terms

To factor the trinomial, $2x^2 + 7x + 6$ this operation is reversed. Hence:

Factors of 2

$$2x^2 + 7x + 6 = (\Box x \quad \Box)(\Box x \quad \Box)$$

Factors of 6

We need to fill in the blanks so that the product of the first terms in the binomials is $2x^2$ and the product of the last terms in the binomials is 6. Furthermore, the factors of $2x^2$ and 6 must be chosen so that the sum of the products of the inner terms and outer terms equals $7x$.

To produce the product $2x^2$ we might try the factors $2x$ and x within the binomials:

$$(2x \quad \Box)(x \quad \Box)$$

To produce a product of 6, the remaining terms in the binomials must either both be positive or both be negative. To produce a positive middle term, we will try positive factors of 6 in the remaining blanks until the correct product is found. The possibilities are $1 \cdot 6, 2 \cdot 3, 3 \cdot 2$, and $6 \cdot 1$.

$(2x + 1)(x + 6) = 2x^2 + 12x + 1x + 6 = 2x^2 + 13x + 6$	Wrong middle term
$(2x + 2)(x + 3) = 2x^2 + 6x + 2x + 6 = 2x^2 + 8x + 6$	Wrong middle term
$(2x + 3)(x + 2) = 2x^2 + 4x + 3x + 6 = 2x^2 + 7x + 6$	Correct!
$(2x + 6)(x + 1) = 2x^2 + 2x + 6x + 6 = 2x^2 + 8x + 6$	Wrong middle term

The correct factorization of $2x^2 + 7x + 6$ is $(2x + 3)(x + 2)$. ✔

As this example shows, we factor a trinomial of the form $ax^2 + bx + c$ by shuffling the factors of a and c within the binomials until the correct product is obtained. However, sometimes it is not necessary to test all the possible combinations of factors. In the previous example, the GCF of the original trinomial is 1. Therefore, any binomial factor that shares a common factor *greater than 1* does not need to be considered. In this case the possibilities $(2x + 2)(x + 3)$ and $(2x + 6)(x + 1)$ cannot work.

$$\underbrace{(2x + 2)}(x + 3) \qquad \underbrace{(2x + 6)}(x + 1).$$

Common
factor of 2

Common
factor of 2

The steps to factor a trinomial by the trial-and-error method are outlined in the following box.

Trial-and-Error Method to Factor $ax^2 + bx + c$

1. Factor out the GCF.
2. List all pairs of positive factors of a and pairs of positive factors of c. Consider the reverse order for either list of factors.
3. Construct two binomials of the form:

Factors of a

$$(\Box x \quad \Box)(\Box x \quad \Box)$$

Factors of c

Test each combination of factors and signs until the correct product is found. If no combination of factors produces the correct product, the trinomial cannot be factored further and is called a prime polynomial.

Before we begin our next example, keep these two important guidelines in mind.

- For any factoring problem you encounter, always factor out the GCF from all terms first.
- To factor a trinomial, write the trinomial in the form $ax^2 + bx + c$.

---

example 1

Factoring a Trinomial by the Trial-and-Error Method

Factor the trinomial by the trial-and-error method: $10x^2 - 9x - 1$

Solution:

$10x^2 - 9x - 1$

Step 1: Factor out the GCF from all terms. In this case, the GCF is 1.

The trinomial is written in the form $ax^2 + bx + c$.

To factor $10x^2 - 9x - 1$, two binomials must be constructed in the form:

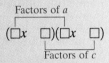

Factors of 10

$$(\Box x \quad \Box)(\Box x \quad \Box)$$

Factors of -1

Step 2: To produce the product $10x^2$, we might try $5x$ and $2x$ or $10x$ and $1x$. To produce a product of -1, we will try the factors $1(-1)$ and $-1(1)$.

Step 3: Construct all possible binomial factors using different combinations of the factors of $10x^2$ and -1.

$(5x + 1)(2x - 1) = 10x^2 - 5x + 2x - 1 = 10x^2 - 3x - 1$　　Wrong middle term

$(5x - 1)(2x + 1) = 10x^2 + 5x - 2x - 1 = 10x^2 + 3x - 1$　　Wrong middle term

Because the numbers 1 and -1 did not produce the correct trinomial when coupled with $5x$ and $2x$, try using $10x$ and $1x$.

$(10x - 1)(1x + 1) = 10x^2 + 10x - 1x - 1 = 10x^2 + 9x - 1$　　Wrong middle term

$(10x + 1)(1x - 1) = 10x^2 - 10x + 1x - 1 = 10x^2 - 9x - 1$　　Correct!

Hence $10x^2 - 9x - 1 = (10x + 1)(x - 1)$.

2. Identifying the Signs When Using the Trial-and-Error Method

In Example 1, the factors of -1 must have opposite signs to produce a negative product. Therefore, one binomial factor is a sum and one is a difference. Determining the correct signs is an important aspect of factoring trinomials. We suggest the following guidelines:

Sign Rules for the Trial-and-Error Method

Given the trinomial $ax^2 + bx + c$, $(a > 0)$ the signs can be determined as follows:

1. If c is positive, then the signs in the binomials must be the same (either both positive or both negative). The correct choice is determined by the middle term. If the middle term is positive, then both signs must be positive. If the middle term is negative, then both signs must be negative.

c is positive
$$20x^2 + 43x + 21$$
$$(4x + 3)(5x + 7)$$
Same signs

c is positive
$$20x^2 - 43x + 21$$
$$(4x - 3)(5x - 7)$$
Same signs

2. If c is negative, then the signs in the binomial must be different. The middle term in the trinomial determines which factor gets the positive sign and which gets the negative sign.

c is negative
$$x^2 + 3x - 28$$
$$(x + 7)(x - 4)$$
Different signs

c is negative
$$x^2 - 3x - 28$$
$$(x - 7)(x + 4)$$
Different signs

example 2

Factoring a Trinomial

Factor the trinomial: $8y^2 + 13y - 6$

Classroom Activity 4.3A

Solution:

$8y^2 + 13y - 6$ **Step 1:**   The GCF is 1.

$(\Box y \ \Box)(\Box y \ \Box)$

Factors of 8	Factors of 6
$1 \cdot 8$	$1 \cdot 6$
$2 \cdot 4$	$2 \cdot 3$
	$3 \cdot 2$ ⎫ (reverse order)
	$6 \cdot 1$ ⎭

Step 2: List the positive factors of 8 and positive factors of 6. Consider the reverse order in one list of factors.

$(2y \ \ 1)(4y \ \ 6)$
$(2y \ \ 2)(4y \ \ 3)$
$(2y \ \ 3)(4y \ \ 2)$
$(2y \ \ 6)(4y \ \ 1)$
$(1y \ \ 1)(8y \ \ 6)$
$(1y \ \ 3)(8y \ \ 2)$

Step 3: Construct all possible binomial factors using different combinations of the factors of 8 and 6.

Without regard to signs, these factorizations cannot work because the terms in a binomial share a common factor greater than 1.

Test the remaining factorizations. Keep in mind that to produce a product of -6, the signs within the parentheses must be opposite (one positive and one negative). Also, the sum of the products of the inner terms and outer terms must be combined to form $13y$.

$(1y \ \ 6)(8y \ \ 1)$ *Incorrect.*   Wrong middle term. Regardless of signs, the product of inner terms, $48y$, and the product of outer terms, $1y$, cannot be combined to form the middle term $13y$.

$(1y \ \ 2)(8y \ \ 3)$ *Correct.*   The terms $16y$ and $3y$ can be combined to form the middle term $13y$, provided the signs are applied correctly. We require $+16y$ and $-3y$.

Hence, the correct factorization is $(y + 2)(8y - 3)$.

Check: $(y + 2)(8y - 3) = 8y^2 - 3y + 16y - 6$
$= 8y^2 + 13y - 6$ ✔

3. Factoring Trinomials with a Leading Coefficient of 1

If a trinomial has a leading coefficient of 1, the factoring process simplifies significantly. Consider the trinomial $x^2 + bx + c$. To produce a leading term of x^2, we can construct binomials of the form $(x + \quad)(x + \quad)$. The remaining terms may be satisfied by two numbers p and q whose product is c and whose sum is b:

$$
\overbrace{}^{\text{Factors of } c}
$$

$$(x + p)(x + q) = x^2 + px + qx + pq = x^2 + \underbrace{(p + q)}x + \underbrace{pq}$$
$$\text{Sum} = b \quad \text{Product} = c$$

This process is demonstrated in the next example.

---

example 3

Factoring a Trinomial with a Leading Coefficient of 1

Factor the trinomial: $x^2 - 10x + 16$

Solution:

$x^2 - 10x + 16$	Factor out the GCF from all terms. In this case, the GCF is 1.
$= (x \quad)(x \quad)$	The trinomial is written in the form $x^2 + bx + c$. To form the product x^2, use the factors x and x.
	Next look for two numbers whose product is 16 and whose sum is -10. Because the middle term is negative, we will consider only the negative factors of 16.

Factors of 16	Sum
$-1(-16)$	$-1 + (-16) = -17$
$-2(-8)$	$-2 + (-8) = -10$
$-4(-4)$	$-4 + (-4) = -8$

The numbers are -2 and -8.

Hence $x^2 - 10x + 16 = (x - 2)(x - 8)$.

Check: $(x - 2)(x - 8) = x^2 - 8x - 2x + 16$
$= x^2 - 10x + 16$ ✔

---

example 4

Factoring Trinomials with a Leading Coefficient of 1

Factor.

a. $t^2 + 34t + 33$ b. $c^2 - 7cd - 30d^2$

Solution:

a. $t^2 + 34t + 33$ Factor out the GCF from all terms. In this case, the GCF is 1.

$= (t\quad)(t\quad)$ To complete the factorization, we need two numbers whose product is 33 and whose sum is 34. The
$= (t + 1)(t + 33)$ numbers are 1 and 33.

Check: $(t + 1)(t + 33) = t^2 + 33t + t + 33$
$= t^2 + 34t + 33$ ✔

b. $c^2 - 7cd - 30d^2$ Factor out the GCF from all terms. In this case, the GCF is 1.

$= (c\quad d)(c\quad d)$ The presence of two variables c and d, does not change the factoring process. We will still look
$= (c - 10d)(c + 3d)$ for two numbers whose product is -30 and whose sum is -7. The numbers are -10 and 3. These will be the coefficients on the d terms.

Check: $(c - 10d)(c + 3d) = c^2 + 3cd - 10cd - 30d^2$
$= c^2 - 7cd - 30d^2$ ✔

4. Greatest Common Factor and Factoring Trinomials

Remember that the first step in any factoring problem is to remove the GCF. By removing the GCF, the remaining terms of the trinomial will be simpler and may have smaller coefficients.

---

example 5

Factoring a Trinomial by the Trial-and-Error Method

Factor the trinomial by the trial-and-error method: $40x^3 - 104x^2 + 10x$

Classroom Activity 4.3B

Solution:

$40x^3 - 104x^2 + 10x$

$= 2x(20x^2 - 52x + 5)$ **Step 1:** The GCF is $2x$.

$= 2x(\Box x\quad\Box)(\Box x\quad\Box)$ **Step 2:** List the factors of 20 and factors of 5. Consider the reverse order in one list of factors.

Tip: Notice when the GCF, $2x$, is removed from the original trinomial, the new trinomial has smaller coefficients. This makes the factoring process simpler. It is easier to list the factors of 20 and 5 rather than the factors of 40 and 10.

Factors of 20	Factors of 5
$1 \cdot 20$	$1 \cdot 5$
$2 \cdot 10$	$5 \cdot 1$
$4 \cdot 5$	

Step 3: Construct all possible binomial factors using different combinations of the factors of 20 and factors of 5. The signs in the parentheses must both be negative.

$$= 2x(1x - 1)(20x - 5)$$

$$= 2x(2x - 1)(10x - 5)$$

Incorrect. The binomials contain a GCF greater than 1.

$$= 2x(4x - 1)(5x - 5)$$

$$= 2x(1x - 5)(20x - 1)$$

Incorrect. Wrong middle term.

$$2x(x - 5)(20x - 1)$$
$$= 2x(20x^2 - 1x - 100x + 5)$$
$$= 2x(20x^2 - 101x + 5)$$

$$= 2x(4x - 5)(5x - 1)$$

Incorrect. Wrong middle term.

$$2x(4x - 5)(5x - 1)$$
$$= 2x(20x^2 - 4x - 25x + 5)$$
$$= 2x(20x^2 - 29x + 5)$$

$$= 2x(2x - 5)(10x - 1)$$

Correct.

$$2x(2x - 5)(10x - 1)$$
$$= 2x(20x^2 - 2x - 50x + 5)$$
$$= 2x(20x^2 - 52x + 5)$$
$$= 40x^3 - 104x^2 + 10x$$

The correct factorization is $2x(2x - 5)(10x - 1)$.

Often it is easier to factor a trinomial when the leading coefficient is positive. If the leading coefficient is negative, consider factoring out the opposite of the GCF.

example 6

Factoring a Trinomial by the Trial-and-Error Method

Factor the trinomial by the trial-and-error method: $-w^2 - 7w + 18$

Solution:

$-w^2 - 7w + 18$

$$= -1(w^2 + 7w - 18)$$

Factor out -1. The resulting trinomial has a leading coefficient of 1.

Avoiding Mistakes

Do not forget to write the GCF as part of the final answer.

$$= -1[(w \quad)(w \quad)]$$

$$= -1[(w + 9)(w - 2)]$$

To complete the factorization, we need two numbers whose product is -18 and whose sum is 7. The numbers are 9 and -2.

Check: $-1[(w + 9)(w - 2)] = -1(w^2 - 2w + 9w - 18)$
$$= -1(w^2 + 7w - 18)$$
$$= -w^2 - 7w + 18 \ ✔$$

Note that not every trinomial is factorable by the methods presented here.

example 7   Factoring a Trinomial by the Trial-and-Error Method

Factor the trinomial by the trial-and-error method: $2p^2 - 8p + 3$

Solution:

$2p^2 - 8p + 3$ **Step 1:**   The GCF is 1.

$(1p \ \square)(2p \ \square)$ **Step 2:**   List the factors of 2 and the factors of 3.

Factors of 2	Factors of 3	**Step 3:** Construct all possible binomial factors using different combinations of the factors of 2 and 3. Because the third term in the trinomial is positive, both signs in the binomial must be the same. Because the middle term coefficient is negative, both signs will be negative.
$1 \cdot 2$	$1 \cdot 3$	
	$3 \cdot 1$	

$$(p - 1)(2p - 3) = 2p^2 - 3p - 2p + 3$$
$$= 2p^2 - 5p + 3 \qquad \textit{Incorrect.} \quad \text{Wrong middle term.}$$

$$(p - 3)(2p - 1) = 2p^2 - p - 6p + 3$$
$$= 2p^2 - 7p + 3 \qquad \textit{Incorrect.} \quad \text{Wrong middle term.}$$

Because none of the combinations of factors results in the correct product, we say that the trinomial $2p^2 - 8p + 3$ is prime and cannot be factored.

---

section 4.3   PRACTICE EXERCISES

For Exercises 1–10, factor out the greatest common factor.   4.1

1. $7a^9 + 28a^3$ $7a^3(a^6 + 4)$ 2. $r^4 - 9r^5$ $r^4(1 - 9r)$

3. $12w^2 - 4w$ $4w(3w - 1)$ 4. $15x^2 + 3x$ $3x(5x + 1)$

5. $21a^2b^2 + 12ab^2 - 15a^2b$ $3ab(7ab + 4b - 5a)$

6. $5uv^2 - 10u^2v + 25u^2v^2$ $5uv(v - 2u + 5uv)$

7. $4y(y - 1) - 5(y - 1)$ $(y - 1)(4y - 5)$

8. $7(n + 5) + 2n(n + 5)$ $(n + 5)(7 + 2n)$

9. $p(4p - 9) + 4(4p - 9)$ $(4p - 9)(p + 4)$

10. $q^2(6q + 1) - 3(6q + 1)$ $(6q + 1)(q^2 - 3)$

For Exercises 11–16, factor by grouping.   4.1

11. $st - 2t + 3s - 6$ $(s - 2)(t + 3)$

12. $3ab + 3b^2 + 2a^2 + 2ab$ $(a + b)(3b + 2a)$

13. $5x - 10 - xy + 2y$ $(x - 2)(5 - y)$

14. $m^2 - 2mn - mn + 2n^2$ $(m - 2n)(m - n)$

15. $5t^2 - 10t + 30t - 60$ $5(t - 2)(t + 6)$

16. $6p^2 + 24p - 12p - 48$ $6(p + 4)(p - 2)$

For Exercises 17–20, assume a, b, and c represent positive integers.

17. When factoring a polynomial of the form $ax^2 + bx - c$, should the signs of the binomials be both positive, both negative, or different?
Different

18. When factoring a polynomial of the form $ax^2 - bx - c$, should the signs of the binomials be both positive, both negative, or different?
Different

19. When factoring a polynomial of the form $ax^2 - bx + c$, should the signs of the binomials be both positive, both negative, or different?
Both negative

20. When factoring a polynomial of the form $ax^2 + bx + c$, should the signs of the binomials be both positive, both negative, or different?

Both positive

For Exercises 21–24, complete the factorization.

21. $x^2 + x - 56 = (x - 7)($ $)$ $(x + 8)$

22. $y^2 + y - 30 = (y - 5)($ $)$ $(y + 6)$

23. $x^2 - x - 56 = (x + 7)($ $)$ $(x - 8)$

24. $y^2 - y - 30 = (y + 5)($ $)$ $(y - 6)$

25. What is a prime polynomial? ❖

26. How do you determine if a trinomial is prime? ❖

For Exercises 27–64, factor the trinomial using the trial-and-error method.

27. $2y^2 - 3y - 2$
$(2y + 1)(y - 2)$

28. $2w^2 + 5w - 3$
$(2w - 1)(w + 3)$

29. $9x^2 - 12x + 4$
$(3x - 2)^2$

30. $3n^2 + 13n + 4$
$(3n + 1)(n + 4)$

31. $2a^2 + 7a + 6$
$(2a + 3)(a + 2)$

32. $8b^2 - 6b - 9$
$(4b + 3)(2b - 3)$

33. $6t^2 + 7t - 3$
$(2t + 3)(3t - 1)$

34. $4p^2 - 9p + 2$
$(4p - 1)(p - 2)$

35. $4m^2 - 20m + 25$
$(2m - 5)^2$

36. $16r^2 + 24r + 9$
$(4r + 3)^2$

37. $5c^2 - c + 2$
Prime

38. $7s^2 + 2s + 9$
Prime

39. $6x^2 - 19xy + 10y^2$
$(2x - 5y)(3x - 2y)$

40. $15p^2 + pq - 2q^2$
$(3p - q)(5p + 2q)$

41. $12m^2 + 11mn - 5n^2$
$(4m + 5n)(3m - n)$

42. $4a^2 + 5ab - 6b^2$
$(4a - 3b)(a + 2b)$

43. $6r^2 + rs - 2s^2$
$(3r + 2s)(2r - s)$

44. $18x^2 - 9xy - 2y^2$
$(6x + y)(3x - 2y)$

45. $4s^2 - 8st + t^2$
Prime

46. $6u^2 - 10uv + 5v^2$
Prime

47. $x^2 + 7x - 18$
$(x + 9)(x - 2)$

48. $y^2 - 6y - 40$
$(y + 4)(y - 10)$

49. $a^2 - 10a - 24$
$(a - 12)(a + 2)$

50. $b^2 + 6b - 7$
$(b + 7)(b - 1)$

51. $r^2 + 5r - 24$
$(r + 8)(r - 3)$

52. $t^2 + 20t + 100$
$(t + 10)^2$

53. $w^2 - 14w + 49$
$(w - 7)^2$

54. $h^2 - 11h + 10$
$(h - 10)(h - 1)$

55. $k^2 + 5k + 4$
$(k + 4)(k + 1)$

56. $u^2 + 9u - 22$
$(u + 11)(u - 2)$

57. $v^2 - 4v + 1$
Prime

58. $x^2 + 5x + 2$
Prime

59. $m^2 - 13mn + 40n^2$
$(m - 8n)(m - 5n)$

60. $r^2 - rs - 42s^2$
$(r - 7s)(r + 6s)$

61. $a^2 + 9ab + 8b^2$
$(a + 8b)(a + b)$

62. $y^2 - yz - 12z^2$
$(y - 4z)(y + 3z)$

63. $x^2 + 9xy + 20y^2$
$(x + 5y)(x + 4y)$

64. $p^2 - 13pq + 36q^2$
$(p - 9q)(p - 4q)$

65. Is the expression $(3x + 6)(x - 5)$ factored completely? Explain why or why not. ❖

66. Is the expression $(5x + 1)(4x - 12)$ factored completely? Explain why or why not. ❖

For Exercises 67–76, first factor out the GCF. Then factor by using the trial-and-error method if possible.

67. $5d^3 + 3d^2 - 10d$
$d(5d^2 + 3d - 10)$

68. $3y^3 - y^2 + 12y$
$y(3y^2 - y + 12)$

69. $4b^3 - 4b^2 - 80b$
$4b(b - 5)(b + 4)$

70. $2w^2 + 20w + 42$
$2(w + 7)(w + 3)$

71. $x^2y^2 - 13xy^2 + 30y^2$
$y^2(x - 3)(x - 10)$

72. $p^2q^2 - 14pq^2 + 33q^2$
$q^2(p - 3)(p - 11)$

73. $-a^2 - 15a + 34$
$-1(a + 17)(a - 2)$

74. $-j^2 - 7j - 10$
$-1(j + 2)(j + 5)$

75. $-2u^2 + 28u - 90$
$-2(u - 5)(u - 9)$

76. $-3v^2 - 3v + 90$
$-3(v - 5)(v + 6)$

For Exercises 77–84, write the polynomial in descending order. Then factor the polynomial.

77. $10x + 1 + 16x^2$
$(8x + 1)(2x + 1)$

78. $k^2 + 16 - 8k$
$(k - 4)^2$

79. $1 + c^2 - 2c$
$(c - 1)^2$

80. $3 - 14t + 16t^2$
$(8t - 3)(2t - 1)$

81. $20z - 18 - 2z^2$
$-2(z - 9)(z - 1)$

82. $25t - 5t^2 - 30$
$-5(t - 2)(t - 3)$

83. $42 - 13q + q^2$
$(q - 7)(q - 6)$

84. $-5w - 24 + w^2$
$(w - 8)(w + 3)$

section

4.4   FACTORING PERFECT SQUARE TRINOMIALS AND THE DIFFERENCE OF SQUARES

1.   Factoring Perfect Square Trinomials

Recall from Section 3.6 that the square of a binomial always results in a perfect square trinomial:

$$(a + b)^2 = (a + b)(a + b) = a^2 + ab + ab + b^2 = a^2 + 2ab + b^2$$
$$(a - b)^2 = (a - b)(a - b) = a^2 - ab - ab + b^2 = a^2 - 2ab + b^2$$

For example, $(3x + 5)^2 = (3x)^2 + 2(3x)(5) + (5)^2 = 9x^2 + 30x + 25$

$$a = 3x \quad b = 5$$

To factor the trinomial $9x^2 + 30x + 25$, the grouping method or the trial-and-error method can be used. However, if we recognize that the trinomial is a perfect square trinomial, we can use one of the following patterns to reach a quick solution.

Factored Form of a Perfect Square Trinomial

$$a^2 + 2ab + b^2 = (a + b)^2$$
$$a^2 - 2ab + b^2 = (a - b)^2$$

Checking for a Perfect Square Trinomial

1. Check if the first and third terms are both perfect squares with positive coefficients.
2. If this is the case, identify a and b, and determine if the middle term equals $2ab$.

example 1

Factoring Perfect Square Trinomials

Factor the trinomials completely.

a. $x^2 + 14x + 49$ b. $25y^2 - 20y + 4$
c. $18c^3 - 48c^2d + 32cd^2$ d. $5w^2 + 50w + 45$

Classroom Activity 4.4A

Solution:

a. $x^2 + 14x + 49$ The GCF is 1.

- The first and third terms are positive.

Perfect squares

$$x^2 + 14x + 49$$

- The first term is a perfect square: $x^2 = (x)^2$.

- The third term is a perfect square: $49 = (7)^2$.

$= (x)^2 + 2(x)(7) + (7)^2$

$= (x + 7)^2$

- The middle term is twice the product of x and 7: $14x = 2(x)(7)$

Hence, the trinomial is in the form $a^2 + 2ab + b^2$, where $a = x$ and $b = 7$.

Factor as $(a + b)^2$.

b. $25y^2 - 20y + 4$

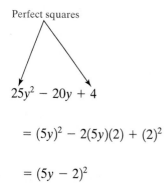

Perfect squares

$25y^2 - 20y + 4$

$= (5y)^2 - 2(5y)(2) + (2)^2$

$= (5y - 2)^2$

The GCF is 1.

- The first and third terms are positive.

- The first term is a perfect square: $25y^2 = (5y)^2$.

- The third term is a perfect square: $4 = (2)^2$.

- The middle term is twice the product of $5y$ and 2: $20y = 2(5y)(2)$

Factor as $(a - b)^2$.

c. $18c^3 - 48c^2d + 32cd^2$

$= 2c(9c^2 - 24cd + 16d^2)$

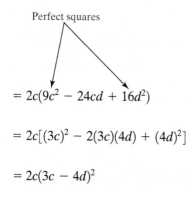

Perfect squares

$= 2c(9c^2 - 24cd + 16d^2)$

$= 2c[(3c)^2 - 2(3c)(4d) + (4d)^2]$

$= 2c(3c - 4d)^2$

The GCF is $2c$.

- The first and third terms are positive.

- The first term is a perfect square: $9c^2 = (3c)^2$.

- The third term is a perfect square: $16d^2 = (4d)^2$.

- The middle term is twice the product of $3c$ and $4d$: $24cd = 2(3c)(4d)$

Factor as $(a - b)^2$.

d. $5w^2 + 50w + 45$

$= 5(w^2 + 10w + 9)$

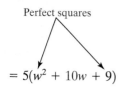

Perfect squares

$= 5(w^2 + 10w + 9)$

The GCF is 5.

The first and third terms are perfect squares.

$$w^2 = (w)^2 \qquad \text{and} \qquad 9 = (3)^2$$

However, the middle term is not 2 times the product of w and 3. Therefore, this is not a perfect square trinomial.

$$10w \neq 2(w)(3)$$

$$= 5(w + 9)(w + 1) \qquad \text{To factor, use either the grouping method or the trial-and-error method.}$$

Tip: To help you identify a perfect square trinomial, it is recommended that you familiarize yourself with the first several perfect squares.

$1 \cdot 1 = 2$	$4 \cdot 4 = 16$	$7 \cdot 7 = 49$	$10 \cdot 10 = 100$	$13 \cdot 13 = 169$
$2 \cdot 2 = 4$	$5 \cdot 5 = 25$	$8 \cdot 8 = 64$	$11 \cdot 11 = 121$	$14 \cdot 14 = 196$
$3 \cdot 3 = 9$	$6 \cdot 6 = 36$	$9 \cdot 9 = 81$	$12 \cdot 12 = 144$	$15 \cdot 15 = 225$

If you do not recognize that a trinomial is a perfect square trinomial, you may still use either the trial-and-error method or the grouping method to factor the trinomial.

2. Factoring a Difference of Squares

Up to this point, we have learned several methods of factoring, including:

- Factoring out the greatest common factor from a polynomial
- Factoring a four-term polynomial by grouping
- Recognizing and factoring perfect square trinomials
- Factoring trinomials by the grouping method or by the trial-and-error method

Next, we will learn how to factor binomials that fit the pattern of a difference of squares. Recall from Section 3.6 that the product of two conjugates results in a difference of squares:

$$(a + b)(a - b) = a^2 - b^2$$

Therefore, to factor a difference of squares, the process is reversed. Identify a and b and construct the conjugate factors.

Factored Form of a Difference of Squares

$$a^2 - b^2 = (a + b)(a - b)$$

In addition to recognizing numbers that are perfect squares, it is helpful to recognize that a variable expression is a perfect square if its exponent is a multiple of 2. For example:

Perfect Squares

$$x^2 = (x)^2$$
$$x^4 = (x^2)^2$$
$$x^6 = (x^3)^2$$
$$x^8 = (x^4)^2$$
$$x^{10} = (x^5)^2$$

example 2

Factoring Differences of Squares

Factor the binomials.

a. $y^2 - 25$ b. $49s^2 - 4t^4$ c. $18w^2z - 2z$

Solution:

a. $y^2 - 25$ The binomial is a difference of squares.

$= (y)^2 - (5)^2$ Write in the form: $a^2 - b^2$, where $a = y, b = 5$.

$= (y + 5)(y - 5)$ Factor as $(a + b)(a - b)$.

<u>Check</u>: $(y + 5)(y - 5) = y^2 - 5y + 5y - 25$
$$= y^2 - 25 \checkmark$$

b. $49s^2 - 4t^4$ The binomial is a difference of squares.

$= (7s)^2 - (2t^2)^2$ Write in the form $a^2 - b^2$, where $a = 7s$ and $b = 2t^2$.

$= (7s + 2t^2)(7s - 2t^2)$ Factor as $(a + b)(a - b)$.

<u>Check</u>: $(7s + 2t^2)(7s - 2t^2) = 49s^2 - 14st^2 + 14st^2 - 4t^4$
$$= 49s^2 - 4t^4 \checkmark$$

c. $18w^2z - 2z$ The GCF is $2z$.

$= 2z(9w^2 - 1)$ $(9w^2 - 1)$ is a difference of squares.

$= 2z[(3w)^2 - (1)^2]$ Write in the form: $a^2 - b^2$, where $a = 3w$, $b = 1$.

$= 2z(3w + 1)(3w - 1)$ Factor as $(a + b)(a - b)$.

<u>Check</u>: $2z(3w + 1)(3w - 1) = 2z[9w^2 - 3w + 3w - 1]$
$$= 2z[9w^2 - 1]$$
$$= 18w^2z - 2z \checkmark$$

3.   Analyzing a Sum of Squares

Suppose a and b share no common factors. Then the difference of squares $a^2 - b^2$ can be factored as $(a + b)(a - b)$. However, the sum of squares $a^2 + b^2$ cannot be factored over the real numbers. To see why, consider the expression $a^2 + b^2$. The factored form would require two binomials of the form:

$$(a \quad b)(a \quad b) \stackrel{?}{=} a^2 + b^2$$

If all possible combinations of signs are considered, none produces the correct product.

$(a + b)(a - b) = a^2 - b^2$ Wrong sign

$(a + b)(a + b) = a^2 + 2ab + b^2$ Wrong middle term

$(a - b)(a - b) = a^2 - 2ab + b^2$ Wrong middle term

After exhausting all possibilities, we see that if a and b share no common factors, then the sum of squares $a^2 + b^2$ is a prime polynomial.

Classroom Activity 4.4B

4. Factoring Using Multiple Methods

Some factoring problems require more than one method of factoring. In general, when factoring a polynomial, be sure to factor completely.

example 3   **Factoring Polynomials**

Factor completely.

a. $w^4 - 16$ b. $4x^3 + 4x^2 - 25x - 25$ c. $8p^3 + 24p^2q + 18pq^2$

Solution:

a. $w^4 - 16$ The GCF is 1. $w^4 - 16$ is a difference of squares.

$= (w^2)^2 - (4)^2$ Write in the form: $a^2 - b^2$, where $a = w^2$, $b = 4$.

$= (w^2 + 4)(w^2 - 4)$ Factor as $(a + b)(a - b)$.

$= (w^2 + 4)(w + 2)(w - 2)$ Note that $w^2 - 4$ can be factored further as a difference of squares. (The binomial $w^2 + 4$ is a sum of squares and cannot be factored further.)

b. $4x^3 + 4x^2 - 25x - 25$ The GCF is 1.

$= 4x^3 + 4x^2 - 25x - 25$ The polynomial has four terms. Factor by grouping.

$= 4x^2(x + 1) - 25(x + 1)$

$= (x + 1)(4x^2 - 25)$ $4x^2 - 25$ is a difference of squares.

$= (x + 1)(2x + 5)(2x - 5)$

c. $8p^3 + 24p^2q + 18pq^2$ The GCF is $2p$.

$= 2p(4p^2 + 12pq + 9q^2)$ $4p^2 + 12pq + 9q^2$ is a perfect square trinomial.

$= 2p(2p + 3q)^2$

---

section 4.4   PRACTICE EXERCISES

For Exercises 1–10, factor the polynomials.   4.1–4.3

1. $3x^2 + x - 10$
    $(3x - 5)(x + 2)$

2. $6a^2b + 3a^3b$
    $3a^2b(2 + a)$

3. $x^2yz^2 + 6y^2z + yz$
    $yz(x^2z + 6y + 1)$

4. $2x^2 - x - 1$
    $(2x + 1)(x - 1)$

5. $12x^2 - 34x + 10$
    $2(3x - 1)(2x - 5)$

6. $3x^2 - 3xy - 6y^2$
    $3(x - 2y)(x + y)$

7. $ax + ab - 6x - 6b$
    $(x + b)(a - 6)$

8. $2xy - 3x - 4y + 6$
    $(2y - 3)(x - 2)$

9. $x^2 + 6x + 9$
    $(x + 3)^2$

10. $y^2 - 4y + 4$
    $(y - 2)^2$

11.   What perfect square trinomial factors to $(2x + 3)^2$? $4x^2 + 12x + 9$

12. What perfect square trinomial factors to $(3k + 5)^2$? $9k^2 + 30k + 25$

13. What perfect square trinomial factors to $(6h - 1)^2$? $36h^2 - 12h + 1$

14. What perfect square trinomial factors to $(4y - 5)^2$? $16y^2 - 40y + 25$

15. a. Identify which trinomial is a perfect square trinomial:

$$x^2 + 4x + 4 \quad \text{or} \quad x^2 + 5x + 4$$
$x^2 + 4x + 4$

  b. Factor both of these trinomials.
  $(x + 2)^2; (x + 4)(x + 1)$

16. a. Identify which trinomial is a perfect square trinomial:

$$x^2 + 13x + 36 \quad \text{or} \quad x^2 + 12x + 36$$
$x^2 + 12x + 36$

  b. Factor both of these trinomials.
  $(x + 9)(x + 4); (x + 6)^2$

 17. a. Identify which trinomial is a perfect square trinomial:

$$4x^2 - 25x + 25 \quad \text{or} \quad 4x^2 - 20x + 25$$
$4x^2 - 20x + 25$

  b. Factor both of these trinomials.
  $(x - 5)(4x - 5); (2x - 5)^2$

18. a. Identify which trinomial is a perfect square trinomial:

$$9x^2 + 12x + 4 \quad \text{or} \quad 9x^2 + 15x + 4$$
$9x^2 + 12x + 4$

  b. Factor both of these trinomials.
  $(3x + 2)^2; (3x + 1)(3x + 4)$

For Exercises 19–40, factor the trinomials, if possible.

19. $y^2 - 10y + 25$
$(y - 5)^2$

20. $t^2 - 16t + 64$
$(t - 8)^2$

21. $m^2 + 6m + 9$
$(m + 3)^2$

22. $n^2 + 18n + 81$
$(n + 9)^2$

23. $r^2 - 2r + 36$
Prime

24. $s^2 - 4s + 100$
Prime

25. $49q^2 - 28q + 4$
$(7q - 2)^2$

26. $64y^2 - 80y + 25$
$(8y - 5)^2$

27. $9p^2 + 42p + 49$
$(3p + 7)^2$

28. $4x^2 + 36x + 81$
$(2x + 9)^2$

29. $25h^2 + 50h + 16$
$(5x + 2)(5x + 8)$

30. $4w^2 - 20w + 9$
$(2x - 1)(2x - 9)$

31. $16a^2 + 8ab + b^2$
$(4a + b)^2$

32. $25m^2 + 10mn + n^2$
$(5m + n)^2$

33. $16q^2 + 40qr + 25r^2$
$(4q + 5r)^2$

34. $u^2 - 2uv + v^2$
$(u - v)^2$

35. $a^2 + 2ab + b^2$
$(a + b)^2$

36. $49h^2 - 14hk + k^2$
$(7h - k)^2$

37. $k^2 - k + \dfrac{1}{4}$ $\left(k - \dfrac{1}{2}\right)^2$

38. $v^2 + \dfrac{2}{3}v + \dfrac{1}{9}$ $\left(v + \dfrac{1}{3}\right)^2$

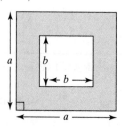 39. $9x^2 + x + \dfrac{1}{36}$ $\left(3x + \dfrac{1}{6}\right)^2$ 40. $4y^2 - y + \dfrac{1}{16}$ $\left(2y - \dfrac{1}{4}\right)^2$

41. What binomial factors as $(x - 5)(x + 5)$? $x^2 - 25$

42. What binomial factors as $(n - 3)(n + 3)$? $n^2 - 9$

43. What binomial factors as $(2w - 3)(2w + 3)$?
$4w^2 - 9$

44. What binomial factors as $(7y - 4)(7y + 4)$?
$49y^2 - 16$

For Exercises 45–64, factor the binomials, if possible.

45. $x^2 - 36$ $(x - 6)(x + 6)$

46. $r^2 - 81$ $(r - 9)(r + 9)$

47. $w^2 - 100$ $(w - 10)(w + 10)$

48. $t^2 - 49$ $(t - 7)(t + 7)$

49. $4a^2 - 121b^2$
$(2a - 11b)(2a + 11b)$

50. $9x^2 - y^2$
$(3x - y)(3x + y)$

51. $49m^2 - 16n^2$
$(7m - 4n)(7m + 4n)$

52. $100a^2 - 49b^2$
$(10a - 7b)(10a + 7b)$

53. $9q^2 + 16$ Prime

54. $36 + s^2$ Prime

55. $c^6 - 25$
$(c^3 - 5)(c^3 + 5)$

56. $z^6 - 4$
$(z^3 - 2)(z^3 + 2)$

57. $25 - 16t^2$
$(5 - 4t)(5 + 4t)$

58. $64 - h^2$
$(8 - h)(8 + h)$

59. $p^2 - \dfrac{1}{9}$ $\left(p - \dfrac{1}{3}\right)\left(p + \dfrac{1}{3}\right)$

60. $q^2 - \dfrac{1}{36}$ $\left(q - \dfrac{1}{6}\right)\left(q + \dfrac{1}{6}\right)$

61. $m^2 + \dfrac{100}{81}$ Prime

62. $n^2 + \dfrac{25}{4}$ Prime

63. $\dfrac{4}{9} - w^2$ $\left(\dfrac{2}{3} - w\right)\left(\dfrac{2}{3} + w\right)$

64. $\dfrac{16}{25} - x^2$ $\left(\dfrac{4}{5} - x\right)\left(\dfrac{4}{5} + x\right)$

65. a. Write a polynomial that represents the area of the shaded region in the figure. $a^2 - b^2$

  b. Factor the expression from part (a).
  $(a - b)(a + b)$

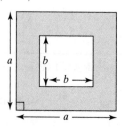

Figure for Exercise 65

66. a. Write a polynomial that represents the area of the shaded region in the figure. $g^2 - h^2$

    b. Factor the expression from part (a).
      $(g - h)(g + h)$

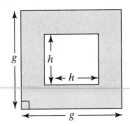

Figure for Exercise 66

For Exercises 67–88, factor the polynomials completely.

67. $3w^2 - 27$
    $3(w - 3)(w + 3)$

68. $6y^2 - 6$
    $6(y - 1)(y + 1)$

69. $50p^4 - 2$
    $2(5p^2 - 1)(5p^2 + 1)$

70. $18q^2 - 98n^2$
    $2(3q - 7n)(3q + 7n)$

71. $2x^2 + 24x + 72$ $2(x + 6)^2$

72. $3x^2y - 66xy + 363y$ $3y(x - 11)^2$

73. $2t^3 - 10t^2 - 2t + 10$ $2(t - 5)(t - 1)(t + 1)$

74. $9a^3 + 27a^2 - 4a - 12$ $(a + 3)(3a - 2)(3a + 2)$

75. $100y^4 + 25x^2$ $25(4y^4 + x^2)$

76. $36a^2 + 9b^4$ $9(4a^2 + b^4)$

77. $4a^2b - 40ab^2 + 100b^3$ $4b(a - 5b)^2$

78. $18u^2 + 24uv + 8v^2$ $2(3u + 2v)^2$

 79. $2x^3 + 3x^2 - 2x - 3$ $(2x + 3)(x - 1)(x + 1)$

80. $3x^3 + x^2 - 12x - 4$ $(3x + 1)(x - 2)(x + 2)$

 81. $81y^4 - 16$
    $(3y - 2)(3y + 2)(9y^2 + 4)$

82. $u^4 - 256$
    $(u - 4)(u + 4)(u^2 + 16)$

83. $81k^2 + 30k + 1$
    $(27k + 1)(3k + 1)$

84. $9h^2 - 15h + 4$
    $(3h - 4)(3h - 1)$

85. $k^3 + 4k^2 - 9k - 36$ $(k + 4)(k - 3)(k + 3)$

86. $w^3 - 2w^2 - 4w + 8$ $(w - 2)^2(w + 2)$

87. $4m^{14} - 20m^7 + 25$
    $(2m^7 - 5)^2$

88. $9n^{12} + 24n^6 + 16$
    $(3n^6 + 4)^2$

EXPANDING YOUR SKILLS

For Exercises 89–100, factor the difference of squares.

89. $0.36x^2 - 0.01$
    $(0.6x - 0.1)(0.6x + 0.1)$

90. $0.81p^2 - 0.25q^2$
    $(0.9p - 0.5q)(0.9p + 0.5q)$

91. $\dfrac{1}{4}w^2 - \dfrac{1}{9}v^2$
    $\left(\dfrac{1}{2}w - \dfrac{1}{3}v\right)\left(\dfrac{1}{2}w + \dfrac{1}{3}v\right)$

92. $\dfrac{4}{9}c^2 - \dfrac{9}{16}d^2$
    $\left(\dfrac{2}{3}c - \dfrac{3}{4}d\right)\left(\dfrac{2}{3}c + \dfrac{3}{4}d\right)$

93. $(y - 3)^2 - 9$
    $y(y - 6)$

94. $(x - 2)^2 - 4$
    $x(x - 4)$

95. $(2p + 1)^2 - 36$
    $(2p - 5)(2p + 7)$

96. $(4q + 3)^2 - 25$
    $8(2q - 1)(q + 2)$ $(-a + 4)(a + 14)$

97. $16 - (t + 2)^2$
    $(-t + 2)(t + 6)$ or $-1(t - 2)(t + 6)$

98. $81 - (a + 5)^2$ or
    $-1(a - 4)(a + 14)$

99. $100 - (2b - 5)^2$
    $(-2b + 15)(2b + 5)$ or
    $-1(2b - 15)(2b + 5)$

100. $49 - (3k - 7)^2$
    $3k(-3k + 14)$ or
    $-3k(3k - 14)$

---

chapter 4   MIDCHAPTER REVIEW: "FACTORING STRATEGY"

1. What is meant by a prime factor?   A prime factor cannot be factored further.

2. What is the first step in factoring any polynomial? Factor out the GCF.

3. When factoring a binomial, what pattern can you look for?   Look for the difference of squares: $a^2 - b^2$.

4. When factoring a trinomial what pattern do you look for first before using the grouping method or trial-and-error method?   Look for a perfect square trinomial: $a^2 + 2ab + b^2$ or $a^2 - 2ab + b^2$.

5. Are factorable polynomials factored completely in one step?   Not all polynomials factor completely in one step. For example: $2x^4 - 32$.

Factoring Strategy

1. Factor out the GCF (Section 4.1).
2. Identify whether the polynomial has two terms, three terms, or more than three terms.
3. If the polynomial has two terms, determine if it fits the pattern for a difference of squares (Section 4.4).
4. If the polynomial has three terms, check first for a perfect square trinomial (Section 4.4). Otherwise, factor the trinomial with the grouping method or the trial-and-error method (Sections 4.2 or 4.3).
5. If the polynomial has more than three terms, try factoring by grouping (Section 4.1).
6. Be sure to factor the polynomial completely.
7. Check by multiplying.

For Exercises 6–37, factor the polynomial completely using the factoring strategy.

6. $6x^2 - 21x - 45$
   $3(2x + 3)(x - 5)$

7. $20y^2 - 14y + 2$
   $2(5y - 1)(2y - 1)$

8. $5a^2bc^3 - 7abc^2$
   $abc^2(5ac - 7)$

9. $8a^2 - 50$
   $2(2a - 5)(2a + 5)$

10. $t^2 + 2t - 63$
    $(t + 9)(t - 7)$

11. $b^2 + 2b - 80$
    $(b + 10)(b - 8)$

12. $ab + ay - b^2 - by$
    $(b + y)(a - b)$

13. $6x^3y^4 + 3x^2y^5$
    $3x^2y^4(2x + y)$

14. $14u^2 - 11uv + 2v^2$
    $(7u - 2v)(2u - v)$

15. $9p^2 - 36pq + 4q^2$
    Prime

16. $4q^2 - 8q - 6$
    $2(2q^2 - 4q - 3)$

17. $9w^2 + 3w - 15$
    $3(3w^2 + w - 5)$

18. $9m^2 + 16n^2$
    Prime

19. $5b^2 - 30b + 45$
    $5(b - 3)^2$

20. $6r^2 + 11r + 3$
    $(3r + 1)(2r + 3)$

21. $4s^2 + 4s - 15$
    $(2s - 3)(2s + 5)$

22. $16a^4 - 1$
    $(2a - 1)(2a + 1)(4a^2 + 1)$

23. $p^3 + p^2c - 9p - 9c$
    $(p + c)(p - 3)(p + 3)$

24. $81u^2 - 90uv + 25v^2$
    $(9u - 5v)^2$

25. $4x^2 + 16$
    $4(x^2 + 4)$

26. $2ax - 6ay + 4bx - 12by$ $2(x - 3y)(a + 2b)$

27. $8m^3 - 10m^2 - 3m$ $m(4m + 1)(2m - 3)$

28. $21x^4y + 41x^3y + 10x^2y$ $x^2y(3x + 5)(7x + 2)$

29. $2m^4 - 128$ $2(m^2 - 8)(m^2 + 8)$

30. $8uv - 6u + 12v - 9$
    $(4v - 3)(2u + 3)$

31. $4t^2 - 20t + st - 5s$
    $(t - 5)(4t + s)$

32. $12x^2 - 12x + 3$
    $3(2x - 1)^2$

33. $p^2 + 2pq + q^2$
    $(p + q)^2$

34. $6n^3 + 5n^2 - 4n$
    $n(2n - 1)(3n + 4)$

35. $4k^3 + 4k^2 - 3k$
    $k(2k - 1)(2k + 3)$

36. $64 - y^2$
    $(8 - y)(8 + y)$

37. $36b - b^3$
    $b(6 - b)(6 + b)$

Concepts

1. Factoring the Sum and Difference of Cubes
2. Factoring Binomials: A Summary
3. General Factoring Summary

4.5 FACTORING THE SUM AND DIFFERENCE OF CUBES AND GENERAL FACTORING SUMMARY

1. Factoring the Sum and Difference of Cubes

In Section 4.4, you learned that a binomial $a^2 - b^2$ is a difference of squares and can be factored as $(a - b)(a + b)$. Furthermore, if a and b share no common factors, then a sum of squares $a^2 + b^2$ is not factorable over the real numbers. In this section we will learn that both a difference of cubes, $a^3 - b^3$, and a sum of cubes $a^3 + b^3$ are factorable.

Factoring a Sum and Difference of Cubes

Sum of Cubes: $a^3 + b^3 = (a + b)(a^2 - ab + b^2)$

Difference of Cubes: $a^3 - b^3 = (a - b)(a^2 + ab + b^2)$

Multiplication can be used to confirm the formulas for factoring a sum or difference of cubes:

$$(a + b)(a^2 - ab + b^2) = a^3 - \cancel{a^2b} + \cancel{ab^2} + \cancel{a^2b} - \cancel{ab^2} + b^3 = a^3 + b^3 ✔$$

$$(a - b)(a^2 + ab + b^2) = a^3 + \cancel{a^2b} + \cancel{ab^2} - \cancel{a^2b} - \cancel{ab^2} - b^3) = a^3 - b^3 ✔$$

To help you remember the formulas for factoring a sum or difference of cubes, keep the following guidelines in mind.

- The factored form is the product of a binomial and a trinomial.
- The first and third terms in the trinomial are the squares of the terms within the binomial factor.
- Without regard to signs, the middle term in the trinomial is the product of terms in the binomial factor.

Square the first term of the binomial.　　　Product of terms in the binomial

$$x^3 + 8 = (x)^3 + (2)^3 = (x + 2)[(x)^2 - (x)(2) + (2)^2]$$

Square the last term of the binomial.

- The sign within the binomial factor is the same as the sign of the original binomial.
- The first and third terms in the trinomial are always positive.
- The sign of the middle term in the trinomial is opposite the sign within the binomial.

Same sign　　　Positive

$$x^3 + 8 = (x)^3 + (2)^3 = (x + 2)[(x)^2 - (x)(2) + (2)^2]$$

Opposite signs

To help you recognize a sum or difference of cubes, we recommend that you familiarize yourself with the first several perfect cubes:

Perfect Cube	Perfect Cube
$1 = (1)^3$	$216 = (6)^3$
$8 = (2)^3$	$343 = (7)^3$
$27 = (3)^3$	$512 = (8)^3$
$64 = (4)^3$	$729 = (9)^3$
$125 = (5)^3$	$1000 = (10)^3$

It is also helpful to recognize that a variable expression is a perfect cube if its exponent is a multiple of 3. For example:

Perfect Cube

$$x^3 = (x)^3$$
$$x^6 = (x^2)^3$$
$$x^9 = (x^3)^3$$
$$x^{12} = (x^4)^3$$

---

example 1

Factoring a Sum of Cubes

Factor: $w^3 + 64$

Solution:

$w^3 + 64$	w^3 and 64 are perfect cubes.
$= (w)^3 + (4)^3$	Write as $a^3 + b^3$, where $a = w$, $b = 4$.
$a^3 + b^3 = (a + b)(a^2 - ab + b^2)$	Apply the formula for a sum of cubes.

$$(w)^3 + (4)^3 = (w + 4)[(w)^2 - (w)(4) + (4)^2]$$
$$= (w + 4)(w^2 - 4w + 16) \qquad \text{Simplify.}$$

Check: $(w + 4)(w^2 - 4w + 16) = w^3 - 4w^2 + 16w + 4w^2 - 16w + 64$
$$= w^3 + 64 \checkmark$$

---

example 2

Factoring a Difference of Cubes

Factor: $27p^3 - q^6$

Classroom Activity 4.5A

Solution:

$27p^3 - q^6$	$27p^3$ and q^6 are perfect cubes.
$(3p)^3 - (q^2)^3$	Write as $a^3 - b^3$, where $a = 3p$, $b = q^2$.
$a^3 - b^3 = (a - b)(a^2 + ab + b^2)$	Apply the formula for a difference of cubes.

$$(3p)^3 - (q^2)^3 = (3p - q^2)[(3p)^2 + (3p)(q^2) + (q^2)^2]$$
$$= (3p - q^2)(9p^2 + 3pq^2 + q^4) \qquad \text{Simplify.}$$

Check: $(3p - q^2)(9p^2 + 3pq^2 + q^4)$
$$= 27p^3 + 9p^2q^2 + 3pq^4 - 9p^2q^2 - 3pq^4 - q^6$$
$$= 27p^3 - q^6 \checkmark$$

2.   Factoring Binomials: A Summary

After removing the GCF, the next step in any factoring problem is to recognize what type of pattern it follows. Exponents that are divisible by 2 are perfect squares and those divisible by 3 are perfect cubes. The formulas for factoring binomials are summarized in the following box:

Factoring Binomials

1.   Difference of Squares: $a^2 - b^2 = (a + b)(a - b)$
2.   Difference of Cubes: $a^3 - b^3 = (a - b)(a^2 + ab + b^2)$
3.   Sum of Cubes: $a^3 + b^3 = (a + b)(a^2 - ab + b^2)$

---

example 3

Factoring Binomials

Factor completely: a. $27y^3 + 1$ b. $m^2 - \dfrac{1}{4}$

c. $3y^4 - 48$ d. $z^6 - 8w^3$

Classroom Activity 4.5B

Solution:

a. $27y^3 + 1$

Sum of cubes: $27y^3 = (3y)^3$
and $1 = (1)^3$.

$= (3y)^3 + (1)^3$

Write as $a^3 + b^3$, where $a = 3y$ and $b = 1$.

$= (3y + 1)((3y)^2 - (3y)(1) + (1)^2)$

Apply the formula
$a^3 + b^3 = (a + b)(a^2 - ab + b^2)$.

$= (3y + 1)(9y^2 - 3y + 1)$

Simplify.

b. $m^2 - \dfrac{1}{4}$

Difference of squares

$= (m)^2 - \left(\dfrac{1}{2}\right)^2$

Write as $a^2 + b^2$, where $a = m$ and $b = \frac{1}{2}$.

$= \left(m + \dfrac{1}{2}\right)\left(m - \dfrac{1}{2}\right)$

Apply the formula $a^2 - b^2 = (a + b)(a - b)$.

c. $3y^4 - 48$

$= 3(y^4 - 16)$

Factor out the GCF. The binomial is a difference of squares.

$= 3[(y^2)^2 - (4)^2]$

Write as $a^2 - b^2$, where $a = y^2$ and $b = 4$.

$$= 3(y^2 + 4)(y^2 - 4)$$

Apply the formula
$a^2 - b^2 = (a + b)(a - b)$.

$y^2 + 4$ is a sum of squares and cannot be factored.

$$= 3(y^2 + 4)(y + 2)(y - 2)$$

$y^2 - 4$ is a difference of squares and can be factored further.

d. $z^6 - 8w^3$

Difference of cubes: $z^6 = (z^2)^3$
and $8w^3 = (2w)^3$

$$= (z^2)^3 - (2w)^3$$

Write as $a^3 - b^3$, where $a = z^2$ and $b = 2w$.

$$= (z^2 - 2w)[(z^2)^2 + (z^2)(2w) + (2w)^2]$$

Apply the formula $a^3 - b^3 = (a - b)(a^2 + ab + b^2)$.

$$= (z^2 - 2w)(z^4 + 2z^2w + 4w^2)$$

Simplify.

Each of the factorizations in Example 3 can be checked by multiplying.

---

example 4

Factoring Binomials

Factor the binomial $x^6 - y^6$ as

a.   A difference of cubes
b.   A difference of squares

Solution:

a. $x^6 - y^6$

Difference of cubes

$$= (x^2)^3 - (y^2)^3$$

Write as $a^3 - b^3$, where $a = x^2$ and $b = y^2$.

$$= (x^2 - y^2)[(x^2)^2 + (x^2)(y^2) + (y^2)^2]$$

Apply the formula $a^3 - b^3 = (a - b)(a^2 + ab + b^2)$

$$= (x^2 - y^2)(x^4 + x^2y^2 + y^2)$$

Factor $x^2 - y^2$ as a difference of squares.

$$= (x + y)(x - y)(x^4 + x^2y^2 + y^2)$$

b. $x^6 - y^6$

Difference of squares

$$= (x^3)^2 - (y^3)^2$$

Write as $a^2 - b^2$, where $a = x^3$ and $b = y^3$.

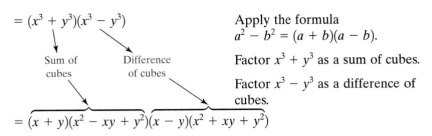

$$= (x^3 + y^3)(x^3 - y^3)$$

Sum of cubes

Difference of cubes

Apply the formula
$a^2 - b^2 = (a + b)(a - b)$.

Factor $x^3 + y^3$ as a sum of cubes.

Factor $x^3 - y^3$ as a difference of cubes.

$$= (x + y)(x^2 - xy + y^2)(x - y)(x^2 + xy + y^2)$$

Notice that the expressions x^6 and y^6 are both perfect squares and perfect cubes because the exponents are both multiples of 2 and of 3. Consequently, $x^6 - y^6$ can be factored initially as either the difference of squares or as the difference of cubes. In such a case, it is recommended that you factor the expression as a difference of squares first because it factors more completely into polynomials of lower degree. Hence:

$$x^6 - y^6 = (x + y)(x^2 - xy + y^2)(x - y)(x^2 + xy + y^2)$$

3. General Factoring Summary

To factor a polynomial, remember always to look for the greatest common factor first. Then identify the number of terms and the type of factoring problem the polynomial represents. A general factoring strategy is given below:

Factoring Strategy

1. Factor out the GCF (Section 4.1).
2. Identify whether the polynomial has two terms, three terms, or more than three terms.
3. If the polynomial has two terms, determine if it fits the pattern for

   - A difference of squares $a^2 - b^2 = (a - b)(a + b)$ (Section 4.4)
   - A difference of cubes $a^3 - b^3 = (a - b)(a^2 + ab + b^2)$ (Section 4.5)
   - A sum of cubes $a^3 + b^3 = (a + b)(a^2 - ab + b^2)$ (Section 4.5)

4. If the polynomial has three terms, check first for a perfect square trinomial (Section 4.4). Otherwise, factor the trinomial with the grouping method or the trial-and-error method (Sections 4.2 or 4.3).
5. If the polynomial has more than three terms, try factoring by grouping (Section 4.1).
6. Be sure to factor the polynomial completely.
7. Check by multiplying.

example 5

Factoring Polynomials

Factor out the GCF and identify the number of terms and type of factoring pattern represented by the polynomial. Then factor the polynomial completely.

a. $abx^2 - 3ax + 5bx - 15$ b. $20y^2 - 110y - 210$
c. $4p^3 + 20p^2 + 25p$ d. $w^3 + 1000$
e. $t^3 - 25t$

Solution:

a. $abx^2 - 3ax + 5bx - 15$ The GCF is 1. The polynomial has four
$abx^2 - 3ax \mid + 5bx - 15$ terms. Therefore, factor by grouping.
$\quad = ax(bx - 3) + 5(bx - 3)$
$\quad = (bx - 3)(ax + 5)$

Check: $(bx - 3)(ax + 5) = abx^2 + 5bx - 3ax - 15$ ✔

b. $20y^2 - 110y - 210$ The GCF is 10. The polynomial has three terms.
$\quad = 10(2y^2 - 11y - 21)$ The trinomial is not a perfect square trinomial.
$\quad = 10(2y + 3)(y - 7)$ Use either the grouping method or the trial-
                               and-error method.

Check: $10(2y + 3)(y - 7) = 10(2y^2 - 14y + 3y - 21)$
$\qquad\qquad\qquad\qquad\quad = 10(2y^2 - 11y - 21)$
$\qquad\qquad\qquad\qquad\quad = 20y^2 - 110y - 210$ ✔

c. $4p^3 + 20p^2 + 25p$ The GCF is p. The polynomial has three terms
$\quad = p(4p^2 + 20p + 25)$ and is a perfect square trinomial, $a^2 + 2ab + b^2$,
                            where $a = 2p$ and $b = 5$.

$\quad = p(2p + 5)^2$ Apply the formula $a^2 + 2ab + b^2 = (a + b)^2$.

Check: $p(2p + 5)^2 = p[(2p)^2 + 2(2p)(5) + (5)^2]$
$\qquad\qquad\qquad\quad = p(4p^2 + 20p + 25)$
$\qquad\qquad\qquad\quad = p^3 + 20p^2 + 25p$ ✔

d. $w^3 + 1000$ The GCF is 1. The polynomial has two
$\quad = (w)^3 + (10)^3$ terms. The binomial is a sum of cubes,
                        $a^3 + b^3$, where $a = w$ and $b = 10$.

$\quad = (w + 10)(w^2 - 10w + 100)$ Apply the formula
                              $a^3 + b^3 = (a + b)(a^2 - ab + b^2)$.

Check: $(w + 10)(w^2 - 10w + 100) = w^3 - \cancel{10w^2} + \cancel{100w} + \cancel{10w^2} - \cancel{100w} + 1000$
$\qquad\qquad\qquad\qquad\qquad\qquad\qquad = w^3 + 100$ ✔

e. $t^3 - 25t$ The GCF is t. The polynomial has two terms. The
$\quad = t(t^2 - 25)$ binomial is a difference of squares, $a^2 - b^2$, where
$\quad = t[(t)^2 - (5)^2]$ $a = t$ and $b = 5$.

$\quad = t(t + 5)(t - 5)$ Apply the formula $a^2 - b^2 = (a + b)(a - b)$.

Check: $t(t + 5)(t - 5) = t(t^2 - \cancel{5t} + \cancel{5t} - 25)$
$\qquad\qquad\qquad\qquad = t(t^2 - 25)$
$\qquad\qquad\qquad\qquad = t^3 - 25t$ ✔

section 4.5   PRACTICE EXERCISES

1. Multiply the polynomials: $(x - y)(x^2 + xy + y^2)$
   3.6 $x^3 - y^3$

2. Multiply the polynomials: $(x + y)(x^2 - xy + y^2)$
   3.6 $x^3 + y^3$

3. What trinomial multiplied by $(x - 2)$ gives a difference of cubes? $x^2 + 2x + 4$

4. What trinomial multiplied by $(p + 3)$ give a sum of cubes? $p^2 - 3p + 9$

5. Write a binomial that when multiplied by $(4x^2 - 2x + 1)$ produces a sum of cubes. $2x + 1$

6. Write a binomial that when multiplied by $(9y^2 + 15y + 25)$ produces a difference of cubes.
   $3y - 5$

7. Identify the expressions that are perfect cubes:

   $\{x^3, 8, 9, y^6, a^4, b^2, 3p^3, 27q^3, w^{12}, r^3s^6\}$
   $x^3, 8, y^6, 27q^3, w^{12}, r^3s^6$

8. Identify the expressions that are perfect cubes:

   $\{z^9, -81, 30, 8, 6x^3, y^{15}, 27a^3, b^2, p^3q^2, -1\}$
   $z^9, 8, y^{15}, 27a^3, -1$

9. How do you determine if a binomial is a sum of cubes?   If the binomial is of the form $a^3 + b^3$.

10. How do you determine if a binomial is a difference of cubes?   If the binomial is of the form $a^3 - b^3$.

11. From memory, write the formula to factor a sum of cubes:

$$a^3 + b^3 = \underline{(a + b)(a^2 - ab + b^2)}$$

12. From memory, write the formula to factor a difference of cubes:

$$a^3 - b^3 = \underline{(a - b)(a^2 + ab + b^2)}$$

For Exercises 13–28, factor the sums and differences of cubes.

13. $y^3 - 8$
    $(y - 2)(y^2 + 2y + 4)$

14. $x^3 + 27$
    $(x + 3)(x^2 - 3x + 9)$

15. $1 - p^3$
    $(1 - p)(1 + p + p^2)$

16. $q^3 + 1$
    $(q + 1)(q^2 - q + 1)$

17. $w^3 + 64$
    $(w + 4)(w^2 - 4w + 16)$

18. $8 - t^3$
    $(2 - t)(4 + 2t + t^2)$

19. $1000a^3 + 27$
    $(10a + 3)(100a^2 - 30a + 9)$

20. $216b^3 - 125$
    $(6b - 5)(36b^2 + 30b + 25)$

21. $n^3 - \dfrac{1}{8}$
    $\left(n - \dfrac{1}{2}\right)\left(n^2 + \dfrac{1}{2}n + \dfrac{1}{4}\right)$

22. $\dfrac{8}{27} + m^6$
    $\left(\dfrac{2}{3} + m^2\right)\left(\dfrac{4}{9} - \dfrac{2}{3}m^2 + m^4\right)$

23. $a^3 + b^6$
    $(a + b^2)(a^2 - ab^2 + b^4)$

24. $u^6 - v^3$
    $(u^2 - v)(u^4 + u^2v + v^2)$

25. $x^9 + 64y^3$
    $(x^3 + 4y)(x^6 - 4x^3y + 16y^2)$

26. $125w^3 - z^9$
    $(5w - z^3)(25w^2 + 5wz^3 + z^6)$

27. $25m^{12} + 16$ Prime

28. $36p^6 + 49q^4$ Prime

29. From memory, write the formula to factor a difference of squares.

$$a^2 - b^2 = \underline{(a - b)(a + b)}$$

30. Write a short paragraph explaining a strategy to factor binomials.   ❖

For Exercises 31–48, factor the binomials completely, if possible.

31. $x^4 - 4$
    $(x^2 - 2)(x^2 + 2)$

32. $b^4 - 25$
    $(b^2 - 5)(b^2 + 5)$

33. $a^2 + 9$
    Prime

34. $w^2 + 36$
    Prime

35. $t^3 + 64$
    $(t + 4)(t^2 - 4t + 16)$

36. $u^3 + 27$
    $(u + 3)(u^2 - 3u + 9)$

37. $g^3 - 4$
    Prime

38. $h^3 - 25$
    Prime

39. $4b^3 + 108$
    $4(b + 3)(b^2 - 3b + 9)$

40. $3c^3 - 24$
    $3(c - 2)(c^2 + 2c + 4)$

41. $5p^2 - 125$
    $5(p - 5)(p + 5)$

42. $2q^4 - 8$
    $2(q^2 - 2)(q^2 + 2)$

43. $\dfrac{1}{64} - 8h^3$
    $(\frac{1}{4} - 2h)(\frac{1}{16} + \frac{1}{2}h + 4h^2)$

44. $\dfrac{1}{125} + k^6$
    $(\frac{1}{5} + k^2)(\frac{1}{25} - \frac{1}{5}k^2 + k^4)$

45. $x^4 - 16$
    $(x - 2)(x + 2)(x^2 + 4)$

46. $p^4 - 81$
    $(p - 3)(p + 3)(p^2 + 9)$

47. $q^6 - 64$ $(q - 2)(q^2 + 2q + 4)$
    $(q + 2)(q^2 - 2q + 4)$

48. $a^6 - 1$ $(a - 1)(a^2 + a + 1)$
    $(a + 1)(a^2 - a + 1)$

For Exercises 49–68, factor completely using the factoring strategy.

49. $4b + 16$
    $4(b + 4)$

50. $2a^2 - 162$
    $2(a - 9)(a + 9)$

51. $y^2 + 4y + 3$
    $(y + 3)(y + 1)$

52. $6w^2 - 6w$
    $6w(w - 1)$

53. $16z^4 - 81$
    $(2z + 3)(2z - 3)(4z^2 + 9)$

54. $3t^2 + 13t + 4$
    $(3t + 1)(t + 4)$

55. $5r^3 + 5$ $5(r + 1)(r^2 - r + 1)$

56. $3ac + ad - 3bc - bd$ $(3c + d)(a - b)$

57. $7p^2 - 29p + 4$
    $(7p - 1)(p - 4)$

58. $3q^2 - 9q - 12$
    $3(q - 4)(q + 1)$

59. $-2x^2 + 8x - 8$
    $-2(x - 2)^2$

60. $18a^2 + 12a$
    $6a(3a + 2)$

61. $54 - 2y^3$
    $2(3 - y)(9 + 3y + y^2)$

62. $4t^2 - 100$
    $4(t - 5)(t + 5)$

63. $4t^2 - 31t - 8$
    $(4t + 1)(t - 8)$

64. $10c^2 + 10c + 10$
    $10(c^2 + c + 1)$

65. $2xw - 10x + 3yw - 15y$ $(w - 5)(2x + 3y)$

❖ See Additional Answers Appendix   🖊 Writing   ⬅ Translating Expression   📐 Geometry   🖩 Scientific Calculator   Video

66. $x^3 + 0.001$ $(x + 0.1)(x^2 - 0.1x + 0.01)$

67. $4q^2 - 9$ $(2p - 3)(2p + 3)$

68. $64 + 16k + k^2$ $(8 + k)^2$

■ Expanding Your Skills

For Exercises 69–72, factor the sum and difference of cubes.

69. $\dfrac{64}{125}p^3 - \dfrac{1}{8}q^3$ ❖

70. $\dfrac{1}{1000}r^3 + \dfrac{8}{27}s^3$ ❖

71. $a^{12} + b^{12}$
$(a^4 + b^4)(a^8 - a^4b^4 + b^8)$

72. $a^9 - b^9$
$(a - b)(a^2 + ab + b^2)(a^6 + a^3b^3 + b^6)$

Use Exercises 73–76, to investigate the relationship between division and factoring.

73.   a. Use long division to divide $x^3 - 8$ by $(x - 2)$.

    b. Factor $x^3 - 8$.   The quotient is $x^2 + 2x + 4$.
    $(x - 2)(x^2 + 2x + 4)$

74.   a. Use long division to divide $y^3 + 27$ by $(y + 3)$.

    b. Factor $y^3 + 27$.   The quotient is $y^2 - 3y + 9$.
    $(y + 3)(y^2 - 3y + 9)$

75.   a. Use long division to divide $m^3 + 1$ by $(m + 1)$.   The quotient is $m^2 - m + 1$.

    b. Factor $m^3 + 1$. $(m + 1)(m^2 - m + 1)$

76.   a. Use long division to divide $n^3 - 64$ by $(n - 4)$.

    b. Factor $n^3 - 64$.   The quotient is $n^2 + 4n + 16$.
    $(n - 4)(n^2 + 4n + 16)$

Concepts

1. Definition of a Quadratic Equation
2. Zero Product Rule
3. Solving Quadratic Equations
4. Solving Higher Degree Polynomial Equations
5. Applications of Quadratic Equations
6. Pythagorean Theorem

section 4.6   Zero Product Rule

1. Definition of a Quadratic Equation

In Section 2.1 we solved linear equations in one variable. These are equations of the form $ax + b = 0$ $(a \neq 0)$. A linear equation in one variable is sometimes called a first-degree polynomial equation because the highest degree of all its terms is 1. A second-degree polynomial equation in one variable is called a quadratic equation.

Definition of a Quadratic Equation in One Variable

If a, b, and c are real numbers such that $a \neq 0$, then a **quadratic equation** is an equation that can be written in the form

$$ax^2 + bx + c = 0$$

The following equations are quadratic because they can each be written in the form $ax^2 + bx + c = 0$, $(a \neq 0)$.

$$-4x^2 + 4x = 1 \qquad x(x - 2) = 3 \qquad (x - 4)(x + 4) = 9$$
$$-4x^2 + 4x - 1 = 0 \qquad x^2 - 2x = 3 \qquad x^2 - 16 = 9$$
$$x^2 - 2x - 3 = 0 \qquad x^2 - 25 = 0$$
$$x^2 + 0x - 25 = 0$$

2. Zero Product Rule

One method for solving a quadratic equation is to factor the equation and apply the zero product rule. The **zero product rule** states that if the product of two factors is zero, then one or both of its factors is zero.

Zero Product Rule

$$\text{If } ab = 0, \text{ then } a = 0 \text{ or } b = 0$$

For example, the quadratic equation $x^2 - x - 12 = 0$ can be written in factored form as $(x - 4)(x + 3) = 0$. By the zero product rule, one or both factors must be zero. Hence, either $x - 4 = 0$ or $x + 3 = 0$. Therefore, to solve the quadratic equation, set each factor equal to zero and solve for x.

$$(x - 4)(x + 3) = 0 \qquad \text{Apply the zero product rule.}$$

$x - 4 = 0$	or	$x + 3 = 0$ Set each factor equal to zero.
$x = 4$	or	$x = -3$ Solve each equation for x.

3. Solving Quadratic Equations

Quadratic equations, like linear equations, arise in many applications in mathematics, science, and business. The following steps summarize the factoring method for solving a quadratic equation.

Steps for Solving a Quadratic Equation by Factoring

1. Write the equation in the form: $ax^2 + bx + c = 0$.
2. Factor the equation completely.
3. Apply the zero product rule. That is, set each factor equal to zero and solve the resulting equations.

Note: The solution(s) found in Step 3 may be checked by substitution into the original equation.

---

example 1

Classroom Activity 4.6A

Solving Quadratic Equations

Solve the quadratic equations.

a. $2x^2 - 9x = 5$ b. $4x^2 + 24x = 0$ c. $5x(5x + 2) = 10x + 9$

Solution:

a. $2x^2 - 9x = 5$

$2x^2 - 9x - 5 = 0$	Write the equation in the form $ax^2 + bx + c = 0$.
$(2x + 1)(x - 5) = 0$	Factor the polynomial completely.
$2x + 1 = 0 \quad \text{or} \quad x - 5 = 0$	Set each factor equal to zero.
$2x = -1 \quad \text{or} \qquad x = 5$	Solve each equation.
$x = -\dfrac{1}{2} \quad \text{or} \qquad x = 5$	The solutions are $x = -\frac{1}{2}$ or $x = 5$.

$$\underline{\text{Check: } x = -\frac{1}{2}} \qquad \underline{\text{Check: } x = 5}$$

$$2x^2 - 9x = 5 \qquad 2x^2 - 9x = 5$$

$$2\left(-\frac{1}{2}\right)^2 - 9\left(-\frac{1}{2}\right) \stackrel{?}{=} 5 \qquad 2(5)^2 - 9(5) \stackrel{?}{=} 5$$

$$2\left(\frac{1}{4}\right) + \frac{9}{2} \stackrel{?}{=} 5 \qquad 2(25) - 45 \stackrel{?}{=} 5$$

$$\frac{1}{2} + \frac{9}{2} \stackrel{?}{=} 5 \qquad 50 - 45 \stackrel{?}{=} 5 \ ✔$$

$$\frac{10}{2} \stackrel{?}{=} 5 \ ✔$$

b. $4x^2 + 24x = 0$ The equation is already in the form $ax^2 + bx + c = 0$ (Note that $c = 0$).

$\qquad = 4x(x + 6) = 0$ Factor completely.

$\quad 4x = 0 \quad$ or $\quad x + 6 = 0$ Set each factor equal to zero.

$\qquad x = 0 \quad$ or $\qquad x = -6$ The solutions are $x = 0$ or $x = -6$.

$\underline{\text{Check: } x = 0} \qquad \underline{\text{Check: } x = -6}$

$$4x^2 + 24x = 0 \qquad 4x^2 + 24x = 0$$

$$4(0)^2 + 24(0) \stackrel{?}{=} 0 \qquad 4(-6)^2 + 24(-6) \stackrel{?}{=} 0$$

$$0 + 0 \stackrel{?}{=} 0 \qquad 4(36) - 144 \stackrel{?}{=} 0$$

$$0 = 0 \ ✔ \qquad 144 - 144 \stackrel{?}{=} 0$$

$$0 = 0 \ ✔$$

c. $\qquad 5x(5x + 2) = 10x + 9$

$\qquad 25x^2 + 10x = 10x + 9$ Clear parentheses.

$25x^2 + 10x - 10x - 9 = 0$ Set the equation equal to zero.

$\qquad 25x^2 - 9 = 0$ The equation is in the form $ax^2 + bx + c = 0$ (Note that $b = 0$).

$\qquad (5x - 3)(5x + 3) = 0$ Factor completely.

$5x - 3 = 0 \quad$ or $\quad 5x + 3 = 0$ Set each factor equal to zero.

$\quad 5x = 3 \quad$ or $\qquad 5x = -3$ Solve each equation.

$$\frac{5x}{5} = \frac{3}{5} \quad \text{or} \quad \frac{5x}{5} = \frac{-3}{5}$$

$\quad x = \frac{3}{5} \quad$ or $\qquad x = -\frac{3}{5}$ The solutions are $x = \frac{3}{5}$ or $x = -\frac{3}{5}$.

Check: $x = \dfrac{3}{5}$ Check: $x = -\dfrac{3}{5}$

$$5x(5x + 2) = 10x + 9$$

$$5(\tfrac{3}{5})[5(\tfrac{3}{5}) + 2] \overset{?}{=} 10(\tfrac{3}{5}) + 9$$

$$3(3 + 2) \overset{?}{=} 6 + 9$$

$$3(5) \overset{?}{=} 15$$

$$15 = 15 \checkmark$$

$$5x(5x + 2) = 10x + 9$$

$$5(-\tfrac{3}{5})[5(-\tfrac{3}{5}) + 2] \overset{?}{=} 10(-\tfrac{3}{5}) + 9$$

$$-3(-3 + 2) \overset{?}{=} -6 + 9$$

$$-3(-1) \overset{?}{=} 3$$

$$3 = 3 \checkmark$$

4.   Solving Higher Degree Polynomial Equations

The zero product rule can be used to solve higher degree polynomial equations provided the equations can be set to zero and written in factored form.

example 2   **Solving Higher Degree Polynomial Equations**

Solve the equations.

a. $-6(y + 3)(y - 5)(2y + 7) = 0$ b. $w^3 + 5w^2 - 9w - 45 = 0$

Classroom Activity 4.6B   **Solution:**

a. $-6(y + 3)(y - 5)(2y + 7) = 0$ The equation is already in factored form and equal to zero.

Set each factor equal to zero. Solve each equation for y.

$-6 \not= 0$ or $y + 3 = 0$ or $y - 5 = 0$ or $2y + 7 = 0$

No solution, $y = -3$ or $y = 5$ or $y = -\dfrac{7}{2}$

Notice that when the constant factor is set equal to zero, the result is a contradiction $-6 = 0$. The constant factor does not produce a solution to the equation. Therefore, the only solutions are $y = -3$, $y = 5$, and $y = -\tfrac{7}{2}$. Each solution can be checked in the original equation.

b. $w^3 + 5w^2 - 9w - 45 = 0$ This is a higher degree polynomial equation.

$$w^3 + 5w^2 - 9w - 45 = 0$$

The equation is already set equal to zero. Now factor.

$$w^2(w + 5) - 9(w + 5) = 0$$
$$(w + 5)(w^2 - 9) = 0$$

Because there are four terms, try factoring by grouping.

$$(w + 5)(w - 3)(w + 3) = 0$$

$w^2 - 9$ is a difference of squares and can be factored further.

$$w + 5 = 0 \quad \text{or} \quad w - 3 = 0 \quad \text{or} \quad w + 3 = 0$$

Set each factor equal to zero.

$$w = -5 \quad \text{or} \quad w = 3 \quad \text{or} \quad w = -3$$

Solve each equation.

Each solution checks in the original equation.

5. Applications of Quadratic Equations

example 3

Using a Quadratic Equation in an Application

The base of a triangle is 3 m more than the height. The area is 35 m². Find the base and height of the triangle.

Solution:

Let x represent the height of the triangle.

Then $x + 3$ represents the base (Figure 4-1).

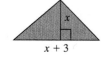

$x + 3$

Figure 4-1

To set up an equation to solve for x, use $A = \frac{1}{2}bh$.

$$\text{Area} = \frac{1}{2}(\text{base})(\text{height})$$

Verbal equation

$$35 = \frac{1}{2}(x + 3)(x)$$

Algebraic equation

$$2 \cdot 35 = 2 \cdot \frac{1}{2}(x + 3)(x)$$

Multiply both sides by 2 to clear fractions.

$$70 = (x + 3)(x)$$

$$70 = x^2 + 3x$$ Clear parentheses.

$$0 = x^2 + 3x - 70$$ Write the equation in the form $ax^2 + bx + c = 0$.

$$0 = (x + 10)(x - 7)$$ Factor the equation.

$$x + 10 = 0 \quad \text{or} \quad x - 7 = 0$$ Set each factor equal to zero.

$$x \cancel{=} -10 \quad \text{or} \quad x = 7$$ Because x represents the height of a triangle, reject the negative solution.

The variable x represents the height of the triangle. Therefore, the height is 7 m.

The expression $x + 3$ represents the base of the triangle. Therefore, the base is 10 m.

---

example 4

Using Translations to Set up a Quadratic Equation

The product of two consecutive integers is 48 more than the larger integer. Find the integers.

Classroom Activity 4.6C

Solution:

Let x represent the first (smaller) integer.

Then $x + 1$ represents the second (larger) integer.   Label the variables.

(First integer)(second integer) = (second integer) + 48   Verbal model

$$x(x + 1) = (x + 1) + 48$$ Algebraic equation

$$x^2 + x = x + 49$$ Simplify.

$$x^2 + x - x - 49 = 0$$ Set the equation equal to zero.

$$x^2 - 49 = 0$$

$$(x - 7)(x + 7) = 0$$ Factor.

$$x - 7 = 0 \quad \text{or} \quad x + 7 = 0$$ Set each factor equal to zero.

$$x = 7 \quad \text{or} \quad x = -7$$ Solve for x.

Recall that x represents the smaller integer. Therefore, there are two possibilities for the pairs of consecutive integers.

If $x = 7$, then the larger integer is $x + 1$ or $7 + 1 = 8$.

If $x = -7$, then the larger integer is $x + 1$ or $-7 + 1 = -6$.

The integers are 7 and 8 or -7 and -6.

Tip: To check your answer in Example 4, verify that each pair of integers satisfies the requirements that the product of integers is equal to 48 more than the larger integer:

Product	Larger Integer $+48$
$(7)(8) = 56$	$8 + 48 = 56$
$(-7)(-6) = 42$	$-6 + 48 = 42$

example 5

Using a Quadratic Equation in an Application

A stone is dropped off a 64-ft cliff and falls into the ocean below. The height of the stone above sea level is given by the equation

$h = -16t^2 + 64$ where h is the stone's height in feet, and t is the time in seconds.

Find the time required for the stone to hit the water.

Solution:

When the stone hits the water, its height is zero. Therefore, substitute $h = 0$ into the equation.

$h = -16t^2 + 64$ The equation is quadratic.

$0 = -16t^2 + 64$ Substitute $h = 0$.

$0 = -16(t^2 - 4)$ Factor out the GCF.

$0 = -16(t - 2)(t + 2)$ Factor as a difference of squares.

$-16 \neq 0$ or $t - 2 = 0$ or $t + 2 = 0$ Set each factor to zero.

No solution, $t = 2$ or $t \neq -2$ Solve for t.

The negative value of t is rejected because the stone cannot fall for a negative time. Therefore, the stone hits the water after 2 seconds.

In Example 5, we can analyze the path of the stone as it falls from the cliff. Compute the height values at various times between 0 and 2 seconds (Table 4-1 and Table 4-2). The ordered pairs can be graphed where t is used in place of x and h is used in place of y.

Table 4-1				Table 4-2	
Time, t (s)	Height, h (ft)			Time, t (s)	Height, h (ft)
0.0		$\longrightarrow h = -16(0.0)^2 + 64 = 64 \longrightarrow$		0.0	64
0.5		$\longrightarrow h = -16(0.5)^2 + 64 = 60 \longrightarrow$		0.5	60
1.0		$\longrightarrow h = -16(1.0)^2 + 64 = 48 \longrightarrow$		1.0	48
1.5		$\longrightarrow h = -16(1.5)^2 + 64 = 28 \longrightarrow$		1.5	28
2.0		$\longrightarrow h = -16(2.0)^2 + 64 = 0 \longrightarrow$		2.0	0

The graph of the height of the stone versus time is shown in Figure 4-2. From the graph, we can verify that the stone hits the water after 2 s.

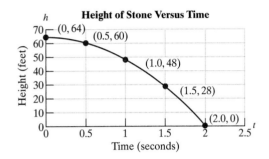

Figure 4-2

6. Pythagorean Theorem

Recall that a right triangle is a triangle that contains a 90° angle. Furthermore, the sum of the squares of the two legs (the shorter sides) of a right triangle equals the square of the hypotenuse (the longest side). This important fact is known as the Pythagorean theorem. The Pythagorean theorem is an enduring landmark of mathematical history from which many mathematical ideas have been built. Although the theorem is named after Pythagoras (sixth century B.C.E.), a Greek mathematician and philosopher, it is thought that the ancient Babylonians were familiar with the principle more than a thousand years earlier.

For the right triangle shown in Figure 4-3, the **Pythagorean theorem** is stated as:

$$a^2 + b^2 = c^2.$$

In this formula, a and b are the legs of the right triangle and c is the hypotenuse. Notice that the hypotenuse is the longest side of the right triangle and is opposite the 90° angle.

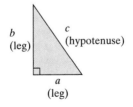

Figure 4-3

example 6

Applying the Pythagorean Theorem

Show that the lengths of the sides of the right triangle in the figure satisfy the Pythagorean theorem.

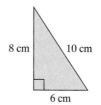

Solution:

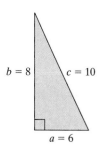

Label the triangle.

$$a^2 + b^2 = c^2 \qquad \text{Apply the Pythagorean theorem.}$$

$$(6)^2 + (8)^2 \overset{?}{=} (10)^2 \qquad a = 6, b = 8, c = 10$$

$$36 + 64 \overset{?}{=} 100$$

$$100 = 100 \; ✔ \qquad \text{The triangle is a right triangle.}$$

example 7

Using a Quadratic Equation in an Application

A 13-ft board is used as a ramp to unload furniture off a loading platform. If the distance between the top of the board and the ground is 7 ft less than the distance between the bottom of the board and the base of the platform, find both distances.

Classroom Activity 4.6D

Solution:

Let x represent the distance between the bottom of the board and the base of the platform. Then $x - 7$ represents the distance between the top of the board and the ground (Figure 4-4).

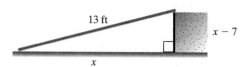

Figure 4-4

$$a^2 + b^2 = c^2$$ Pythagorean theorem

$$x^2 + (x - 7)^2 = (13)^2$$

Avoiding Mistakes

Recall that the square of a binomial results in a perfect square trinomial.

$$(a - b)^2 = a^2 - 2ab + b^2$$

Don't forget the middle term.

$$x^2 + [(x)^2 - 2(x)(7) + (7)^2] = 169$$

$$x^2 + x^2 - 14x + 49 = 169$$

$$2x^2 - 14x + 49 = 169$$ Combine *like* terms.

$$2x^2 - 14x + 49 - 169 = 0$$ Set the equation equal to zero.

$$2x^2 - 14x - 120 = 0$$ Write the equation in the form $ax^2 + bx + c = 0$.

$$2(x^2 - 7x - 60) = 0$$ Factor.

$$2(x - 12)(x + 5) = 0$$

$$2 \neq 0 \quad \text{or} \quad x - 12 = 0 \quad \text{or} \quad x + 5 = 0$$ Set each factor equal to zero.

$$x = 12 \quad \text{or} \quad x \neq -5$$ Solve both equations for x.

Recall that x represents the distance between the bottom of the board and the base of the platform. We reject the negative value of x because a distance cannot be negative. Therefore, the distance between the bottom of the board and the base of the platform is 12 ft. The distance between the top of the board and the ground is $x - 7 = 5$ ft.

---

section 4.6 PRACTICE EXERCISES

For Exercises 1–10, factor completely. 4.1–4.4

1. $4x - 2 + 2bx - b$
   $(2x - 1)(2 + b)$
2. $6a - 8 - 3ab + 4b$
   $(3a - 4)(2 - b)$
3. $4b^2 - 44b + 120$
   $4(b - 5)(b - 6)$
4. $8u^2v^2 - 4uv$
   $4uv(2uv - 1)$
5. $16w^2 - 1$
   $(4w - 1)(4w + 1)$
6. $3x^2 + 10x - 8$
   $(3x - 2)(x + 4)$
7. $12k + 16$
   $4(3k + 4)$
8. $3h^2 - 75$
   $3(h - 5)(h + 5)$
9. $2y^2 + 3y - 44$
   $(2y + 11)(y - 4)$
10. $4x^2 + 16y^2$
    $4(x^2 + 4y^2)$

For Exercises 11–18, identify the polynomials as linear, quadratic, or neither.

11. $4 - 5x$
    Linear
12. $5x^3 + 2$
    Neither
13. $3x - 6x^2$
    Quadratic
14. $1 - x + 2x^2$
    Quadratic
15. $7x^4 + 8$
    Neither
16. $3x + 2$
    Linear

17. $6x^2 - 7x - 2$
    Quadratic
18. $4x^2 - 1$
    Quadratic

19. State the zero product rule.
    If $ab = 0$, then $a = 0$ or $b = 0$.

For Exercises 20–27, solve the equations using the zero product rule.

20. $(x - 5)(x + 1) = 0$
    $x = 5, x = -1$
21. $(x + 3)(x - 1) = 0$
    $x = -3, x = 1$
22. $(3x - 2)(3x + 2) = 0$
    $x = \frac{2}{3}, x = -\frac{2}{3}$
23. $(2x - 7)(2x + 7) = 0$
    $x = \frac{7}{2}, x = -\frac{7}{2}$
24. $2(x - 7)(x - 7) = 0$
    $x = 7$
25. $3(x + 5)(x + 5) = 0$
    $x = -5$
26. $x(x - 4)(2x + 3) = 0$
    $x = 0, x = 4, x = -\frac{3}{2}$
27. $x(3x + 1)(x + 1) = 0$
    $x = 0, x = -\frac{1}{3}, x = -1$

---

28. For a quadratic equation of the form $ax^2 + bx + c = 0$, what must be done before applying the zero product rule?

The polynomial must be factored completely.

For Exercises 29–40, solve the equations.

29. $p^2 - 2p - 15 = 0$

$p = 5, p = -3$

30. $y^2 - 7y - 8 = 0$

$y = 8, y = -1$

31. $z^2 + 10z - 24 = 0$

$z = -12, z = 2$

32. $w^2 - 10w + 16 = 0$

$w = 8, w = 2$

33. $2q^2 - 7q - 4 = 0$

34. $4x^2 - 11x - 3 = 0$ ❖

35. $0 = 9x^2 - 4$ ❖

36. $4a^2 - 49 = 0$ ❖

37. $2k^2 - 28k + 96 = 0$

$k = 6, k = 8$

38. $0 = 2t^2 + 20t + 50$

$t = -5$

39. $0 = 2m^3 - 5m^2 - 12m$ $m = 0, m = -\dfrac{3}{2}, m = 4$

40. $3n^3 + 4n^2 + n = 0$ $n = 0, n = -\dfrac{1}{3}, n = -1$

41. What are the requirements needed to use the zero product rule to solve a quadratic equation or higher degree polynomial equation? ❖

For Exercises 42–63, solve the equations.

42. $x^2 + 10x = 24$

$x = -12, x = 2$

43. $x^2 - 10x = -16$

$x = 8, x = 2$

44. $9d^2 = 4$ $d = \dfrac{2}{3}, d = -\dfrac{2}{3}$

45. $4p^2 = 49$ $p = \dfrac{7}{2}, p = -\dfrac{7}{2}$

46. $2(c^2 - 14c) = -96$

$c = 6, c = 8$

47. $2(q^2 + 10q) = -50$

$q = -5$

48. $12x = 2x^3 - 5x^2$ ❖

49. $-x = 3x^3 + 4x^2$ ❖

50. $3(a^2 + 2a) = 2a^2 - 9$ ❖

51. $9(k - 1) = -4k^2$ ❖

52. $2n(n + 2) = 6$

$n = -3, n = 1$

53. $3p(p - 1) = 18$

$p = 3, p = -2$

54. $27q^2 = 9q$ $q = 0, q = \dfrac{1}{3}$

55. $21w^2 = 14w$ $w = 0, w = \dfrac{2}{3}$

56. $3(c^2 - 2c) = 0$

$c = 0, c = 2$

57. $2(4d^2 + d) = 0$

$d = 0, d = -\dfrac{1}{4}$

58. $y^3 - 3y^2 - 4y + 12 = 0$

$y = 3, y = -2, y = 2$

59. $t^3 + 2t^2 - 16t - 32$ $t = -2, t = 4, t = -4$

60. $(x - 1)(x + 2) = 18$ $x = -5, x = 4$

61. $(w + 5)(w - 3) = 20$ $w = -7, w = 5$

62. $(p + 2)(p + 3) = 1 - p$ $p = -5, p = -1$

63. $(k - 6)(k - 1) = -k - 2$ $k = 4, k = 2$

64. If eleven is added to the square of a number, the result is sixty. Find all such numbers.

The numbers are 7 and −7.

65. If a number is added to two times its square, the result is thirty-six. Find all such numbers. ❖

66. If twelve is added to six times a number, the result is twenty-eight less than the square of the number. Find all such numbers.

The numbers are 10 and −4.

67. The square of a number is equal to twenty more than the number. Find all such numbers.

The numbers are 5 and −4.

68. The product of two consecutive odd integers is sixty-three. Find all such integers.

The numbers are −9 and −7 or 7 and 9.

69. The product of two consecutive even integers is forty-eight. Find all such integers.

The numbers are 6 and 8 or −8 and −6.

70. The sum of the squares of two consecutive integers is one more than ten times the larger number. Find all such integers.

The numbers are 5 and 6 or −1 and 0.

71. The sum of the squares of two consecutive integers is nine less than ten times the sum of the integers. Find all such integers.

The numbers are 0 and 1 or 9 and 10.

72. The length of a rectangular room is 5 yd more than the width. If 300 yd^2 of carpeting cover the room, what are the dimensions of the room?

The room is 15 yd by 20 yd.

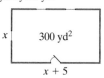

Figure for Exercise 72

73. The width of a rectangular painting is 2 in. less than the length. The area is 120 in.^{2} Find the length and width.

The painting has length 12 in. and width 10 in.

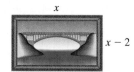

Figure for Exercise 73

74. The width of a rectangular slab of concrete is 3 m less than the length. If the area is 28 m^2,

a. What are the dimensions of the rectangle? ❖

b. What is the perimeter of the rectangle? ❖

75. The width of a rectangular picture is 7 in. less than the length. If the area of the picture is 78 in.^{2},

a. What are the dimensions of the rectangle? ❖

b. What is the perimeter of the rectangle? ❖

❖ See Additional Answers Appendix   Writing   Translating Expression   Geometry   Scientific Calculator   Video

76. The base of a triangle is 1 ft less than twice the height. The area is 14 ft². Find the base and height of the triangle.
    The base is 7 ft and the height is 4 ft.

77. The height of a triangle is 5 cm less than 3 times the base. The area is 125 cm². Find the base and height of the triangle.
    The base is 10 cm and the height is 25 cm.

78. In a physics experiment, a ball is dropped off a 144-ft platform. The height of the ball above the ground is given by the equation

    $h = -16t^2 + 144$ where h is the ball's height in feet and t is the time in seconds after the ball is dropped ($t \geq 0$).

    Find the time required for the stone to hit the ground. (*Hint*: Let $h = 0$)   3 s

79. A stone is dropped off a 64-ft cliff. The height of the stone above the ground is given by the equation

    $h = -16t^2 + 64$ where h is the stone's height in feet, and t is the time in seconds after the stone is dropped ($t \geq 0$).

    Find the time required for the stone to hit the ground.   2 s

80. An object is shot straight up into the air from ground level with initial speed of 24 ft/s. The height of the object (in feet) is given by the equation

    $h = -16t^2 + 24t$ where t is the time in seconds after launch ($t \geq 0$).

    Find the time(s) when the object is at ground level.   0 s and 1.5 s

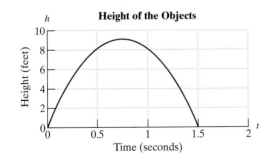

Figure for Exercise 80

81. A rocket is launched straight up into the air from the ground with initial speed of 64 ft/s. The height of the rocket (in feet) is given by the equation

    $h = -16t^2 + 64t$ where t is the time in seconds after launch ($t \geq 0$).

    Find the time(s) when the ball is at ground level.
    0 s and 4 s

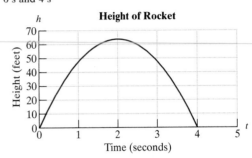

Figure for Exercise 81

82. Draw a right triangle and label the sides with the words *leg* and *hypotenuse*. ❖

83. State the Pythagorean theorem.   Given a right triangle with legs a and b and hypotenuse c, then $a^2 + b^2 = c^2$.

For Exercises 84–87, use the Pythagorean theorem to determine whether the triangle could be a right triangle.

84. Yes

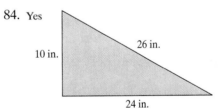

85.

Yes

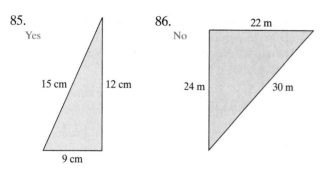

15 cm

12 cm

9 cm

86.

No

22 m

24 m

30 m

87.

No

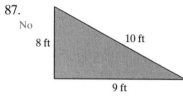

8 ft

10 ft

9 ft

88. Darcy holds the end of a kite string 3 ft (1 yd) off the ground and wants to estimate the height of the kite. Her friend Jenna is 24 yd away from her, standing directly under the kite as shown in the figure. If Darcy has 30 yd of string out, find the height of the kite (ignore the sag in the string). 19 yd

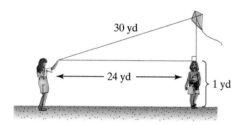

30 yd

24 yd

1 yd

Figure for Exercise 88

89. A 17-ft ladder rests against the side of a house. The distance between the top of the ladder and the ground is 7 ft more than the distance between the base of the ladder and the bottom of the house. Find both distances.

The bottom of the ladder is 8 ft from the house. The distance from the top of the ladder to the ground is 15 ft.

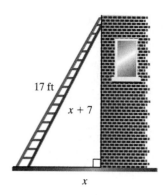

17 ft

$x + 7$

x

Figure for Exercise 89

90. Two boats leave a marina. One travels east, and the other travels south. After 30 min, the second boat has traveled 1 mile farther than the first boat and the distance between the boats is 5 miles. Find the distance each boat traveled. ❖

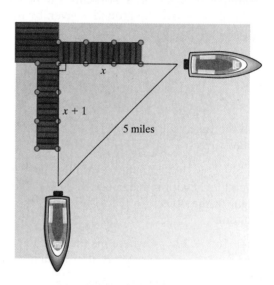

x

$x + 1$

5 miles

Figure for Exercise 90

91. One leg of a right triangle is 4 m less than the hypotenuse. The other leg is 2 m less than the hypotenuse. Find the length of the hypotenuse.
10 m

92. The longer leg of a right triangle is 1 cm less than twice the shorter leg. The hypotenuse is 1 cm greater than twice the shorter leg. Find the length of the shorter leg.
8 cm

■ EXPANDING YOUR SKILLS

93. The formula

$$N = \frac{x(x-3)}{2}$$

gives the number of diagonals, *N*, for a polygon with *x* sides, where $x \geq 3$.

a. Find the number of diagonals for a four-sided polygon.   Two diagonals

b. Find the number of diagonals for a five-sided polygon.   Five diagonals

c. Find the number of sides of a polygon if the polygon has 35 diagonals.   Ten sides

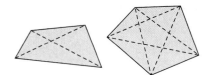

Figure for Exercise 93

94. A cardboard box is to be constructed from a square piece of cardboard by cutting out 2-in. squares from the corners and folding up the sides. If the volume of the box is 128 in.³, find the original dimensions of the square piece of cardboard.   The original square was 12 in. by 12 in.

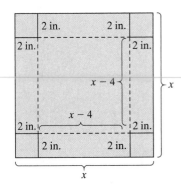

Figure for Exercise 94

---

section

4.7   CONNECTIONS TO GRAPHING: *x*- AND *y*-INTERCEPTS

Concepts

1. **Definition of *x*- and *y*-Intercepts**

2. **Determining *x*- and *y*-Intercepts from a Graph**

3. **Determining *x*- and *y*-Intercepts from an Equation**

4. **Applications of *x*- and *y*-Intercepts**

1. Definition of *x*- and *y*-Intercepts

For many applications of graphing it is advantageous to know the points where a graph crosses the *x*- or *y*-axis. These points are called the *x*- and *y*-intercepts.

In Figure 4-5, the *x*-intercept is at the point $(-3, 0)$ and the *y*-intercept is at $(0, 2)$. In Figure 4-6, there are three *x*-intercepts, one at $(-4, 0)$, one at $(1, 0)$, and one at $(5, 0)$. The *y*-intercept is at $(0, 20)$.

Notice that any point on the *x*-axis must have a *y*-coordinate of zero. Similarly, any point on the *y*-axis must have an *x*-coordinate of zero.

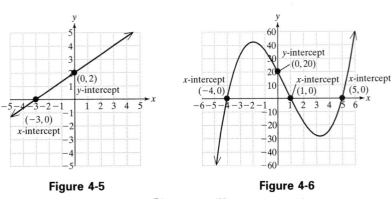

Figure 4-5   **Figure 4-6**

---

*Definition of *x*- and *y*-Intercepts

An ***x*-intercept** of an equation is a point $(a, 0)$ where the graph intersects the *x*-axis.

A ***y*-intercept** of an equation is a point $(0, b)$ where the graph intersects the *y*-axis.

*In some applications, an *x*-intercept is defined as the *x-coordinate* of a point of intersection that a graph makes with the *x*-axis. For example, if an *x*-intercept is at the point $(5, 0)$, it is sometimes stated simply as 5 (the *y*-coordinate is assumed to be zero). Similarly, a *y*-intercept is sometimes defined as the *y-coordinate* of a point of intersection that a graph makes with the *y*-axis. For example, if a *y*-intercept is at the point $(0, -3)$, it may be stated simply as -3 (the *x*-coordinate is assumed to be zero).

---

2. Determining *x*- and *y*-Intercepts from a Graph

example 1   **Determining the *x*- and *y*-Intercepts from a Graph**

Identify the *x*- and *y*-intercepts from the graphs.

a.
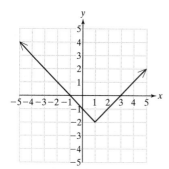

b.
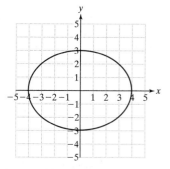

c.
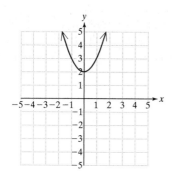

d.
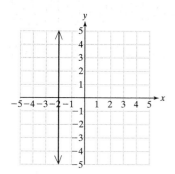

Classroom Activity 4.7A

Solution:

a. *x*-intercepts: $(-1, 0)$ and $(3, 0)$ b. *x*-intercepts: $(-4, 0)$ and $(4, 0)$

   y-intercept: $(0, -1)$ *y*-intercepts: $(0, -3)$ and $(0, 3)$

c. *x*-intercept: None   d. *x*-intercept: $(-2, 0)$

   y-intercept: $(0, 2)$ *y*-intercept: None

3.   Determining x- and y-Intercepts from an Equation

The x- and y-intercepts can be determined directly from the graph of an equation. However, if the graph is not given, the x- and y-intercepts may be determined from the equation. Because an x-intercept has a y-coordinate of zero, we can find an x-intercept by substituting $y = 0$ into the equation and solving for x. Similarly, the x-coordinate of a y-intercept is zero. Therefore, we can find a y-intercept by substituting $x = 0$ into the equation and solving for y.

Steps to Find the x- and y-Intercepts from an Equation

Given an equation in x and y,

1. Find the x-intercept(s) by substituting $y = 0$ into the equation and solving for x.
2. Find the y-intercept(s) by substituting $x = 0$ into the equation and solving for y.

example 2

Finding the x- and y-Intercepts of an Equation

Find the x- and y-intercepts.

a. $2x - 3y = 6$ b. $y = x^2 + 3x - 10$

Classroom Activity 4.7B

Solution:

a.   <u>To find the x-intercept, substitute $y = 0$:</u>

$$2x - 3y = 6$$

$2x - 3(0) = 6$ Substitute $y = 0$.

$\quad\quad 2x = 6$ Solve the equation for x (the equation is linear).

$\dfrac{2x}{2} = \dfrac{6}{2}$ Solve for x.

$\quad\quad x = 3$

The x-intercept is $(3, 0)$.

<u>To find the y-intercept, substitute $x = 0$:</u>

$$2x - 3y = 6$$

$2(0) - 3y = 6$ Substitute $x = 0$.

$\quad\quad -3y = 6$ Solve the equation for y (the equation is linear).

$\dfrac{-3y}{-3} = \dfrac{6}{-3}$ Solve for y.

$\quad\quad y = -2$

The y-intercept is $(0, -2)$.

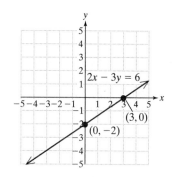

Figure 4-7

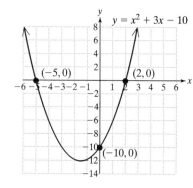

Figure 4-8

b.   To find the *x*-intercept, substitute $y = 0$:

$$y = x^2 + 3x - 10$$

$$0 = x^2 + 3x - 10 \qquad \text{Substitute } y = 0 \text{ (the equation is quadratic).}$$

$$0 = (x + 5)(x - 2) \qquad \text{Factor.}$$

$$x + 5 = 0 \quad \text{or} \quad x - 2 = 0 \qquad \text{Set each factor equal to zero.}$$

$$x = -5 \quad \text{or} \quad x = 2 \qquad \text{Solve each equation.}$$

The *x*-intercepts are $(-5, 0)$ and $(2, 0)$.

To find the *y*-intercept, substitute $x = 0$:

$$y = x^2 + 3x - 10$$

$$y = (0)^2 + 3(0) - 10 \qquad \text{Substitute } x = 0.$$

$$y = -10 \qquad \text{Simplify.}$$

The *y*-intercept is $(0, -10)$.

Notice that the equation $2x - 3y = 6$ is linear because it is in the form $ax + by = c$. The equation $y = x^2 + 3x - 10$ is nonlinear (the presence of the term x^2 makes it impossible to write the equation in the form $ax + by = c$). The graphs of the equations $2x - 3y = 6$ and $y = x^2 + 3x - 10$ are shown in Figures 4-7 and 4-8. The *x*- and *y*-intercepts of each equation can be confirmed from the graph.

4.   Applications of *x*- and *y*-Intercepts

example 3

Interpreting the *x*- and *y*-Intercepts in an Application

The guarantee period for an $80 truck battery is 4 years. If the battery does not last the entire 4 years, the manufacturer will pay the customer a refund for a portion of the loss. The amount the company will pay the customer, *y*, is given by the equation:

$$y = 80 - 20x \quad 0 \le x \le 4 \qquad \text{where } y \text{ is in dollars, and } x \text{ is the time in years that the battery lasts.}$$

a.   If a battery lasts for only 1 year, how much will the company refund to the customer?

b.   If the battery lasts for $3\frac{1}{2}$ years, how much will the company refund to the customer?

c.   Find the *y*-intercept of this equation and interpret the meaning of the *y*-intercept in the context of this problem.

d.   Find the *x*-intercept of this equation and interpret the meaning of the *x*-intercept in the context of this problem.

Solution:

a. $y = 80 - 20x$

$y = 80 - 20(1)$ Substitute $x = 1$.

$y = 60$

If a battery lasts only 1 year, the company will pay the customer $60.

b. $y = 80 - 20x$

$y = 80 - 20(3.5)$ Substitute $x = 3.5$.

$y = 80 - 70$

$y = 10$

If a battery lasts only $3\frac{1}{2}$ years, the company will pay the customer $10.

c. $y = 80 - 20x$

$y = 80 - 20(0)$ To find the *y*-intercept, substitute $x = 0$.

$y = 80$

The *y*-intercept is (0, 80). The *y*-intercept indicates that if a battery lasts 0 years, the company will give a full refund of $80.

d. $y = 80 - 20x$

$0 = 80 - 20x$ To find the *x*-intercept, substitute $y = 0$.

$-80 = -20x$ Solve for *x*.

$4 = x$

The *x*-intercept is (4, 0). The *x*-intercept indicates that if a battery lasts 4 years, the company will pay the customer $0. That is, if the battery lasts for the full warranty period, the company does not have to issue a refund.

The answers to parts (a)–(d) in Example 3 can be verified from the graph of the equation, $y = 80 - 20x$ (Figure 4-9). Notice that the longer a battery lasts, the less money the company must refund to the customer. After 4 years ($x = 4$), the company pays nothing ($y = 0$) to the customer. On the other hand, if a battery is dead at the time of purchase ($x = 0$), the company gives a full refund ($y = 80$).

Refund Amount Versus Time a Battery Lasts

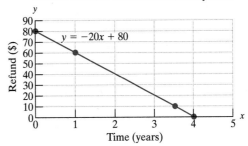

Figure 4-9

example 4

Interpreting the *x*- and *y*-Intercepts in an Application

A rock is dropped off a 144-ft cliff. The height of the rock, *y*, as it falls is given by

$$y = -16x^2 + 144$$ where *y* is measured in feet and *x* is the time in seconds after the rock is released ($0 \le x \le 3$).

a.   Use the equation to find the *y*-intercept and interpret the meaning of the *y*-intercept in the context of this problem.

b.   Use the equation to find the *x*-intercept(s) and interpret the meaning of the *x*-intercept in the context of this problem.

Classroom Activity 4.7C

Solution:

a. $y = -16x^2 + 144$

$y = -16(0)^2 + 144$ To find the *y*-intercept, substitute $x = 0$.

$y = 144$

The *y*-intercept is (0, 144) and means that at time of release ($x = 0$), the height of the rock is 144 ft.

b. $y = -16x^2 + 144$

$0 = -16x^2 + 144$ To find the *x*-intercept(s), substitute $y = 0$.

$0 = -16(x^2 - 9)$ Factor. Begin by factoring out -16.

$0 = -16(x - 3)(x + 3)$ Factor $x^2 - 9$ as a difference of squares.

$0 \not= -16$ or $0 = x - 3$ or $0 = x + 3$ Set each factor equal to zero.

$x = 3$ or $x = -3$ Solve for *x*.

A negative value of *x* corresponds to a time before the rock was dropped. Hence only the positive value of *x* makes sense in the context of this problem. The *x*-intercept is (3, 0) and means that 3 seconds after the rock was dropped, the rock had a height of 0 ft (the rock hit the ground). See Figure 4-10.

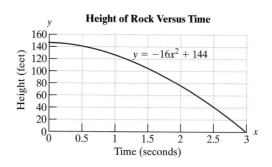

Figure 4-10

Calculator Connections

The *x*- and *y*-intercepts of the equation $2x - 3y = 6$ were found in Example 2a using algebraic methods. The *x*-intercept is $(3, 0)$ and the *y*-intercept is $(-2, 0)$. Students are urged to master the algebraic methods to study equations and their related graphs. However, a graphing calculator is an excellent verification tool to confirm the results.

To graph the equation $2x - 3y = 6$ on a calculator, it is necessary to write the equation with the *y*-variable isolated. Then, enter the equation into the calculator and graph the equation. For example:

$$2x - 3y = 6$$
$$-3y = -2x + 6$$
$$\frac{-3y}{-3} = \frac{-2x}{-3} + \frac{6}{-3}$$
$$y = \frac{2}{3}x - 2$$

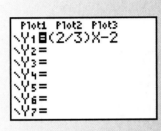

Tip: A "friendly" window on most calculators may give the exact coordinates of the *x*-intercept of this equation (Figure 4-11). A friendly window is set so that the coordinates of each pixel are terminating decimals. A friendly window can be defined by using a *ZDecimal* option or a comparable feature given in your user's manual.

Many calculators have a *Trace* feature that moves the cursor along the graph. The coordinates of the points along the graph are updated as the cursor moves.

To verify that the *x*-intercept is $(3, 0)$, move the cursor to the point where the line crosses the *x*-axis. Depending on the viewing window, the coordinates will be *close* to $x = 3$ and $y = 0$. Similarly, to verify that the *y*-intercept is $(0, -2)$, move the cursor to the point where the line crosses the *y*-axis. Depending on the viewing window, the coordinates will be *close* to $x = 0$ and $y = -2$.

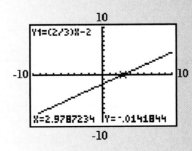

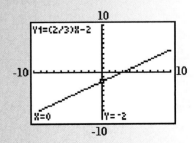

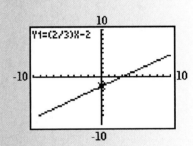

Figure 4-11

In Example 2b, we found the x- and y-intercepts of the equation $y = x^2 + 3x - 10$ algebraically. The x-intercepts are $(-5, 0)$ and $(2, 0)$, and the y-intercept is $(0, -10)$. A graphing calculator can be used to confirm these results.

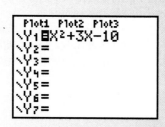
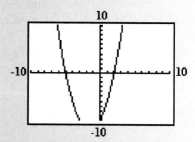

The x-intercepts appear to be at $(-5, 0)$ and $(2, 0)$ as expected. The y-intercept is difficult to see in the standard viewing window. Therefore, the viewing window must be set to accommodate values of y less than -10. Use a *Window* or *Range* option (see your user's manual) to change the minimum and maximum x- and y-values.

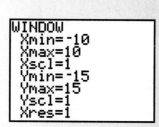
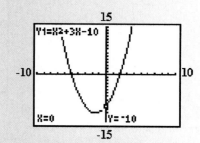

Calculator Exercises

For each of the following equations, use algebraic methods to find the x- and y-intercepts. Then use a graphing calculator to verify your results.

1. $y = 2x - 4$ ❖
2. $y = -3x + 6$ ❖
3. $3x + 4y = 6$ ❖
4. $-2x + 5y = -4$ ❖
5. $y = x^2 + 6x - 7$ ❖
6. $y = x^2 - 1$ ❖
7. $y = 2x^2 - 2x - 12$ ❖
8. $y = x^2 - 7x - 18$ ❖
9. $y = x - 15$ ❖

---

section 4.7   PRACTICE EXERCISES

1. Factor.   4.2, 4.3
    $2x^2 - 2x - 144$
    $2(x - 9)(x + 8)$

2. Factor.   4.4
    $t^2 - 9$
    $(t - 3)(t + 3)$

3. Solve.   4.6
    $x = 9, x = -8$
    $x(2x - 2) = 144$

4. Solve.   4.6
    $t = 3, t = -3$
    $2t^2 = 18$

5. A parallelogram has a base that is 2 ft less than twice its height. The area is 144 ft². Find the base and height of the parallelogram.   4.6
    The base is 16 ft, and the height is 9 ft.

6. A rock is dropped off of a bridge 144 ft high. The height of the rock is given by the equation

$$h = -16t^2 + 144$$ where h is the rock's height in feet and t is the time in seconds $(t \geq 0)$.

Find the time required for the rock to hit the water.   <u>4.6</u>   3 s

 7. State the definition of an x-intercept.   An x-intercept is a point $(a, 0)$ where a graph intersects the x-axis.

 8. Given a two-variable equation in x and y, explain how to find an x-intercept. Substitute $y = 0$, and solve for x.

 9. Given a two-variable equation in x and y, explain how to find a y-intercept. Substitute $x = 0$, and solve for y.

 10. State the definition of a y-intercept.   A y-intercept is a point $(0, b)$ where a graph intersects the y-axis.

For Exercises 11–16, identify the x- and y-intercepts from the graphs.

11. 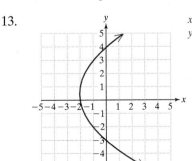 x-intercept: $(0, 0)$; y-intercepts: $(0, 0)(0, -3)$

12. 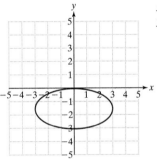 x-intercept: none; y-intercept: $(0, 4)$

13. 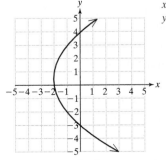 x-intercept: $(-2, 0)$; y-intercepts: $(0, 4)(0, -3)$

14. 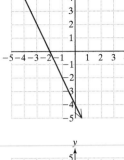 x-intercept: $(-2, 0)$; y-intercept: $(0, -4)$

 15. 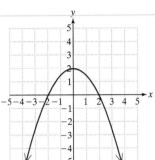 x-intercepts: $(2, 0)(-2, 0)$; y-intercept: $(0, 2)$

16. 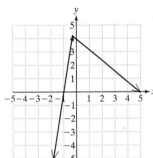 x-intercepts: $(-1, 0)(5, 0)$; y-intercept: $(0, 4)$

For Exercises 17–30, find the x- and y-intercepts.

 17. $2x - 4y = 8$
x-intercept: $(4, 0)$; y-intercept: $(0, -2)$

18. $6x + 3y = 12$
x-intercept: $(2, 0)$; y-intercept: $(0, 4)$

19. $y = 5x + 10$
x-intercept: $(-2, 0)$; y-intercept: $(0, 10)$

20. $y = 4x - 12$
x-intercept: $(3, 0)$; y-intercept: $(0, -12)$

21. $y = x^2 - 16$ x-intercepts: $(4, 0)(-4, 0)$; y-intercept: $(0, -16)$

22. $y = x^2 - 25$ x-intercepts: $(5, 0)(-5, 0)$; y-intercept: $(0, -25)$

 23. $y = x^2 - 7x + 10$
x-intercepts: $(5, 0)(2, 0)$; y-intercept: $(0, 10)$

24. $y = x^2 + 7x + 6$ ❖

25. $y = x^2 + 4x + 4$
x-intercept: $(-2, 0)$; y-intercept: $(0, 4)$

26. $y = x^2 - 6x + 9$
x-intercept: $(3, 0)$; y-intercept: $(0, 9)$

27. $y = x^3 + 2x^2 - 15x$ ❖

28. $y = x^3 - 10x^2 + 16x$ ❖

29. $y = x^3 - x^2 - 16x + 16$ x-intercepts: $(4, 0)(-4, 0)(1, 0)$; y-intercept: $(0, 16)$

30. $y = x^3 + 3x^2 - 4x - 12$ x-intercepts: $(2, 0)(-2, 0)(-3, 0)$; y-intercept: $(0, -12)$

❖ See Additional Answers Appendix   Writing   Translating Expression   Geometry   Scientific Calculator    Video

31. A child throws a ball straight up from ground level. The height of the ball, y, is given by

$$y = -16x^2 + 32x$$ where y is given in feet and x is the time in seconds ($x \geq 0$).

A graph of the equation is shown.

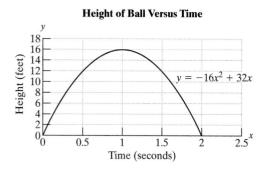

Height of Ball Versus Time

$y = -16x^2 + 32x$

Figure for Exercise 31

a. Use the equation to find the y-intercept. Verify your answer from the graph. (0, 0)

b. What does the y-intercept mean in the context of this problem? 0 s after release, the ball is at ground level.

c. Use the equation to find the x-intercepts. (*Hint*: Substitute $y = 0$, and solve the equation for x.) Verify your answer from the graph. (0, 0)(2, 0)

d. What do the x-intercepts mean in the context of this problem? At 0 s and 2 s after release, the ball is at ground level.

32. A toy rocket is shot upward from a 128-ft platform. The height of the rocket, y, is given by the equation

$$y = -16x^2 + 112x + 128$$ where y is measured in feet, and x is the time in seconds ($x \geq 0$).

A graph of the equation is shown.

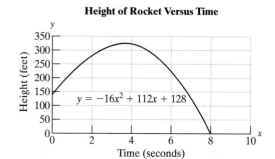

Height of Rocket Versus Time

$y = -16x^2 + 112x + 128$

Figure for Exercise 32

 ❖ See Additional Answers Appendix

a. Use the equation to find the y-intercept. Verify your answer from the graph. (0, 128)

b. What does the y-intercept mean in the context of this problem? 0 s after launch the rocket's height is 128 ft.

c. Use the equation to find the x-intercept. (*Hint*: Substitute $y = 0$, and solve the equation for x.) (8, 0)

d. What does the x-intercept mean in the context of this problem? 8 s after launch, the rocket is at ground level.

33. The guarantee period for a car air-conditioning unit is 3 years. If the air-conditioner does not last the entire 3 years, the dealer will pay the customer a refund for a portion of the loss. The amount, y, that the dealer will refund to the customer is given by the equation

$$y = 1500 - 500x \quad 0 \leq x \leq 3$$ where y is in dollars, and x is the time in years that the air-conditioner lasts.

The graph of the equation is shown.

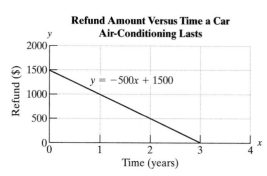

Refund Amount Versus Time a Car Air-Conditioning Lasts

$y = -500x + 1500$

Figure for Exercise 33

a. If an air-conditioner lasts for only 1 year, how much will the dealer refund to the customer? Verify your answer from the graph. ❖

b. If the air-conditioner lasts for 2 years, how much will the dealer refund to the customer? ❖

c. Find the y-intercept of this equation and interpret the meaning of the y-intercept in the context of this problem. ❖

d. Find the x-intercept of this equation and interpret the meaning of the x-intercept in the context of this problem. ❖

Writing   Translating Expression   Geometry   Scientific Calculator   Video

34. The guarantee period for a stereo system is 2 years. If the stereo does not last the entire 2 years, the store will pay the customer a refund for a portion of the loss. The amount, *y*, that the store will refund to the customer is given by the equation

$$y = -400x + 800 \quad 0 \le x \le 2$$

where *y* is in dollars, and *x* is the time in years that the stereo lasts.

The graph of the equation is shown.

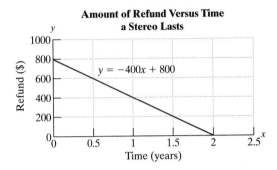

Amount of Refund Versus Time a Stereo Lasts

$y = -400x + 800$

Figure for Exercise 34

a. If a stereo lasts for only 1 year, how much will the store refund to the customer? Verify your answer from the graph. ❖

b. If the stereo lasts for $1\frac{1}{2}$ years, how much will the store refund to the customer? ❖

c. Find the *y*-intercept of this equation and interpret the meaning of the *y*-intercept in the context of this problem. ❖

d. Find the *x*-intercept of this equation and interpret the meaning of the *x*-intercept in the context of this problem. ❖

35. A stadium holds 75,000 seats. If a concert promoter charges *x* dollars per ticket for a concert, then the total number of tickets sold, *N*, can be expressed as

$$N = 75{,}000 - 500x \quad (x \ge 0)$$

where *x* is the price per ticket.

The graph of the equation is shown. The variable *N* is used in place of *y*.

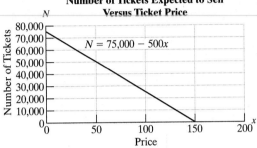

Number of Tickets Expected to Sell Versus Ticket Price

$N = 75{,}000 - 500x$

Figure for Exercise 35

a. If the ticket price is $20, how many tickets can the promoter expect to sell? Verify your answer from the graph.   65,000 tickets

b. If the ticket price is $50, how many tickets can the promoter expect to sell?   50,000 tickets

c. Find the *N*-intercept of this equation and interpret the meaning of the *N*-intercept in the context of this problem.   (0, 75000) If tickets cost $0 (free), the promoter will sell 75,000 tickets.

d. Find the *x*-intercept of this equation and interpret the meaning of the *x*-intercept in the context of this problem.   (150, 0) If tickets cost $150, the promoter will sell 0 tickets.

36. A certain sports arena holds 20,000 seats. If a sports franchise charges *x* dollars per ticket for a game, then the total number of tickets sold, *N*, can be approximated by

$$N = 20{,}000 - 400x \quad (x \ge 0)$$

where *x* is the price per ticket.

The graph of the equation is shown. The variable N is used in place of y.

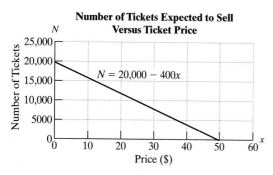

Number of Tickets Expected to Sell Versus Ticket Price

$$N = 20{,}000 - 400x$$

Figure for Exercise 36

a. If the ticket price is \$10, how many tickets can the promoter expect to sell? Verify your answer from the graph.   16,000 tickets

b. If the ticket price is \$40, how many tickets can the promoter expect to sell?   4000 tickets

c. Find the N-intercept of this equation and interpret the meaning of the N-intercept in the context of this problem.   (0, 20000) If tickets cost \$0 (free), the promoter will "sell" 20,000 tickets.

d. Find the x-intercept of this equation and interpret the meaning of the x-intercept in the context of this problem.   (50, 0) If tickets cost \$50, the promoter will sell 0 tickets.

EXPANDING YOUR SKILLS

37. Draw a graph of a line with x-intercept $(-3, 0)$. (Answers may vary) ❖

38. Draw a graph of a line with x-intercept $(4, 0)$. (Answers may vary) ❖

39. Draw a graph of a line with y-intercept $(0, -4)$. (Answers may vary) ❖

40. Draw a graph of a line with y-intercept $(0, 0)$. (Answers may vary) ❖

---

chapter 4   SUMMARY

SECTION 4.1—GREATEST COMMON FACTOR AND FACTORING BY GROUPING

KEY CONCEPTS:

The greatest common factor (GCF) is the greatest factor common to all terms of a polynomial. To factor out the GCF from a polynomial, use the distributive property.

A four-term polynomial may be factorable by grouping.

Steps to Factoring by Grouping

1. Identify and factor out the GCF from all four terms.
2. Factor out the GCF from the first pair of terms. Factor out the GCF or its opposite from the second pair of terms.
3. If the two terms share a common binomial factor, factor out the binomial factor.

EXAMPLES:

Factor out the GCF:

$$3x^2(a + b) - 6x(a + b)$$
$$= 3x(a + b)x - 3x(a + b)(2)$$
$$= 3x(a + b)(x - 2)$$

Factor by Grouping:

$$60xa - 30xb - 80ya + 40yb$$
$$= 10[6xa - 3xb - 8ya + 4yb]$$
$$= 10[3x(2a - b) - 4y(2a - b)]$$

$$= 10[(2a - b)(3x - 4y)]$$

KEY TERMS:

factoring   greatest common factor
factoring by grouping   (GCF)
                              prime factorization

SECTION 4.2—FACTORING TRINOMIALS: GROUPING METHOD

KEY CONCEPTS:

Grouping Method for Factoring Trinomials of the Form $ax^2 + bx + c$ **(where $a \neq 0$)**

1. Factor out the GCF from all terms.
2. Find the product ac.
3. Find two integers whose product is ac and whose sum is b. (If no pair of integers can be found, then the trinomial is prime.)
4. Rewrite the middle term (bx) as the sum of two terms whose coefficients are the numbers found in Step 3.
5. Factor the polynomial by grouping.

KEY TERMS:

factoring trinomials by grouping
prime polynomial

EXAMPLES:

Factor:

$$10y^2 + 35y - 20$$
$$= 5(2y^2 + 7y - 4)$$
$$ac = (2)(-4) = -8$$

Find two integers whose product is -8 and whose sum is 7. The numbers are 8 and -1.

$$5[2y^2 + 8y - 1y - 4]$$
$$= 5[2y(y + 4) - 1(y + 4)]$$
$$= 5(y + 4)(2y - 1)$$

SECTION 4.3—FACTORING TRINOMIALS: TRIAL-AND-ERROR METHOD

KEY CONCEPTS:

Trial-and-Error Method for Factoring Trinomials in the Form $ax^2 + bx + c$

1. Factor out the GCF from all terms.
2. List the pairs of factors of a and the pairs of factors of c. Consider the reverse order in either list.
3. Construct two binomials of the form

4. Test each combination of factors and signs until the product forms the correct trinomial.
5. If no combination of factors produces the correct product, then the trinomial is prime.

EXAMPLES:

Factor:

$$10y^2 + 35y - 20$$
$$= 5(2y^2 + 7y - 4)$$

The pairs of factors of 2 are: $2 \cdot 1$
The pairs of factors of -4 are:

$$-1 \cdot 4 \qquad 1(-4)$$
$$-2 \cdot 2 \qquad 2(-2)$$
$$-4 \cdot 1 \qquad 4(-1)$$

$(2y - 1)(y + 4) = 2y^2 + 7y - 4$	Yes
$(2y - 2)(y + 2) = 2y^2 + 2y - 4$	No
$(2y - 4)(y + 1) = 2y^2 - 2y - 4$	No
$(2y + 1)(y - 4) = 2y^2 - 7y - 4$	No

KEY TERMS:

factoring trinomials by the trial-and-error method

$(2y + 2)(y - 2) = 2y^2 - 2y - 4$ No

$(2y + 4)(y - 1) = 2y^2 + 2y - 4$ No

SECTION 4.4—FACTORING PERFECT SQUARE TRINOMIALS AND THE DIFFERENCE OF SQUARES

KEY CONCEPTS:

The factored form of a perfect square trinomial is the square of a binomial:

$$a^2 + 2ab + b^2 = (a + b)^2$$
$$a^2 - 2ab + b^2 = (a - b)^2$$

Difference of Squares

$$a^2 - b^2 = (a + b)(a - b)$$

KEY TERMS:

factoring a difference of squares
factoring a perfect square trinomial

EXAMPLES:

Factor:

$$9w^2 - 30wz + 25z^2$$
$$= (3w)^2 - 2(3w)(5z) + (5z)^2$$

$$= (3w - 5z)^2$$

Factor:

$$25z^2 - 4y^2$$
$$= (5z + 2y)(5z - 2y)$$

SECTION 4.5—FACTORING THE SUM AND DIFFERENCE OF CUBES AND GENERAL FACTORING SUMMARY

KEY CONCEPTS:

Factoring a Difference of Cubes

$$a^3 - b^3 = (a - b)(a^2 + ab + b^2)$$

Factoring a Sum of Cubes

$$a^3 + b^3 = (a + b)(a^2 - ab + b^2)$$

KEY TERMS:

difference of cubes
factoring strategy
sum of cubes

EXAMPLES:

Factor:

$$m^3 - 64$$
$$= (m)^3 - (4)^3$$

This is a difference of cubes: $a = m$ and $b = 4$. Apply the formula:

$$m^3 - 64 = (m)^3 - (4)^3$$
$$= (m - 4)(m^2 + 4m + 16)$$

Factor:

$$x^6 + 8y^3$$
$$= (x^2)^3 + (2y)^3$$

This is a sum of cubes: $a = x^2$ and $b = 2y$. Apply the formula:

$$x^6 + 8y^3 = (x^2)^3 + (2y)^3$$
$$= (x^2 + 2y)(x^4 - 2x^2y + 4y^2)$$

SECTION 4.6—ZERO PRODUCT RULE

KEY CONCEPTS:

An equation of the form: $ax^2 + bx + c = 0$, where $a \neq 0$ is a quadratic equation.

The zero product rule states that if $ab = 0$, then either $a = 0$ or $b = 0$. The zero product rule can be used to solve a quadratic equation or a higher degree polynomial equation that is factored and set to zero.

KEY TERMS:

Pythagorean theorem
quadratic equation
steps for solving quadratic equations by factoring
zero product rule

EXAMPLES:

The equation $2x^2 - 17x + 30 = 0$ is a *quadratic equation*.

Solve:

$$3w(w - 4)(2w + 1) = 0$$

$3w = 0$ or $w - 4 = 0$ or $2w + 1 = 0$

$w = 0$ or $w = 4$ or $w = -\dfrac{1}{2}$

Solve: $\quad 4x^2 = 34x - 60$

$4x^2 - 34x + 60 = 0$
$2(2x^2 - 17x + 30) = 0$
$2(2x - 5)(x - 6) = 0$

$\cancel{2} \neq 0$ or $2x - 5 = 0$ or $x - 6 = 0$

$\qquad\qquad\qquad x = \dfrac{5}{2}$ or $x = 6$

SECTION 4.7—CONNECTIONS TO GRAPHING: x- AND y-INTERCEPTS

KEY CONCEPTS:

An x-intercept of an equation is a point $(a, 0)$ where the graph intersects the x-axis.

A y-intercept of an equation is a point $(0, b)$ where the graph intersects the y-axis.

EXAMPLES:

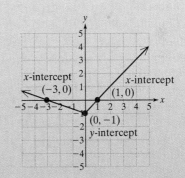

Finding x- and y-Intercepts from an Equation

1. To find an x-intercept from an equation, substitute $y = 0$ and solve for x.

Find the x- and y-intercepts of $3x - y = 3$:

1. To find the x-intercept, substitute $y = 0$

$$3x - y = 3$$

$$3x - (0) = 3$$

$$3x = 3$$

$$x = 1 \qquad \text{The } x\text{-intercept is } (1, 0)$$

2. To find a y-intercept from an equation, substitute $x = 0$ and solve for y.

KEY TERMS:

x-intercept
y-intercept

2. To find the y-intercept, substitute $x = 0$

$$3x - y = 3$$

$$3(0) - y = 3$$

$$0 - y = 3$$

$$-y = 3$$

$$y = -3 \qquad \text{The } y\text{-intercept is } (0, -3)$$

---

chapter 4   REVIEW EXERCISES

Section 4.1

For Exercises 1–6, identify the greatest common factor between each pair of terms.

1. $24, 18$ 6

2. $16x^2, 20x^3$ $4x^2$

3. $15a^2b^4, 22ab^5$ ab^4

4. $3(x + 5), x(x + 5)$ $x + 5$

5. $2c^3(3c - 5), 4c(3c - 5)$
   $2c(3c - 5)$

6. $-2wyz, -4xyz$
   $-2yz$ or $2yz$

For Exercises 7–14, factor out the greatest common factor.

7. $6x^2 + 2x^3 - 8x$
   $2x(3x + x^2 - 4)$

8. $11w^3 - 44w^2$
   $11w^2(w - 4)$

9. $32y^2 - 48$
   $16(2y^2 - 3)$

10. $5a^3 + 9a^2 + 2a$
    $a(5a^2 + 9a + 2)$

11. $-t^2 + 5t$
    $t(-t + 5)$ or $-t(t - 5)$

12. $-6u^2 - u$
    $u(-6u - 1)$ or $-u(6u + 1)$

13. $3b(b + 2) - 7(b + 2)$ $(b + 2)(3b - 7)$

14. $2(5x + 9) + 8x(5x + 9)$ $2(5x + 9)(1 + 4x)$

For Exercises 15–20, factor by grouping.

15. $7w^2 + 14w + wb + 2b$ $(w + 2)(7w + b)$

16. $b^2 - 2b + 5b - 10$ $(b - 2)(b + 5)$

17. $x^2 - 6x - 4x + 24$ $(x - 6)(x - 4)$

18. $18p^2 + 12pq - 3p - 2q$ $(3p + 2q)(6p - 1)$

19. $60y^2 - 45y - 12y + 9$ $3(4y - 3)(5y - 1)$

20. $6a - 3a^2 - 2ab + a^2b$ $a(2 - a)(3 - b)$

Section 4.2

For Exercises 21–26, find a pair of integers whose product and sum are given.

21. Product: -6 sum: -5 $-6, 1$

22. Product: 12 sum: 13 $12, 1$

23. Product: 24 sum: 11 $8, 3$

24. Product: -60 sum: 17 $20, -3$

25. Product: -5 sum: 4 $5, -1$

26. Product: 15 sum: -8 $-3, -5$

For Exercises 27–40, factor the trinomial using the grouping method.

27. $3c^2 - 5c - 2$
    $(c - 2)(3c + 1)$

28. $4y^2 + 13y + 3$
    $(y + 3)(4y + 1)$

29. $2t^2 + 11st + 12s^2$ $(t + 4s)(2t + 3s)$

30. $4x^3 + 17x^2 - 15x$ $x(x + 5)(4x - 3)$

31. $w^3 + 4w^2 - 5w$
    $w(w + 5)(w - 1)$

32. $p^2 - 8pq + 15q^2$
    $(p - 3q)(p - 5q)$

33. $40v^2 + 22v - 6$ $2(4v + 3)(5v - 1)$

34. $40s^2 + 30s - 100$
    $10(4s - 5)(s + 2)$

35. $x^2 + 9x - 22$
    $(x - 2)(x + 11)$

36. $y^2 - 9y + 8$ $(y - 8)(y - 1)$

37. $a^3b - 10a^2b^2 + 24ab^3$
    $ab(a - 6b)(a - 4b)$

38. $2z^6 + 8z^5 - 42z^4$
    $2z^4(z + 7)(z - 3)$

39. $3m + 9m^2 - 2$
    $(3m - 1)(3m + 2)$

40. $10 + 6p^2 + 19p$
    $(3p + 2)(2p + 5)$

Section 4.3

For Exercises 41–44, let a, b, and c represent positive integers.

41. When factoring a polynomial of the form $ax^2 - bx - c$, should the signs of the binomials be both positive, both negative, or different?
Different

42. When factoring a polynomial of the form $ax^2 - bx + c$, should the signs of the binomials be both positive, both negative, or different?
Both negative

43. When factoring a polynomial of the form $ax^2 + bx + c$, should the signs of the binomials be both positive, both negative, or different?
Both positive

44. When factoring a polynomial of the form $ax^2 + bx - c$, should the signs of the binomials be both positive, both negative, or different?
Different

For Exercises 45–58, factor the trinomial using the trial-and-error method.

45. $2y^2 - 5y - 12$
$(2y + 3)(y - 4)$

46. $4w^2 - 5w - 6$
$(4w + 3)(w - 2)$

47. $2p^2 - 4p - 48$
$2(p - 6)(p + 4)$

48. $3c^2 + 18c - 21$
$3(c + 7)(c - 1)$

49. $10z^2 + 29z + 10$
$(2z + 5)(5z + 2)$

50. $8z^2 + 6z - 9$
$(4z - 3)(2z + 3)$

51. $2p^2 - 5p + 1$
Prime

52. $5r^2 - 3r + 7$
Prime

53. $10w^2 - 60w - 270$
$10(w - 9)(w + 3)$

54. $3y^2 - 18y - 48$
$3(y - 8)(y + 2)$

55. $9c^2 - 30cd + 25d^2$ $(3c - 5d)^2$

56. $121m^2 + 154mn + 49n^2$ $(11m + 7n)^2$

57. $v^4 - 2v^2 - 3$
$(v^2 + 1)(v^2 - 3)$

58. $x^4 + 7x^2 + 10$
$(x^2 + 5)(x^2 + 2)$

Section 4.4

For Exercises 59–64, determine if the trinomial is a perfect square trinomial. If it is, factor the trinomial. If the trinomial is not a perfect square trinomial, explain why.

59. $4x^2 - 20x + 25$
$(2x - 5)^2$

60. $y^2 + 12y + 36$
$(y + 6)^2$

61. $c^2 - 6c + 9$
$(c - 3)^2$

62. $9b^2 + 6b + 1$
$(3b + 1)^2$

63. $t^2 + 8t + 49$ ❖

64. $k^2 - 10k + 64$ ❖

For Exercises 65–72, determine if the binomial is a difference of two squares. If it is, factor the binomial. If the binomial is not a difference of squares, explain why.

65. $a^2 - 49$
$(a - 7)(a + 7)$

66. $d^2 - 64$
$(d - 8)(d + 8)$

67. $h - 25$ Not a difference of squares. h is not a perfect square.

68. $c - 9$ Not a difference of squares. c is not a perfect square.

69. $100 - 81t^2$
$(10 - 9t)(10 + 9t)$

70. $4 - 25k^2$
$(2 - 5k)(2 + 5k)$

71. $x^2 + 16$
This is a sum of squares.

72. $y^2 + 121$
This is a sum of squares.

For Exercises 73–78, factor completely.

73. $2c^4 - 18$
$2(c^2 - 3)(c^2 + 3)$

74. $72x^2 - 2y^2$
$2(6x - y)(6x + y)$

75. $8x^2 + 24x + 18$
$2(2x + 3)^2$

76. $48t^2 - 24t + 3$
$3(4t - 1)^2$

77. $p^3 + 3p^2 - 16p - 48$ $(p + 3)(p - 4)(p + 4)$

78. $4k - 8 - k^3 + 2k^2$
$(k - 2)(2 - k)(2 + k)$ or $-1(k - 2)^2(2 + k)$

Section 4.5

79. Write the formula for factoring the sum of cubes: $a^3 + b^3$ $(a + b)(a^2 - ab + b^2)$

80. Write the formula for factoring the difference of cubes: $a^3 - b^3$ $(a - b)(a^2 + ab + b^2)$

For Exercises 81–88, factor the sums and differences of cubes.

81. $z^3 - w^3$
$(z - w)(z^2 + zw + w^2)$

82. $r^3 + s^3$
$(r + s)(r^2 - rs + s^2)$

83. $64 + a^3$
$(4 + a)(16 - 4a + a^2)$

84. $125 - b^3$
$(5 - b)(25 + 5b + b^2)$

85. $p^6 + 8$
$(p^2 + 2)(p^4 - 2p^2 + 4)$

86. $q^6 - \dfrac{1}{27}$
$\left(q^2 - \dfrac{1}{3}\right)\left(q^4 + \dfrac{1}{3}q^2 + \dfrac{1}{9}\right)$

87. $6x^3 - 48$
$6(x - 2)(x^2 + 2x + 4)$

88. $7y^3 + 7$
$7(y + 1)(y^2 - y + 1)$

Match the polynomials in Exercises 89–91 with its factored form.

89. $a^2 - b^2$ x.

90. $a^3 - b^3$ v.

91. $a^3 + b^3$ vi.

i. $(a + b)^2$
ii. $(a - b)^3$
iii. $(a - b)(a^2 + b^2)$
iv. $(a + b)(a^2 + b^2)$
v. $(a - b)(a^2 + ab + b^2)$
vi. $(a + b)(a^2 - ab + b^2)$
vii. $(a + b)^3$
viii. $(a - b)(a^2 - ab + b^2)$
ix. $(a + b)(a^2 + ab + b^2)$
x. $(a - b)(a + b)$

For Exercises 92–97, factor the binomials completely.

92. $y^2 - 81$
$(y - 9)(y + 9)$

93. $216w^3 - 1$
$(6w - 1)(36w^2 + 6w + 1)$

94. $4a^2 + b^2$
Prime

95. $128 + 2v^6$
$2(4 + v^2)(16 - 4v^2 + v^4)$

96. $p^6 + 1$ $(p^2 + 1)(p^4 - p^2 + 1)$

97. $q^6 - 1$ $(q - 1)(q^2 + q + 1)$
$(q + 1)(q^2 - q + 1)$

❖ See Additional Answers Appendix   ✎ Writing   ⬌ Translating Expression   Geometry   Scientific Calculator   Video

For Exercises 98–107, factor completely.

98. $6y^2 - 11y - 2$
$(6y + 1)(y - 2)$

99. $3p^2 - 6p + 3$
$3(p - 1)^2$

100. $x^3 - 36x$
$x(x - 6)(x + 6)$

101. $k^2 - 13k + 42$
$(k - 7)(k - 6)$

102. $7ac - 14ad - bc + 2bd$ $(c - 2d)(7a - b)$

103. $q^4 - 64q$
$q(q - 4)(q^2 + 4q + 16)$

104. $8h^2 + 20$
$4(2h^2 + 5)$

105. $2t^2 + t + 3$
Prime

106. $m^2 - 8m$
$m(m - 8)$

107. $x^3 + 4x^2 - x - 4$ $(x + 4)(x + 1)(x - 1)$

Section 4.6

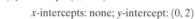 108. For which of the following equations can the zero product rule be applied directly? Explain.

$(x - 3)(2x + 1) = 0$ or $(x - 3)(2x + 1) = 6$

$(x - 3)(2x + 1) = 0$ can be solved by the zero product rule because it is a product of factors set equal to zero.

For Exercises 109–122, solve the equation using the zero product rule.

109. $(4x - 1)(3x + 2) = 0$ $x = \dfrac{1}{4}, x = -\dfrac{2}{3}$

110. $(a - 9)(2a - 1) = 0$ $a = 9, a = \dfrac{1}{2}$

111. $3w(w + 3)(5w + 2) = 0$ $w = 0, w = -3, w = -\dfrac{2}{5}$

112. $6u(u - 7)(4u - 9) = 0$ $u = 0, u = 7, u = \dfrac{9}{4}$

113. $7k^2 - 9k - 10 = 0$ $k = -\dfrac{5}{7}, k = 2$

114. $4h^2 - 23h - 6 = 0$ $h = -\dfrac{1}{4}, h = 6$

115. $q^2 - 144 = 0$
$q = 12, q = -12$

116. $r^2 = 25$
$r = 5, r = -5$

117. $5v^2 - v = 0$ $v = 0, v = \dfrac{1}{5}$

118. $x(x - 6) = -8$
$x = 4, x = 2$

119. $36t^2 + 60t = -25$ $t = -\dfrac{5}{6}$

120. $9s^2 + 12s = -4$ $s = -\dfrac{2}{3}$

121. $3(y^2 + 4) = 20y$ $y = \dfrac{2}{3}, y = 6$

122. $2(p^2 - 66) = -13p$ $p = \dfrac{11}{2}, p = -12$

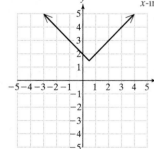 123. The base of a parallelogram is 1 ft more than twice the height. If the area is 78 ft^2 what are the base and height of the parallelogram?
The height is 6 ft, and the base is 13 ft.

124. A ball is tossed into the air from ground level with initial speed of 16 ft/s. The height of the ball is given by the equation

$$h = -16x^2 + 16x \quad (x \geq 0)$$ where h is the ball's height in feet and x is the time in seconds

Find the time(s) when the ball is at ground level.
The ball is at ground level at 0 and 1 s.

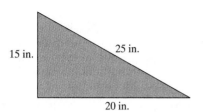 125. Using the Pythagorean theorem determine whether the triangle could be a right triangle.
Yes

Figure for Exercise 125

126. A right triangle has one leg that is 2 ft more than the other leg. The hypotenuse is 2 ft less than twice the shorter leg. Find the length of all sides of the triangle.
The legs are 6 ft and 8 ft; the hypotenuse is 10 ft.

127. If the square of a number is subtracted from 60, the result is -4. Find all such numbers.
The numbers are -8 and 8.

128. The product of two consecutive integers is 44 more than 14 times their sum.
The numbers are 29 and 30, or -2 and -1.

129. The base of a triangle is 1 m more than twice the height. If the area of the triangle is 18 m^2, find the base and height.
The height is 4 m and the base is 9 m.

Section 4.7

For Exercises 130–137, find the x- and y-intercepts.

130. $5x + 2y = 15$
x-intercept: $(3, 0)$; y-intercept: $\left(0, \dfrac{15}{2}\right)$

131. $x - 4y = 8$
x-intercept: $(8, 0)$; y-intercept $(0, -2)$

132. $y = 3x - 2$
x-intercept: $\left(\dfrac{2}{3}, 0\right)$; y-intercept: $(0, -2)$

133. $y = -\dfrac{1}{2}x + 4$
x-intercept: $(8, 0)$; y-intercept: $(0, 4)$

134. $y = x^2 - 9$ x-intercepts: $(3, 0)(-3, 0)$; y-intercept: $(0, -9)$

135. $y = x^2 - x - 12$
x-intercepts: $(4, 0)(-3, 0)$; y-intercept: $(0, -12)$

136.
x-intercepts: none; y-intercept: $(0, 2)$

✎ Writing   ⟷ Translating Expression   Geometry   Scientific Calculator   Video

137.

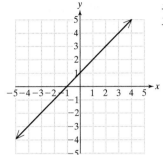

x-intercept: $(-1, 0)$;
y-intercept: $(0, 1)$

138.   Given $2x - 5y = 10$

   a.   Find the x- and y-intercepts.   ❖

   b.   Use the intercepts to graph the line.   ❖

139.   Given $4x - 8y = 0$

   a.   Find the x- and y-intercepts.   ❖

   b.   Use the intercepts to graph the line.   ❖

140.   Chocolate chip cookies are made for a large bake sale. The profit, y, (in dollars) depends on the number of cookies produced, x, according to the equation

$$y = -\frac{1}{1200}x(x - 840)$$

The graph of the equation is shown.

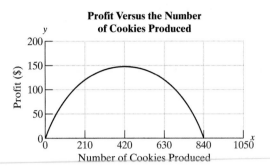

Profit Versus the Number of Cookies Produced

Figure for Exercise 140

   a.   Use the equation to find the x-intercepts. Verify your answer from the graph. $(0, 0)(840, 0)$

   b.   Interpret the x-intercepts in context of the problem.   When 0 cookies or 840 cookies are produced, the profit is $0.

---

chapter 4   TEST

1.   Factor out the GCF: $15x^4 - 3x + 6x^3$
   $3x(5x^3 - 1 + 2x^2)$

2.   Factor by grouping: $7a - 35 - a^2 + 5a$
   $(a - 5)(7 - a)$

3.   Factor the trinomial: $6w^2 - 43w + 7$
   $(6w - 1)(w - 7)$

4.   Factor the difference of squares: $169 - p^2$
   $(13 - p)(13 + p)$

5.   Factor the perfect square trinomial:
   $q^2 - 16q + 64$ $(q - 8)^2$

6.   Factor the sum of cubes: $8 + t^3$
   $(2 + t)(4 - 2t + t^2)$

For Exercises 7–16, factor completely.

7. $3a^2 + 27ab + 54b^2$ $3(a + 6b)(a + 3b)$

8. $c^4 - 1$ $(c - 1)(c + 1)(c^2 + 1)$

9. $xy - 7x + 3y - 21$ $(y - 7)(x + 3)$

10. $49 + p^2$ Prime

11. $-10u^2 + 30u - 20$ $-10(u - 2)(u - 1)$

12. $12t^2 - 75$ $3(2t - 5)(2t + 5)$

13. $5y^2 - 50y + 125$ $5(y - 5)^2$

❖ See Additional Answers Appendix   Writing   Translating Expression   Geometry   Scientific Calculator   Video

14. $21q^2 + 14q$ $7q(3q + 2)$

15. $2x^3 + x^2 - 8x - 4$ $(2x + 1)(x - 2)(x + 2)$

16. $y^3 - 125$ $(y - 5)(y^2 + 5y + 25)$

For Exercises 17–20, solve the equation.

17. $(2x - 3)(x + 5) = 0$ $x = \dfrac{3}{2}, x = -5$

18. $x^2 - 7x = 0$ $x = 0, x = 7$

19. $x^2 - 6x = 16$ $x = 8, x = -2$

20. $x(5x + 4) = 1$ $x = \dfrac{1}{5}, x = -1$

21. A tennis court has area of 312 yd². If the length is 2 yd more than twice the width, find the dimensions of the court. The tennis court is 12 yd by 26 yd.

22. The hypotenuse of a right triangle is 2 ft less than three times the shorter leg. The longer leg is 3 ft less than three times the shorter leg. Find the length of the shorter leg.   The shorter leg is 5 ft.

23. Find the x- and y-intercepts and match the equation with the graph.

   a. $2x - 3y = 18$ x-intercept: $(9, 0)$; y-intercept: $(0, -6)$ ii

   b. $y = x^2 - 4x - 5$
   x-intercepts: $(5, 0)(-1, 0)$; y-intercept: $(0, -5)$ i

   i.

   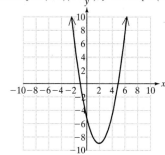

   ii.

   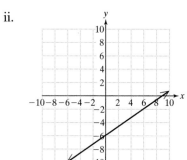

   Figure for Exercise 23

24. A baker sells loaves of sweet bread at a flea market. The profit, y, (in dollars) depends on the number of loaves produced, x, according to the equation,

$$y = -\frac{3}{260}x(x - 260)$$

The graph of the equation is shown here.

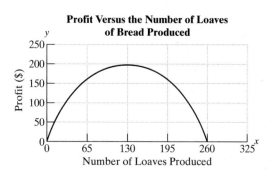

Figure for Exercise 24

a. Use the equation to find the x-intercepts. Verify your answer from the graph. $(0, 0)(260, 0)$

b. Interpret the x-intercepts in context of the problem.   The baker makes no profit if 0 loaves or 260 loaves of bread are produced.

c. Why do you think there are two values where the profit is zero? (Answers may vary.)   If the baker produces too many loaves, the baker floods the market and cannot sell them all.

CUMULATIVE REVIEW EXERCISES, CHAPTERS 1–4

For Exercises 1–2, simplify completely.

1. $\dfrac{|4 - 25 \div (-5) \cdot 2|}{\sqrt{8^2 + 6^2}}$ $\dfrac{7}{5}$

2. $-\dfrac{2}{3} - \dfrac{1}{3}(3^2 + \sqrt{81})$ $-\dfrac{20}{3}$

3. Solve for x: $-3.5 - 2.5x = 1.5(x - 3)$ $x = 0.25$

4. Solve for t: $5 - 2(t + 4) = 3t + 12$ $t = -3$

5. Solve for y: $3x - 2y = 8$ $\dfrac{8 - 3x}{-2}$ or $\dfrac{3x - 8}{2}$

6. The circumference of a circular fountain is 50 ft. Find the radius of the fountain. Round to the nearest tenth of a foot.   The radius is 8.0 ft.

7. A child's piggy bank has \$3.80 in quarters, dimes, and nickels. The number of nickels is two more than the number of quarters. The number of dimes is three less than the number of quarters. Find the number of each type of coin in the bank.   There are 10 quarters, 12 nickels, and 7 dimes.

8. Solve the inequality. Graph the solution and write the solution set in interval notation.   ❖

$$-\dfrac{5}{12}x \le \dfrac{5}{3}$$

For Exercises 9–11, perform the indicated operations.

9. $2\left(\dfrac{1}{3}y^3 - \dfrac{3}{2}y^2 - 7\right) - \left(\dfrac{2}{3}y^3 + \dfrac{1}{2}y^2 + 5y\right)$ $-\dfrac{7}{2}y^2 - 5y - 14$

10. $(4p^2 - 5p - 1)(2p - 3)$ $8p^3 - 22p^2 + 13p + 3$

11. $(2w - 7)^2$ $4w^2 - 28w + 49$

12. Divide using long division:
$(r^4 + 2r^3 - 5r + 1) \div (r - 3)$ $r^3 + 5r^2 + 15r + 40 + \dfrac{121}{r - 3}$

For Exercises 13–15, simplify the expressions. Write the final answer using positive exponents only.

13. $\dfrac{c^{12}c^{-5}}{c^3}$ c^4

14. $\left(\dfrac{2a^2b^{-3}}{c}\right)^{-2}$ $\dfrac{b^6c^2}{4a^4}$

15. $\left(\dfrac{1}{2}\right)^0 - \left(\dfrac{1}{4}\right)^{-2}$ -15

16. Divide. Write the final answer in scientific notation: $\dfrac{8.0 \times 10^{-3}}{5.0 \times 10^{-6}}$ 1.6×10^3

For Exercises 17–22, factor completely.

17. $w^4 - 16$ $(w - 2)(w + 2)(w^2 + 4)$

18. $2ax + 10bx - 3ya - 15yb$ $(a + 5b)(2x - 3y)$

19. $4a^2 - 12a + 9$ $(2a - 3)^2$

20. $4x^2 - 8x - 5$ $(2x - 5)(2x + 1)$

21. $y^3 - 27$ $(y - 3)(y^2 + 3y + 9)$

22. $p^6 + q^6$ $(p^2 + q^2)(p^4 - p^2q^2 + q^4)$

For Exercises 23–24, solve the equation.

23. $4x(2x - 1)(x + 5) = 0$ $x = 0, x = \dfrac{1}{2}, x = -5$

24. $x(x + 2) = 35$ $x = -7, x = 5$

25. Given the equation $y = -x^2 + 4$
    a. Is the equation linear or quadratic?   ❖
    b. Complete the table.
    c. Sketch the curve defined by the ordered pairs in the table.   ❖
    d. Identify the x-intercepts.   ❖
    e. Identify the y-intercept.   ❖

x	y
-3	-5
-2	0
-1	3
0	4
1	3
2	0
3	-5

Table for Exercise 25

26. A truck used in a business has an initial value of $24,000. The truck's value decreases over the years as shown in the graph.

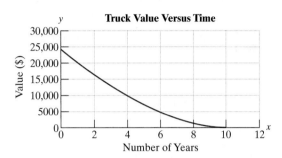

Truck Value Versus Time

Figure for Exercise 26

a. Use the graph to estimate the truck's value after 4 years. $10,000

b. Use the graph to estimate the truck's value after 6 years. $5000

c. What is the y-intercept and what does it mean in the context of this problem? (0, 24000)
The y-intercept represents the initial value of the truck, $24,000.

d. What is the x-intercept and what does it mean in the context of this problem?   (10, 0)
The x-intercept indicates that after 10 years the value of the truck is $0.

RATIONAL EXPRESSIONS

The equation $d = rt$ gives the fundamental relationship among the variables: distance, rate, and time. To find the average rate of travel, we can use the rational equation:

$$r = \frac{d}{t}$$

When making a round trip, the average rate for the round trip is given by

$$r = \frac{2d}{t_1 + t_2}$$

where the distance is doubled, and t_1 and t_2 are the times for the original and return flights, respectively.

When traveling by air, the variables t_1 and t_2 may be different for each flight due to variations in wind speed. Furthermore, when traveling east to west or west to east the difference in the arrival time and departure time does not necessarily give the total travel time. Accommodations must be made for differences in time zones. For more information visit citydist and timezones at

www.mhhe.com/miller_oneill

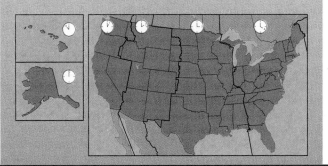

section

5.1 Introduction to Rational Expressions

1. Definition of a Rational Expression

In Section 1.1, we defined a rational number as the ratio of two integers, $\frac{p}{q}$, where $q \neq 0$.

$$\text{Examples of rational numbers:} \quad \frac{2}{3}, -\frac{1}{5}, 9$$

In a similar way, we define a **rational expression** as the ratio of two polynomials, $\frac{p}{q}$, where $q \neq 0$.

$$\text{Examples of rational expressions:} \quad \frac{3x-6}{x^2-4}, \quad \frac{3}{4}, \quad \frac{6r^5+2r}{7}, \quad x$$

2. Evaluating Rational Expressions

example 1

Evaluating Rational Expressions

Evaluate the rational expression (if possible) for the given values of x: $\dfrac{12}{x-3}$

a. $x = 0$ b. $x = 1$ c. $x = -3$ d. $x = 3$

Solution:

a. $\dfrac{12}{x-3}$

$\dfrac{12}{(0)-3}$ Substitute $x = 0$.

$= -4$

b. $\dfrac{12}{x-3}$

$\dfrac{12}{(1)-3}$ Substitute $x = 1$.

$= -6$

c. $\dfrac{12}{x-3}$

$\dfrac{12}{(-3)-3}$ Substitute $x = -3$.

$= \dfrac{12}{-6}$

$= -2$

d. $\dfrac{12}{x - 3}$

$\dfrac{12}{(3) - 3}$ Substitute $x = 3$.

$= \dfrac{12}{0}$ (undefined)   Recall that division by zero is undefined.

As with fractions, the denominator of a rational expression may not equal zero. If a value of the variable makes the denominator of a rational expression equal to zero, we say that the rational expression is undefined for that value of the variable. For example, the rational expression, $12/(x - 3)$ is undefined for $x = 3$.

3.   Domain of a Rational Expression

In Example 1(d), the expression $12/(x - 3)$ is undefined for $x = 3$. The fact that a rational expression may be defined for some values of the variable but not for others leads us to an important concept called the domain of an expression.

Informal Definition of the Domain of an Algebraic Expression

Given an algebraic expression, the **domain** of the expression is the set of real numbers that when substituted for the variable makes the expression result in a real number.

Note: For a rational expression, the domain is all real numbers except those that make the denominator zero.

According to this definition, the domain of the expression $12/(x - 3)$ is the set of real numbers *excluding* $x = 3$. This can be stated in set-builder notation as

$$\{x \,|\, x \neq 3\}$$

To find the domain of a rational expression we must identify and exclude the values of the variable that make the denominator zero.

Steps to Find the Domain of a Rational Expression

1.   Set the denominator equal to zero and solve the resulting equation.
2.   The domain is the set of real numbers *excluding* the values found in Step 1.

example 2

Finding the Domain of Rational Expressions

Find the domain of the expressions. Write the answers in set-builder notation.

a. $\dfrac{y - 3}{2y + 7}$ b. $\dfrac{-5}{x}$ c. $\dfrac{a + 10}{a^2 - 25}$ d. $\dfrac{2x^3 + 5}{x^2 + 9}$

Solution:

a. $\dfrac{y - 3}{2y + 7}$

$2y + 7 = 0$ Set the denominator equal to zero.

$2y = -7$ Solve the equation.

$\dfrac{2y}{2} = \dfrac{-7}{2}$

$y = -\dfrac{7}{2}$ The domain is the set of real numbers except $-\frac{7}{2}$.

Domain: $\{y \mid y \neq -\frac{7}{2}\}$

b. $\dfrac{-5}{x}$

$x = 0$ Set the denominator equal to zero.

Domain: $\{x \mid x \neq 0\}$ The domain is the set of real numbers except 0.

c. $\dfrac{a + 10}{a^2 - 25}$

$a^2 - 25 = 0$ Set the denominator equal to zero. The equation is quadratic.

$(a - 5)(a + 5) = 0$ Factor the equation.

$a - 5 = 0$ or $a + 5 = 0$ Set each factor equal to zero.

$a = 5$ or $a = -5$ The domain is the set of real numbers except 5 and -5.

Domain: $\{a \mid a \neq 5 \text{ and } a \neq -5\}$

d. $\dfrac{2x^3 + 5}{x^2 + 9}$

The quantity x^2 cannot be negative for any real number, x, so the denominator $x^2 + 9$ cannot equal zero. Therefore, no numbers are excluded from the domain. The domain is the set of all real numbers.

4.   Reducing Rational Expressions

In many cases it is advantageous to reduce a fraction to lowest terms. The same is true for rational expressions.

The method for reducing rational expressions mirrors the process for reducing fractions. In each case, factor the numerator and denominator. Common factors in the numerator and denominator form a ratio of 1 and can be reduced.

Reducing a fraction: $\dfrac{21}{35} \xrightarrow{\text{factor}} \dfrac{3 \cdot \overset{1}{\cancel{7}}}{5 \cdot \cancel{7}} = \dfrac{3}{5} \cdot (1) = \dfrac{3}{5}$

Reducing a rational expression: $\dfrac{2x - 6}{x^2 - 9} \xrightarrow{\text{factor}} \dfrac{2(\overset{1}{\cancel{x - 3}})}{(x + 3)(\cancel{x - 3})} = \dfrac{2}{(x + 3)}(1) = \dfrac{2}{x + 3}$

Informally, to reduce a rational expression we reduce common factors whose ratio is 1. Formally, this is accomplished by applying the fundamental principle of rational expressions.

Fundamental Principle of Rational Expressions

Let p, q, and r represent polynomials. Then

$$\frac{pr}{qr} = \frac{p}{q} \text{ for } q \neq 0 \text{ and } r \neq 0$$

---

example 3

Reducing a Rational Expression

Given the expression $\dfrac{2p - 14}{p^2 - 49}$

a.   Factor the numerator and denominator.
b.. Determine the domain of the expression and write the domain in set-builder notation.
c.   Reduce the expression.

Solution:

Classroom Activity 5.1A

a. $\dfrac{2p - 14}{p^2 - 49}$ Factor out the GCF in the numerator.

$= \dfrac{2(p - 7)}{(p + 7)(p - 7)}$ Factor the denominator as a difference of squares.

⬢ **Avoiding Mistakes**

The domain of a rational expression is always determined *before* reducing the expression.

b. $(p + 7)(p - 7) = 0$ To find the domain restrictions, set the denominator equal to zero. The equation is quadratic.

$p + 7 = 0$ or $p - 7 = 0$ Set each factor equal to 0.

$p = -7$ or $p = 7$ The domain is all real numbers except -7 and 7.

Domain: $\{p \mid p \neq -7,\ p \neq 7\}$

c. $\dfrac{2(\cancel{p-7})}{(p+7)(\cancel{p-7})}$ Reduce common factors whose ratio is 1.

$= \dfrac{2}{p+7}$ (provided $p \neq 7$ and $p \neq -7$)

In Example 3, it is important to note that the expressions

$$\dfrac{2p-14}{p^2-49} \quad \text{and} \quad \dfrac{2}{p+7}$$

are equal for all values of p that make each expression a real number. Therefore,

$$\dfrac{2p-14}{p^2-49} = \dfrac{2}{p+7}$$

for all values of p except $p = 7$ and $p = -7$. (At $p = 7$ and $p = -7$, the original expression is undefined.) This is why the domain of an expression is always determined before the expression is reduced.

From this point forward, we will write statements of equality between two rational expressions under the assumption that they are equal for all values of the variable for which each expression is defined.

example 4   **Reducing Rational Expressions**

Reduce the rational expressions.

a. $\dfrac{18a^4}{9a^5}$ b. $\dfrac{2c-8}{10c^2-80c+160}$

Solution:

a. $\dfrac{18a^4}{9a^5}$

$= \dfrac{2 \cdot 3 \cdot 3 \cdot a \cdot a \cdot a \cdot a}{3 \cdot 3 \cdot a \cdot a \cdot a \cdot a \cdot a}$ Factor the numerator and denominator.

$= \dfrac{2 \cdot \overset{1}{(\cancel{3 \cdot 3 \cdot a \cdot a \cdot a \cdot a})}}{(\cancel{3 \cdot 3 \cdot a \cdot a \cdot a \cdot a}) \cdot a}$ Reduce common factors.

$= \dfrac{2}{a}$

Tip: To reduce rational expressions, we can also use the property of exponents for dividing expressions of the same base. Recall, for integers, m and n, and $a \neq 0$.

$$\frac{a^m}{a^n} = a^{m-n}$$

In Example 4(a), we could have subtracted exponents to simplify the expression

$$\frac{18a^4}{9a^5} = \frac{18}{9} \cdot \frac{a^4}{a^5} = 2a^{4-5} = 2a^{-1} = \frac{2}{a}$$

b. $\dfrac{2c - 8}{10c^2 - 80c + 160}$

$= \dfrac{2(c - 4)}{10(c^2 - 8c + 16)}$ Factor out the GCF.

$= \dfrac{2(c - 4)}{10(c - 4)^2}$ The denominator is a perfect square trinomial.

$= \dfrac{\overset{1}{\cancel{2}}(\overset{1}{\cancel{c - 4}})}{\cancel{2} \cdot 5(\cancel{c - 4})(c - 4)}$ Reduce common factors whose ratio is 1.

$= \dfrac{1}{5(c - 4)}$

◆ Avoiding Mistakes

The fundamental principle of rational expressions indicates that common factors in the numerator and denominator may be reduced.

$$\frac{pr}{qr} = \frac{p}{q} \cdot \frac{r}{r} = \frac{p}{q}(1) = \frac{p}{q}$$

This property is based on the identity property of multiplication, so reducing or canceling applies only to *factors* (remember that factors are multiplied). Therefore, terms that are added or subtracted cannot be reduced or canceled. For example:

$$\frac{3x}{3y} = \frac{\overset{1}{\cancel{3}} \cdot x}{\cancel{3} \cdot y} = \frac{x}{y} \qquad \text{however,}$$

↑
reduce
common factor

$$\frac{x + 3}{y + 3} \quad \text{cannot be reduced}$$

↑
cannot reduce
common terms

The objective of reducing a rational expression is to create an equivalent expression that is simpler to use. Consider the rational expression from Example 4(b) in its original form and in its reduced form. If we choose an arbitrary value of c from the domain of the original expression and substitute that value into each expression, we see that the reduced form is easier to evaluate. For example, substitute $c = 3$:

	Original Expression	**Simplified Expression**
	$\dfrac{2c - 8}{10c^2 - 80c + 160}$	$\dfrac{1}{5(c - 4)}$
Substitute $c = 3$	$= \dfrac{2(3) - 8}{10(3)^2 - 80(3) + 160}$	$= \dfrac{1}{5(3 - 4)}$
	$= \dfrac{6 - 8}{10(9) - 240 + 160}$	$= \dfrac{1}{5(-1)}$
	$= \dfrac{-2}{90 - 240 + 160}$	$= -\dfrac{1}{5}$
	$= \dfrac{-2}{10} \quad \text{or} \quad -\dfrac{1}{5}$	

5.   Reducing a Ratio of -1

When two factors are identical in the numerator and denominator, they form a ratio of 1 and can be reduced. Sometimes we encounter two factors that are opposites and form a ratio of -1. For example:

Reduced Form   **Details/Notes**

$\dfrac{-5}{5} = -1$ The ratio of a number and its opposite is -1.

$\dfrac{100}{-100} = -1$ The ratio of a number and its opposite is -1.

$\dfrac{x+7}{-x-7} = -1$ $\dfrac{x+7}{-x-7} = \dfrac{x+7}{-1(x+7)} = \dfrac{\cancel{x+7}^{1}}{-1(\cancel{x+7})} = \dfrac{1}{-1} = -1$

factor out -1

$\dfrac{2-x}{x-2} = -1$ $\dfrac{2-x}{x-2} = \dfrac{-1(-2+x)}{x-2} = \dfrac{-1(\cancel{x-2})}{\cancel{x-2}} = \dfrac{-1}{1} = -1$

Recognizing factors that are opposites is useful when reducing rational expressions.

---

example 5   **Reducing Rational Expressions**

Reduce the rational expressions.

a. $\dfrac{3c-3d}{d-c}$ b. $\dfrac{5-y}{y^2-25}$

Classroom Activity 5.1B   **Solution:**

a. $\dfrac{3c-3d}{d-c}$

$= \dfrac{3(c-d)}{d-c}$ Factor the numerator and denominator.

Notice that $(c-d)$ and $(d-c)$ are opposites and form a ratio of -1.

$= \dfrac{3(\cancel{c-d})^{-1}}{\cancel{d-c}}$ <u>Details:</u> $\dfrac{3(c-d)}{d-c} = \dfrac{3(c-d)}{-1(-d+c)} = \dfrac{3(c-d)}{-1(c-d)} = \dfrac{3}{-1} = -3$

$= 3(-1)$

$= -3$

b. $\dfrac{5 - y}{y^2 - 25}$

$= \dfrac{5 - y}{(y - 5)(y + 5)}$ Factor the numerator and denominator.

Notice that $5 - y$ and $y - 5$ are opposites and form a ratio of -1.

$= \dfrac{\overset{-1}{\cancel{5 - y}}}{\cancel{(y - 5)}(y + 5)}$ Details: $\dfrac{5 - y}{(y - 5)(y + 5)} = \dfrac{-1(-5 + y)}{(y - 5)(y + 5)}$

$= \dfrac{-1(y - 5)}{(y - 5)(y + 5)} = \dfrac{-1}{y + 5}$

$= \dfrac{-1}{y + 5}$ or $\dfrac{1}{-(y + 5)}$ or $-\dfrac{1}{y + 5}$

Tip: It is important to recognize that a rational expression may be written in several equivalent forms, particularly when a negative factor is present. For example, since two numbers with opposite signs form a negative quotient, the number $-\frac{3}{4}$ can be written as $\frac{-3}{4}$ or as $\frac{3}{-4}$. The $-$ sign can be written in the numerator, in the denominator, or out in front of the fraction.

For this reason, the expression in Example 5(b) can be written in a variety of forms.

$\dfrac{-1}{y + 5}$ or $\dfrac{1}{-(y + 5)}$ or $-\dfrac{1}{y + 5}$

---

section 5.1   PRACTICE EXERCISES

1. a. ^{p} What is a rational number?
    A number $\frac{p}{q}$, where p and q are integers and $q \neq 0$.
    b. ^{q} What is a rational expression?
    An expression $\frac{p}{q}$, where p and q are polynomials and $q \neq 0$.

2. ^{q} a. Write an example of a rational number. (Answers will vary.) For example: $\frac{2}{3}$

    b. Write an example of a rational expression. (Answers will vary.) For example: $\frac{3x^2 + 1}{2x + 5}$

For Exercises 3–10, substitute the given number into the expression and simplify (if possible).

3. $\dfrac{1}{x - 6}$ let $x = -2$ $-\frac{1}{8}$ 4. $\dfrac{1}{x + 1}$ let $x = 4$ $\frac{1}{5}$

5. $\dfrac{w - 10}{w + 6}$ let $w = 0$ $-\frac{5}{3}$ 6. $\dfrac{w - 4}{2w + 8}$ let $w = 0$ $-\frac{1}{2}$

7. $\dfrac{y - 8}{(y + 1)(2y - 1)}$ let $y = 8$ 0

8. $\dfrac{y + 3}{(3y + 2)(y - 9)}$ let $y = -3$ 0

9. $\dfrac{(a - 7)(a + 1)}{(a - 2)(a + 5)}$ let $a = 2$ Undefined

10. $\dfrac{(a + 4)(a + 1)}{(a - 4)(a - 1)}$ let $a = 1$ Undefined

❖ See Additional Answers Appendix   ✎ Writing   ⬌ Translating Expression   Geometry   Scientific Calculator   Video

11. A bicyclist rides 24 miles against a wind and returns 24 miles with the same wind. His average speed for the return trip traveling with the wind is 8 mph faster than his speed going out against the wind. If x represents the bicyclist's speed going out against the wind, then the total time, t, required for the round trip is given by

$$t = \frac{24}{x} + \frac{24}{x + 8}$$ where $x > 0$ and t is measured in hours.

a. Find the time required for the round trip if the cyclist rides 12 mph against the wind.
   $3\frac{1}{5}$ h or 3.2 h

b. Find the time required for the round trip if the cyclist rides 24 mph against the wind.
   $1\frac{3}{4}$ h or 1.75 h

12. The manufacturer of mountain bikes has a fixed cost of $52,000, plus a variable cost of $140 per bike. The average cost per bike, y (in dollars), is given by the equation:

$$y = \frac{56{,}000 + 140x}{x}$$ where x represents the number of bikes produced and $x \geq 0$.

a. Find the average cost per bike if the manufacturer produces 1000 bikes. $196

b. Find the average cost per bike if the manufacturer produces 2000 bikes. $168

c. Find the average cost per bike if the manufacturer produces 10,000 bikes. $145.60

For Exercises 13–18, write the domain of the expression in set-builder notation.

13. $\dfrac{5}{k + 2}$ $\{k \mid k \neq -2\}$

14. $\dfrac{-3}{h - 4}$ $\{h \mid h \neq 4\}$

15. $\dfrac{x + 5}{(x - 5)(x + 8)}$
    $\{x \mid x \neq 5, x \neq -8\}$

16. $\dfrac{4y + 1}{(y + 7)(y + 3)}$
    $\{y \mid y \neq -7, y \neq -3\}$

17. $\dfrac{b + 12}{(b + 2)(b + 3)(2b - 5)}$ $\left\{ b \mid b \neq -2, b \neq -3, b \neq \dfrac{5}{2} \right\}$

18. $\dfrac{c - 11}{(2c + 1)(c - 6)(c + 1)}$ $\left\{ c \mid c \neq -\dfrac{1}{2}, c \neq 6, c \neq -1 \right\}$

19. Construct a rational expression that is undefined for $x = 2$. (Answers will vary.)   For example: $\dfrac{1}{x - 2}$

20. Construct a rational expression that is undefined for $x = 5$. (Answers will vary.)   For example: $\dfrac{1}{x - 5}$

21. Construct a rational expression that is undefined for $x = -3$ and $x = 7$. (Answers will vary.)   For example: $\dfrac{1}{(x + 3)(x - 7)}$

22. Construct a rational expression that is undefined for $x = -1$ and $x = 4$. (Answers will vary.)   For example: $\dfrac{1}{(x + 1)(x - 4)}$

23. Substitute $x = 4$ in the expressions in parts (a) and (b). Compare your answers.

   a. $\dfrac{5x + 5}{x^2 - 1}$ $\dfrac{5}{3}$ b. $\dfrac{5}{x - 1}$ $\dfrac{5}{3}$

24. Substitute $x = 3$ in the expressions in parts (a) and (b). Compare your answers.

   a. $\dfrac{2x^2 - 4x - 6}{2x^2 - 18}$ b. $\dfrac{x + 1}{x + 3}$ $\dfrac{2}{3}$
      Undefined

25. Substitute $x = -1$ in the expressions in parts (a) and (b). Compare your answers.

   a. $\dfrac{3x^2 - 2x - 1}{6x^2 - 7x - 3}$ $\dfrac{2}{5}$ b. $\dfrac{x - 1}{2x - 3}$ $\dfrac{2}{5}$

26. Substitute $x = 4$ in the expressions in parts (a) and (b). Compare your answers.

   a. $\dfrac{(x + 5)^2}{x^2 + 6x + 5}$ $\dfrac{9}{5}$ b. $\dfrac{x + 5}{x + 1}$ $\dfrac{9}{5}$

For Exercises 27–38, reduce the expression.

27. $\dfrac{7b^2}{21b}$ $\dfrac{b}{3}$

28. $\dfrac{15c^3}{3c^5}$ $\dfrac{5}{c^2}$

29. $\dfrac{18st^5}{12st^3}$ $\dfrac{3}{2}t^2$

30. $\dfrac{20a^4b^2}{25ab^2}$ $\dfrac{4}{5}a^3$

31. $\dfrac{-24x^2y^5z}{8xy^4z^3}$ $-\dfrac{3xy}{z^2}$

32. $\dfrac{60rs^4t^2}{-12r^4s^2t^3}$ $-\dfrac{5s^2}{r^3t}$

33. $\dfrac{3(y + 2)}{6(y + 2)}$ $\dfrac{1}{2}$

34. $\dfrac{8(x - 1)}{4(x - 1)}$ 2

35. $\dfrac{(p-3)(p+5)}{(p+5)(p+4)}$ $\dfrac{p-3}{p+4}$

36. $\dfrac{(c+4)(c-1)}{(c+4)(c+2)}$ $\dfrac{c-1}{c+2}$

58. $\dfrac{3x-3y}{a^2x-a^2y+b^2x-b^2y}$ $\dfrac{3}{a^2+b^2}$

37. $\dfrac{(m+11)}{4(m+11)(m-11)}$ $\dfrac{1}{4(m-11)}$

38. $\dfrac{(n-7)}{9(n+2)(n-7)}$ $\dfrac{1}{9(n+2)}$

59. $\dfrac{5x^3+4x^2-45x-36}{x^2-9}$ $5x+4$

For Exercises 39–48:

a. Factor both the numerator and denominator.
b. Write the domain in set-builder notation.
c. Reduce the expression.

60. $\dfrac{x^2-1}{ax^3-bx^2-ax+b}$ $\dfrac{1}{ax-b}$

61. $\dfrac{2x^2-xy-3y^2}{2x^2-11xy+12y^2}$ $\dfrac{x+y}{x-4y}$ 62. $\dfrac{2c^2+cd-d^2}{5c^2+3cd-2d^2}$ $\dfrac{2c-d}{5c-2d}$

39. $\dfrac{3y+6}{6y+12}$ ❖

40. $\dfrac{8x-8}{4x-4}$ ❖

63. What is the relationship between $x-2$ and $2-x$? They are opposites.

41. $\dfrac{t^2-1}{t+1}$ ❖

42. $\dfrac{r^2-4}{r-2}$ ❖

64. What is the relationship between $w+p$ and $-w-p$? They are opposites.

43. $\dfrac{7w}{21w^2-35w}$ ❖

44. $\dfrac{12a^2}{24a^2-18a}$ ❖

For Exercises 65–72, reduce the expression involving a ratio of -1.

45. $\dfrac{9x^2-4}{6x+4}$ ❖

46. $\dfrac{8b-20}{4b^2-25}$ ❖

65. $\dfrac{x-5}{5-x}$ -1

66. $\dfrac{8-p}{p-8}$ -1

47. $\dfrac{a^2+3a-10}{a^2+a-6}$ ❖

48. $\dfrac{t^2+3t-10}{t^2+t-20}$ ❖

67. $\dfrac{-4-y}{4+y}$ -1

68. $\dfrac{z+10}{-z-10}$ -1

For Exercises 49–62, reduce the rational expressions.

69. $\dfrac{3y-6}{12-6y}$ $-\dfrac{1}{2}$

70. $\dfrac{4q-4}{12-12q}$ $-\dfrac{1}{3}$

49. $\dfrac{y^2+6y+9}{2y^2+y-15}$ $\dfrac{y+3}{2y-5}$

50. $\dfrac{h^2+h-6}{h^2+2h-8}$ $\dfrac{h+3}{h+4}$

71. $\dfrac{x^2-x-12}{16-x^2}$ $-\dfrac{x+3}{4+x}$

72. $\dfrac{49-b^2}{b^2-10b+21}$ $-\dfrac{7+b}{b-3}$

51. $\dfrac{3x^2+7x-6}{x^2+7x+12}$ $\dfrac{3x-2}{x+4}$

52. $\dfrac{x^2-5x-14}{2x^2-x-10}$ $\dfrac{x-7}{2x-5}$

EXPANDING YOUR SKILLS

53. $\dfrac{5q^2+5}{q^4-1}$ $\dfrac{5}{(q+1)(q-1)}$

54. $\dfrac{4t^2+16}{t^4-16}$ $\dfrac{4}{(t-2)(t+2)}$

For Exercises 73–76, factor and reduce the expressions.

55. $\dfrac{ac-ad+2bc-2bd}{2ac+ad+4bc+2bd}$ $\dfrac{c-d}{2c+d}$

73. $\dfrac{w^3-8}{w^2+2w+4}$ $w-2$

74. $\dfrac{y^3+27}{y^2-3y+9}$ $y+3$

56. $\dfrac{3pr-ps-3qr+qs}{3pr-ps+3qr-qs}$ $\dfrac{p-q}{p+q}$

75. $\dfrac{z^2-16}{z^3-64}$ $\dfrac{z+4}{z^2+4z+16}$

76. $\dfrac{x^2-25}{x^3+125}$ $\dfrac{x-5}{x^2-5x+25}$

57. $\dfrac{49p^2-28pq+4q^2}{14p-4q}$ $\dfrac{7p-2q}{2}$

 See Additional Answers Appendix   Writing   Translating Expression   Geometry   Scientific Calculator   Video

section

5.2   MULTIPLICATION AND DIVISION OF RATIONAL EXPRESSIONS

1.   Multiplication of Rational Expressions

Recall from Section R.1 that to multiply fractions, we multiply the numerators and multiply the denominators. The same is true for multiplying rational expressions.

Multiplication of Rational Expressions

Let p, q, r, and s represent polynomials, such that $q \neq 0, s \neq 0$. Then,

$$\frac{p}{q} \cdot \frac{r}{s} = \frac{pr}{qs}$$

For example:

Multiply the Fractions	Multiply the Rational Expressions
$\dfrac{2}{3} \cdot \dfrac{5}{7} = \dfrac{10}{21}$	$\dfrac{2x}{3y} \cdot \dfrac{5z}{7} = \dfrac{10xz}{21y}$

Sometimes it is possible to reduce a ratio of common factors to 1 *before* multiplying. To do so, we must first factor the numerators and denominators of each fraction.

$$\frac{15}{14} \cdot \frac{21}{10} = \frac{3 \cdot \overset{1}{\cancel{5}}}{2 \cdot \cancel{7}} \cdot \frac{3 \cdot \overset{1}{\cancel{7}}}{2 \cdot \cancel{5}} = \frac{9}{4}$$

The same process is also used to multiply rational expressions.

Steps to Multiply Rational Expressions

1.   Factor the numerators and denominators of all rational expressions.
2.   Reduce the ratios of common factors to 1.
3.   Multiply the remaining factors in the numerator and multiply the remaining factors in the denominator.

---

example 1   **Multiplying Rational Expressions**

a. $\dfrac{5a^2b}{2} \cdot \dfrac{6a}{10b}$ b. $\dfrac{3c - 3d}{6c} \cdot \dfrac{2}{c^2 - d^2}$ c. $\dfrac{35 - 5x}{5x + 5} \cdot \dfrac{x^2 + 5x + 4}{x^2 - 49}$

Classroom Activity 5.2A

Solution:

a. $\dfrac{5a^2b}{2} \cdot \dfrac{6a}{10b}$

$$= \frac{5 \cdot a \cdot a \cdot b}{2} \cdot \frac{2 \cdot 3 \cdot a}{2 \cdot 5 \cdot b} \qquad \text{Factor.}$$

$$= \frac{\overset{1}{\cancel{5}} \cdot a \cdot a \cdot \overset{1}{\cancel{b}}}{2} \cdot \frac{\overset{1}{\cancel{2}} \cdot 3 \cdot a}{\cancel{2} \cdot \cancel{5} \cdot \cancel{b}} \qquad \text{Reduce.}$$

$$= \frac{3a^3}{2} \qquad \text{Multiply remaining factors.}$$

b. $\dfrac{3c - 3d}{6c} \cdot \dfrac{2}{c^2 - d^2}$

$$= \frac{3(c - d)}{2 \cdot 3 \cdot c} \cdot \frac{2}{(c - d)(c + d)} \qquad \text{Factor.}$$

$$= \frac{\overset{1}{\cancel{3}}(\overset{1}{\cancel{c - d}})}{\cancel{2} \cdot \cancel{3} \cdot c} \cdot \frac{\overset{1}{\cancel{2}}}{(\cancel{c - d})(c + d)} \qquad \text{Reduce.}$$

$$= \frac{1}{c(c + d)}$$

> **⬢ Avoiding Mistakes**
>
> If all factors in the numerator reduce to a ratio of 1, do not forget to write the factor of 1 in the numerator.

c. $\dfrac{35 - 5x}{5x + 5} \cdot \dfrac{x^2 + 5x + 4}{x^2 - 49}$

$$= \frac{5(7 - x)}{5(x + 1)} \cdot \frac{(x + 4)(x + 1)}{(x - 7)(x + 7)} \qquad \begin{array}{l}\text{Factor the numerators and denominators}\\ \text{completely.}\end{array}$$

$$= \frac{\overset{1}{\cancel{5}}(\overset{-1}{\cancel{7 - x}})}{\cancel{5}(\cancel{x + 1})} \cdot \frac{(x + 4)(\overset{1}{\cancel{x + 1}})}{(\cancel{x - 7})(x + 7)} \qquad \begin{array}{l}\text{Reduce the ratios of common factors to 1}\\ \text{or } -1.\end{array}$$

$$= \frac{-1(x + 4)}{x + 7}$$

$$= \frac{-(x + 4)}{x + 7} \qquad \text{or} \qquad \frac{x + 4}{-(x + 7)} \qquad \text{or} \qquad -\frac{x + 4}{x + 7}$$

2. Division of Rational Expressions

Recall that to divide fractions, multiply the first fraction by the reciprocal of the second.

$$\frac{21}{10} \div \frac{49}{15} \xrightarrow[\text{of the second fraction}]{\text{multiply by the reciprocal}} \frac{21}{10} \cdot \frac{15}{49} \xrightarrow{\text{factor}} \frac{3 \cdot \overset{1}{\cancel{7}}}{2 \cdot \cancel{5}} \cdot \frac{3 \cdot \cancel{5}}{\cancel{7} \cdot 7} \overset{\text{reduce}}{=} \frac{9}{14}$$

The same process is used to divide rational expressions.

Division of Rational Expressions

Let p, q, r, and s represent polynomials, such that $q \neq 0$, $r \neq 0$, $s \neq 0$. Then,

$$\frac{p}{q} \div \frac{r}{s} = \frac{p}{q} \cdot \frac{s}{r} = \frac{ps}{qr}$$

---

example 2 **Dividing Rational Expressions**

Divide.

a. $\dfrac{5t - 15}{2} \div \dfrac{t^2 - 9}{10}$ b. $\dfrac{p^2 - 11p + 30}{10p^2 - 250} \div \dfrac{30p - 5p^2}{2p + 4}$ c. $\dfrac{\dfrac{3x}{4y}}{\dfrac{5x}{6y}}$

Solution:

a. $\dfrac{5t - 15}{2} \div \dfrac{t^2 - 9}{10}$

$= \dfrac{5t - 15}{2} \cdot \dfrac{10}{t^2 - 9}$ Multiply the first fraction by the reciprocal of the second.

$= \dfrac{5(t - 3)}{2} \cdot \dfrac{2 \cdot 5}{(t - 3)(t + 3)}$ Factor each polynomial.

$= \dfrac{5\overset{1}{\cancel{(t - 3)}}}{\underset{1}{\cancel{2}}} \cdot \dfrac{\cancel{2} \cdot 5}{\cancel{(t - 3)}(t + 3)}$ Reduce common factors.

$= \dfrac{25}{t + 3}$

b. $\dfrac{p^2 - 11p + 30}{10p^2 - 250} \div \dfrac{30p - 5p^2}{2p + 4}$

$= \dfrac{p^2 - 11p + 30}{10p^2 - 250} \cdot \dfrac{2p + 4}{30p - 5p^2}$

Multiply the first fraction by the reciprocal of the second.

Factor the trinomial.
$p^2 - 11p + 30 = (p - 5)(p - 6)$

Factor out the GCF.
$2p + 4 = 2(p + 2)$

Factor out the GCF. Then factor the difference of squares.
$10p^2 - 250 = 10(p^2 - 25)$
$\qquad\qquad\quad = 2 \cdot 5(p - 5)(p + 5)$

$= \dfrac{(p - 5)(p - 6)}{2 \cdot 5(p - 5)(p + 5)} \cdot \dfrac{2(p + 2)}{5p(6 - p)}$

Factor out the GCF.
$30p - 5p^2 = 5p(6 - p)$

Tip: $(p - 6)$ and $(6 - p)$ are opposites and form a ratio of -1.

$\dfrac{p - 6}{6 - p} = \dfrac{p - 6}{-1(-6 + p)}$
$\qquad = \dfrac{p - 6}{-1(p - 6)} = -1$

$= \dfrac{\overset{1}{\cancel{(p - 5)}}\,\overset{-1}{\cancel{(p - 6)}}}{\overset{}{\cancel{2} \cdot 5\cancel{(p - 5)}(p + 5)}} \cdot \dfrac{\overset{1}{\cancel{2}}(p + 2)}{5p\cancel{(6 - p)}}$

Reduce common factors.

$= -\dfrac{(p + 2)}{25p(p + 5)}$

Tip: A fraction with one or more rational expressions in its numerator or denominator is called a **complex fraction**.

$\dfrac{\dfrac{3x}{4y}}{\dfrac{5x}{6y}}$

c. $\dfrac{\dfrac{3x}{4y}}{\dfrac{5x}{6y}}$ ⟵ ————— This fraction bar denotes division ($\div$).

$= \dfrac{3x}{4y} \div \dfrac{5x}{6y}$

$= \dfrac{3x}{4y} \cdot \dfrac{6y}{5x}$

Multiply by the reciprocal of the second fraction.

$= \dfrac{3 \cdot \overset{1}{\cancel{x}}}{\cancel{2} \cdot 2 \cdot \cancel{y}} \cdot \dfrac{\overset{1}{\cancel{2}} \cdot 3 \cdot \overset{1}{\cancel{y}}}{5 \cdot \cancel{x}}$

Reduce common factors.

$= \dfrac{9}{10}$

3. Performing the Order of Operations

Sometimes multiplication and division of rational expressions appear in the same problem. In such a case, apply the order of operations by multiplying or dividing in order from left to right.

---

example 3

Multiplying and Dividing Rational Expressions

Perform the indicated operations.

$$\frac{4}{c^2 - 9} \div \frac{6}{c - 3} \cdot \frac{3c}{8}$$

Classroom Activity 5.2B

Solution:

In this example, division occurs first, before multiplication. Parentheses may be inserted to reinforce the proper order.

$$\left(\frac{4}{c^2 - 9} \div \frac{6}{c - 3}\right) \cdot \frac{3c}{8}$$

$$= \left(\frac{4}{c^2 - 9} \cdot \frac{c - 3}{6}\right) \cdot \frac{3c}{8}$$
Multiply the first fraction by the reciprocal of the second.

$$= \left(\frac{2 \cdot 2}{(c - 3)(c + 3)} \cdot \frac{c - 3}{2 \cdot 3}\right) \cdot \frac{3 \cdot c}{2 \cdot 2 \cdot 2}$$
Now that each operation is written as multiplication, factor the polynomials and reduce the common factors.

$$= \frac{\overset{1}{\cancel{2}} \cdot \overset{1}{\cancel{2}}}{(\cancel{c - 3})(c + 3)} \cdot \frac{\overset{1}{\cancel{c - 3}}}{2 \cdot \cancel{3}} \cdot \frac{\overset{1}{\cancel{3}} \cdot c}{\cancel{2} \cdot \cancel{2} \cdot 2}$$

$$= \frac{c}{4(c + 3)}$$
Simplify.

---

4. Converting Units of Measurement

Converting units of measurement can be accomplished using a process similar to multiplying rational expressions. Consider a board for instance that is 30 in. long (Figure 5-1). Because 1 ft equals 12 in., the expressions

$$\frac{1 \text{ ft}}{12 \text{ in.}} \quad \text{and} \quad \frac{12 \text{ in.}}{1 \text{ ft}}$$

each form a ratio of two equal lengths. These ratios are called **conversion factors**.

To convert 30 in. to an equivalent distance in feet, multiply by the appropriate conversion factor. In this case, multiply by 1 ft/12 in. so that the units of inches will cancel and the units of feet will remain in the numerator.

$$\left(\frac{30 \text{ in.}}{1}\right)\left(\frac{1 \text{ ft}}{12 \text{ in.}}\right) = \frac{30}{12} \text{ ft} = \frac{5}{2} \text{ ft} = 2\tfrac{1}{2} \text{ ft}$$

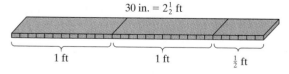

30 in. = $2\tfrac{1}{2}$ ft

1 ft   1 ft $\tfrac{1}{2}$ ft

Figure 5-1

Before we present further examples of unit conversion, review the following list of units of measurement and their abbreviations.

Length		**Volume/Capacity**		**Mass**	
in.	inch	pt	pint	mg	milligram
ft	foot	gal	gallon	g	gram
yd	yard	mL	milliliter	kg	kilogram
mm	millimeter	L	liter		
cm	centimeter	cc	cubic centimeter	**Force**	
m	meter		(also written as cm^3)	lb	pound
km	kilometer				

Table 5-1 provides the relationship among several units of measurement. This table is provided for reference purposes only and does not need to be memorized.

Table 5-1

Length	Area	Volume	Mass/Force
1 mile = 5280 ft	1 in.^{2} = 6.452 cm^2	1 pt = 2 cups	1 g = 1000 mg
1 mile = 1.609 km	1 yd^2 = 9 ft^2	1 qt = 2 pt	1 kg = 1000 g
1 in. = 2.54 cm	1 m^2 = 10.76 ft^2	1 gal = 4 qt	*1 kg = 2.2 lb
1 m = 3.281 ft	1 mile^2 = 640 acres	1 L = 1000 cm^3	1 lb = 16 oz
1 light-year = 9.46 × 10^{15} m		1 gal = 3.785 L	
1 m = 1000 mm		1 ft^3 = 7.481 gal	
1 m = 100 cm		1 L = 1000 ml	
1 km = 1000 m		1 ml = 1 cc	

*A kilogram (kg) is a unit of mass, whereas a pound (lb) is a measure of the force of gravity acting on a mass. Thus, a 1-kg mass produces 2.2 lb of force under the influence of earth's gravity.

example 4

Converting Units of Measurement

a. Find the mass in kilograms of a 150-lb person. (Round to the nearest tenth of a kilogram)
b. Find the mass in grams of a 150-lb person.

Solution:

a. From Table 5-1, we know that for measurements taken on earth, 1 kg = 2.2 lb. Multiply 150 lb by 1 kg/2.2 lb so that the units of lb will cancel, leaving only the units of kg in the numerator.

$$(150 \text{ lb})\left(\frac{1 \text{ kg}}{2.2 \text{ lb}}\right) \approx 68.2 \text{ kg}$$

b. From part (a), we know that 150 lb ≈ 68.2 kg. From Table 5-1, we know that 1 kg = 1000 g. Multiply 68.2 kg by 1000 g/1 kg so that the units of kg cancel.

$$(68.2 \text{ kg})\left(\frac{1000 \text{ g}}{1 \text{ kg}}\right) = 68{,}200 \text{ g}$$

Sometimes it is necessary to use more than one conversion factor as shown in the next example.

example 5   **Converting Units of Measurement**

At its cruising altitude of approximately 225 miles above the earth, the Space Shuttle orbits the globe in roughly 90 min. Its average speed is approximately 25,000 ft/s. Convert 25,000 ft/s to miles per hour. (*Source:* National Aeronautics and Space Administration)

Classroom Activity 5.2C

Solution:

The equivalent units needed for this problem are

$$1 \text{ mile} = 5280 \text{ ft}$$

$$1 \text{ min} = 60 \text{ s}$$

$$1 \text{ h} = 60 \text{ min}$$

$$\left(25{,}000\frac{\text{ft}}{\text{s}}\right)\left(\frac{1 \text{ mile}}{5280 \text{ ft}}\right)\left(\frac{60 \text{ s}}{1 \text{ min}}\right)\left(\frac{60 \text{ min}}{1 \text{ h}}\right) = 17{,}045\frac{\text{miles}}{\text{h}}$$

Converts feet to miles   Converts seconds to minutes   Converts minutes to hours

example 6   **Converting Units of Measurement**

A room measures 180 ft². If carpeting costs \$12.99/yd², find the cost to carpet the room.

Solution:

Because the cost of carpeting is given by the square yard, we must determine the area of the room in square yards. From Table 5-1, 1 yd² = 9 ft². Hence:

$$(180 \text{ ft}^2)\left(\frac{1 \text{ yd}^2}{9 \text{ ft}^2}\right) = 20 \text{ yd}^2$$

At \$12.99/yd², the cost is $(20 \text{ yd}^2)(\$12.99/\text{yd}^2) = \$259.80$.

Tip: To understand why 1 yd^2 = 9 ft^2, recall that the area of a square is given by $A = s^2$, where s represents the length of a side. The area of a square that is 1 yd × 1 yd is 1 yd^2. The same square measures 3 ft by 3 ft for an area of 9 ft^2.

Area = 1 yd^2

1 yd

1 yd

Area = 9 ft^2

3 ft

3 ft

Hence, 1 yd^2 = 9 ft^2.

---

section 5.2 PRACTICE EXERCISES

For Exercises 1–6, write the domain in set-builder notation and reduce each rational expression. 5.1

1. $\dfrac{(x+2)(x-1)}{(x-3)(x+2)}$ $\{x \mid x \neq 3, x = -2\}, \dfrac{x-1}{x-3}$

2. $\dfrac{(y+6)}{(y-1)(y+6)}$ $\{y \mid y \neq 1, y \neq -6\}, \dfrac{1}{y-1}$

3. $\dfrac{a^2-4}{a^2-4a+4}$ $\{a \mid a \neq 2\}, \dfrac{a+2}{a-2}$

4. $\dfrac{b^2+10b+25}{b^2-25}$ $\{b \mid b \neq 5, b \neq -5\}, \dfrac{b+5}{b-5}$

5. $\dfrac{12t-6}{3-6t}$ $\left\{t \mid t \neq \dfrac{1}{2}\right\}, -2$

6. $\dfrac{15p-10}{8-12p}$ $\left\{p \mid p \neq \dfrac{2}{3}\right\}, -\dfrac{5}{4}$

For Exercises 7–14, multiply or divide the fractions.

7. $\dfrac{3}{5} \cdot \dfrac{1}{2}$ $\dfrac{3}{10}$

8. $\dfrac{6}{7} \cdot \dfrac{5}{12}$ $\dfrac{5}{14}$

9. $\dfrac{3}{4} \div \dfrac{3}{8}$ 2

10. $\dfrac{18}{5} \div \dfrac{2}{5}$ 9

11. $6 \cdot \dfrac{5}{12}$ $\dfrac{5}{2}$

12. $\dfrac{7}{25} \cdot 5$ $\dfrac{7}{5}$

13. $\dfrac{\frac{21}{4}}{\frac{7}{5}}$ $\dfrac{15}{4}$

14. $\dfrac{\frac{9}{2}}{\frac{3}{4}}$ 6

For Exercises 15–50, multiply or divide as indicated.

15. $\dfrac{4x-24}{20x} \cdot \dfrac{5x}{8}$ $\dfrac{x-6}{8}$

16. $\dfrac{5a+20}{a} \cdot \dfrac{3a}{10}$ $\dfrac{3(a+4)}{2}$

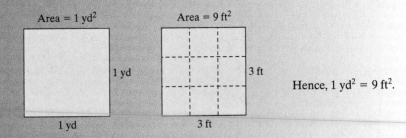 17. $\dfrac{3y+18}{y^2} \cdot \dfrac{4y}{6y+36}$ $\dfrac{2}{y}$

18. $\dfrac{2p-4}{6p} \cdot \dfrac{4p^2}{8p-16}$ $\dfrac{p}{6}$

19. $\dfrac{10}{2-a} \cdot \dfrac{a-2}{16}$ $-\dfrac{5}{8}$

20. $\dfrac{b-3}{6} \cdot \dfrac{20}{3-b}$ $-\dfrac{10}{3}$

21. $\dfrac{b^2-a^2}{a-b} \cdot \dfrac{a}{a^2-ab}$ $-\dfrac{b+a}{a-b}$

22. $\dfrac{(x-y)^2}{x^2+xy} \cdot \dfrac{x}{y-x}$ $-\dfrac{x-y}{x+y}$

23. $\dfrac{4a+12}{6a-18} \div \dfrac{3a+9}{5a-15}$ $\dfrac{10}{9}$

24. $\dfrac{8b-16}{3b+3} \div \dfrac{5b-10}{2b+2}$ $\dfrac{16}{15}$

25. $\dfrac{3x-21}{6x^2-42x} \div \dfrac{7}{12x}$ $\dfrac{6}{7}$

26. $\dfrac{4a^2-4a}{9a-9} \div \dfrac{5}{12a}$ $\dfrac{16a^2}{15}$

27. $\dfrac{y^2+5y-36}{y^2-2y-8} \cdot \dfrac{y+2}{y-6}$ $\dfrac{y+9}{y-6}$

28. $\dfrac{z^2-11z+28}{z-1} \cdot \dfrac{z+1}{z^2-6z-7}$ $\dfrac{z-4}{z-1}$

29. $\dfrac{t^2+4t-5}{t^2+7t+10} \cdot \dfrac{t+4}{t-1}$ $\dfrac{t+4}{t+2}$

30. $\dfrac{p^2-3p+2}{p^2-4p+3} \cdot \dfrac{p+1}{p-2}$ $\dfrac{p+1}{p-3}$

31. $\dfrac{m^2-n^2}{9} \div \dfrac{3n-3m}{27m}$ $-m(m+n)$

32. $\dfrac{9-b^2}{15b+15} \div \dfrac{b-3}{5b}$ $-\dfrac{b(3+b)}{3(b+1)}$

❖ See Additional Answers Appendix ✎ Writing ⬌ Translating Expression ▱ Geometry ▦ Scientific Calculator ▭ Video

33. $\dfrac{3p + 4q}{p^2 + 4pq + 4q^2} \div \dfrac{4}{p + 2q}$ $\dfrac{3p + 4q}{4(p + 2q)}$

34. $\dfrac{x^2 + 2xy - 3y^2}{2x - y} \div \dfrac{x + 3y}{5}$ $\dfrac{5(x - y)}{2x - y}$

35. $(w + 3) \cdot \dfrac{w}{2w^2 + 5w - 3}$ $\dfrac{w}{2w - 1}$

36. $\dfrac{5t + 1}{5t^2 - 31t + 6} \cdot (t - 6)$ $\dfrac{5t + 1}{5t - 1}$

37. $\dfrac{\dfrac{5t - 10}{12}}{\dfrac{4t - 8}{8}}$ $\dfrac{5}{6}$

38. $\dfrac{\dfrac{6m + 6}{5}}{\dfrac{3m + 3}{10}}$ 4

39. $\dfrac{q + 1}{5q^2 - 28q - 12} \cdot (5q + 2)$ $\dfrac{q + 1}{q - 6}$

40. $(r - 5) \cdot \dfrac{4r}{2r^2 - 7r - 15}$ $\dfrac{4r}{2r + 3}$

41. $\dfrac{2a^2 + 13a - 24}{8a - 12} \div (a + 8)$ $\dfrac{1}{4}$

42. $\dfrac{3y^2 + 20y - 7}{5y + 35} \div (3y - 1)$ $\dfrac{1}{5}$

43. $(5t - 1) \div \dfrac{5t^2 + 9t - 2}{3t + 8}$ $\dfrac{3t + 8}{t + 2}$

44. $(2q - 3) \div \dfrac{2q^2 + 5q - 12}{q - 7}$ $\dfrac{q - 7}{q + 4}$

45. $\dfrac{x^2 + 2x - 3}{x^2 - 3x + 2} \cdot \dfrac{x^2 + 2x - 8}{x^2 + 4x + 3}$ $\dfrac{x + 4}{x + 1}$

46. $\dfrac{y^2 + y - 12}{y^2 - y - 20} \cdot \dfrac{y^2 + y - 30}{y^2 - 2y - 3}$ $\dfrac{y + 6}{y + 1}$

47. $\dfrac{\dfrac{w^2 - 6w + 9}{8}}{\dfrac{9 - w^2}{4w + 12}}$ $-\dfrac{w - 3}{2}$

48. $\dfrac{\dfrac{p^2 - 6p + 8}{24}}{\dfrac{16 - p^2}{6p + 6}}$ $\dfrac{(p - 2)(p + 1)}{4(4 + p)}$

49. $\dfrac{k^2 + 3k + 2}{k^2 + 5k + 4} \div \dfrac{k^2 + 5k + 6}{k^2 + 10k + 24}$ $\dfrac{k + 6}{k + 3}$

50. $\dfrac{4h^2 - 5h + 1}{h^2 + h - 2} \div \dfrac{6h^2 - 7h + 2}{2h^2 + 3h - 2}$ $\dfrac{4h - 1}{3h - 2}$

 For Exercises 51–66, use conversion factors to solve the problem.

51. A certain commercial airliner flies at 36,000 ft. Convert 36,000 ft to miles. (Round the answer to the nearest tenth of a mile.)   6.8 miles

52. The Space Shuttle orbits the earth at approximately 225 miles above its surface. Convert 225 miles to feet.   1,188,000 ft

53. A desktop has dimensions of 150 cm by 76 cm. Convert these measurements to meters.
1.5 m by 0.76 m

54. A legal envelope measures 10.4 cm by 24.1 cm. Convert the length and width to millimeters.
104 mm by 241 mm

55. The external fuel tank on the Space Shuttle feeds fuel at a rate of 227,000 L/minute. Convert 227,000 L to gallons. (Round to the nearest ten gallons.)   59,970 gal

56. How many gallons can a tank hold if the capacity of the tank is 108 ft³? (Round to the nearest tenth of a gallon.)   807.9 gal

57. How many miles in 1 light-year? Assume that light travels at 6.71×10^8 mph. (A light-year is the distance light travels in 1 year.)
Approximately 5.88×10^{12} miles

58. A remote-controlled toy race car can travel at nearly 2 mph. Convert 2 mph to feet per second. (Round to one decimal place.)   2.9 ft/s

59. A child weighing 92 lb weighs herself on a metric scale. What is the equivalent measure in kilograms? (Round to the nearest tenth of a kilogram.)   41.8 kg

60. A person weighing 180 lb weighs himself on a metric scale. What is the equivalent measure in kilograms? (Round to the nearest tenth of a kilogram.)   81.8 kg

61. Material for a dress costs $5.49/yd². Find the cost per square foot. (Round to the nearest cent.) $0.61/ft²

62. Floor tile sells for $2.59/ft². How much would it cost to tile a 25-yd² room? (Round to the nearest cent.) $582.75

63. A bottle of soda contains 2.5 L of fluid. Convert this volume to milliliters.   2500 mL

64. A mug contains 425 mL of coffee. Convert this volume to liters.   0.425 L

65. A student walks 1550 m to school. Convert this distance to kilometers.   1.55 km

66. A baby has a mass of 4250 g at birth. Convert this mass to kilograms.   4.25 kg

▊ EXPANDING YOUR SKILLS

For Exercises 67–72, multiply or divide as indicated.

67. $\dfrac{b^3 - 3b^2 + 4b - 12}{b^4 - 16} \cdot \dfrac{3b^2 + 5b - 2}{3b^2 - 10b + 3} \div \dfrac{3}{6b - 12}$ 2

68. $\dfrac{x^2 - 25}{3x^2 + 3xy} \cdot \dfrac{x^2 + 4x + xy + 4y}{x^2 + 9x + 20} \div \dfrac{x - 5}{x}$ $\dfrac{1}{3}$

69. $\dfrac{a^2 - 5a}{a^2 + 7a + 12} \div \dfrac{a^3 - 7a^2 + 10a}{a^2 + 9a + 18} \div \dfrac{a + 6}{a + 4}$ $\dfrac{1}{a - 2}$

70. $\dfrac{t^2 + t - 2}{t^2 + 5t + 6} \div \dfrac{t - 1}{t} \div \dfrac{5t - 5}{t + 3}$ $\dfrac{t}{5(t - 1)}$

71. $\dfrac{p^3 - q^3}{p - q} \cdot \dfrac{p + q}{2p^2 + 2pq + 2q^2}$ $\dfrac{p + q}{2}$

72. $\dfrac{r^3 + s^3}{r - s} \div \dfrac{r^2 + 2rs + s^2}{r^2 - s^2}$ $r^2 - rs + s^2$

Concepts

1. **Equivalent Rational Expressions**

2. **Writing Equivalent Fractions**

3. **Least Common Denominator**

4. **Writing Fractions with the Least Common Denominator**

section
5.3   LEAST COMMON DENOMINATOR

1. Equivalent Rational Expressions

In Sections 5.1 and 5.2, we learned how to reduce, multiply, and divide rational expressions. Our next goal is to add and subtract rational expressions. As with fractions, rational expressions may be added or subtracted only if they have the same denominator. Therefore, we must first learn how to identify a common denominator between two or more rational expressions. Then we must learn how to convert a rational expression into an **equivalent rational expression** with the indicated denominator.

Using the identity property of multiplication, we know that for $q \neq 0$ and $r \neq 0$,

$$\frac{p}{q} = \frac{p}{q} \cdot 1 = \frac{p}{q} \cdot \frac{r}{r} = \frac{pr}{qr}$$

This principle is used to convert a rational expression into an equivalent expression with a different denominator. For example, $\frac{1}{2}$ can be converted into an equivalent expression with a denominator of 12 as follows:

$$\frac{1}{2} = \frac{1}{2} \cdot \frac{6}{6} = \frac{1 \cdot 6}{2 \cdot 6} = \frac{6}{12}$$

In this example, we multiplied $\frac{1}{2}$ by a convenient form of 1. The ratio $\frac{6}{6}$ was chosen so that the product produced a new denominator of 12. Notice that multiplying $\frac{1}{2}$ by $\frac{6}{6}$ is equivalent to multiplying the numerator and denominator of the original expression by 6. In general, if the numerator and denominator of a rational expression are multiplied by the same quantity, the value of the expression remains unchanged.

 ❖ See Additional Answers Appendix   Writing   Translating Expression   Geometry   Scientific Calculator   Video

2. Writing Equivalent Fractions

example 1

Creating Equivalent Fractions

Convert each expression into an equivalent expression with the indicated denominator.

a. $\dfrac{5}{x} = \dfrac{}{xyz}$

b. $\dfrac{7}{5p^2} = \dfrac{}{20p^6}$

c. $\dfrac{w}{w+5} = \dfrac{}{(w+5)(w-2)}$

d. $\dfrac{1}{5} = \dfrac{}{x^2 - x - 12}$

Classroom Activity 5.3A

Solution:

a. $\dfrac{5}{x} = \dfrac{}{xyz}$

To convert $\frac{5}{x}$ to an equivalent fraction with a denominator of xyz, multiply both numerator and denominator by the missing factor of yz.

$\dfrac{5 \cdot yz}{x \cdot yz} = \dfrac{5yz}{xyz}$

b. $\dfrac{7}{5p^2} = \dfrac{}{20p^6}$

Multiply the numerator and denominator of the fraction by the missing factor of $4p^4$.

$\dfrac{7 \cdot 4p^4}{5p^2 \cdot 4p^4} = \dfrac{28p^4}{20p^6}$

Tip: Notice that in Example 1(c) we multiplied the polynomials in the numerator but left the denominator in factored form. This convention is followed because when we add and subtract rational expressions in the next section, the terms in the numerators must be combined.

c. $\dfrac{w}{w+5} = \dfrac{}{(w+5)(w-2)}$

Multiply numerator and denominator by the missing factor of $(w-2)$.

$\dfrac{w}{w+5} = \dfrac{w \cdot (w-2)}{(w+5) \cdot (w-2)}$ or $\dfrac{w^2 - 2w}{(w+5)(w-2)}$

d. $\dfrac{1}{5} = \dfrac{}{x^2 - x - 12}$

$\dfrac{1}{5} = \dfrac{}{(x-4)(x+3)}$

Factor the denominator of the second expression. The factors missing from the denominator of the first expression are $(x-4)$ and $(x+3)$.

Multiply numerator and denominator by $(x-4)(x+3)$.

$\dfrac{1}{5} = \dfrac{1 \cdot (x-4)(x+3)}{5 \cdot (x-4)(x+3)} = \dfrac{x^2 + 3x - 4x - 12}{5(x-4)(x+3)} = \dfrac{x^2 - x - 12}{5(x-4)(x+3)}$

3.   Least Common Denominator

Recall from Section R.1 that to add or subtract fractions, the fractions must have a common denominator. The same is true for rational expressions. In this section, we present a method to find the least common denominator of two rational expressions.

The **least common denominator (LCD)** of two or more rational expressions is defined as the least common multiple of the denominators. For example, consider the fractions $\frac{1}{20}$ and $\frac{1}{8}$. By inspection, you can probably see that the least common denominator is 40. To understand why, find the prime factorization of both denominators:

$$20 = 2^2 \cdot 5 \quad \text{and} \quad 8 = 2^3$$

A common multiple of 20 and 8 must be a multiple of 5, a multiple of 2^2, and a multiple of 2^3. However, any number that is a multiple of $2^3 = 8$ is automatically a multiple of $2^2 = 4$. Therefore, it is sufficient to construct the least common denominator as the product of unique prime factors, in which each factor is raised to its highest power.

$$\text{The LCD of } \frac{1}{20} \text{ and } \frac{1}{8} \text{ is } 2^3 \cdot 5 = 40.$$

Steps to Find the Least Common Denominator of Two or More Rational Expressions

1.   Factor all denominators completely.
2.   The LCD is the product of unique factors from the denominators, in which each factor is raised to the highest power to which it appears in any denominator.

---

example 2

Finding the Least Common Denominator of Rational Expressions

Find the LCD of the following sets of rational expressions.

a. $\dfrac{5}{14}; \dfrac{3}{49}; \dfrac{1}{8}$ b. $\dfrac{5}{3x^2z}; \dfrac{7}{x^5y^3}$ c. $\dfrac{a+b}{a^2-25}; \dfrac{1}{2a-10}$

d. $\dfrac{x-5}{x^2-2x}; \dfrac{1}{x^2-4x+4}$

Classroom Activity 5.3B

Solution:

a. $\dfrac{5}{14}; \dfrac{3}{49}; \dfrac{1}{8}$

$= \dfrac{5}{2 \cdot 7}; \dfrac{3}{7^2}; \dfrac{1}{2^3};$ **Step 1:** Factor the denominators.

The LCD is $2^3 \cdot 7^2 = 392$.   **Step 2:** The LCD is the product of unique factors, each raised to its highest power.

b. $\dfrac{5}{3x^2z}; \dfrac{7}{x^5y^3}$

$= \dfrac{5}{3x^2z}; \dfrac{7}{x^5y^3}$

Step 1: The denominators are already factored.

The LCD is $3x^5y^3z$.

Step 2: The LCD is the product of unique factors, each raised to its highest power.

c. $\dfrac{a+b}{a^2-25}; \dfrac{1}{2a-10}$

$= \dfrac{a+b}{(a-5)(a+5)}; \dfrac{1}{2(a-5)}$

Step 1: Factor the denominators.

The LCD is $2(a-5)(a+5)$.

Step 2: The LCD is the product of unique factors, each raised to its highest power.

d. $\dfrac{x-5}{x^2-2x}; \dfrac{1}{x^2-4x+4}$

$= \dfrac{x-5}{x(x-2)}; \dfrac{1}{(x-2)^2}$

Step 1: Factor the denominators.

The LCD is $x(x-2)^2$.

Step 2: The LCD is the product of unique factors, each raised to its highest power.

4. Writing Fractions with the Least Common Denominator

To add or subtract two rational expressions, the expressions must have the same denominator. Therefore, we must first practice the skill of converting each rational expression into an equivalent expression with the LCD as its denominator. The process is as follows: Identify the LCD for the two expressions. Then, multiply the numerator and denominator of each fraction by the factors from the LCD that are missing from the original denominators.

example 3

Converting to the Least Common Denominator

Find the LCD of each pair of rational expressions. Then convert each expression to an equivalent fraction with the denominator equal to the LCD.

a. $\dfrac{3}{2ab}; \dfrac{6}{5a^2}$ b. $\dfrac{4}{x+1}; \dfrac{7}{x-4}$ c. $\dfrac{w+2}{w^2-w-12}; \dfrac{1}{w^2-9}$

Solution:

a. $\dfrac{3}{2ab}; \dfrac{6}{5a^2}$

The LCD is $10a^2b$.

$$\dfrac{3}{2ab} = \dfrac{3 \cdot 5a}{2ab \cdot 5a} = \dfrac{15}{10a^2b}$$

The first expression is missing the factor $5a$ from the denominator.

$$\dfrac{6}{5a^2} = \dfrac{6 \cdot 2b}{5a^2 \cdot 2b} = \dfrac{12b}{10a^2b}$$

The second expression is missing the factor $2b$ from the denominator.

b. $\dfrac{4}{x+1}; \dfrac{7}{x-4}$

The LCD is $(x+1)(x-4)$.

$$\dfrac{4}{x+1} = \dfrac{4(x-4)}{(x+1)(x-4)} = \dfrac{4x-16}{(x+1)(x-4)}$$

The first expression is missing the factor $(x-4)$ from the denominator.

$$\dfrac{7}{x-4} = \dfrac{7(x+1)}{(x-4)(x+1)} = \dfrac{7x+7}{(x-4)(x+1)}$$

The second expression is missing the factor $(x+1)$ from the denominator.

c. $\dfrac{w+2}{w^2-w-12}; \dfrac{1}{w^2-9}$

To find the LCD, factor each denominator.

$$\dfrac{w+2}{(w-4)(w+3)}; \dfrac{1}{(w-3)(w+3)}$$

The LCD is $(w-4)(w+3)(w-3)$.

$$\dfrac{w+2}{(w-4)(w+3)} = \dfrac{(w+2)(w-3)}{(w-4)(w+3)(w-3)}$$

The first expression is missing the factor $(w-3)$ from the denominator.

$$= \dfrac{w^2-w-6}{(w-4)(w+3)(w-3)}$$

$$\dfrac{1}{(w-3)(w+3)} = \dfrac{1(w-4)}{(w-3)(w+3)(w-4)}$$

The second expression is missing the factor $(w-4)$ from the denominator.

$$= \dfrac{w-4}{(w-3)(w+3)(w-4)}$$

---

example 4

Converting to the Least Common Denominator

Find the LCD of the expressions

$$\dfrac{3}{x-7} \quad \text{and} \quad \dfrac{1}{7-x}$$

Solution:

Notice that the expressions $x - 7$ and $7 - x$ are opposites and differ by a factor of -1. Therefore, we may use either $x - 7$ or $7 - x$ as a common denominator. Each case is detailed in the following conversions.

Converting to the Denominator $x - 7$

$$\frac{3}{x - 7}, \frac{1}{7 - x}$$

$$\frac{1}{7 - x} = \frac{(-1)1}{(-1)(7 - x)}$$
Multiply the *second* rational expression by the ratio $\frac{-1}{-1}$ to change its denominator to $x - 7$.

$$= \frac{-1}{-7 + x}$$
Apply the distributive property.

$$= \frac{-1}{x - 7}$$

Converting to the Denominator $7 - x$

$$\frac{3}{x - 7}, \frac{1}{7 - x}$$

Tip: In Example 4, the expressions

$$\frac{3}{x - 7} \quad \text{and} \quad \frac{1}{7 - x}$$

have opposite factors in the denominators. In such a case, you do *not* need to include both factors in the LCD.

$$= \frac{(-1)3}{(-1)(x - 7)};$$
Multiply the first rational expression by the ratio $\frac{-1}{-1}$ to change its denominator to $7 - x$.

$$= \frac{-3}{-x + 7}$$
Apply the distributive property.

$$= \frac{-3}{7 - x}$$

---

section 5.3   PRACTICE EXERCISES

For Exercises 1–4, write the domain in set-builder notation and reduce.   5.1

1. $\dfrac{3x + 3}{5x^2 - 5}$ $\{x \mid x \neq 1, x \neq -1\};$ $\dfrac{3}{5(x - 1)}$

2. $\dfrac{x + 2}{x^2 - 3x - 10}$ $\{x \mid x \neq -2, x \neq 5\};$ $\dfrac{1}{x - 5}$

3. $\dfrac{t^2 - 3t - 4}{t^2 + 4t + 3}$ $\{t \mid t \neq -3, t \neq -1\};$ $\dfrac{t - 4}{t + 3}$

4. $\dfrac{c^2 - 9}{c^2 - 3c}$ $\{c \mid c \neq 0, c \neq 3\};$ $\dfrac{c + 3}{c}$

For Exercises 5–8, multiply or divide as indicated.   5.2

5. $\dfrac{a + 3}{a + 7} \cdot \dfrac{a^2 + 3a - 10}{a^2 + a - 6}$ $\dfrac{a + 5}{a + 7}$

6. $\dfrac{16y^2}{9y + 36} \div \dfrac{8y^3}{3y + 12}$ $\dfrac{2}{3y}$

7. $\dfrac{6(a + 2b)}{2(a - 3b)} \cdot \dfrac{4(a + 3b)(a - 3b)}{9(a + 2b)(a - 2b)}$ $\dfrac{4(a + 3b)}{3(a - 2b)}$

8. $\dfrac{5b^2 + 6b + 1}{b^2 + 5b + 6} \div (5b + 1)$ $\dfrac{b + 1}{(b + 2)(b + 3)}$

9. How many liters can a 15-gal gas tank hold? Round to the nearest tenth of a liter (1 gal = 3.785 L).   5.2   56.8 L

10. How many quarts can a 2-L bottle of soda hold? Round to the nearest tenth of a quart (1 gal = 4 qt).   _5.2_   2.1 qt

For Exercises 11–26, convert the expressions into equivalent expressions with the indicated denominator.

11. $\dfrac{6}{7} = \dfrac{36}{42}$

12. $\dfrac{4}{9} = \dfrac{32}{72}$

13. $\dfrac{2}{13} = \dfrac{6}{39}$

14. $\dfrac{1}{8} = \dfrac{8}{64}$

15. $\dfrac{3}{p^2q} = \dfrac{15p}{5p^3q}$

16. $\dfrac{2}{3rs} = \dfrac{12s^2}{18rs^3}$

17. $\dfrac{2x}{yz} = \dfrac{12xyz^3}{6y^2z^4}$

18. $\dfrac{8a}{b^2c} = \dfrac{16ab^2c^4}{2b^4c^5}$

19. $\dfrac{w+6}{w-7} = \dfrac{w^2+8w+12}{(w-7)(w+2)}$

20. $\dfrac{z-1}{z+1} = \dfrac{z^2-4z+3}{(z+1)(z-3)}$

21. $\dfrac{6}{x-3} = \dfrac{-6}{3-x}$

22. $\dfrac{2}{a-9} = \dfrac{-2}{9-a}$

23. $\dfrac{t+3}{t+4} = \dfrac{5t^2+13t-6}{5t^2+18t-8}$

24. $\dfrac{w-4}{w-2} = \dfrac{2w^2-5w-12}{2w^2-w-6}$

25. $\dfrac{p-12}{p^2+7p} = \dfrac{p^2-10p-24}{p^3+9p^2+14p}$

26. $\dfrac{2y+3}{2y+1} = \dfrac{6y^3+5y^2-6y}{6y^3-y^2-2y}$

27. Which of the expressions are equivalent to
$$-\dfrac{5}{x-3}?$$

a. $\dfrac{-5}{x-3}$ b. $\dfrac{5}{-x+3}$

c. $\dfrac{5}{3-x}$ d. $\dfrac{5}{-(x-3)}$

a, b, c, d

28. Which of the expressions are equivalent to
$$\dfrac{4-a}{6}?$$

a. $\dfrac{a-4}{-6}$ b. $\dfrac{a-4}{6}$

c. $\dfrac{-(4-a)}{-6}$ d. $-\dfrac{a-4}{6}$

a, c, d

29. Explain why the least common denominator of $\dfrac{1}{x^3}, \dfrac{1}{x^5}$, and $\dfrac{1}{x^4}$ is x^5. x^5 is the lowest power of x that has x^3, x^5, x^4 as factors.

30. Explain why the least common denominator of $\dfrac{2}{y^3}, \dfrac{9}{y^6}$, and $\dfrac{4}{y^5}$ is y^6. y^6 is the lowest power of y that has y^3, y^6, and y^5 as factors.

31. Explain why the least common denominator of
$$\dfrac{1}{x+3} \quad \text{and} \quad \dfrac{3}{x-2}$$
is $(x+3)(x-2)$.   The product of unique factors is $(x+3)(x-2)$.

32. Explain why the least common denominator of
$$\dfrac{7}{y-8} \quad \text{and} \quad \dfrac{3}{y+1}$$
is $(y-8)(y+1)$.   The product of unique factors is $(y-8)(y+1)$.

33. Explain why a common denominator of
$$\dfrac{b+1}{b-1} \quad \text{and} \quad \dfrac{b}{1-b}$$
could be either $(b-1)$ or $(1-b)$.   Because $(b-1)$ and $(1-b)$ are opposites, they differ by a factor of -1.

34. Explain why a common denominator of
$$\dfrac{1}{6-t} \quad \text{and} \quad \dfrac{t}{t-6}$$
could be either $(6-t)$ or $(t-6)$.   Because $(6-t)$ and $(t-6)$ are opposites, they differ by a factor of -1.

For Exercises 35–52, identify the LCD.

35. $\dfrac{4}{15}, \dfrac{5}{9}$ 45

36. $\dfrac{7}{12}, \dfrac{1}{18}$ 36

37. $\dfrac{3}{16}, \dfrac{1}{4}$ 16

38. $\dfrac{1}{2}, \dfrac{11}{12}$ 12

39. $\dfrac{1}{7}, \dfrac{2}{9}$ 63

40. $\dfrac{2}{3}, \dfrac{5}{8}$ 24

41. $\dfrac{1}{3x^2y}, \dfrac{8}{9xy^3}$ $9x^2y^3$

42. $\dfrac{5}{2a^4b^2}, \dfrac{1}{8ab^3}$ $8a^4b^3$

43. $\dfrac{6}{w^2}, \dfrac{7}{y}$ w^2y

44. $\dfrac{2}{r}, \dfrac{3}{s^2}$ rs^2

45. $\dfrac{p}{(p+3)(p-1)}, \dfrac{2}{(p+3)(p+2)}$ $(p+3)(p-1)(p+2)$

46. $\dfrac{6}{(q+4)(q-4)}, \dfrac{q^2}{(q+1)(q+4)}$ $(q+4)(q-4)(q+1)$

47. $\dfrac{7}{3t(t+1)}, \dfrac{10t}{9(t+1)^2}$ $9t(t+1)^2$

48. $\dfrac{13x}{15(x-1)^2}, \dfrac{5}{3x(x-1)}$ $15x(x-1)^2$

49. $\dfrac{y}{y^2-4}, \dfrac{3y}{y^2+5y+6}$ $(y-2)(y+2)(y+3)$

50. $\dfrac{4}{w^2-3w+2}, \dfrac{w}{w^2-4}$ $(w-1)(w-2)(w+2)$

51. $\dfrac{17x}{8x-4}, \dfrac{22}{3x-6x^2}$
$12x(2x-1)$ or $12(1-2x)$

52. $\dfrac{6}{9y^2-1}, \dfrac{y-3}{9-27y}$
$9(3y-1)(3y+1)$ or $9(1-3y)(3y+1)$

For Exercises 53–62, find the LCD. Then convert each expression to an equivalent expression with the denominator equal to the LCD.

53. $\dfrac{5}{6a^2b}, \dfrac{a}{12b}$ $\dfrac{10}{12a^2b}, \dfrac{a^3}{12a^2b}$

54. $\dfrac{x}{15y^2}, \dfrac{y}{5xy}$ $\dfrac{x^2}{15xy^2}, \dfrac{3y^2}{15xy^2}$

55. $\dfrac{6}{m+4}, \dfrac{3}{m-1}$ ❖

56. $\dfrac{3}{n-5}, \dfrac{7}{n+2}$ ❖

57. $\dfrac{6}{(w+3)(w-8)}, \dfrac{w}{(w-8)(w+1)}$ ❖

58. $\dfrac{t}{(t+2)(t+12)}, \dfrac{18}{(t-2)(t+2)}$ ❖

59. $\dfrac{6p}{p^2-4}, \dfrac{3}{p^2+4p+4}$ $\dfrac{6p^2+12p}{(p-2)(p+2)^2}, \dfrac{3p-6}{(p-2)(p+2)^2}$

60. $\dfrac{5}{q^2-6q+9}, \dfrac{q}{q^2-9}$ $\dfrac{5q+15}{(q-3)^2(q+3)}, \dfrac{q^2-3q}{(q-3)^2(q+3)}$

61. $\dfrac{1}{a-4}, \dfrac{a}{4-a}$
$\dfrac{1}{a-4}, \dfrac{-a}{a-4}$ or $\dfrac{-1}{4-a}, \dfrac{a}{4-a}$

62. $\dfrac{3b}{2b-5}, \dfrac{2b}{5-2b}$
$\dfrac{3b}{2b-5}, \dfrac{-2b}{2b-5}$ or $\dfrac{-3b}{5-2b}, \dfrac{2b}{5-2b}$

⬛ EXPANDING YOUR SKILLS

For Exercises 63–66, find the LCD. Then convert each expression to an equivalent expression with the denominator equal to the LCD.

63. $\dfrac{z}{z^2+9z+14}, \dfrac{-3z}{z^2+10z+21}, \dfrac{5}{z^2+5z+6}$ ❖

64. $\dfrac{6}{w^2-3w-4}, \dfrac{1}{w^2+6w+5}, \dfrac{-9w}{w^2+w-20}$ ❖

65. $\dfrac{3}{p^3-8}, \dfrac{p}{p^2-4}, \dfrac{5p}{p^2+2p+4}$ ❖

66. $\dfrac{7}{q^3+125}, \dfrac{q}{q^2-25}, \dfrac{12}{q^2-5q+25}$ ❖

Concepts

1. Addition and Subtraction of Rational Expressions with the Same Denominator

2. Addition and Subtraction of Rational Expressions with Different Denominators

3. Performing the Order of Operations

5.4   ADDITION AND SUBTRACTION OF RATIONAL EXPRESSIONS

1.   Addition and Subtraction of Rational Expressions with the Same Denominator

To add or subtract rational expressions, the expressions must have the same denominator. As with fractions, we add or subtract rational expressions with the same denominator by combining the terms in the numerator and then writing the result over the common denominator. Then, if possible, we reduce the expression to lowest terms.

Addition and Subtraction of Rational Expressions

Let p, q, and r represent polynomials where $q \neq 0$. Then,

1. $\dfrac{p}{q} + \dfrac{r}{q} = \dfrac{p+r}{q}$

2. $\dfrac{p}{q} - \dfrac{r}{q} = \dfrac{p-r}{q}$

---

example 1

Adding and Subtracting Rational Expressions with a Common Denominator

Add or subtract as indicated.

a. $\dfrac{1}{12} + \dfrac{7}{12}$ b. $\dfrac{2}{5p} - \dfrac{7}{5p}$

c. $\dfrac{2}{3d+5} + \dfrac{7d}{3d+5}$ d. $\dfrac{x^2}{x-3} - \dfrac{-5x+24}{x-3}$

Classroom Activity 5.4A

Solution:

a. $\dfrac{1}{12} + \dfrac{7}{12}$ The fractions have the same denominator.

$= \dfrac{1+7}{12}$ Add the terms in the numerators and write the result over the common denominator.

$= \dfrac{8}{12}$ Simplify.

$= \dfrac{2}{3}$ Reduce.

b. $\dfrac{2}{5p} - \dfrac{7}{5p}$ The rational expressions have the same denominator.

$= \dfrac{2 - 7}{5p}$ Subtract the terms in the numerators and write the result over the common denominator.

$= \dfrac{-5}{5p}$

$= \dfrac{-\overset{1}{\cancel{5}}}{\cancel{5}p}$ Reduce.

$= -\dfrac{1}{p}$

c. $\dfrac{2}{3d + 5} + \dfrac{7d}{3d + 5}$ The rational expressions have the same denominator.

$= \dfrac{2 + 7d}{3d + 5}$ Add the terms in the numerators and write the result over the common denominator.

$= \dfrac{7d + 2}{3d + 5}$ Because the numerator and denominator share no common factors, the expression is in lowest terms.

d. $\dfrac{x^2}{x - 3} - \dfrac{-5x + 24}{x - 3}$ The rational expressions have the same denominator.

Subtract the terms in the numerators and write the result over the common denominator.

$= \dfrac{x^2 - (-5x + 24)}{x - 3}$

$= \dfrac{x^2 + 5x - 24}{x - 3}$ Simplify the numerator.

$= \dfrac{(x + 8)(x - 3)}{(x - 3)}$ Factor the numerator and denominator to determine if the rational expression can be reduced.

$= \dfrac{(x + 8)\cancel{(x - 3)}^{\,1}}{\cancel{(x - 3)}}$ Reduce common factors.

$= x + 8$

Avoiding Mistakes

When subtracting rational expressions, use parentheses to group the terms in the numerator that follow the subtraction sign. This will help you remember to apply the distributive property.

2. Addition and Subtraction of Rational Expressions with Different Denominators

To add or subtract two rational expressions with unlike denominators, we must convert the expressions to equivalent expressions with the same denominator. For example, consider adding

$$\dfrac{1}{10} + \dfrac{12}{5y}$$

The LCD is $10y$. For each expression, identify the factors from the LCD that are missing from the denominator. Then multiply the numerator and denominator of the expression by the missing factor(s).

$$\underbrace{\frac{1}{10}}_{\substack{\text{missing} \\ y}} + \underbrace{\frac{12}{5y}}_{\substack{\text{missing} \\ 2}}$$

$$= \frac{1 \cdot y}{10 \cdot y} + \frac{12 \cdot 2}{5y \cdot 2}$$

$$= \frac{y}{10y} + \frac{24}{10y} \qquad \text{The rational expressions now have the same denominators.}$$

$$= \frac{y + 24}{10y} \qquad \text{Add the numerators.}$$

After successfully adding or subtracting two rational expressions, always check to see if the final answer is reduced. If necessary, factor the numerator and denominator and reduce common factors. The expression

$$\frac{y + 24}{10y}$$

is in lowest terms because the numerator and denominator do not share any common factors.

Steps to Add or Subtract Rational Expressions

1. Factor the denominators of each rational expression.
2. Identify the LCD.
3. Rewrite each rational expression as an equivalent expression with the LCD as its denominator.
4. Add or subtract the numerators and write the result over the common denominator.
5. Simplify and reduce.

---

example 2

Adding and Subtracting Rational Expressions with Unlike Denominators

Add or subtract as indicated.

a. $\dfrac{4}{7k} - \dfrac{3}{k^2}$ b. $\dfrac{2q - 4}{3} - \dfrac{q + 1}{2}$ c. $\dfrac{1}{x - 5} + \dfrac{-10}{x^2 - 25}$

Solution:

a. $\dfrac{4}{7k} - \dfrac{3}{k^2}$

Step 1: The denominators are already factored.

Step 2: The LCD is $7k^2$.

$= \dfrac{4 \cdot k}{7k \cdot k} - \dfrac{3 \cdot 7}{k^2 \cdot 7}$ **Step 3:** Write each expression with the LCD.

$= \dfrac{4k}{7k^2} - \dfrac{21}{7k^2}$

$= \dfrac{4k - 21}{7k^2}$ **Step 4:** Subtract the numerators and write the result over the LCD.

Step 5: The expression is in lowest terms because the numerator and denominator share no common factors.

b. $\dfrac{2q - 4}{3} - \dfrac{q + 1}{2}$

Step 1: The denominators are already factored.

Step 2: The LCD is 6.

$= \dfrac{2(2q - 4)}{2 \cdot 3} - \dfrac{3(q + 1)}{3 \cdot 2}$ **Step 3:** Write each expression with the LCD.

$= \dfrac{2(2q - 4) - 3(q + 1)}{6}$ **Step 4:** Subtract the numerators and write the result over the LCD.

$= \dfrac{4q - 8 - 3q - 3}{6}$

$= \dfrac{q - 11}{6}$ **Step 5:** The expression is in lowest terms because the numerator and denominator share no common factors.

c. $\dfrac{1}{x - 5} + \dfrac{-10}{x^2 - 25}$

$= \dfrac{1}{x - 5} + \dfrac{-10}{(x - 5)(x + 5)}$ **Step 1:** Factor the denominators.

Step 2: The LCD is $(x - 5)(x + 5)$.

$= \dfrac{1(x + 5)}{(x - 5)(x + 5)} + \dfrac{-10}{(x - 5)(x + 5)}$ **Step 3:** Write each expression with the LCD.

$$= \frac{1(x+5) + (-10)}{(x-5)(x+5)}$$

Step 4: Add the numerators and write the result over the LCD.

$$= \frac{x+5-10}{(x-5)(x+5)}$$

$$= \frac{\overset{1}{\cancel{x-5}}}{\cancel{(x-5)}(x+5)}$$

Step 5: Simplify and reduce.

$$= \frac{1}{x+5}$$

example 3

Adding and Subtracting Rational Expressions with Different Denominators

Add or subtract as indicated.

$$\frac{p+2}{p-1} - \frac{2}{p+6} - \frac{14}{p^2+5p-6}$$

Solution:

$$\frac{p+2}{p-1} - \frac{2}{p+6} - \frac{14}{p^2+5p-6}$$

$$= \frac{p+2}{p-1} - \frac{2}{p+6} - \frac{14}{(p-1)(p+6)}$$

Step 1: Factor the denominators.

Step 2: The LCD is $(p-1)(p+6)$.

Step 3: Write each expression with the LCD.

$$= \frac{(p+2)(p+6)}{(p-1)(p+6)} - \frac{2(p-1)}{(p+6)(p-1)} - \frac{14}{(p-1)(p+6)}$$

$$= \frac{(p+2)(p+6) - 2(p-1) - 14}{(p-1)(p+6)}$$

Step 4: Combine the numerators and write the result over the LCD.

$$= \frac{p^2 + 6p + 2p + 12 - 2p + 2 - 14}{(p - 1)(p + 6)}$$

Step 5: Clear parentheses in the numerator.

$$= \frac{p^2 + 6p}{(p - 1)(p + 6)}$$

Combine *like* terms.

$$= \frac{p(p + 6)}{(p - 1)(p + 6)}$$

Factor the numerator to determine if the expression is in lowest terms.

$$= \frac{p\overset{1}{\cancel{(p + 6)}}}{(p - 1)\cancel{(p + 6)}}$$

Reduce common factors.

$$= \frac{p}{p - 1}$$

When the denominator of two rational expressions are opposites, we can produce identical denominators by multiplying one of the expressions by the ratio $\frac{-1}{-1}$. This is demonstrated in the next example.

---

example 4

Adding Rational Expressions with Unlike Denominators

Add the rational expressions.

$$\frac{1}{d - 7} + \frac{5}{7 - d}$$

Classroom Activity 5.4B

Solution:

$$\frac{1}{d - 7} + \frac{5}{7 - d}$$

The expressions $d - 7$ and $7 - d$ are opposites and differ by a factor of -1. Therefore, multiply the numerator and denominator of *either* expression by -1 to obtain a common denominator.

$$= \frac{1}{d - 7} + \frac{(-1)5}{(-1)(7 - d)}$$

Note that $-1(7 - d) = -7 + d$ or $d - 7$.

$$= \frac{1}{d - 7} + \frac{-5}{d - 7}$$

Simplify.

$$= \frac{1 + (-5)}{d - 7}$$

Add the terms in the numerators and write the result over the common denominator.

$$= \frac{-4}{d - 7}$$

3. Performing the Order of Operations

example 5

Applying the Order of Operations

Simplify the expression: $\left(\dfrac{2a}{bc} - \dfrac{4}{ac}\right)\left(\dfrac{ab}{2}\right)$

Solution:

$\left(\dfrac{2a}{bc} - \dfrac{4}{ac}\right)\left(\dfrac{ab}{2}\right)$

The order of operations indicates that we simplify expressions inside parentheses first. For the two rational expressions in the first parentheses the LCD is abc.

$= \left(\dfrac{2a \cdot a}{bc \cdot a} - \dfrac{4 \cdot b}{ac \cdot b}\right)\left(\dfrac{ab}{2}\right)$ Subtract the expressions in first parentheses by writing the expressions with a common denominator.

$= \left(\dfrac{2a^2 - 4b}{abc}\right)\left(\dfrac{ab}{2}\right)$

$= \dfrac{2(a - 2b)}{abc} \cdot \dfrac{ab}{2}$ To multiply the rational expressions, factor and reduce common factors.

$= \dfrac{\overset{1}{\cancel{2}}(a - 2b)}{\cancel{ab} \cdot c} \cdot \dfrac{\overset{1}{\cancel{ab}}}{\cancel{2}}$ Reduce.

$= \dfrac{a - 2b}{c}$

example 6

Using Rational Expressions in Translations

Translate the English phrase into a mathematical expression. Then simplify by combining the rational expressions.

The difference of the reciprocal of x and the quotient of x and 3.

Solution:

The difference of the reciprocal of x and the quotient of x and 3.

The difference of

$\left(\dfrac{1}{x}\right) - \left(\dfrac{x}{3}\right)$

the reciprocal of x the quotient of x and 3

$$\frac{1}{x} - \frac{x}{3}$$ The LCD is $3x$.

$$= \frac{3 \cdot 1}{3 \cdot x} - \frac{x \cdot x}{3 \cdot x}$$ Write each expression over the LCD.

$$= \frac{3 - x^2}{3x}$$ Subtract the numerators.

section 5.4   PRACTICE EXERCISES

1. For the rational expression   <u>5.1</u>

   $$\frac{x^2 - 4x - 5}{x^2 - 7x + 10}$$

   a. Find the value of the expression (if possible) when $x = 0, 1, -1, 2,$ and 5.
   $-\frac{1}{2}, -2, 0,$ undefined, undefined
   b. Factor the denominator and identify the domain. Write the domain in set-builder notation. $(x - 5)(x - 2); \{x \mid x \ne 5, x \ne 2\}$

   c. Reduce the expression. $\dfrac{x + 1}{x - 2}$

2. For the rational expression   <u>5.1</u>

   $$\frac{a^2 + a - 2}{a^2 - 4a - 12}$$

   a. Find the value of the expression (if possible) when $x = 0, 1, -2, 2,$ and 6.
   $\frac{1}{6}, 0,$ undefined, $-\frac{1}{4},$ undefined
   b. Factor the denominator and identify the domain. Write the domain in set-builder notation. $(a - 6)(a + 2); \{a \mid a \ne 6, a \ne -2\}$

   c. Reduce the expression. $\dfrac{a - 1}{a - 6}$

For Exercises 3–4, multiply or divide as indicated.   <u>5.2</u>

3. $\dfrac{2b^2 - b - 3}{2b^2 - 3b - 9} \cdot \dfrac{4b - 12}{2b - 3} \div \dfrac{b^2 - 1}{4b + 6}$ $\dfrac{8}{b - 1}$

4. $(t - 6) \div \dfrac{5t - 30}{6t - 1} \cdot \dfrac{10t - 25}{2t^2 - 3t - 5}$ $\dfrac{6t - 1}{t + 1}$

5. How many cups are in $\frac{1}{2}$ qt of ice cream? (2 cups = 1 pt, 2 pt = 1 qt, 4 qt = 1 gal)   <u>5.2</u>
   2 cups

6. How many quarts can a 1-L bottle of soda hold? Round to the nearest tenth of a quart. (4 qt = 1 gal, 1 gal = 3.786 L)   1.1 qt

For Exercises 7–20, add or subtract the expressions with like denominators as indicated.

7. $\dfrac{7}{8} + \dfrac{3}{8}$ $\dfrac{5}{4}$

8. $\dfrac{1}{3} + \dfrac{7}{3}$ $\dfrac{8}{3}$

9. $\dfrac{9}{16} - \dfrac{3}{16}$ $\dfrac{3}{8}$

10. $\dfrac{14}{15} - \dfrac{4}{15}$ $\dfrac{2}{3}$

11. $\dfrac{5a}{a + 2} - \dfrac{3a - 4}{a + 2}$ 2

12. $\dfrac{2b}{b - 3} - \dfrac{b - 9}{b - 3}$ $\dfrac{b + 9}{b - 3}$

13. $\dfrac{5c}{c + 6} + \dfrac{30}{c + 6}$ 5

14. $\dfrac{12}{2 + d} + \dfrac{6d}{2 + d}$ 6

15. $\dfrac{5}{t - 8} - \dfrac{2t + 1}{t - 8}$ $\dfrac{-2(t - 2)}{t - 8}$

16. $\dfrac{7p + 1}{2p + 1} - \dfrac{p - 4}{2p + 1}$ $\dfrac{6p + 5}{2p + 1}$

17. $\dfrac{10}{3x - 7} - \dfrac{5}{3x - 7}$ $\dfrac{5}{3x - 7}$

18. $\dfrac{8}{2w - 1} - \dfrac{4}{2w - 1}$ $\dfrac{4}{2w - 1}$

19. $\dfrac{m^2}{m + 5} + \dfrac{10m + 25}{m + 5}$ $m + 5$

20. $\dfrac{k^2}{k - 3} - \dfrac{6k - 9}{k - 3}$ $k - 3$

 For Exercises 21–22, find an expression that represents the perimeter of the figure (assume that $x > 0, y > 0,$ and $t > 0$).

21.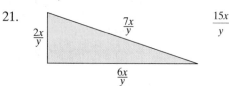

22.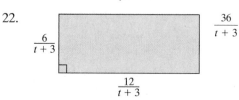

For Exercises 23–28, find the least common denominator. Then convert each expression to an equivalent expression with the denominator equal to the LCD.   5.3

23. $\dfrac{4}{5xy^3}, \dfrac{2x}{15y^2}$ $\dfrac{12}{15xy^3}, \dfrac{2x^2y}{15xy^3}$

24. $\dfrac{5}{3a^2b}, \dfrac{-7}{6b^2}$ $\dfrac{10b}{6a^2b^2}, \dfrac{-7a^2}{6a^2b^2}$

25. $\dfrac{z+7}{3z-9}, \dfrac{z-6}{z-3}$ ❖

26. $\dfrac{3w+4}{w-2}, \dfrac{w-3}{2w-4}$ ❖

27. $\dfrac{6}{a^2-b^2}, \dfrac{2a}{a^3+a^2b}$ $\dfrac{6a^2}{(a-b)(a+b)a^2}, \dfrac{2a^2-2ab}{(a-b)(a+b)a^2}$

28. $\dfrac{7x}{x^2+2xy+y^2}, \dfrac{3}{x^2+xy}$ $\dfrac{7x^2}{x(x+y)^2}, \dfrac{3x+3y}{x(x+y)^2}$

For Exercises 29–44, add or subtract the expressions with unlike denominators as indicated.

29. $\dfrac{5}{a+1} + \dfrac{4}{3a+3}$ ❖

30. $\dfrac{2}{c-4} + \dfrac{1}{5c-20}$ ❖

31. $\dfrac{k}{k^2-9} - \dfrac{4}{k-3}$ ❖

32. $\dfrac{7}{h+5} - \dfrac{2h-3}{h^2-25}$ ❖

33. $\dfrac{3a-7}{6a+10} - \dfrac{10}{3a^2+5a}$ ❖

34. $\dfrac{k+2}{8k} - \dfrac{3-k}{12k}$ ❖

35. $\dfrac{10}{3x-7} + \dfrac{5}{7-3x}$ ❖

36. $\dfrac{8}{2w-1} + \dfrac{4}{1-2w}$ ❖

37. $\dfrac{4n}{n-8} - \dfrac{2n-1}{8-n}$ ❖

38. $\dfrac{m}{m-2} - \dfrac{3m+1}{2-m}$ ❖

39. $\dfrac{5}{x} + \dfrac{3}{x+2}$ $\dfrac{2(4x+5)}{x(x+2)}$

40. $\dfrac{6}{y-1} + \dfrac{9}{y}$ $\dfrac{3(5y-3)}{y(y-1)}$

41. $\dfrac{4w}{w^2+2w-3} + \dfrac{2}{1-w}$ $\dfrac{2(w-3)}{(w+3)(w-1)}$

42. $\dfrac{z-23}{z^2-z-20} - \dfrac{2}{5-z}$ $\dfrac{3}{z+4}$

43. $\dfrac{3a-8}{a^2-5a+6} + \dfrac{a+2}{a^2-6a+8}$ $\dfrac{4a-13}{(a-3)(a-4)}$

44. $\dfrac{3b+5}{b^2+4b+3} + \dfrac{-b+5}{b^2+2b-3}$ $\dfrac{2b}{(b+1)(b-1)}$

For Exercises 45–46, find an expression that represents the perimeter of the figure (assume that $x > 0$ and $t > 0$).

45.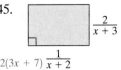

$\dfrac{2}{x+3}$ $\dfrac{1}{x+2}$ $\dfrac{2(3x+7)}{(x+3)(x+2)}$

46.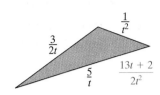

$\dfrac{3}{2t}$ $\dfrac{1}{t^2}$ $\dfrac{5}{t}$ $\dfrac{13t+2}{2t^2}$

47. Let a number be represented by n. Write the reciprocal of n. $\dfrac{1}{n}$

48. Write the reciprocal of the sum of a number and 6. $\dfrac{1}{n+6}$

49. Let a number be represented by p. Write the quotient of 12 and p. $\dfrac{12}{p}$

50. Write the quotient of 5 and the sum of a number and 2. $\dfrac{5}{n+2}$

For Exercises 51–54, translate the English phrases into algebraic expressions. Then simplify by combining the rational expressions.

51. The sum of a number and the quantity seven times the reciprocal of the number. $n + \left(7 \cdot \dfrac{1}{n}\right) = \dfrac{n^2+7}{n}$

52. The sum of a number and the quantity five times the reciprocal of the number. $n + \left(5 \cdot \dfrac{1}{n}\right) = \dfrac{n^2+5}{n}$

53. The difference of the reciprocal of n and the quotient of 2 and n. $\dfrac{1}{n} - \dfrac{2}{n} = -\dfrac{1}{n}$

54. The difference of the reciprocal of m and the quotient of $3m$ and 7. $\dfrac{1}{m} - \dfrac{3m}{7} = \dfrac{7-3m^2}{7m}$

For Exercises 55–58, simplify by applying the order of operations.

55. $\left(\dfrac{2}{k+1} + 3\right)\left(\dfrac{k+1}{4k+7}\right)$ $\dfrac{3k+5}{4k+7}$

56. $\left(\dfrac{p+1}{3p+4}\right)\left(\dfrac{1}{p+1} + 2\right)$ $\dfrac{2p+3}{3p+4}$

57. $\left(\dfrac{1}{10a} - \dfrac{b}{10a^2}\right) \div \left(\dfrac{1}{10} - \dfrac{b}{10a}\right)$ $\dfrac{1}{a}$

58. $\left(\dfrac{1}{2m} + \dfrac{n}{2m^2}\right) \div \left(\dfrac{1}{4} + \dfrac{n}{4m}\right)$ $\dfrac{2}{m}$

■ EXPANDING YOUR SKILLS

For Exercises 59–64, perform the indicated operations.

59. $\dfrac{-3}{w^3 + 27} - \dfrac{1}{w^2 - 9}$ ❖   60. $\dfrac{m}{m^3 - 1} + \dfrac{1}{(m - 1)^2}$ ❖

61. $\dfrac{2p}{p^2 + 5p + 6} - \dfrac{p + 1}{p^2 + 2p - 3} + \dfrac{3}{p^2 + p - 2}$ ❖

62. $\dfrac{3t}{8t^2 + 2t - 1} - \dfrac{5t}{2t^2 - 9t - 5} + \dfrac{2}{4t^2 - 21t + 5}$ ❖

63. $\dfrac{3m}{m^2 + 3m - 10} + \dfrac{5}{4 - 2m} - \dfrac{1}{m + 5}$ ❖

64. $\dfrac{2n}{3n^2 - 8n - 3} + \dfrac{1}{6 - 2n} - \dfrac{3}{3n + 1}$ ❖

---

chapter 5   MIDCHAPTER REVIEW

In Sections 5.1–5.4 we learned how to reduce, add, subtract, multiply, and divide rational expressions. The procedure for each operation is different and it takes considerable practice to determine the correct method to apply for a given problem. The following review exercises give you the opportunity to practice the specific techniques for simplifying rational expressions.

1. a. Write the steps to reduce a rational expression. Factor the numerator and denominator completely, then reduce factors whose ratio is 1 or −1.
   b. Reduce.
   $$\frac{a^2 - 4a}{a^2 - 11a + 28}   \frac{a}{a - 7}$$

2. a. Write the steps to add or subtract rational expressions. ❖
   b. Add or subtract as indicated.
   $$\frac{1}{y} - \frac{3}{y^2 + 3y} + \frac{y}{3y + 9}$$ ❖

3. a. Write the steps to multiply rational expressions. ❖
   b. Multiply.
   $$\frac{2x^2 - 5x - 3}{x^2 - 9} \cdot \frac{x^2 + 6x + 9}{10x + 5}$$ ❖

4. a. Write the steps to divide rational expressions. ❖
   b. Divide.
   $$\frac{c^2 + 5c + 6}{c^2 + c - 2} \div \frac{c}{c - 1}$$ ❖

5. Summarize which operations involve finding an LCD. Adding and subtracting rational expressions

6. Summarize which operations *do not* involve finding an LCD. Multiplying, dividing, and reducing rational expressions

7. Identify the operations on rational expressions in which the first step is to factor the numerator and denominator. Multiplying, dividing, and reducing rational expressions.

8. Identify the operations on rational expressions in which the first step is to factor only the denominator. Adding and subtracting rational expressions.

For Exercises 9–20, perform the indicated operation.

9. Reduce. $\dfrac{6a^2b^3}{72ab^7c}$ $\dfrac{a}{12b^4c}$

10. Subtract. $\dfrac{2a}{a + b} - \dfrac{b}{a - b} - \dfrac{-4ab}{a^2 - b^2}$ $\dfrac{2a - b}{a - b}$

11. Divide. $\dfrac{p^2 + 10pq + 25q^2}{p^2 + 6pq + 5q^2} \div \dfrac{10p + 50q}{2p^2 - 2q^2}$ $\dfrac{p - q}{5}$

12. Add. $\dfrac{3k - 8}{k - 5} + \dfrac{k - 12}{k - 5}$ 4

13. Reduce. $\dfrac{20x^2 + 10x}{4x^3 + 4x^2 + x}$ $\dfrac{10}{2x + 1}$

❖ See Additional Answers Appendix   🖋 Writing   ⬅➡ Translating Expression   📐 Geometry   🖩 Scientific Calculator   Video

14. Multiply.

$$\frac{w^2 - 81}{w^2 + 10w + 9} \cdot \frac{w^2 + w + 2zw + 2z}{w^2 - 9w + zw - 9z} \quad \frac{w + 2z}{w + z}$$

15. Divide. $\dfrac{h^2 - 49}{h + 1} \div \dfrac{h + 7}{h^2 - 1}$ $(h - 7)(h - 1)$

16. Reduce. $\dfrac{xy + 7x + 5y + 35}{x^2 + ax + 5x + 5a}$ $\dfrac{y + 7}{x + a}$

17. Subtract. $\dfrac{a}{a^2 - 9} - \dfrac{3}{6a - 18}$ $\dfrac{1}{2(a + 3)}$

18. Add. $\dfrac{4}{y^2 - 36} + \dfrac{2}{y^2 - 4y - 12}$ $\dfrac{2(3y + 10)}{(y - 6)(y + 6)(y + 2)}$

19. Multiply. $(t^2 + 5t - 24)\left(\dfrac{t + 8}{t - 3}\right)$ $(t + 8)^2$

20. Reduce. $\dfrac{b^2 + 5b - 14}{b - 2}$ $b + 7$

Concepts

1. Simplifying Complex Fractions (Method I)
2. Simplifying Complex Fractions (Method II)

5.5   COMPLEX FRACTIONS

1.   Simplifying Complex Fractions (Method I)

A **complex fraction** is a fraction whose numerator or denominator contains one or more rational expressions. For example,

$$\frac{\dfrac{1}{ab}}{\dfrac{2}{b}} \quad \text{and} \quad \frac{1 + \dfrac{3}{4} - \dfrac{1}{6}}{\dfrac{1}{2} + \dfrac{1}{3}}$$

are complex fractions.

Two methods will be presented to simplify complex fractions. The first method (Method I) follows the order of operations to simplify the numerator and denominator separately before dividing. The process is summarized as follows.

> ### Steps to Simplify a Complex Fraction (Method I)
>
> 1. Add or subtract expressions in the numerator to form a single fraction. Add or subtract expressions in the denominator to form a single fraction.
> 2. Divide the rational expressions from Step 1 by multiplying the numerator of the complex fraction by the reciprocal of the denominator of the complex fraction.
> 3. Simplify and reduce if possible.

example 1   **Simplifying Complex Fractions (Method I)**

Simplify the expression.

$$\frac{\dfrac{1}{ab}}{\dfrac{2}{b}}$$

Solution:

Step 1: The numerator and denominator of the complex fraction are already single fractions.

$$\frac{\dfrac{1}{ab}}{\dfrac{2}{b}} \longleftarrow \text{This fraction bar denotes division } (\div).$$

$$= \frac{1}{ab} \div \frac{2}{b}$$

$$= \frac{1}{ab} \cdot \frac{b}{2} \qquad \textbf{Step 2:} \quad \text{Multiply the numerator of the complex fraction by the reciprocal of } \tfrac{2}{b}, \text{ which is } \tfrac{b}{2}.$$

$$= \frac{1}{a \cdot \cancel{b}} \cdot \frac{\overset{1}{\cancel{b}}}{2} \qquad \textbf{Step 3:} \quad \text{Reduce common factors and simplify.}$$

$$= \frac{1}{2a}$$

Sometimes it is necessary to simplify the numerator and denominator of a complex fraction before the division can be performed. This is illustrated in the next example.

example 2   **Simplifying Complex Fractions (Method I)**

Simplify the expression.

$$\frac{1 + \dfrac{3}{4} - \dfrac{1}{6}}{\dfrac{1}{2} + \dfrac{1}{3}}$$

Solution:

$$\frac{1 + \dfrac{3}{4} - \dfrac{1}{6}}{\dfrac{1}{2} + \dfrac{1}{3}}$$

Step 1: Combine fractions in the numerator and denominator separately.

$$= \frac{\dfrac{12}{12} + \dfrac{9}{12} - \dfrac{2}{12}}{\dfrac{3}{6} + \dfrac{2}{6}}$$

The LCD in the numerator is 12. The LCD in the denominator is 6.

$$= \frac{\dfrac{19}{12}}{\dfrac{5}{6}}$$

Form a single fraction in the numerator and in the denominator.

$$= \frac{19}{\overset{2}{\cancel{12}}} \cdot \frac{\overset{1}{\cancel{6}}}{5}$$

Step 2: Multiply by the reciprocal of $\frac{5}{6}$, which is $\frac{6}{5}$.

$$= \frac{19}{10}$$

Step 3: Simplify and reduce.

---

example 3

Simplifying Complex Fractions (Method I)

Simplify the expression.

$$\frac{\dfrac{1}{x} + \dfrac{1}{y}}{x - \dfrac{y^2}{x}}$$

Classroom Activity 5.5A

Solution:

$$\frac{\dfrac{1}{x} + \dfrac{1}{y}}{x - \dfrac{y^2}{x}}$$

The LCD in the numerator is xy. The LCD in the denominator is x.

$$= \frac{\dfrac{1 \cdot y}{x \cdot y} + \dfrac{1 \cdot x}{y \cdot x}}{\dfrac{x \cdot x}{1 \cdot x} - \dfrac{y^2}{x}}$$

Rewrite the expressions using common denominators.

$$= \frac{\dfrac{y}{xy} + \dfrac{x}{xy}}{\dfrac{x^2}{x} - \dfrac{y^2}{x}}$$

$$= \frac{\dfrac{y+x}{xy}}{\dfrac{x^2-y^2}{x}}$$ Form single fractions in the numerator and denominator.

$$= \frac{y+x}{xy} \cdot \frac{x}{x^2-y^2}$$ Multiply by the reciprocal of the denominator.

$$= \frac{\overset{1}{\cancel{y+x}}}{xy} \cdot \frac{\overset{1}{\cancel{x}}}{(x+y)(x-y)}$$ Factor and reduce. Note that $(y+x) = (x+y)$.

$$= \frac{1}{y(x-y)}$$ Simplify.

2. Simplifying Complex Fractions (Method II)

We will now simplify the expressions from Examples 2 and 3 again using a second method to simplify complex fractions (Method II). Recall that multiplying the numerator and denominator of a rational expression by the same quantity does not change the value of the expression. This is the basis for Method II.

Steps to Simplifying a Complex Fraction (Method II)

1. Multiply the numerator and denominator of the complex fraction by the LCD of *all* individual fractions within the expression.
2. Apply the distributive property and simplify the numerator and denominator.
3. Reduce if necessary.

example 4 **Simplifying Complex Fractions (Method II)**

Simplify the expression.

$$\frac{1 + \dfrac{3}{4} - \dfrac{1}{6}}{\dfrac{1}{2} + \dfrac{1}{3}}$$

Solution:

$$\frac{1 + \dfrac{3}{4} - \dfrac{1}{6}}{\dfrac{1}{2} + \dfrac{1}{3}}$$ The LCD of the expressions: $1, \frac{3}{4}, \frac{1}{6}, \frac{1}{2}$, and $\frac{1}{3}$ is 12.

$$= \frac{12\left(1 + \dfrac{3}{4} - \dfrac{1}{6}\right)}{12\left(\dfrac{1}{2} + \dfrac{1}{3}\right)}$$

Step 1: Multiply the numerator and denominator of the complex fraction by 12.

$$= \frac{12 \cdot 1 + 12 \cdot \dfrac{3}{4} - 12 \cdot \dfrac{1}{6}}{12 \cdot \dfrac{1}{2} + 12 \cdot \dfrac{1}{3}}$$

Step 2: Apply the distributive property.

$$= \frac{12 \cdot 1 + \overset{3}{\cancel{12}} \cdot \dfrac{3}{\cancel{4}} - \overset{2}{\cancel{12}} \cdot \dfrac{1}{\cancel{6}}}{\overset{6}{\cancel{12}} \cdot \dfrac{1}{\cancel{2}} + \overset{4}{\cancel{12}} \cdot \dfrac{1}{\cancel{3}}}$$

Simplify each term.

$$= \frac{12 + 9 - 2}{6 + 4}$$

Simplify.

$$= \frac{19}{10}$$

Step 3: The expression is already reduced.

---

example 5

Simplifying a Complex Fraction (Method II)

Simplify the expression.
$$\frac{\dfrac{1}{x} + \dfrac{1}{y}}{x - \dfrac{y^2}{x}}$$

Solution:

$$\frac{\dfrac{1}{x} + \dfrac{1}{y}}{x - \dfrac{y^2}{x}}$$

The LCD of the expressions: $\frac{1}{x}, \frac{1}{y}, x$, and $\frac{y^2}{x}$ is xy.

$$= \frac{xy\left(\dfrac{1}{x} + \dfrac{1}{y}\right)}{xy\left(x - \dfrac{y^2}{x}\right)}$$

Step 1: Multiply numerator and denominator of the complex fraction by xy.

$$= \frac{\cancel{xy} \cdot \dfrac{1}{\cancel{x}} + x\cancel{y} \cdot \dfrac{1}{\cancel{y}}}{xy \cdot x - \cancel{x}y \cdot \dfrac{y^2}{\cancel{x}}}$$

Step 2: Apply the distributive property and simplify each term.

$$= \frac{y + x}{x^2y - y^3}$$

$$= \frac{y + x}{y(x^2 - y^2)}$$

Step 3: Factor completely and reduce common factors.

$$= \frac{\cancel{y + x}}{y\cancel{(x + y)}(x - y)}$$

Note that $(y + x) = (x + y)$.

$$= \frac{1}{y(x - y)}$$

---

example 6   **Simplifying a Complex Fraction (Method II)**

Simplify the expression. $\dfrac{\dfrac{1}{k + 1} - 1}{\dfrac{1}{k + 1} + 1}$

Solution:

$$\frac{\dfrac{1}{k + 1} - 1}{\dfrac{1}{k + 1} + 1}$$

The LCD of $\dfrac{1}{k + 1}$ and 1 is $(k + 1)$.

$$= \frac{(k + 1)\left(\dfrac{1}{k + 1} - 1\right)}{(k + 1)\left(\dfrac{1}{k + 1} + 1\right)}$$

Step 1: Multiply numerator and denominator of the complex fraction by $(k + 1)$.

$$= \frac{(k + 1)\left(\dfrac{1}{k + 1} - 1\right)}{(k + 1)\left(\dfrac{1}{k + 1} + 1\right)}$$

Step 2: Apply the distributive property.

$$= \frac{1 - (k + 1)}{1 + (k + 1)}$$

Simplify.

$$= \frac{1 - k - 1}{1 + k + 1}$$

$$= \frac{-k}{k + 2}$$

Step 3: The expression is already reduced.

section 5.5 PRACTICE EXERCISES

For Exercises 1–4, write the domain in set-builder notation and reduce the expression. 5.1

1. $\dfrac{(c - 2)(c + 3)}{(c + 1)(c - 2)}$
$\{c \mid c \neq -1, c \neq 2\}; \dfrac{c + 3}{c + 1}$

2. $\dfrac{y(2y + 9)}{y^2(2y + 9)}$
$\left\{y \mid y \neq 0, y = -\dfrac{9}{2}\right\}; \dfrac{1}{y}$

3. $\dfrac{6x + 12}{3x^2 - 12}$
$\{x \mid x \neq 2, x \neq -2\}; \dfrac{2}{x - 2}$

4. $\dfrac{a + 5}{2a^2 + 7a - 15}$
$\left\{a \mid a \neq \dfrac{3}{2}, a = -5\right\}; \dfrac{1}{2a - 3}$

For Exercises 5–12, perform the indicated operations. 5.2–5.4

5. $\dfrac{2}{w - 2} + \dfrac{3}{w}$ $\dfrac{5w - 6}{w(w - 2)}$

6. $\dfrac{6}{5} - \dfrac{3}{5k - 10}$ $\dfrac{3(2k - 5)}{5(k - 2)}$

7. $\dfrac{p^2 + 2p}{2p - 1} \cdot \dfrac{10p^2 - 5p}{12p^3 + 24p^2}$ $\dfrac{5}{12}$

8. $\dfrac{x^2 - 2xy + y^2}{x^4 - y^4} \div \dfrac{3x^2y - 3xy^2}{x^2 + y^2}$ $\dfrac{1}{3xy(x + y)}$

9. $\left(\dfrac{5}{t + 4}\right) \div \left(\dfrac{1}{t - 4} - \dfrac{2}{t^2 - 16}\right)$ $\dfrac{5(t - 4)}{t + 2}$

10. $\left(\dfrac{b}{b + 1} + 1\right) \div \left(\dfrac{2b + 1}{b - 1}\right)$ $\dfrac{b - 1}{b + 1}$

11. $\left(\dfrac{1}{z} - \dfrac{1}{2z}\right) \div \left(\dfrac{1}{2} + \dfrac{1}{2z}\right)$ $\dfrac{1}{z + 1}$

12. $\left(\dfrac{2}{3a^2} - \dfrac{3}{b}\right) \div \left(\dfrac{5}{ab} - 4\right)$ $\dfrac{2b - 9a^2}{3a(5 - 4ab)}$

 13. Mount Rainier National Park in Washington is approximately 235,612 acres. How many square miles is this? Round to the nearest square mile. (1 mile^2 = 640 acres) 5.2   368 square miles

 14. Yellowstone National Park in Wyoming is approximately 3468 miles^2. How many acres is this? (1 mile^2 = 640 acres) 5.2   2,219,520 acres

For Exercises 15–18, translate the English phrases into algebraic expressions. Then simplify the expressions. 5.4

$\dfrac{\frac{1}{2} + \frac{2}{3}}{5} = \dfrac{7}{30}$ 15. The sum of one-half and two-thirds, divided by five.

16. The quotient of ten and the difference of two-fifths and one-fourth. $\dfrac{10}{\frac{2}{5} - \frac{1}{4}} = \dfrac{200}{3}$

17. The quotient of three and the sum of two-thirds and three-fourths. ❖

18. The difference of three-fifths and one-half, divided by four. $\dfrac{\frac{3}{5} - \frac{1}{2}}{4} = \dfrac{1}{40}$

For Exercises 19–38, simplify the complex fractions.

19. $\dfrac{\dfrac{1}{8} + \dfrac{4}{3}}{\dfrac{1}{2} - \dfrac{5}{12}}$ $\dfrac{35}{2}$

20. $\dfrac{\dfrac{8}{9} - \dfrac{1}{3}}{\dfrac{7}{6} + \dfrac{1}{9}}$ $\dfrac{10}{23}$

21. $\dfrac{\dfrac{1}{h} + \dfrac{1}{k}}{\dfrac{1}{hk}}$ $k + h$

22. $\dfrac{\dfrac{1}{b} + 1}{\dfrac{1}{b}}$ $1 + b$

23. $\dfrac{\dfrac{n + 1}{n^2 - 9}}{\dfrac{2}{n + 3}}$ $\dfrac{n + 1}{2(n - 3)}$

24. $\dfrac{\dfrac{5}{k - 5}}{\dfrac{k + 1}{k^2 - 25}}$ $\dfrac{5(k + 5)}{k + 1}$

25. $\dfrac{2 + \dfrac{1}{x}}{4 + \dfrac{1}{x}}$ $\dfrac{2x + 1}{4x + 1}$

26. $\dfrac{6 + \dfrac{6}{k}}{1 + \dfrac{1}{k}}$ 6

27. $\dfrac{\dfrac{m}{7} - \dfrac{7}{m}}{\dfrac{1}{7} + \dfrac{1}{m}}$ $m - 7$

28. $\dfrac{\dfrac{2}{p} + \dfrac{p}{2}}{\dfrac{p}{3} - \dfrac{3}{p}}$ $\dfrac{3(4 + p^2)}{2(p - 3)(p + 3)}$

29. $\dfrac{\dfrac{1}{5} - \dfrac{1}{y}}{\dfrac{7}{10} + \dfrac{1}{y^2}}$ $\dfrac{2y(y - 5)}{7y^2 + 10}$

30. $\dfrac{\dfrac{1}{m^2} + \dfrac{2}{3}}{\dfrac{1}{m} - \dfrac{5}{6}}$ $\dfrac{2(3 + 2m^2)}{m(6 - 5m)}$

31. $\dfrac{\dfrac{8}{a + 4} + 2}{\dfrac{12}{a + 4} - 2}$ $-\dfrac{a + 8}{a - 2}$ or $\dfrac{a + 8}{2 - a}$

32. $\dfrac{\dfrac{2}{w + 1} + 3}{\dfrac{3}{w + 1} + 4}$ $\dfrac{3w + 5}{4w + 7}$

33. $\dfrac{1 - \dfrac{4}{t^2}}{1 - \dfrac{2}{t} - \dfrac{8}{t^2}}$ $\dfrac{t - 2}{t - 4}$

34. $\dfrac{1 - \dfrac{9}{p^2}}{1 - \dfrac{1}{p} - \dfrac{6}{p^2}}$ $\dfrac{p + 3}{p + 2}$

 ❖ See Additional Answers Appendix   Writing   Translating Expression   Geometry   Scientific Calculator   Video

35. $\dfrac{\dfrac{1}{z^2 - 9} + \dfrac{2}{z + 3}}{\dfrac{3}{z - 3}}$ $\dfrac{2z - 5}{3(z + 3)}$

36. $\dfrac{\dfrac{5}{w^2 - 25} - \dfrac{3}{w + 5}}{\dfrac{4}{w - 5}}$ $\dfrac{-3w + 20}{4(w + 5)}$

37. $\dfrac{\dfrac{2}{x - 1} + 2}{\dfrac{2}{x + 1} - 2}$ $\dfrac{x + 1}{-(x - 1)}$ or $\dfrac{x + 1}{1 - x}$

38. $\dfrac{\dfrac{1}{y - 3} + 1}{\dfrac{2}{y + 3} - 1}$ $\dfrac{(y + 3)(y - 2)}{(y + 1)(y - 3)}$

39. In electronics, resistors oppose the flow of current. For two resistors in parallel, the total resistance is given by

$$R = \dfrac{1}{\dfrac{1}{R_1} + \dfrac{1}{R_2}}$$

a. Find the total resistance if $R_1 = 2\ \Omega$ (ohms) and $R_2 = 3\ \Omega$. $\dfrac{6}{5}\ \Omega$

b. Find the total resistance if $R_1 = 10\ \Omega$ and $R_2 = 15\ \Omega$. $6\ \Omega$

40. Suppose that Joelle makes a round trip to a location that is d miles away. If the average rate going to the location is r_1 and the average rate on the return trip is given by r_2, the average rate of the entire trip, R, is given by

$$R = \dfrac{2d}{\dfrac{d}{r_1} + \dfrac{d}{r_2}}$$

a. Find the average rate of a trip to a destination 30 miles away when the average rate going there was 60 mph and the average rate returning home was 45 mph. (Round to the nearest tenth of a mile per hour.) 51.4 mph average

b. Find the average rate of a trip to a destination that is 50 miles away if the driver travels at the same rates as in part (a). (Round to the nearest tenth of a mile per hour.) 51.4 mph

c. Compare your answers from parts (a) and (b) and explain the results in the context of the problem. Because the rates going to and leaving from the destination are the same, the average rate is unchanged. The average rate is not affected by the distance traveled.

▣ EXPANDING YOUR SKILLS

For Exercises 41–43, simplify the complex fractions. (*Hint*: Use the order of operations and begin with the fraction on the lower right.)

41. $1 + \dfrac{1}{1 + 1}$ $\dfrac{3}{2}$

42. $1 + \dfrac{1}{1 + \dfrac{1}{1 + 1}}$ $\dfrac{5}{3}$

43. $1 + \dfrac{1}{1 + \dfrac{1}{1 + \dfrac{1}{1 + 1}}}$ $\dfrac{8}{5}$

---

section

5.6   RATIONAL EQUATIONS

Concepts

1. **Definition of a Rational Equation**
2. **Clearing Fractions**
3. **Solving Rational Equations**
4. **Solving Formulas Involving Rational Equations**

1. Definition of a Rational Equation

Thus far we have studied two specific types of equations in one variable: linear equations and quadratic equations. Recall,

$ax + b = 0$, where $a \neq 0$, is a **linear equation**

$ax^2 + bx + c = 0$, where $a \neq 0$, is a **quadratic equation**.

We will now study another type of equation called a rational equation.

Definition of a Rational Equation

An equation with one or more rational expressions is called a **rational equation**.

The following equations are rational equations:

$$\frac{y}{2} + \frac{y}{4} = 6 \qquad \frac{1}{x} + \frac{1}{3} = \frac{5}{6} \qquad \frac{6}{t^2 - 7t + 12} + \frac{2t}{t - 3} = \frac{3t}{t - 4}$$

2. Clearing Fractions

To understand the process of solving a rational equation, first review the process of clearing fractions from Section 2.4.

example 1

Solving a Rational Equation

Solve. $\quad \dfrac{y}{2} + \dfrac{y}{4} = 6$

Solution:

$$\frac{y}{2} + \frac{y}{4} = 6 \qquad \text{The LCD of all terms in the equation is 4.}$$

$$4\left(\frac{y}{2} + \frac{y}{4}\right) = 4(6) \qquad \text{Multiply both sides of the equation by 4 to clear fractions.}$$

$$4 \cdot \frac{y}{2} + 4 \cdot \frac{y}{4} = 4(6) \qquad \text{Apply the distributive property.}$$

$$2y + y = 24 \qquad \text{Clear fractions.}$$

$$3y = 24 \qquad \text{Solve the resulting equation (linear).}$$

$$y = 8$$

Check: $\quad \dfrac{y}{2} + \dfrac{y}{4} = 6$

$$\frac{(8)}{2} + \frac{(8)}{4} \stackrel{?}{=} 6$$

$$4 + 2 \stackrel{?}{=} 6$$

$$6 = 6 \checkmark$$

3. Solving Rational Equations

The same process of clearing fractions is used to solve rational equations when variables are present in the denominator.

example 2

Solving a Rational Equation

Solve the equation. $\dfrac{1}{x} + \dfrac{1}{3} = \dfrac{5}{6}$

Solution:

$$\frac{1}{x} + \frac{1}{3} = \frac{5}{6}$$ 　　　The LCD of all expressions is $6x$.

$$6x\left(\frac{1}{x} + \frac{1}{3}\right) = 6x\left(\frac{5}{6}\right)$$ 　　　Multiply by the LCD.

$$6x \cdot \frac{1}{x} + 6x \cdot \frac{1}{3} = 6x \cdot \frac{5}{6}$$ 　　　Apply the distributive property.

$$6 + 2x = 5x$$ 　　　Clear fractions.

$$6 = 3x$$ 　　　Solve the resulting equation (linear).

$$x = 2$$

$$\underline{\text{Check:}} \quad \frac{1}{x} + \frac{1}{3} = \frac{5}{6}$$

$$\frac{1}{(2)} = \frac{1}{3} \overset{?}{=} \frac{5}{6}$$

$$\frac{3}{6} + \frac{2}{6} \overset{?}{=} \frac{5}{6}$$

$$\frac{5}{6} = \frac{5}{6} \checkmark$$

example 3

Solving a Rational Equation

Solve the equation. $1 + \dfrac{3a}{a-2} = \dfrac{6}{a-2}$

Solution:

$$1 + \frac{3a}{a-2} = \frac{6}{a-2}$$ 　　　The LCD of all expressions is $a - 2$.

$$(a-2)\left(1 + \frac{3a}{a-2}\right) = (a-2)\left(\frac{6}{a-2}\right)$$ 　　　Multiply by the LCD.

$$(a-2)1 + (a-2)\left(\frac{3a}{a-2}\right) = (a-2)\left(\frac{6}{a-2}\right)$$ 　　　Apply the distributive property.

$$a - 2 + 3a = 6$$

$$4a - 2 = 6$$

$$4a = 8$$

$$a = 2$$

Solve the resulting equation (linear).

Check: $1 + \dfrac{3a}{a-2} = \dfrac{6}{a-2}$

$1 + \dfrac{3(2)}{(2)-2} \overset{?}{=} \dfrac{6}{(2)-2}$

$1 + \dfrac{6}{0} \overset{?}{=} \dfrac{6}{0}$

The denominator is 0 when $a = 2$.

Because the value $a = 2$ makes the denominator zero in one (or more) of the rational expressions within the equation, the equation is undefined for $a = 2$. No other potential solutions exist for the equation,

$$1 + \frac{3a}{a-2} = \frac{6}{a-2};$$

hence, it has no solution.

Examples 1–3 show that the steps to solve a rational equation mirror the process of clearing fractions from Section 2.4. However, there is one significant difference. The solutions of a rational equation must not make the denominator equal to zero for any expression within the equation. When $a = 2$ is substituted into the expression

$$\frac{3a}{a-2} \qquad \text{or} \qquad \frac{6}{a-2}$$

the denominator is zero and the expression is undefined. Hence, $a = 2$ cannot be a solution to the equation

$$1 + \frac{3a}{a-2} = \frac{6}{a-2}$$

The steps to solve a rational equation are summarized as follows.

Steps to Solve a Rational Equation

1. Factor the denominators of all rational expressions.
2. Identify the LCD of all expressions in the equation.
3. Multiply both sides of the equation by the LCD.
4. Solve the resulting equation.
5. Check potential solutions in the original equation.

example 4   **Solving Rational Equations**

Solve the equations.

a. $1 - \dfrac{4}{p} = -\dfrac{3}{p^2}$ b. $\dfrac{6}{t^2 - 7t + 12} + \dfrac{2t}{t - 3} = \dfrac{3t}{t - 4}$

Classroom Activity 5.6A

Solution:

a. $\qquad 1 - \dfrac{4}{p} = -\dfrac{3}{p^2}$ **Step 1:**   The denominators are already factored.

Step 2:   The LCD of all expressions is p^2.

$p^2\left(1 - \dfrac{4}{p}\right) = p^2\left(-\dfrac{3}{p^2}\right)$ **Step 3:**   Multiply by the LCD.

$p^2(1) - p^2\left(\dfrac{4}{p}\right) = p^2\left(-\dfrac{3}{p^2}\right)$ Apply the distributive property.

$p^2 - 4p = -3$ **Step 4:**   Solve the resulting equation (quadratic).

$p^2 - 4p + 3 = 0$ Set the equation equal to zero and factor.

$(p - 3)(p - 1) = 0$

$p - 3 = 0 \quad \text{or} \quad p - 1 = 0$ Set each factor equal to zero.

$p = 3 \quad \text{or} \quad p = 1$ **Step 5:** $\underline{\text{Check: } p = 3} \qquad \underline{\text{Check: } p = 1}$

$$1 - \dfrac{4}{p} = -\dfrac{3}{p^2} \qquad 1 - \dfrac{4}{p} = -\dfrac{3}{p^2}$$

$$1 - \dfrac{4}{(3)} \overset{?}{=} -\dfrac{3}{(3)^2} \qquad 1 - \dfrac{4}{(1)} \overset{?}{=} -\dfrac{3}{(1)^2}$$

$$\dfrac{3}{3} - \dfrac{4}{3} \overset{?}{=} -\dfrac{3}{9} \qquad 1 - 4 \overset{?}{=} -3$$

Both solutions $p = 3$ and $p = 1$ check. $-\dfrac{1}{3} = -\dfrac{1}{3} \checkmark \qquad -3 = -3 \checkmark$

b. $\dfrac{6}{t^2 - 7t + 12} + \dfrac{2t}{t - 3} = \dfrac{3t}{t - 4}$

$\dfrac{6}{(t - 3)(t - 4)} + \dfrac{2t}{t - 3} = \dfrac{3t}{t - 4}$ **Step 1:**   Factor the denominators.

Step 2:   The LCD is $(t - 3)(t - 4)$.

Step 3: Multiply by the LCD on both sides.

$$(t-3)(t-4)\left(\frac{6}{(t-3)(t-4)} + \frac{2t}{t-3}\right) = (t-3)(t-4)\left(\frac{3t}{t-4}\right)$$

$$\cancel{(t-3)}\cancel{(t-4)}\left(\frac{6}{\cancel{(t-3)}\cancel{(t-4)}}\right) + \cancel{(t-3)}(t-4)\left(\frac{2t}{\cancel{t-3}}\right) = (t-3)\cancel{(t-4)}\left(\frac{3t}{\cancel{t-4}}\right)$$

$$6 + 2t(t-4) = 3t(t-3)$$

$6 + 2t^2 - 8t = 3t^2 - 9t$ **Step 4:** Solve the resulting equation.

$0 = 3t^2 - 2t^2 - 9t + 8t - 6$

$0 = t^2 - t - 6$

$0 = (t-3)(t+2)$

Because the resulting equation is quadratic, set the equation equal to zero and factor.

$t - 3 = 0$ or $t + 2 = 0$ Set each factor equal to zero.

     $t = 3$ or $t = -2$

Step 5: Check the potential solutions in the original equation.

Check: $t = 3$ Check: $t = -2$

$t = 3$ cannot be a solution to the equation because it will make the denominator zero in the original equation.

$$\frac{6}{t^2 - 7t + 12} + \frac{2t}{t-3} = \frac{3t}{t-4}$$

$$\frac{6}{(-2)^2 - 7(-2) + 12} + \frac{2(-2)}{(-2) - 3} \stackrel{?}{=} \frac{3(-2)}{(-2) - 4}$$

$$\frac{6}{t^2 - 7t + 12} + \frac{2t}{t-3} = \frac{3t}{t-4}$$

$$\frac{6}{4 + 14 + 12} + \frac{-4}{-5} \stackrel{?}{=} \frac{-6}{-6}$$

$$\frac{6}{(3)^2 - 7(3) + 12} + \frac{2(3)}{(3) - 3} \stackrel{?}{=} \frac{3(3)}{(3) - 4}$$

$$\frac{6}{30} + \frac{4}{5} \stackrel{?}{=} 1$$

$$\frac{6}{0} + \frac{6}{0} \stackrel{?}{=} \frac{9}{-1}$$

$$\frac{1}{5} + \frac{4}{5} = 1 \checkmark$$

zero in the denominator

$t = -2$ is a solution.

The only solution is $t = -2$.

---

example 5   **Solving a Rational Equation**

Ten times the reciprocal of a number is added to four. The result is equal to the quotient of twenty-two and the number. Find the number.

Solution:

Let x represent the number.

$$\underset{\substack{\uparrow \\ \text{is added} \\ \text{to four}}}{4} + \underset{\substack{\uparrow \\ \text{the reciprocal} \\ \text{of a number}}}{10\underset{\substack{\text{10} \\ \text{times}}}{}\left(\frac{1}{x}\right)} \underset{\substack{\uparrow \\ \text{the result} \\ \text{is equal to}}}{=} \underset{\substack{\uparrow \\ \text{the quotient of} \\ \text{22 and the number}}}{\frac{22}{x}}$$

$4 + \dfrac{10}{x} = \dfrac{22}{x}$

Step 1: The denominators are already factored.

Step 2: The LCD is x.

$x\left(4 + \dfrac{10}{x}\right) = x\left(\dfrac{22}{x}\right)$

Step 3: Multiply both sides by the LCD.

$4x + 10 = 22$

Apply the distributive property.

$4x = 22 - 10$

Step 4: Solve the resulting equation (linear).

$4x = 12$

$x = 3$ is a potential solution.

Step 5: Substituting $x = 3$ into the original equation verifies that it is a solution.

The number is 3.

4. Solving Formulas Involving Rational Equations

example 6

Solving Formulas Involving Rational Equations

Solve for b: $h = \dfrac{2A}{B + b}$

Classroom Activity 5.6B

Solution:

To solve for b, we must clear fractions so that b appears in the numerator.

$h = \dfrac{2A}{B + b}$ The LCD is $B + b$.

$h(B + b) = \left(\dfrac{2A}{B + b}\right) \cdot (B + b)$ Multiply both sides of the equation by the LCD.

$hB + hb = 2A$ Apply the distributive property.

$$hb = 2A - hB$$ Subtract hB from both sides to isolate the b term.

$$\frac{\cancel{h}b}{\cancel{h}} = \frac{2A - hB}{h}$$ Divide by h.

$$b = \frac{2A - hB}{h}$$

Tip: The solution to Example 6 can be written in several forms. The quantity

$$\frac{2A - hB}{h}$$

can be left as a single rational expression or can be split into two fractions and reduced. Any of the following forms is a valid representation of b:

$$b = \frac{2A - hB}{h} = \frac{2A}{h} - \frac{hB}{h} = \frac{2A}{h} - B$$

---

section 5.6   PRACTICE EXERCISES

For Exercises 1–6, perform the indicated operations. 5.2–5.5

1. $\dfrac{2}{x - 3} - \dfrac{3}{x^2 - x - 6}$ $\dfrac{2x + 1}{(x - 3)(x + 2)}$

2. $\dfrac{\dfrac{3}{a} + \dfrac{5}{2a}}{1 + \dfrac{2}{a + 2}}$ $\dfrac{11(a + 2)}{2a(a + 4)}$

3. $\dfrac{t^2 - 5t + 6}{t^2 - 5t - 6} \div \dfrac{t^2 - 4}{t^2 + 2t + 1}$ $\dfrac{(t - 3)(t + 1)}{(t - 6)(t + 2)}$

4. $\dfrac{2y}{y - 3} + \dfrac{4}{y^2 - 9}$ $\dfrac{2(y + 2)(y + 1)}{(y - 3)(y + 3)}$

5. $\dfrac{h - \dfrac{1}{h}}{\dfrac{1}{5} - \dfrac{1}{5h}}$ $5(h + 1)$

6. $\dfrac{w - 4}{w^2 - 9} \cdot \dfrac{w - 3}{w^2 - 8w + 16}$ $\dfrac{1}{(w + 3)(w - 4)}$

7. The distance between St. Johns, Newfoundland, Canada and Gander, Newfoundland, Canada is 154 miles. Convert this distance to kilometers. Round to the nearest tenth of a kilometer. (1 mile = 1.609 km)   5.2   247.8 km

8. The distance between Beijing, China, and London, England, is 8161 km. Convert this distance to miles. Round to the nearest mile. (1 mile = 1.609 km)   5.2   5072 miles

For Exercises 9–14, solve the equations by first clearing the fractions. 2.4

9. $\dfrac{1}{3}z + \dfrac{2}{3} = -2z + 10$ $z = 4$

10. $\dfrac{5}{2} + \dfrac{1}{2}b = 5 - \dfrac{1}{3}b$ $b = 3$

11. $\dfrac{3}{2}p + \dfrac{1}{3} = \dfrac{2p - 3}{4}$ $p = -\dfrac{13}{12}$

12. $\dfrac{5}{3} - \dfrac{1}{6}k = \dfrac{3k + 5}{4}$ $k = \dfrac{5}{11}$

 See Additional Answers Appendix   ✏ Writing   ⬌ Translating Expression   🌐 Geometry   🖩 Scientific Calculator   📼 Video

13. $\dfrac{2x - 3}{4} + \dfrac{9}{10} = \dfrac{x}{5}$ $x = -\dfrac{1}{2}$

14. $\dfrac{4y + 2}{3} - \dfrac{7}{6} = -\dfrac{y}{6}$ $y = \dfrac{1}{3}$

15. For the equation.

$$\dfrac{1}{w} - \dfrac{1}{2} = -\dfrac{1}{4}$$

   a. Identify the LCD of all of the denominators of the equation. $4w$

   b. Solve the equation. $w = 4$

16. For the equation.

$$\dfrac{3}{z} - \dfrac{4}{5} = -\dfrac{1}{5}$$

   a. Identify the LCD of all of the denominators of the equation. $5z$

   b. Solve the equation. $z = 5$

17. For the equation.

$$\dfrac{x + 1}{x^2 + 2x - 3} = \dfrac{1}{x + 3} - \dfrac{1}{x - 1}$$

   a. Identify the LCD of all of the denominators of the equation. $(x + 3)(x - 1)$

   b. Solve the equation. $x = -5$

18. For the equation.

$$\dfrac{10}{x - 2} - \dfrac{40}{x^2 + x - 6} = \dfrac{12}{x + 3}$$

   a. Identify the LCD of all of the denominators of the equation. $(x - 2)(x + 3)$

   b. Solve the equation. $x = 7$

For Exercises 19–34, solve the equations.

19. $1 - \dfrac{2}{y} = \dfrac{3}{y^2}$ $y = 3, y = -1$

20. $1 - \dfrac{2}{m} = \dfrac{8}{m^2}$ $m = 4, m = -2$

21. $\dfrac{a + 1}{a} = 1 + \dfrac{a - 2}{2a}$ $a = 4$

22. $\dfrac{7b - 4}{5b} = \dfrac{9}{5} - \dfrac{4}{b}$ $b = 8$

23. $\dfrac{w}{5} - \dfrac{w + 3}{w} = -\dfrac{3}{w}$ $w = 5; (w = 0$ does not check.)

24. $\dfrac{t}{12} + \dfrac{t + 3}{3t} = \dfrac{1}{t}$ $t = -4; (t = 0$ does not check.)

25. $\dfrac{2}{m + 3} = \dfrac{5}{4m + 12} - \dfrac{3}{8}$ $m = -5$

26. $\dfrac{2}{4n - 4} - \dfrac{7}{4} = \dfrac{-3}{n - 1}$ $n = 3$

27. $\dfrac{p}{p - 4} - 5 = \dfrac{4}{p - 4}$ No solution; ($p = 4$ does not check.)

28. $\dfrac{-5}{q + 5} = \dfrac{q}{q + 5} + 2$ No solution; ($q = -5$ does not check.)

29. $\dfrac{2t}{t + 2} - 2 = \dfrac{t - 8}{t + 2}$ $t = 4$

30. $\dfrac{4w}{w - 3} - 3 = \dfrac{3w - 1}{w - 3}$ $w = 5$

31. $\dfrac{2x}{x + 4} - \dfrac{8}{x - 4} = \dfrac{2x^2 + 32}{x^2 - 16}$ No solution; ($x = -4$ does not check.)

32. $\dfrac{4x}{x + 3} - \dfrac{12}{x - 3} = \dfrac{4x^2 + 36}{x^2 - 9}$ No solution; ($x = -3$ does not check.)

33. $\dfrac{x}{x + 6} = \dfrac{72}{x^2 - 36} + 4$ $x = 4; (x = -6$ does not check.)

34. $\dfrac{y}{y + 4} = \dfrac{32}{y^2 - 16} + 3$ $y = 2; (y = -4$ does not check.)

35. The reciprocal of a number is added to three. The result is the quotient of 25 and the number. Find the number. The number is 8.

36. The difference of three and the reciprocal of a number is equal to the quotient of 20 and the number. Find the number. The number is 7.

37. If a number added to five is divided by the difference of the number and two, the result is three-fourths. Find the number. The number is −26.

38. If twice a number added to three is divided by the number plus one, the result is three-halves. Find the number. The number is −3.

For Exercises 39–48, solve for the indicated variable.

39. $K = \dfrac{ma}{F}$ for m $m = \dfrac{FK}{a}$

40. $K = \dfrac{ma}{F}$ for a $a = \dfrac{FK}{m}$

41. $K = \dfrac{IR}{E}$ for E $E = \dfrac{IR}{K}$

42. $K = \dfrac{IR}{E}$ for R $R = \dfrac{KE}{I}$

43. $I = \dfrac{E}{R + r}$ for R $R = \dfrac{E - Ir}{I}$ or $R = \dfrac{E}{I} - r$

44. $I = \dfrac{E}{R + r}$ for r $r = \dfrac{E - IR}{I}$ or $r = \dfrac{E}{I} - R$

45. $h = \dfrac{2A}{B + b}$ for B $B = \dfrac{2A - hb}{h}$ or $B = \dfrac{2A}{h} - b$

46. $\dfrac{C}{\pi r} = 2$ for r $r = \dfrac{C}{2\pi}$

47. $\dfrac{V}{\pi h} = r^2$ for h $h = \dfrac{V}{r^2\pi}$

48. $\dfrac{V}{lw} = h$ for w $w = \dfrac{V}{lh}$

section

5.7   Applications of Rational Equations and Proportions

1.   Proportions

Recall from Section 2.6 that a **proportion** is an equation that relates two ratios. That is, for $b \neq 0$ and $d \neq 0$, the equation $\frac{a}{b} = \frac{c}{d}$ is a proportion. A proportion can be solved by multiplying both sides by the LCD and clearing fractions.

---

example 1   **Solving a Proportion**

Solve the proportion.

$$\frac{3}{11} = \frac{123}{w}$$

Solution:

$$\frac{3}{11} = \frac{123}{w} \qquad \text{The LCD is } 11w.$$

$$11w\left(\frac{3}{11}\right) = 11w\left(\frac{123}{w}\right) \qquad \text{Multiply by the LCD and clear fractions.}$$

$$3w = 11 \cdot 123 \qquad \text{Solve the resulting equation (linear).}$$

$$3w = 1353$$

$$\frac{3w}{3} = \frac{1353}{3}$$

$$w = 451$$

$$\underline{\text{Check}}: w = 451$$

$$\frac{3}{11} = \frac{123}{w}$$

$$\frac{3}{11} \stackrel{?}{=} \frac{123}{(451)}$$

Reduce

$$\frac{3}{11} = \frac{3}{11} \checkmark$$

Tip: Recall from Section 2.6 we learned that the cross products of a proportion are equal. That is, for $b \neq 0$ and $d \neq 0$, the proportion $\dfrac{a}{b} = \dfrac{c}{d}$ is equivalent to $ad = bc$.

Some rational equations are proportions and can be solved by equating the cross products. Consider the proportion from Example 1:

$$\frac{3}{11} \bowtie \frac{123}{w}$$

$$3 \cdot w = 11 \cdot 123 \qquad \text{Equate the cross products.}$$

$$3w = 1353 \qquad \text{Solve the resulting equation.}$$

$$\frac{3w}{3} = \frac{1353}{3}$$

$$w = 451$$

2. Applications of Proportions

example 2

Using a Proportion in an Application

The population of Alabama in 1997 was approximately 4.2 million. At that time, Alabama had seven representatives in the U.S. House of Representatives. In 1997, North Carolina had a population of approximately 7.2 million. If representation in the House is based on population in equal proportions for each state, how many representatives would North Carolina be expected to have?

Classroom Activity 5.6A

Solution:

Let x represent the number of representatives for North Carolina.

Set up a proportion by writing two equivalent ratios.

$$\boxed{\frac{\text{Population of Alabama}}{\text{Number of representatives}}} \rightarrow \frac{4.2}{7} = \frac{7.2}{x} \leftarrow \boxed{\frac{\text{Population of North Carolina}}{\text{Number of representatives}}}$$

$$\frac{4.2}{7} = \frac{7.2}{x}$$

Tip: The equation from Example 2 could have been solved by first equating the cross products:

$$\frac{4.2}{7} \bowtie \frac{7.2}{x}$$

$$4.2x = (7.2)(7)$$

$$4.2x = 50.4$$

$$x = 12$$

$$7x \cdot \frac{4.2}{7} = 7x \cdot \frac{7.2}{x} \qquad \text{Multiply by the LCD, } 7x.$$

$$4.2x = (7.2)(7) \qquad \text{Solve the resulting equation (linear).}$$

$$4.2x = 50.4$$

$$\frac{4.2x}{4.2} = \frac{50.4}{4.2}$$

$$x = 12 \qquad \text{We would expect North Carolina to have had 12 representatives in 1997.}$$

example 3

Using a Proportion in an Application

In a large lecture class in chemistry, the ratio of men to women is 3 to 1. If there are 124 more men than women, find the number of women.

Solution:

Let x represent the number of women.

Then $x + 124$ represents the number of men.

Set up a proportion by writing two equivalent ratios.

$$\boxed{\dfrac{\text{Number of men}}{\text{Number of women}}} \begin{array}{c} \rightarrow \\ \rightarrow \end{array} \dfrac{3}{1} = \dfrac{x + 124}{x} \begin{array}{c} \leftarrow \\ \leftarrow \end{array} \boxed{\dfrac{\text{Number of men}}{\text{Number of women}}}$$

$$x \cdot \frac{3}{1} = x \cdot \left(\frac{x + 124}{x}\right) \qquad \text{Multiply by the LCD, } x.$$

$$3x = x + 124 \qquad \text{Solve the resulting equation (linear).}$$

$$2x = 124$$

$$x = 62$$

There are 62 women.

Because $x + 124 = 62 + 124 = 186$, there are 186 men.

3. Distance, Rate, Time Applications

In Section 2.6, we presented applications involving the relationship among the variables distance, rate, and time. Recall that $d = rt$.

example 4

Using a Rational Equation in a Distance, Rate, Time Application

A small plane flies 440 miles with the wind from Memphis, Tennessee, to Oklahoma City, Oklahoma. In the same amount of time, the plane flies 340 miles against the wind from Oklahoma City to Little Rock, Arkansas (see Figure 5-2). If the wind speed is 30 mph, find the speed of the plane in still air.

Figure 5-2

Solution:

Let x represent the speed of the plane in still air.

Organize the given information in a chart.

	Distance	Rate	Time
With the Wind	440	$x + 30$	$\dfrac{440}{x + 30}$
Against the Wind	340	$x - 30$	$\dfrac{340}{x - 30}$

Because $d = rt$, then $t = \dfrac{d}{r}$

The plane travels with the wind for the same amount of time as it travels against the wind, so we can equate the two expressions for time.

$$\begin{pmatrix} \text{Time with} \\ \text{the wind} \end{pmatrix} = \begin{pmatrix} \text{time against} \\ \text{the wind} \end{pmatrix}$$

$$\frac{440}{x + 30} = \frac{340}{x - 30} \qquad \text{The LCD is } (x + 30)(x - 30).$$

Tip: The equation

$$\frac{440}{x + 30} = \frac{340}{x - 30}$$

is a proportion. The fractions can also be cleared by equating the cross products.

$$\frac{440}{x + 30} \diagup\!\!\!\!\diagdown \frac{340}{x - 30}$$

$$440(x - 30) = 340(x + 30)$$

$$\cancel{(x + 30)}(x - 30) \cdot \frac{440}{\cancel{x + 30}} = (x + 30)\cancel{(x - 30)} \cdot \frac{340}{\cancel{x - 30}}$$

$$440(x - 30) = 340(x + 30)$$

$$440x - 13{,}200 = 340x + 10{,}200 \qquad \text{Solve the resulting linear equation.}$$

$$100x = 23{,}400$$

$$x = 234$$

The plane's speed in still air is 234 mph.

---

example 5   **Using a Rational Equation in a Distance, Rate, Time Application**

A motorist drives 100 miles between two cities in a bad rainstorm. For the return trip in sunny weather, she averages 10 mph faster and takes $\frac{1}{2}$ hour less time. Find the average speed of the motorist in the rainstorm and in sunny weather.

Solution:

Let x represent the motorist's speed during the rain.

Then $x + 10$ represents the speed in sunny weather.

	Distance	Rate	Time
Trip during Rainstorm	100	x	$\dfrac{100}{x}$
Trip during Sunny Weather	100	$x + 10$	$\dfrac{100}{x + 10}$

Because $d = rt$, then $t = \dfrac{d}{r}$

Because the same distance is traveled in $\frac{1}{2}$ h less time, the difference between the time of the trip during the rainstorm and the time during sunny weather is $\frac{1}{2}$ h.

$$\left(\begin{matrix}\text{Time during}\\\text{the rainstorm}\end{matrix}\right) - \left(\begin{matrix}\text{time during}\\\text{sunny weather}\end{matrix}\right) = \left(\frac{1}{2}\text{ h}\right) \qquad \text{Verbal model}$$

$$\frac{100}{x} - \frac{100}{x + 10} = \frac{1}{2} \qquad \text{Mathematical equation}$$

$$2x(x + 10)\left(\frac{100}{x} - \frac{100}{x + 10}\right) = 2x(x + 10)\left(\frac{1}{2}\right) \qquad \text{Multiply by the LCD.}$$

$$2x(x + 10)\left(\frac{100}{x}\right) - 2x(x + 10)\left(\frac{100}{x + 10}\right) = 2x(x + 10)\left(\frac{1}{2}\right) \qquad \text{Apply the distributive property.}$$

$$200(x + 10) - 200x = x(x + 10) \qquad \text{Clear fractions.}$$

$$200x + 2000 - 200x = x^2 + 10x \qquad \text{Solve the resulting equation (quadratic).}$$

$$2000 = x^2 + 10x$$

$$0 = x^2 + 10x - 2000 \qquad \text{Set the equation equal to zero.}$$

$$0 = (x - 40)(x + 50) \qquad \text{Factor.}$$

$$x = 40 \qquad \text{or} \qquad x = -50$$

Because a rate of speed cannot be negative, reject $x = -50$. Therefore, the speed of the motorist in the rainstorm is 40 mph. Because $x + 10 = 40 + 10 = 50$, the average speed for the return trip in sunny weather is 50 mph.

Avoiding Mistakes

The equation

$$\frac{100}{x} - \frac{100}{x + 10} = \frac{1}{2}$$

is not a proportion because the left-hand side has more than one fraction. Do not try to multiply the cross products. Instead, multiply by the LCD to clear fractions.

4. "Work" Problems

The next example demonstrates how work rates are related to a portion of a job that can be completed in one unit of time.

example 6

Classroom Activity 5.7C

Using a Rational Equation in a Work Problem

A new printing press can print the morning edition in 2 hours, whereas the old printer required 4 hours. How long would it take to print the morning edition if both printers were working together?

Solution:

Let x represent the time required for both printers working together to complete the job.

One method to approach this problem is to determine the portion of the job that each printer can complete in 1 hour and extend that rate to the portion of the job completed in x hours.

- The old printer can perform the job in 4 hours. Therefore, it completes $\frac{1}{4}$ of the job in 1 hour and $\frac{1}{4}x$ jobs in x hours.
- The new printer can perform the job in 2 hours. Therefore, it completes $\frac{1}{2}$ of the job in 1 hour and $\frac{1}{2}x$ jobs in x hours.

	Work Rate	Time	Portion of Job Completed
Old Printer	$\dfrac{1 \text{ job}}{4 \text{ h}}$	x hours	$\dfrac{1}{4}x$
New Printer	$\dfrac{1 \text{ job}}{2 \text{ h}}$	x hours	$\dfrac{1}{2}x$

The sum of the portions of the job completed by each printer must equal one whole job.

$$\begin{pmatrix} \text{Portion of job} \\ \text{completed by} \\ \text{old printer} \end{pmatrix} + \begin{pmatrix} \text{portion of job} \\ \text{completed by} \\ \text{new printer} \end{pmatrix} = \begin{pmatrix} 1 \\ \text{whole} \\ \text{job} \end{pmatrix}$$

$\dfrac{1}{4}x + \dfrac{1}{2}x = 1$ The LCD is 4.

$4\left(\dfrac{1}{4}x + \dfrac{1}{2}x\right) = 4(1)$ Multiply by the LCD.

$4 \cdot \dfrac{1}{4}x + 4 \cdot \dfrac{1}{2}x = 4 \cdot 1$ Apply the distributive property.

$x + 2x = 4$ Solve the resulting linear equation.

$3x = 4$

$x = \dfrac{4}{3}$ or $x = 1\dfrac{1}{3}$ The time required to print the morning edition using both printers is $1\frac{1}{3}$ h.

section 5.7   PRACTICE EXERCISES

For Exercises 1–8, determine whether each of the following is an equation or an expression. If it is an equation, solve it. If it is an expression, perform the indicated operation.   5.2–5.6

1. $\dfrac{b}{5} + 3 = 9$ Expression; $\dfrac{m^2 + m + 2}{(m-1)(m+3)}$

Equation; $b = 30$

2. $\dfrac{m}{m-1} - \dfrac{2}{m+3}$

3. $\dfrac{2}{a+5} + \dfrac{5}{a^2 - 25}$ ❖

4. $\dfrac{n}{2n+2} + \dfrac{5}{4n+4}$ ❖

5. $\dfrac{3y+6}{20} \div \dfrac{4y+8}{8}$ Expression; $\dfrac{3}{10}$

6. $\dfrac{z^2 + z}{24} \cdot \dfrac{8}{z+1}$ Expression; $\dfrac{z}{3}$

7. $\dfrac{3}{p+3} = \dfrac{12p+19}{p^2 + 7p + 12} - \dfrac{5}{p+4}$ Equation; $p = 2$

8. $\dfrac{\dfrac{1}{t^2} + \dfrac{2}{3}}{\dfrac{1}{t} - \dfrac{5}{6}}$ Expression; $\dfrac{2(3+2t^2)}{t(6-5t)}$

For Exercises 9–18, solve the proportions.

9. $\dfrac{5}{3} = \dfrac{a}{8}$ $a = \dfrac{40}{3}$

10. $\dfrac{b}{14} = \dfrac{3}{8}$ $b = \dfrac{21}{4}$

11. $\dfrac{2}{1.9} = \dfrac{x}{38}$ $x = 40$

12. $\dfrac{16}{1.3} = \dfrac{30}{p}$ $p = 2.4375$

13. $\dfrac{y+1}{2y} = \dfrac{2}{3}$ $y = 3$

14. $\dfrac{w-2}{4w} = \dfrac{1}{6}$ $w = 6$

15. $\dfrac{9}{2z-1} = \dfrac{3}{z}$ $z = -1$

16. $\dfrac{1}{t} = \dfrac{1}{4-t}$ $t = 2$

17. $\dfrac{8}{9a-1} = \dfrac{5}{3a+2}$ $a = 1$

18. $\dfrac{4p+1}{3} = \dfrac{2p-5}{6}$ $p = -\dfrac{7}{6}$

19. Charles' law describes the relationship between the temperature and volume of a gas held at a constant pressure.

$$\dfrac{V_i}{V_f} = \dfrac{T_i}{T_f}$$

a.   Solve the equation for V_f. $V_f = \dfrac{V_i T_f}{T_i}$

b.   Solve the equation for T_f. $T_f = \dfrac{T_i V_f}{V_i}$

20. The relationship between the area, height, and base of a triangle is given by the proportion

$$\dfrac{A}{b} = \dfrac{h}{2},$$ where A is area, b is the base, and h is the height.

a.   Solve the equation for A. $A = \dfrac{hb}{2}$

b.   Solve the equation for b. $b = \dfrac{2A}{h}$

21. Jennifer shot a 51 on nine holes of golf. At this rate, what can she expect her score to be if she plays all 18 holes?   102

22. A liquid plant food is prepared by using 3 oz for each gallon of water. At this rate, how many ounces of plant food are required for 5 gal of water?   15 oz

23. Geoff has a garden that is 5 ft in length by 3 ft in width in the front yard. He would like a garden with the dimensions in the same proportion in the backyard. If he has a length of 8 ft available, how wide should he make the garden?   4.8 ft

24. Cooking oatmeal requires 1 cup of water for every $\frac{1}{2}$ cup of oats. How many cups of water will be required for $\frac{3}{4}$ cup of oats?   1.5 cups

25. A map has a scale of 75 miles/in. If two cities measure 3 in. apart, how many miles does this represent?   225 miles

26. A map has a scale of 50 miles/in. If two cities measure 6 in. apart, how many miles does this represent?   300 miles

27. A boat travels 54 miles upstream against the current in the same amount of time it takes to travel 66 miles downstream with the current. If the boat has the speed of 20 mph in still water, what is the speed of the current? (Use $t = \frac{d}{r}$ to complete the table.)   The speed of the current is 2 mph.

	Distance	Rate	Time
With the current (downstream)			
Against the current (upstream)			

Table for Exercise 27

❖ See Additional Answers Appendix   ✎ Writing   ⬌ Translating Expression   Geometry   Scientific Calculator   Video

28. A fisherman travels 9 miles downstream with the current in the same time that he travels 3 miles upstream against the current. If the speed of the current is 6 mph, what is the speed at which the fisherman travels in still water? The speed of the boat in still water is 12 mph.

29. A plane flies 630 miles with the wind in the same time that it takes to fly 455 miles against the wind. If this plane flies at the rate of 217 mph in still air, what is the speed of the wind? (Use $t = \frac{d}{r}$ to complete the table.) The wind is 35 mph.

	Distance	Rate	Time
With the wind			
Against the wind			

Table for Exercise 29

30. A plane flies 370 miles with the wind in the same time that it takes to fly 290 miles against the wind. If the speed of the wind is 20 mph, what is the speed of the plane in still air? The plane travels 165 mph in still air.

31. One motorist travels 15 mph faster than another. The faster driver can cover 360 miles in the same time as the slower driver covers 270 miles. What are the speeds of the two motorists? The speeds are 45 mph and 60 mph.

32. A train can travel 325 miles in the same time an express bus can travel 200 miles. If the speed of the bus is 25 mph slower than the speed of the train, what is the speed of the train? The train travels 65 mph.

33. If it takes a person 2 h to paint a room, what fraction of the room would be painted in 1 h? $\frac{1}{2}$ of the room

34. If it takes a copier 3 h to complete a job, what fraction of the job would be completed in 1 h? $\frac{1}{3}$ of the job

35. If the cold-water faucet is left on, the sink will fill in 10 min. If the hot-water faucet is left on, the sink will fill in 12 min. How long would it take the sink to fill if both faucets are left on? $5\frac{5}{11} (5.\overline{45})$ minutes

36. The CUT-IT-OUT lawn mowing company consists of two people: Tina and Bill. If Tina cuts a lawn by herself, she can do it in 4 h. If Bill cuts the same lawn himself, it takes him an hour longer than Tina. How long would it take them if they worked together? $2\frac{2}{9} (2.\overline{2})$ h

37. A manuscript needs to be printed. One printer can do the job in 50 min and another printer can do the job in 40 min. How long would it take if both printers were used? $22\frac{2}{9} (22.\overline{2})$ min

38. A pump can empty a small pond in 4 h. Another more efficient pump can do the job in 3 h. How long would it take to empty the pond if both pumps were used? $1\frac{5}{7}$ (approximately 1.7) h

39. Tim and Al are bricklayers. Tim can construct an outdoor grill in 5 days. If Al helps Tim, they can build it in only 2 days. How long would it take Al to build the grill alone? $3\frac{1}{3} (3.\overline{3})$ days

40. Norma is a new and inexperienced secretary. It takes her 3 h to prepare a mailing. If her boss helps her, the mailing can be completed in 1 h. How long would it take the boss to do the job by herself? $1\frac{1}{2} (1.5)$ h

▪ EXPANDING YOUR SKILLS

A polygon is a closed figure with many sides. Two polygons are similar if their corresponding sides are proportional. For Exercises 41–44, assume that the polygons are similar.

41. $\triangle ABC$ is similar to $\triangle DEF$.
    a. Find the length of side $\overline{EF}$. 4 cm
    b. Find the length of side $\overline{DF}$. 5 cm

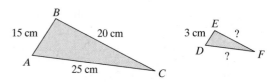

Figures for Exercise 41

42. Figure $ABCD$ is similar to Figure $EFGH$.
    a. Find the length of side $\overline{EH}$. 10 in.
    b. Find the length of side $\overline{AB}$. 13.5 in.
    c. Find the length of side $\overline{BC}$. 18 in.

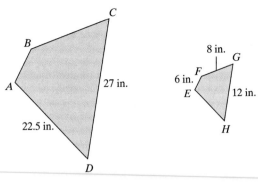

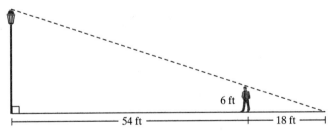

Figures for Exercise 42

43. A 6-ft tall man standing 54 ft from a light post casts an 18-ft shadow. What is the height of the light post?   The light post is 24 ft high.

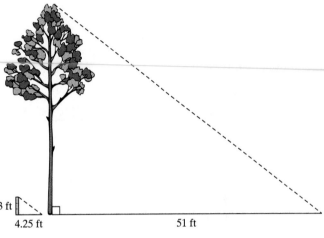

44. For a science project at school, a student must measure the height of a tree. The student measures the length of the shadow of the tree and then measures the length of the shadow cast by a yardstick. Use similar triangles to find the height of the tree.   The tree is 36 ft high.

Figure for Exercise 44

Figure for Exercise 43

---

chapter 5   SUMMARY

SECTION 5.1—INTRODUCTION TO RATIONAL EXPRESSIONS

KEY CONCEPTS:

A rational expression is a ratio of the form $\frac{p}{q}$ where p and q are polynomials and $q \neq 0$.

The domain* of an algebraic expression is the set of real numbers that when substituted for the variable makes the expression result in a real number.

*For a rational expression the domain is all real numbers except those that make the denominator zero.

EXAMPLES:

$\dfrac{x + 2}{x^2 - 5x - 14}$ is a rational expression.

Find the Domain: $\dfrac{x + 2}{x^2 - 5x - 14}$

$= \dfrac{x + 2}{(x + 2)(x - 7)}$ Factor the denominator.

The domain is $\{x \mid x \neq -2, x \neq 7\}$.

Reducing a Rational Expression:

Factor the numerator and denominator completely and reduce factors whose ratio is equal to 1 or to -1. A rational expression written in lowest terms will still have the same restrictions on the domain as the original expression.

KEY TERMS:

domain of a rational expression
fundamental principle of rational expressions
rational expression

Reduce: $\dfrac{x + 2}{x^2 - 5x - 14}$

$$= \dfrac{\overset{1}{\cancel{x + 2}}}{\cancel{(x + 2)}(x - 7)} \qquad \text{Factor and reduce.}$$

$$= \dfrac{1}{x - 7}$$

SECTION 5.2—MULTIPLICATION AND DIVISION OF RATIONAL EXPRESSIONS

KEY CONCEPTS:

Multiplying Rational Expressions:

Factor the numerator and denominator completely. Then reduce factors whose ratio is 1 or -1.

EXAMPLES:

Multiply:

$$\frac{b^2 - a^2}{a^2 - 2ab + b^2} \cdot \frac{a^2 - 3ab + 2b^2}{2a + 2b}$$

$$= \frac{\overset{-1}{\cancel{(b - a)}}\overset{1}{\cancel{(b + a)}}}{\cancel{(a - b)}\cancel{(a - b)}} \cdot \frac{(a - 2b)\overset{1}{\cancel{(a - b)}}}{2\cancel{(a + b)}}$$

$$= -\frac{a - 2b}{2} \qquad \text{or} \qquad \frac{2b - a}{2}$$

Dividing Rational Expressions:

Multiply the first expression by the reciprocal of the second expression. That is, for $q \neq 0$, $r \neq 0$, and $s \neq 0$,

$$\frac{p}{q} \div \frac{r}{s} = \frac{p}{q} \cdot \frac{s}{r}$$

Divide:

$$\frac{2c^2 d^5}{15e^4} \div \frac{6c^4 d^3}{20e}$$

$$= \frac{2c^2 d^5}{15e^4} \cdot \frac{20e}{6c^4 d^3}$$

$$= \frac{40c^2 d^5 e}{90c^4 d^3 e^4}$$

$$= \frac{4d^2}{9c^2 e^3}$$

Converting Between Two Units of Measure:

Multiply by the conversion factor.

KEY TERMS:

complex fraction
conversion factor
division of rational expressions
multiplication of rational expressions

Convert:

15.7 miles to kilometers. Round to the nearest hundredth of a kilometer and use the fact that 1 mile = 1.609 km.

$$(15.7 \, \cancel{\text{miles}})\left(1.609 \frac{\text{km}}{\cancel{\text{mile}}}\right) \approx 25.3 \text{ km}$$

SECTION 5.3—THE LEAST COMMON DENOMINATOR

KEY CONCEPTS:

Converting a Rational Expression to an Equivalent Expression with a Different Denominator:

Multiply numerator and denominator of the rational expression by the missing factors necessary to create the desired denominator.

EXAMPLES:

Convert $\dfrac{-3}{x-2}$ to an equivalent expression with the indicated denominator:

$$\frac{-3}{x-2} = \frac{}{5x^2 - 20}$$

$$\frac{-3}{x-2} = \frac{}{5(x^2 - 4)} \qquad \text{Factor.}$$

$$\frac{-3}{x-2} = \frac{}{5(x-2)(x+2)}$$

Multiply numerator and denominator by the missing factors from the denominator.

$$\frac{-3 \cdot 5(x+2)}{(x-2) \cdot 5(x+2)} = \frac{-15x - 30}{5(x-2)(x+2)}$$

Finding the Least Common Denominator (LCD) of Two or More Rational Expressions:

1. Factor all denominators completely.
2. The LCD is the product of unique factors from the denominators, where each factor is raised to its highest power.

KEY TERMS:

equivalent rational expressions
least common denominator (LCD)

Identify the LCD:

$$\frac{1}{8x^3 y^2 z}; \quad \frac{5}{6xy^4}$$

1. Write the denominators as a product of prime factors:

$$\frac{1}{2^3 x^3 y^2 z}; \quad \frac{5}{2 \cdot 3xy^4}$$

2. The LCD is $2^3 3 x^3 y^4 z$ or $24 x^3 y^4 z$

SECTION 5.4—ADDITION AND SUBTRACTION OF RATIONAL EXPRESSIONS

KEY CONCEPTS:

To add or subtract rational expressions, the expressions must have the same denominator.

Steps to Add or Subtract Rational Expressions:

1. Factor the denominators of each rational expression.
2. Identify the LCD.
3. Rewrite each rational expression as an equivalent expression with the LCD as its denominator.
4. Add or subtract the numerators and write the result over the common denominator.
5. Simplify and reduce.

KEY TERMS:

addition of rational expressions
subtraction of rational expressions

EXAMPLES:

Subtract: $\dfrac{c}{c^2 - c - 12} - \dfrac{1}{2c - 8}$

$$= \dfrac{c}{(c - 4)(c + 3)} - \dfrac{1}{2(c - 4)}$$

The LCD is $2(c - 4)(c + 3)$.

$$= \dfrac{2c}{2(c - 4)(c + 3)} - \dfrac{1(c + 3)}{2(c - 4)(c + 3)}$$

$$= \dfrac{2c - (c + 3)}{2(c - 4)(c + 3)}$$

$$= \dfrac{2c - c - 3}{2(c - 4)(c + 3)} = \dfrac{c - 3}{2(c - 4)(c + 3)}$$

SECTION 5.5—COMPLEX FRACTIONS

KEY CONCEPTS:

Complex fractions can be simplified by using Method I or Method II.

Method I

1. Simplify the numerator and denominator of the complex fraction separately to form a single fraction in the numerator and a single fraction in the denominator.
2. Perform the division represented by the complex fraction. (Multiply the numerator of the complex fraction by the reciprocal of the denominator of the complex fraction.)
3. Simplify and reduce if possible.

EXAMPLES:

Simplify Using Method I:

$$\dfrac{1 - \dfrac{4}{w^2}}{1 - \dfrac{1}{w} - \dfrac{6}{w^2}}$$

$$\dfrac{1 - \dfrac{4}{w^2}}{1 - \dfrac{1}{w} - \dfrac{6}{w^2}} = \dfrac{\dfrac{w^2}{w^2} - \dfrac{4}{w^2}}{\dfrac{w^2}{w^2} - \dfrac{w}{w^2} - \dfrac{6}{w^2}}$$

$$= \dfrac{\dfrac{w^2 - 4}{w^2}}{\dfrac{w^2 - w - 6}{w^2}} = \dfrac{w^2 - 4}{w^2} \cdot \dfrac{w^2}{w^2 - w - 6}$$

$$= \dfrac{(w - 2)(w + 2)}{w^2} \cdot \dfrac{w^2}{(w - 3)(w + 2)}$$

$$= \dfrac{w - 2}{w - 3}$$

Method II

1. Multiply the numerator and denominator of the complex fraction by the LCD of all individual fractions within the expression.
2. Apply the distributive property and simplify the result.
3. Reduce if necessary.

KEY TERM:

complex fraction

Simplify Using Method II:

$$\frac{1 - \dfrac{4}{w^2}}{1 - \dfrac{1}{w} - \dfrac{6}{w^2}}$$

$$= \frac{w^2\left(1 - \dfrac{4}{w^2}\right)}{w^2\left(1 - \dfrac{1}{w} - \dfrac{6}{w^2}\right)} = \frac{w^2 - 4}{w^2 - w - 6}$$

$$= \frac{(w - 2)(w + 2)}{(w - 3)(w + 2)} = \frac{w - 2}{w - 3}$$

SECTION 5.6—RATIONAL EQUATIONS

KEY CONCEPTS:

An equation with one or more rational expressions is called a rational equation.

Steps to Solve a Rational Equation

1. Factor the denominators of all rational expressions.
2. Identify the LCD of all expressions in the equation.
3. Multiply both sides of the equation by the LCD.
4. Solve the resulting equation.
5. Check each potential solution in the original equation.

KEY TERMS:

linear equation
quadratic equation
rational equation

EXAMPLES:
Solve:

$$\frac{1}{w} - \frac{1}{2w - 1} = \frac{-2w}{2w - 1}$$

The LCD is $w(2w - 1)$.

$$w(2w - 1)\frac{1}{w} - w(2w - 1)\frac{1}{2w - 1}$$
$$= w(2w - 1)\frac{-2w}{2w - 1}$$

$$(2w - 1)1 - w(1) = w(-2w)$$

$$2w - 1 - w = -2w^2 \qquad \text{Quadratic equation}$$

$$2w^2 + w - 1 = 0$$

$$(2w - 1)(w + 1) = 0$$

$$w \neq \tfrac{1}{2} \quad \text{or} \quad w = -1$$
$$\text{Does not check.} \qquad\qquad \text{Checks.}$$

Solve for I:

$$q = \frac{VQ}{I}$$

$$I \cdot q = \frac{VQ}{I} \cdot I$$

$$Iq = VQ$$

$$I = \frac{VQ}{q}$$

SECTION 5.7—APPLICATIONS OF RATIONAL EXPRESSIONS

KEY CONCEPTS:

An equation that equates two ratios is called a proportion:

$$\frac{a}{b} = \frac{c}{d} \quad (b \neq 0, d \neq 0)$$

The cross products of a proportion are equal. $ad = bc$.

Application of Proportions.

Applications of Rational Equations.

1. Applications involving $d = rt$
    (Distance = rate $\cdot$ time)

EXAMPLES:

Solve:

$$\frac{3}{5} = \frac{45}{n}$$

$$3 \cdot n = 5 \cdot 45$$

$$3n = 225$$

$$n = 75$$

A 90-g serving of a particular ice-cream contains 10 g of fat. How much fat does 400 g of the same ice-cream contain?

$$\frac{10 \text{ g fat}}{90 \text{ g ice-cream}} = \frac{x \text{ grams fat}}{400 \text{ g ice-cream}}$$

$$\frac{10}{90} = \frac{x}{400}$$

$$10 \cdot 400 = 90 \cdot x$$

$$4000 = 90x$$

$$x = \frac{4000}{90} = \frac{400}{9} \approx 44.4 \text{ g}$$

Solve:

Two cars travel from Los Angeles to Las Vegas. One car travels an average of 8 mph faster than the other car. If the faster car travels 189 miles in the same time as the slower car travels 165 miles, what is the average speed of each car?

Let r represent the speed of the slower car.
Let $r + 8$ represent the speed of the faster car.

	Distance	Rate	Time
Slower car	165	r	$\dfrac{165}{r}$
Faster car	189	$r + 8$	$\dfrac{189}{r + 8}$

$$\frac{165}{r} = \frac{189}{r+8}$$

$$165(r+8) = 189r$$

$$165r + 1320 = 189r$$

$$1320 = 24r$$

$$55 = r$$

The slower car travels at 55 mph and the faster car travels $55 + 8 = 63$ mph.

2. Application involving work.

Solve:

Beth and Cecelia have a housecleaning business. Beth can clean a particular house in 5 h by herself. Cecelia can clean the same house in 4 h. How long would it take if they cleaned the house together?

Let x be the number of hours it takes for both Beth and Cecelia to clean the house.

Beth can clean $\frac{1}{5}$ of the house in an hour and $\frac{x}{5}$ of the house in x hours.

Cecelia can clean $\frac{1}{4}$ of the house in an hour and $\frac{x}{4}$ of the house in x hours.

$$\frac{x}{5} + \frac{x}{4} = 1 \qquad \text{Together they clean one whole house.}$$

$$20\left(\frac{x}{5} + \frac{x}{4}\right) = (1)20$$

$$4x + 5x = 20$$

$$9x = 20$$

$$x = \frac{20}{9}, \text{ or } 2\frac{2}{9} \text{ h working together.}$$

KEY TERM:

proportion

chapter 5   REVIEW EXERCISES

Section 5.1

1. For the rational expression $\dfrac{t-2}{t+9}$

   a. Evaluate the expression (if possible) for
   $t = 0, 1, 2, -3, -9$ $-\dfrac{2}{9}, -\dfrac{1}{10}, 0, -\dfrac{5}{6},$ undefined
   b. Write the domain of the expression in set-builder notation. $\{t \mid t \neq -9\}$

2. For the rational expression $\dfrac{k+1}{k-5}$

   a. Evaluate the expression for $k = 0, 1, 5,$
   $-1, -2$ $-\dfrac{1}{5}, -\dfrac{1}{2},$ undefined, $0, \dfrac{1}{7}$
   b. Write the domain of the expression in set-builder notation. $\{k \mid k \neq 5\}$

3. Which of the rational expressions are equal to -1 for all values of x for which the expressions are defined?   a, c, d

   a. $\dfrac{2-x}{x-2}$ b. $\dfrac{x-5}{x+5}$

   c. $\dfrac{-x-7}{x+7}$ d. $\dfrac{x^2-4}{4-x^2}$

For Exercises 4–13, write the domain in set-builder notation. Then reduce the expression.

4. $\dfrac{x-3}{(2x-5)(x-3)}$ ❖

5. $\dfrac{h+7}{(3h+1)(h+7)}$ ❖

6. $\dfrac{4a^2+7a-2}{a^2-4}$ ❖

7. $\dfrac{2w^2+11w+12}{w^2-16}$ ❖

8. $\dfrac{z^2-4z}{8-2z}$ ❖

9. $\dfrac{15-3k}{2k^2-10k}$ ❖

10. $\dfrac{2b^2+4b-6}{4b+12}$ ❖

11. $\dfrac{3m^2-12m-15}{9m+9}$ ❖

12. $\dfrac{n+3}{n^2+6n+9}$ ❖

13. $\dfrac{p+7}{p^2+14p+49}$ ❖

Section 5.2

For Exercises 14–25, multiply or divide as indicated.

14. $\dfrac{3y^3}{3y-6} \cdot \dfrac{y-2}{y}$ y^2

15. $\dfrac{2u+10}{u} \cdot \dfrac{u^3}{4u+20}$ $\dfrac{u^2}{2}$

16. $\dfrac{11}{v-2} \cdot \dfrac{2v^2-8}{22}$ $v+2$

17. $\dfrac{8}{x^2-25} \cdot \dfrac{3x+15}{16}$ $\dfrac{3}{2(x-5)}$

18. $\dfrac{4c^2+4c}{c^2-25} \div \dfrac{8c}{c^2-5c}$ $\dfrac{c(c+1)}{2(c+5)}$

19. $\dfrac{q^2-5q+6}{2q+4} \div \dfrac{2q-6}{q+2}$ $\dfrac{q-2}{4}$

20. $\left(\dfrac{-2t}{t+1}\right)(t^2-4t-5)$ $-2t(t-5)$

21. $(s^2-6s+8)\left(\dfrac{4s}{s-2}\right)$ $4s(s-4)$

22. $\dfrac{\dfrac{a^2+5a+1}{7a-7}}{\dfrac{a^2+5a+1}{a-1}}$ $\dfrac{1}{7}$

23. $\dfrac{\dfrac{n^2+n+1}{n^2-4}}{\dfrac{n^2+n+1}{n+2}}$ $\dfrac{1}{n-2}$

24. $\dfrac{5h^2-6h+1}{h^2-1} \div \dfrac{16h^2-9}{4h^2+7h+3} \cdot \dfrac{3-4h}{30h-6}$ $-\dfrac{1}{6}$

25. $\dfrac{3m-3}{6m^2+18m+12} \cdot \dfrac{2m^2-8}{m^2-3m+2} \div \dfrac{m+3}{m+1}$ $\dfrac{1}{m+3}$

26. One paper clip has approximately 2.5 g of mass. What is the weight in pounds of 200 paper clips? Round to the nearest pound. (1 kg = 2.2 lb, 1 kg = 1000 g)   1.1 lb

27. A medicine is sold in bottles containing 0.25 L of fluid. How many cubic centimeters (cc) is this? (1 L = 1000 mL, 1 mL = 1 cc)   250 cc

Section 5.3

For Exercises 28–33, convert each expression into an equivalent expression with the indicated denominator.

28. $\dfrac{x+1}{x-2} = \dfrac{}{5x-10}$
$5x+5$

29. $\dfrac{y+2}{y-3} = \dfrac{}{2y-6}$
$2y+4$

30. $\dfrac{6}{w} = \dfrac{}{w^2-4w}$
$6w-24$

31. $\dfrac{2}{r} = \dfrac{}{r^2+3r}$
$2r+6$

32. $\dfrac{s-2}{s+4} = \dfrac{}{s^2-16}$
s^2-6s+8

33. $\dfrac{u+1}{u+6} = \dfrac{}{u^2-36}$
u^2-5u-6

For Exercises 34–41, identify the LCD.

34. $\dfrac{2}{a^2bc^2}, \dfrac{5}{ab^3}$ $a^2b^3c^2$

35. $\dfrac{6x}{y^2z}, \dfrac{3}{xy^2z^4}$ xy^2z^4

36. $\dfrac{5}{p+2}, \dfrac{p}{p-4}$
$(p+2)(p-4)$

37. $\dfrac{6}{q}, \dfrac{1}{q+8}$ $q(q+8)$

38. $\dfrac{8}{m^2-16}, \dfrac{7}{m^2-m-12}$ $(m-4)(m+4)(m+3)$

39. $\dfrac{6}{n^2-9}, \dfrac{5}{n^2-n-6}$ $(n-3)(n+3)(n+2)$

40. $\dfrac{4}{2t-5}, \dfrac{5}{5-2t}$
$2t-5$ or $5-2t$

41. $\dfrac{-2}{3k-1}, \dfrac{6}{1-3k}$
$3k-1$ or $1-3k$

42. State two possible LCDs that could be used to add the fractions. $c-2$ or $2-c$

$$\dfrac{7}{c-2} + \dfrac{4}{2-c}$$

43. State two possible LCDs that could be used to subtract the fractions: $3-x$ or $x-3$

$$\dfrac{10}{3-x} - \dfrac{5}{x-3}$$

Section 5.4

For Exercises 44–55, add or subtract as indicated.

44. $\dfrac{h+3}{h+1} + \dfrac{h-1}{h+1}$ 2

45. $\dfrac{b-6}{b-2} + \dfrac{b+2}{b-2}$ 2

46. $\dfrac{a^2}{a-5} - \dfrac{25}{a-5}$ $a+5$

47. $\dfrac{x^2}{x+7} - \dfrac{49}{x+7}$ $x-7$

48. $\dfrac{y}{y^2-81} + \dfrac{2}{9-y}$ ❖

49. $\dfrac{3}{4-t^2} + \dfrac{t}{2-t}$ ❖

50. $\dfrac{4}{3m} - \dfrac{1}{m+2}$ ❖

51. $\dfrac{5}{2r+12} - \dfrac{1}{r}$ ❖

52. $\dfrac{4p}{p^2+6p+5} - \dfrac{3p}{p^2+5p+4}$ $\dfrac{p}{(p+4)(p+5)}$

53. $\dfrac{3q}{q^2+7q+10} - \dfrac{2q}{q^2+6q+8}$ $\dfrac{q}{(q+5)(q+4)}$

54. $\dfrac{1}{h} + \dfrac{h}{2h+4} - \dfrac{2}{h^2+2h}$ $\dfrac{1}{2}$

55. $\dfrac{x}{3x+9} - \dfrac{3}{x^2+3x} + \dfrac{1}{x}$ $\dfrac{1}{3}$

Section 5.5

For Exercises 56–63, simplify the complex fractions.

56. $\dfrac{\dfrac{a-4}{3}}{\dfrac{a-2}{3}}$ $\dfrac{a-4}{a-2}$

57. $\dfrac{\dfrac{z+5}{z}}{\dfrac{z-5}{3}}$ $\dfrac{3(z+5)}{z(z-5)}$

58. $\dfrac{\dfrac{2-3w}{2}}{\dfrac{2}{w}-3}$ $\dfrac{w}{2}$

59. $\dfrac{\dfrac{2}{y}+6}{\dfrac{3y+1}{4}}$ $\dfrac{8}{y}$

60. $\dfrac{\dfrac{y}{x}-\dfrac{x}{y}}{\dfrac{1}{x}+\dfrac{1}{y}}$ $y-x$

61. $\dfrac{\dfrac{b}{a}-\dfrac{a}{b}}{\dfrac{1}{b}-\dfrac{1}{a}}$ $-(b+a)$

62. $\dfrac{\dfrac{6}{p+2}+4}{\dfrac{8}{p+2}-4}$ $\dfrac{2p+7}{2p}$

63. $\dfrac{\dfrac{25}{k+5}+5}{\dfrac{5}{k+5}-5}$ $-\dfrac{k+10}{k+4}$

Section 5.6

For Exercises 64–71, solve the equations.

64. $\dfrac{2}{x} + \dfrac{1}{2} = \dfrac{1}{4}$ $x=-8$

65. $\dfrac{1}{y} + \dfrac{3}{4} = \dfrac{1}{4}$ $y=-2$

66. $\dfrac{2}{h-2} + 1 = \dfrac{h}{h+2}$ $h=0$

67. $\dfrac{w}{w-1} = \dfrac{3}{w+1} + 1$ $w=2$

 ❖ See Additional Answers Appendix   ✎ Writing   ↔ Translating Expression   Geometry   Scientific Calculator   Video

68. $\dfrac{t+1}{3} - \dfrac{t-1}{6} = \dfrac{1}{6}$ $t = -2$

69. $\dfrac{4p-4}{p^2+5p-14} + \dfrac{2}{p+7} = \dfrac{1}{p-2}$ $p = 3$

70. $\dfrac{1}{z+2} = \dfrac{4}{z^2-4} - \dfrac{1}{z-2}$ No solution; ($z = 2$ does not check.)

71. $\dfrac{y+1}{y+3} = \dfrac{y^2-11y}{y^2+y-6} - \dfrac{y-3}{y-2}$ $y = -11, y = 1$

72. Four times a number is added to 5. The sum is then divided by 6. The result is $\frac{7}{2}$. Find the number.   The number is 4.

73. Solve the formula $\dfrac{V}{h} = \dfrac{\pi r^2}{3}$ for h. $h = \dfrac{3V}{\pi r^2}$

74. Solve the formula $\dfrac{A}{b} = \dfrac{h}{2}$ for b. $b = \dfrac{2A}{h}$

Section 5.7

For Exercises 75–76, solve the proportions.

75. $\dfrac{m+2}{8} = \dfrac{m}{3}$

    $m = \dfrac{6}{5}$

76. $\dfrac{a-3}{4a} = \dfrac{2}{a+7}$

    $a = 7, a = -3$

77. A bag of popcorn states that it contains 4 g of fat per serving. If a serving is 2 oz, how many grams of fat are in a 6-oz bag?   12 g

78. Bud goes 10 mph faster on his Harley Davidson motorcycle than Ed goes on his Honda motorcycle. If Bud travels 105 miles in the same time that Ed travels 90 miles, what are the rates of the two bikers?   Ed travels at 60 mph, and Bud travels at 70 mph.

79. There are two pumps set up to fill a small swimming pool. One pump takes 24 min by itself to fill the pool, but the other takes 56 min by itself. How long would it take if both pumps work together?   Together the pumps would fill the pool in 16.8 min.

---

chapter 5   TEST

For Exercises 1–2,

a. Write the domain in set-builder notation.

b. Reduce the rational expression.

1. $\dfrac{5(x-2)(x+1)}{30(2-x)}$

    a. $\{x \mid x \neq 2\}$

    b. $-\dfrac{x+1}{6}$

2. $\dfrac{7a^2 - 42a}{a^3 - 4a^2 - 12a}$

    a. $\{a \mid a \neq 0, a \neq 6, a \neq -2\}$

    b. $\dfrac{7}{a+2}$

3. Identify the rational expressions that are equal to -1 for all values of x for which the expression is defined.   b, c, d

    a. $\dfrac{x+4}{x-4}$

    b. $\dfrac{7-2x}{2x-7}$

    c. $\dfrac{9x^2+16}{-9x^2-16}$

    d. $-\dfrac{x+5}{x+5}$

For Exercises 4–9, perform the indicated operation.

4. $\dfrac{2}{y^2 + 4y + 3} + \dfrac{1}{3y + 9}$ $\dfrac{y + 7}{3(y + 3)(y + 1)}$

5. $\dfrac{9 - b^2}{5b + 15} \div \dfrac{b - 3}{b + 3}$ $-\dfrac{b + 3}{5}$

6. $\dfrac{w^2 - 4w}{w^2 - 8w + 16} \cdot \dfrac{w - 4}{w^2 + w}$ $\dfrac{1}{w + 1}$

7. $\dfrac{t}{t - 2} - \dfrac{8}{t^2 - 4}$ $\dfrac{t + 4}{t + 2}$

8. $\dfrac{1}{x + 4} + \dfrac{2}{x^2 + 2x - 8} + \dfrac{x}{x - 2}$
$\dfrac{x(x + 5)}{(x + 4)(x - 2)}$

9. $\dfrac{1 - \dfrac{4}{m}}{m - \dfrac{16}{m}}$ $\dfrac{1}{m + 4} \cdot \dfrac{m}{m}$

10. In 1998, Eddie Cheever won the Indianapolis 500 with an average speed of 145.155 mph. Convert this speed to feet per second. (Round to one decimal place.) (1 mile = 5280 ft)   212.9 ft/s

For Exercises 11–12, solve the equation.

11. $\dfrac{p}{p - 1} + \dfrac{1}{p} = \dfrac{p^2 + 1}{p^2 - p}$ $p = 2$

12. $\dfrac{3}{c - 2} - \dfrac{1}{c + 1} = \dfrac{7}{c^2 - c - 2}$ $c = 1$

13. Solve the formula $\dfrac{C}{2} = \dfrac{A}{r}$ for r. $r = \dfrac{2A}{C}$

14. If $\frac{3}{2}$ is added to the reciprocal of a number the result is $\frac{2}{5}$ times the reciprocal of that number. Find the number.   The number is $-\dfrac{2}{5}$.

15. Solve the proportion.
$$\dfrac{y + 7}{-4} = \dfrac{1}{4}$$ $y = -8$

16. A recipe for vegetable soup calls for $\frac{1}{2}$ cup of carrots for six servings. How many cups of carrots are needed to prepare 15 servings? $1\frac{1}{4}$ (1.25) cups of carrots

17. A motorboat can travel 28 miles downstream in the same amount of time as it can travel 18 miles upstream. Find the speed of the current if the boat can travel 23 mph in still water.   The speed of the current is 5 mph.

18. Two printers working together can complete a job in 2 h. If one printer requires 6 h to do the job alone, how many hours would the second printer need to complete the job alone?   It would take the second printer 3 h to do the job working alone.

CUMULATIVE REVIEW EXERCISES, CHAPTERS 1–5

For Exercises 1–2, simplify completely.

1. $\left(\dfrac{1}{2}\right)^{-4} + 2^4$ 32

2. $|3 - 5| + |-2 + 7|$ 7

3. Which of the following are rational numbers and which are irrational numbers?   ❖
$$\sqrt{4}, \sqrt{5}, \sqrt{9}, \sqrt{16}, \sqrt{20}, \sqrt{49}$$

4. Solve for y: $\dfrac{1}{2} - \dfrac{3}{4}(y - 1) = \dfrac{5}{12}$ $y = \dfrac{10}{9}$

5. Solve the inequality. Graph the solution set and express the solution in interval notation:
$-3(x - 5) - 2 < -2x + 5$ ❖

6. Complete the table.   ❖

Set-Builder Notation	Graph	Interval Notation
$\{x \mid x \geq -1\}$	⟶	
	⟶	$(-\infty, 5)$

Table for Exercise 6

7. The perimeter of a rectangular swimming pool is 104 m. The length is 1 m more than twice the width. Find the length and width.   ❖

8. The height of a triangle is 2 in. less than the base. The area is 40 in.^{2} Find the base and height of the triangle.   ❖

9. Find the values of the vertical angles.
   $x = 10$; the angles are $37°$.

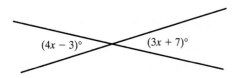

$(4x - 3)°$ $(3x + 7)°$

Figure for Exercise 9

10. The mass of a proton is approximately 1.66×10^{-24} g. What is the mass of 6.02×10^{23} protons? 9.9932×10^{-1} g or 0.99932 g.

11. Show that the lengths of the sides of the right triangle satisfy the Pythagorean theorem.   ❖

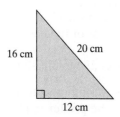

16 cm   20 cm

12 cm

Figure for Exercise 11

For Exercises 12–13, simplify the expressions. Write the final answer with positive exponents only.

12. $\dfrac{(x^{-1})^2 x^5}{x^3}$ 1

13. $\left(\dfrac{4x^{-1}y^{-2}}{z^4}\right)^{-2} (2y^{-1}z^3)^3$ $\dfrac{x^2 y z^{17}}{2}$

14. The length and width of a rectangle are given in terms of x.

   a. Write a polynomial that represents the perimeter of the rectangle. $6x + 4$

   b. Write a polynomial that represents the area of the rectangle. $2x^2 + x - 3$

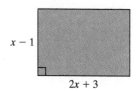

$x - 1$

$2x + 3$

Figure for Exercise 14

15. Perform the indicated operation: $(5x - 3)^2$
    $25x^2 - 30x + 9$

16. Factor completely: $25x^2 - 30x + 9$
    $(5x - 3)^2$

17. Divide.

$$\frac{8a^2b^4 - 2ab^3 + a^3b^2}{2ab^2} \qquad 4ab^2 - b + \frac{a^2}{2}$$

For Exercises 18–20, factor completely.

18. $27x^2 - 75y^2$ ❖

19. $10cd + 5d - 6c - 3$ ❖

20. $x^2 - x - 20$ ❖

21. What is the domain of the expression.

$$\frac{x + 3}{(x - 5)(2x + 1)} \qquad \left\{ x \mid x \neq 5, x \neq -\frac{1}{2} \right\}$$

22. Reduce.

$$\frac{x^2 - 9}{x^2 + 8x + 15} \qquad \frac{x - 3}{x + 5}$$

For Exercises 23–24, perform the indicated operations.

23. $\dfrac{2x - 6}{x^2 - 16} \div \dfrac{10x^2 - 90}{x^2 - x - 12}$ $\dfrac{1}{5(x + 4)}$

24. $\dfrac{x^2 - 6x}{x - 3} - \dfrac{-9}{x - 3}$ $x - 3$

25. Simplify.

$$\frac{\dfrac{3}{4} - \dfrac{1}{x}}{\dfrac{1}{3} - \dfrac{1}{4x}} \qquad -3$$

26. Solve.

$$\frac{7}{y^2 - 4} = \frac{3}{y - 2} + \frac{2}{y + 2} \qquad y = 1$$

27. Solve the proportion.

$$\frac{2b - 5}{6} = \frac{4b}{7} \qquad b = -\frac{7}{2}$$

28. A small boat can sail at a rate of 18 km per hour in still water. If the boat can go 63 km downstream with the current in the same amount of time it takes to go 45 km upstream against the current, what is the speed of the current?   ❖

29. Given the equation $5x - y = 10$.

    a. Is the equation linear or nonlinear? ❖

    b. Complete the table.

x	y
0	−10
2	0
1	−5

    Table for Exercise 29

    c. Sketch the graph passing through the points given in the table. ❖

30. Given the equation $y = x^2 - 25$.

    a. Is the equation linear or nonlinear?   nonlinear

    b. Find the y-intercept. $(0, -25)$

    c. Find the x-intercept(s). $(5, 0)$ and $(-5, 0)$

GRAPHING LINEAR EQUATIONS IN TWO VARIABLES

6

In Chapter 6 we investigate linear relationships and their graphs. In mathematical applications, a linear relationship may be given by a fixed value plus a constant rate of increase or decrease. For example:

Suppose 8 inches of snow is on the ground at 12:00 noon. A blizzard dumps more snow at a rate of $\frac{3}{4}$ in. per hour for nine hours. The depth of the snow, d (in inches), is given by:

$$d = \frac{3}{4}x + 8$$

where x is the number of hours after the storm began, $(0 \le x \le 9)$.

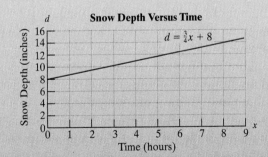

Snow Depth Versus Time

$d = \frac{3}{4}x + 8$

This and other linear equations may be analyzed and graphed using the techniques presented in Chapter 6. For more information visit xyplot at

www.mhhe.com/miller_oneill

section

6.1   LINEAR EQUATIONS IN TWO VARIABLES

1. Solutions to Linear Equations in Two Variables

Recall from Section 2.8 that a **linear equation in two variables** x and y can be written in the form $ax + by = c$, where a and b are not both zero. A **solution** to such an equation is an ordered pair (x, y) that makes the equation a true statement. The set of all solutions to a linear equation is called the solution set.

2. Graphing Linear Equations by Plotting Points

The solution set of a linear equation in two variables can be represented by a line in a rectangular coordinate system. To graph a line, a minimum of two points must be determined. To find an ordered pair solution to a linear equation, we can pick any arbitrary number to substitute for x and then solve for y. Similarly, we can pick any arbitrary number to substitute for y and then solve for x. This process is reviewed in the next example.

example 1

Graphing a Linear Equation in Two Variables

Graph the equation $x - 2y = 8$.

Solution:

It is necessary to find two ordered pairs that are solutions to the equation $x - 2y = 8$. Pick arbitrary values for x or y such as those shown in the table. Complete the table to find the corresponding ordered pairs.

x	y
2	
	−1

Let $x = 2$:

$x - 2y = 8$

$(2) - 2y = 8$

$-2y = 8 - 2$

$-2y = 6$

$y = -3$

Let $y = -1$:

$x - 2y = 8$

$x - 2(-1) = 8$

$x + 2 = 8$

$x = 6$

x	y
2	−3
6	−1

The ordered pairs $(2, -3)$ and $(6, -1)$ are solutions to the linear equation $x - 2y = 8$. The line through these points represents all solutions to the equation (Figure 6-1).

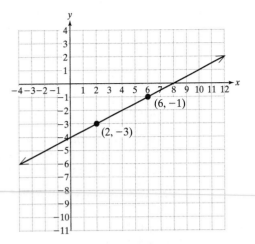

Figure 6-1

A third point can be found as a check point. Choose another arbitrary value for x or y such as $x = 0$:

x	y
2	−3
6	−1
0	

<u>Let $x = 0$:</u>

$$x - 2y = 8$$
$$(0) - 2y = 8$$
$$-2y = 8$$
$$y = -4$$

x	y
2	−3
6	−1
0	−4

Notice that the check point $(0, -4)$ also lies on the line (Figure 6-2). Because the three points "line up," we can be reasonably sure that all three ordered pairs were computed correctly.

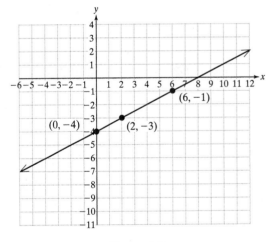

Figure 6-2

example 2

Graphing a Linear Equation in Two Variables

Graph the line $y = -\dfrac{1}{3}x + 1$.

Solution:

Because the y-variable is isolated in the equation, it is easy to substitute a value for x and simplify the right-hand side to find y. Since any number for x can be picked, choose numbers that are multiples of 3 that will simplify easily when multiplied by $-\frac{1}{3}$.

x	y
3	
0	
−3	

$$y = -\dfrac{1}{3}x + 1$$

Let $x = 3$:

$$y = -\dfrac{1}{3}(3) + 1$$
$$y = -1 + 1$$
$$y = 0$$

Let $x = 0$:

$$y = -\dfrac{1}{3}(0) + 1$$
$$y = 0 + 1$$
$$y = 1$$

Let $x = -3$:

$$y = -\dfrac{1}{3}(-3) + 1$$
$$y = 1 + 1$$
$$y = 2$$

x	y
3	0
0	1
−3	2

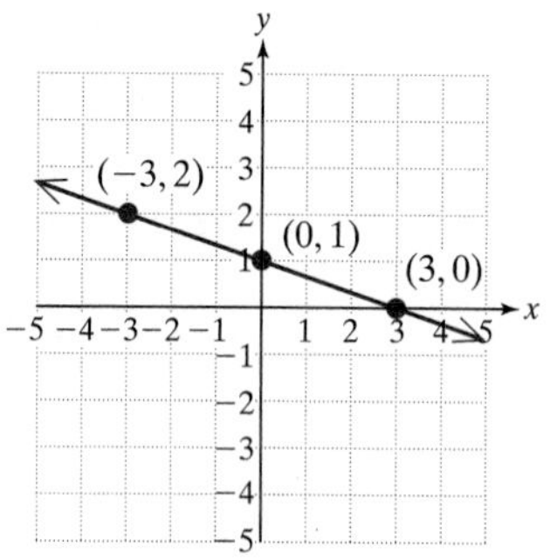

Figure 6-3

The line through the three ordered pairs $(3, 0)$, $(0, 1)$, and $(-3, 2)$ is shown in Figure 6-3. The line represents the set of all solutions to the equation $y = -\frac{1}{3}x + 1$.

3. *x*- and *y*-Intercepts

In Section 4.7 we defined the x- and y-intercepts of a graph. These are the points where the graph crosses the x- and y-axes, respectively. From Example 2, we see that the x-intercept is at the point $(3, 0)$ and the y-intercept is at the point $(0, 1)$. Notice that a y-intercept is a point on the y-axis and must have an x-coordinate of 0. Likewise, an x-intercept is a point on the x-axis and has a y-coordinate of 0.

Definition of *x*- and *y*-Intercepts

An **x-intercept** of an equation is a point $(a, 0)$ where the graph intersects the x-axis.

A **y-intercept** of an equation is a point $(0, b)$ where the graph intersects the y-axis.

In some applications, an *x*-intercept is defined as the *x*-coordinate of a point of intersection that a graph makes with the *x*-axis. For example, if an *x*-intercept is at the point $(3, 0)$, it is sometimes stated simply as 3 (the *y*-coordinate is assumed to be 0). Similarly, a *y*-intercept is sometimes defined as the *y*-coordinate of a point of intersection that a graph makes with the *y*-axis. For example, if a *y*-intercept is at the point $(0, 7)$, it may be stated simply as 7 (the *x*-coordinate is assumed to be 0).

Although any two points may be used to graph a line, in some cases it is convenient to use the *x*- and *y*-intercepts of the line. To find the *x*- and *y*-intercepts of any two-variable equation in *x* and *y*, follow these steps:

Steps to Finding the *x*- and *y*-Intercepts from an Equation

Given an equation in *x* and *y*,

1. Find the *x*-intercept(s) by substituting $y = 0$ into the equation and solving for *x*.
2. Find the *y*-intercept(s) by substituting $x = 0$ into the equation and solving for *y*.

---

example 3   Finding the *x*- and *y*-Intercepts of a Line

Given the equation $-3x + 2y = 8$,

a. Find the *x*-intercept.
b. Find the *y*-intercept.
c. Graph the equation.

Solution:

a. To find the *x*-intercept, substitute $y = 0$.

$$-3x + 2y = 8$$
$$-3x + 2(0) = 8$$
$$-3x = 8$$
$$\frac{-3x}{-3} = \frac{8}{-3}$$
$$x = -\frac{8}{3}$$

The *x*-intercept is $\left(-\frac{8}{3}, 0\right)$.

b. To find the *y*-intercept, substitute $x = 0$.

$$-3x + 2y = 8$$
$$-3(0) + 2y = 8$$
$$2y = 8$$
$$y = 4$$

The *y*-intercept is $(0, 4)$.

c. The line through the ordered pairs $\left(-\frac{8}{3}, 0\right)$ and $(0, 4)$ is shown in Figure 6-4. Note that the point $\left(-\frac{8}{3}, 0\right)$ can be written as $\left(-2\frac{2}{3}, 0\right)$.
The line represents the set of all solutions to the equation $-3x + 2y = 8$.

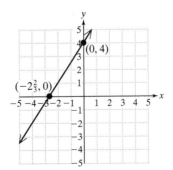

Figure 6-4

example 4

Finding the *x*- and *y*-Intercepts of a Line

Given the equation: $4x + 5y = 0$

a. Find the *x*-intercept.
b. Find the *y*-intercept.
c. Graph the line.

Solution:

a. To find the *x*-intercept, substitute $y = 0$.

$$4x + 5y = 0$$
$$4x + 5(0) = 0$$
$$4x \qquad = 0$$
$$x = 0$$

The *x*-intercept is $(0, 0)$.

b. To find the *y*-intercept, substitute $x = 0$.

$$4x + 5y = 0$$
$$4(0) + 5y = 0$$
$$5y = 0$$
$$y = 0$$

The *y*-intercept is $(0, 0)$.

c. Because the *x*-intercept and the *y*-intercept are the same point (the origin), one or more additional points are needed to graph the line. In Table 6-1, we have arbitrarily selected additional values for *x* and *y* to find two more points on the line.

Table 6-1

x	*y*
−5	
	2

Let $x = -5$:
$$4x + 5y = 0$$
$$4(-5) + 5y = 0$$
$$-20 + 5y = 0$$
$$5y = 20$$
$$y = 4$$

Let $y = 2$:
$$4x + 5y = 0$$
$$4x + 5(2) = 0$$
$$4x + 10 = 0$$
$$4x = -10$$
$$x = -\frac{10}{4}$$
$$x = -\frac{5}{2} \text{ or } x = -2\tfrac{1}{2}$$

x	*y*
−5	4
$-\frac{5}{2}$	2

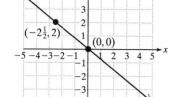

The line through the ordered pairs $(0, 0)$, $(-5, 4)$, and $\left(-\frac{5}{2}, 2\right)$ is shown in Figure 6-5. Note that the point $\left(-\frac{5}{2}, 2\right)$ can be written as $\left(-2\tfrac{1}{2}, 2\right)$.

The line represents the set of all solutions to the equation $4x + 5y = 0$.

Figure 6-5

4.   Applications of *x*- and *y*-Intercepts

In many applications of mathematics, variables are chosen with a meaningful letter to reflect the quantity being represented. In the next example, the letters *a* and *T* are used instead of *x* and *y* to represent altitude and temperature.

example 5

Interpreting Intercepts in an Application

Under certain conditions, the temperature of the air changes linearly with increasing altitude. For example, anyone who has ever hiked in the mountains might know that temperatures tend to become cooler at higher elevations. Scientists can use the following equation to predict the temperature of the air (measured in degrees Celsius, °C) according to the elevation above sea level (measured in meters). (*Source:* U.S. National Oceanic and Atmospheric Administration)

$$T = -0.01a + 20$$ where *a* is the altitude (in meters) and *T* is the temperature in degrees Celsius.

a.   Find the temperature at 1000 m above sea level.
b.   Find the *a*-intercept and the *T*-intercept for this equation and graph the line. (Graph altitude on the horizontal axis and temperature on the vertical axis.)
c.   Interpret the meaning of the *a*-intercept and the *T*-intercept in the context of this problem.

Classroom Activity 6.1A

Solution:

It is important to realize in this example, that the variable *a* is being used in place of *x*. Similarly, the variable *T* is being used in place of *y*.

a. $T = -0.01a + 20$

$T = -0.01(1000) + 20$ To find the temperature at 1000 meters, substitute *a* = 1000.

$T = -10 + 20$

$T = 10$

At 1000 m above sea level, the temperature is 10°C.

b.   To find the *a*-intercept, substitute *T* = 0.

$T = -0.01a + 20$

$0 = -0.01a + 20$

$-20 = -0.01a$

$$\frac{-20}{-0.01} = \frac{-0.01a}{-0.01}$$

$2000 = a$

The *a*-intercept is (2000, 0).

To find the *T*-intercept, substitute *a* = 0.

$T = -0.01a + 20$

$T = -0.01(0) + 20$

$T = 0 + 20$

$T = 20$

The *T*-intercept is (0, 20).

Using the intercepts, we can graph the line $T = -0.01a + 20$ (Figure 6-6).

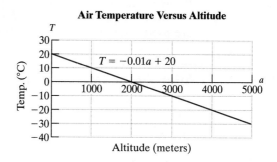

Air Temperature Versus Altitude

Figure 6-6

c. The a-intercept $(2000, 0)$ indicates that 2000 m above sea level, the temperature is 0°C.

The T-intercept $(0, 20)$ indicates that 0 m above sea level, the temperature is 20°C.

5. Horizontal and Vertical Lines

Recall that a linear equation can be written in the form of $ax + by = c$, where a and b are not both zero. However, if a or b is 0 then the line is either parallel to the x-axis (horizontal) or parallel to the y-axis (vertical).

Definitions of Vertical and Horizontal Lines

1. A **vertical line** is a line that can be written in the form, $x = k$, where k is a constant.
2. A **horizontal line** is a line that can be written in the form, $y = k$, where k is a constant.

example 6

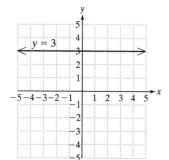

Figure 6-7

Graphing a Horizontal Line

Graph the line $y = 3$.

Solution:

Because this equation is in the form $y = k$, the line is horizontal and must cross the y-axis at $y = 3$ (Figure 6-7).

Alternative Solution:

Create a table of values for the equation $y = 3$. The choice for the y-coordinate must be 3, but x can be any real number.

Tip: Notice that a horizontal line has a y-intercept, but does not have an x-intercept (unless the horizontal line is the x-axis itself).

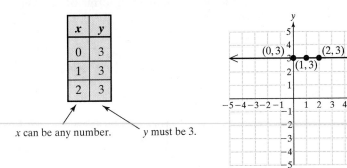

x	y
0	3
1	3
2	3

x can be any number. y must be 3.

---

example 7

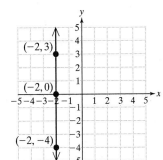

Figure 6-8

Graphing a Vertical Line

Graph the line $4x = -8$.

Solution:

Because the equation does not have a y-variable, we can solve the equation for x.

$$4x = -8 \quad \text{is equivalent to} \quad x = -2$$

This equation is in the form $x = k$, indicating that the line is vertical and must cross the x-axis at $x = -2$ (Figure 6-8).

Alternative Solution:

Create a table of values for the equation $x = -2$. The choice for the x-coordinate must be -2 but y can be any real number.

Tip: Notice that a vertical line has an x-intercept but does not have a y-intercept (unless the vertical line is the y-axis itself).

Classroom Activity 6.1B

x	y
-2	0
-2	3
-2	-4

x must be -2. y can be any number.

Calculator Connections

To graph a linear equation, most calculators require that the y-variable be isolated so that y is expressed in terms of x. To graph the line from Example 1, we have

$$x - 2y = 8$$
$$-2y = -x + 8$$
$$\frac{-2y}{-2} = \frac{-x}{-2} + \frac{8}{-2}$$
$$y = \frac{1}{2}x - 4$$

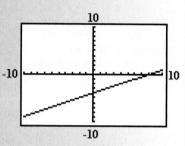

A *Table* feature can be used to verify ordered-pair solutions to an equation in two variables. Consider the equation $y = \frac{1}{2}x - 4$ from Example 1. The table shows y-values corresponding to selected x-values from 0 to 6.

Graphing Calculator Exercises

For each of the following equations, use algebraic methods to graph the line. Then use a graphing calculator to verify the results.

1. $y = 3x - 2$ ❖
2. $y = -2x + 5$ ❖
3. $-x + y = 0$ ❖
4. $-2x + 5y = 0$ ❖
5. $y = 8$ ❖
6. $-5x - 4y = 8$ ❖

---

section 6.1 PRACTICE EXERCISES

1. Determine whether the ordered pairs are solutions to the equation $x + 2y = 6$.

   a. $(0, 3)$ Yes
   b. $(1, 2)$ No
   c. $(-4, 5)$ Yes
   d. $(8, -1)$ Yes

2. Determine whether the ordered pairs are solutions to the equation $y = -\frac{3}{4}x - \frac{1}{4}$.

   a. $\left(0, -\frac{1}{4}\right)$ Yes
   b. $(1, -1)$ Yes
   c. $(4, -3)$ No
   d. $\left(-\frac{1}{3}, 0\right)$ Yes

For Exercises 3–8, identify the equations as linear or nonlinear. Recall that a linear equation in two variables x and y can be written in the form: $ax + by = c$, where a and b are not both zero.

3. $2x - 3y = 8$ Linear
4. $y = -5x - \frac{1}{2}$ Linear
5. $2y^2 = 8$ Nonlinear
6. $x^2 + y^2 = 25$ Nonlinear
7. $\frac{1}{3}x = 6 - \frac{1}{4}y$ Linear
8. $x = -3$ Linear

9. Explain why it is recommended that three points be used to graph a line when only two are needed. The third point can be used as a check point.

 See Additional Answers Appendix   Writing   Translating Expression   Geometry   Scientific Calculator   Video

For Exercises 10–25, graph the lines by making a table of at least three ordered pairs and plotting the points.

10. $x = y + 2$ ❖

11. $x - y = 4$ ❖

12. $-3x + y = -6$ ❖

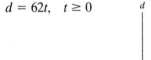 13. $2x - 5y = 10$ ❖

14. $y = 4x$ ❖

15. $y = -2x$ ❖

16. $y = -\frac{1}{2}x + 3$ ❖

17. $y = \frac{1}{4}x - 2$ ❖

18. $x + y = 0$ ❖

19. $-x + y = 0$ ❖

20. $2x + 3y = 8$ ❖

21. $4x - 5y = 15$ ❖

22. $y = 0.75x + 0.25$ ❖

23. $y = -0.8x - 1.2$ ❖

24. $50x - 40y = 200$ ❖

25. $-30x - 20y = 60$ ❖

26. The distance, d, (in miles) a car travels is related to the amount of time, t, (in hours) traveled according to the equation

$$d = 62t, \quad t \geq 0$$

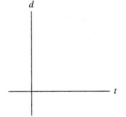

a. Graph the line using t on the horizontal axis and d on the vertical axis. ❖

b. Use the equation to find the distance the car will travel in 2 h. ❖

c. Use the equation to find the distance the car will travel in 3 h and 15 min. ❖

27. The perimeter, P, (in meters) of a square is related to the length of a side, x, (in meters) by the equation

$$P = 4x, \quad x \geq 0$$

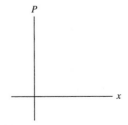

a. Graph the line using x on the horizontal axis and P on the vertical axis. ❖

b. Use the equation to find the perimeter of a square whose sides are 2.2 m. ❖

c. Use the equation to find the perimeter of a square whose sides are 16.5 m. ❖

28. Explain why every line does not have both an x- and a y-intercept.   A horizontal line may not have an x-intercept. A vertical line may not have a y-intercept.

29. The x-intercept is on which axis? x-axis

30. The y-intercept is on which axis? y-axis

31. Which of the lines will have only one intercept?

a. $2x - 3y = 6$ b. $x = 5$

c. $2y = 8$ d. $-x + y = 0$
b, c, d

32. Which of the lines will have only one intercept?

a. $y = 2$ b. $x + y = 0$

c. $2x - 10 = 2$ d. $x + 4y = 8$
a, b, c

For Exercises 33– 54, find the x- and y-intercepts (if they exist), and graph the line.

33. $5x + y = 15$ ❖

34. $x - 3y = -9$ ❖

35. $y = \frac{2}{3}x - 1$ ❖

36. $y = -\frac{3}{4}x + 2$ ❖

37. $x - 3 = y$ ❖

38. $2x + 8 = y$ ❖

39. $-3x + y = 0$ ❖

40. $2x - 2 = 0$ ❖

41. $5y = 8$ ❖

42. $y + 9 = 7$ ❖

43. $x - 3 = -2$ ❖

44. $4x = 5$ ❖

45. $25y = 10x + 100$ ❖

46. $20x = -40y + 200$ ❖

47. $1.2x - 2.4y = 3.6$ ❖

48. $-8.1x - 10.8y = 16.2$ ❖

49. $x = 2y$ ❖

50. $x = -5y$ ❖

51. $2 = 4 + x$ ❖

52. $3 = 7 + y$ ❖

53. $10 = 5y$ ❖

54. $-6 = -6x$ ❖

55. An automobile depreciates immediately after purchase. The equation $V = -5000n + 30{,}000$ determines the value of a certain automobile over the first 5 years of ownership, where V is the value of the automobile (in dollars) and n is the number of years of ownership. See figure.

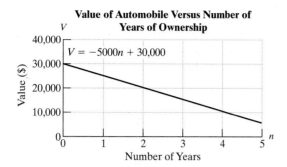

Figure for Exercise 55

a. Find the V-intercept for this equation.
$(0, 30{,}000)$

b. Interpret the meaning of the V-intercept in terms of the value of the automobile and the number of years since its purchase.   At the time of purchase ($n = 0$), the value of the car is $30,000.

c. Find the value of the automobile after 3 years. $15,000

56. The weekly pay for an employee working overtime can be expressed by the equation $P = 18x + 480$, where P is the gross pay in dollars (before taxes and deductions) and x is the number of hours of overtime.

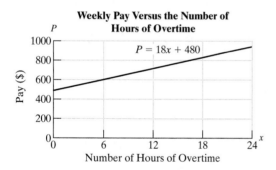

Figure for Exercise 56

a. Find the P-intercept for this equation. $(0, 480)$

b. Interpret the meaning of the P-intercept in terms of gross pay and the number of hours of overtime.   If an employee works 0 hours overtime, the gross pay will be $480.

c. Find the weekly pay for a person who works 6 h overtime. $588

57. Graph the lines on the same set of coordinate axes, and identify any similarities and differences.

a. $y = 2x$ ❖

b. $y = 2x + 3$ ❖

c. $y = 2x - 2$ ❖

58. Graph the lines on the same set of coordinate axes, and identify any similarities and differences.

a. $x + y = 0$ ❖

b. $x + y = 3$ ❖

c. $x + y = -1$ ❖

59. True or False. If the statement is false, rewrite it to be true.

a. The line $x = 3$ is horizontal.   False, $x = 3$ is vertical.

b. The line $y = -4$ is horizontal.   True

60. True or False. If the statement is false, rewrite it to be true.

a. A line parallel to the y-axis is vertical.   True

b. A line perpendicular to the x-axis is vertical.   True

For Exercises 61–72, identify the equation as representing a vertical line or a horizontal line. Then graph the line.

61. $x = 3$ ❖   62. $y = -1$ ❖

63. $-2y = 8$ ❖   64. $5x = 20$ ❖

65. $x + 3 = 7$ ❖   66. $y - 8 = -13$ ❖

67. $3y = 0$ ❖   68. $5x = 0$ ❖

69. $2x + 7 = 10$ ❖ 70. $-3y + 2 = 9$ ❖

71. $9 = 3 + 4y$ ❖   72. $7 = -2x - 5$ ❖

73. The average time per person spent watching television in the United States remained relatively constant between the years 1994 and 1999.

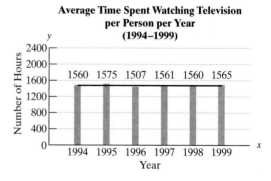

Figure for Exercise 73

(*Source:* Veronis, Suhler & Associates Inc., *Communications Industry Report*)

❖ See Additional Answers Appendix   Writing   Translating Expression   Geometry   Scientific Calculator   Video

a. Find the average of the amount of time spent watching television per year from 1994 to 1999. That is, find the average of: 1560, 1575, 1507, 1561, 1560, and 1565 h. Round to the nearest hour.   1555 h

b. Because the time spent watching television was approximately constant during this time period, an appropriate linear equation is one that represents a horizontal line. Mathematically, a horizontal line is represented by an equation of the form, $y = k$, where k is a constant. Write an equation of a horizontal line to represent these data. (*Hint*: Use the average value found in part (a) for the constant k.) $y = 1555$

74. The batting average of the leader in the American League was relatively constant for the years 1991 through 1997.

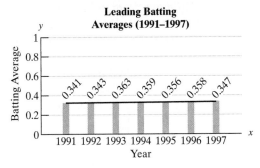

**Leading Batting
Averages (1991–1997)**

Figure for Exercise 74

a. Find the average of the batting averages for the 7 years given. That is, find the average of: 0.341, 0.343, 0.363, 0.359, 0.356, 0.358, and 0.347. Round to 3 decimal places.   0.352

b. Because the batting average appears to be constant during this period, an appropriate linear equation is that of a horizontal line $y = k$, where k is a constant. Write an equation of a horizontal line to represent these data. Use the average value found in part (a) as the constant k. $y = 0.352$

EXPANDING YOUR SKILLS

75. Write an equation representing the x-axis. $y = 0$

76. Write an equation representing the y-axis. $x = 0$

77. Write an equation of the line parallel to the x-axis and passing through the point $(0, 2)$. $y = 2$

78. Write an equation of the line parallel to the y-axis and passing through the point $(4, 0)$. $x = 4$

79. Write an equation of the line perpendicular to the x-axis and passing through the point $(3, -2)$.
$x = 3$

80. Write an equation of the line perpendicular to the y-axis and passing through the point $(-1, 5)$.
$y = 5$

81. Under what conditions will the x- and y-intercepts of the equation $ax + by = c$ be at the origin?
When $c = 0$, the x- and y-intercepts are at the origin.

section

6.2   SLOPE OF A LINE

1. Introduction to Slope

The x- and y-intercepts represent the points where a line crosses the x- and y-axes. Another important feature of a line is its slope. Geometrically, the slope of a line measures the "steepness" of the line. For example, two ski runs are depicted by the lines in Figure 6-9.

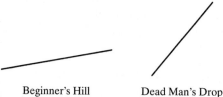

Beginner's Hill   Dead Man's Drop

Figure 6-9

By visual inspection, Dead Man's Drop is "steeper" than Beginner's Hill. To measure the slope of a line quantitatively, consider two points on the line. The slope of the line is the ratio of the vertical change (change in y) between the two points and the horizontal change (change in x). As a memory device, we might think of the slope of a line as "rise over run." See Figure 6-10.

$$\text{Slope} = \frac{\text{change in } y}{\text{change in } x} = \frac{\text{rise}}{\text{run}}$$

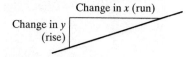

Change in x (run)

Change in y (rise)

Figure 6-10

To move from point A to point B on Beginner's Hill, rise 2 ft and move to the right 6 ft (Figure 6-11).

To move from point A to point B on Dead Man's Drop, rise 12 ft and move to the right 6 ft (Figure 6-12).

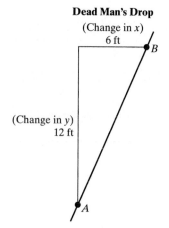

Dead Man's Drop
(Change in x)
6 ft
B
(Change in y)
12 ft
A

Figure 6-12

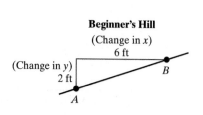

Beginner's Hill
(Change in x)
6 ft
B
(Change in y)
2 ft
A

Figure 6-11

$$\text{Slope} = \frac{\text{change in } y}{\text{change in } x} = \frac{2 \text{ ft}}{6 \text{ ft}} = \frac{1}{3}$$

$$\text{Slope} = \frac{\text{change in } y}{\text{change in } x} = \frac{12 \text{ ft}}{6 \text{ ft}} = \frac{2}{1} = 2$$

The slope of Dead Man's Drop is greater than the slope of Beginner's Hill, confirming the observation that Dead Man's Drop is steeper. On Dead Man's Drop, there is a 12-ft change in elevation for every 6 ft of horizontal distance (a 2:1 ratio). On Beginner's Hill there is only a 2-ft change in elevation for every 6 ft of horizontal distance (a 1:3 ratio).

example 1

Finding Slope in an Application

Find the slope of the ramp up the stairs.

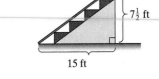

Solution:

$$\text{Slope} = \frac{\text{change in } y}{\text{change in } x} = \frac{7\frac{1}{2} \text{ ft}}{15 \text{ ft}}$$

$$\frac{7\frac{1}{2}}{15} = \frac{\frac{15}{2}}{\frac{15}{1}} \qquad \text{Write the mixed number as an improper fraction.}$$

$$\frac{15}{2} \cdot \frac{1}{15} = \frac{1}{2} \qquad \text{Multiply by the reciprocal and simplify.}$$

The slope is $\frac{1}{2}$.

2. Slope Formula

The slope of a line may be found using any two points on the line—call these points (x_1, y_1) and (x_2, y_2). The change in y between the points can be found by taking the difference of the y-values: $y_2 - y_1$. The change in x can be found by taking the difference of the x-values in the same order: $x_2 - x_1$ (Figure 6-13).

The slope of a line is often symbolized by the letter m and is given by the following formula.

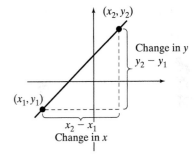

Figure 6-13

Definition of the Slope of a Line

The **slope** of a line passing through the distinct points (x_1, y_1) and (x_2, y_2) is

$$m = \frac{y_2 - y_1}{x_2 - x_1} \quad \text{provided } x_2 - x_1 \neq 0$$

example 2

Finding the Slope of a Line through Two Points

Find the slope of the line through the points $(-1, 3)$ and $(-4, -2)$.

Solution:

To use the slope formula, first label the coordinates of each point and then substitute the coordinates into the slope formula.

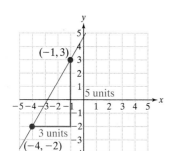

$$
\begin{array}{ccc}
(-1, 3) & \text{and} & (-4, -2) \\
(x_1, y_1) & & (x_2, y_2)
\end{array}
\qquad \text{Label the points.}
$$

$$
m = \frac{y_2 - y_1}{x_2 - x_1} = \frac{(-2) - (3)}{(-4) - (-1)} \qquad \text{Apply the slope formula.}
$$

$$
= \frac{-5}{-3}; \quad \text{hence, } m = \frac{5}{3} \qquad \text{Simplify and reduce.}
$$

The slope of the line can be verified from the graph (Figure 6-14).

Figure 6-14

Tip: The slope formula is not dependent on which point is labeled (x_1, y_1) and which point is labeled (x_2, y_2). In Example 2, reversing the order in which the points are labeled results in the same slope.

$$
\begin{array}{ccc}
(-1, 3) & \text{and} & (-4, -2) \\
(x_2, y_2) & & (x_1, y_1)
\end{array}
\qquad \text{Label the points.}
$$

then

$$
m = \frac{(3) - (-2)}{(-1) - (-4)} = \frac{5}{3} \qquad \text{Apply the slope formula.}
$$

3. Positive, Negative, Zero, and Undefined Slopes

The value of the slope of a line may be positive, negative, zero, or undefined.

Lines that increase, or rise, from left to right have a positive slope.
Lines that decrease, or fall, from left to right have a negative slope.
Horizontal lines have a slope of zero.
Vertical lines have an undefined slope.

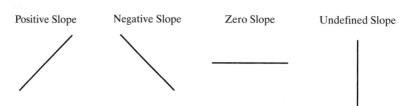

Positive Slope   Negative Slope   Zero Slope   Undefined Slope

example 3

Finding the Slope of a Line Between Two Points

Find the slope of the line passing through the points $(-5, 0)$ and $(2, -3)$.

Solution:

$$
\begin{array}{cc}
(-5, 0) \quad \text{and} \quad (2, -3) \\
(x_1, y_1) \qquad\qquad (x_2, y_2)
\end{array}
\qquad \text{Label the points.}
$$

$$
m = \frac{y_2 - y_1}{x_2 - x_1} = \frac{(-3) - (0)}{(2) - (-5)} \qquad \text{Apply the slope formula.}
$$

$$
= \frac{-3}{7} \quad \text{or} \quad -\frac{3}{7} \qquad \text{Simplify.}
$$

By graphing the points $(-5, 0)$ and $(2, -3)$, we can verify that the slope is $-\frac{3}{7}$ (Figure 6-15). Notice that the line slopes downward from left to right.

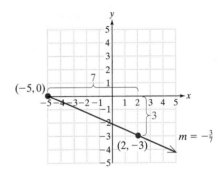

Figure 6-15

example 4

Determining the Slope of a Horizontal and Vertical Line

a. Find the slope of the line passing through the points $(2, -1)$ and $(2, 4)$.
b. Find the slope of the line passing through the points $\left(\frac{8}{3}, \frac{4}{5}\right)$ and $\left(\frac{5}{6}, \frac{4}{5}\right)$.

Classroom Activity 6.2A

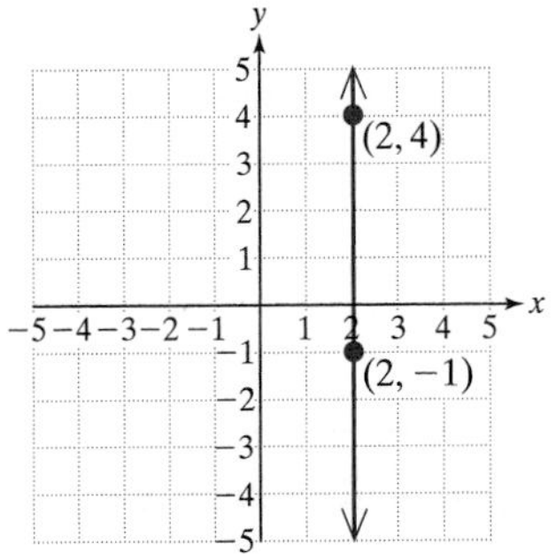

Solution:

a.
$$
\begin{array}{cc}
(2, -1) \quad \text{and} \quad (2, 4) \\
(x_1, y_1) \qquad\qquad (x_2, y_2)
\end{array}
\qquad \text{Label the points.}
$$

$$
m = \frac{y_2 - y_1}{x_2 - x_1} = \frac{(4) - (-1)}{(2) - (2)} \qquad \text{Apply the slope formula.}
$$

$$
m = \frac{5}{0} \quad \text{Undefined}
$$

Because the slope, m, is undefined, we expect the points to form a vertical line as shown in Figure 6-16.

Figure 6-16

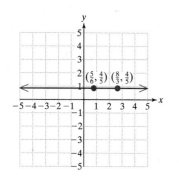

Figure 6-17

b. $\left(\dfrac{8}{3}, \dfrac{4}{5}\right)$ and $\left(\dfrac{5}{6}, \dfrac{4}{5}\right)$

$\quad (x_1, y_1) \qquad\qquad\quad (x_2, y_2)$ Label the points.

$m = \dfrac{y_2 - y_1}{x_2 - x_1} = \dfrac{\left(\frac{4}{5}\right) - \left(\frac{4}{5}\right)}{\left(\frac{5}{6}\right) - \left(\frac{8}{3}\right)}$ Apply the slope formula.

$m = \dfrac{\frac{4}{5} - \frac{4}{5}}{\left(\frac{5}{6}\right) - \left(\frac{16}{6}\right)} = \dfrac{0}{-\frac{11}{6}} = 0$

Because the slope is 0, we expect the points to form a horizontal line as shown in Figure 6-17.

4. Parallel and Perpendicular Lines

Lines that do not intersect are called *parallel lines*. Parallel lines have the same slope and different *y*-intercepts (Figure 6-18).

Lines that intersect at a right angle are *perpendicular lines*. If two lines are perpendicular then the slope of one line is the opposite of the reciprocal of the slope of the other line (provided neither line is vertical) (Figure 6-19).

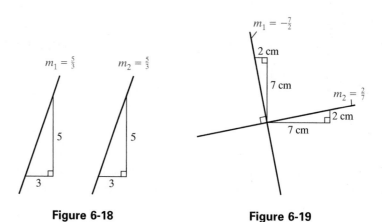

Figure 6-18   **Figure 6-19**

Slopes of Parallel Lines

If m_1 and m_2 represent the **slopes of two parallel** (nonvertical) **lines,** then

$$m_1 = m_2$$

Slopes of Perpendicular Lines

If $m_1 \neq 0$ and $m_2 \neq 0$ represent the **slopes of two perpendicular lines,** then

$$m_1 = -\dfrac{1}{m_2} \quad \text{or} \quad m_2 = -\dfrac{1}{m_1}$$

or equivalently, $m_1 m_2 = -1$.

example 5

Determining the Slope of Parallel and Perpendicular Lines

Suppose a given line has a slope of $-\frac{1}{4}$.

a.   Find the slope of a line parallel to the given line.
b.   Find the slope of a line perpendicular to the given line.

Classroom Activity 6.2B

Solution:

a.   Parallel lines must have the same slope. The slope of a line parallel to the given line is: $m = -\frac{1}{4}$.

b.   Perpendicular lines must have opposite and reciprocal slopes. The slope of a line perpendicular to the given line is: $m = +\frac{4}{1}$ or simply, $m = 4$.

5.   Applications of Slope

In many applications, the interpretation of slope refers to the *rate of change* of the y-variable to the x-variable.

example 6

Interpreting Slope in an Application

Mario earns $10.00/h working for a landscaping company. Shannelle earns $15.00/h working for an in-home nursing agency. Figure 6-20 shows their total earnings versus the number of hours they work.

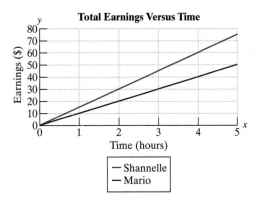

Figure 6-20

a.   Find the slope of the line representing Mario's earnings.
b.   Find the slope of the line representing Shannelle's earnings.

Solution:

a. After 1 hour, Mario earns $10. After 2 hours, he earns $20, and so on. For each 1-hour change in time, there is a $10 increase in wages. The slope of the line representing Mario's earnings is $10/h.
b. After 1 hour, Shannelle earns $15. After 2 hours, she earns $30, and so on. For each 1-hour change in time, there is a $15 increase in wages. The slope of the line representing Shannelle's earnings is $15/h.

---

example 7

Interpreting Slope in an Application

Figure 6-21 depicts the annual median income for males in the United States between 1990 and 1996. The trend is approximately linear. Find the slope of the line and interpret the meaning of the slope in the context of this problem.

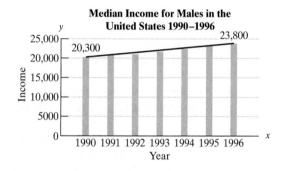

Figure 6-21

Source: U.S. Department of the Census.

Classroom Activity 6.2C

Solution:

To determine the slope we need to know two points on the line. From the graph, the median income for males in 1990 was $20,300. This corresponds to the ordered pair (1990, 20,300). In 1996, the median income was $23,800. This corresponds to the ordered pair (1996, 23,800).

$$\underset{(x_1, y_1)}{(1990, 20{,}300)} \quad \text{and} \quad \underset{(x_2, y_2)}{(1996, 23{,}800)} \qquad \text{Label the points.}$$

$$m = \frac{y_2 - y_1}{x_2 - x_1} = \frac{(23{,}800) - (20{,}300)}{(1996) - (1990)} \qquad \text{Apply the slope formula.}$$

$$m = \frac{3500}{6} \approx 583.33 \qquad \text{Simplify.}$$

The slope indicates that the median income for males in the United States increased at a rate of approximately $583.33 per year between 1990 and 1996.

section 6.2 PRACTICE EXERCISES

For Exercises 1–8, find the *x*- and *y*-intercepts (if they exist). Then graph the lines. <u>6.1</u>

1. $x - 3y = 6$ ❖

2. $y = -3$ ❖

3. $x - 5 = 2$ ❖

4. $y = \dfrac{2}{3}x$ ❖

5. $2y - 3 = 0$ ❖

6. $4x + y = 8$ ❖

7. $2x = 4y$ ❖

8. $-1 = 2x + 3$ ❖

For Exercises 9–12, fill in the blank with the appropriate term: zero, negative, positive, or undefined.

9. The slope of a line parallel to the *y*-axis is <u>undefined</u>.

10. The slope of a horizontal line is <u>zero</u>.

11. The slope of a line that rises from left to right is <u>positive</u>.

12. The slope of a line that falls from left to right is <u>negative</u>.

For Exercises 13–20, label the lines as having a positive, negative, zero, or undefined slope.

13.
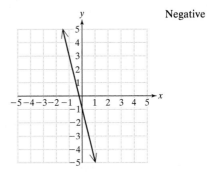
Negative

14.
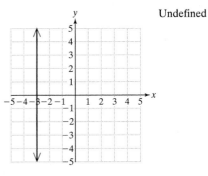
Undefined

15.
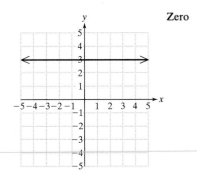
Zero

16.
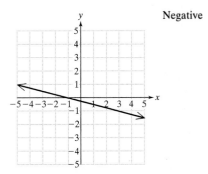
Negative

17.
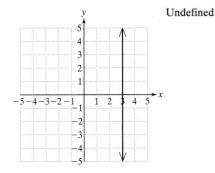
Undefined

18.
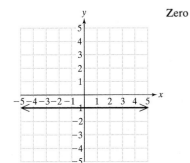
Zero

19.

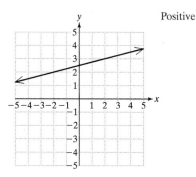

Positive

20.

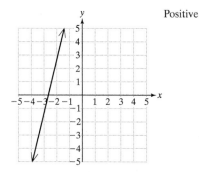

Positive

21. Point A is located 3 units up and 4 units to the right of point B. What is the slope of the line passing through the points? $\frac{3}{4}$

22. Point A is located 2 units up and 5 units to the left of point B. What is the slope of the line passing through the points? $-\frac{2}{5}$

23. Point A is located 1 unit down and 2 units to the right of point B. What is the slope of the line passing through the points? $-\frac{1}{2}$

24. Point A is located 3 units down and 2 units to the left of point B. What is the slope of the line passing through the points? $\frac{3}{2}$

25. Point A is located 3 units up and 3 units to the left of point B. What is the slope of the line passing through the points? -1

26. Point A is located 2 units down and 2 units to the left of point B. What is the slope of the line passing through the points? 1

27. Point A is located 5 units to the right of point B. What is the slope of the line passing through the points? Zero

28. Point A is located 3 units down from point B. What is the slope of the line passing through the points? Undefined

For Exercises 29–46, find the slope of the line that passes through the two points.

29. $(2, 4)$ and $(-1, -1)$ $\frac{5}{3}$

30. $(0, 4)$ and $(3, 0)$ $-\frac{4}{3}$

31. $(-2, 3)$ and $(-1, 0)$ -3

32. $(-3, -4)$ and $(1, -5)$ $-\frac{1}{4}$

33. $(5, 3)$ and $(-2, 3)$ Zero

34. $(0, -1)$ and $(-4, -1)$ Zero

35. $(2, -7)$ and $(2, 5)$ Undefined

36. $(-4, 3)$ and $(-4, -4)$ Undefined

37. $\left(\frac{1}{2}, \frac{3}{5}\right)$ and $\left(\frac{1}{4}, -\frac{4}{5}\right)$ $\frac{28}{5}$

38. $\left(-\frac{2}{7}, \frac{1}{3}\right)$ and $\left(\frac{8}{7}, -\frac{5}{6}\right)$ $-\frac{49}{60}$

39. $(3\frac{3}{4}, -1\frac{1}{4})$ and $(-5, 6\frac{1}{2})$ $-\frac{31}{35}$

40. $(-6\frac{7}{8}, 5\frac{2}{5})$ and $(-10, 4\frac{1}{10})$ $\frac{52}{125}$

41. $(6.8, -3.4)$ and $(-3.2, 1.1)$ -0.45

42. $(-3.15, 8.25)$ and $(6.85, -4.25)$ -1.25

43. $(-5.50, 1.75)$ and $(-1.50, -4.80)$ -1.6375

44. $(11.2, 8.4)$ and $(-3.8, -11.6)$ $1.\overline{3}$

45. $(1994, 35{,}000)$ and $(2000, 24{,}000)$ -1833.3 or $-\frac{5500}{3}$

46. $(1988, 4.65)$ and $(1998, 9.25)$ 0.46

47. Graph the line through the point $(0, 3)$ having a slope of $\frac{3}{4}$. ❖

48. Graph the line through the point $(0, -2)$ having a slope of $\frac{2}{3}$. ❖

49. Graph the line through the point $(0, 3)$ having a slope of $-\frac{3}{4}$. ❖

50. Graph the line through the point $(0, -2)$ having a slope of $-\frac{2}{3}$. ❖

51. Graph the line through the point $(-4, 1)$ having a slope of -2. ❖

52. Graph the line through the point $(3, 2)$ having a slope of -3. ❖

❖ See Additional Answers Appendix   Writing   Translating Expression   Geometry   Scientific Calculator   Video

53. In 1980, there were 304 thousand male inmates in federal and state prisons. By 1996, the number increased to 1069 thousand.

Let *x* represent the year and let *y* represent the number of male inmates in federal and state prisons (in thousands).

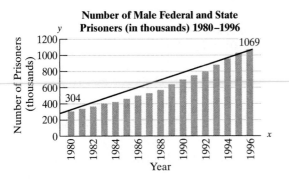

Number of Male Federal and State Prisoners (in thousands) 1980–1996

Figure for Exercise 53

(*Source:* U.S. Bureau of Statistics)

a. Using the ordered pairs (1980, 304) and (1996, 1069), find the slope of the line. $\frac{765}{16}$

b. Interpret the meaning of the slope in the context of this problem. The number of male inmates increased by 765 thousand in 16 years, or approximately 47.8 thousand inmates per year.

54. In 1980, there were 12 thousand female inmates in federal and state prisons. By 1996, the number increased to 70 thousand.

Let *x* represent the year and let *y* represent the number of female inmates in federal and state prisons (in thousands).

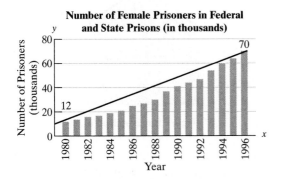

Number of Female Prisoners in Federal and State Prisons (in thousands)

Figure for Exercise 54

(*Source:* U.S. Bureau of Statistics)

a. Using the ordered pairs (1980, 12) and (1996, 70), find the slope of the line. $\frac{29}{8}$

b. Interpret the meaning of the slope in the context of this problem. The number of female inmates increased by 29 thousand in 8 years, or approximately 3.6 thousand inmates per year.

55. In Exercises 53 and 54 you found the slope of the lines representing the rise in male and female federal and state prisoners per year.

a. Which group, males or females, had the larger rate of increase in prisoners per year? Males had a larger rate of increase.

b. For each year between 1980 and 1996, the number of female prisoners was less than the number of male prisoners. Based on the slopes of the two lines, will the number of female prisoners ever "catch up" to the number of male prisoners if these linear trends continue? No, the females will not catch up because their rate of increase is less than the males' rate of increase.

56. The following graph depicts the median income for females in the United States between 1990 and 1996.

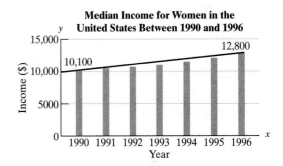

Median Income for Women in the United States Between 1990 and 1996

Figure for Exercise 56

(*Source:* U.S. Bureau of the Census)

a. Find the slope of the line and interpret the meaning of the slope in the context of this problem. $m = 450$. The median income for women in the United States increased by \$450/year.

b. Compare the slopes for the rise in median income per year for women and for men (see Example 7). Based on the slopes of the two lines, will the median income for women ever catch-up to the median income for men if these linear trends continue? Explain. No, the rate of increase in women's median income is less than the rate of increase in men's median income.

For Exercises 57–62, answers may vary.

57. a. Draw a line with a positive slope. ❖

b. Draw a line parallel to the line in part (a). ❖

58. a. Draw a line with a negative slope. ❖

b. Draw a line parallel to the line in part (a). ❖

❖ See Additional Answers Appendix  Writing Translating Expression Geometry Scientific Calculator Video

59. a. Draw a line with a negative slope. ❖

    b. Draw a line perpendicular to the line in part (a). ❖

60. a. Draw a line with a positive slope. ❖

    b. Draw a line perpendicular to the line in part (a). ❖

61. a. Draw a line with an undefined slope. ❖

    b. Draw a line perpendicular to the line in part (a). ❖

62. a. Draw a line with a slope of 0. ❖

    b. Draw a line parallel to the line in part (a). ❖

63. Suppose a given line has a slope of -2.

    a. What is the slope of a line parallel to the given line? -2

    b. What is the slope of a line perpendicular to the given line? $\frac{1}{2}$

64. Suppose a given line has a slope of $\frac{2}{3}$.

    a. What is the slope of a line parallel to the given line? $\frac{2}{3}$

    b. What is the slope of a line perpendicular to the given line? $-\frac{3}{2}$

65. Suppose a given line has a slope of 0.

    a. What is the slope of a line parallel to the given line? 0

    b. What is the slope of a line perpendicular to the given line? Undefined

66. Suppose a line has an undefined slope.

    a. What is the slope of a line parallel to the given line? Undefined

    b. What is the slope of a line perpendicular to the given line? 0

For Exercises 67–74, find the slopes of the lines l_1 and l_2 determined by the two given points. Then identify whether l_1 and l_2 are parallel, perpendicular, or neither.

67. l_1: (2, 4) and (−1, −2)
    l_2: (1, 7) and (0, 5)   Parallel

68. l_1: (0, 0) and (−2, 4)
    l_2: (1, −5) and (−1, −1)   Parallel

69. l_1: (1, 9) and (0, 4)
    l_2: (5, 2) and (10, 1)   Perpendicular

70. l_1: (3, −4) and (−1, −8)
    l_2: (5, −5) and (−2, 2)   Perpendicular

71. l_1: (4, 4) and (0, 3)
    l_2: (1, 7) and (−1, −1)   Neither

72. l_1: (3, 5) and (−2, −5)
    l_2: (2, 0) and (−4, −3)   Neither

73. l_1: (3.1, 6.3) and (3.1, −5.7)
    l_2: (1.2, 4.7) and (1.2, −5.3)   Parallel

74. l_1: (4.5, −6.7) and (−2.3, −6.7)
    l_2: (−2.2, −6.7) and (−1.4, −6.7)   Parallel

75. Determine the pitch (slope) of the roof. $\pm\frac{1}{3}$

Figure for Exercise 75

76. Find the height, y, so that the pitch (slope) of the garage roof is $\frac{1}{4}$. $\frac{9}{2}$ ft or $4\frac{1}{2}$ ft

Figure for Exercise 76

77. The distance, d, (in miles) between a lightning strike and an observer is given by the equation:

    $d = 0.2t$ where t is the time (in seconds) between seeing lightning and hearing thunder.

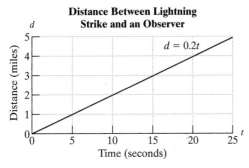

Distance Between Lightning Strike and an Observer

$d = 0.2t$

Distance (miles) / Time (seconds)

Figure for Exercise 77

a. If an observer counts 5 s between seeing lightning and hearing thunder, how far away was the lightning strike? 1 mile

b. If an observer counts 10 s between seeing lightning and hearing thunder, how far away was the lightning strike? 2 miles

c. If an observer counts 15 s between seeing lightning and hearing thunder, how far away was the lightning strike? 3 miles

d. What is the slope of the line? Interpret the meaning of the slope in the context of this problem. $m = 0.2$; The distance between a lightning strike and an observer increases by 0.2 miles for every additional second between seeing lightning and hearing thunder.

78. Jorge is paid by the hour according to the equation

$P = 11.50x$ P is his total pay (in dollars) and x is the number of hours worked

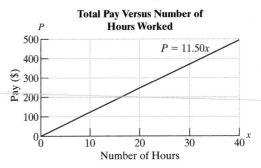

Total Pay Versus Number of Hours Worked

$P = 11.50x$

Pay ($) / Number of Hours

Figure for Exercise 78

a. How much money will Jorge earn if he works 20 h? $230.00

b. How much money will Jorge earn if he works 21 h? $241.50

c. How much money will Jorge earn if he works 22 h? $253.00

d. What is the slope of the line? Interpret the meaning of the slope in the context of this problem. $m = 11.5$; Jorge's pay increases $11.50 for each additional hour worked.

EXPANDING YOUR SKILLS

79. Find the slope between the points $(a + b, 4m - n)$ and $(a - b, m + 2n)$. $\dfrac{3m - 3n}{2b}$

80. Find the slope between the points $(3c - d, s + t)$ and $(c - 2d, s - t)$. $\dfrac{2t}{2c + d}$

81. Find the x-intercept of the line $ax + by = c$. $\left(\dfrac{c}{a}, 0\right)$

82. Find the y-intercept of the line $ax + by = c$. $\left(0, \dfrac{c}{b}\right)$

83. Find another point on the line that contains the point $(2, -1)$ and has a slope of $\frac{2}{5}$. For example: $(7, 1)$

84. Find another point on the line that contains the point $(-3, 4)$ and has a slope of $\frac{1}{4}$. For example: $(1, 5)$

section

6.3 SLOPE-INTERCEPT FORM OF A LINE

1. Slope-Intercept Form of a Line

In Section 6.1, we learned that an equation of the form: $ax + by = c$ (where a and b are not both zero) represents a line in a rectangular coordinate system. An equation of a line written in this way is said to be in **standard form**. In this section, we will learn a new form, called **slope-intercept form**, which is useful in determining the slope and *y*-intercept of a line.

Let $(0, b)$ represent the *y*-intercept of a line. Let (x, y) represent any other point on the line. Then the slope of the line can be found as follows:

Let $(0, b)$ represent (x_1, y_1) and let (x, y) represent (x_2, y_2). Apply the slope formula.

$$m = \frac{(y_2 - y_1)}{(x_2 - x_1)} \rightarrow m = \frac{y - b}{x - 0} \qquad \text{Apply the slope formula.}$$

$$m = \frac{y - b}{x} \qquad \text{Simplify.}$$

$$mx = \left(\frac{y - b}{x}\right)x \qquad \text{Multiply by } x \text{ to clear fractions.}$$

$$mx = y - b$$

$$mx + b = y - b + b \qquad \text{To isolate } y, \text{ add } b \text{ to both sides.}$$

$$mx + b = y \qquad \text{or} \qquad y = mx + b \qquad \text{The equation is in slope-intercept form.}$$

Slope-Intercept Form of a Line
___

$y = mx + b$ is the slope-intercept form of a line.

m is the slope and the point $(0, b)$ is the *y*-intercept.

For example, the equation, $y = -3x + 1$ is written in slope-intercept form. By inspection, the slope of the line is -3 and the *y*-intercept is $(0, 1)$.

2. Identifying the Slope and *y*-Intercept of a Line

example 1

Identifying the Slope and *y*-Intercept of a Line

Given the line $-5x - 2y = 6$,

a. Write the slope-intercept form of the line.
b. Identify the slope and *y*-intercept.

Solution:

a.   Write the equation in slope-intercept form, $y = mx + b$, by solving for y.

$$-5x - 2y = 6$$

$$-2y = 5x + 6 \qquad \text{Add } 5x \text{ to both sides.}$$

$$\frac{-2y}{-2} = \frac{5x + 6}{-2} \qquad \text{Divide both sides by } -2.$$

$$y = \frac{5x}{-2} + \frac{6}{-2}$$

$$y = -\frac{5}{2}x - 3 \qquad \text{Slope-intercept form.}$$

b.   The slope is $m = -\frac{5}{2}$ and the y-intercept is $(0, -3)$.

3.   Graphing a Line from Its Slope and y-Intercept

Slope-intercept form is a useful tool to graph a line. The y-intercept is a known point on the line. The slope indicates the direction of the line and can be used to find a second point. Using slope-intercept form to graph a line is demonstrated in the next example.

example 2

Classroom Activity 6.3A

Tip: Example 2 illustrates that for a line with a slope of $-\frac{5}{2}$ we can start at the y-intercept and either move down 5 units and to the right 2 units or up 5 units and to the left 2 units to plot a second point.

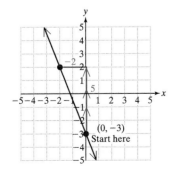

Figure 6-23

Graphing a Line Using the Slope and y-Intercept

Graph the line $y = -\frac{5}{2}x - 3$ by using the slope and y-intercept.

Solution:

First plot the y-intercept, $(0, -3)$.

The slope, $m = -\frac{5}{2}$ can be written as

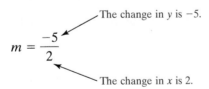

$$m = \frac{-5}{2}$$

The change in y is -5.

The change in x is 2.

To find a second point on the line, start at the y-intercept and move down 5 units and to the right 2 units. Then draw the line through the two points (Figure 6-22).

Similarly, the slope can be written as

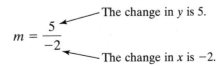

$$m = \frac{5}{-2}$$

The change in y is 5.

The change in x is -2.

To find a second point, start at the y-intercept and move up 5 units and to the left 2 units. Then draw the line through the two points (Figure 6-23).

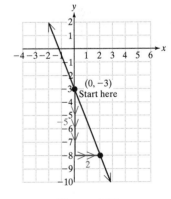

Figure 6-22

4. Determining Whether Two Lines Are Parallel, Perpendicular, or Neither

example 3

Determining If Two Lines are Parallel, Perpendicular, or Neither

Use the slope-intercept form of the following equations to determine if the lines l_1 and l_2 are parallel, perpendicular, or neither.

$$l_1: -x + 2y = 2 \quad \text{and} \quad l_2: 2x + y = -4$$

Solution:

Write each line in slope-intercept form.

$l_1: -x + 2y = 2$ $l_2: 2x + y = -4$

$\qquad 2y = x + 2$ $\qquad y = -2x - 4$ The slope is -2.

$\qquad \dfrac{2y}{2} = \dfrac{x}{2} + \dfrac{2}{2}$

$\qquad y = \dfrac{1}{2}x + 1$ The slope is $\frac{1}{2}$.

Since $\frac{1}{2}$ is the opposite of the reciprocal of -2, then the lines l_1 and l_2 are perpendicular.

Tip: To confirm that the lines l_1 and l_2 are perpendicular, show that $m_1 m_2 = -1$. If $m_1 = \frac{1}{2}$ and $m_2 = -2$, then $m_1 m_2 = \frac{1}{2}(-2) = -1$ as desired.

5. Writing an Equation of a Line Using Slope-Intercept Form

The slope-intercept form of a line can be used to write an equation of a line when the slope is known and a point on the line is known.

example 4

Writing an Equation of a Line Using Slope-Intercept Form

Write an equation of the line whose slope is $\frac{2}{3}$ and whose y-intercept is $(0, 8)$.

Solution:

The slope is given as $m = \frac{2}{3}$ and the y-intercept $(0, b)$ is given as $(0, 8)$. Substitute the values $m = \frac{2}{3}$ and $b = 8$, into the slope-intercept form of a line.

$$y = mx + b$$
$$y = \frac{2}{3}x + 8$$

In Example 4, both the slope and y-intercept were given in the problem, and the values of m and b could be substituted directly into slope-intercept form. What if the y-intercept is not known and a different point is given instead? Example 5 illustrates this case.

example 5   **Writing an Equation of a Line Using Slope-Intercept Form**

Write an equation of the line having a slope of 3 and passing through the point $(-2, -4)$. Write the answer in slope-intercept form.

Solution:

To find an equation of a line using slope-intercept form, it is necessary to find the value of m and b. The slope is given in the problem as $m = 3$. Therefore, the slope-intercept form becomes

$$y = mx + b$$
$$y = 3x + b$$

Because the point $(-2, -4)$ is on the line, it is a solution to the equation. Therefore, to find b, substitute the values of x and y from the ordered pair $(-2, -4)$ and solve the resulting equation.

$$y = 3x + b$$
$$-4 = 3(-2) + b \qquad \text{Substitute } y = -4 \text{ and } x = -2.$$
$$-4 = -6 + b \qquad \text{Simplify.}$$
$$2 = b \qquad \text{Solve for } b. \qquad\qquad y = mx + b$$

Now with m and b known, the slope-intercept form is: $y = 3x + 2$.

Tip: The equation from Example 5 can be checked by graphing the line $y = 3x + 2$. The slope $m = 3$ can be written as $m = \frac{3}{1}$. Therefore, to graph the line, start at the y-intercept $(0, 2)$ and move up 3 units and to the right 1 unit (Figure 6-24).
   The graph verifies that the line passes through the point $(-2, -4)$ as it should.

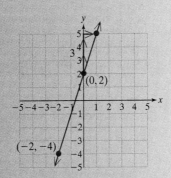

Figure 6-24

6. Writing an Equation of a Line Through Two Points

The slope-intercept form of a line can also be used to write an equation of a line when two points on the line are known.

example 6

Writing an Equation of a Line Through Two Points

Write an equation of the line passing through the points $(-3, 0)$ and $(3, -4)$. Write the answer in slope-intercept form.

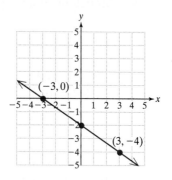

Figure 6-25

Solution:

Given two points on a line, the slope can be found with the slope formula.

$$\begin{array}{cc} (-3, 0) & \text{and} & (3, -4) \\ (x_1, y_1) & & (x_2, y_2) \end{array} \quad \text{Label the points.}$$

$$m = \frac{y_2 - y_1}{x_2 - x_1} = \frac{(-4) - (0)}{(3) - (-3)} = \frac{-4}{6} = -\frac{2}{3}$$

With the slope $m = -\frac{2}{3}$, the slope-intercept form of the line is: $y = -\frac{2}{3}x + b$.

To find b, substitute the values of x and y from *either* ordered pair and then solve the resulting equation for b. We will use the point $(-3, 0)$.

$$y = -\frac{2}{3}x + b$$

$$0 = -\frac{2}{3}(-3) + b \qquad \text{Substitute } y = 0 \text{ and } x = -3.$$

$$0 = 2 + b \qquad \text{Simplify.}$$

$$-2 = b \qquad \text{Solve for } b.$$

Tip: The value of b could also have been found by using the point $(3, -4)$:

$$y = -\frac{2}{3}x + b$$

$$-4 = -\frac{2}{3}(3) + b$$

$$-4 = -2 + b$$

$$-2 = b$$

The slope-intercept form is $y = -\frac{2}{3}x - 2$.

A sketch of the line shows that the line passes through the points $(-3, 0)$ and $(3, -4)$ as desired (Figure 6-25).

7. Writing an Equation of a Line Parallel or Perpendicular to Another Line

example 7

Writing an Equation of a Line Parallel to Another Line

Write an equation of the line passing through the point $(1, 2)$ and parallel to the line $y = 3x - 4$. Write the answer in slope-intercept form.

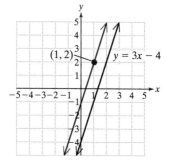

Figure 6-26

Solution:

Figure 6-26 shows the line $y = 3x - 4$ (pictured in black) and a line parallel to it (pictured in blue) that passes through the point $(1, 2)$. The given line, $y = 3x - 4$, is written in slope-intercept form and its slope is easily identified as 3. The line parallel to the given line must also have a slope of 3.

The slope-intercept form of the line we want to find is

$$y = 3x + b$$

Because the point $(1, 2)$ is on the line, it is a solution to the equation. Therefore, to find b, substitute the values of x and y from the ordered pair $(1, 2)$, and solve the resulting equation.

$$y = 3x + b$$
$$2 = 3(1) + b$$
$$2 = 3 + b$$
$$-1 = b$$

The slope-intercept form of the line is $y = 3x - 1$.

---

example 8

Writing an Equation of a Line Perpendicular to Another Line

Write an equation of the line passing through the point $(-1, 3)$ and perpendicular to the line $2x + y = 1$. Write the answer in slope-intercept form.

Classroom Activity 6.3B

Solution:

The equation $2x + y = 1$ can be written in slope-intercept form as: $y = -2x + 1$. The slope of this line is -2. Therefore, the slope of a line perpendicular to $y = -2x + 1$ is $\frac{1}{2}$ (the opposite of the reciprocal of -2).

The slope-intercept form of the line we want to find is $y = \frac{1}{2}x + b$.

Because the point $(-1, 3)$ is on the line, it is a solution to the equation. To find b, substitute the values of x and y from the ordered pair $(-1, 3)$, and solve the resulting equation.

$$y = \frac{1}{2}x + b$$

$$3 = \frac{1}{2}(-1) + b \qquad \text{Substitute } y = 3 \text{ and } x = -1.$$

$$3 = -\frac{1}{2} + b \qquad \text{Simplify.}$$

$$3 + \frac{1}{2} = b \qquad \text{Solve for } b.$$

$$\frac{6}{2} + \frac{1}{2} = b \qquad \text{Get a common denominator.}$$

$$\frac{7}{2} = b$$

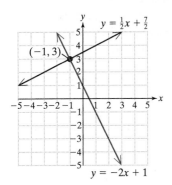

Figure 6-27

The slope-intercept form of the line is $y = \frac{1}{2}x + \frac{7}{2}$. A sketch will verify that the line $y = \frac{1}{2}x + \frac{7}{2}$ passes through the point $(-1, 3)$ and is perpendicular to the line $y = -2x + 1$ (Figure 6-27).

8. Different Forms of Linear Equations: A Summary

A linear equation can be written in several different forms as summarized in Table 6-2.

Table 6-2

Form	Example	Comments
Standard Form $ax + by = c$	$4x + 2y = 8$	a and b must not *both* be zero.
Horizontal Line $y = k$ (k is constant)	$y = 4$	The slope is zero and the y-intercept is $(0, k)$.
Vertical Line $x = k$ (k is constant)	$x = -1$	The slope is undefined and the x-intercept is $(k, 0)$.
Slope-Intercept Form $y = mx + b$ Slope $= m$ y-intercept is $(0, b)$	$y = -3x + 7$ Slope $= -3$ y-intercept is $(0, 7)$	Solving a linear equation for y results in slope-intercept form. The coefficient of the x-term is the slope, and the constant defines the location of the y-intercept.

Although standard form and slope-intercept form can be used to express an equation of a line, often the slope-intercept form is used to give a unique representation of the line. For example, the following linear equations are all written in standard form, yet they each define the same line:

$$2x + 5y = 10$$

$$-4x - 10y = -20$$

$$6x + 15y = 30$$

$$\frac{2}{5}x + y = 2$$

The line can be written uniquely in slope-intercept form as: $y = -\frac{2}{5}x + 2$.

Although it is important to understand and apply slope-intercept form, it is not necessarily applicable to all problems. The following example illustrates how a little common sense and ingenuity may lead to a simple solution.

example 9

Writing an Equation of a Line

Write an equation of the line passing through the point $(-3, 2)$ and perpendicular to the x-axis.

Solution:

Because the line is perpendicular to the x-axis, the line must be vertical. Recall that all vertical lines can be written in the form $x = k$, where k is a constant. A quick sketch can determine the value of the constant (Figure 6-28).

Because the line must pass through a point whose x-coordinate is -3, the equation of the line is $x = -3$.

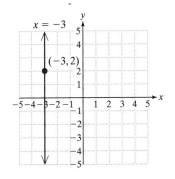

Figure 6-28

Tip: Vertical lines cannot be written in slope-intercept form because the slope is undefined.

Calculator Connections

In Example 8, we found an equation $y = \frac{1}{2}x + \frac{7}{2}$ that represents the line through the point $(-1, 3)$ and perpendicular to the line $y = -2x + 1$. The equations and graphs of both lines are shown here.

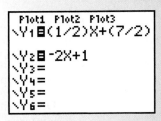

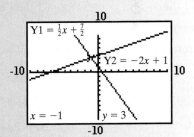

Notice that the lines do not appear perpendicular in the calculator display. That is, they do not appear to form a right angle at the point of intersection. Because many calculators have a rectangular screen, the standard viewing window is elongated in the horizontal direction. To eliminate this distortion, try using a *ZSquare* option. This feature will set the viewing window so that equal distances on the display denote an equal number of units on the graph.

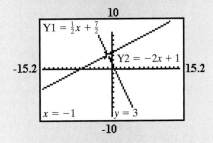

Calculator Exercises

For each pair of lines, determine if the lines are parallel, perpendicular, or neither. Then use a square viewing window to graph the lines on a graphing calculator to verify your results (see Example 3).

1. $x + y = 1$ ❖   2. $3x + y = -2$ ❖   3. $2x - y = 4$ ❖

   $x - y = -3$ $6x + 2y = 6$ $3x + 2y = 4$

4. Graph the lines: $y = x + 1$ and $y = 0.99x + 3$.
   Are these lines parallel? Explain. The lines may appear parallel; however, they are not parallel because the slopes are different.

5. Graph the lines $y = -2x - 1$ and $y = -2x - 0.99$.
   Are these lines the same? Explain. The lines may appear to coincide on a graph; however, they are not the same line because the y-intercepts are different.

section 6.3   PRACTICE EXERCISES

For Exercises 1–8

a. Find the x- and y-intercepts (if they exist).   6.1

b. Graph the line.

c. Determine the slope of the line (if it exists).   6.2

1. $x - 5y = 10$ ❖   2. $3x + y = -12$ ❖

3. $3y = -9$ ❖   4. $2 + y = 5$ ❖

5. $-4x = 6y$ ❖   6. $2x - 2y = 3$ ❖

7. $-x + 3 = 8$ ❖   8. $5x = 20$ ❖

For Exercises 9–20, write each equation in slope-intercept form, if possible. Then identify the slope and y-intercept, if they exist.

9. $2x - 5y = 4$ ❖   10. $3x + 2y = 9$ ❖

11. $3x - y = 5$ 12. $7x - 3y = -6$ ❖
$y = 3x - 5; m = 3; y$-intercept $(0, -5)$

13. $x + y = 6$ $y = -x + 6$; 14. $x - y = 1$ $y = x - 1$;
$m = -1; y$-intercept $(0, 6)$ $m = 1; y$-intercept $(0, -1)$

15. $x + 6 = 8$ ❖   16. $-4 + x = 1$ ❖

17. $1 - 8y = 2$ 18. $1 - y = 9$ $y = -8$;
$y = -\frac{1}{4}; m = 0; y$-intercept $\left(0, -\frac{1}{4}\right)$ $m = 0; y$-intercept $(0, -8)$

19. $3y - 2x = 0$ ❖   20. $5x = 6y$ ❖

21. Graph the line through the point $(0, 2)$, having a slope of -4. ❖

22. Graph the line through the point $(0, -1)$, having a slope of -3. ❖

23. Graph the line through the point $(0, -5)$, having a slope of $\frac{3}{2}$. ❖

24. Graph the line through the point $(0, 3)$, having a slope of $-\frac{1}{4}$. ❖

For Exercises 25–32, draw a line as indicated. Answers may vary.

25. Draw a line with a positive slope and a positive y-intercept. ❖

26. Draw a line with a positive slope and a negative y-intercept. ❖

27. Draw a line with a negative slope and a negative y-intercept. ❖

28. Draw a line with a negative slope and positive y-intercept. ❖

29. Draw a line with a zero slope and a positive y-intercept. ❖

30. Draw a line with a zero slope and a negative y-intercept. ❖

31. Draw a line with undefined slope and a negative x-intercept. ❖

32. Draw a line with undefined slope and a positive x-intercept. ❖

For Exercises 33–52, write each equation in slope-intercept form (if possible) and graph the line.

33. $x - 2y = 6$ ❖   34. $5x - 2y = 2$ ❖

35. $2x + y = 9$ ❖   36. $-6x + y = 8$ ❖

❖ See Additional Answers Appendix   ✎ Writing   ⬌ Translating Expression   Geometry   Scientific Calculator   Video

37. $3y + 6x = -1$ ❖

38. $-4y - 2x = -7$ ❖

39. $2x = -4y + 6$ ❖

40. $3x = y - 7$ ❖

41. $x + y = 0$ ❖

42. $x - y = 0$ ❖

43. $0.2x - 0.5y = 0.1$ ❖

44. $1.5x + 2.5y = 5.0$ ❖

45. $5y = 9x$ ❖

46. $-2x = 5y$ ❖

47. $3y + 2 = 0$ ❖

48. $1 + 5y = 6$ ❖

49. $3x + 1 = 7$ ❖

50. $-2x - 5 = 1$ ❖

51. $\frac{1}{2}x + \frac{1}{4}y = \frac{1}{2}$ ❖

52. $\frac{1}{3}x - \frac{1}{6}y = \frac{1}{2}$ ❖

For Exercises 53–58, let m_1 and m_2 represent the slopes of two lines. Determine if the lines are parallel, perpendicular, or neither.

53. $m_1 = -2, m_2 = \frac{1}{2}$

Perpendicular

54. $m_1 = \frac{2}{3}, m_2 = \frac{3}{2}$

Neither

55. $m_1 = 1, m_2 = \frac{4}{4}$

Parallel

56. $m_1 = \frac{3}{4}, m_2 = -\frac{8}{6}$

Perpendicular

57. $m_1 = \frac{2}{7}, m_2 = -\frac{2}{7}$

Neither

58. $m_1 = 5, m_2 = 5$

Parallel

For Exercises 59–70, write an equation of the line given the following information. Write the answer in slope-intercept form if possible.

59. The slope is $-\frac{1}{3}$ and the y-intercept is $(0, 2)$. $y = -\frac{1}{3}x + 2$

60. The slope is $\frac{2}{3}$ and the y-intercept is $(0, -1)$. $y = \frac{2}{3}x - 1$

61. The slope is 2 and the line passes through the point $(-3, 4)$. $y = 2x + 10$

62. The slope is -1 and the line passes through the point $(-1, 4)$. $y = -x + 3$

63. The slope is 0 and the line passes through the point $(4, -1)$. $y = -1$

64. The slope is 0 and the line passes through the point $(0, 9)$. $y = 9$

65. The line passes through the points $(-1, -5)$ and $(2, 1)$. $y = 2x - 3$

66. The line passes through the points $(-1, 6)$ and $(1, 2)$. $y = -2x + 4$

67. The line passes through the points $(-3, 1)$ and $(-3, 4)$. $x = -3$

68. The line passes through the points $(2, 4)$ and $(2, -5)$. $x = 2$

69. The line has x- and y-intercepts $(3, 0)$ and $(0, -1)$. $y = \frac{1}{3}x - 1$

70. The line has x- and y-intercepts $(-5, 0)$ and $(0, -2)$. $y = -\frac{2}{5}x - 2$

71. The line passes through the point $(2, 1)$ and is parallel to the line $x = -3$ (see the figure). $x = 2$

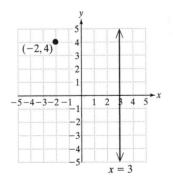

Figure for Exercise 71

72. The line passes through the point $(-2, 4)$ and is parallel to the line $x = 3$ (see the figure). $x = -2$

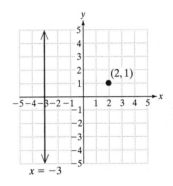

Figure for Exercise 72

73. The line passes through the point $(0, 4)$ and is perpendicular to the line $x = 1$. (*Hint*: Sketch the line first.) $y = 4$

74. The line passes through the point $\left(\frac{1}{2}, \frac{2}{3}\right)$ and is perpendicular to the line $x = 4$. (*Hint*: Sketch the line first.) $y = \frac{2}{3}$

75. The line passes through the point $(5, 2)$ and is parallel to the line $y = \frac{1}{2}$. (*Hint*: Sketch the line first.) $y = 2$

76. The line passes through the point $(-7, \frac{1}{3})$ and is parallel to the line $y = -1$. (*Hint*: Sketch the line first.) $y = \frac{1}{3}$

77. The line passes through the point $(3, 3)$ and is perpendicular to the line $y = 0$. (*Hint*: Sketch the line first.) $x = 3$

78. The line passes through the point $(6, -1)$ and is perpendicular to the line $y = 5$. (*Hint*: Sketch the line first.) $x = 6$

79. The cost for a rental car is $49.95 per day plus a flat fee of $31.95 for insurance. The equation, $C = 49.95x + 31.95$ represents the total cost, C, (in dollars) to rent the car for x days.

    a. Identify the slope. Interpret the meaning of the slope in the context of this problem.
    $m = 49.95$. The cost increases $49.95 per day.
    b. Identify the C-intercept. Interpret the meaning of the C-intercept in the context of this problem. $(0, 31.95)$. The cost to rent the car for 0 days is $31.95.
    c. Use the equation to determine how much it would cost to rent the car for 1 week. $381.60

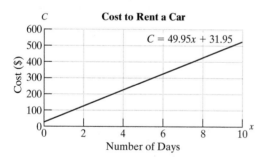

Figure for Exercise 79

80. A phone bill is determined each month by a $16.95 flat fee plus $0.10/min of long distance. The equation $C = 0.10x + 16.95$ represents the total monthly cost, C, for x minutes of long distance.

    a. Identify the slope. Interpret the meaning of the slope in the context of this problem.
    $m = 0.10$. The cost increases $0.10/min of long distance.
    b. Identify the C-intercept. Interpret the meaning of the C-intercept in the context of this problem. $(0, 16.95)$. The monthly bill is $16.95 if 0 minutes of long distance is used.

c. Use the equation to determine the total cost of 234 min of long distance. $40.35

Phone Bill Cost Versus Number of Minutes of Long Distance

$C = 0.10x + 16.95$

Figure for Exercise 80

■ EXPANDING YOUR SKILLS

For Exercises 81–88, write an equation of the line given the following information. Write the answer in slope-intercept form if possible.

81. The line passes through the point $(-5, 1)$ and is perpendicular to the line $y = \frac{1}{3}x + 1$. $y = -3x - 14$

82. The line passes through the point $(2, 2)$ and is perpendicular to the line $y = \frac{1}{2}x - 4$. $y = -2x + 6$

83. The line passes through the point $(3, 3)$ and is parallel to the line $x - 3y = 8$. $y = \frac{1}{3}x + 2$

84. The line passes through the point $(-2, -5)$ and is parallel to the line $2x + 3y = -2$. $y = -\frac{2}{3}x - \frac{19}{3}$

85. The line passes through the point $(0, -2)$ and is perpendicular to the line $-5x + y = 1$. $y = -\frac{1}{5}x - 2$

86. The line passes through the point $(4, 0)$ and is perpendicular to the line $x - 2y = 6$. $y = -2x + 8$

 88. The line passes through the point $(1.25, -5.5)$ and is parallel to the line $y = -2.5x - 7.5$.
$y = -2.5x - 2.375$

 87. The line passes through the point $(-2.3, 4.1)$ and is parallel to the line $20.4x - 10.2y = 30.6$.
$y = 2x + 8.7$

---

chapter 6   MIDCHAPTER REVIEW

For Exercises 1–6, identify whether the equation is in slope-intercept form or standard form.

1. $y = 3x + 8$
   Slope-intercept form
2. $4x + 9y = 8$
   Standard form
3. $-2x - 8y = 3$
   Standard form
4. $y = 6x - 12$
   Slope-intercept form
5. $\frac{2}{3}x - \frac{3}{4}y = 2$
   Standard form
6. $x - 6 = y$
   Slope-intercept form

For Exercises 7–12, identify which equations represent horizontal lines and which represent vertical lines.

7. $y = 6$ Horizontal
8. $x = 2$ Vertical
9. $x = 5$ Vertical
10. $y = -1$ Horizontal
11. $-2y + 9 = 0$
    Horizontal
12. $5x - 3 = 0$ Vertical

13. Graph the linear equation using three different methods as indicated: $4x + 3y = 6$.

    a. Complete the table, and graph the line through the points. ❖

x	y
2	
	-2
-3	

    Table for Exercise 13

    b. Graph the line by plotting the x- and y-intercepts. ❖

    c. Write the equation in slope-intercept form, and graph the line using the slope and y-intercept. ❖

    d. State any advantages or disadvantages of each method for graphing this linear equation. ❖

14. Graph the linear equation using three different methods as indicated: $y = \frac{1}{3}x + 5$.

    a. Complete the table, and graph the line through the points. ❖

x	y
-3	
-2	
-6	

    Table for Exercise 14

    b. Graph the line by plotting the x- and y-intercepts. ❖

    c. Write the equation in slope-intercept form, and graph the line using the slope and y-intercept. ❖

    d. State any advantages or disadvantages of each method for graphing this linear equation. ❖

15. Graph the linear equation using three different methods as indicated: $4x + y = 0$.

    a. Complete the table, and graph the line through the points. ❖

x	y
1	
2	
-4	

    Table for Exercise 15

    b. Graph the line by plotting the x- and y-intercepts. ❖

❖ See Additional Answers Appendix   🖊 Writing   ⬄ Translating Expression   Geometry   Scientific Calculator   Video

c. Write the equation in slope-intercept form, and graph the line using the slope and y-intercept. ❖

d. State any advantages or disadvantages of each method for graphing this linear equation. ❖

16. Graph the linear equation using three different methods as indicated: $y = \frac{3}{4}$.

a. Complete the table, and graph the line through the points. ❖

x	y
0	
1	
2	

Table for Exercise 16

b. Graph the line by plotting the x- and y-intercepts if possible. ❖

c. Write the equation in slope-intercept form, and graph the line using the slope and y-intercept. ❖

d. State any advantages or disadvantages of each method for graphing this linear equation. ❖

17. Graph the linear equation using three different methods as indicated: $2x - 4 = 0$.

a. Complete the table, and graph the line through the points. ❖

x	y
	0
	1
	2

Table for Exercise 17

b. Graph the line by plotting the x- and y-intercepts if possible. ❖

c. Write the equation in slope-intercept form if possible, and graph the line using the slope and y-intercept. ❖

d. State any advantages or disadvantages of each method for graphing this linear equation. ❖

18. Graph the lines on the same set of coordinate axes and state any similarities and differences.

a. $y = 2x + 1$ ❖

b. $y = 3x + 1$ ❖

c. $y = 4x + 1$ ❖

19. Graph the lines on the same set of coordinate axes, and state any similarities and differences.

a. $y = \frac{3}{4}x - 1$ ❖

b. $y = \frac{1}{2}x - 1$ ❖

c. $y = \frac{1}{4}x - 1$ ❖

20. Graph the lines on the same set of coordinate axes, and state any similarities and differences.

a. $y = -\frac{2}{3}x$ ❖

b. $y = -\frac{2}{3}x + 3$ ❖

c. $y = -\frac{2}{3}x - 1$ ❖

21. Graph the lines on the same set of coordinate axes, and state any similarities and differences.

a. $y = 4x$ ❖

b. $y = 4x - 3$ ❖

c. $y = 4x + 3$ ❖

❖ See Additional Answers Appendix   Writing   Translating Expression   Geometry   Scientific Calculator   Video

1. Point-Slope Formula

In Section 6.3, the slope-intercept form of a line was used as a tool to construct an equation of a line. Another useful tool to determine an equation of a line is the point-slope formula. The point-slope formula can be derived from the slope formula as follows:

Suppose a line passes through a given point (x_1, y_1) and has slope m. If (x, y) is any other point on the line, then:

$$m = \frac{y - y_1}{x - x_1}$$ Slope formula

$$m(x - x_1) = \frac{y - y_1}{x - x_1}(x - x_1)$$ Clear fractions.

$$m(x - x_1) = y - y_1$$

or

$$y - y_1 = m(x - x_1)$$ Point-slope formula

Point-Slope Formula

The **point-slope formula** is given by

$$y - y_1 = m(x - x_1)$$

where m is the slope of the line and (x_1, y_1) is a known point on the line.

2. Writing an Equation of a Line Using the Point-Slope Formula

To illustrate the use of the point-slope formula for finding an equation of a line, we will repeat Example 5 from Section 6.3.

example 1

Writing an Equation of a Line Using the Point-Slope Formula

Use the point-slope formula to find an equation of the line having a slope of 3 and passing through the point $(-2, -4)$. Write the final answer in slope-intercept form.

Solution:

The slope of the line is given: $m = 3$.

A point on the line is given: $(x_1, y_1) = (-2, -4)$.

The point-slope formula:

$$y - y_1 = m(x - x_1)$$

$$y - (-4) = 3[x - (-2)]$$ Substitute $m = 3$, $x_1 = -2$, and $y_1 = -4$.

$$y + 4 = 3(x + 2)$$ Simplify. Because the final answer is required in slope-intercept form, simplify the equation and solve for y.

$$y + 4 = 3x + 6$$ Apply the distributive property.

$$y = 3x + 6 - 4$$ Subtract 4 from both sides.

$$y = 3x + 2$$ Slope-intercept form

The answer to Example 1 is the same as the answer to Example 5 in the previous section. This example shows that either slope-intercept form or the point-slope formula may be used to find an equation of a line when a slope and a point are known.

3. Writing an Equation of a Line Through Two Points

example 2

Writing an Equation of a Line Through Two Points

Use the point-slope formula to find an equation of the line passing through the points $(-2, 5)$ and $(4, -1)$. Write the final answer in slope-intercept form.

Solution:

Given two points on a line, the slope can be found with the slope formula.

$$(-2, 5) \quad \text{and} \quad (4, -1)$$
$$(x_1, y_1) \qquad\qquad (x_2, y_2) \quad \text{Label the points.}$$

$$m = \frac{y_2 - y_1}{x_2 - x_1} = \frac{(-1) - (5)}{(4) - (-2)} = \frac{-6}{6} = -1$$

To apply the point-slope formula, use the slope, $m = -1$ and either given point. We will choose the point $(-2, 5)$ as (x_1, y_1).

$$y - y_1 = m(x - x_1)$$

$$y - 5 = -1[x - (-2)]$$ Substitute $m = -1$, $x_1 = -2$, $y_1 = 5$.

$$y - 5 = -1(x + 2)$$ Simplify.

$$y - 5 = -x - 2$$

$$y = -x + 3$$

Tip: The point-slope formula can be applied using either given point for (x_1, y_1). In Example 2, using the point $(4, -1)$ for (x_1, y_1) produces the same result.

$$y - y_1 = m(x - x_1)$$

$$y - (-1) = -1(x - 4)$$

$$y + 1 = -x + 4$$

$$y = -x + 3$$

4. Writing an Equation of a Line Parallel or Perpendicular to Another Line

example 3   Writing an Equation of a Line Parallel to Another Line

Use the point-slope formula to find an equation of the line passing through the point $(-1, 0)$ and parallel to the line $y = -4x + 3$. Write the final answer in slope-intercept form.

Solution:

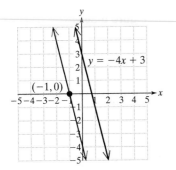

Figure 6-29

Figure 6-29 shows the line $y = -4x + 3$ (pictured in black) and a line parallel to it (pictured in blue) that passes through the point $(-1, 0)$. The equation of the given line, $y = -4x + 3$, is written in slope-intercept form, and its slope is easily identified as -4. The line parallel to the given line must also have a slope of -4.

Apply the point-slope formula using $m = -4$ and the point $(x_1, y_1) = (-1, 0)$.

$$y - y_1 = m(x - x_1)$$
$$y - 0 = -4[x - (-1)]$$
$$y = -4(x + 1)$$
$$y = -4x - 4$$

example 4   Writing an Equation of a Line Perpendicular to Another Line

Use the point-slope formula to find an equation of the line passing through the point $(-3, 1)$ and perpendicular to the line $3x + y = -2$. Write the final answer in slope-intercept form.

Classroom Activity 6.4A

Solution:

The given line can be written in slope-intercept form as $y = -3x - 2$. The slope of this line is -3. Therefore, the slope of a line perpendicular to the given line is $\frac{1}{3}$.

Apply the point-slope formula with $m = \frac{1}{3}$, and $(x_1, y_1) = (-3, 1)$.

$y - y_1 = m(x - x_1)$	Point-slope formula
$y - (1) = \frac{1}{3}[x - (-3)]$	Substitute $m = \frac{1}{3}$, $x_1 = -3$, and $y_1 = 1$.
$y - 1 = \frac{1}{3}(x + 3)$	To write the final answer in slope-intercept form, simplify the equation and solve for y.
$y - 1 = \frac{1}{3}x + 1$	Apply the distributive property.
$y = \frac{1}{3}x + 2$	Add 1 to both sides.

A sketch of the perpendicular lines $y = \frac{1}{3}x + 2$ and $y = -3x - 2$ is shown in Figure 6-30. Notice that the line $y = \frac{1}{3}x + 2$ passes through the point $(-3, 1)$.

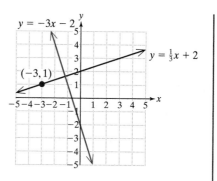

Figure 6-30

5. Different Forms of Linear Equations: A Summary

A linear equation can be written in several different forms as summarized in Table 6-3.

Table 6-3

Form	Example	Comments
Standard Form $ax + by = c$	$4x + 2y = 8$	a and b must not both be zero.
Horizontal Line $y = k$ (k is constant)	$y = 4$	The slope is zero and the y-intercept is $(0, k)$.
Vertical Line $x = k$ (k is constant)	$x = -1$	The slope is undefined and the x-intercept is $(k, 0)$.
Slope-Intercept Form $y = mx + b$ the slope is m y-intercept is $(0, b)$	$y = -3x + 7$ Slope $= -3$ y-intercept is $(0, 7)$	Solving a linear equation for y results in slope-intercept form. The coefficient of the x-term is the slope, and the constant defines the location of the y-intercept.
Point-Slope Formula $y - y_1 = m(x - x_1)$	$m = -3$ $(x_1, y_1) = (4, 2)$ $y - 2 = -3(x - 4)$	This formula is typically used to build an equation of a line when a point on the line is known and the slope of the line is known.

Although it is important to understand and apply slope-intercept form and the point-slope formula, they are not necessarily applicable to all problems, particularly when dealing with a horizontal or vertical line.

example 5

Writing an Equation of a Line

Find an equation of the line passing through the point $(2, -4)$ and parallel to the x-axis.

Solution:

Because the line is parallel to the x-axis, the line must be horizontal. Recall that all horizontal lines can be written in the form $y = k$, where k is a constant. A quick sketch can help find the value of the constant. See Figure 6-31.

Because the line must pass through a point whose y-coordinate is -4, then the equation of the line must be $y = -4$.

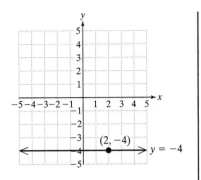

Figure 6-31

---

section 6.4   PRACTICE EXERCISES

For Exercises 1–8, graph the equations.   6.1

1. $4x + 3y = 9$ ❖

2. $2x - 3y = -3$ ❖

3. $y = -2x$ ❖

4. $3 - y = 9$ ❖

5. $3x - 1 = 5$ ❖

6. $x = -1$ ❖

7. $y = -4$ ❖

8. $y = \dfrac{4}{5}x$ ❖

For Exercises 9–14, match the form or formula on the left with its name on the right.   6.3

9. $x = k$ iv

10. $y = mx + b$ v

11. $m = \dfrac{y_2 - y_1}{x_2 - x_1}$ vi

12. $y - y_1 = m(x - x_1)$ ii

13. $y = k$ iii

14. $ax + by = c$ i

i.   Standard form

ii. Point-slope formula

iii. Horizontal line

iv. Vertical line

v.   Slope-intercept form

vi. Slope formula

For Exercises 15–20, find the slope of the line that passes through the given points.   6.2

15. $(1, -3)$ and $(2, 6)$ 9

16. $(2, -4)$ and $(-2, 4)$ -2

17. $(0, 0)$ and $(-1, 0)$ 0

18. $(-2, 5)$ and $(5, 5)$ 0

19. $(6.1, 2.5)$ and $(6.1, -1.5)$ Undefined

20. $(0, 5.4)$ and $(0, 9.3)$ Undefined

For Exercises 21–46, use the point-slope formula (if possible) to write an equation of the line given the following information. Write the final answer in slope-intercept form if possible.

21. The slope is 3 and the line passes through the point $(-2, 1)$. $y = 3x + 7$

22. The slope is -2 and the line passes through the point $(1, -5)$. $y = -2x - 3$

23. The slope is $\frac{1}{4}$ and the line passes through the point $(-8, 6)$. $y = \dfrac{1}{4}x + 8$

24. The slope is $\frac{2}{5}$ and the line passes through the point $(-5, 4)$. $y = \dfrac{2}{5}x + 6$

25. The slope is 4.1 and the line passes through the point $(5.3, -2.2)$. $y = 4.1x - 23.93$

26. The slope is -3.6 and the line passes through the point $(10.0, 8.2)$. $y = -3.6x + 44.2$

27. The slope is 0 and the line passes through the point $(3, -2)$. $y = -2$

28. The slope is 0 and the line passes through the point $(0, 5)$. $y = 5$

29. The line passes through the points $(-2, -6)$ and $(1, 0)$. $y = 2x - 2$

30. The line passes through the points $(-2, 5)$ and $(0, 1)$. $y = -2x + 1$

31. The line passes through the points $(1, -3)$ and $(-7, 2)$. $y = -\dfrac{5}{8}x - \dfrac{19}{8}$

❖ See Additional Answers Appendix   ✎ Writing   ⬌ Translating Expression   🌐 Geometry   🖩 Scientific Calculator   Video

32. The line passes through the points $(0, -4)$ and $(-1, -3)$. $y = -x - 4$

33. The line passes through the points $(2.2, 3.1)$ and $(12.2, -5.3)$. $y = -0.84x + 4.948$

34. The line passes through the points $(4.75, -2.50)$ and $(-0.25, 6.75)$. $y = -1.85x + 6.2875$

35. The line passes through the point $(3, 1)$ and is parallel to the line $y = -4$. See the figure. $y = 1$

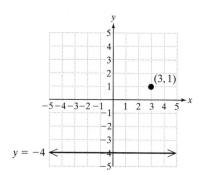

Figure for Exercise 35

36. The line passes through the point $(-1, 1)$ and is parallel to the line $y = 2$. See the figure. $y = 1$

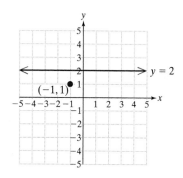

Figure for Exercise 36

37. The line passes through the point $(2, 6)$ and is perpendicular to the line $y = 1$. (*Hint*: Sketch the line first.) $x = 2$

38. The line passes through the point $(0, 3)$ and is perpendicular to the line $y = -5$. (*Hint*: Sketch the line first.) $x = 0$

39. The line passes through the point $(\frac{5}{2}, \frac{1}{2})$ and is parallel to the line $x = 4$. $x = \frac{5}{2}$

40. The line passes through the point $(-6, \frac{2}{3})$ and is parallel to the line $x = -2$. $x = -6$

41. The line passes through the point $(2, 2)$ and is perpendicular to the line $x = 0$. $y = 2$

42. The line passes through the point $(5, -2)$ and is perpendicular to the line $x = 0$. $y = -2$

43. The slope is undefined and the line passes through the point $(-6, -3)$. $x = -6$

44. The slope is undefined and the line passes through the point $(2, -1)$. $x = 2$

45. The line passes through the points $(-4, 0)$ and $(-4, 3)$. $x = -4$

46. The line passes through the points $(1, 3)$ and $(1, -4)$. $x = 1$

47. The following table represents the median selling price, y, of new privately owned one-family houses sold in the Midwest from 1980 to 1995. Let x represent the number of years after 1980. Let y represent price in thousands of dollars.

Year		Price in ($1000)
1980	$x = 0$	67
1985	$x = 5$	84
1990	$x = 10$	108
1995	$x = 15$	142

Table for Exercise 47

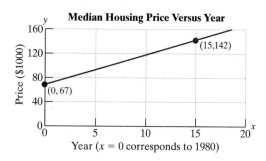

Figure for Exercise 47

Source: U.S. Bureau of Census.

a. Find the slope of the line between the points $(0, 67)$ and $(15, 142)$. 5

b. Find an equation of the line between the points $(0, 67)$ and $(15, 142)$. Write the answer in slope-intercept form. $y = 5x + 67$

c. Use the equation from part (b) to estimate the median price of a one-family house sold in the Midwest in the year 2005. The median price of a one-family house in 2005 would be $192,000.

48. The following table represents the percentage of females, y, who smoked for selected years. Let x represent the number of years after 1965. Let y represent percentage of women who smoked.

Year		Percentage
1965	$x = 0$	33.9
1975	$x = 10$	32.1
1985	$x = 20$	27.9
1995	$x = 30$	23.5

Table for Exercise 48

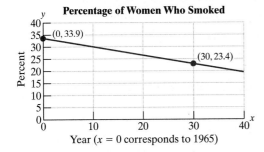

Percentage of Women Who Smoked

Year ($x = 0$ corresponds to 1965)

Figure for Exercise 48

Source: U.S. National Center for Health Statistics.

a. Find the slope of the line between the points $(0, 33.9)$ and $(30, 23.4)$. -0.35

b. Find an equation of the line between the points $(0, 33.9)$ and $(30, 23.5)$. Write the answer in slope-intercept form. $y = -0.35x + 33.9$

c. Use the equation from part (b) to estimate the percentage of women who smoked in the year 2000. In 2000, 21.65% of women smoked.

49. Consider the line having a slope of 2 and a y-intercept of $(0, -4)$:

a. Write an equation of the line using slope-intercept form. $y = 2x - 4$

b. Write an equation of the line using the point-slope formula. $y = 2x - 4$

c. Compare your answers to parts (a) and (b). The equations are the same.

50. Consider the line having a slope of -3 and a y-intercept of $(0, 1)$:

a. Write an equation of the line using slope-intercept form. $y = -3x + 1$

b. Write an equation of the line using the point-slope formula. $y = -3x + 1$

c. Compare your answers to parts (a) and (b). The equations are the same.

51. Consider the line having a slope of $-\frac{5}{2}$ and passing through the point $(4, -7)$:

a. Write an equation of the line using slope-intercept form. $y = -\frac{5}{2}x + 3$

b. Write an equation of the line using the point-slope formula. $y = -\frac{5}{2}x + 3$

c. Compare your answers to parts (a) and (b). The equations are the same.

52. Consider the line having a slope of $-\frac{1}{4}$ and passing through the point $(-5, 1)$:

a. Write an equation of the line using slope-intercept form. $y = -\frac{1}{4}x - \frac{1}{4}$

b. Write an equation of the line using the point-slope formula. $y = -\frac{1}{4}x - \frac{1}{4}$

c. Compare your answers to parts (a) and (b). The equations are the same.

■ EXPANDING YOUR SKILLS

For Exercises 53–60, use the point-slope formula to write an equation of the line given the following information. Write the final answer in slope-intercept form if possible.

53. The line passes through the point $(-5, 2)$ and is perpendicular to the line $y = \frac{1}{2}x + 3$. $y = -2x - 8$

54. The line passes through the point $(-2, -2)$ and is perpendicular to the line $y = \frac{1}{3}x - 5$. $y = -3x - 8$

55. The line passes through the point $(4, 4)$ and is parallel to the line $3x - y = 6$. $y = 3x - 8$

56. The line passes through the point $(-1, -7)$ and is parallel to the line $5x + y = -5$. $y = -5x - 12$

57. The line passes through the point $(0, -6)$ and is perpendicular to the line $-5x + y = 4$. $y = -\frac{1}{5}x - 6$

58. The line passes through the point $(0, -8)$ and is perpendicular to the line $2x - y = 5$. $y = -\frac{1}{2}x - 8$

Concepts

1. Definition of Dependent and Independent Variables
2. Interpreting a Linear Equation in Two Variables
3. Writing a Linear Equation Using Observed Data Points
4. Writing a Linear Model Given a Fixed Value and a Rate of Change

6.5 CONNECTIONS TO GRAPHING: APPLICATIONS OF LINEAR EQUATIONS

1. Definition of Dependent and Independent Variables

Linear equations can often be used to describe (or model) the relationship between two variables in a real-world event. In an xy-coordinate system, the variable being predicted by the mathematical equation is called the **dependent variable** (or response variable) and is represented by y. The variable used to make the prediction is called the **independent variable** (or predictor variable) and is represented by x.

2. Interpreting a Linear Equation in Two Variables

---

example 1 **Interpreting a Linear Equation**

The cost, y, of a speeding ticket (in dollars) is given by $y = 10x + 100$, where $x > 0$ is the number of miles per hour over the speed limit.

a. Which variable is the dependent variable?
b. Which is the independent variable?
c. What is the slope of the line?
d. Interpret the meaning of the slope in terms of cost and the number of miles per hour over the speed limit.
e. Graph the line.

Solution:

a. The dependent variable is the cost of the speeding ticket and is represented by y. The cost of the ticket *depends* on the number of miles per hour over the speed limit.
b. The independent variable is the number of miles over the speed limit and is represented by x.
c. The line is written in slope-intercept form where $m = 10$.
d. The slope $m = 10$ or $\frac{10}{1}$ indicates that there is a $10 increase in the cost of the speeding ticket for every 1 mph over the speed limit.

e.

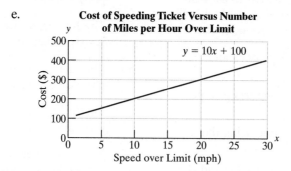

Cost of Speeding Ticket Versus Number of Miles per Hour Over Limit

---

example 2

Interpreting a Linear Equation

The total number of crimes in the United States decreased from 1994 to 1997 (Figure 6-32). The decrease followed a linear trend and can be represented by the linear equation

$y = -2.6x + 45$ where y is the number of crimes measured in millions and x is the number of years since 1994.

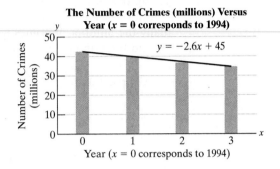

The Number of Crimes (millions) Versus Year ($x = 0$ corresponds to 1994)

Figure 6-32

Source: U.S. Federal Bureau of Investigation.

a. Which is the dependent variable?
b. Which is the independent variable?
c. Use the equation to predict the number of crimes in 1995.
d. What is the y-intercept of the line? Interpret the meaning of the y-intercept in terms of the number of crimes and the year.
e. What is the slope of the line? Interpret the meaning of the slope in terms of the number of crimes and the year.
f. Is it possible for this linear trend to continue indefinitely?
g. From the equation, determine the value of the x-intercept. Interpret the meaning of the x-intercept in terms of the number of crimes and the year. Realistically, is it possible for this linear trend to continue to its x-intercept?

Classroom Activity 6.5A

Solution:

a. The number of crimes is the dependent variable and is represented by y.
b. The number of years since 1994 is the independent variable and is represented by x.
c. Because $x = 0$ represents the year 1994, then $x = 1$ represents 1995. Substitute $x = 1$ into the linear equation.

$$y = -2.6x + 45$$

$$= -2.6(1) + 45 \qquad \text{Substitute } x = 1.$$

$$= -2.6 + 45$$

$$= 42.4$$

The number of crimes in the year 1995 was approximately 42.4 million.

d. To find the *y*-intercept, substitute $x = 0$.

$$y = -2.6(0) + 45$$
$$= 45$$

The *y*-intercept is $(0, 45)$ and indicates that in 1994, the number of crimes was approximately 45 million.

e. From the slope-intercept form of the line, $y = -2.6x + 45$, the slope is -2.6. This indicates that the number of crimes decreased by 2.6 million per year during this time period.

f. It is not possible for the linear trend to continue indefinitely because eventually the number of crimes would reduce to a negative number.

g. To find the *x*-intercept, substitute $y = 0$.

$$y = -2.6x + 45$$
$$0 = -2.6x + 45$$
$$-45 = -2.6x$$
$$\frac{-45}{-2.6} = \frac{-2.6x}{-2.6}$$
$$17.3 \approx x$$

The *x*-intercept is $(17.3, 0)$ and indicates that approximately 17.3 years after 1994 (the year 2011), the number of crimes will be 0. Although the concept of having zero crimes committed in the United States is appealing, it is not realistic. This shows that the linear trend will not continue indefinitely.

3. Writing a Linear Equation Using Observed Data Points

example 3

Writing a Linear Equation from Observed Data Points

The average amount of time per year that a person in the United States spent listening to the radio decreased between 1994 and 1999 (Figure 6-33).

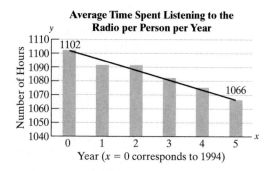

Figure 6-33

Source: Veronis, Suhler Associates, Inc., "Communication Industry Report."

Let x represent the number of years since 1994. Let y represent the average number of hours spent listening to the radio.

a. Find a linear equation that represents the average time an individual spent listening to the radio versus the number of years since 1994.
b. Use the linear equation found in part (a) to predict the amount of time spent listening to the radio in the year 1998. Round to the nearest hour.

Classroom Activity 6.5B

Solution:

a. From the graph, two data values are given. In the year 1994, ($x = 0$), the average number of hours spent listening to the radio per person was 1102. This can be written as the ordered pair (0, 1102). Similarly the ordered pair (5, 1066) indicates that 5 years later, the number of hours spent listening to the radio was 1066.

Using the points (0, 1102) and (5, 1066), find the slope of the line. Once the slope is known, either slope-intercept form or the point-slope formula can be used to find a linear equation. Because the point (0, 1102) is the y-intercept of the line, we will use slope-intercept form.

$$(0, 1102) \quad \text{and} \quad (5, 1066)$$
$$(x_1, y_1) \qquad\qquad (x_2, y_2) \qquad \text{Label the points.}$$

$$m = \frac{y_2 - y_1}{x_2 - x_1} = \frac{(1066) - (1102)}{(5) - (0)} \qquad \text{Apply the slope formula.}$$

$$= \frac{-36}{5} \quad \text{or} \quad m = -7.2 \qquad$$ The slope indicates that there has been a 7.2-hour decrease per year in the average number of hours spent listening to the radio.

With $m = -7.2$ and the y-intercept (0, 1102), we have

$$y = mx + b$$
$$y = -7.2x + 1102$$

b. The value $x = 0$ represents the year 1994. Because the year 1998 is 4 years after 1994, substitute $x = 4$ into the linear equation.

$$y = -7.2x + 1102$$
$$= -7.2(4) + 1102 \qquad \text{Substitute } x = 4.$$
$$= -28.8 + 1102$$
$$= 1073.2 \approx 1073 \qquad \text{Round to the nearest hour.}$$

The average time spent listening to the radio per person in the United States during the year 1998 is estimated to be 1073 h.

4. Writing a Linear Model Given a Fixed Value and a Rate of Change

Another way to look at the equation $y = mx + b$ is to identify the term mx as the variable term and the term b as the constant term. The value of the term mx will change with the value of x (this is why the slope, m, is called a rate of change).

However, the term b will remain constant regardless of the value of x. With these ideas in mind, we can write a linear equation if the rate of change and the constant are known.

---

example 4

Finding a Linear Equation

A stack of posters to advertise a school play costs $9.95 plus $0.50 per poster at the printer.

a. Write a linear equation to compute the cost, y, of buying x posters.
b. Use the equation to compute the cost of 125 posters.

Classroom Activity 6.5C

Solution:

a. The constant cost is $9.95. The variable cost is $0.50 per poster. If m is replaced with 0.50 and b is replaced with 9.95, the equation is

$$y = 0.50x + 9.95 \qquad \text{where } y \text{ represents the total cost of buying } x \text{ posters.}$$

b. Because x represents the number of posters, substitute $x = 125$.

$$y = 0.50(125) + 9.95$$
$$= 62.5 + 9.95$$
$$= 72.45$$

The total cost of buying 125 posters is $72.45.

---

example 5

Finding a Linear Equation

A small word-processing business has a fixed monthly cost of $6000 (this includes rent, utilities, and salaries of its employees). The business has a variable cost of $5 per project (this includes mostly paper and computer supplies).

a. Write a linear equation to compute the total cost, y, for 1 month if x projects are completed.
b. Use the equation to compute the cost to run the business for 1 month if 800 projects are completed.

Solution:

a. $6000 is the constant (fixed) monthly cost. The variable cost is $5 per project. If the slope, m, is replaced with 5, and b is replaced with 6000, then the equation $y = mx + b$ becomes

$$y = 5x + 6000 \qquad \text{where } y \text{ represents the total cost of completing } x \text{ projects in 1 month}$$

b. Because x represents the number of projects, substitute $x = 800$.

$$y = 5(800) + 6000$$
$$= 4000 + 6000$$
$$= 10,000$$

The total cost of completing 800 projects is $10,000.

Calculator Connections

In Example 2, the equation $y = -2.6x + 45$ was used to represent the number of crimes in the United States, y, (in millions) versus the number of years, x, after 1994. The equation is based on data between 1994 and 1997 that correspond to x-values between 0 and 3. Therefore, to graph the equation on a graphing calculator, the viewing window can be set for x between 0 and 3. Furthermore, the window must accommodate y-values up to 45 to show the y-intercept of the equation.

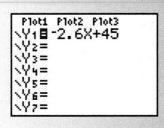
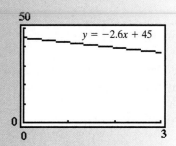

An *Eval* feature can be used to find solutions to an equation by evaluating the value of y for user-defined values of x. For example, the ordered pair (2, 39.8) indicates that in 1996, there were 39.8 million crimes in the United States.

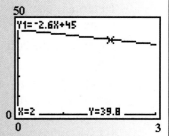

In Example 3, a linear equation was found relating the average amount of time, y, (in hours) that Americans spent listening to the radio x years after 1994.

$$y = -7.2x + 1102$$

Try graphing the equation $y = -7.2x + 1102$ on an appropriate viewing window. Then evaluate the equation at $x = 4$ to confirm that the equation passes through the point (4, 1073.2). This is consistent with the information given in Example 3.

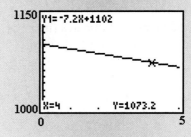

Calculator Exercises

Use a graphing calculator to graph the lines on an appropriate viewing window. Evaluate the equation at the given values of x.

1. $y = -4.6x + 27.1$ at $x = 3$ ❖

2. $y = -3.6x - 42.3$ at $x = 0$ ❖

3. $y = 40x + 105$ at $x = 6$ ❖

4. $y = 20x - 65$ at $x = 8$ ❖

section 6.5 PRACTICE EXERCISES

 1. The electric bill charge for a certain utility company is \$0.095 per kilowatt-hour. The total cost, y, depends on the number of kilowatt-hours, x, according to the equation

$$y = 0.095x \quad x \geq 0$$

a. Determine the cost of using 1000 kilowatt-hours. ❖

b. Determine the cost of using 2000 kilowatt-hours. ❖

c. What is the y-intercept of the equation? Interpret the meaning of the y-intercept in the context of this problem. ❖

d. What is the slope of the equation? Interpret the meaning of the slope in the context of this problem. ❖

e. Graph the equation. ❖

 2. The following graph depicts the rise in the number of jail inmates in the United States from 1987 to 1997. Two linear equations are given: one to describe the number of female inmates and one to describe the number of male inmates by year.

Let y represent the number of inmates (in thousands). Let x represent the number of years since 1987.

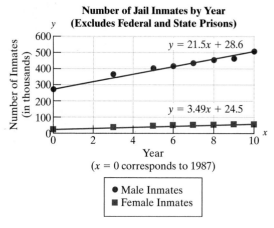

Number of Jail Inmates by Year
(Excludes Federal and State Prisons)

Figure for Exercise 2

Source: U.S. Bureau of Justice Statistics

a. What is the slope of the line representing the number of female inmates? Interpret the meaning of the slope in the context of this problem. $m = 3.49$. The number of female inmates has increased by 3.49 thousand per year between 1987 and 1997.

b. What is the slope of the line representing the number of male inmates? Interpret the meaning of the slope in the context of this problem. $m = 21.5$. The number of male inmates has increased by 21.5 thousand per year between 1987 and 1997.

c. Which group, males or females, has the largest slope? What does this imply about the rise in the number of male and female prisoners? Males. The number of male inmates is increasing at a faster rate than the number of female inmates.

3. Mammography is a diagnostic test used to screen for breast cancer. It is very important in the early detection of breast cancer because it can find tumors and irregularities in the breast tissue while they are still too small to be felt. The percentage of women over 40 who have had a mammogram within the previous 2 years differs according to education level.

Years of School Completed	Percentage of Women Over 40 in the U.S. Who Had Had a Mammogram Within the Previous 2 Years.		
	1990 ($x = 0$)	1993 ($x = 3$)	1994 ($x = 4$)
less than 12	36.4	46.4	48.2
12 years	52.7	59.0	61.0
more than 12	62.8	69.5	69.7

Table for Exercise 3

Source: U.S. National Center for Health Statistics.

The following equations model the percentage of women over 40, y, who had a mammogram within the previous 2 years. In each equation, x represents the number of years since 1990.

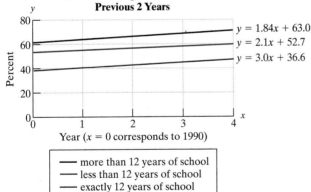

Percentage of U.S. Women over 40 Who Had a Mammogram Within the Previous 2 Years

$y = 1.84x + 63.0$
$y = 2.1x + 52.7$
$y = 3.0x + 36.6$

— more than 12 years of school
— less than 12 years of school
— exactly 12 years of school

Figure for Exercise 3

For each educational level, the percentage of women over 40 who had a mammogram within the previous 2 years increased linearly between 1990 and 1994.

a. Notice that women with the highest education level (more than 12 years) show the largest percentage who had mammograms. Which group has the smallest percentage who had mammograms? The women with less than 12 years of education

b. Based on the linear equations, which line has the greatest slope? The line depicting the women with less than 12 years of education

c. Which line has the smallest slope? The line depicting the women with more than 12 years of education

d. Which group has the greatest rate of increase in the percentage of women who had a mammogram within 2 years? The women with less than 12 years of education

e. Which group has the smallest rate of increase in the percentage of women who had a mammogram within two years? The women with more than 12 years of education

4. The percentage of women over 40 who have had a mammogram within the previous 2 years differs according to economic status.

Economic Status	Percentage of Women Over 40 in the U.S. Who Had Had a Mammogram Within the Previous 2 Years.		
	1990 ($x = 0$)	1993 ($x = 3$)	1994 ($x = 4$)
Below poverty level	28.7	41.6	43.3
At or above poverty level	54.8	62.8	64.2

Table for Exercise 4

Source: U.S. National Center for Health Statistics.

The following equations model the percentage of women over 40, *y*, who had a mammogram within the previous 2 years. In each equation, *x* represents the number of years since 1990.

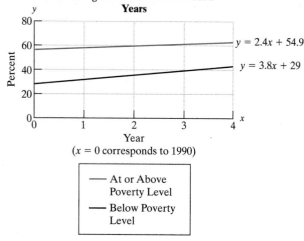

Percentage of U.S. Women over 40 Who Had a Mammogram Within the Previous 2 Years

$y = 2.4x + 54.9$

$y = 3.8x + 29$

Year
(*x* = 0 corresponds to 1990)

—— At or Above Poverty Level

—— Below Poverty Level

Figure for Exercise 4

For each economic group, the percentage of women over 40 who had a mammogram within the last 2 years increased linearly between 1990 and 1994.

a. Which economic group shows a higher percentage of women who had mammograms?
The women at or above the poverty level

b. Based on the linear equations, which line has the greater slope? The line depicting the women below the poverty level

c. Which group has the greatest rate of increase in the percentage of women who had a mammogram within two years? The women below the poverty level

5. The average daily temperature in January for cities along the eastern seaboard of the United States and Canada generally decreases for cities farther north. A city's latitude in the northern hemisphere is a measure of how far north it is on the globe.

City	x Latitude (°N)	y Average Daily Temperature, (°F)
Jacksonville, FL	30.3	52.4
Miami, FL	25.8	67.2
Atlanta, GA	33.8	41.0
Baltimore, MD	39.3	31.8
Boston, MA	42.3	28.6
Atlantic City, NJ	39.4	30.9
New York, NY	40.7	31.5
Portland, ME	43.7	20.8
Charlotte, NC	35.2	39.3
Norfolk, VA	36.9	39.1

Table for Exercise 5

The average temperature, *y*, (measured in degrees Fahrenheit) can be described by the equation

$y = -2.333x + 124.0$ where *x* is the latitude of the city.

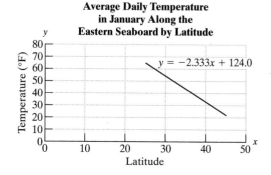

Average Daily Temperature in January Along the Eastern Seaboard by Latitude

$y = -2.333x + 124.0$

Figure for Exercise 5

Source: U.S. National Oceanic and Atmospheric Administration.

a. Which variable is the dependent variable?
y, temperature

b. Which variable is the independent variable?
x, latitude

c. Use the equation to predict the average daily temperature in January for Philadelphia, PA, whose latitude is 40.0°N. Round to one decimal place. 30.7°

d. Use the equation to predict the average daily temperature in January for Edmundston, New Brunswick, Canada, whose latitude is 47.4°N. Round to one decimal place. 13.4°

e.   What is the slope of the line? Interpret the meaning of the slope in terms of the latitude and temperature. $m = -2.333$. The average temperature in January decreases 2.333° per 1° of latitude.

f.   From the equation, determine the value of the *x*-intercept. Round to one decimal place. Interpret the meaning of the *x*-intercept in terms of latitude and temperature. (53.2, 0). At 53.2° latitude, the average temperature in January is 0°.

6. The water bill charge for a certain utility company is $4.20 per 1000 gallons used. The total cost, *y*, depends on the number of thousands of gallons of water, *x*, according to the equation

$$y = 4.20x \quad x \geq 0$$

a.   Determine the cost of using 3000 gallons. (*Hint*: *x* = 3.) ❖

b.   Determine the cost of using 5000 gallons. ❖

c.   What is the *y*-intercept of the equation? Interpret the meaning of the *y*-intercept in the context of this problem. ❖

d.   What is the slope of the equation? Interpret the meaning of the slope in the context of this problem. ❖

e.   Graph the equation. ❖

7. The average amount of time per year that a person in the United States spent reading newspapers decreased between 1994 and 1999.

Let *x* represent the number of years since 1994. Let *y* represent the average time (hours) spent reading newspapers.

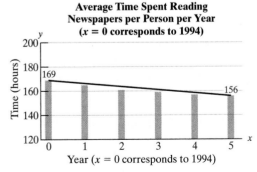

Average Time Spent Reading Newspapers per Person per Year
(x = 0 corresponds to 1994)

Figure for Exercise 7
Source: Veronis, Suhler & Associates Inc., "Communications Industry Report"

a.   Find a linear equation that represents the time spent reading newspapers versus the year. (*Hint*: See Example 3.) $y = -\dfrac{13}{5}x + 169$

b.   Use the linear equation found in part (a) to predict the amount of time spent reading newspapers in the year 2000.   153.4 h

8. The average length of stay for community hospitals has been decreasing in the United States from 1980 to 1997.

Let *x* represent the number of years since 1980. Let *y* represent the average length of a hospital stay in days.

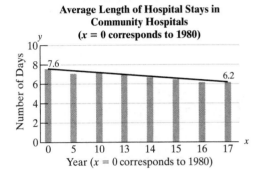

Average Length of Hospital Stays in Community Hospitals
(x = 0 corresponds to 1980)

Figure for Exercise 8
Source: U.S. National Center for Health Statistics

a.   Find a linear equation that relates the average length of hospital stays versus the year. (*Hint*: See Example 3.) Round the slope to three decimal places. $y = -0.082x + 7.6$

b.   Use the linear equation found in part (a) to predict the average length of stay in community hospitals in the year 2000. $y = 5.96$

9. The figure depicts a relationship between a person's height, *y*, (in inches) and the length of the person's arm, *x* (measured in inches from shoulder to wrist).

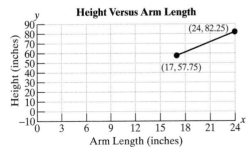

Height Versus Arm Length

(24, 82.25)
(17, 57.75)

Figure for Exercise 9

a. Use the points (17, 57.75) and (24, 82.25) to find a linear equation relating height to arm length. ❖

b. What is the slope of the line? Interpret the slope in the context of this problem. ❖

c. Use the equation from part (a) to estimate the height of a person whose arm length is 21.5 in. ❖

10. In a certain city, the time required to commute to work, y, (in minutes) by car is related linearly to the distance traveled, x (in miles).

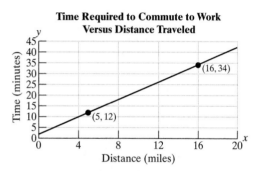

Time Required to Commute to Work Versus Distance Traveled

Figure for Exercise 10

a. Use the points (5, 12) and (16, 34) to find a linear equation relating the commute time to work to the distance traveled. $y = 2x + 2$

b. What is the slope of the line? Interpret the slope in the context of this problem.
$m = 2$. For each additional mile, the time is increased by 2 min.

c. Use the equation from part (a) to find the time required to commute to work for a motorist who lives 18 miles away. 38 min

11. The cost to rent a car, y, for 1 day is $20 plus $0.25 per mile.

a. Write a linear equation to compute the cost, y, of driving a car x miles for 1 day. (*Hint:* See Example 4) $y = 0.25x + 20$

b. Use the equation to compute the cost of driving 258 miles in the rental car. $84.50

12. A phone bill is determined each month by a $18.95 flat fee plus $0.08 per minute of long distance.

a. Write a linear equation to compute the monthly cost of a phone bill, y, if x minutes of long distance are used. (*Hint:* See Example 4) $y = 0.08x + 18.95$

b. Use the equation to compute the phone bill for a month in which 1 h and 27 min of long distance was used. $25.91

13. A tennis instructor charges a student $25 per lesson plus a one-time court fee of $20.

a. Write a linear equation to compute the total cost, y, for x tennis lessons. $y = 25x + 20$

b. What is the total cost to a student who takes 20 tennis lessons? $520.00

14. The cost to rent a 10 ft by 10 ft storage space is $90 per month plus a nonrefundable deposit of $105.

a. Write a linear equation to compute the cost, y, of renting a 10 ft by 10 ft space for x months. $y = 90x + 105$

b. What is the cost of renting such a storage space for 1 year (12 months)? $1185.00

15. A business has a fixed monthly cost of $1200. In addition, the business has a variable cost of $35 for each item produced.

a. Write a linear equation to compute the total cost, y, for 1 month if x items are produced. $y = 35x + 1200$

b. Use the equation to compute the cost for 1 month if 100 items are produced. $4700.00

16. An air-conditioning and heating company has a fixed monthly cost of $5000. Furthermore, each service call costs the company $25.

a. Write a linear equation to compute the total cost, y, for 1 month if x service calls are made. $y = 25x + 5000$

b. Use the equation to compute the cost for 1 month if 150 service calls are made. $8750.00

17. A bakery that specializes in bread rents a booth at a flea market. The daily cost to rent the booth is $100. Each loaf of bread costs the bakery $0.80 to produce.

    a. Write a linear equation to compute the total cost, y, for 1 day if x loaves of bread are produced. $y = 0.8x + 100$

    b. Use the equation to compute the cost for 1 day if 200 loaves of bread are produced.
    $260.00

18. A beverage company rents a booth at an art show to sell lemonade. The daily cost to rent a booth is $35. Each lemonade costs $0.50 to produce.

    a. Write a linear equation to compute the total cost, y, for 1 day if x lemonades are produced. $y = 0.5x + 35$

    b. Use the equation to compute the cost for 1 day if 350 lemonades are produced.
    $210.00

---

chapter 6 — SUMMARY

SECTION 6.1—LINEAR EQUATIONS IN TWO VARIABLES

KEY CONCEPTS:

A linear equation in two variables can be written in the form $ax + by = c$, where a, b, and c are real numbers and a and b are not both zero.

A solution to a linear equation in x and y is an ordered pair (x, y) that satisfies the equation. The set of all solutions to a linear equation can be represented by a line in a rectangular coordinate system.

An x-intercept of an equation is a point $(a, 0)$ where the graph intersects the x-axis.

A y-intercept of an equation is a point $(0, b)$ where the graph intersects the y-axis.

EXAMPLES:

Graph the Equation: $-2x + y = 4$

At least two points are needed to graph a line. To find a point on the line, choose a value for either x or y and solve for the other variable.

x	y
0	4
-2	0
-1	2

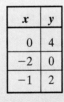

Find the x- and y-Intercepts: $x + 3y = 5$

x-intercept: $x + 3(0) = 5$

$$x = 5 \quad (5, 0)$$

y-intercept: $(0) + 3y = 5$

$$3y = 5$$

$$y = \frac{5}{3} \quad \left(0, \frac{5}{3}\right)$$

 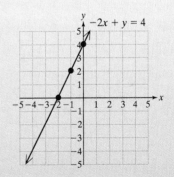

A vertical line can be written in the form $x = k$.

A horizontal line can be written in the form $y = k$.

KEY TERMS:

x-intercept
y-intercept
horizontal line
solution to a linear equation in two variables
vertical line

$x = 3$ is a
Vertical Line

$y = 3$ is a
Horizontal Line

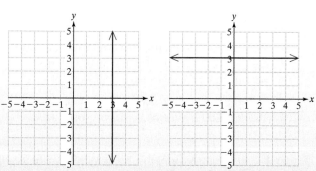

SECTION 6.2—SLOPE OF A LINE

KEY CONCEPTS:

The slope, m, of a line between two points (x_1, y_1) and (x_2, y_2) is given by

$$m = \frac{y_2 - y_1}{x_2 - x_1} \quad \text{or} \quad \frac{\text{change in } y}{\text{change in } x}$$

The slope of a line may be positive, negative, zero, or undefined.

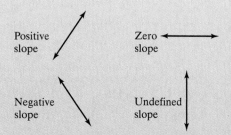

If m_1 and m_2 represent the slopes of two parallel (nonvertical) lines, then $m_1 = m_2$.

If $m_1 \neq 0$ and $m_2 \neq 0$ represent the slopes of two perpendicular lines, then

$$m_1 = -\frac{1}{m_2} \quad \text{or} \quad m_2 = -\frac{1}{m_1}$$

or equivalently, $m_1 m_2 = -1$.

KEY TERMS:

slope
slope formula
slope of a horizontal line
slope of a vertical line
slopes of parallel lines
slopes of perpendicular lines

EXAMPLES:

Find the Slope of the Line Between $(1, -3)$ and $(-3, 7)$

$$m = \frac{7 - (-3)}{-3 - 1} = \frac{10}{-4} = -\frac{5}{2}$$

Determine Whether the Lines are Parallel, Perpendicular, or Neither. The slopes of two lines are given.

a. $m_1 = -7$ and $m_2 = -7$ Parallel

b. $m_1 = -\dfrac{1}{5}$ and $m_2 = 5$ Perpendicular

c. $m_1 = -\dfrac{3}{2}$ and $m_2 = -\dfrac{2}{3}$ Neither

SECTION 6.3—SLOPE-INTERCEPT FORM OF A LINE

KEY CONCEPTS:

The slope-intercept form of a line is

$$y = mx + b$$

where m is the slope of the line and $(0, b)$ is the y-intercept.

Slope-intercept form is used to identify the slope and y-intercept of a line when the equation is given.

Slope-intercept form can also be used to graph a line.

Slope-intercept form can be used to construct an equation of a line given a point on the line and the slope of the line.

KEY TERMS:

slope-intercept form of a line
standard form of a line

EXAMPLES:

Find the Slope and *y*-Intercept:

$$7x - 2y = 4$$

$$-2y = -7x + 4 \qquad \text{Solve for } y.$$

$$y = \frac{7}{2}x - 2$$

The slope is $\frac{7}{2}$. The y-intercept is $(0, -2)$.

Graph the Line:

$$y = \frac{7}{2}x - 2$$

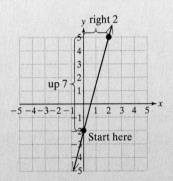

Find an Equation of the Line Passing Through the Point (2, −3) and Having Slope *m* = −4.

Using slope-intercept form:

$$y = mx + b$$

$$-3 = (-4)(2) + b$$

$$-3 = -8 + b$$

$$5 = b \qquad \Rightarrow \quad y = -4x + 5$$

SECTION 6.4—POINT-SLOPE FORMULA

KEY CONCEPTS:

The point-slope formula is used primarily to construct an equation of a line given a point and a slope.

Equations of Lines—A Summary:

Standard form: $ax + by = c$
Horizontal line: $y = k$
Vertical line: $x = k$
Slope intercept form: $y = mx + b$
Point-slope formula: $y - y_1 = m(x - x_1)$

KEY TERM:

point-slope formula

EXAMPLES:

Find an Equation of the Line Passing Through the Point (6, −4) and Having a Slope of $-\frac{1}{2}$.

Label the given information: $m = -\frac{1}{2}$ and $(6, -4) = (x_1, y_1)$

$$y - y_1 = m(x - x_1)$$
$$y - (-4) = -\tfrac{1}{2}(x - 6)$$
$$y + 4 = -\tfrac{1}{2}x + 3$$
$$y = -\tfrac{1}{2}x - 1$$

SECTION 6.5—CONNECTIONS TO GRAPHING: APPLICATIONS OF LINEAR EQUATIONS

KEY CONCEPTS:

Linear equations can often be used to describe or model the relationship between variables in a real world event. In such applications, the slope may be interpreted as a rate of change.

EXAMPLES:

The number of drug-related arrests for a small city has been growing approximately linearly since 1980.

Let y represent the number of drug arrests, and let x represent the number of years after 1980.

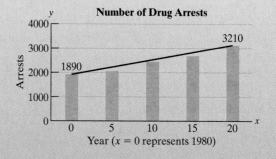

a. Use the ordered pairs (0, 1890) and (20, 3210) to find an equation of the line shown in the graph.

$$m = \frac{y_2 - y_1}{x_2 - x_1} = \frac{3210 - 1890}{20 - 0}$$
$$= \frac{1320}{20} = 66$$

The slope is 66 indicating that the number of drug arrests is increasing at a rate of 66 per year.
$m = 66$, and the y-intercept is $(0, 1890)$. Hence:

$$y = mx + b \implies y = 66x + 1890$$

KEY TERMS:
dependent variable
independent variable

b. Use the equation in part (a) to predict the number of drug-related arrests in the year 2010. (The year 2010 is 30 years after 1980. Hence $x = 30$)

$$y = 66(30) + 1890$$

$$y = 3870$$

The number of drug arrests is predicted to be 3870 by the year 2010.

chapter 6   REVIEW EXERCISES

Section 6.1

1. Is the ordered pair $(-2, 3)$ a solution to the equation $2x + y = -1$?   Yes

2. Is the ordered pair $(1, -5)$ a solution to the equation $5x - y = 0$?   No

3. Is the ordered pair $(-5, -2)$ a solution to the equation $y - 2 = 4$?   No

4. Is the ordered pair $(4, -8)$ a solution to the equation $3x = 12$?   Yes

5. a. Write an example of a linear equation in two variables.   For example: $3x + 2y = 6$

    b. Write an example of a two-variable equation that is not linear.   For example: $5x^2 + 8y^2 = 16$

For Exercises 6–9, find the x- and y-intercepts, if they exist. x-intercept: $\left(-\frac{1}{2}, 0\right)$; y-intercept: $\left(0, \frac{1}{4}\right)$

6. $-4x + 8y = 2$
7. $2x + y = 0$
x-intercept: $(0, 0)$; y-intercept: $(0, 0)$

8. $6y = -24$
9. $5x + 1 = 11$
x-intercept: none; y-intercept: $(0, -4)$ x-intercept: $(2, 0)$; y-intercept: none

For Exercises 10–15, graph the equations by plotting at least three points.

10. $x - 5y = 10$ ❖   11. $y = -\dfrac{3}{2}$ ❖

12. $y = 3x$ ❖   13. $y = \dfrac{1}{4}x - 2$ ❖

14. $x - 7 = 0$ ❖   15. $6x + y = 0$ ❖

16. Write an equation of a line that has only an x-intercept. (Answers may vary.)   For example: $x = 3$

17. Write an equation of a line that has only a y-intercept. (Answers may vary.)   For example: $y = -2$

18. Write an equation of a line that has two distinct x- and y-intercepts. (Answers may vary.)
For example: $5x - 2y = 10$

Section 6.2

19. Draw a line with a positive slope.   ❖

20. Draw a line with a negative slope.   ❖

21. Draw a line with the slope of 0.   ❖

22. Draw a line with an undefined slope.   ❖

23. What is the slope of this ladder leaning up against a wall? $m = \dfrac{12}{5}$

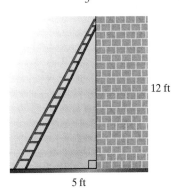

12 ft

5 ft

Figure for Exercise 23

24. Point A is located 4 units down and 2 units to the right of point B. What is the slope of the line through points A and B? -2

25. Determine the slope of the line that passes through the points $(7, -9)$ and $(-5, -1)$. $-\frac{2}{3}$

26. Determine the slope of the line that has x- and y-intercepts of $(-1, 0)$ and $(0, 8)$. 8

27. Determine the slope of the line that passes through the points $(3, 0)$ and $(3, -7)$. Undefined

28. Determine the slope of the horizontal line $y = -1$. 0

29. A given line has a slope of -5.
    a. What is the slope of a line parallel to the given line? -5
    b. What is the slope of a line perpendicular to the given line? $\frac{1}{5}$

30. A given line has a slope of 0.
    a. What is the slope of a line parallel to the given line? 0
    b. What is the slope of a line perpendicular to the given line? Undefined

For Exercises 31–34, find the slopes of the lines l_1 and l_2 from the two given points. Then determine whether l_1 and l_2 are parallel, perpendicular or neither.

31. l_1: $(3, 7)$ and $(0, 5)$ $m_1 = \frac{2}{3}$
    l_2: $(6, 3)$ and $(-3, -3)$ $m_2 = \frac{2}{3}$; parallel

32. l_1: $(-2, 1)$ and $(-1, 9)$ $m_1 = 8$
    l_2: $(0, -6)$ and $(2, 10)$ $m_2 = 8$; parallel

33. l_1: $(0, \frac{5}{6})$ and $(2, 0)$ $m_1 = -\frac{5}{12}$
    l_2: $(0, \frac{6}{5})$ and $(-\frac{1}{2}, 0)$ $m_2 = \frac{12}{5}$; perpendicular

34. l_1: $(1, 1)$ and $(1, -8)$ $m_1 =$ undefined
    l_2: $(4, -5)$ and $(7, -5)$ $m_2 = 0$; perpendicular

Section 6.3

For Exercises 35–42, write each equation in slope-intercept form. Identify the slope and the y-intercept, and graph the line.

35. $5x - 2y = 10$ ❖   36. $3x + 4y = 12$ ❖

37. $2x + y = -3$ ❖   38. $3x - y = 4$ ❖

39. $x - 3y = 0$ ❖   40. $5y - 8 = 4$ ❖

41. $2y = -5$ ❖   42. $y - x = 0$ ❖

For Exercises 43–50, write an equation of the line given the following information. Write the final answer in slope-intercept form.

43. The slope is $-\frac{4}{3}$ and the y-intercept is $(0, -1)$. $y = -\frac{4}{3}x - 1$

44. The slope is 0 and the y-intercept is $(0, 2)$. $y = 2$

45. The slope is -3 and the line passes through the point $(-6, 2)$. $y = -3x - 16$

46. The line passes through the points $(-5, 7)$ and $(3, 1)$. $y = -\frac{3}{4}x + \frac{13}{4}$

47. The line passes through the point $(4, 17)$ and is parallel to the line $y = 5$. $y = 17$

48. The line has x- and y-intercepts $(-1, 0)$ and $(0, -9)$, respectively. $y = -9x - 9$

49. The line passes through the point $(8, 12)$ and is parallel to the line $y = 4x - 2$. $y = 4x - 20$

50. The line passes through the point $(-6, 4)$ and is perpendicular to the line $2x - 3y = 6$. $y = -\frac{3}{2}x - 5$

Section 6.4

51. Write a linear equation in two variables in slope-intercept form. (Answers may vary.)
    For example: $y = 3x + 2$

52. Write a linear equation in two variables in standard form. (Answers may vary.)
    For example: $5x + 2y = -4$

53. Write the slope formula to find the slope of the line between the point (x_1, y_1) and (x_2, y_2). ❖

54. Write the point-slope formula. $y - y_1 = m(x - x_1)$

55. Write an equation of a vertical line (answers may vary). For example: $x = 6$

56. Write an equation of a horizontal line (answers may vary). For example: $y = -5$

For Exercises 57–62, use the point-slope formula to write an equation of a line given the following information. Write the answer in slope-intercept form.

57. The slope is -6 and the line passes through the point $(-1, 8)$. $y = -6x + 2$

58. The slope is $\frac{2}{3}$ and the line passes through the point $(5, 5)$. $y = \frac{2}{3}x + \frac{5}{3}$

❖ See Additional Answers Appendix  Writing Translating Expression Geometry Scientific Calculator Video

59. The line passes through the points $(0, -4)$ and $(8, -2)$. $\quad y = \frac{1}{4}x - 4$

60. The line passes through the points $(2, -5)$ and $(8, -5)$. $\quad y = -5$

61. The line passes through the point $(5, 12)$ and is perpendicular to the line $y = -\frac{5}{6}x - 3$. $\quad y = \frac{6}{5}x + 6$

62. The line passes through the point $(-6, 7)$ and is parallel to the line $4x - y = 0$. $\quad y = 4x + 31$

Section 6.5

63. The number of reported property crimes decreased from 1994 to 1997. This decrease followed a linear trend and can be represented by the equation

$$y = -1.8x + 32.9 \qquad \text{where } y \text{ is the number}$$

of crimes measured in millions and x is the number of years after 1994.

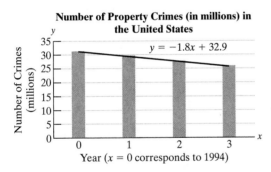

Number of Property Crimes (in millions) in the United States

Figure for Exercise 63

Source: U.S. Federal Bureau of Investigation

a. Which variable is the dependent variable?
   y, the number of crimes
b. Which variable is the independent variable?
   x, the year where $x = 0$ corresponds to 1994
c. Use the equation to predict the number of reported property crimes in the year 2000.
   22,100,000
d. What is the slope of the line? Interpret the meaning of the slope in the context of the problem. $m = -1.8$. Between 1994 and 1997 the number of reported property crimes dropped by an average of 1.8 million per year.
e. From the equation, determine the value of the y-intercept. Interpret the meaning of the y-intercept in the context of the problem.
   $(0, 32.9)$. In the year 1994 32.9 million property crimes were reported.

64. The number of robberies in the United States in 1994 was approximately 1.3 million. By 1997, the number dropped to approximately 0.9 million.
   Let $x = 1$ represent the year 1994. Let y represent the number of robberies (in millions).

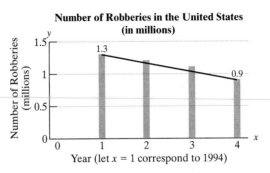

Number of Robberies in the United States (in millions)

Figure for Exercise 64

Source: U.S. Federal Bureau of Investigation

a. Using the ordered pairs $(1, 1.3)$ and $(4, 0.9)$, find the slope of the line. Round to two decimal places. $\quad m = -\frac{2}{15}$ or -0.13
b. Interpret the meaning of the slope in the context of this problem. The number of robberies decreased by an average of 0.13 million per year between 1994 and 1999.
c. Write an equation that represents the number of robberies versus the year.
   $y = -0.13x + 1.43$
d. Use the equation in part (c) to approximate the number of robberies in 1998 $(x = 5)$.
   780,000 robberies

65. During the 1990s, consumer debt grew linearly. The amount of money (in \$billions) that U.S. consumers had in outstanding automobile loans for the years 1992–1997 is expressed in the graph.
   Let x represent the number of years since 1992. Let y represent the total debt in auto loans (in \$billions).

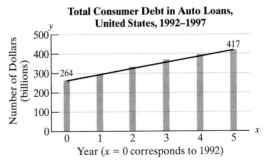

Total Consumer Debt in Auto Loans, United States, 1992–1997

Figure for Exercise 65

Source: Federal Reserve Board.

a. Find a linear equation that represents the total debt in auto loans, y, versus the year, x.
$y = 30.6x + 264$

b. Use the linear equation found in part (a) to predict the debt in the year 2000.
$508.8 billion

66. A water purification company charges $20 per month and a $55 installation fee.

a. Write a linear equation to compute the total cost, y, of renting this system for x months.
$y = 20x + 55$

b. Use the equation from part (a) to determine the total cost to rent the system for 9 months.
$235

67. A small cleaning company has a fixed monthly cost of $700 and a variable cost of $8 per service call.

a. Write a linear equation to compute the total cost, y, of making x service calls in one month. $y = 8x + 700$

b. Use the equation from part (a) to determine the total cost of making 80 service calls.
$1340

chapter 6   TEST

1. Determine which of the ordered pairs are solutions to the equation: $x - 6y = -8$.

    a. $(2, 1)$ b. $(4, -2)$ c. $(4, 2)$
       No   No   Yes

2. Find the x-intercept and the y-intercept of the line $-4x + 3y = 6$. x-intercept: $\left(-\frac{3}{2}, 0\right)$; y-intercept: $(0, 2)$

3. What is the average slope of the hill? $\frac{2}{5}$

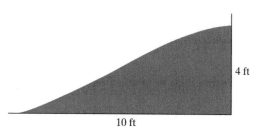

4 ft

10 ft

Figure for Exercise 3

4. a. Find the slope of the line that passes through the points $(-2, 0)$ and $(-5, -1)$. $\frac{1}{3}$

    b. Find the slope of the line $4x - 3y = 9$. $\frac{4}{3}$

5. a. What is the slope of a line parallel to the line $x + 4y = -16$? $-\frac{1}{4}$

    b. What is the slope of a line perpendicular to the line $x + 4y = -16$? 4

6. a. What is the slope of the line $x = 5$? Undefined

    b. What is the slope of the line $y = -3$? 0

For Exercises 7–10, find the x- and y-intercepts if they exist, and graph the lines.

7. $y = 8x + 2$ ❖   8. $2x + 9y = 0$ ❖

9. $x - 3 = 0$ ❖   10. $12y = -4$ ❖

11. Determine whether the lines l_1 and l_2 are parallel, perpendicular, or neither.

    l_1: $2y = 3x - 3$ l_2: $4x = -6y + 1$
    Perpendicular

12. Write an equation of the line that has y-intercept $\left(0, \frac{1}{2}\right)$ and slope $\frac{1}{4}$. Write the equation in slope-intercept form. $y = \frac{1}{4}x + \frac{1}{2}$

13. Write an equation of the line that passes through the points $(2, 8)$, and $(4, 1)$. Write the equation in slope-intercept form. $y = -\frac{7}{2}x + 15$

14. Write an equation of the line that passes through the point $(2, -6)$ and is parallel to the x-axis.
    $y = -6$

15. Write an equation of the line that passes through the point $(3, 0)$ and is parallel to the line $2x + 6y = -5$. $y = -\frac{1}{3}x + 1$

16. Write an equation of the line that passes through the point $(-3, -1)$ and is perpendicular to the line $x + 3y = 9$. $y = 3x + 8$

17. Hurricane Floyd dumped rain at an average rate of $\frac{3}{4}$ in./h on Southport, North Carolina. Further inland, in Lumberton, North Carolina, the storm dropped $\frac{1}{2}$ in. of rain per hour. The following graph depicts the total amount of rainfall (in

inches) versus the time (in hours) for both locations in North Carolina.

a. What is the slope of the line representing the rainfall for Southport? $\frac{3}{4}$

b. What is the slope of the line representing the rainfall for Lumberton? $\frac{1}{2}$

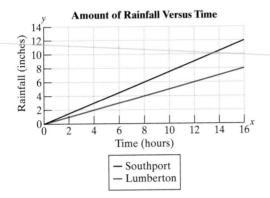

Amount of Rainfall Versus Time

Figure for Exercise 17

18. To attend a State Fair, the cost is $10 per person to cover exhibits and musical entertainment. There is an additional cost of $1.50 per ride.

a. Write an equation that gives the total cost, y, of visiting the State Fair and going on x rides. $y = 1.5x + 10$

b. Use the equation from part (a) to determine the cost of going to the State Fair and going on 10 rides. $25

19. The winner of the women's 100-m freestyle swimming event in the 1912 Olympics was Fanny

Durack of Australia. Her time was 82.2 s. Since that time the winning time for this event has dropped. By the 1996 Olympics, the winning time posted by Le Jingyi of China was 54.5 s.

Let x represent the number of years since 1912. Let y represent the winning time in seconds for the women's 100-m freestyle event.

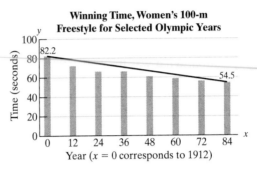

Winning Time, Women's 100-m Freestyle for Selected Olympic Years

Figure for Exercise 19

a. Find the slope of the line shown in the graph. Interpret the meaning of the slope in the context of this problem. Round to two decimal places. $m = -0.33$. The winning time has dropped by an average of 0.33 s/year between 1912 and 1996.

b. Use the slope and y-intercept to find an equation of the line that represents the winning time, y, versus the number of years, x, since 1912. $y = -0.33x + 82.2$

c. Use the equation from part (b) to estimate the winning time in the women's 100-m freestyle event in the year 1948. 70.32 s

CUMULATIVE REVIEW EXERCISES, CHAPTERS 1–6

1. Simplify.

$$\frac{\sqrt{12 - 9 \div 3}}{|14 - 20|} \quad \frac{1}{2}$$

2. Determine whether the radical expression represents a rational number or an irrational number.

a. $\sqrt{9 - 2^3}$ b. $\sqrt{2^2 + 3^2}$
   Rational   Irrational

3. Simplify.

$$\frac{1}{5} - \left(\frac{3}{5}\right)^2 \div \left(-\frac{3}{10}\right) \quad \frac{7}{5}$$

4. Solve for x: $-3x + 2 = 14$ $x = -4$

5. Solve for x. Graph the solution set and express the solution in interval notation: ❖

$$-3x + 2 > 14$$

 ❖ See Additional Answers Appendix    Writing   Translating Expression   Geometry   Scientific Calculator   Video

6. Solve for x: $5(x - 2) + 7 = 2(x + 1) + 3x - 5$
    All real numbers

7. Explain what is meant by the term *complementary angles*. Explain what is meant by the term *supplementary angles*. Two angles are complementary if their sum is 90°. Two angles are supplementary if their sum is 180°.

8. One angle is 9° less than twice another angle. The angles are complementary. Find the measure of each angle.   The angles are 33° and 57°

9. The population of Alaska in 1997 was 609,000. This represents a 10.7% increase from the 1990 population. What was the population in 1990? Round to the nearest whole unit.   550,136 people

10. Simplify.
    $$\left(\frac{5a^7b^4}{10a^3b^5}\right)^2 \quad \frac{a^8}{4b^2}$$

11. Simplify.
    $$\left(\frac{1}{5}\right)^{-1} + \left(\frac{1}{3}\right)^{-2} + \left(\frac{1}{2}\right)^0 \quad 15$$

12. Divide and write the final answer in scientific notation.
    $$\frac{6.0 \times 10^{-8}}{1.2 \times 10^{-3}} \quad 5.0 \times 10^{-5}$$

13. Perform the indicated operations:
    a. $(x + 2)^2$ $x^2 + 4x + 4$
    b. $-3(4x + 1)$ $-12x - 3$
    c. $(x + 2)^2 - 3(4x + 1)$ $x^2 - 8x + 1$

14. Divide using long division:
    $(3x^4 - 2x^3 + 5x - 7) \div (x - 2)$
    $3x^3 + 4x^2 + 8x + 21 + \dfrac{35}{x - 2}$

For Exercises 15 and 16, factor completely:

15. $16x^2y - 28xy + 6y$ $2y(2x - 3)(4x - 1)$

16. $16b^4 - 81a^4$ $(2b - 3a)(2b + 3a)(4b^2 + 9a^2)$

17. Solve for t: $9t^2 - 30t = -25$ $t = \dfrac{5}{3}$

18. In a rectangular room the length of the room is 2 yd more than three times the width. Find the dimensions of the room if the area is 33 yd².
    The width is 3 yd and the length is 11 yd.

19. Divide.
    $$\frac{2a^2b^{12}}{6ab^3c} \div \frac{5b}{12ac^4} \quad \frac{4a^2b^8c^3}{5}$$

20. What is the domain of the expression?
    $$\frac{x^2}{x^2 - 9} \quad \{x \mid x \neq 3, x \neq -3\}$$

21. Subtract.
    $$\frac{c^2}{c - 5} - \frac{2c + 15}{c - 5} \quad c + 3$$

22. Simplify the complex fraction.
    $$\frac{c + \dfrac{1}{d}}{d + \dfrac{1}{c}} \quad \frac{c}{d}$$

23. The quotient of 6 and 2 more than a number is equal to the quotient of 4 and the number. Find the number.   The number is 4.

24. Solve the equation.
    $$\frac{1}{y^2 - 1} = \frac{1}{y + 1} + \frac{1}{y - 1} \quad y = \frac{1}{2}$$

25. Given the equation: $3y = -4x + 6$
    a. Is the equation linear or nonlinear?   ❖
    b. What is the x-intercept?   ❖
    c. What is the y-intercept?   ❖
    d. Write the equation in slope-intercept form.   ❖
    e. What is the slope of the line?   ❖
    f. Graph the line.   ❖

26. Find an equation of the line passing through the point $(2, -3)$ and having a slope of $\frac{1}{2}$. Write the final answer in slope-intercept form. $y = \dfrac{1}{2}x - 4$

❖ See Additional Answers Appendix   <img_1> Writing   ⬌ Translating Expression   Geometry   Scientific Calculator   Video

SYSTEMS OF LINEAR EQUATIONS IN TWO VARIABLES

A system of linear equations can be used to solve an application where two variables are subject to two linear constraints. For example, the speed of a plane is influenced by the plane's still air speed, p, and by the speed of the wind, w. When traveling with the wind, the net speed of the plane is $p + w$. When traveling against the wind, the net speed is $p - w$.

Suppose a plane travels 1500 miles in 3 h with a tail wind, and 1200 miles in 3 h against a head wind. Using the relationship $d = rt$, we have two linear constraints:

$$\text{Distance} = \text{rate} \times \text{time}$$
$$1500 = (p + w)3$$
$$1200 = (p - w)3$$

Using techniques presented in this chapter, we find that $p = 450$ and $w = 50$. Therefore, the plane's still air speed is 450 mph, and the speed of the wind is 50 mph. For more information visit systemsolve *at*

www.mhhe.com/miller_oneill

1. Introduction to Systems of Linear Equations in Two Variables

A linear equation in two variables has an infinite number of solutions that form a line in a rectangular coordinate system. Two or more linear equations form a **system of linear equations**. For example:

$$x - 3y = -5 \qquad\qquad y = \tfrac{1}{4}x - \tfrac{3}{4} \qquad\qquad 5a + b = 4$$
$$2x + 4y = 10 \qquad -2x + 8y = -6 \qquad\qquad -10a - 2b = 8$$

2. Determining Solutions to a System of Linear Equations

A **solution to a system of linear equations** is an ordered pair that is a solution to *each* individual linear equation.

---

example 1

Determining Solutions to a System of Linear Equations

Determine whether the ordered pairs are solutions to the system.

$$x + y = 4$$
$$-2x + y = -5$$

a. $(3, 1)$ b. $(0, 4)$

Solution:

a.   Substitute the ordered pair $(3, 1)$ into both equations:

$$x + y = 4 \longrightarrow (3) + (1) \overset{?}{=} 4 \; ✔$$
$$-2x + y = -5 \longrightarrow -2(3) + (1) \overset{?}{=} -5 \; ✔$$

Because the ordered pair $(3, 1)$ is a solution to each equation, it is a solution to the *system* of equations.

b.   Substitute the ordered pair $(0, 4)$ into both equations.

$$x + y = 4 \longrightarrow (0) + (4) \overset{?}{=} 4 \; ✔$$
$$-2x + y = -5 \longrightarrow -2(0) + (4) \overset{?}{=} -5 \qquad \text{False}$$

Because the ordered pair $(0, 4)$ is not a solution to the second equation, it is *not* a solution to the system of equations.

A solution to a system of two linear equations may be interpreted graphically as a point of intersection between the two lines. Using slope-intercept form to graph the lines from Example 1, we have

$$x + y = 4 \longrightarrow y = -x + 4$$
$$-2x + y = -5 \longrightarrow y = 2x - 5$$

Notice that the lines intersect at $(3, 1)$ (Figure 7-1).

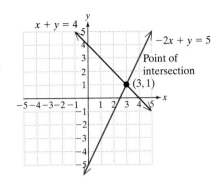

Figure 7-1

3.   Dependent and Inconsistent Systems of Linear Equations

When two lines are drawn in a rectangular coordinate system, three geometric relationships are possible:

1.   Two lines may intersect at *exactly one point*.

2.   Two lines may intersect at *no point*. This occurs if the lines are parallel.

3.   Two lines may intersect at *infinitely many points* along the line. This occurs if the equations represent the same line (the lines are coinciding).

If a system of linear equations has one or more solutions, the system is said to be **consistent**. If a linear equation has no solution, it is said to be **inconsistent**.

If two equations represent the same line, then all points along the line are solutions to the system of equations. In such a case, the system is characterized as a **dependent system**. An **independent system** is one in which the two equations represent different lines.

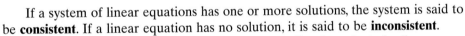

Solutions to Systems of Linear Equations in Two Variables

One Unique Solution	No Solution	Infinitely Many Solutions

One point of intersection	Parallel lines	Coinciding lines
System is consistent.	System is inconsistent.	System is consistent.
System is independent.	System is independent.	System is dependent.

4. Solving Systems of Linear Equations by Graphing

example 2

Solving a System of Linear Equations by Graphing

Solve the system by graphing both linear equations and finding the point(s) of intersection.

$$x - 2y = -2$$
$$-3x + 2y = 6$$

Solution:

To graph each equation, write the equation in slope-intercept form: $y = mx + b$.

$$x - 2y = -2 \qquad\qquad -3x + 2y = 6$$
$$-2y = -x - 2 \qquad\qquad 2y = 3x + 6$$
$$\frac{-2y}{-2} = \frac{-x}{-2} - \frac{2}{-2} \qquad\qquad \frac{2y}{2} = \frac{3x}{2} + \frac{6}{2}$$
$$y = \frac{1}{2}x + 1 \qquad\qquad y = \frac{3}{2}x + 3$$

From their slope-intercept forms, we see that the lines have different slopes, indicating that the lines are different and nonparallel. Therefore, the lines must intersect at exactly one point. Graph the lines to find that point (Figure 7-2).

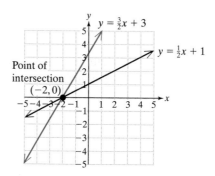

Figure 7-2

The point $(-2, 0)$ appears to be the point of intersection. This can be confirmed by substituting $x = -2$ and $y = 0$ into both equations.

$$x - 2y = -2 \longrightarrow (-2) - 2(0) \overset{?}{=} -2 \; ✔$$
$$-3x + 2y = 6 \longrightarrow -3(-2) + 2(0) \overset{?}{=} 6 \; ✔$$

The solution is $(-2, 0)$.

> **Tip:** In Example 2, the lines could also have been graphed by using the x- and y-intercepts or by using a table of points. However, the advantage of writing the equations in slope-intercept form is that we can compare the slopes and y-intercepts of each line.
>
> 1. If the slopes differ, the lines are different and nonparallel and must cross in exactly one point.
> 2. If the slopes are the same and the y-intercepts are different, the lines are parallel and will not intersect.
> 3. If the slopes are the same and the y-intercepts are the same, the two equations represent the same line.

example 3

Solving a System of Equations by Graphing

Solve the system by graphing.

$$-x + 3y = -6$$
$$6y = 2x + 6$$

Solution:

To graph the line, write each equation in slope-intercept form.

$$-x + 3y = -6 \qquad\qquad 6y = 2x + 6$$
$$3y = x - 6$$
$$\frac{3y}{3} = \frac{x}{3} - \frac{6}{3} \qquad \frac{6y}{6} = \frac{2x}{6} + \frac{6}{6}$$
$$y = \frac{1}{3}x - 2 \qquad\qquad y = \frac{1}{3}x + 1$$

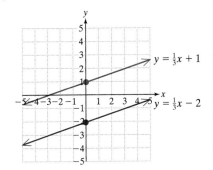

Because the lines have the same slope, but different y-intercepts, they are parallel (Figure 7-3). Two parallel lines do not intersect, which implies that the system has no solution. The system is inconsistent.

Figure 7-3

example 4

Solving a System of Linear Equations by Graphing

Solve the system by graphing.

$$x + 4y = 8$$
$$y = -\frac{1}{4}x + 2$$

Classroom Activity 7.1A

Solution:

Write the first equation in slope-intercept form. The second equation is already in slope-intercept form.

$$x + 4y = 8 \qquad\qquad y = -\frac{1}{4}x + 2$$

$$4y = -x + 8$$

$$\frac{4y}{4} = \frac{-x}{4} + \frac{8}{4}$$

$$y = -\frac{1}{4}x + 2$$

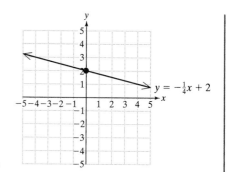

Figure 7-4

Notice that the slope-intercept forms of the two lines are identical. Therefore, the equations represent the same line (Figure 7-4). The system is dependent, and the solution to the system of equations is the set of all points on the line.

Because the ordered pairs in the solution set cannot all be listed, we can write the solution in set-builder notation. $\{(x, y) \mid y = -\frac{1}{4}x + 2\}$. This may be read as "the set of all ordered pairs, (x, y) such that the ordered pairs satisfy the equation $y = -\frac{1}{4}x + 2$."

In Chapter 2 we solved word problems using one linear equation and one variable. In this chapter, we will use two linear equations and two variables to solve similar problems.

---

example 5

Using a System of Linear Equations in an Application

The sum of two numbers is 5 and their difference is -9. Find the numbers.

Classroom Activity 7.1B

Solution:

Let x represent one number.
Let y represent the other number.

The sum of the two numbers is 5: $x + y = 5$

The difference is -9: $x - y = -9$

To solve the system, graph each line and find the point(s) of intersection (Figure 7-5).

$$x + y = 5 \qquad\qquad x - y = -9$$

$$y = -x + 5 \qquad\qquad -y = -x - 9$$

$$\frac{-y}{-1} = \frac{-x}{-1} - \frac{9}{-1}$$

$$y = x + 9$$

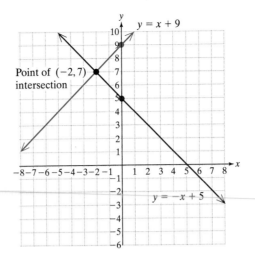

Figure 7-5

The point of intersection of the two graphs appears to be $(-2, 7)$. The solution can be confirmed by substituting $x = -2$ and $y = 7$ into both equations.

$$x + y = 5 \longrightarrow (-2) + (7) \stackrel{?}{=} 5 \checkmark$$

$$x - y = -9 \longrightarrow (-2) - (7) \stackrel{?}{=} -9 \checkmark$$

Because the ordered pair $(-2, 7)$ is a solution to both equations, it is a solution to the system of equations.

Hence the numbers are -2 and 7.

Calculator Connections

The solution to a system of equations can be found by using either a *Trace* feature or an *Intersect* feature on a graphing calculator to find the point of intersection between two curves.

For example, consider the system:

$$-2x + y = 6$$

$$5x + y = -1$$

First graph the equations together on the same viewing window. Recall that to enter the equations into the calculator, the equations must be written with the y-variable isolated.

isolate y

$$-2x + y = 6 \longrightarrow y = 2x + 6$$

$$5x + y = -1 \longrightarrow y = -5x - 1$$

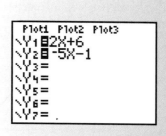

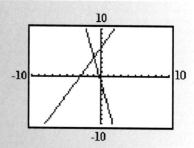

By inspection of the graph, it appears that the solution is $(-1, 4)$. The *Trace* option on the calculator may come close to $(-1, 4)$ but may not show the exact solution (Figure 7-6). However, an *Intersect* feature on a graphing calculator may provide the exact solution (Figure 7-7). See your user's manual for further details.

Using *Trace*

Using *Intersect*

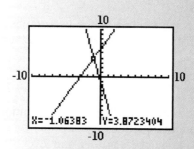

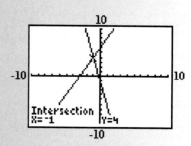

Figure 7-6

Figure 7-7

Calculator Exercises

Use a graphing calculator to graph each linear equation on the same viewing window. Use a *Trace* or *Intersect* feature to find the point(s) of intersection.

1. $y = 2x - 3$

$y = -4x + 9$ ❖

2. $y = -\dfrac{1}{2}x + 2$

$y = \dfrac{1}{3}x - 3$ ❖

3. $x + y = 4$ Example 1

$-2x + y = -5$ ❖

4. $x - 2y = -2$ Example 2

$-3x + 2y = 6$ ❖

5. $-x + 3y = -6$ Example 3

$6y = 2x + 6$ ❖

6. $x + 4y = 8$ Example 4

$y = -\dfrac{1}{4}x + 2$ ❖

section 7.1   PRACTICE EXERCISES

For Exercises 1–6, determine if the given point is a solution to the system.

1. $3x - y = 7$ $(2, -1)$
   $x - 2y = 4$ Yes

2. $x - y = 3$ $(4, 1)$
   $x + y = 5$ Yes

3. $2x - 3y = 12$ $(0, 4)$
   $3x + 4y = 12$ No

4. $x - 2y = 6$ $(9, -1)$
   $x + 3y = 6$ No

5. $3x - 6y = 9$ $\left(4, \dfrac{1}{2}\right)$
   $x - 2y = 3$ Yes

6. $x - y = 4$ $(6, 2)$
   $3x - 3y = 12$ Yes

7. Graph each system of equations.
   a. $y = 2x - 3$ b. $y = 2x + 1$
   $y = 2x + 5$ ❖ $y = 4x - 5$ ❖
   c. $y = 3x - 5$
   $y = 3x - 5$ ❖

For Exercises 8–18, determine which system of equations (a, b, or c) makes the statement true. (*Hint*: Refer to the graphs from Exercise 7.)

8. The lines are parallel.   a   a. $y = 2x - 3$
                           $y = 2x + 5$

9. The lines are coinciding.   c

10. The lines intersect at exactly one point.   b   b. $y = 2x + 1$
                                 $y = 4x - 5$

11. The system is inconsistent.   a

12. The system is dependent.   c   c. $y = 3x - 5$
                              $y = 3x - 5$

13. The lines have the same slope but different y-intercepts.   a

14. The lines have the same slope and same y-intercept.   c

15. The lines have different slopes.   b

16. The system has exactly one solution.   b

17. The system has infinitely many solutions.   c

18. The system has no solution.   a

For Exercises 19–22, match the graph of the system of equations with the appropriate description of the solution.

19.   b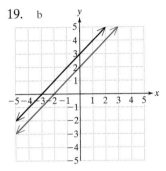

     a. The solution is $(1, 3)$.
     b. No solution.
     c. There are infinitely many solutions.
     d. The solution is $(0, 0)$.

20.   c   21.   d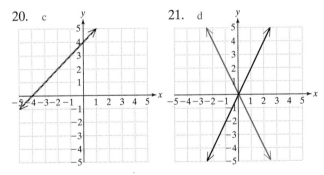

22.   a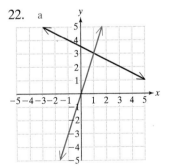

For Exercises 23–48, solve the systems by graphing. Identify each system as consistent or inconsistent. Identify each system as dependent or independent.

23. $y = -x + 4$ ❖
    $y = x - 2$

24. $y = 3x + 2$ ❖
    $y = 2x$

25. $2x + y = 0$ ❖
    $3x + y = 1$

26. $x + y = -1$ ❖
    $2x - y = -5$

27. $2x + y = 6$ ❖
    $x = 1$

28. $4x + 3y = 9$ ❖
    $x = 3$

29. $-6x - 3y = 0$ ❖
    $4x + 2y = 4$

30. $2x - 6y = 12$ ❖
    $-3x + 9y = 12$

31. $-2x + y = 3$ ❖
    $6x - 3y = -9$

32. $x + 3y = 0$ ❖
    $-2x - 6y = 0$

33. $y = 6$ ❖
    $2x + 3y = 12$

34. $y = -2$ ❖
    $x - 2y = 10$

35. $-5x + 3y = -9$ ❖
    $y = \dfrac{5}{3}x - 3$

36. $4x + 2y = 6$ ❖
    $y = -2x + 3$

37. $x = 4 + y$ ❖
    $3y = -3x$

38. $3y = 4x$ ❖
    $x - y = -1$

39. $-x + y = 3$ ❖
    $4y = 4x + 6$

40. $x - y = 4$ ❖
    $3y = 3x + 6$

41. $x = 4$ ❖
    $2y = 4$

42. $-3x = 6$ ❖
    $y = 2$

43. $2x + 3y = 8$ ❖
    $-4x - 6y = 6$

44. $4x + 4y = 8$ ❖
    $5x + 5y = 5$

45. $2x + y = 4$ ❖
    $4x - 2y = -4$

46. $6x + 6y = 3$ ❖
    $2x - y = 4$

47. $y = 0.5x + 2$ ❖
    $-x + 2y = 4$

48. $3x - 4y = 6$ ❖
    $-6x + 8y = -12$

49. Two tennis instructors have two different fee schedules. Owen charges $25 per lesson plus a one-time court fee of $20 at the tennis club. Joan charges $30 per lesson, but does not require a court fee. The total cost, y, depends on the number of lessons, x, according to the equations

    Owen: $y = 25x + 20$

    Joan: $y = 30x$

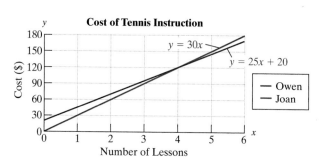

Figure for Exercise 49

From the graph, determine the number of lessons for which the total cost is the same for both instructors.   4 lessons will cost $120 for each instructor.

50. The cost to rent a 10 ft by 10 ft storage space is different for two different storage companies. The Storage Bin charges $90 per month plus a nonrefundable deposit of $120. AAA Storage charges $110 per month with no deposit. The total cost, y, to rent a 10 ft by 10 ft space depends on the number of months, x, according to the equations

    The Storage Bin: $y = 90x + 120$

    AAA Storage: $y = 110x$

❖ See Additional Answers Appendix   🖉 Writing   ⬌ Translating Expression   Geometry   Scientific Calculator   Video

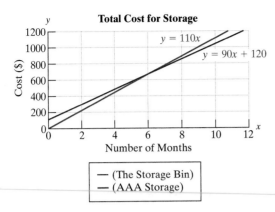

Figure for Exercise 50

From the graph, determine the number of months required for which the cost to rent space is equal for both companies.   For 6 months, the cost is $660 for each company.

51. The following graph depicts the number of persons in the United States living with AIDS between 1993 and 1998 by race or ethnicity. Estimate the year in which the number of black (not Hispanic) persons living with AIDS and the number of white (not Hispanic) persons living with AIDS was approximately the same.   1997

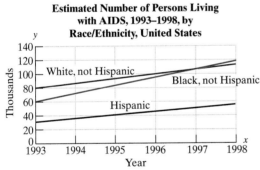

Figure for Exercise 51

Source: Centers for Disease Control and Prevention.

52. The following graph depicts the number of men and number of women in the United States living with AIDS, between 1993 and 1998. If the trends continue, will the number of women living with AIDS ever equal the number of men living with AIDS? Explain why or why not. No, these lines will not intersect in the future because the rate of increase of women with AIDS is less than the rate of increase of men with AIDS.

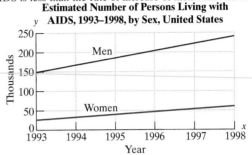

Figure for Exercise 52

Source: Centers for Disease Control and Prevention.

EXPANDING YOUR SKILLS

53. Write a system of linear equations whose solution is $(2, 1)$.   For example: $4x + y = 9$
$-2x - y = -5$

54. Write a system of linear equations whose solution is $(1, 4)$.   For example: $x - 3y = -11$
$3x + y = 7$

55. One equation in a system of linear equations is $x + y = 4$. Write a second equation such that the system will have no solution. (Answers may vary.)   For example: $2x + 2y = 1$

56. One equation in a system of linear equations is $x - y = 3$. Write a second equation such that the system will have infinitely many solutions. (Answers may vary.)   For example: $3x - 3y = 9$

section

7.2 SUBSTITUTION METHOD

1. Introduction to the Substitution Method

In the previous section we used the graphing method to find the solution set to a system of equations. However, sometimes it is difficult to determine the solution using this method because of limitations in the accuracy of the graph. This is particularly true when the coordinates of a solution are not integer values or when the solution is a point not sufficiently close to the origin. Identifying the coordinates of the point $\left(\frac{3}{17}, -\frac{23}{9}\right)$ or $(-251, 8349)$ for example, might be difficult from a graph.

In this section and the next, we will cover two algebraic methods to solve a system of equations that do not require graphing. The first method we present here is called the **substitution method**. For example, consider the following system of equations:

$$x + y = 4$$
$$-2x + y = -5$$

The first step in the substitution process is to isolate one of the variables in one of the equations. For instance, solving the first equation for x yields: $x = 4 - y$. Then, because x is equal to $4 - y$, the expression $4 - y$ may be substituted for x in the second equation. This leaves the second equation in terms of y only.

First equation: $\qquad x + y = 4 \longrightarrow x = \underline{4 - y}$

Second equation: $\qquad -2x + y = -5$

$\qquad\qquad -2(4 - y) + y = -5$ $\qquad$ This equation now contains only one variable.

$\qquad\qquad -8 + 2y + y = -5$ $\qquad$ Solve the resulting equation.

$\qquad\qquad -8 + 3y = -5$

$\qquad\qquad 3y = -5 + 8$

$\qquad\qquad 3y = 3$

$\qquad\qquad y = 1$

To find x, substitute $y = 1$ back into the expression

$$x = 4 - y$$
$$x = 4 - (1)$$
$$x = 3$$

The ordered pair $(3, 1)$ can be checked in the original equations to verify the answer.

$$x + y = 4 \longrightarrow (3) + (1) = 4 \; \checkmark$$
$$-2x + y = -5 \longrightarrow -2(3) + (1) = -5 \; \checkmark$$

The solution is $(3, 1)$.

2. Solving a System of Linear Equations by the Substitution Method

Solving a System of Equations by the Substitution Method

1. Isolate one of the variables from one equation.
2. Substitute the quantity found in Step 1 into the other equation.
3. Solve the resulting equation.
4. Substitute the value found in Step 3 back into the equation in Step 1 to find the value of the remaining variable.
5. Check the solution in both equations and write the answer as an ordered pair.

example 1

Solving a System of Linear Equations Using the Substitution Method

Solve the system by using the substitution method.

$$3x + 5y = 17$$
$$2x - y = -6$$

Classroom Activity 7.2A

Solution:

The y-variable in the second equation is the easiest variable to isolate because its coefficient is -1.

$$3x + 5y = 17$$
$$2x - y = -6 \longrightarrow -y = -2x - 6$$
$$y = 2x + 6$$

Step 1: Solve the second equation for y.

$$3x + 5(2x + 6) = 17$$

Step 2: Substitute the quantity $2x + 6$ for y in the other equation.

⬡ **Avoiding Mistakes**

Do not substitute $y = 2x + 6$ into the same equation from which it came. This mistake will result in an identity:

$$2x - y = -6$$
$$2x - (2x + 6) = -6$$
$$2x - 2x - 6 = -6$$
$$-6 = -6$$

$$3x + 10x + 30 = 17$$
$$13x + 30 = 17$$
$$13x = 17 - 30$$
$$13x = -13$$
$$x = -1$$

Step 3: Solve for x.

$$y = 2x + 6$$
$$y = 2(-1) + 6$$
$$y = -2 + 6$$
$$y = 4$$

Step 4: Substitute $x = -1$ into the expression $y = 2x + 6$.

Step 5: The ordered pair $(-1, 4)$ can be checked in the original equations to verify the answer.

$$3x + 5y = 17 \longrightarrow 3(-1) + 5(4) = 17 \longrightarrow -3 + 20 = 17 ✔$$
$$2x - y = -6 \longrightarrow 2(-1) - (4) = -6 \longrightarrow -2 - 4 = -6 ✔$$

The solution is $(-1, 4)$.

3.   Systems of Linear Equations with No Solution or Infinitely Many Solutions

Recall from Section 7.1, that a system of linear equations may represent two parallel lines. In such a case there is no solution to the system.

example 2

Solving a System of Linear Equations Using the Substitution Method

Solve the system by using the substitution method.

$$2x + 3y = 6$$
$$y = -\tfrac{2}{3}x + 4$$

Solution:

$$2x + 3y = 6$$
$$y = -\tfrac{2}{3}x + 4$$

Step 1: The variable y is already isolated in the second equation.

$$2x + 3(-\tfrac{2}{3}x + 4) = 6$$

Step 2: Substitute $y = -\tfrac{2}{3}x + 4$ from the second equation into the first equation.

$$2x - 2x + 12 = 6$$
$$12 = 6$$

Step 3: Solve the resulting equation.

The equation results in a contradiction. There are no values of x and y that will make 12 equal to 6. Therefore, there is no solution, and the system is inconsistent.

Tip: The answer to Example 2 can be verified by writing each equation in slope-intercept form and graphing the lines.

Equation 1	**Equation 2**

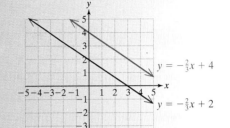

$$2x + 3y = 6 \qquad\qquad y = -\frac{2}{3}x + 4$$

$$3y = -2x + 6$$

$$\frac{3y}{3} = \frac{-2x}{3} + \frac{6}{3}$$

$$y = -\frac{2}{3}x + 2$$

The equations indicate that the lines have the same slope but different y-intercepts. Therefore, the lines must be parallel. There is no point of intersection, indicating that the system has no solution.

Recall that a system of two linear equations may represent the same line. In such a case, the solution is the set of all points on the line.

---

example 3

Solving a System of Linear Equations Using Substitution

Solve the system by using the substitution method.

$$\frac{1}{2}x - \frac{1}{4}y = 1$$

$$6x - 3y = 12$$

Solution:

$$\frac{1}{2}x - \frac{1}{4}y = 1 \qquad$$ To make the first equation easier to work with, we have the option of clearing fractions.

$$6x - 3y = 12$$

$$\frac{1}{2}x - \frac{1}{4}y = 1 \xrightarrow{\text{multiply by 4}} 4\left(\frac{1}{2}x\right) - 4\left(\frac{1}{4}y\right) = 4(1) \longrightarrow 2x - y = 4$$

Now the system becomes:

$$2x - y = 4 \qquad$$ The y-variable in the first equation is the easiest to isolate because its coefficient is -1.

$$6x - 3y = 12$$

$$2x - y = 4 \xrightarrow{\text{solve for } y} -y = -2x + 4 \rightarrow y = 2x - 4 \qquad$$ **Step 1:** Isolate one of the variables.

$$6x - 3y = 12$$

$$6x - 3(2x - 4) = 12$$

Step 2: Substitute $y = 2x - 4$ from the first equation into the second equation.

$$6x - 6x + 12 = 12$$

Step 3: Solve the resulting equation.

$$12 = 12$$

Because the equation produces an identity, we know that x can be any real number. Substituting any real number into the expression $y = 2x - 4$ produces an ordered pair on the line $y = 2x - 4$. Hence the solution set to the system of equations is the set of all ordered pairs on the line $y = 2x - 4$. This can be written as $\{(x, y) \mid y = 2x - 4\}$. The system is dependent.

Tip: The solution to Example 3 can be verified by writing each equation in slope-intercept form and graphing the lines.

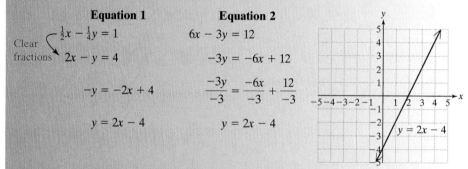

Equation 1	Equation 2
$-\frac{1}{2}x - \frac{1}{4}y = 1$	$6x - 3y = 12$
Clear fractions $\searrow$ $2x - y = 4$	$-3y = -6x + 12$
$-y = -2x + 4$	$\frac{-3y}{-3} = \frac{-6x}{-3} + \frac{12}{-3}$
$y = 2x - 4$	$y = 2x - 4$

Notice that the slope-intercept forms for both equations are identical. The equations represent the same line, indicating that the system is dependent. Each point on the line is a solution to the system of equations.

4. Solutions to Systems of Linear Equations: A Summary

The following summary reviews the three different geometric relationships between two lines and the solutions to the corresponding systems of equations.

Classroom Activity 7.2B

Solutions to a System of Two Linear Equations

1. The lines may intersect at one point (yielding one unique solution).
2. The lines may be parallel and intersect at no point (yielding no solution). This is detected algebraically when a contradiction (false statement) is obtained (for example: $0 = -3$ and $12 = 6$).
3. The lines may be the same and intersect at all points on the line (yielding an infinite number of solutions). This is detected algebraically when an identity is obtained (for example: $0 = 0$ and $12 = 12$).

5. Applications of the Substitution Method

| example 4 | **Using the Substitution Method** |

One number is 3 more than $\frac{1}{4}$ of another. Their sum is 133. Find the numbers.

Solution:

Let x represent one number.
Let y represent the other number.

One number is 3 more than $\frac{1}{4}$ of another. $\rightarrow$ $x = \dfrac{1}{4}y + 3$

Their sum is 133. $\longrightarrow$ $x + y = 133$

$$\left(\frac{1}{4}y + 3\right) + y = 133$$

Step 1: Notice that x is already isolated in the first equation.

Step 2: Substitute $x = \frac{1}{4}y + 3$ into the second equation.

$$\frac{5}{4}y + 3 = 133$$

Step 3: Solve the resulting equation.

$$\frac{5}{4}y = 130$$

$$\frac{4}{5} \cdot \frac{5}{4}y = 130 \cdot \frac{4}{5}$$

$$y = 104$$

$$x = \frac{1}{4}y + 3$$

Step 4: Substitute $y = 104$ into the expression $x = \frac{1}{4}y + 3$.

$$x = \frac{1}{4}(104) + 3$$

$$x = 26 + 3$$

$$x = 29$$

Tip: Check that the numbers 104 and 29 meet the conditions of Example 4.

- $\frac{1}{4}$ of 104 is 26. Three more than 26 is 29. ✔
- The sum of the numbers should be 133:
  $29 + 104 = 133$ ✔

One number is 104 and the other is 29.

example 5   Using the Substitution Method in a Geometry Application

Two angles are supplementary. One angle is 15° more than twice the other angle. Find the two angles.

Classroom Activity 7.2C

Solution:

Let x represent one angle.
Let y represent the other.

The sum of supplementary angles is 180° ⟶ $x + y = 180$

One angle is 15° more than twice the other angle ⟶ $x = 2y + 15$

$$x + y = 180$$

$$x = 2y + 15$$ **Step 1:** The x-variable in the second equation is already isolated.

$$(2y + 15) + y = 180$$ **Step 2:** Substitute $x = 2y + 15$ from the second equation into the first equation.

$$2y + 15 + y = 180$$ **Step 3:** Solve the resulting equation.

$$3y + 15 = 180$$

$$3y = 165$$

$$\frac{3y}{3} = \frac{165}{3}$$

$$y = 55$$

Tip: Check that the angles 55° and 125° meet the conditions of Example 5.

- Because 55° + 125° = 180°, the angles are supplementary. ✔
- The angle 125° is 15° more than twice 55°: 125° = 2(55°) + 15°. ✔

$$x = 2y + 15$$

$$x = 2(55) + 15$$ **Step 4:** Substitute $y = 55$ into the expression $x = 2y + 15$.

$$x = 110 + 15$$

$$x = 125$$

One angle is 55° and the other is 125°.

---

section 7.2   PRACTICE EXERCISES

For Exercises 1–6, write each pair of lines in slope-intercept form. Then identify whether the lines intersect in exactly one point or if the lines are parallel or coinciding.   7.1

1. $2x - y = 4$ $y = 2x - 4$

   $-2y = -4x + 8$
   $y = 2x - 4$; coinciding lines

2. $x - 2y = 5$ $y = \frac{1}{2}x - \frac{5}{2}$

   $3x = 6y + 15$

   $y = \frac{1}{2}x - \frac{5}{2}$; coinciding lines

3. $2x + 3y = 6$ $y = -\frac{2}{3}x + 2$

   $x - y = 5$
   $y = x - 5$; intersecting lines

4. $x - y = -1$ $y = x + 1$

   $x + 2y = 4$

   $y = -\frac{1}{2}x + 2$; intersecting lines

5. $2x = \frac{1}{2}y + 2$

   $y = 4x - 4$

   $4x - y = 13$
   $y = 4x - 13$; parallel lines

6. $4y = 3x$ $y = \frac{3}{4}x$

   $3x - 4y = 15$

   $y = \frac{3}{4}x - \frac{15}{4}$; parallel lines

 See Additional Answers Appendix   Writing   Translating Expression   Geometry   Scientific Calculator   Video

For Exercises 7–14, solve each system using the substitution method.

7. $3x + 2y = -3$
   $y = 2x - 12$ $(3, -6)$

8. $4x - 3y = -19$
   $y = -2x + 13$ $(2, 9)$

9. $x = -4y + 16$
   $3x + 5y = 20$ $(0, 4)$

10. $x = -y + 3$
    $-2x + y = 6$ $(-1, 4)$

11. $3x + 5y = 7$
    $y = -\dfrac{3}{5}x + 3$ No solution

12. $4x - 3y = -28$
    $y = \dfrac{4}{3}x + 3$ No solution

13. $x = \dfrac{6}{5}y + 3$
    Infinitely many solutions
    $5x - 6y = 15$
    $\{(x, y) | x = \frac{6}{5}y + 3\}$

14. $x = \dfrac{1}{2}y + 5$
    Infinitely many solutions
    $4x - 2y = 20$
    $\{(x, y) | x = \frac{1}{2}y + 5\}$

15. Given the system:
    $$4x - 2y = -6$$
    $$3x + y = 8$$
    a. Which variable from which equation is easiest to isolate and why?   *y in the second equation is easiest to solve for because its coefficient is 1.*
    b. Solve the system using the substitution method. $(1, 5)$

16. Given the system:
    $$x - 5y = 2$$
    $$11x + 13y = 22$$
    a. Which variable from which equation is easiest to isolate and why?   *x in the first equation is easiest to solve for because its coefficient is 1.*
    b. Solve the system using the substitution method. $(2, 0)$

For Exercises 17–46, solve each system using the substitution method.

17. $4x - y = -1$
    $2x + 4y = 13$ $\left(\dfrac{1}{2}, 3\right)$

18. $5x - 3y = -2$
    $10x - y = 1$ $\left(\dfrac{1}{5}, 1\right)$

19. $x - 3y = -1$
    $2x = 4y + 2$ $(5, 2)$

20. $3x + y = 1$
    $2y = x + 9$ $(-1, 4)$

21. $-2x + 5y = 5$
    $x - 4y = -10$ $(10, 5)$

22. $3x - 7y = -2$
    $2x + y = 27$ $(11, 5)$

23. $3x + 2y = -1$ No solution
    $\dfrac{3}{2}x + y = 4$

24. $5x - 2y = 6$ No solution
    $-\dfrac{5}{2}x + y = 5$

25. $10x - 30y = -10$ Infinitely many solutions
    $2x - 6y = -2$
    $\{(x, y) | y = \frac{1}{3}x + \frac{1}{3}\}$

26. $3x + 6y = 6$ Infinitely many solutions
    $-6x - 12y = -12$
    $\{(x, y) | -\frac{1}{2}x + 1\}$

27. $2x + y = 3$
    $y = -7$ $(5, -7)$

28. $-3x = 2y + 23$
    $x = -1$ $(-1, -10)$

29. $x + 2y = -2$
    $4x = -2y - 17$ $\left(-5, \dfrac{3}{2}\right)$

30. $x + y = 1$
    $2x - y = -2$ $\left(-\dfrac{1}{3}, \dfrac{4}{3}\right)$

31. $y = -\dfrac{1}{2}x - 4$
    $y = 4x - 13$ $(2, -5)$

32. $y = \dfrac{2}{3}x - 3$
    $y = 6x - 19$ $(3, -1)$

33. $y = -2x + 1$ No solution
    $y - 4 = -2(x + 3)$

34. $x = 6y + 12$ No solution
    $x - 3 = 6(y + 2)$

35. $3x + 2y = 4$
    $2x - 3y = -6$ $(0, 2)$

36. $4x + 3y = 4$
    $-2x + 5y = -2$ $(1, 0)$

37. $y = 0.25x + 1$ Infinitely many solutions
    $-x + 4y = 4$ $\{(x, y) | y = 0.25x + 1\}$

38. $y = 0.75x - 3$ Infinitely many solutions
    $-3x + 4y = -12$ $\{(x, y) | y = 0.75x - 3\}$

39. $11x + 6y = 17$
    $5x - 4y = 1$ $(1, 1)$

40. $3x - 8y = 7$
    $10x - 5y = 45$ $(5, 1)$

41. $x + 2y = 4$
    $4y = -2x - 8$ No solution

42. $-y = x - 6$ No solution
       $2x + 2y = 4$

43. $\frac{1}{3}(2x + y) = 1$ $(-1, 5)$
    $x + y = 4$

44. $2(x - y) = 4$ $(3, 1)$
    $3x + y = 10$

45. $\frac{a}{3} + \frac{b}{2} = -4$ $(-6, -4)$
    $a - 3b = 6$

46. $a - 2b = -5$ $(-1, 2)$
    $\frac{2a}{3} + \frac{b}{3} = 0$

For Exercises 47–56, set up a system of linear equations and solve for the indicated quantities.

47. Two numbers have a sum of 106. One number is 10 less than the other. Find the numbers.
    The numbers are 48 and 58.

48. Two positive numbers have a difference of 8. The larger number is 2 less than 3 times the smaller number. Find the numbers. The numbers are 13 and 5.

49. Two angles are supplementary. One angle is 15° more than 10 times the other angle. Find the measure of each angle.   The angles are 165° and 15°.

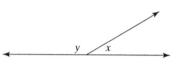

Figure for Exercise 49

50. Two angles are complementary. One angle is 1° less than 6 times the other angle. Find the measure of each angle.
    The angles are 13° and 77°.

51. Two angles are complementary. One angle is 10° more than 3 times the other angle. Find the measure of each angle.   The angles are 70° and 20°.

Figure for Exercise 50

52. Two angles are supplementary. One angle is 5° less than twice the other angle. Find the measure of each angle.   The angles are $118\frac{1}{3}°$ and $61\frac{2}{3}°$.

53. In a right triangle, one of the acute angles is 6° less than the other acute angle. Find the measure of each acute angle.   The angles are 42° and 48°.

Figure for Exercise 53

54. In a right triangle, one of the acute angles is 9° less than twice the other acute angle. Find the measure of each acute angle.
    The angles are 57° and 33°.

55. At a ballpark the total cost of a soft drink and a hot dog is $2.50. The price of the hot dog is $1.00 more than the cost of the soft drink. Find the cost of a soft drink and the cost of a hot dog.
    A hot dog costs $1.75 and a drink costs $0.75.

56. Ray played two rounds of golf for a total score of 154. If his score in the second round is 10 more than his score in the first round, find the scores for each round.   The score was 72 on the first round and 82 on the second round.

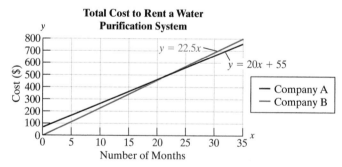

57. Two water purification systems are priced differently. Company A charges a $55 installation fee plus $20 per month. Company B charges $22.50 per month with no installation fee. The total cost, y, depends on the number of months, x, according to the following equations:

    Company A: $y = 20x + 55$

    Company B: $y = 22.50x$

Total Cost to Rent a Water Purification System

Figure for Exercise 57

a. Use the graph to determine how many months are required for the cost to rent from Company A to equal the cost to rent from Company B.   ❖

b. Use the substitution method to solve the system of equations. Interpret the answer in terms of the number of months and the total cost of renting a water purification system. ❖

c. Which company is more expensive if the water system is rented for more than 22 months? Which is more expensive if the water system is rented for less than 22 months? ❖

58. Two rental car companies use different fee rates. Company A charges $20 per day and $0.20 per mile. Company B charges 0.30 per mile with no daily fee. If a car is rented for one day, then the total cost, y, depends on the number of miles driven, x, according to the following equations:

Company A: $y = 0.20x + 20$

Company B: $y = 0.30x$

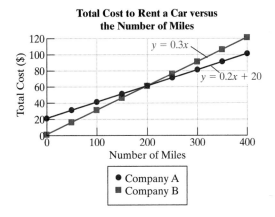

Total Cost to Rent a Car versus the Number of Miles

Figure for Exercise 58

a. Use the graph to determine how many miles would be required for the cost to rent from Company A to equal the cost to rent from Company B.   200 miles

b. Use the substitution method to solve the system of equations. Interpret the answer in terms of the number of miles driven and the total cost.   For 200 miles, the rental companies both charge $60.

c. Which car rental company is more expensive if a car is rented for one day and driven for more than 200 miles?   Company B is more expensive if traveling over 200 miles.

EXPANDING YOUR SKILLS

59. The following system of equations is dependent and has infinitely many solutions. Find three ordered pairs that are solutions to the system of equations.

$$y = 2x + 3$$
$$-4x + 2y = 6$$ For example: $(0, 3)(1, 5)(-1, 1)$

60. The following system of equations is dependent and has infinitely many solutions. Find three ordered pairs that are solutions to the system of equations.

$$y = -x + 1$$
$$2x + 2y = 2$$ For example: $(0, 1)(1, 0)(-1, 2)$

Concepts

1. Introduction to the Addition Method

2. Solving a System of Linear Equations by the Addition Method

3. Systems of Linear Equations with No Solution or Infinitely Many Solutions

4. Summary of Methods for Solving Linear Equations in Two Variables

section

7.3   ADDITION METHOD

1.   Introduction to the Addition Method

Thus far in Chapter 7 we have used the graphing method and the substitution method to solve a system of linear equations in two variables. In this section, we present another algebraic method to solve a system of linear equations, called the **addition method** (sometimes called the elimination method).

example 1　Using the Addition Method in an Application

The sum of two numbers is -2. The difference of the smaller number and the larger number is -6. Find the numbers.

Solution:

Let x represent the smaller number.
Let y represent the larger number.

The sum of the numbers is -2. ⟶ $x + y = -2$

The difference of the smaller and the larger is -6. → $x - y = -6$

Notice that the coefficients of the y-variables are opposites:

$$x + 1y = -2 \quad \text{Coefficient is 1.}$$
$$x - 1y = -6 \quad \text{Coefficient is } -1.$$

Because the coefficients of the y-variables are opposites, we can add the two equations to eliminate the y-variable.

$$
\begin{array}{r}
x + y = -2 \\
\underline{x - y = -6} \\
2x \quad\;\; = -8
\end{array}
$$
← After adding the equations, we have one equation and one variable.

$2x = -8$ 　　Solve the resulting equation.

$x = -4$

Tip: Notice that the value $x = -4$ could have been substituted into the second equation, to obtain the same value for y.

$x - y = -6$
$(-4) - y = -6$
$-y = -6 + 4$
$-y = -2$
$y = 2$

To find the value of y, substitute $x = -4$ into *either* of the original equations.

$x + y = -2$ 　　First equation
$(-4) + y = -2$
$y = -2 + 4$
$y = 2$

The numbers are -4 and 2.

Check that the numbers -4 and 2 meet the conditions of this problem.

1. The sum of the numbers is -2. 　　　　　　$-4 + 2 = -2$ ✔
2. The difference of the smaller and the larger is -6. 　$-4 - 2 = -6$ ✔

2.　Solving a System of Linear Equations by the Addition Method

It is important to note that the addition method works on the premise that the two equations have *opposite* values for the coefficients of one of the variables. Sometimes it is necessary to manipulate the original equations to create two coefficients that are opposites. This is accomplished by multiplying one or both equations by an appropriate constant. The process is outlined as follows.

Solving a System of Equations by the Addition Method

1. Write both equations in standard form: $ax + by = c$.
2. Clear fractions or decimals (optional).
3. Multiply one or both equations by nonzero constants to create opposite coefficients for one of the variables.
4. Add the equations from Step 3 to eliminate one variable.
5. Solve for the remaining variable.
6. Substitute the known value from Step 5 into one of the original equations to solve for the other variable.
7. Check the solution in both equations.

example 2

Solving a System of Linear Equations Using the Addition Method

Solve the system using the addition method.

$$3x + 5y = 17$$
$$2x - y = -6$$

Solution:

$3x + 5y = 17$ **Step 1:** Both equations are already written in standard form.

$2x - y = -6$ **Step 2:** There are no fractions or decimals.

Notice that neither the coefficients of x nor the coefficients of y are opposites. However, multiplying the second equation by 5 creates the term $-5y$ in the second equation. This is the opposite of the term $+5y$ in the first equation.

$$3x + 5y = 17$$
$$2x - y = -6 \xrightarrow[\text{Multiply by 5}]{}$$

$$\begin{aligned} 3x + 5y &= 17 \\ 10x - 5y &= -30 \\ \hline 13x \phantom{{}+5y} &= -13 \end{aligned}$$

Step 3: Multiply the second equation by 5.

Step 4: Add the equations.

$$13x = -13$$

$$x = -1$$

Step 5: Solve the equation.

$$3x + 5y = 17 \qquad \text{First equation}$$

$$3(-1) + 5y = 17$$

$$-3 + 5y = 17$$

$$5y = 20$$

$$y = 4$$

Step 6: Substitute the known value of x into one of the original equations.

The solution is $(-1, 4)$.

Step 7: Check the solution in both original equations.

<u>Check:</u>

$3x + 5y = 17 \longrightarrow 3(-1) + 5(4) \stackrel{?}{=} 17 \longrightarrow -3 + 20 = 17$ ✔

$2x - y = -6 \longrightarrow 2(-1) - (4) \stackrel{?}{=} -6 \longrightarrow -2 - 4 = -6$ ✔

---

example 3

Solving a System of Linear Equations Using the Addition Method

Solve the system using the addition method.

$$3(x - 10) = 7y + 11$$
$$-2(x - y) = 2x - 18$$

Classroom Activity 7.3A

Solution:

Step 1: Write the equations in standard form.

     Clear parentheses.   Write as $ax + by = c$.

$3(x - 10) = 7y + 11 \longrightarrow 3x - 30 = 7y + 11 \longrightarrow 3x - 7y = 41$

$-2(x - y) = 2x - 18 \longrightarrow -2x + 2y = 2x - 18 \longrightarrow -4x + 2y = -18$

Step 2: There are no fractions or decimals.

Notice that neither the coefficients of x nor the coefficients of y are opposites. However, it is possible to change the coefficients of x to 12 and -12 (notice that 12 is the LCM, of 3 and 4). This is accomplished by multiplying the first equation by 4 and the second equation by 3.

Step 3: Opposite coefficients of x.

$3x - 7y = 41 \xrightarrow{\text{Multiply by 4}} 12x - 28y = 164$

$-4x + 2y = -18 \xrightarrow{\text{Multiply by 3}} \underline{-12x + 6y = -54}$

$-22y = 110$

Step 4: Add the equations.

$-22y = 110$

$\dfrac{-22y}{-22} = \dfrac{110}{-22}$

$y = -5$

Step 5: Solve the resulting equation.

$$3x - 7y = 41 \quad \text{First equation}$$
$$3x - 7(-5) = 41$$
$$3x + 35 = 41$$
$$3x = 6$$
$$x = 2$$

The solution is $(2, -5)$.

Check:

$$3(x - 10) = 7y + 11 \longrightarrow 3(2 - 10) \stackrel{?}{=} 7(-5) + 11 \longrightarrow -24 = -24 \checkmark$$
$$-2(x - y) = 2x - 18 \longrightarrow -2[2 - (-5)] \stackrel{?}{=} 2(2) - 18 \longrightarrow -14 = -14 \checkmark$$

Step 6: Substitute the known value of y into one of the original equations.

Step 7: Check the solution in the original equations.

Tip: When using the addition method, it makes no difference which variable is eliminated. In Example 3, we eliminated x. However, we could easily have eliminated y by changing the coefficients of y to -14 and 14. This would be accomplished by multiplying the first equation by 2 and the second equation by 7.

$$3x - 7y = 41 \xrightarrow{\text{Multiply by 2}} 6x - 14y = 82$$
$$-4x + 2y = -18 \xrightarrow[\text{Multiply by 7}]{} \frac{-28x + 14y = -126}{-22x \qquad = -44}$$

Because $-22x = -44$, then $x = 2$. Substituting $x = 2$ into either original equation yields $y = -5$.

example 4

Solving a System of Linear Equations Using the Addition Method

Solve the system using the addition method.

$$34x - 22y = 4$$
$$17x - 88y = -19$$

Solution:

The equations are already in standard form. There are no fractions or decimals to clear.

$$34x - 22y = 4 \xrightarrow{\hspace{3cm}} 34x - 22y = 4$$
$$17x - 88y = -19 \xrightarrow[\text{Multiply by } -2]{} \frac{-34x + 176y = 38}{154y = 42}$$

$$154y = 42$$

$$\frac{154y}{154} = \frac{42}{154}$$

$$y = \frac{3}{11}$$

To find the value of x, we normally substitute y into one of the original equations and solve for x. However, because the value of y is a fraction, we may choose to repeat the addition method again, this time eliminating y and solving for x.

Multiply by -4

$$34x - 22y = 4 \longrightarrow -136x + 88y = -16$$
$$17x - 88y = -19 \longrightarrow \underline{17x - 88y = -19}$$
$$-119x = -35$$

$$-119x = -35 \qquad \text{Solve for } x.$$

$$\frac{-119x}{-119} = \frac{-35}{-119}$$

$$x = \frac{5}{17} \qquad \text{Reduce.}$$

The solution is $\left(\frac{5}{17}, \frac{3}{11}\right)$. These values can be checked in the original equations:

$$34x - 22y = 4 \qquad\qquad 17x - 88y = -19$$

$$34\left(\frac{5}{17}\right) - 22\left(\frac{3}{11}\right) \stackrel{?}{=} 4 \qquad 17\left(\frac{5}{17}\right) - 88\left(\frac{3}{11}\right) \stackrel{?}{=} -19$$

$$10 - 6 = 4 \; ✔ \qquad\qquad 5 - 24 = -19 \; ✔$$

3. Systems of Linear Equations with No Solution or Infinitely Many Solutions

example 5 **Solving a System of Linear Equations**

Solve the system using the addition method.

$$2x - 5y = 10$$

$$\frac{1}{2}x = 1 + \frac{5}{4}y$$

Solution:

$$2x - 5y = 10 \longrightarrow 2x - 5y = 10$$

$$\frac{1}{2}x = 1 + \frac{5}{4}y \longrightarrow \frac{1}{2}x - \frac{5}{4}y = 1 \qquad \textbf{Step 1:} \quad \text{Write the equations in standard form.}$$

Step 2: Multiply both sides of the second equation by 4 to clear fractions.

$$\frac{1}{2}x - \frac{5}{4}y = 1 \longrightarrow 4\left(\frac{1}{2}x - \frac{5}{4}y\right) = 4(1) \longrightarrow 2x - 5y = 4$$

Now the system becomes

$$2x - 5y = 10$$
$$2x - 5y = 4$$

To make either the *x*-coefficients or *y*-coefficients opposites, multiply either equation by −1.

$$2x - 5y = 10 \xrightarrow{\text{Multiply by } -1} -2x + 5y = -10 \qquad \textbf{Step 3:} \text{ Create opposite coefficients.}$$

$$2x - 5y = 4 \longrightarrow \underline{2x - 5y = 4}$$

$$0 = -6 \qquad \textbf{Step 4:} \text{ Add the equations.}$$

Because the equation results in a contradiction, there is no solution, and the system is inconsistent. Writing each line in slope-intercept form verifies that the lines are parallel (Figure 7-8).

$$2x - 5y = 10 \xrightarrow{\text{Slope-intercept form}} y = \frac{2}{5}x - 2$$

$$\frac{1}{2}x = 1 + \frac{5}{4}y \xrightarrow{\text{Slope-intercept form}} y = \frac{2}{5}x - \frac{4}{5}$$

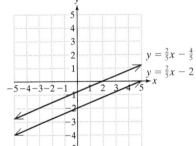

Figure 7-8

---

example 6

Solving a System of Linear Equations

Solve the system by the addition method.

$$3x - y = 4$$
$$2y = 6x - 8$$

Solution:

$$3x - y = 4 \longrightarrow 3x - y = 4 \qquad \textbf{Step 1:} \text{ Write the equations in standard form.}$$

$$2y = 6x - 8 \longrightarrow -6x + 2y = -8 \qquad \textbf{Step 2:} \text{ There are no fractions or decimals.}$$

Notice that the equations differ exactly by a factor of -2, which indicates that these two equations represent the same line. Multiply the first equation by 2 to create opposite coefficients for the variables.

Multiply by 2

$$3x - y = 4 \xrightarrow{\hspace{2cm}} 6x - 2y = 8 \qquad \textbf{Step 3:} \quad \text{Create opposite}$$
$$-6x + 2y = -8 \qquad\qquad \underline{-6x + 2y = -8} \qquad\qquad \text{coefficients.}$$
$$0 = 0 \qquad\qquad \textbf{Step 4:} \quad \text{Add the equations.}$$

Because the resulting equation is an identity, the original equations represent the same line. This can be confirmed by writing each equation in slope-intercept form.

$$3x - y = 4 \longrightarrow -y = -3x + 4 \longrightarrow y = 3x - 4$$
$$-6x + 2y = -8 \longrightarrow 2y = 6x - 8 \longrightarrow y = 3x - 4$$

The solution is the set of all points on the line, or equivalently, $\{(x, y) | y = 3x - 4\}$.

Classroom Activity 7.3B

4. Summary of Methods for Solving Linear Equations in Two Variables

If no method of solving a system of linear equations is specified, you may use the method of your choice. However, we recommend the following guidelines:

1. If one of the equations is written with a variable isolated, the substitution method is a good choice. For example:

$$2x + 5y = 2 \qquad\qquad \text{or} \qquad\qquad y = \frac{1}{3}x - 2$$
$$x = y - 6 \qquad\qquad\qquad\qquad x - 6y = 9$$

2. If both equations are written in standard form, $ax + by = c$, where none of the variables has coefficients of 1 or -1, then the elimination method is a good choice.

$$4x + 5y = 12$$
$$5x + 3y = 15$$

3. If both equations are written in standard form, $ax + by = c$, and at least one variable has a coefficient of 1 or -1, then either the substitution method or the addition method is a good choice.

---

section 7.3 PRACTICE EXERCISES

For Exercises 1–4, check to see if the given ordered pair is a solution to the system. 7.1

1. $x + y = 8$ $(5, 3)$ Yes

    $y = x - 2$

2. $x = y + 1$ $(3, 2)$ No

    $-x + 2y = 0$

3. $3x + 2y = 14$ $(5, -2)$ No

    $5x - 2y = 29$

 See Additional Answers Appendix   Writing   Translating Expression   Geometry   Scientific Calculator   Video

4. $\quad x = 2y - 11 \quad (-9, 1)$ No

$\quad -x + 5y = 23$

For Exercises 5–6, answer as true or false, and explain why.

5. Given the system

$$5x - 4y = 1$$

$$7x - 2y = 5$$

a. To eliminate the y-variable using the addition method, multiply the second equation by 2. False, multiply by -2.

b. To eliminate the x-variable, multiply the first equation by 7 and the second equation by -5. True

6. Given the system

$$3x + 5y = -1$$

$$9x - 8y = -26$$

a. To eliminate the x-variable using the addition method, multiply the first equation by -3. True

b. To eliminate the y-variable, multiply the first equation by 8 and the second equation by -5. False, multiply the second equation by 5.

7. Given the system

$$3x - 4y = 2$$

$$17x + y = 35$$

a. Which variable, x or y, is easier to eliminate using the addition method? Explain. y would be easier.

b. Solve the system using the addition method. $(2, 1)$

8. Given the system

$$-2x + 5y = -15$$

$$6x - 7y = 21$$

a. Which variable, x or y, is easier to eliminate using the addition method? Explain. x would be easier.

b. Solve the system using the addition method. $(0, -3)$

9. In solving a system of equations, suppose you get the statement $0 = 5$. How many solutions will the system have? What can you say about the graphs of these equations? The system will have no solution. The lines are parallel.

10. In solving a system of equations, suppose you get the statement $0 = 0$. How many solutions will the system have? What can you say about the graphs of these equations? There are infinitely many solutions. The lines coincide.

11. In solving a system of equations, suppose you get the statement $3 = 3$. How many solutions will the system have? What can you say about the graphs of these equations? There are infinitely many solutions. The lines coincide.

12. In solving a system of equations, suppose you get the statement $2 = -5$. How many solutions will the system have? What can you say about the graphs of these equations? The system will have no solution. The lines are parallel.

For Exercises 13–32, solve the systems using the addition method.

13. $\quad x + 2y = 8$
$(2, 3)$
$\quad 5x - 2y = 4$

14. $\quad 2x - 3y = 11$
$(4, -1)$
$\quad -4x + 3y = -19$

15. $\quad a + b = 3$
$(5, -2)$
$\quad 3a + b = 13$

16. $\quad -2u + 6v = 10$
$(4, 3)$
$\quad -2u + v = -5$

17. $\quad -3x + y = 1$
$(0, 1)$
$\quad -6x - 2y = -2$

18. $\quad 5p - 2q = 4$
$(2, 3)$
$\quad 3p + q = 9$

19. $\quad 3x - 5y = 13$
$(1, -2)$
$\quad x - 2y = 5$

20. $\quad 7a + 2b = -1$
$(1, -4)$
$\quad 3a - 4b = 19$

21. $\quad -2x + y = -5$
No solution
$\quad 8x - 4y = 12$

22. $\quad x - 3y = 2$
No solution
$\quad -5x + 15y = 10$

23. $\quad x + 2y = 2$
Infinitely many solutions
$\quad -3x - 6y = -6$
$\{(x, y) | y = -\frac{1}{2}x + 1\}$

24. $\quad 4x - 3y = 6$
Infinitely many solutions
$\quad -12x + 9y = -18$
$\{(x, y) | y = \frac{4}{3}x - 2\}$

25. $\quad 3a + 2b = 11$
$(1, 4)$
$\quad 7a - 3b = -5$

26. $\quad 4y + 5z = -2$
$(2, -2)$
$\quad 5y - 3z = 16$

27. $\quad 3x - 5y = 7$
$(-1, -2)$
$\quad 5x - 2y = -1$

28. $\quad 4s + 3t = 9$
$(0, 3)$
$\quad 3s + 4t = 12$

29. $\quad 2(x + 1) = -3y + 9 \quad (2, 1)$

$\quad 3x - 10 = -4y$

30. $\quad -3(x - 2) + 7y = 5 \quad (5, 2)$

$$5y = 2x$$

31. $\quad 4x - 5y = 0$
No solution
$\quad 8(x - 1) = 10y$

32. $\quad y = 2x + 1$
No solution
$\quad -3(2x - y) = 0$

See Additional Answers Appendix Writing Translating Expression Geometry Scientific Calculator Video

For Exercises 33–50, solve the system by either the addition method or the substitution method.

33. $5x - 2y = 4$
    (2, 3)
    $y = -3x + 9$

34. $-x = 8y + 5$
    (−5, 0)
    $4x - 3y = -20$

35. $x + y = 6$ $\left(\frac{7}{2}, \frac{5}{2}\right)$
    $x - y = 1$

36. $x + y = 2$ $\left(\frac{5}{2}, -\frac{1}{2}\right)$
    $x - y = 3$

37. $3x = 5y - 9$ $\left(\frac{1}{3}, 2\right)$
    $2y = 3x + 3$

38. $10x - 5 = 3y$ $\left(\frac{1}{2}, 0\right)$
    $2x - 3y = 1$

39. $y = -5x + 1$ Infinitely many solutions
    $\{(x, y)|y = -5x + 1\}$
    $15x - 3 = -3y$

40. $4x + 5y = -2$ Infinitely many solutions
    $\{(x, y)|y = -\frac{4}{5}x - \frac{2}{5}\}$
    $8x = -10y - 4$

41. $x + 2y = 4$ $\left(\frac{2}{3}, \frac{5}{3}\right)$
    $x - y = -1$

42. $-3x + y = 1$ (0, 1)
    $-6x - 2y = -2$

43. $8x - 16y = 24$
    No solution
    $2x - 4y = 0$

44. $y = -\frac{1}{2}x - 5$
    No solution
    $2x + 4y = -8$

45. $\frac{m}{2} + \frac{n}{5} = \frac{13}{10}$ (1, 4)
    $3(m - n) = m - 10$

46. $\frac{a}{4} - \frac{3b}{2} = \frac{15}{2}$ (0, −5)
    $\frac{1}{5}(a + 2b) = -2$

47. $2(p - 3q) = p + 4$ (4, 0)
    $3p + 8 = 5p - q$

48. $m - 3n = 10$ (4, −2)
    $3(m + 4n) = -12$

49. $9a - 2b = 8$ Infinitely many solutions
    $\{(a, b)|b = \frac{9}{2}a - 4\}$
    $6(3a + 1) = 4b + 22$

50. $a = 5 + 2b$ Infinitely many solutions
    $\{(a, b)|b = \frac{1}{2}a - \frac{5}{2}\}$
    $3(a - 2b) = 15$

For Exercises 51–56, set up a system of linear equations, and solve for the indicated quantities.

51. The sum of two positive numbers is 26. Their difference is 14. Find the numbers.
    The numbers are 20 and 6.

52. The difference of two positive numbers is 2. The sum of the numbers is 36. Find the numbers.
    The numbers are 17 and 19.

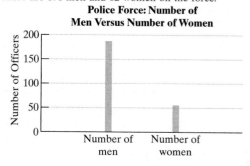 53. Eight times the smaller of two numbers plus 2 times the larger number is 44. Three times the smaller number minus 2 times the larger number is zero. Find the numbers.   The numbers are 4 and 6.

54. Six times the smaller of two numbers minus the larger number is −9. Ten times the smaller number plus five times the larger number is 5. Find the numbers.   The numbers are −1 and 3.

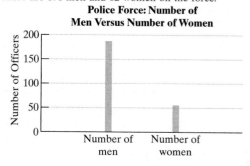 55. The number of calories in a piece of cake is 20 less than 3 times the number of calories in a scoop of ice cream. Together, the cake and ice cream have 460 Calories. How many calories are in each?
    The cake has 340 Calories, and the ice cream has 120 Calories.

56. A police force has 240 officers. If there are 116 more men than women, find the number of men and the number of women on the force.
    There are 178 men and 62 women on the force.

Police Force: Number of Men Versus Number of Women

Figure for Exercise 56

 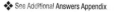

For Exercises 57–60, use the addition method to eliminate the *x*-variable to solve for *y*. Then use the addition
method to eliminate the *y*-variable to solve for *x* (see
Example 4).

57. $2x + 3y = 6$ $\left(\frac{21}{5}, -\frac{4}{5}\right)$ 58. $6x + 6y = 8$ $\left(\frac{7}{9}, \frac{5}{9}\right)$

$x - y = 5$ $9x - 18y = -3$

59. $2x - 5y = 4$ $\left(\frac{8}{9}, -\frac{4}{9}\right)$ 60. $6x - 5y = 7$ $\left(\frac{7}{16}, -\frac{7}{8}\right)$

$3x - 3y = 4$ $4x - 6y = 7$

62. The solution to the following system of linear
equations is $(1, 2)$. Find A and B.

$$Ax + 3y = 8 \quad A = 2, B = -4$$

$$x + By = -7$$

63. The solution to the following system of linear
equations is $(-3, 4)$. Find A and B.

$$4x + Ay = -32 \quad A = -5, B = 2$$

$$Bx + 6y = 18$$

EXPANDING YOUR SKILLS

 61. Explain why a system of linear equations cannot
have exactly two solutions.   A line would have to
"bend" to have two points of intersection. This is not possible.

---

chapter 7   MIDCHAPTER REVIEW

 1. Given the system

$$x = -2y + 5$$

$$2x - 4y = 10$$

a. Which method would you choose to solve
the system, the substitution method or the
addition method? Explain your choice.

b. Solve the system. $(5, 0)$

 2. Given the system

$$3x - 2y = 22$$

$$5x + 2y = 10$$

a. Which method would you choose to solve
the system, the substitution method or the
addition method? Explain your choice.

b. Solve the system. $(4, -5)$

 3. Given the system

$$3x - 6y = 30$$

$$2x + 3y = -22$$

a. Which method would you choose to solve
the system, the substitution method or the
addition method? Explain your choice.

b. Solve the system. $(-2, -6)$

 4. Given the system

$$-2x + y = -14$$

$$4x - 2y = 28$$

a. Which method would you choose to solve
the system, the substitution method or the
addition method? Explain your choice.

b. Solve the system.
Infinitely many solutions; $\{(x, y)|y = 2x - 14\}$

 5. Given the system

$$y = 0.4x - 0.3$$

$$-4x + 10y = 20$$

a. Which method would you choose to solve
the system, the substitution method or the
addition method? Explain your choice.

b. Solve the system.   No solution

For Exercises 6–15, first simplify the equation by clearing parentheses and collecting like terms. Then choose
a method to solve the system.

6. $u - v = 3$ $\left(\frac{1}{2}, -\frac{5}{2}\right)$ 7. $5v + 6w = -11$
$6u - 4v = 13$ $(-1, -1)$
                                            $3v + w = -4$

8. $-8y = x - 1$ $\left(-5, \dfrac{3}{4}\right)$

$-x + 2y = \dfrac{13}{2}$

9. $1 + x = 10y$

$-x + 3y = -\dfrac{9}{5}$

$\left(3, \dfrac{2}{5}\right)$

10. $3x + y - 7 = x - 4$ $(5, -7)$

$3x - 4y + 4 = -6y + 5$

11. $7y - 8y - 3 = -3x + 4$ $(2, -1)$

$5(2x - y) - 12 = 13$

12. $\dfrac{1}{2}x + \dfrac{1}{6}y = \dfrac{1}{3}$ No solution

$x = \dfrac{1}{6} - \dfrac{1}{3}y$

13. $\dfrac{1}{3}x + \dfrac{1}{2}y = \dfrac{2}{3}$ $(2, 0)$

$y = \dfrac{2}{3}x - \dfrac{4}{3}$

14. $x + y = 3200$ $(2200, 1000)$

$0.06x + 0.04y = 172$

15. $x + y = 4500$ $(3300, 1200)$

$0.07x + 0.05y = 291$

For Exercises 16–21, set up a system of linear equations and solve for the indicated quantities.

16. One number is 4 more than another number. The sum of the numbers is 56. Find the numbers. The numbers are 30 and 26.

17. The sum of two numbers is 75. One number is 3 less than twice the other number. Find the numbers. The numbers are 49 and 26.

18. The sum of Denisha's age and her sister's age is 48. If Denisha is 4 years older than her sister, find each of their ages. Denisha is 26 and her sister is 22.

19. In a right triangle, one acute angle is 10° less than 4 times the other acute angle. Find the measure of each acute angle. The angles are 70° and 20°.

20. Two angles are complementary. The difference of the larger angle and the smaller angle is 14°. Find the measure of each angle. The angles are 52° and 38°.

21. Two angles are supplementary. The measure of one angle is 11 times the measure of the other angle. Find the measure of each angle. The angles are 165° and 15°.

Concepts

1. Applications Involving Cost

2. Applications Involving Principal and Interest

3. Applications Involving Mixtures

4. Applications Involving Distance, Rate, and Time

section

7.4 Applications of Linear Equations in Two Variables

1. Applications Involving Cost

In Sections 2.5 and 2.6, we solved several applied problems by setting up a linear equation in one variable. When solving an application that involves two unknowns, sometimes it is convenient to use a system of linear equations in two variables.

example 1 Using a System of Linear Equations Involving Cost

At a movie theater a couple buys one large popcorn and two drinks for $5.75. A group of teenagers buys two large popcorns and five drinks for $13.00. Find the cost of one large popcorn and the price of one drink.

Solution:

Let x represent the cost of one large popcorn.
Let y represent the cost of one drink.

$$\left(\begin{array}{c}\text{Cost of 1}\\\text{large popcorn}\end{array}\right) + \left(\begin{array}{c}\text{cost of 2}\\\text{drinks}\end{array}\right) = \left(\begin{array}{c}\text{total}\\\text{cost}\end{array}\right) \longrightarrow x + 2y = 5.75$$

$$\left(\begin{array}{c}\text{Cost of 2}\\\text{large popcorns}\end{array}\right) + \left(\begin{array}{c}\text{cost of 5}\\\text{drinks}\end{array}\right) = \left(\begin{array}{c}\text{total}\\\text{cost}\end{array}\right) \longrightarrow 2x + 5y = 13.00$$

To solve this system, we may either use the substitution method or the addition method. We will use the substitution method by solving for x in the first equation.

$x + 2y = 5.75 \longrightarrow x = -2y + 5.75$	Isolate x in the first equation.
$2x + 5y = 13.00$	

$2(-2y + 5.75) + 5y = 13.00$	Substitute $x = -2y + 5.75$ into the other equation.
$-4y + 11.50 + 5y = 13.00$	
$y + 11.50 = 13.00$	Solve for y.
$y = 13.00 - 11.50$	
$y = 1.50$	

$x = -2y + 5.75$	
$x = -2(1.50) + 5.75$	Substitute $y = 1.50$ into the expression $x = -2y + 5.75$.
$x = -3.00 + 5.75$	
$x = 2.75$	

The cost of one large popcorn is $2.75 and the cost of one drink is $1.50.

Check by verifying that the solutions meet the specified conditions.

1 popcorn + 2 drinks = 1($2.75) + 2($1.50) = $5.75 ✔

2 popcorns + 5 drinks = 2($2.75) + 5($1.50) = $13.00 ✔

2. Applications Involving Principal and Interest

In Section 2.6 we solved problems in which money is invested in two accounts at different rates. We determined how the principal was to be divided to produce a specified amount of interest. Remember, that when investing there are two sources of money: the principal amount invested and the interest earned. Recall that simple interest is computed by the formula, $I = Prt$. If the time of the investment is 1 year, we have $I = Pr$.

example 2   Using a System of Linear Equations Involving Investments

Joanne has a total of $6000 to deposit in two accounts. One account earns 3.5% simple interest and the other earns 2.5% simple interest. If the total amount of interest at the end of 1 year is $195, find the amount she deposited in each account.

Solution:

Let x represent the principal deposited in the 2.5% account.
Let y represent the principal deposited in the 3.5% account.

	2.5% Account	3.5% Account	Total
Principal	x	y	6000
Interest	$0.025x$	$0.035y$	195

Each row of the table yields an equation in x and y:

$$\begin{pmatrix} \text{Principal} \\ \text{invested} \\ \text{at } 2.5\% \end{pmatrix} + \begin{pmatrix} \text{principal} \\ \text{invested} \\ \text{at } 3.5\% \end{pmatrix} = \begin{pmatrix} \text{total} \\ \text{principal} \end{pmatrix} \longrightarrow x + y = 6000$$

$$\begin{pmatrix} \text{Interest} \\ \text{earned} \\ \text{at } 2.5\% \end{pmatrix} + \begin{pmatrix} \text{interest} \\ \text{earned} \\ \text{at } 3.5\% \end{pmatrix} = \begin{pmatrix} \text{total} \\ \text{interest} \end{pmatrix} \longrightarrow 0.025x + 0.035y = 195$$

We will choose the addition method to solve the system of equations. First multiply the second equation by 1000 to clear decimals.

Multiply by -25

$$x + y = 6000 \longrightarrow \quad x + y = 6000 \quad \longrightarrow -25x - 25y = -150,000$$
$$0.025x + 0.035y = 195 \longrightarrow 25x + 35y = 195,000 \longrightarrow \underline{\quad 25x + 35y = \quad 195,000}$$

Multiply by 1000

$$10y = 45,000$$

$$10y = 45,000 \qquad \text{After eliminating the } x\text{-variable, solve for } y.$$

$$\frac{10y}{10} = \frac{45,000}{10}$$

$$y = 4500 \qquad \text{The amount invested in the 3.5\% account is \$4500.}$$

$$x + y = 6000 \qquad \text{Substitute } y = 4500 \text{ into the equation } x + y = 6000.$$

$$x + 4500 = 6000$$

$$x = 1500 \qquad \text{The amount invested in the 2.5\% account is \$1500.}$$

Joanne deposited $1500 in the 2.5% account and $4500 in the 3.5% account.

To check the solution, verify that the conditions of the problem have been met.

1. The sum of $1500 and $4500 is $6000 as desired. ✔

2. The interest earned on $1500 at 2.5% is: 0.025($1500) = $37.5
    The interest earned on $4500 at 3.5% is: 0.035($4500) = $157.5
    Total interest: $195.00 ✔

3. Applications Involving Mixtures

example 3

Using a System of Linear Equations in a Mixture Application

A 10% alcohol solution is mixed with a 40% alcohol solution to produce 30 L of a 20% alcohol solution. Find the number of liters of 10% solution and the number of liters of 40% solution required for this mixture.

Classroom Activity 7.4A

Solution:

Each solution contains a percentage of alcohol plus some other mixing agent such as water. Before we set up a system of equations to model this situation, it is helpful to have background understanding of the problem. In Figure 7-9, the liquid depicted in blue is pure alcohol and the liquid shown in gray is the mixing agent (such as water). Together these liquids form a solution. (Realistically the mixture may not separate as shown, but this image may be helpful for your understanding.)

Let x represent the number of liters of 10% solution.
Let y represent the number of liters of 40% solution.

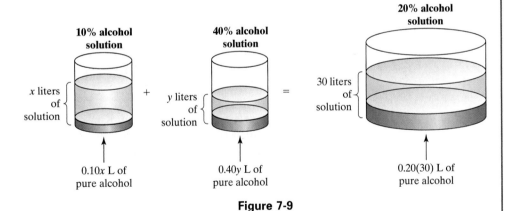

Figure 7-9

The information given in the statement of the problem can be organized in a chart.

	10% Alcohol	40% Alcohol	20% Alcohol
Number of liters of solution	x	y	30
Number of liters of pure alcohol	0.10x	0.40y	0.20(30) = 6

From the first row, we have

$$\begin{pmatrix} \text{Amount of} \\ 10\% \text{ solution} \end{pmatrix} + \begin{pmatrix} \text{amount of} \\ 40\% \text{ solution} \end{pmatrix} = \begin{pmatrix} \text{total amount} \\ \text{of } 20\% \text{ solution} \end{pmatrix} \longrightarrow x + y = 30$$

From the second row, we have

$$\begin{pmatrix} \text{Amount of} \\ \text{pure alcohol} \\ \text{solution 1} \end{pmatrix} + \begin{pmatrix} \text{amount of} \\ \text{pure alcohol} \\ \text{solution 2} \end{pmatrix} = \begin{pmatrix} \text{total amount of} \\ \text{pure alcohol} \end{pmatrix} \longrightarrow 0.10x + 0.40y = 6$$

We will solve the system with the addition method by first clearing decimals.

$$
\begin{array}{ccccc}
x + y = 30 & \longrightarrow & x + y = 30 & \xrightarrow{\text{Multiply by } -1} & -x - y = -30 \\
0.10x + 0.40y = 6 & \xrightarrow{\text{Multiply by 10}} & x + 4y = 60 & \longrightarrow & x + 4y = 60 \\
& & & & \overline{3y = 30}
\end{array}
$$

$3y = 30$ \qquad After eliminating the x-variable, solve for y.

$y = 10$ \qquad 10 L of 40% solution is needed.

$x + y = 30$ \qquad Substitute $y = 10$ into either of the original equations.

$x + (10) = 30$

$x = 20$ \qquad 20 L of 10% solution is needed.

10 L of 40% solution must be mixed with 20 L of 10% solution.

4. Applications Involving Distance, Rate, and Time

Using a System of Linear Equations in a Distance, Rate, Time Application

A plane travels with a tail wind from Kansas City, Missouri, to Denver, Colorado, a distance of 600 miles in 2 hours. The return trip against a head wind takes 3 hours. Find the speed of the plane in still air and find the speed of the wind.

Solution:

Let x represent the speed of the plane in still air.
Let y represent the speed of the wind.

Notice that when the plane travels with the wind, the net speed is $x + y$. When the plane travels against the wind, the net speed is $x - y$.

The information given in the problem can be organized in a chart.

	Distance	Rate	Time
With a tail wind	600	$x + y$	2
Against a head wind	600	$x - y$	3

To set up two equations in x and y, recall that $d = rt$.

From the first row, we have

$$\begin{pmatrix} \text{Distance} \\ \text{with the wind} \end{pmatrix} = \begin{pmatrix} \text{rate with} \\ \text{the wind} \end{pmatrix} \begin{pmatrix} \text{time traveled} \\ \text{with the wind} \end{pmatrix} \longrightarrow 600 = (x + y) \cdot 2$$

From the second row, we have

$$\begin{pmatrix} \text{Distance} \\ \text{against the wind} \end{pmatrix} = \begin{pmatrix} \text{rate against} \\ \text{the wind} \end{pmatrix} \begin{pmatrix} \text{time traveled} \\ \text{against the wind} \end{pmatrix} \longrightarrow 600 = (x - y) \cdot 3$$

Using the distributive property to clear parentheses, produces the following system:

$$2x + 2y = 600$$
$$3x - 3y = 600$$

The coefficients of the y-variable can be changed to 6 and -6 by multiplying the first equation by 3 and the second equation by 2.

$$
\begin{array}{l}
2x + 2y = 600 \xrightarrow{\text{Multiply by 3}} 6x + 6y = 1800 \\
3x - 3y = 600 \xrightarrow{\text{Multiply by 2}} \underline{6x - 6y = 1200} \\
\hphantom{3x - 3y = 600 \xrightarrow{\text{Multiply by 2}} } 12x \hphantom{ - 6y} = 3000
\end{array}
$$

$$12x = 3000$$

$$\frac{12x}{12} = \frac{3000}{12}$$

$$x = 250 \qquad \text{The speed of the plane in still air is 250 mph.}$$

Tip: To create opposite coefficients on the y-variables, we could have divided the first equation by 2 and divided the second equation by 3:

$$
\begin{array}{l}
2x + 2y = 600 \xrightarrow{\text{Divide by 2}} x + y = 300 \\
3x - 3y = 600 \xrightarrow{\text{Divide by 3}} \underline{x - y = 200} \\
\hphantom{3x - 3y = 600 \xrightarrow{\text{Divide by 3}} } 2x \hphantom{ - y} = 500
\end{array}
$$

$$x = 250$$

$$2x + 2y = 600 \qquad \text{Substitute } x = 250 \text{ into the first equation.}$$

$$2(250) + 2y = 600$$

$$500 + 2y = 600$$

$$2y = 100$$

$$y = 50 \qquad \text{The speed of the wind is 50 mph.}$$

The speed of the plane in still air is 250 mph. The speed of the wind is 50 mph.

section 7.4 PRACTICE EXERCISES

For Exercises 1–4, solve each system by three different methods: 7.1–7.3

    a. Graphing method

    b. Substitution method

    c. Addition method

1. $-2x + y = 6$
    $2x + y = 2$ $(-1, 4)$

2. $x - y = 2$
    $x + y = 6$ $(4, 2)$

3. $y = -2x + 6$
    $4x - 2y = 8$ $\left(\frac{5}{2}, 1\right)$

4. $2x = y + 4$
    $4x = 2y + 8$ Infinitely many solutions; $\{(x, y) \mid y = 2x - 4\}$

For Exercises 5–36, set up a system of linear equations in two variables to solve for the unknown quantities.

5. Two angles are complementary. One angle is 10° less than 9 times the other. Find the measure of each angle. 7.2 The angles are 80° and 10°.

6. Two angles are supplementary. One angle is 9° more than twice the other angle. Find the measure of each angle. 7.2 The angles are 123° and 57°.

7. In the 1994 Super Bowl, the Dallas Cowboys scored four more points than twice the number of points scored by the Buffalo Bills. If the total number of points scored by both teams was 43, find the number of points scored by each individual team. Dallas scored 30 points, and Buffalo scored 13 points.

8. In the 1973 Super Bowl, the Miami Dolphins scored twice as many points as the Washington Redskins. If the total number of points scored by both teams was 21, find the number of points scored by each individual team. Miami scored 14 points, and Washington scored 7 points.

9. Kent bought three tapes and 2 CDs for $62.50. Demond bought one tape and four CDs for $72.50. Find the cost of one tape and the cost of one CD. Tapes are $10.50 each, and CDs are $15.50 each.

10. Tanya bought three adult tickets and one child's ticket to a movie for $23.00. Li bought two adult tickets and five children's tickets for $30.50. Find the cost of one adult ticket and the cost of one children's ticket. Adult tickets cost $6.50, and children's tickets cost $3.50.

11. Linda bought 100 shares of a technology stock and 200 shares of a mutual fund for $3800. Her sister, Sandie, bought 300 shares of technology stock and 50 shares of a mutual fund for $5350. Find the cost per share of the technology stock, and the cost per share of the mutual fund. Technology stock costs $16 per share, and the mutual fund costs $11 per share.

12. Two videos and three DVDs can be rented for $19.15. Four videos and one DVD can be rented for $17.35. Find the cost to rent one video and the cost to rent one DVD. It costs $3.29 to rent a video and $4.19 to rent a DVD.

13. Shanelle invested $10,000 and at the end of 1 year she received $805 in interest. She invested part of the money in an account earning 10% simple interest and the remaining money in an account earning 7% simple interest. How much did she invest in each account?

	10% Account	7% Account	Total
Principal invested			
Interest earned			

Table for Exercise 13

Shanelle invested $3500 in the 10% account and $6500 in the 7% account.

14. $2000 more is invested in an account earning 12% simple interest than in an account earning 8% simple interest. If the total interest at the end of 1 year is $1240, how much was invested in each account?

	12% Account	8% Account	Total
Principal invested			
Interest earned			

Table for Exercise 14

$7000 was invested in the 12% account and $5000 was invested in the 8% account.

15. Janise invested twice as much money in an account earning 7% simple interest as she did in an account earning 4% simple interest. If the total interest at the end of 1 year was $720, how much did Janise invest in each account? Janise invested $8000 in the 7% account and $4000 in the 4% account.

16. Mario invested 4 times as much money in an account earning 9% simple interest as he did in an account earning 5% simple interest. If he received $1435 in total interest after 1 year, how much was invested in each account? Mario invested $14,000 in the 9% account and $3500 in the 5% account.

17. How much 50% disinfectant solution must be mixed with a 40% disinfectant solution to produce 25 gal of a 46% disinfectant solution?

	50% Mixture	40% Mixture	46% Mixture
Amount of solution			
Amount of disinfectant			

Table for Exercise 17

15 gallons of the 50% mixture should be mixed with 10 gal of the 40% mixture.

18. How many gallons of 20% antifreeze solution and a 10% antifreeze solution must be mixed to obtain 40 gal of a 16% antifreeze solution?

	20% Mixture	10% Mixture	16% Mixture
Amount of solution			
Amount of antifreeze			

Table for Exercise 18

24 gallons of the 20% mixture should be mixed with 16 gal of the 10% mixture.

19. How much 45% disinfectant solution must be mixed with a 30% disinfectant solution to produce 20 gal of a 39% disinfectant solution? ❖

20. How many gallons of a 25% antifreeze solution and a 15% antifreeze solution must be mixed to obtain 15 gal of a 23% antifreeze solution? ❖

21. It takes a boat 2 h to go 16 miles downstream with the current, and 4 h to return against the current. Find the speed of the boat in still water and the speed of the current.

	Distance	Rate	Time
Downstream			
Return			

Table for Exercise 21

The speed of the boat in still water is 6 mph, and the speed of the current is 2 mph.

22. A boat takes $1\frac{1}{2}$ h to go 12 miles upstream against the current. It can go 24 miles downstream with the current in the same amount of time. Find the speed of the current and the speed of the boat in still water.

	Distance	Rate	Time
Downstream			
Upstream			

Table for Exercise 22

The speed of the boat is 12 mph, and the speed of the current is 4 mph.

23. A plane can fly 800 miles with the wind in $2\frac{1}{2}$ h. It takes the same amount of time to fly 700 miles against the wind. What is the speed of the plane in still air and the speed of the wind? The speed of the plane in still air is 300 mph, and the wind is 20 mph.

24. A plane flies 600 miles with the wind in $2\frac{1}{2}$ h. The return trip takes $3\frac{1}{3}$ h. What is the speed of the wind and the speed of the plane in still air? The speed of the plane in still air is 210 mph, and the wind is 30 mph.

25. Debi has $2.80 in a collection of dimes and nickels. The number of nickels is five more than the number of dimes. Find the number of each type of coin.

	Dimes	Nickels	Total
Number of coins			
Value of coins			

Table for Exercise 25

There are 17 dimes and 22 nickels.

26. A child is collecting state quarters and new $1 coins. If she has a total of 25 coins, and the number of quarters is nine more than the number of dollar coins, how many of each type of coin does she have? She has eight $1 coins and 17 quarters.

27. In the 1961–1962 NBA basketball season, Wilt Chamberlain of the Philadelphia Warriors made 2432 baskets. Some of the baskets were free-throws (worth 1 point each) and some were field goals

(worth 2 points each). The number of field goals was 762 more than the number of free-throws.

a. How many field goals did he make and how many free-throws did he make? 835 free-throws and 1597 field goals

b. What was the total number of points scored? 4029 points

c. If Wilt Chamberlain played 80 games during this season, what was the average number of points per game? Approximately 50 points per game

28. In the 1971–1972 NBA basketball season, Kareem Abdul-Jabbar of the Milwaukee Bucks made 1663 baskets. Some of the baskets were free-throws (worth 1 point each) and some were field goals (worth 2 points each). The number of field goals he scored was 151 more than twice the number of free-throws.

a. How many field goals did he make and how many free-throws did he make? 504 free-throws and 1159 field goals

b. What was the total number of points scored? 2822 points

c. If Kareem Abdul-Jabbar played 81 games during this season, what was the average number of points per game? Approximately 35 points per game

29. A small plane can fly 350 miles with a tailwind in $1\frac{3}{4}$ hours. In the same amount of time the same plane can travel only 210 miles with a headwind. What is the speed of the plane in still air and the speed of the wind? The speed of the plane in still air is 160 mph, and the wind is 40 mph.

30. A plane takes 2 h to travel 1000 miles with the wind. It can travel only 880 miles against the wind in the same time. Find the speed of the wind and the speed of the plane in still air. The speed of the plane in still air is 470 mph, and the wind is 30 mph.

31. At the holidays, Erica likes to sell a candy/nut mixture to her neighbors. She wants to combine candy that costs $1.80 per pound with nuts that costs $1.20 per pound. If Erica needs 20 lb of mixture that will sell for $1.56 per pound, how many pounds of candy and how many pounds of nuts should she use? 12 pounds of candy should be mixed with 8 lb of nuts.

32. Mary Lee's natural food store sells a combination of teas. The most popular is a mixture of a tea that sells for $3.00 per pound with one that sells for $4.00

per pound. If she needs 40 lb of tea that will sell for $3.65 per pound, how many pounds of each tea should she use? 14 pounds of $3 per pound tea should be mixed with 26 lb of $4 per pound tea.

33. A total of $60,000 is invested in two accounts, one that earns 5.5% simple interest, and one that earns 6.5% simple interest. If the total interest at the end of 1 year is $3750, find the amount invested in each account. $15,000 is invested in the 5.5% account, and $45,000 is invested in the 6.5% account.

34. Jacques borrows a total of $15,000. Part of the money is borrowed from a bank that charges 12% simple interest per year. Jacques borrows the remaining part of the money from his sister and promises to pay her 7% simple interest per year. If Jacques total interest for the year is $1475, find the amount he borrowed from each source. Jacques borrowed $8500 at 12% and $6500 at 7%.

35. Miracle-Gro All-Purpose Plant Food contains 15% nitrogen. Green Light Super Bloom contains 12% nitrogen. How much Miracle-Gro and how much Green Light fertilizer must be mixed to obtain 60 oz of a mixture that is 13% nitrogen? 20 oz of Miracle-Gro should be mixed with 40 oz of Green Light.

36. A textile manufacturer wants to combine a mixture of 20% dye with a mixture that is 50% dye to form 200 gal of a mixture that is 42.5% dye. How much of the 20% and 50% dye mixtures should he use? 50 gal of 20% dye should be mixed with 150 gal of 50% dye.

EXPANDING YOUR SKILLS

37. In a survey conducted among 500 college students, 340 said that the campus lacked adequate lighting. If $\frac{4}{5}$ of the women and $\frac{1}{2}$ of the men said that they thought the campus lacked adequate lighting, how many men and how many women were in the survey? There were 300 women and 200 men in the survey.

38. A thousand people were surveyed in southern California, and 445 said that they worked out at least three times a week. If $\frac{1}{2}$ of the women and $\frac{3}{8}$ of the men said that they worked out at least three times a week, how many men and how many women were in the survey? There were 560 women and 440 men in the survey.

39. During a 1-hour television program, there were 22 commercials. Some commercials were 15 seconds (s), and some were 30 s long. Find the number of 15-s commercials and the number of 30-s commercials if the total playing time for commercials was 9.5 min. There are six 15-s commercials and sixteen 30-s commercials.

 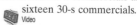

Concepts

section

7.5

CONNECTIONS TO GRAPHING: LINEAR INEQUALITIES IN TWO VARIABLES

1. Introduction to Linear Inequalities in Two Variables

A **linear inequality in two variables** x and y is an inequality that can be written in one of the following forms: $ax + by < c$, $ax + by > c$, $ax + by \leq c$, or $ax + by \geq c$.

A solution to a linear inequality in two variables is an ordered pair that makes the inequality true. For example, solutions to the inequality $x + y < 3$ are ordered pairs (x, y) such that the sum of the x- and y-coordinates is less than 3. There are an infinite number of solutions to this inequality, and therefore it is convenient to express the solution set as a graph.

To graph a linear inequality in two variables, we will use a process called the **test point method**. The basis of the test point method is to graph the related equation, which in this case will be a line in the xy-plane. Then test ordered pairs above and below the line to determine which region in the xy-plane represents the solution set. The process is demonstrated in the next example.

---

example 1

Graphing a Linear Inequality in Two Variables

Graph the inequality: $2x + y \leq 3$

Solution:

$2x + y \leq 3 \xrightarrow[\substack{\text{Set up the} \\ \text{related equation.}}]{} 2x + y = 3$

Step 1: Set up the related equation.

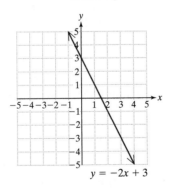

Step 2: Graph the related equation (Figure 7-10). The slope-intercept form is $y = -2x + 3$.

$y = -2x + 3$

Figure 7-10

Step 3: Notice that Figure 7-10 shows the xy-plane divided into three regions: the region below the line, the region above the line, and the line itself. To find the solution set, we will select a test point from each region. Then substitute the point into the original inequality to determine if it makes the inequality a true statement. Select a test point above the line such as $(4, 3)$ and select a test point below the line such as $(0, 0)$.

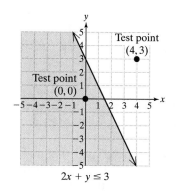

$2x + y \leq 3$

Figure 7-11

Test point above: $(4, 3)$

$$2x + y \leq 3$$
$$2(4) + (3) \overset{?}{\leq} 3$$
$$8 + 3 \overset{?}{\leq} 3$$
$$11 \overset{?}{\leq} 3 \text{ False}$$

Test point below: $(0, 0)$

$$2x + y \leq 3$$
$$2(0) + (0) \overset{?}{\leq} 3$$
$$0 + 0 \overset{?}{\leq} 3$$
$$0 \overset{?}{\leq} 3 \text{ True } \checkmark$$

Shade below the line.

A test point below the line makes the inequality a true statement, the region below the line is therefore contained in the solution set. Next, we consider the line itself. Because the original inequality contains equality ($\leq$), all points on the line $2x + y = 3$ must be solutions. Hence the line itself is also included. The solution set is all points below and including the line $2x + y = 3$ (Figure 7-11).

Tip: A test point on the line itself is not needed to determine whether the line is included in the solution set. Instead, we know that if the inequality sign contains equality ($\leq$ or $\geq$) then the line *is included*. If the inequality is strict ($<$ or $>$), then the line is *not included* in the solution set.

To test the regions above and below the line, be sure to select points *not* on the line itself. It is also important to understand that the test points are *arbitrary*. Any points above and below the line may be used.

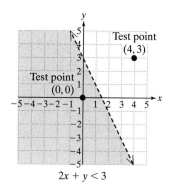

$2x + y < 3$

Figure 7-12

In Example 1, the line $2x + y = 3$ is included in the solution set. We show that this boundary is included in the solution set by drawing the line as a solid line.

If the original inequality had been a strict inequality, $2x + y < 3$, or $2x + y > 3$, then a dashed line would be used to signify that the boundary is *not* included.

The "boundary" of the inequality $2x + y < 3$ is graphed with a dashed line (Figure 7-12).

2.   Test Point Method

The test point method to graph linear inequalities in two variables is summarized as follows:

Test Point Method: Summary

1. Set-up the related equation.
2. Graph the related equation from Step 1. The equation will be a boundary line in the *xy*-plane.
   - If the original inequality is a strict inequality, $<$ or $>$, then the line is not part of the solution set. Therefore, graph the boundary as a *dashed line.*
   - If the original inequality uses $\leq$ or $\geq$ then the line is part of the solution set. Graph the boundary as a *solid line.*
3. From each region above and below the line, select an ordered pair as a test point and substitute it into the original inequality.
   - If a test point makes the inequality true, then that region is part of the solution set.

3. Graphing Linear Inequalities in Two Variables

example 2 **Graphing a Linear Inequality in Two Variables**

Graph the inequality: $4x - 2y > 6$.

Classroom Activity 7.5A

Solution:

$$4x - 2y > 6 \xrightarrow[\substack{\text{Set up the} \\ \text{related equation.}}]{} 4x - 2y = 6$$
Step 1: Set up the related equation.

Step 2: Graph the related equation (Figure 7-13). Use a dashed line because the original inequality is $>$. (The line is not included in the solution set.)

Slope-intercept form can be used to graph the line:

$$4x - 2y = 6$$
$$-2y = -4x + 6$$
$$\frac{-2y}{-2} = \frac{-4x}{-2} + \frac{6}{-2}$$
$$y = 2x - 3$$

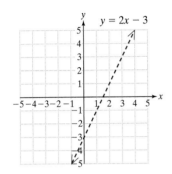

Figure 7-13

Tip: To graph the related equation, you may either create a table of points, or you may use the slope-intercept form of the line.

Table of points   Slope-intercept form

$y = 2x - 3$

x	y
0	-3
$\frac{3}{2}$	0
3	3

Step 3: Select test points above and below the line. For the point above, we have arbitrarily selected $(0, 0)$. For the point below, we have selected $(2, -1)$. Test each point in the original inequality (Figure 7-14).

Test point above: $(0, 0)$
$$4x - 2y > 6$$
$$4(0) - 2(0) \overset{?}{>} 6$$
$$0 - 0 \overset{?}{>} 6$$
$$0 \overset{?}{>} 6 \text{ False}$$

Test point below: $(2, -1)$
$$4x - 2y > 6$$
$$4(2) - 2(-1) \overset{?}{>} 6$$
$$8 + 2 \overset{?}{>} 6$$
$$10 > 6 \text{ True } ✔$$

Shade below the line.

All points strictly *below* the line are solutions to the inequality $4x - 2y > 6$.

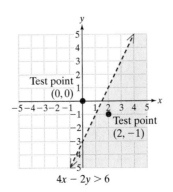

$4x - 2y > 6$

Figure 7-14

Tip: An inequality can also be graphed by first solving the inequality for y. Then,

*   Shade *below* the line if the inequality is of the form $y < mx + b$ or $y \le mx + b$.
*   Shade *above* the line if the inequality is of the form $y > mx + b$ or $y \ge mx + b$.

From Example 2, we have

$$4x - 2y > 6$$

$$-2y > -4x + 6$$

$$\frac{-2y}{-2} < \frac{-4x}{-2} + \frac{6}{-2} \qquad \begin{array}{l}\text{Reverse the} \\ \text{inequality sign.}\end{array}$$

$$y < 2x - 3 \qquad \text{Shade below the line.}$$

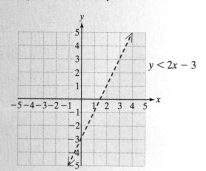

---

example 3

Graphing a Linear Inequality in Two Variables

Graph the inequality: $2x > -4$

Solution:

$$2x > -4 \longrightarrow 2x = -4 \qquad \textbf{Step 1:} \quad \text{Set up the related equation.}$$

$$x = -2 \qquad \textbf{Step 2:} \quad \text{Graph the line. This is the equation of the vertical line, } x = -2 \text{ (Figure 7-15). Because the original inequality is strict } (>), \text{ a dashed line is used to indicate that the line itself is not part of the solution set.}$$

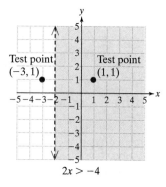

Figure 7-15

Step 3: Select test points on the right and left sides of the line. On the left side we will choose $(-3, 1)$, and on the right side we will choose $(1, 1)$ (Figure 7-16). Test each point in the original inequality.

Test point left: $(-3, 1)$

$$2x > -4$$
$$2(-3) \overset{?}{>} -4$$
$$-6 \overset{?}{>} -4 \text{ False}$$

Test point right: $(1, 1)$

$$2x > -4$$
$$2(1) \overset{?}{>} -4$$
$$2 \overset{?}{>} -4 \text{ True } \checkmark$$

Shade the region to the right.

The solution set is the set of all ordered pairs strictly to the right of the line $x = -2$.

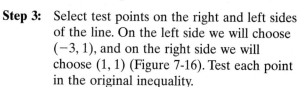

Figure 7-16

4. Systems of Linear Inequalities in Two Variables

Sometimes more than one inequality can be used to define a region in the xy-plane.

---

example 4   **Sketching a System of Linear Inequalities in Two Variables**

Sketch the region bounded by the inequalities.

$$x > 0$$
$$y > 0$$
$$x + y < 5$$

Solution:

The inequality $x > 0$ represents the set of points to the right of the y-axis.

The inequality $y > 0$ represents the set of points above the x-axis.

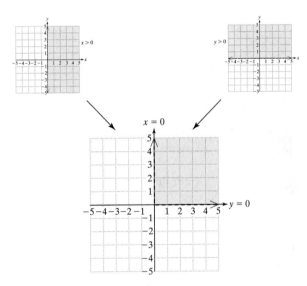

Figure 7-17

The region bounded by the first two inequalities is the first quadrant (Figure 7-17).

The third inequality can be sketched by using the test point method.

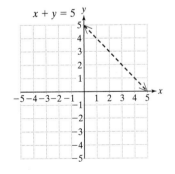

Figure 7-18

$x + y < 5 \longrightarrow x + y = 5$ **Step 1:** Set up the related equation.

$\qquad\qquad y = -x + 5$ **Step 2:** Graph the related equation using a dashed line (Figure 7-18).

Step 3: Select test points above and below the line. For the point above, we will choose $(4, 4)$. For the point below, we will choose $(1, 1)$. Test each point in the original inequality.

Test point above: $(4, 4)$ **Test point below:** $(1, 1)$

$$x + y < 5$$
$$(4) + (4) \overset{?}{<} 5$$
$$8 \overset{?}{<} 5 \text{ False}$$

$$x + y < 5$$
$$(1) + (1) \overset{?}{<} 5$$
$$2 \overset{?}{<} 5 \text{ True } ✔$$

Shade below the line.

The region bounded by all three inequalities is the region in the first quadrant below the line $x + y = 5$ (Figure 7-19).

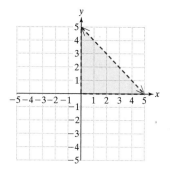

Figure 7-19

---

example 5

Sketching a System of Linear Inequalities

Graph the region bounded by the inequalities.

$$y \geq \tfrac{1}{2}x + 1$$
$$x + y < 1$$

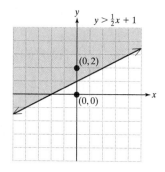

Figure 7-20

Solution:

Set up the related equations and write each line in slope-intercept form to graph the lines (Figures 7-20 and 7-21).

$$y \geq \frac{1}{2}x + 1 \longrightarrow y = \frac{1}{2}x + 1$$

Test point above: $(0, 2)$ **Test point below:** $(0, 0)$

$$2 \overset{?}{\geq} \frac{1}{2}(0) + 1$$
$$2 \overset{?}{\geq} 1 \qquad \text{True } ✔$$

$$0 \overset{?}{\geq} \frac{1}{2}(0) + 1$$
$$0 \overset{?}{\geq} 1 \qquad \text{False}$$

Shade above the line.

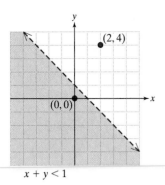

$x + y < 1$

Figure 7-21

$x + y < 1 \longrightarrow x + y = 1$

$$y = -x + 1$$

Test point above: $(2, 4)$ **Test point below:** $(0, 0)$

$4 \overset{?}{<} -(2) + 1$ $0 \overset{?}{<} -(0) + 1$

$4 \overset{?}{<} -1$ False $0 \overset{?}{<} 1$ True ✔

Shade below the line.

The region bounded by the inequalities is the region above the line $y = \frac{1}{2}x + 1$ and below the line $y = -x + 1$ (Figure 7-22).

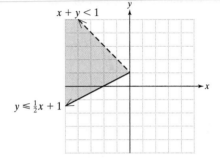

Figure 7-22

---

section 7.5   PRACTICE EXERCISES

For Exercises 1–4, graph the equations using the slope and the y-intercept. 6.3, 7.1

1. $y = 5x + 1$ ❖

2. $y = \frac{3}{5}x + 2$ ❖

3. $y = -\frac{4}{3}x - 2$ ❖

4. $y = -3x - 4$ ❖

For Exercises 5–8, graph the equations using the x- and y-intercepts. 4.7, 6.1

5. $-3x + 2y = 12$ ❖

6. $5x - y = 15$ ❖

7. $x + 3y = -9$ ❖

8. $2x - 4y = -16$ ❖

For Exercises 9–12, graph the horizontal and vertical lines. 6.1

9. $x = -3$ ❖

10. $y = -\frac{1}{4}$ ❖

11. $x = \frac{1}{2}$ ❖

12. $y = 5$ ❖

 13. When is a solid line used in the graph of a linear inequality in two variables? When the inequality symbol is ≤ or ≥

❖ See Additional Answers Appendix   Writing   Translating Expression   Geometry   Scientific Calculator   Video

14. When is a dashed line used in the graph of a linear inequality in two variables?
   When the inequality symbol is < or >

15. When graphing a linear inequality in two variables, how do you determine which side of the boundary line to shade?   Test a point in each region.
   Shade the region that yields a true statement.

16. What does the shaded region represent in the graph of a linear inequality in two variables?
   All of the points in the shaded region are solutions to the inequality.

17. Which is the graph of $-2x - y \le 2$?   a

   a.
   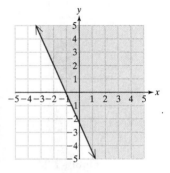

   b.
   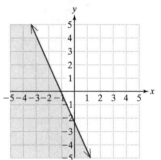

   c.
   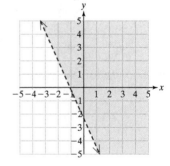

18. Which is the graph of $-3x + y > -1$?   c

   a.
   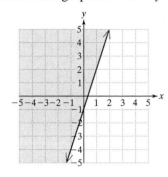

   b.
   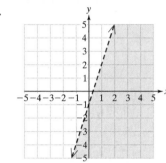

   c.
   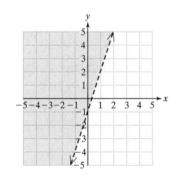

For Exercises 19–24, graph the linear inequalities. Then write three ordered pairs that are solutions to the inequality.

19. $y \ge -x + 5$ ❖
20. $y \le 2x - 1$ ❖
21. $y < 4x$ ❖
22. $y > -5x$ ❖
23. $3x + 7y \le 14$ ❖
24. $5x - 6y \ge 18$ ❖

For Exercises 25–38, graph the linear inequalities.

25. $x - y > 6$ ❖
26. $x + y < 5$ ❖
27. $x \ge -1$ ❖
28. $x \le 6$ ❖

29. $y < 3$ ❖

30. $y > -3$ ❖

31. $y \le -\frac{3}{4}x + 2$ ❖

32. $y \ge \frac{2}{3}x + 1$ ❖

33. $y - 2x > 0$ ❖

34. $y + 3x < 0$ ❖

35. $x \le 0$ ❖

36. $y \le 0$ ❖

37. $y \ge 0$ ❖

38. $x \ge 0$ ❖

39. a. Describe the solutions to the inequality $x + y > 4$. Find three solutions to the inequality (answers will vary). ❖

  b. Describe the solutions to the equation $x + y = 4$. Find three solutions to the equation (answers will vary). ❖

  c. Describe the solutions to the inequality $x + y < 4$. Find three solutions to the inequality (answers will vary). ❖

40. a. Describe the solutions to the inequality $x + y < 3$. Find three solutions to the inequality (answers will vary). ❖

  b. Describe the solutions to the equation $x + y = 3$. Find three solutions to the equation (answers will vary). ❖

  c. Describe the solutions to the inequality $x + y > 3$. Find three solutions to the inequality (answers will vary). ❖

EXPANDING YOUR SKILLS

41. Describe the region bounded by the inequalities $x > 0$ and $y > 0$. This is Quadrant I.

42. Describe the region bounded by the inequalities $x < 0$ and $y > 0$. This is Quadrant II.

For Exercises 43–52, sketch the region bounded by the inequalities.

43. $x \ge 0$ ❖
$y \ge 0$
$x + y \le 4$

44. $x \le 0$ ❖
$y \ge 0$
$-x + y \ge 3$

45. $y < 2$ ❖
$y > -3$

46. $x \le 4$ ❖
$x \ge -1$

47. $y < x$ ❖
$y \ge 0$
$x < 4$

48. $y > x$ ❖
$x \ge 0$
$y < 3$

49. $x \le 2$ ❖
$y \le 2$
$x \ge -2$
$y \ge -2$

50. $x > -1$ ❖
$y > -1$
$x < 1$
$y < 1$

51. $2x + y < 3$ ❖
$y \ge x + 3$

52. $x + y \le 3$ ❖
$y - x > 0$

chapter 7   Summary

Section 7.1—Introduction to Systems of Linear Equations

Key Concepts:

A system of two linear equations can be solved by graphing.

A solution to a system of linear equations is an ordered pair that satisfies each equation in the system. Graphically, this represents a point of intersection of the lines.

Examples:

Solve by graphing:

$$x + y = 3$$
$$2x - y = 0$$

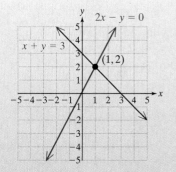

The solution is $(1, 2)$.

Solve by graphing:

$$3x - 2y = 2$$
$$-6x + 4y = 4$$

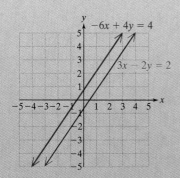

There is no solution. The system is inconsistent.

There may be one solution, infinitely many solutions, or no solution.

One solution	Many solutions	No solution
Consistent	Consistent	Inconsistent
Independent	Dependent	Independent

A system of equations is consistent if there is at least one solution. A system is inconsistent if there is no solution.

A linear system in x and y is dependent if two equations represent the same line. The solution set is the set of all points on the line.

If two linear equations represent different lines, then the system of equation is independent.

KEY TERMS:

consistent system
dependent system
inconsistent system
independent system
solution to a system of linear equations
system of linear equations

Solve by graphing:

$$x + 2y = 2$$
$$-3x - 6y = -6$$

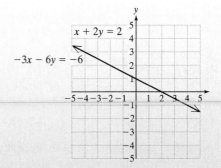

The system is dependent, and the solution set consists of all points on the line, given by
$$\{(x, y) \mid y = -\tfrac{1}{2}x + 1\}$$

SECTION 7.2—SUBSTITUTION METHOD

KEY CONCEPTS:

Steps to Solve a System of Equations by the Substitution Method:

1. Isolate one of the variables from one equation.
2. Substitute the quantity found in Step 1 into the other equation.
3. Solve the resulting equation.
4. Substitute the value found in Step 3 back into the equation in Step 1 to find the remaining variable.
5. Check the solution in both equations and write the answer as an ordered pair.

EXAMPLES:

Solve by the substitution method:

$$x + 4y = -11$$
$$3x - 2y = -5$$

Isolate x in the first equation: $x = -4y - 11$
Substitute into the second equation.

$$3(-4y - 11) - 2y = -5 \qquad \text{Solve the equation.}$$

$$-12y - 33 - 2y = -5$$
$$-14y = 28$$
$$y = -2$$

$$\begin{aligned} & \qquad\qquad\qquad\quad \text{Substitute} \\ x = -4y - 11 & \qquad\qquad y = -2. \\ x = -4(-2) - 11 & \qquad\qquad \text{Solve for } x. \\ x = -3 & \end{aligned}$$

The solution is $(-3, -2)$ and checks in both original equations.

An inconsistent system has no solution and is detected algebraically by a contradiction (such as $0 = 3$).

Solve by the substitution method:

$$3x + y = 4$$

$$-6x - 2y = 2$$

Isolate y in the first equation: $y = -3x + 4$. Substitute into the second equation.

$$-6x - 2(-3x + 4) = 2$$

$$-6x + 6x - 8 = 2$$

$$-8 = 2 \qquad \text{Contradiction}$$

The system is inconsistent and has no solution.

If two linear equations represent the same line, the system is dependent. This is detected algebraically by an identity (such as $0 = 0$).

Solve by the substitution method:

$$y = x + 2 \qquad y \text{ is already isolated.}$$

$$x - y = -2$$

$$x - (x + 2) = -2 \qquad \text{Substitute } y = x + 2 \text{ into}$$

$$x - x - 2 = -2 \qquad \text{the second equation.}$$

$$-2 = -2 \qquad \text{Identity}$$

The system is dependent. The solution set is all points on the line $y = x + 2$ or $\{(x, y) \mid y = x + 2\}$.

KEY TERM:

substitution method

SECTION 7.3—ADDITION METHOD

KEY CONCEPTS:

Solving a System of Linear Equations by the Addition Method:

1. Write both equations in standard form: $ax + by = c$.
2. Clear fractions or decimals (optional).
3. Multiply one or both equations by a nonzero constant to create opposite coefficients for one of the variables.
4. Add the equations to eliminate one variable.
5. Solve for the remaining variable.
6. Substitute the known value into one of the original equations to solve for the other variable.
7. Check the solution in both equations.

KEY TERM:

addition method

EXAMPLES:

Solve by using the addition method:

$$5x = -4y - 7 \qquad \text{Write the first equation in}$$

$$6x - 3y = 15 \qquad \text{standard form.}$$

$$5x + 4y = -7 \xrightarrow{\text{Multiply by 3}} 15x + 12y = -21$$

$$6x - 3y = 15 \xrightarrow[\text{Multiply by 4}]{} \underline{24x - 12y = 60}$$

$$39x \qquad\quad = 39$$

$$x = 1$$

$$5x = -4y - 7$$

$$5(1) = -4y - 7$$

$$5 = -4y - 7$$

$$12 = -4y$$

$$-3 = y \qquad \text{The solution is } (1, -3) \text{ and checks in both original equations.}$$

Solve by using the addition method:

$$x + y = 5 \xrightarrow{\text{Multiply by } -2} \begin{array}{r} -2x - 2y = -10 \\ 2x + 2y = 4 \\ \hline 0 = -6 \end{array}$$

$$2x + 2y = 4$$

The contradiction $0 = -6$ implies that there is no solution. The system is inconsistent.

Solve by using the addition method:

$$\frac{1}{2}x + \frac{3}{4}y = 4 \xrightarrow{\text{Multiply by } 4} \begin{array}{r} 2x + 3y = 16 \\ -2x - 3y = -16 \\ \hline 0 = 0 \end{array}$$

$$-2x - 3y = -16$$

The identity $0 = 0$ implies that the two linear equations represent the same line. The solution set is the set of all points on the line or $\{(x, y) | 2x + 3y = 16\}$. The system is dependent.

SECTION 7.4—APPLICATIONS OF LINEAR EQUATIONS IN TWO VARIABLES

EXAMPLES:

A riverboat travels 36 miles with the current to a marina in 2 h. The return trip takes 3 h against the current. Find the speed of the current and the speed of the boat in still water.

Let x represent the speed of the boat in still water. Let y represent the speed of the current.

	Distance	Rate	Time
Against current	36	$x - y$	3
With current	36	$x + y$	2

Distance = (rate)(time)

$$36 = (x - y) \cdot 3$$
$$36 = (x + y) \cdot 2$$

EXAMPLES:

Diane invests $15,000 more in an account earning 8% simple interest than in an account earning 5% simple interest. If the total interest after 1 year is $1850, how much was invested in each account?

	8%	5%	Total
Principal	x	y	
Interest	$0.08x$	$0.05y$	1850

$$x = y + 15{,}000$$
$$0.08x + 0.05y = 1850$$

$$36 = 3x - 3y \xrightarrow{\text{Multiply by 2}} 72 = 6x - 6y$$

$$36 = 2x + 2y \xrightarrow[\text{Multiply by 3}]{} \underline{108 = 6x + 6y}$$

$$180 = 12x$$

$$15 = x$$

$$36 = 2(15) + 2y$$

$$36 = 30 + 2y$$

$$6 = 2y$$

$$3 = y$$

The speed of the boat in still water is 15 mph, and the speed of the current is 3 mph.

Substitute $x = y + 15{,}000$ into the second equation:

$$0.08(y + 15{,}000) + 0.05y = 1850$$

$$0.08y + 1200 + 0.05y = 1850$$

$$0.13y + 1200 = 1850$$

$$0.13y = 650$$

$$y = \frac{650}{0.13},$$

$$y = 5000$$

$$x = y + 15{,}000$$

$$x = 5000 + 15{,}000$$

$$x = 20{,}000$$

The amount invested in the 8% account is $20,000 and the amount invested at 5% is $5000.

SECTION 7.5—CONNECTIONS TO GRAPHING: LINEAR INEQUALITIES IN TWO VARIABLES

KEY CONCEPTS:

A linear inequality in two variables is an inequality of the form: $ax + by < c$, $ax + by > c$, $ax + by \leq c$, or $ax + by \geq c$.

Steps for Using the Test Point Method to Solve a Linear Inequality in Two Variables:

1. Set-up the related *equation*.
2. Graph the related equation. This will be a line in the *xy*-plane.
   - If the original inequality is a strict inequality, $<$ or $>$, then the line is not part of the solution set. Therefore, graph the boundary as a dashed line.

   - If the original inequality uses $\leq$ or $\geq$ then the line is part of the solution set. Therefore, graph the boundary as a solid line.

EXAMPLES:

Graph the inequality:

$$2x - y < 4$$

1. The related equation is $2x - y = 4$
2. Graph the equation $2x - y = 4$ (dashed line)

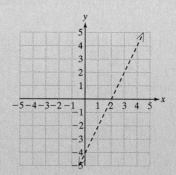

3. From each region above and below the line, select an ordered pair as a test point and substitute it into the original inequality. If a test point makes the inequality true, then that region is part of the solution set.

KEY TERMS:

linear inequality in two variables
test point method

3. *Test Points:*

$(0, 0)$	$2(0) - (0) < 4$	True Shade above.
$(4, -4)$	$2(4) - (-4) < 4$	False Do not shade below.

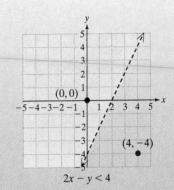

$2x - y < 4$

---

chapter 7   REVIEW EXERCISES

Section 7.1

For Exercises 1–4, determine if the ordered pair is a solution to the system.

1. $x - 4y = -4$ $(4, 2)$

   $x + 2y = 8$ Yes

2. $x - 6y = 6$ $(12, 1)$

   $-x + y = 4$ No

3. $3x + y = 9$ $(1, 3)$

   $y = 3$ No

4. $2x - y = 8$ $(2, -4)$

   $x = 2$ Yes

For Exercises 5–10, identify whether the system represents intersecting lines, parallel lines, or coinciding lines by comparing their slopes and y-intercepts.

5. $y = -\dfrac{1}{2}x + 4$

   $y = x - 1$ Intersecting lines

6. $y = -3x + 4$

   $y = 3x + 4$ Intersecting lines

7. $y = -\dfrac{4}{7}x + 3$ 8. $y = 5x - 3$

Parallel lines   Intersecting lines

$y = -\dfrac{4}{7}x - 5$ $y = \dfrac{1}{5}x - 3$

9. $y = 9x - 2$ 10. $x = -5$

Coinciding lines   Intersecting lines

$9x - y = 2$ $y = 2$

For Exercises 11–18, solve the systems by graphing. Identify whether the system is consistent or inconsistent. Identify whether the system is dependent or independent.

11. $y = -\dfrac{2}{3}x - 2$ ❖

$-x + 3y = -6$

12. $y = -2x - 1$ ❖

$2x + 3y = 5$

13. $4x = -2y + 10$ ❖

$2x + y = 5$

14. $10y = 2x - 10$ ❖

$-x + 5y = -5$

15. $6x - 3y = 9$ ❖   16. $5x + y = -11$ ❖

$y = -1$ $x = -1$

17. $x - 7y = 14$ ❖   18. $y = -5x + 6$ ❖

$-2x + 14y = 14$ $10x + 2y = 6$

19.   The following graph depicts the proportion of AIDS cases by race and ethnicity in the United States between 1985–1998. Estimate the year in which the proportion of AIDS cases attributed to black (not Hispanic) persons equaled the proportion of AIDS cases attributed to white (not Hispanic) persons.   1994

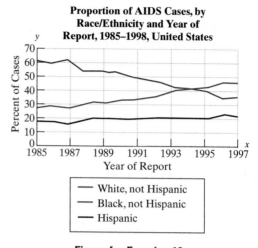

Proportion of AIDS Cases, by Race/Ethnicity and Year of Report, 1985–1998, United States

Figure for Exercise 19

Source: Centers for Disease Control and Prevention.

Section 7.2

For Exercises 20–23, solve the systems using the substitution method.

20. $6x + y = 2$ $\left(\dfrac{2}{3}, -2\right)$ 21. $2x + 3y = -5$ $(-4, 1)$

$y = 3x - 4$ $x = y - 5$

22. $2x + 6y = 10$ No solution

$x = -3y + 6$

23. $4x + 2y = 4$ Infinitely many solutions

$\{(x, y)|y = -2x + 2\}$

$y = -2x + 2$

24.   Given the system:

$x + 2y = 11$

$5x + 4y = 40$

a.   Which variable from which equation is easiest to isolate and why?

x in the first equation is easiest to solve for because its coefficient is 1.

b.   Solve the system using the substitution method. $\left(6, \dfrac{5}{2}\right)$

25.   Given the system:

$4x - 3y = 9$

$2x + y = 12$

a. Which variable from which equation is easiest to isolate and why?
y in the second equation is easiest to solve for because its coefficient is 1.

b. Solve the system using the substitution method. $\left(\frac{9}{2}, 3\right)$

For Exercises 26–29, solve the systems using the substitution method.

26. $3x - 2y = 23$ (5, −4)

$x + 5y = -15$

27. $x + 5y = 20$ (0, 4)

$3x + 2y = 8$

28. $x - 3y = 9$
Infinitely many solutions
$5x - 15y = 45$
$\{(x, y)|y = \frac{1}{3}x - 3\}$

29. $-3x + y = 15$
No solution
$6x - 2y = 12$

30. The difference of two positive numbers is 42. The larger number is 2 more than 6 times the smaller number. Find the numbers. *The numbers are 50 and 8.*

31. In a right triangle, one of the acute angles is 6° less than the other acute angle. Find the measure of each acute angle. *The angles are 42° and 48°.*

32. During the first 13 years of his football career, Jerry Rice scored a total of 166 touchdowns. One touchdown was scored on a kickoff return and the remaining 165 were scored rushing or receiving. The number of receiving touchdowns he scored was 5 more than 15 times the number of rushing touchdowns he scored. How many receiving touchdowns and how many rushing touchdowns did he score?
He scored 155 receiving touchdowns and 10 touchdowns rushing.

33. Two angles are supplementary. One angle is 14° less than two times the other angle. Find the measure of each angle. *The angles are $115\frac{1}{3}°$ and $64\frac{2}{3}°$.*

Section 7.3

34. Explain the process for solving a system of two equations using the addition method. ❖

35. Given the system:

$$3x - 5y = 1$$
$$2x - y = -4$$

a. Which variable, x or y, is easier to eliminate using the addition method?

b. Solve the system using the addition method. (−3, −2)

36. Given the system:

$$9x - 2y = 14$$
$$4x + 3y = 14$$

a. Which variable, x or y, is easier to eliminate using the addition method?

b. Solve the system using the addition method. (2, 2)

For Exercises 37–44, solve the systems using the addition method.

37. $2x + 3y = 1$ (2, −1)

$x - 2y = 4$

38. $x + 3y = 0$ (−6, 2)

$-3x - 10y = -2$

39. $8(x + 1) = -6y + 6$ $\left(-\frac{1}{2}, \frac{1}{3}\right)$

$10x = 9y - 8$

40. $12x = 5(y + 1)$ $\left(\frac{1}{4}, -\frac{2}{5}\right)$

$5y = -1 - 4x$

41. $-4x - 6y = -2$
Infinitely many solutions
$6x + 9y = 3$
$\{(x, y)|y = -\frac{2}{3}x + \frac{1}{3}\}$

42. $-8x - 4y = 16$
No solution
$10x + 5y = 5$

43. $\frac{1}{2}x - \frac{3}{4}y = -\frac{1}{2}$
(−4, −2)
$\frac{1}{3}x + y = -\frac{10}{3}$

44. $0.5x - 0.2y = 0.5$ (1, 0)

$0.4x + 0.7y = 0.4$

45. Given the system:

$$4x + 9y = -7$$
$$y = 2x - 13$$

a. Which method would you choose to solve the system, the substitution method or the addition method? Explain your choice.

b. Solve the system. (5, −3)

46. Given the system:

$$5x - 8y = -2$$
$$3x - y = -5$$

a. Which method would you choose to solve the system, the substitution method or the addition method? Explain your choice.

b. Solve the system. $(-2, -1)$

Section 7.4

47. Miami Metrozoo charges $8.75 for adult admission and $5.75 for each child under 12. The total bill before tax for a school group of 60 people is $369. How many adults and how many children were admitted?
There were 8 adult tickets and 52 childrens tickets sold.

48. Emillo invested $20,000 and at the end of one year, he received $1525 in interest. If he invested part of the money at 5% simple interest and the remaining money at 8% simple interest, how much did he invest in each account?
Emillo invested $2500 in the 5% account and $17,500 in the 8% account.

49. To produce a 16% alcohol solution, a chemist mixes a 20% alcohol solution and a 14% alcohol solution. How much 20% solution and how much 14% solution must be used to produce 15 L of a 16% alcohol solution? 5 liters of the 20% solution should be mixed with 10 L of the 14% solution.

50. A boat travels 80 miles downstream with the current in 4 hours and 80 miles upstream against the current in 5 h. Find the speed of the current and the speed of the boat in still water.
The speed of the boat is 18 mph, and that of the stream is 2 mph.

51. Suzanne has a collection of new quarters and new $1 coins. She has four more quarters than dollar coins and the total value of the coins is $4.75. How many of each coin does Suzanne have? Suzanne has seven quarters and three dollar coins.

52. In a recent election 5700 votes were cast and 3675 voters voted for the winning candidate. If $\frac{5}{8}$ of the women and $\frac{2}{3}$ of the men voted for the winning candidate, how many men and how many women voted? 3000 women and 2700 men

Section 7.5

53. Which graph represents the solution to $x + y \le 4$? a

a.

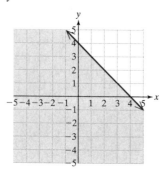

b.

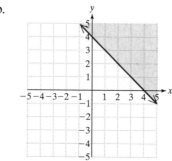

c.

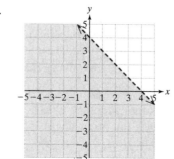

54. Which graph represents the solution to $x > -2$? c

a.

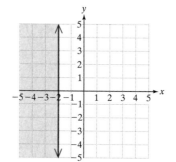

b.

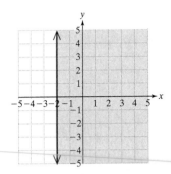

c.

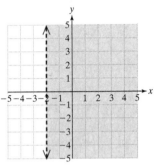

For Exercises 55–58, graph the inequalities. Then write three ordered pairs that are in the solution set (answers may vary).

55. $y < 3x - 1$ ❖

56. $y > -2x + 6$ ❖

57. $4x - 2y \leq 10$ ❖

58. $-2x - 3y \geq 8$ ❖

For Exercises 59–64, graph the inequalities.

59. $x - 5y \geq 0$ ❖

60. $7x - y \leq 0$ ❖

61. $x > 5$ ❖

62. $y < -4$ ❖

63. $x \geq 0$ ❖

64. $y \geq 0$ ❖

For Exercises 65–70, sketch the regions bounded by the inequalities.

65. $x \leq 0$ ❖
     $y \leq 0$
     $x + y > -4$

66. $x \geq 0$ ❖
     $y \leq 0$
     $2x - y < 2$

67. $x > -3$ ❖
    $x < 2$

68. $y \geq -1$ ❖
    $y \leq 5$

69. $x \geq 0$ ❖
     $x \leq 4$
     $y \geq 0$
     $y \leq 6$

70. $x \leq 0$ ❖
     $x \geq -3$
     $y \geq 0$
     $y \leq 4$

---

chapter 7   TEST

1. Write each line in slope-intercept form. Then determine if the lines represent intersecting lines, parallel lines, or coinciding lines.   **Parallel lines**

$$5x + 2y = -6$$

$$-\frac{5}{2}x - y = -3$$

2. a. Solve the system by graphing ❖

$$y = \frac{1}{2}x + \frac{9}{2}$$

$$y = -\frac{1}{3}x + 2$$

b. Check your answer by substituting the ordered pair in both equations.

3. The supply and demand of an item depends on the number of items produced and the price of the item. If the price is high, the demand decreases and there is a surplus in supply. Answer the following questions based on the graph provided.

a. What is the equilibrium price (that is, for what price does supply equals demand)?   **$15**

b. How many items should be produced at the equilibrium price?   **5,000,000 items**

❖ See Additional Answers Appendix   🖋 Writing   ↔ Translating Expression   📐 Geometry   🖩 Scientific Calculator   📼 Video

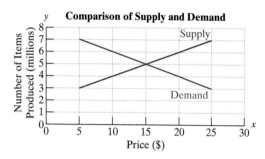

Figure for Exercise 3

4. Solve the system using the substitution method. $(-2, 0)$

$$x = 5y - 2$$
$$2x + y = -4$$

5. In the 1998 WNBA season, the league's leading scorer, Cynthia Cooper from the Houston Comets scored 227 more points than her teammate Sheryl Swoopes. Together they scored a total of 1133 points. How many points did each player score?
Cooper had 680 points, and Swoopes had 453 points.

6. Solve the system using the addition method. $\left(2, -\dfrac{1}{3}\right)$

$$3x - 6y = 8$$
$$2x + 3y = 3$$

7. How many milliliters of a 50% acid solution and how many milliliters of a 20% acid solution must be mixed to produce 36 mL of a 30% acid solution?   12 milliliters of the 50% acid solution should be mixed with 24 mL of the 20% solution.

8. a. How many solutions does a system of two linear equations have if the equations represent parallel lines?   No solution

    b. How many solutions does a system of two linear equations have if the equations represent coinciding lines?   Infinitely many solutions

    c. How many solutions does a system of two linear equations have if the equations represent intersecting lines?   One solution

For Exercises 9–12, solve the systems using any method.

9. $\dfrac{1}{3}x + y = \dfrac{7}{3}$ $(-5, 4)$

$$x = \dfrac{3}{2}y - 11$$

10. $2(x - 6) = y$ No solution

$$2x - \dfrac{1}{2}y = x + 5$$

11. $-0.25x - 0.05y = 0.2$ Infinitely many solutions $\{(x, y)\mid y = -5x - 4\}$

$$10x + 2y = -8$$

12. $3(x + y) = -2y - 7$ $(1, -2)$

$$-3y = 10 - 4x$$

13. In a right triangle, one of the acute angles is 9° less than twice the other acute angle. Find the measure of each acute angle.
The angles are 57° and 33°.

14. Max has 30 coins consisting of nickels, dimes, and quarters. He has 10 nickels and the number of dimes is 1 less than twice the number of quarters.

    a. How many of each type of coin does he have?
13 dimes, 7 quarters, 10 nickels

    b. How much money does he have? $3.55

15. Five thousand dollars less was invested in an account earning 9% simple interest than in an account earning 11% simple interest. If the total interest at the end of 1 year is $1950, how much was invested in each account? What is the total amount invested? $7000 was invested at 9%, $12,000 was invested at 11%, totaling $19,000.

16. A consumer who is tired of paying rising costs for cable television looked into the possibility of purchasing a satellite dish. The cost of a particular satellite dish is $120 plus $26 per month. The cable television is already installed and costs $34 per month. The total cost, y, for either option depends on the number of months of service, x, according to the following equations:

Satellite: $y = 26x + 120$

Cable: $y = 34x$

    a. Use the graph to estimate how many months are required for the cost to use a satellite dish and the cost to use cable to be equal.
15 months

    b. Solve the system of equations by any method. Interpret the meaning of the solution in terms of the number of months and the total cost.   For 15 months of service the cost is $510 for either cable or a satellite dish.

**Comparison Between Cost of Cable and
Cost of a Satellite Dish**

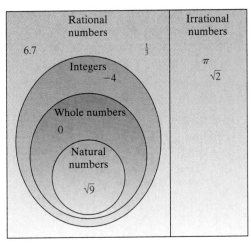

$y = 34x$

$y = 26x + 120$

Number of Months

● Satellite Dish
■ Cable

Figure for Exercise 16

17. A plane travels 880 miles in 2 hours (h) against the wind, and 1000 miles in 2 h with the same wind. Find the speed of the plane in still air and the speed of the wind.
The plane is traveling at 470 mph, and the wind is 30 mph.

18. Graph the inequality: $5x - y > -6$ ❖

19. Sketch the region bounded by the inequalities. ❖

$$y \geq 0$$

$$y \leq -x$$

$$x \geq -3$$

CUMULATIVE REVIEW EXERCISES, CHAPTERS 1–7

1. Simplify.

$$\frac{|2 - 5| + 10 \div 2 + 3}{\sqrt{10^2 - 8^2}} \quad \frac{11}{6}$$

2. Place the numbers in the diagram below: $\{-4, \pi, \sqrt{9}, 6.7, \frac{1}{3}, 0, \sqrt{2}\}$

Rational numbers   Irrational numbers

6.7 $\frac{1}{3}$

Integers
-4

π
$\sqrt{2}$

Whole numbers
0

Natural numbers

$\sqrt{9}$

Figure for Exercise 2

3. Solve for x: $\frac{1}{3}x - \frac{3}{4} = \frac{1}{2}(x + 2)$ $x = -\frac{21}{2}$

4. Solve for a: $-4(a + 3) + 2 = -5(a + 1) + a$
No solution

5. Solve for y: $3x - 2y = 6$ $y = \frac{3}{2}x - 3$

6. Solve for z. Graph the solution set and write the solution in interval notation: ❖

$$-2(3z + 1) \leq 5(z - 3) + 10$$

7. In any triangle, what is the sum of the three inscribed angles?   180°

8. The largest angle in a triangle is 110°. Of the remaining two angles, one is 4° less than the other angle. Find the measure of the three angles.
The angles are 37°, 33°, 110°.

9. Two hikers start at opposite ends of an 18-mile trail and walk toward each other. One hiker walks predominately down hill and averages 2 mph faster than the other hiker. Find the average rate of each hiker if they meet in 3 h.
The rates of the hikers are 2 mph and 4 mph.

10. Simplify. Write the final answer with positive exponents only:

a. $(2c^2d^{-3})(5c^{-4}d^8)$ $\frac{10d^5}{c^2}$ b. $\frac{(x^{-2})^{-3}(x^5)^2}{x^{-2}}$ x^{18}

11. Multiply. Write the answer in scientific notation: $(3.0 \times 10^4)(6.0 \times 10^8)$ 1.8×10^{13}

12. Multiply: $(3y^2 + 2y - 4)(y - 2)$ $3y^3 - 4y^2 - 8y + 8$

13. Subtract: $(\frac{1}{2}w^2 - \frac{3}{5}w + \frac{1}{2}) - (\frac{3}{2}w^2 + \frac{1}{10}w - 2)$
$-w^2 - \frac{7}{10}w + \frac{5}{2}$

14. Divide.

$$\frac{8p^3q - 4p^2q^2 + 16pq^3}{4p^2q^2} \quad \frac{2p}{q} - 1 + \frac{4q}{p}$$

15. Factor completely: $5x^2 - 125$ $5(x - 5)(x + 5)$

16. Factor completely: $5xa - 10xb - 2ya + 4yb$
    $(a - 2b)(5x - 2y)$

17. Solve the equation: $5y(y - 3)(2y + 1) = 0$
    $y = 0, y = 3, y = -\frac{1}{2}$

18. Solve the equation: $t^2 - 10t = -25$
    $t = 5$

19. Given the equation: $y = x^2 - x - 12$

    a. Is the equation linear or nonlinear?   Nonlinear

    b. Find the x- and y-intercepts.
    x-intercepts: $(4, 0)(-3, 0)$; y-intercept: $(0, -12)$

20. Multiply.

$$\frac{20x - 30}{2x + 10} \cdot \frac{x^2 + 8x + 15}{4x^2 - 9} \quad \frac{5(x + 3)}{2x + 3}$$

21. Add.

$$\frac{2a}{a - 3} + \frac{28}{a^2 - 9} + \frac{6}{a + 3} \quad \frac{2(a + 5)(a + 1)}{(a + 3)(a - 3)}$$

22. Solve for a. $a = -5, a = -1$

$$\frac{2a}{a - 3} + \frac{28}{a^2 - 9} + \frac{6}{a + 3} = 0$$

23. Explain the difference between the process used to work Problems 21 and 22.   ❖

24. Joelle can do a job in 4 h. Her husband Bob can do the same job in 3 h. How long will it take them if they work together?
    It would take them $1\frac{5}{7}$ h.

25. The slope of a given line is $-\frac{2}{3}$.

    a. What is the slope of a line parallel to the given line? $-\frac{2}{3}$

    b. What is the slope of a line perpendicular to the given line? $\frac{3}{2}$

26. Find an equation of the line passing through the point $(2, -3)$ and perpendicular to the line $x - 3y = 4$. Write the final answer in slope-intercept form. $y = -3x + 3$

27. Sketch the following equations on the same graph.

    a. $2x + 5y = 10$ ❖

    b. $2y = 4$ ❖

    c. Find the point of intersection and check the solution in each equation.   ❖

28. Solve the system of equations using any method.

$$2x + 5y = 10 \quad (0, 2)$$
$$2y = 4$$

29. a. Graph the line $2x + y = 3$.   ❖

    b. Graph the inequality: $2x + y < 3$.   ❖

    c. Explain the difference between the graphs in parts (a) and (b).   ❖

30. How many gallons of a 15% antifreeze solution should be mixed with a 60% antifreeze solution to produce 60 gal of a 45% antifreeze solution?   ❖

31. Set up a system of two linear equations to solve for x and y. x is 27°; y is 63°

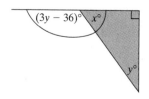

Figure for Exercise 31

32. In 1920, the average speed for the winner of the Indianapolis 500 car race was 88.6 mph. By 1990, the average speed of the winner was 186.0 mph.

    a. Find the slope of the line shown in the figure. Round to one decimal place.   1.4

    b. Interpret the meaning of the slope in the context of this problem.   Each year between 1920 and 1990 the winning speed in the Indianapolis 500 increased on average by 1.4 mph.

Figure for Exercise 32

RADICALS

The area of a triangle can be found if the length of one side (the base) and the corresponding height of the triangle are known.

$$A = \frac{1}{2}bh$$

However, if the height of a triangle is not known, but the lengths of the three sides, a, b, and, c are given, the area of the triangle can be found using Heron's formula:

$$A = \sqrt{s(s - a)(s - b)(s - c)}$$

where

$$s = \frac{1}{2}(a + b + c)$$

Heron's formula and other applications of radicals are studied in this chapter.

The Louvre pyramid, designed by architect I. M. Pei, is a glass structure that serves as the entrance to the Louvre Museum in Paris. Each triangular face is made of glass with dimensions as shown.

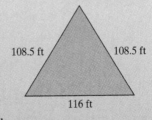

The area of each face can be found using Heron's formula:

$$s = \frac{1}{2}(108.5 + 108.5 + 116) = 166.5$$

$$A = \sqrt{166.5(166.5 - 108.5)(166.5 - 108.5)(166.5 - 116)}$$

$$\approx 5318.4$$

The area of each triangular face required approximately 5318.4 ft^2 of glass.

section

8.1   INTRODUCTION TO ROOTS AND RADICALS

1.   Definition of a Square Root

Recall that to square a number means to multiply the number by itself: $b^2 = b \cdot b$. The reverse operation to squaring a number is to find its square roots. For example, finding a square root of 49 is equivalent to asking: "What number when squared equals 49?"

One obvious answer to this question is 7, because $(7)^2 = 49$. But -7 will also work, because $(-7)^2 = 49$.

Definition of a Square Root

b is a **square root** of a if $b^2 = a$.

---

example 1

Identifying the Square Roots of a Number

Identify the square roots of

a. 9   b. 121   c. 0   d. −4

Tip: All positive real numbers have two real-valued square roots: one positive and one negative. Zero has only one square root, which is 0 itself. Finally, for any negative real number, there are no real-valued square roots.

Solution:

a. 3 is a square root of 9 because $(3)^2 = 9$.
    -3 is a square root of 9 because $(-3)^2 = 9$.

b. 11 is a square root of 121 because $(11)^2 = 121$
    -11 is a square root of 121 because $(-11)^2 = 121$

c. 0 is a square root of 0 because $(0)^2 = 0$.

d. There are no real numbers that when squared will equal a negative number. Therefore, there are no real-valued square roots of -4.

Recall from Section 1.2, that the positive square root of a real number can be denoted with a radical sign, $\sqrt{\ }$.

Notation for Positive and Negative Square Roots

Let a represent a positive real number. Then,

1. $\sqrt{a}$ is the **positive square root** of a. The positive square root is also called the **principal square root**.
2. $-\sqrt{a}$ is the **negative square root** of a.
3. $\sqrt{0} = 0$

2. Finding the Principal Square Root of a Nonnegative Real Number

example 2

Simplifying a Square Root

Simplify the square roots.

a. $\sqrt{36}$ b. $\sqrt{225}$ c. $\sqrt{1}$ d. $\sqrt{\dfrac{9}{4}}$

◆ **Avoiding Mistakes**

a. $\sqrt{36} = 6$ (not -6)
b. $\sqrt{225} = 15$ (not -15)
c. $\sqrt{1} = 1$ (not -1)
d. $\sqrt{\dfrac{9}{4}} = \dfrac{3}{2}$ $\left(\text{not } -\dfrac{3}{2}\right)$

Solution:

a. $\sqrt{36}$ denotes the positive square root of 36. $\sqrt{36} = 6$

b. $\sqrt{225}$ denotes the positive square root of 225. $\sqrt{225} = 15$

c. $\sqrt{1}$ denotes the positive square root of 1. $\sqrt{1} = 1$

d. $\sqrt{\dfrac{9}{4}}$ denotes the positive square root of $\dfrac{9}{4}$. $\sqrt{\dfrac{9}{4}} = \dfrac{3}{2}$

Tip: Before using a calculator to evaluate a square root, try estimating the value first.

$\sqrt{13}$ must be a number between 3 and 4 because $\sqrt{9} < \sqrt{13} < \sqrt{16}$.

$\sqrt{42}$ must be a number between 6 and 7 because $\sqrt{36} < \sqrt{42} < \sqrt{49}$.

The numbers 36, 225, 1, and $\frac{9}{4}$ are **perfect squares** because their square roots are rational numbers. Radicals that cannot be simplified to rational numbers are irrational numbers. Recall that an irrational number cannot be written as a terminating or repeating decimal. For example, the symbol, $\sqrt{13}$, is used to represent the exact value of the square root of 13. The symbol, $\sqrt{42}$, is used to represent the exact value of the square root of 42. These values are irrational numbers but can be approximated by rational numbers by using a calculator.

$$\sqrt{13} \approx 3.605551275 \qquad \sqrt{42} \approx 6.480740698$$

3. Square Roots of Negative Numbers

A negative number cannot have a real number as a square root because no real number when squared is negative. For example, $\sqrt{-25}$ is *not a real number* because there is no real number, b, for which $(b)^2 = -25$.

example 3

Evaluating Square Roots if Possible

Simplify the square roots if possible.

a. $\sqrt{-100}$ b. $-\sqrt{100}$ c. $\sqrt{-64}$

Classroom Activity 8.1A

Solution:

a. $\sqrt{-100}$ Not a real number

Tip: In the expression $-\sqrt{100}$ the negative sign is *outside* the radical.

b. $-\sqrt{100}$

$-1 \cdot \sqrt{100}$ The expression $-\sqrt{100}$ is equivalent to $-1 \cdot \sqrt{100}$.

$-1 \cdot 10 = -10$

c. $\sqrt{-64}$ Not a real number

4. Definition of an *n*th-Root

Finding a square root of a number is the reverse process of squaring a number. This concept can be extended to finding a third root (called a cube root), a fourth root, and in general, an *n*th root.

Definition of an *n*th-Root

b is an **nth-root** of a if $b^n = a$.

The radical sign, $\sqrt{}$, is used to denote the principal square root of a number. The symbol, $\sqrt[n]{}$, is used to denote the principal *n*th-root of a number.

In the expression $\sqrt[n]{a}$, n is called the **index** of the radical, and a is called the **radicand**. For a square root, the index is 2, but it is usually not written ($\sqrt[2]{a}$ is denoted simply as $\sqrt{a}$). A radical with an index of three is called a **cube root**, $\sqrt[3]{a}$.

Definition of $\sqrt[n]{a}$

1. If n is a positive *even* integer and $a > 0$, then $\sqrt[n]{a}$ is the principal (positive) *n*th-root of a.
2. If n is a positive *odd* integer, then $\sqrt[n]{a}$ is the *n*th root of a.
3. If n is any positive integer, then $\sqrt[n]{0} = 0$.

For the purpose of simplifying radicals, it is helpful to know the following patterns:

Perfect cubes	Perfect fourth powers	Perfect fifth powers
$1^3 = 1$	$1^4 = 1$	$1^5 = 1$
$2^3 = 8$	$2^4 = 16$	$2^5 = 32$
$3^3 = 27$	$3^4 = 81$	$3^5 = 243$
$4^3 = 64$	$4^4 = 256$	$4^5 = 1024$
$5^3 = 125$	$5^4 = 625$	$5^5 = 3125$

example 4

Simplifying *n*th-Roots

Simplify the expressions (if possible).

a. $\sqrt[3]{8}$ b. $\sqrt[4]{16}$ c. $\sqrt[5]{32}$ d. $\sqrt[3]{-64}$

e. $\sqrt[3]{\dfrac{125}{27}}$ f. $\sqrt{0.01}$ g. $\sqrt[4]{-81}$

Classroom Activity 8.1B

Solution:

a. $\sqrt[3]{8} = 2$ because $(2)^3 = 8$

b. $\sqrt[4]{16} = 2$ because $(2)^4 = 16$

⬢ **Avoiding Mistakes**

c. $\sqrt[5]{32} = 2$ because $(2)^5 = 32$

When evaluating $\sqrt[n]{a}$, where n is *even*, always choose the principal (positive) root. Hence:

d. $\sqrt[3]{-64} = -4$ because $(-4)^3 = -64$

e. $\sqrt[3]{\dfrac{125}{27}} = \dfrac{5}{3}$ because $\left(\dfrac{5}{3}\right)^3 = \dfrac{125}{27}$

$\sqrt[4]{16} = 2$ (not -2)

$\sqrt{0.01} = 0.1$ (not -0.1)

f. $\sqrt{0.01} = 0.1$ because $(0.1)^2 = 0.01$

Note: $\sqrt{0.01}$ is equivalent to $\sqrt{\dfrac{1}{100}} = \dfrac{1}{10}$, or 0.1.

g. $\sqrt[4]{-81}$ is not a real number because no real number raised to the fourth power equals -81.

Example 4(g) illustrates that an *n*th root of a negative number is not a real number if the index is even because no real number raised to an even power is negative.

5.   Finding Roots of Variable Expressions

Finding an *n*th-root of a variable expression is similar to finding an *n*th-root of a numerical expression. However, for roots with an even index, particular care must be taken to obtain a nonnegative solution.

Definition of $\sqrt[n]{a^n}$

1.   If n is a positive odd integer, then $\sqrt[n]{a^n} = a$
2.   If n is a positive even integer, then $\sqrt[n]{a^n} = |a|$

The absolute value bars are necessary for roots with an even index because the variable, a, may represent a positive quantity or a negative quantity. By using absolute value bars $\sqrt[n]{a^n} = |a|$ is nonnegative and represents the principal *n*th-root of a.

example 5

Simplifying Expressions of the Form $\sqrt[n]{a^n}$

Simplify the expressions.

a. $\sqrt{(-3)^2}$ b. $\sqrt{x^2}$ c. $\sqrt[3]{x^3}$ d. $\sqrt[4]{x^4}$ e. $\sqrt[5]{x^5}$

Solution:

a. $\sqrt{(-3)^2} = |-3| = 3$ Because the index is *even*, absolute value bars are necessary to make the answer nonnegative.

b. $\sqrt{x^2} = |x|$ Because the index is *even*, absolute value bars are necessary to make the answer nonnegative.

c. $\sqrt[3]{x^3} = x$ Because the index is *odd*, no absolute value bars are necessary.

d. $\sqrt[4]{x^4} = |x|$ Because the index is *even*, absolute value bars are necessary to make the answer nonnegative.

e. $\sqrt[5]{x^5} = x$ The index is *odd*, so no absolute value bars are necessary.

If n is an even integer, then $\sqrt[n]{a^n} = |a|$. However, if the variable a is assumed to be nonnegative, then the absolute value bars may be dropped. That is $\sqrt[n]{a^n} = a$ provided $a \geq 0$. In many examples and exercises, we will make the assumption that the variables within a radical expression are positive real numbers. In such a case, the absolute value bars are not needed to evaluate $\sqrt[n]{a^n}$.

It is helpful to become familiar with the patterns associated with perfect squares and perfect cubes involving variable expressions.

The following powers of x are perfect squares:

Perfect squares

$(x^1)^2 = x^2$

$(x^2)^2 = x^4$

$(x^3)^2 = x^6$

$(x^4)^2 = x^8$

$\ldots$

The following powers of x are perfect cubes:

Perfect cubes

$(x^1)^3 = x^3$

$(x^2)^3 = x^6$

$(x^3)^3 = x^9$

$(x^4)^3 = x^{12}$

$\ldots$

Tip: In general, any expression raised to an even power (multiple of 2) is a perfect square.

Tip: In general, any expression raised to a power that is a multiple of 3 is a perfect cube.

example 6

Simplifying *n*th-Roots

Simplify the expressions. Assume that all variables are positive real numbers.

a. $\sqrt{c^6}$ b. $\sqrt[3]{d^{15}}$ c. $\sqrt[4]{y^8}$ d. $\sqrt{a^2b^2}$ e. $\sqrt[3]{64z^6}$

Classroom Activity 8.1C

Solution:

a. $\sqrt{c^6}$ Because 6 is an even number, the expression c^6 is a perfect square.

$\sqrt{c^6} = c^3$ because $(c^3)^2 = c^6$

b. $\sqrt[3]{d^{15}}$ Because 15 is a multiple of 3, the expression d^{15} is a perfect cube.

$\sqrt[3]{d^{15}} = d^5$ because $(d^5)^3 = d^{15}$

c. $\sqrt[4]{y^8}$ Because 8 is a multiple of 4, the expression y^8 is a fourth power

$\sqrt[4]{y^8} = y^2$ because $(y^2)^4 = y^8$

d. $\sqrt{a^2b^2} = ab$ because $(ab)^2 = a^2b^2$

e. $\sqrt[3]{64z^6} = 4z^2$ because $(4z^2)^3 = 64z^6$

6.   Translations Involving *n*th-Roots

It is important to understand the vocabulary and language associated with *n*th-roots. For instance you must be able to distinguish between "squaring a number" and "taking the square *root* of a number." The following example offers practice translating between English form and algebraic form.

example 7

Translating from English Form to Algebraic Form

Translate each English phrase into an algebraic expression.

a. The difference of the square of x and the square root of seven
b. The quotient of one and the cube root of z

Solution:

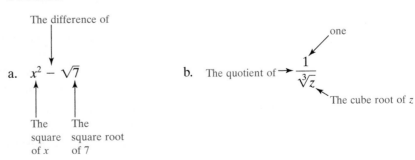

example 8   Translating from Algebraic Form to English Form

Translate each algebraic expression into an English phrase. (Answers may vary.) Assume a and c represent positive real numbers.

a. $7 \cdot \sqrt{10}$ b. $\sqrt{a} + c^2$

Solution:

a.   The product of 7 and the square root of 10
     OR
     Seven times the square root of 10
b.   The sum of the square root of a and the square of c
     OR
     The square root of a plus the square of c.

Tip: The phrase, "the square of c" can also be written as "c squared."

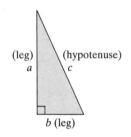

(leg) a (hypotenuse) c

b (leg)

Figure 8-1

7.   Pythagorean Theorem

Recall that the Pythagorean theorem relates the lengths of the three sides of a right triangle (Figure 8-1).

$$a^2 + b^2 = c^2$$

The principal square root can be used to solve for an unknown side of a right triangle if the lengths of the other two sides are known.

example 9   Applying the Pythagorean Theorem

Use the Pythagorean theorem and the definition of the principal square root of a number to find the length of the unknown side.

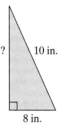

? | 10 in.

8 in.

Solution:

Label the sides of the triangle.

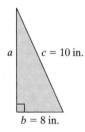

a | $c = 10$ in.

$b = 8$ in.

$a^2 + b^2 = c^2$

$a^2 + (8)^2 = (10)^2$ Apply the Pythagorean theorem.

$a^2 + 64 = 100$ Simplify.

$a^2 = 100 - 64$

$a^2 = 36$ This equation is quadratic. One method for solving the equation is to set the equation equal to zero, factor, and apply the zero product rule. However, we can also use the definition of a square root to solve for a.

$a = \sqrt{36}$ By definition, a must be one of the square roots of 36
$a = 6$ (either 6 or -6). However, because a represents a distance, choose the *positive* (principal) square root of 36.

The third side is 6 in. long.

example 10

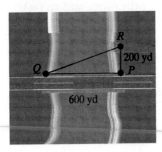

Figure 8-2

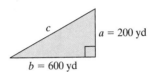

c

$a = 200$ yd

$b = 600$ yd

Classroom Activity 8.1D

Applying the Pythagorean Theorem

A bridge across a river is 600 yd long. A boat ramp is 200 yd due north of point P on the bridge, such that the line segments $\overline{PQ}$ and $\overline{PR}$ form a right angle (Figure 8-2). How far does a kayak travel if it leaves from the boat ramp and paddles to point Q? Use a calculator to round the answer to the nearest yard.

Solution:

Label the triangle:

$$a^2 + b^2 = c^2$$

$(200)^2 + (600)^2 = c^2$ Apply the Pythagorean theorem.

$40{,}000 + 360{,}000 = c^2$ Simplify.

$400{,}000 = c^2$ By definition, c must be one of the square roots of 400,000. Because the value of c is a distance, choose the positive square root of 400,000.

$c = \sqrt{400{,}000}$

$c \approx 632$ Use a calculator to approximate the positive square root of 400,000.

The kayak must travel 632 yd.

Tip: The values $\sqrt{60}$ and $\sqrt{\frac{13}{3}}$ are approximated on the calculator to 10 digits. However, $\sqrt{60}$ and $\sqrt{\frac{13}{3}}$ are actually irrational numbers. Their decimal forms are nonterminating and nonrepeating.

Calculator Connections

A calculator can be used to approximate the value of a radical expression. To evaluate a square root, use the $\sqrt{}$ key. For example, evaluate: $\sqrt{25}$, $\sqrt{60}$, $\sqrt{\frac{13}{3}}$

Scientific Calculator

Enter: 2 5 $\sqrt{}$ Result: [5]

Enter: 6 0 $\sqrt{}$ Result: [7.745966692]

Enter: 1 3 $\div$ 3 $=$ $\sqrt{}$ Result: [2.081665999]

Graphing Calculator

```
√(25)
              5
√(60)
       7.745966692
√(13/3)
       2.081665999
```

To evaluate cube roots, your calculator may have a $\boxed{\sqrt[3]{}}$ key. Otherwise for cube roots and roots of higher index (fourth roots, fifth roots, and so on), try using the $\boxed{\sqrt[x]{y}}$ key or $\boxed{\sqrt[x]{}}$ key. For example, evaluate $\sqrt[3]{64}$, $\sqrt[4]{81}$, and $\sqrt[3]{162}$:

Scientific Calculator

Enter: $\boxed{6}\ \boxed{4}\ \boxed{2^{nd}}\ \boxed{\sqrt[x]{y}}\ \boxed{3}\ \boxed{=}$ **Result:** $\boxed{\qquad 4}$

Enter: $\boxed{8}\ \boxed{1}\ \boxed{2^{nd}}\ \boxed{\sqrt[x]{y}}\ \boxed{4}\ \boxed{=}$ **Result:** $\boxed{\qquad 3}$

Enter: $\boxed{1}\ \boxed{6}\ \boxed{2}\ \boxed{2^{nd}}\ \boxed{\sqrt[x]{y}}\ \boxed{3}\ \boxed{=}$ **Result:** $\boxed{5.451361778}$

Graphing Calculator

On a graphing calculator, the index is usually entered first.

```
3ˣ√(64)
             4
4ˣ√(81)
             3
3ˣ√(162)
     5.451361778
```

Calculator Exercises

Use a calculator to evaluate each expression.

1. $\sqrt{100}$ ❖
2. $\sqrt{110}$ ❖
3. $\sqrt{9}$ ❖
4. $\sqrt{8}$ ❖

5. $\sqrt{4366}$ ❖
6. $\sqrt{86{,}035}$ ❖
7. $\sqrt{\dfrac{1}{2}}$ ❖
8. $\sqrt{\dfrac{3}{5}}$ ❖

9. $\sqrt{0.01}$ ❖
10. $\sqrt{0.25}$ ❖
11. $\sqrt[3]{64}$ ❖
12. $\sqrt[3]{65}$ ❖

13. $\sqrt[3]{1000}$ ❖
14. $\sqrt[3]{1045}$ ❖
15. $\sqrt[4]{16}$ ❖
16. $\sqrt[4]{15}$ ❖

17. $\sqrt[4]{\dfrac{8895}{13}}$ ❖
18. $\sqrt[4]{\dfrac{1{,}094{,}496}{30}}$ ❖
19. $\sqrt[5]{455}$ ❖
20. $\sqrt[5]{680{,}498}$ ❖

---

section 8.1   PRACTICE EXERCISES

1. a. What is the principal square root of 64? 8
   b. What is the negative square root of 64? −8

2. a. What is the principal square root of 144? 12
   b. What is the negative square root of 144? −12

3. Does every number have two square roots? Explain. ❖

4. Which number has only one square root? 0

5. Which of the following are perfect squares?
   {0, 1, 4, 7, 12, 16, 25, 30, 36, 42, 49}
   0, 1, 4, 16, 25, 36, 49

6. Which of the following are perfect squares?
   {50, 64, 72, 81, 95, 100, 121, 140, 144, 169}
   64, 81, 100, 121, 144, 169 ❖

7. Which of the following are perfect cubes?
   {0, 1, 3, 6, 8, 9, 16, 20, 27, 30, 36} 0, 1, 8, 27

8. Which of the following are perfect cubes?
   {42, 60, 64, 90, 111, 125, 133, 150, 216} 64, 125, 216

9. Using the definition of a square root, explain why $\sqrt{-16}$ does not have a real-valued square root. There is no real value of b for which $b^2 = -16$.

10. Using the definition of an nth root, explain why $\sqrt[4]{-16}$ does not have a real-valued fourth root.
    There is no real value of b for which $b^4 = -16$.

11. Does $\sqrt[3]{-27}$ have a real-valued cube root?
    Yes, −3

12. Does $\sqrt[4]{-81}$ have a real-valued fourth root? No

For Exercises 13–40, evaluate the roots, if possible.

13. $\sqrt{169}$ 13

14. $\sqrt{81}$ 9

15. $\sqrt{225}$ 15

16. $\sqrt{625}$ 25

17. $\sqrt[4]{81}$ 3

18. $\sqrt[3]{27}$ 3

19. $\sqrt[5]{1}$ 1

20. $\sqrt[6]{64}$ 2

21. $\sqrt[3]{0}$ 0

22. $\sqrt[4]{0}$ 0

23. $-\sqrt{4}$ −2

24. $-\sqrt{36}$ −6

25. $\sqrt{-4}$
Not a real number

26. $\sqrt{-36}$
Not a real number

27. $\sqrt[4]{-16}$
Not a real number

28. $\sqrt[4]{-1}$
Not a real number

29. $-\sqrt[4]{16}$
−2

30. $-\sqrt[4]{-1}$
Not a real number

31. $\sqrt[6]{-729}$
Not a real number

32. $\sqrt[3]{-64}$
−4

33. $-\sqrt[3]{64}$
−4

34. $-\sqrt[5]{32}$ −2

35. $\sqrt{\dfrac{1}{9}}$ $\dfrac{1}{3}$

36. $\sqrt{\dfrac{25}{16}}$ $\dfrac{5}{4}$

37. $\sqrt{0.25}$ 0.5

38. $\sqrt{0.16}$ 0.4

39. $\sqrt[4]{-256}$
Not a real number

40. $\sqrt[3]{216}$ 6

For Exercises 41–48, estimate the value of each radical. Then use a calculator to approximate the radical to three decimal places.

41. $\sqrt{5}$ 2.236

42. $\sqrt{17}$ 4.123

43. $\sqrt{50}$ 7.071

44. $\sqrt{96}$ 9.798

45. $\sqrt{33}$ 5.745

46. $\sqrt{145}$ 12.042

47. $\sqrt{80}$ 8.944

48. $\sqrt{170}$ 13.038

For Exercises 49–56, estimate the value of each radical. Then use a calculator to approximate the radical to three decimal places.

49. $\sqrt[3]{7}$ 1.913

50. $\sqrt[3]{28}$ 3.037

51. $\sqrt[3]{65}$ 4.021

52. $\sqrt[3]{124}$ 4.987

53. $\sqrt[4]{17}$ 2.031

54. $\sqrt[4]{82}$ 3.009

55. $\sqrt[4]{15}$ 1.968

56. $\sqrt[4]{140}$ 3.440

57. Determine which of the expressions are perfect squares. Then state a rule for determining perfect squares based on the exponent of the expression. ❖

$\{x^2,\ a^3,\ y^4,\ z^5,\ (ab)^6,\ (pq)^7,\ w^8x^8,\ c^9d^9,\ m^{10},\ n^{11}\}$

58. Determine which of the expressions are perfect cubes. Then state a rule for determining perfect cubes based on the exponent of the expression. ❖

$\{a^2,\ b^3,\ c^4,\ d^5,\ e^6,\ (xy)^7,\ (wz)^8,\ (pq)^9,\ t^{10}s^{10},\ m^{11}n^{11},\ u^{12}v^{12}\}$

59. Determine which of the expressions are perfect fourth powers. Then state a rule for determining perfect fourth powers based on the exponent of the expression. ❖

$\{m^2,\ n^3,\ p^4,\ q^5,\ r^6,\ s^7,\ t^8,\ u^9,\ v^{10},\ (ab)^{11},\ (cd)^{12}\}$

60. Determine which of the expressions are perfect fifth powers. Then state a rule for determining perfect fifth powers based on the exponent of the expression. ❖

$\{a^2,\ b^3,\ c^4,\ d^5,\ e^6,\ k^7,\ w^8,\ x^9,\ y^{10},\ z^{11}\}$

For Exercises 61–100, simplify the expressions.

61. $\sqrt{(4)^2}$ 4

62. $\sqrt{(8)^2}$ 8

63. $\sqrt{(-4)^2}$ 4

64. $\sqrt{(-8)^2}$ 8

65. $\sqrt[3]{(5)^3}$ 5

66. $\sqrt[3]{(7)^3}$ 7

67. $\sqrt[3]{(-5)^3}$ −5

68. $\sqrt[3]{(-7)^3}$ −7

69. $\sqrt[4]{(2)^4}$ 2

70. $\sqrt[4]{(10)^4}$ 10

71. $\sqrt[4]{(-2)^4}$ 2

72. $\sqrt[4]{(-10)^4}$ 10

73. $\sqrt[5]{(1)^5}$ 1

74. $\sqrt[5]{(3)^5}$ 3

75. $\sqrt[5]{(-1)^5}$ −1

76. $\sqrt[5]{(-3)^5}$ −3

77. $\sqrt[6]{(5)^6}$ 5

78. $\sqrt[6]{(11)^6}$ 11

79. $\sqrt[6]{(-5)^6}$ 5

80. $\sqrt[6]{(-11)^6}$ 11

81. $\sqrt{a^2}$ $|a|$

82. $\sqrt{b^2}$ $|b|$

83. $\sqrt[3]{y^3}$ y

84. $\sqrt[3]{z^3}$ z

85. $\sqrt[4]{w^4}$ $|w|$

86. $\sqrt[4]{p^4}$ $|p|$

87. $\sqrt[5]{x^5}$ x

88. $\sqrt[5]{y^5}$ y

89. $\sqrt[6]{m^6}$ $|m|$

90. $\sqrt[6]{n^6}$ $|n|$

91. $\sqrt[7]{c^7}$ c

92. $\sqrt[7]{d^7}$ d

93. $\sqrt{(5x)^2}$ $5|x|$

94. $\sqrt{(6w)^2}$ $6|w|$

95. $\sqrt{25x^2}$ $5|x|$

96. $\sqrt{36w^2}$ $6|w|$

97. $\sqrt[3]{(5p^2)^3}$ $5p^2$

98. $\sqrt[3]{(2k^4)^3}$ $2k^4$

99. $\sqrt[3]{125p^6}$ $5p^2$

100. $\sqrt[3]{8k^{12}}$ $2k^4$

For Exercises 101–104, translate the English phrases into an algebraic expression.

101. The sum of the square root of q and the square of p. $\sqrt{q} + p^2$

102. The product of the square root of eleven and the cube of x. $\sqrt{11} \cdot x^3$

103. The quotient of six and the fourth root of x. $\dfrac{6}{\sqrt[4]{x}}$

104. The difference of the square of y and one. $y^2 - 1$

❖ See Additional Answers Appendix   ✎ Writing   ⬌ Translating Expression   Geometry   Scientific Calculator   Video

For Exercises 105–108, translate each algebraic expression into an English phrase (answers may vary).

105. $5 - \sqrt{6}$
The difference of 5 and the square root of 6

106. $x^2 + \sqrt{y}$
The sum of the square of x and the square root of y

107. $4 \cdot \sqrt[3]{x}$
The product of 4 and the cube root of x

108. $\dfrac{\sqrt{6}}{14}$
The quotient of the square root of 6 and 14

For Exercises 109–114, find the length of the third side of each triangle using the Pythagorean theorem. Round to the nearest tenth if necessary.

109.
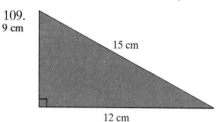
9 cm
15 cm
12 cm

110. 10 in.
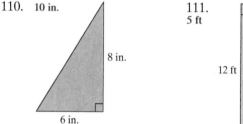
8 in.
6 in.

111.
5 ft
12 ft
13 ft

112. 3 m
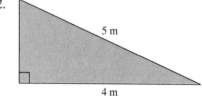
5 m
4 m

113. 6.5 cm
6.9 cm
2.4 cm

114. 11.6 ft
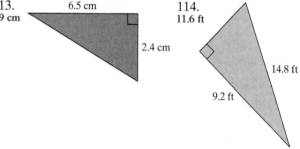
14.8 ft
9.2 ft

115. Find the length of the diagonal of the square tile shown in the figure. Round the answer to the nearest tenth of an inch. **17.0 in.**

12 in.
12 in.

Figure for Exercise 115

116. A baseball diamond is 90 ft on a side. Find the distance between home plate and second base. Round the answer to the nearest tenth of a foot. **127.3 ft**

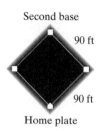
Second base
90 ft
90 ft
Home plate

Figure for Exercise 116

117. On a map, Fresno, California, is 108 miles east of Salinas, California. Reno, Nevada, is 190 miles north of Fresno. Approximate the distance between Reno and Salinas to the nearest mile. **219 miles**

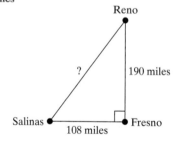
Reno
?
190 miles
Salinas
108 miles
Fresno

Figure for Exercise 117

118. Washington, DC, is north of Richmond, Virginia, a distance of 98 miles. Louisville, Kentucky, is west of Richmond, a distance of 454 miles. How far is it from Washington to Louisville? Round the answer to the nearest mile. **464 miles**

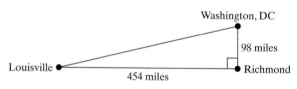
Washington, DC
98 miles
Louisville
454 miles
Richmond

Figure for Exercise 118

119. On a map, the cities Asheville, North Carolina, Roanoke, Virginia, and Greensboro, North Carolina, form a right triangle (see the figure). The distance between Asheville and Roanoke is 300 km. The distance between Roanoke and Greensboro is 134 km. How far is it from Greensboro to Asheville? Round the answer to the nearest kilometer. 268 km

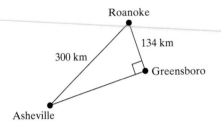

Figure for Exercise 119

120. Jackson, Mississippi, is west of Meridian, Mississippi, a distance of 141 km. Tupelo, Mississippi, is north of Meridian, a distance of 209 km. How far is it from Jackson to Tupelo? Round the answer to the nearest kilometer. 252 km

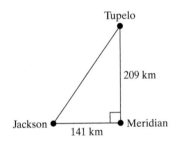

Figure for Exercise 120

121. Before a hurricane, homeowners often apply strips of masking tape over windows to prevent shards of glass from flying into a room.

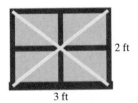

Figure for Exercise 121

a. How much masking tape is needed to tape one diagonal in the window shown in the figure? (Round to the nearest tenth of a foot.) 3.6 ft

b. How much tape is needed to tape both diagonals? 7.2 ft

c. How much tape is needed to tape 20 such windows? 144 ft

EXPANDING YOUR SKILLS

122. For what values of x will $\sqrt{x}$ be a real number?
$x \geq 0$

123. For what values of x will $\sqrt{-x}$ be a real number?
$x \leq 0$

124. A motorist must drive between Frankville and Clayton. Normally the driver takes the route around the mountains by driving through Hamilton (see figure).

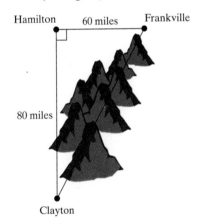

Figure for Exercise 124

a. If the driver goes around the mountains, the roads from Frankville to Hamilton and from Hamilton to Clayton are both highways. The driver can average 60 mph. How long would it take him to travel that route? $2\frac{1}{3}$ h (2 h 20 min)

b. What is the direct distance between Frankville and Clayton? 100 miles

c. If the driver goes directly from Frankville to Clayton through the mountains, he can only average 40 mph. How long would it take him to travel that route? $2\frac{1}{2}$ h (2 h 30 min)

section

8.2 PROPERTIES OF RADICALS

1. Multiplication and Division Properties of Radicals

You may have already recognized certain properties of radicals involving a product or quotient.

Multiplication and Division Properties of Radicals

Let a and b represent real numbers such that $\sqrt[n]{a}$ and $\sqrt[n]{b}$ are both real. Then,

1. $\sqrt[n]{ab} = \sqrt[n]{a} \cdot \sqrt[n]{b}$ **Multiplication property of radicals**

2. $\sqrt[n]{\dfrac{a}{b}} = \dfrac{\sqrt[n]{a}}{\sqrt[n]{b}}$ $b \neq 0$ **Division property of radicals**

The multiplication and division properties of radicals indicate that a product or quotient within a radicand can be written as a product or quotient of radicals, provided the roots are real numbers.

$$\sqrt{100} = \sqrt{25} \cdot \sqrt{4}$$

$$\sqrt{\frac{4}{9}} = \frac{\sqrt{4}}{\sqrt{9}}$$

The reverse process is also true. A product or quotient of radicals can be written as a single radical, provided the roots are real numbers and they have the same indices.

Same index
$$\sqrt{2} \cdot \sqrt{18} = \sqrt{36}$$

Same index
$$\frac{\sqrt[3]{125}}{\sqrt[3]{8}} = \sqrt[3]{\frac{125}{8}}$$

2. Simplified Form of a Radical

In algebra it is customary to simplify radical expressions as much as possible. A radical expression is in simplified form if all of the following conditions are true.

Simplified Form of a Radical

Consider any radical expression where the radicand is written as a product of prime factors. The expression is in **simplified form** if all of the following conditions are met:

1. The radicand has no factor raised to a power greater than or equal to the index.
2. There are no radicals in the denominator of a fraction.
3. The radicand does not contain a fraction.

---

3. Simplifying Radicals Using the Multiplication Property of Radicals

The expression $\sqrt{x^2}$ is not simplified because it fails condition 1. Because x^2 is a perfect square, $\sqrt{x^2}$ is easily simplified.

$$\sqrt{x^2} = x \quad (\text{for } x \geq 0).$$

However, how is an expression such as $\sqrt{x^7}$ simplified? This and many other radical expressions are simplified using the multiplication property of radicals. The following examples illustrate how nth powers can be removed from the radicands of nth-roots.

---

example 1

Using the Multiplication Property to Simplify a Radical Expression

Use the multiplication property of radicals to simplify the expression $\sqrt{x^7}$. Assume $x \geq 0$.

Solution:

The expression $\sqrt{x^7}$ is equivalent to $\sqrt{x^6 \cdot x}$. By applying the multiplication property of radicals, we have

$$\sqrt{x^6 \cdot x} = \sqrt{x^6} \cdot \sqrt{x} \qquad x^6 \text{ is a perfect square because } (x^3)^2 = x^6$$

$$= x^3 \cdot \sqrt{x} \qquad \text{Simplify.}$$

$$= x^3\sqrt{x}$$

In Example 1, the expression x^7 is not a perfect square. Therefore, to simplify $\sqrt{x^7}$, it was necessary to write the expression as the product of the largest perfect square and a remaining or "left-over" factor: $\sqrt{x^7} = \sqrt{x^6 \cdot x}$. This process also applies to simplifying nth-roots as shown in the next example.

---

example 2

Using the Multiplication Property to Simplify Radicals

Use the multiplication property of radicals to simplify the expressions. Assume all variables represent positive real numbers.

a. $\sqrt{a^{15}}$ b. $\sqrt{x^2y^5}$ c. $\sqrt[3]{z^5}$ d. $\sqrt[4]{s^9t^{12}}$

Solution:

The goal is to rewrite each radicand as the product of the largest perfect square (perfect cube, perfect fourth-power, etc.) and a left-over factor.

a. $\sqrt{a^{15}}$

$$= \sqrt{a^{14} \cdot a} \qquad a^{14} \text{ is the largest perfect square in the radicand.}$$

$$= \sqrt{a^{14}} \cdot \sqrt{a} \qquad \text{Apply the multiplication property of radicals.}$$

$$= a^7\sqrt{a} \qquad \text{Simplify.}$$

b. $\sqrt{x^2y^5}$

$\qquad = \sqrt{x^2y^4 \cdot y}$ $\qquad$ x^2y^4 is the largest perfect square in the radicand.

$\qquad = \sqrt{x^2y^4} \cdot \sqrt{y}$ $\qquad$ Apply the multiplication property of radicals.

$\qquad = xy^2\sqrt{y}$ $\qquad$ Simplify.

c. $\sqrt[3]{z^5}$

$\qquad = \sqrt[3]{z^3 \cdot z^2}$ $\qquad$ z^3 is the largest perfect cube in the radicand.

$\qquad = \sqrt[3]{z^3} \cdot \sqrt[3]{z^2}$ $\qquad$ Apply the multiplication property of radicals.

$\qquad = z\sqrt[3]{z^2}$ $\qquad$ Simplify.

d. $\sqrt[4]{s^9t^{12}}$

$\qquad = \sqrt[4]{s^8t^{12} \cdot s}$ $\qquad$ s^8t^{12} is the largest perfect fourth power in the radicand.

$\qquad = \sqrt[4]{s^8t^{12}} \cdot \sqrt[4]{s}$ $\qquad$ Apply the multiplication property of radicals.

$\qquad = s^2t^3\sqrt[4]{s}$ $\qquad$ Simplify.

Each expression in Example 2 involved a radicand that is a product of variable factors. If a numerical factor is present, sometimes it is necessary to factor the coefficient before simplifying the radical.

---

example 3   **Using the Multiplication Property to Simplify Radicals**

Use the multiplication property of radicals to simplify the expressions. Assume all variables represent positive real numbers.

a. $\sqrt{50}$ $\qquad$ b. $\sqrt[3]{80}$ $\qquad$ c. $\sqrt{24a^6}$ $\qquad$ d. $\sqrt[4]{81x^4y^9}$

Classroom Activity 8.2A

Solution:

The goal is to rewrite each radicand as the product of the largest perfect square (perfect cube, perfect fourth-power, etc.) and a left-over factor.

a. Write the radicand as a product of prime factors. From the prime factorization, the largest perfect square is easily identified.

$\qquad \sqrt{50} = \sqrt{5^2 \cdot 2}$ $\qquad$ Factor the radicand. $\qquad$ 2 | 50

$\qquad\qquad\qquad\qquad$ 5^2 is the largest perfect square. $\qquad$ 5 | 25

$\qquad\qquad = \sqrt{5^2} \cdot \sqrt{2}$ $\qquad$ Apply the multiplication $\qquad$ 5 | 5

$\qquad\qquad\qquad\qquad$ property of radicals. $\qquad$ 1

$\qquad\qquad = 5\sqrt{2}$ $\qquad$ Simplify.

b. $\sqrt[3]{80} = \sqrt[3]{2^4 \cdot 5}$ $\qquad$ Write the radicand as a $\qquad$ 2 | 80

$\qquad\qquad\qquad\qquad$ product of prime factors. $\qquad$ 2 | 40

$\qquad\qquad = \sqrt[3]{2^3 \cdot 2 \cdot 5}$ $\qquad$ 2^3 is the largest perfect $\qquad$ 2 | 20

$\qquad\qquad\qquad\qquad$ cube in the radicand. $\qquad$ 2 | 10

$\qquad\qquad = \sqrt[3]{2^3} \cdot \sqrt[3]{2 \cdot 5}$ $\qquad$ Apply the multiplication $\qquad$ 5 | 5

$\qquad\qquad\qquad\qquad$ property of radicals. $\qquad$ 1

$\qquad\qquad = 2\sqrt[3]{10}$ $\qquad$ Simplify.

Avoiding Mistakes

The multiplication property of radicals allows us to simplify a product of factors within a radical. For example:

$$\sqrt{x^2y^2} = \sqrt{x^2} \cdot \sqrt{y^2} = xy$$
(for $x \geq 0$ and $y \geq 0$)

However, this rule does not apply to *terms* that are added or subtracted within the radical. For example,

$$\sqrt{x^2 + y^2} \quad \text{and} \quad \sqrt{x^2 - y^2}$$
cannot be simplified

c. $\sqrt{24a^6} = \sqrt{2^3 \cdot 3 \cdot a^6}$ Write the radicand as a product of prime factors $24 = 2^3 \cdot 3$.

$\phantom{\sqrt{24a^6}} = \sqrt{2^2 a^6 \cdot 2 \cdot 3}$ $2^2 a^6$ is the largest perfect square in the radicand.

$\phantom{\sqrt{24a^6}} = \sqrt{2^2 a^6} \cdot \sqrt{2 \cdot 3}$ Apply the multiplication property of radicals.

$\phantom{\sqrt{24a^6}} = 2a^3\sqrt{6}$ Simplify.

d. $\sqrt[4]{81x^4y^9} = \sqrt[4]{3^4 x^4 y^9}$ Write the radicand as a product of prime factors $81 = 3^4$.

$\phantom{\sqrt[4]{81x^4y^9}} = \sqrt[4]{3^4 x^4 y^8 \cdot y}$ $3^4 x^4 y^8$ is the largest fourth power in the radicand.

$\phantom{\sqrt[4]{81x^4y^9}} = \sqrt[4]{3^4 x^4 y^8} \cdot \sqrt[4]{y}$ Apply the multiplication property of radicals.

$\phantom{\sqrt[4]{81x^4y^9}} = 3xy^2\sqrt[4]{y}$ Simplify.

4. Simplifying Radicals Using the Division Property of Radicals

The division property of radicals indicates that a radical of a quotient can be written as the quotient of the radicals and vice versa, provided all roots are real numbers.

example 4

Simplifying Radicals Using the Division Property of Radicals

Use the division property of radicals to simplify the expressions. Assume all variables represent positive real numbers.

a. $\sqrt{\dfrac{a^5}{a^3}}$ b. $\sqrt[3]{\dfrac{a^6}{b^3}}$ c. $\sqrt{\dfrac{27x^5}{3x}}$ d. $\dfrac{\sqrt[3]{2}}{\sqrt[3]{16}}$

Classroom Activity 8.2B

Solution:

a. $\sqrt{\dfrac{a^5}{a^3}}$ The radical contains a fraction. However, the fraction can be reduced.

$ = \sqrt{a^2}$ Reduce the fraction.

$ = a$ Simplify the radical.

b. $\sqrt[3]{\dfrac{a^6}{b^3}}$ The radical contains an irreducible fraction.

$ = \dfrac{\sqrt[3]{a^6}}{\sqrt[3]{b^3}}$ Apply the division property of radicals.

$ = \dfrac{a^2}{b}$ Simplify.

c. $\sqrt{\dfrac{27x^5}{3x}}$ The fraction within the radicand can be reduced.

$= \sqrt{9x^4}$ Reduce.

$= 3x^2$ Simplify.

d. $\dfrac{\sqrt[3]{2}}{\sqrt[3]{16}}$ Notice that the radicands have a common factor.

$= \sqrt[3]{\dfrac{2}{16}}$ Apply the division property of radicals to write the radicals as a single radical.

$= \sqrt[3]{\dfrac{1}{8}}$ Reduce the fraction.

$= \dfrac{1}{2}$ Simplify.

---

example 5

Simplifying Radicals Using the Division Property of Radicals

Use the division property of radicals to simplify the expression

$$\dfrac{5\sqrt{20}}{2}$$

Avoiding Mistakes

The division property of radicals allows us to reduce a ratio of two radicals, provided they have the same index. For example:

$$\dfrac{\sqrt{15}}{\sqrt{5}} = \sqrt{\dfrac{15}{5}} = \sqrt{3}$$

However, a factor within the radicand cannot be simplified with a factor outside the radicand. For example,

$\dfrac{\sqrt{15}}{5}$ cannot be simplified

Solution:

$\dfrac{5\sqrt{20}}{2} = \dfrac{5\sqrt{2^2 \cdot 5}}{2}$ 2^2 is the largest perfect square in the radicand.

$= \dfrac{5\sqrt{2^2} \cdot \sqrt{5}}{2}$ Apply the multiplication property of radicals.

$= \dfrac{5 \cdot 2\sqrt{5}}{2}$ Simplify the radical.

$= \dfrac{5 \cdot \cancel{2}\sqrt{5}}{\cancel{2}}$ Reduce.

$= 5\sqrt{5}$

Calculator Connections

A calculator can support the multiplication property of radicals. For example, use a calculator to evaluate $\sqrt{50}$ and its simplified form $5\sqrt{2}$.

Scientific Calculator

Enter: 5 0 √   **Result:** 7.071067812

Enter: 2 √ × 5 =   **Result:** 7.071067812

Graphing Calculator

```
√(50)
          7.071067812
5*√(2)
          7.071067812
```

Tip: The decimal approximation for $\sqrt{50}$ and $5\sqrt{2}$ agree for the first 10 digits. This in itself does not make $\sqrt{50} = 5\sqrt{2}$. It is the multiplication property of radicals that guarantees that the expressions are equal.

A calculator can support the division property of radicals. For example, use a calculator to evaluate $\sqrt{\dfrac{21}{5}}$ and its equivalent form $\dfrac{\sqrt{21}}{\sqrt{5}}$

Scientific Calculator

Enter: 2 1 ÷ 5 = √   **Result:** 2.049390153

Enter: 2 1 √ ÷ 5 √ =   **Result:** 2.049390153

Graphing Calculator

```
√(21/5)
          2.049390153
√(21)/√(5)
          2.049390153
```

Tip: The calculator approximations agree to the first 10 digits. However, it is the division property of radicals that guarantees that

$$\sqrt{\frac{21}{5}} = \frac{\sqrt{21}}{\sqrt{5}}$$

Calculator Exercises

Simplify the radical expressions algebraically. Then use a calculator to approximate the original expression and its simplified form.

1. $\sqrt{125}$ ❖   2. $\sqrt{18}$ ❖   3. $\sqrt{288}$ ❖   4. $\sqrt[3]{54}$ ❖   5. $\sqrt[3]{108}$ ❖   6. $\sqrt{180}$ ❖

Use a calculator to find a decimal approximation for both the right- and left-hand sides of the expressions.

7. $\sqrt{\dfrac{40}{3}} = \dfrac{\sqrt{40}}{\sqrt{3}}$ ❖   8. $\sqrt{\dfrac{105}{17}} = \dfrac{\sqrt{105}}{\sqrt{17}}$ ❖   9. $\sqrt[3]{\dfrac{128}{2}} = \dfrac{\sqrt[3]{128}}{\sqrt[3]{2}}$ ❖

❖ See Additional Answers Appendix   🖉 Writing   ⬌ Translating Expression   Geometry   Scientific Calculator   Video

section 8.2 PRACTICE EXERCISES

1. Which of the following are perfect squares?
$\{2, 4, 6, 16, 20, 25, x^2, x^3, x^{15}, x^{20}, x^{25}\}$ **8.1**
4, 16, 25, x^2, x^{20}

2. Which of the following are perfect cubes?
$\{3, 6, 8, 9, 12, 27, y^3, y^8, y^9, y^{12}, y^{27}\}$ **8.1**
8, 27, y^3, y^9, y^{12}, y^{27}

3. Which of the following are perfect fourth powers? $\{4, 16, 20, 25, 81, w^4, w^{16}, w^{20}, w^{25}, w^{81}\}$ **8.1**
16, 81, w^4, w^{16}, w^{20}

4. Which of the following are perfect fifth powers?
$\{5, 10, 25, 32, 243, z^5, z^{10}, z^{25}, z^{32}\}$ **8.1**
32, 243, z^5, z^{10}, z^{25}

For Exercises 5–12, simplify the expressions, if possible. Assume all variables represent positive real numbers. **8.1**

5. $-\sqrt{25}$ -5

6. $\sqrt{-25}$ Not a real number

7. $-\sqrt[3]{27}$ -3

8. $\sqrt[3]{-27}$ -3

9. $\sqrt[4]{a^8}$ a^2

10. $\sqrt[5]{b^{15}}$ b^3

11. $\sqrt{4x^2y^4}$ $2xy^2$

12. $\sqrt{9p^{10}}$ $3p^5$

13. On a map, Seattle, Washington, is 378 km west of Spokane, Washington. Portland, Oregon, is 236 km south of Seattle. Approximate the distance between Portland and Spokane to the nearest kilometer. **8.1**   446 km

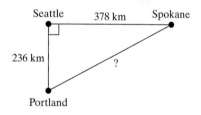

Figure for Exercise 13

14. A new roof is needed on a shed. How many square feet of tar paper are needed to cover the top of the roof? **8.1**   1040 sq. ft.

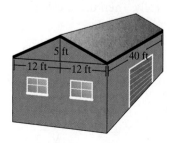

Figure for Exercise 14

For Exercises 15–18, round answers to three decimal places.

15. a. Use a calculator to approximate $\sqrt{80}$.   8.944
    b. Simplify $\sqrt{80}$. $4\sqrt{5}$
    c. Use a calculator to approximate the simplified expression from part (b).   8.944
    d. Compare the answers to parts (a) and (c).
       They are the same.

16. a. Use a calculator to approximate $\sqrt{12}$.   3.464
    b. Simplify $\sqrt{12}$. $2\sqrt{3}$
    c. Use a calculator to approximate the simplified expression from part (b).   3.464
    d. Compare the answers to parts (a) and (c).
       They are the same.

17. a. Use a calculator to approximate $\sqrt{98}$.   9.899
    b. Simplify $\sqrt{98}$. $7\sqrt{2}$
    c. Use a calculator to approximate the simplified expression from part (b).   9.899
    d. Compare the answers to parts (a) and (c).
       They are the same.

18. a. Use a calculator to approximate $\sqrt{60}$.   7.746
    b. Simplify $\sqrt{60}$. $2\sqrt{15}$
    c. Use a calculator to approximate the simplified expression from part (b).   7.746
    d. Compare the answers to parts (a) and (c).
       They are the same.

For Exercises 19–58, use the multiplication property of radicals to simplify the expressions. Assume all variables represent positive real numbers.

19. $\sqrt{18}$ $3\sqrt{2}$

20. $\sqrt{75}$ $5\sqrt{3}$

21. $\sqrt{28}$ $2\sqrt{7}$

22. $\sqrt{40}$ $2\sqrt{10}$

23. $\sqrt{a^5}$ $a^2\sqrt{a}$

24. $\sqrt{b^9}$ $b^4\sqrt{b}$

25. $\sqrt{w^{22}}$ w^{11}

26. $\sqrt{p^{18}}$ p^9

27. $\sqrt{m^4n^5}$ $m^2n^2\sqrt{n}$

28. $\sqrt{c^2d^9}$ $cd^4\sqrt{d}$

29. $\sqrt{x^{13}y^{10}}$ $x^6y^5\sqrt{x}$

30. $\sqrt{u^{10}v^7}$ $u^5v^3\sqrt{v}$

31. $\sqrt{8x^3}$ $2x\sqrt{2x}$

32. $\sqrt{27y^5}$ $3y^2\sqrt{3y}$

33. $\sqrt{16z^3}$ $4z\sqrt{z}$

34. $\sqrt{9y^5}$ $3y^2\sqrt{y}$

35. $-\sqrt{45w^6}$ $-3w^3\sqrt{5}$

36. $-\sqrt{56v^8}$ $-2v^4\sqrt{14}$

37. $-\sqrt{15z^{11}}$ $-z^5\sqrt{15z}$

38. $-\sqrt{6k^{15}}$ $-k^7\sqrt{6k}$

39. $\sqrt{104a^2b^7}$ $2ab^3\sqrt{26b}$

40. $\sqrt{88m^4n^{11}}$ $2m^2n^5\sqrt{22n}$

41. $\sqrt{26pq}$ $\sqrt{26pq}$

42. $\sqrt{15a}$ $\sqrt{15a}$

43. $\sqrt[3]{a^8}$ $a^2\sqrt[3]{a^2}$

44. $\sqrt[3]{8v^3}$ $2v$

45. $\sqrt[3]{16z^3}$ $2z\sqrt[3]{2}$

46. $\sqrt[3]{54t^6}$ $3t^2\sqrt[3]{2}$

47. $\sqrt[4]{16a^5b^6}$ $2ab\sqrt[4]{ab^2}$

48. $\sqrt[4]{81p^9q^{11}}$ $3p^2q^2\sqrt[4]{pq^3}$

49. $\sqrt[5]{u^{10}v^7}$ $u^2v\sqrt[5]{v^2}$

50. $\sqrt[5]{25w^6}$ $w\sqrt[5]{25w}$

51. $\sqrt[5]{64k^5}$ $2k\sqrt[5]{2}$

52. $\sqrt[5]{s^7t^{11}}$ $st^2\sqrt[5]{s^2t}$

53. $\sqrt[4]{m^{11}n^{16}}$ $m^2n^4\sqrt[4]{m^3}$

54. $\sqrt[4]{c^4d^{13}}$ $cd^3\sqrt[4]{d}$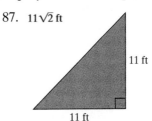

55. $\sqrt[4]{48a^3b^5c^4}$ $2bc\sqrt[4]{3a^3b}$

56. $\sqrt{18xy^4z^3}$ $3y^2z\sqrt{2xz}$

57. $\sqrt{75u^4v^5}$ $5u^2v^2\sqrt{3v}$

58. $\sqrt[3]{96p^5q^2}$ $2p\sqrt[3]{12p^2q^2}$

For Exercises 59–86, use the division property of radicals to simplify the expressions. Assume all variables represent positive real numbers.

59. $\sqrt{\dfrac{3}{16}}$ $\dfrac{\sqrt{3}}{4}$

60. $\sqrt{\dfrac{7}{25}}$ $\dfrac{\sqrt{7}}{5}$

61. $\sqrt{\dfrac{a^4}{b^4}}$ $\dfrac{a^2}{b^2}$

62. $\sqrt{\dfrac{y^6}{z^2}}$ $\dfrac{y^3}{z}$

63. $\sqrt{\dfrac{a^9}{a}}$ (*Hint:* Reduce the radicand first.) a^4

64. $\sqrt{\dfrac{x^5}{x}}$ (*Hint:* Reduce the radicand first.) x^2

65. $\sqrt{\dfrac{9}{36}}$ $\dfrac{1}{2}$

66. $\sqrt{\dfrac{4}{64}}$ $\dfrac{1}{4}$

67. $\sqrt{\dfrac{c^3}{4}}$ $\dfrac{c\sqrt{c}}{2}$

68. $\sqrt{\dfrac{d^5}{9}}$ $\dfrac{d^2\sqrt{d}}{3}$

69. $\sqrt{\dfrac{a^9}{b^4}}$ $\dfrac{a^4\sqrt{a}}{b^2}$

70. $\sqrt{\dfrac{y^3}{z^{10}}}$ $\dfrac{y\sqrt{y}}{z^5}$

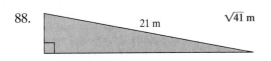 71. $\sqrt{\dfrac{200}{81}}$ $\dfrac{10\sqrt{2}}{9}$

72. $\sqrt{\dfrac{80}{49}}$ $\dfrac{4\sqrt{5}}{7}$

73. $\dfrac{\sqrt{8}}{\sqrt{50}}$ (*Hint:* Write the expression as a single radical and simplify.) $\dfrac{2}{5}$

74. $\dfrac{\sqrt{21}}{\sqrt{12}}$ (*Hint:* Write the expression as a single radical and simplify.) $\dfrac{\sqrt{7}}{2}$

75. $\sqrt{\dfrac{p}{4p^3}}$ $\dfrac{1}{2p}$

76. $\dfrac{\sqrt{9t}}{\sqrt{t^5}}$ $\dfrac{3}{t^2}$

77. $\dfrac{\sqrt[3]{z^4}}{\sqrt[3]{z}}$ z

78. $\dfrac{\sqrt[3]{w^8}}{\sqrt[3]{w^2}}$ w^2

79. $\sqrt[3]{\dfrac{x^2}{27}}$ $\dfrac{\sqrt[3]{x^2}}{3}$

80. $\sqrt[3]{\dfrac{c^2}{8}}$ $\dfrac{\sqrt[3]{c^2}}{2}$

81. $\sqrt[3]{\dfrac{y^5}{27y^3}}$ $\dfrac{\sqrt[3]{y^2}}{3}$

82. $\sqrt[3]{\dfrac{7ac}{64c^4}}$ $\dfrac{\sqrt[3]{7a}}{4c}$

83. $\dfrac{\sqrt[4]{2w^6}}{\sqrt[4]{32w^2}}$ $\dfrac{w}{2}$

84. $\dfrac{\sqrt[4]{81m}}{\sqrt[4]{16m^5}}$ $\dfrac{3}{2m}$

85. $\sqrt[4]{\dfrac{p^7q^5}{p^3}}$ $pq\sqrt[4]{q}$

86. $\sqrt[3]{\dfrac{7a^2b^3}{8c^6}}$ $\dfrac{b\sqrt[3]{7a^2}}{2c^2}$

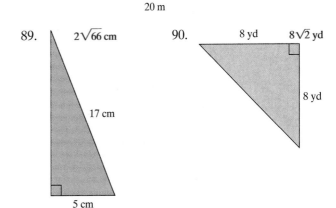

For Exercises 87–90, find the exact length of the third side of each triangle using the Pythagorean theorem. Simplify the radical if possible.

87. $11\sqrt{2}$ ft

   11 ft   11 ft

88. 21 m $\sqrt{41}$ m   20 m

89. $2\sqrt{66}$ cm   17 cm   5 cm

90. 8 yd $8\sqrt{2}$ yd   8 yd

For Exercises 91–114, simplify the expressions. Assume all variables represent positive real numbers.

91. $\sqrt[3]{16a^3}$ $2a\sqrt[3]{2}$

92. $\sqrt[3]{125x^6}$ $5x^2$

93. $\sqrt{16a^3}$ $4a\sqrt{a}$

94. $\sqrt{125x^6}$ $5x^3\sqrt{5}$

95. $\sqrt{\dfrac{4x^3}{y^2}}$ $\dfrac{2x\sqrt{x}}{y}$

96. $\sqrt{\dfrac{9z^5}{w^2}}$ $\dfrac{3z^2\sqrt{z}}{w}$

97. $\sqrt[3]{\dfrac{b^4}{27b}}$ $\dfrac{b}{3}$

98. $\sqrt[3]{\dfrac{k^6}{64}}$ $\dfrac{k^2}{4}$

99. $\sqrt[4]{32}$ $2\sqrt[4]{2}$

100. $\sqrt[4]{64}$ $2\sqrt[4]{4}$

101. $\sqrt{52u^4v^7}$ $2u^2v^3\sqrt{13v}$

102. $\sqrt{44p^8q^{10}}$ $2p^4q^5\sqrt{11}$

103. $\sqrt{216}$ $6\sqrt{6}$

104. $\sqrt{250}$ $5\sqrt{10}$

105. $\sqrt[3]{216}$ 6

106. $\sqrt[3]{250}$ $5\sqrt[3]{2}$

107. $\dfrac{\sqrt{3}}{\sqrt{27}}$ $\dfrac{1}{3}$

108. $\dfrac{\sqrt{5}}{\sqrt{125}}$ $\dfrac{1}{5}$

109. $\sqrt[3]{\dfrac{x^5}{x^2}}$ x

110. $\sqrt[3]{\dfrac{y^{11}}{y^2}}$ y^3

111. $\sqrt[3]{15m^4n^{22}}$ $mn^7\sqrt[3]{15mn}$

112. $\sqrt[3]{20s^{15}t^{11}}$ $s^5t^3\sqrt[3]{20t^2}$

113. $\sqrt[4]{8p^2q}$ $\sqrt[4]{8p^2q}$

114. $\sqrt[4]{6cd^3}$ $\sqrt[4]{6cd^3}$

■ EXPANDING YOUR SKILLS

For Exercises 115–118, simplify the expressions. Assume all variables represent positive real numbers.

115. $\sqrt{(-2-5)^2+(-4+3)^2}$ $5\sqrt{2}$

116. $\sqrt{(-1-7)^2+[1-(-1)]^2}$ $2\sqrt{17}$

117. $\sqrt{x^2+10x+25}$ $x+5$

118. $\sqrt{x^2+6x+9}$ $x+3$

Concepts

1. **Definition of *Like* Radicals**
2. **Addition and Subtraction of Radicals**
3. **Recognizing Un*like* Radicals**

section

8.3   ADDITION AND SUBTRACTION OF RADICALS

1.   Definition of *Like* Radicals

Definition of *Like* Radicals

Two radical expressions are said to be ***like* radicals** if their radical factors have the same index and the same radicand.

The following are pairs of *like* radicals:

Same index

$2x\sqrt{6}$ and $5x\sqrt{6}$ Indices and radicands are the same.

Same radicand

Same index

$-4\sqrt[3]{17y}$ and $\dfrac{1}{2}\sqrt[3]{17y}$ Indices and radicands are the same.

Same radicand

These pairs are *not like* radicals:

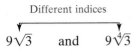

Different indices

$9\sqrt{3}$ and $9\sqrt[4]{3}$ Radicals have different indices.

$8ab\sqrt{5}$ and $ab\sqrt{10}$ Radicals have different radicands.

Different radicands

2. Addition and Subtraction of Radicals

Expressions with radicals can be added or subtracted if they are *like* radicals. To add or subtract *like* radicals, use the distributive property. For example:

$$3\sqrt{7} + 5\sqrt{7} = (3 + 5)\sqrt{7}$$
$$= 8\sqrt{7}$$

$$9\sqrt[3]{2y} - 4\sqrt[3]{2y} = (9 - 4)\sqrt[3]{2y}$$
$$= 5\sqrt[3]{2y}$$

example 1   **Adding and Subtracting Radicals**

Add or subtract the radicals as indicated. Assume all variables represent positive real numbers.

a. $\sqrt{5} + \sqrt{5}$ b. $6\sqrt[3]{15} + 3\sqrt[3]{15} + \sqrt[3]{15}$

c. $\sqrt{xy} - 6\sqrt{xy} + 4\sqrt{xy}$ d. $4\sqrt{2} - 7\sqrt{2}$

Solution:

a. $\sqrt{5} + \sqrt{5}$

$= 1\sqrt{5} + 1\sqrt{5}$ *Note:* $\sqrt{5} = 1\sqrt{5}$.

$= (1 + 1)\sqrt{5}$ Apply the distributive property.

$= 2\sqrt{5}$ Simplify.

b. $6\sqrt[3]{15} + 3\sqrt[3]{15} + \sqrt[3]{15}$ The radicals have the same radicand and same index.

$= 6\sqrt[3]{15} + 3\sqrt[3]{15} + 1\sqrt[3]{15}$ *Note:* $\sqrt[3]{15} = 1\sqrt[3]{15}$.

$= (6 + 3 + 1)\sqrt[3]{15}$ Apply the distributive property.

$= 10\sqrt[3]{15}$

c. $\sqrt{xy} - 6\sqrt{xy} + 4\sqrt{xy}$ The radicals have the same radicand and same index.

$= 1\sqrt{xy} - 6\sqrt{xy} + 4\sqrt{xy}$ *Note:* $\sqrt{xy} = 1\sqrt{xy}$.

$= (1 - 6 + 4)\sqrt{xy}$ Apply the distributive property.

$= -1\sqrt{xy}$ Simplify.

$= -\sqrt{xy}$

⬢ **Avoiding Mistakes**

The process of adding *like* radicals with the distributive property is similar to adding *like* terms. The end result is that the numerical coefficients are added and the radical factor is unchanged.

$\sqrt{5} + \sqrt{5}$

$= 1\sqrt{5} + 1\sqrt{5}$

$= 2\sqrt{5}$ Correct

Be careful: $\sqrt{5} + \sqrt{5} \neq \sqrt{10}$
In general:
$\sqrt{x} + \sqrt{y} \neq \sqrt{x + y}$

d. $4\sqrt{2} - 7\sqrt{2}$ The radicals have the same radicand and same index.

$= (4 - 7)\sqrt{2}$ Apply the distributive property.

$= -3\sqrt{2}$

Sometimes it is necessary to simplify radicals before adding or subtracting.

example 2   **Simplifying Radicals Before Adding or Subtracting**

Add or subtract the radicals as indicated. Assume all variables represent positive real numbers.

a. $\sqrt{20} + 7\sqrt{5}$ b. $\sqrt{50} - \sqrt{8}$

c. $-4\sqrt{3x^2} - x\sqrt{27} + 5x\sqrt{3}$ d. $a\sqrt[3]{8a^4} + 6\sqrt[3]{a^7}$

Classroom Activity 8.3A

Solution:

a. $\sqrt{20} + 7\sqrt{5}$ Because the radicands are different, try simplifying the radicals first.

$= \sqrt{2^2 \cdot 5} + 7\sqrt{5}$ Factor the radicand.

$= 2\sqrt{5} + 7\sqrt{5}$ The terms have *like* radicals.

$= (2 + 7)\sqrt{5}$ Apply the distributive property.

$= 9\sqrt{5}$ Simplify.

b. $\sqrt{50} - \sqrt{8}$ Because the radicands are different, try simplifying the radicals first.

$= \sqrt{5^2 \cdot 2} - \sqrt{2^2 \cdot 2}$ Factor the radicand.

$= 5\sqrt{2} - 2\sqrt{2}$ The terms have *like* radicals.

$= (5 - 2)\sqrt{2}$ Apply the distributive property.

$= 3\sqrt{2}$ Simplify.

c. $-4\sqrt{3x^2} - x\sqrt{27} + 5x\sqrt{3}$ Simplify each radical.

$= -4\sqrt{3x^2} - x\sqrt{3^2 \cdot 3} + 5x\sqrt{3}$ Factor the radicands.

$= -4x\sqrt{3} - 3x\sqrt{3} + 5x\sqrt{3}$ The terms have *like* radicals.

$= (-4 - 3 + 5)x\sqrt{3}$ Apply the distributive property.

$= -2x\sqrt{3}$ Simplify.

d. $a\sqrt[3]{8a^4} + 6\sqrt[3]{a^7}$ Simplify each radical.

$= a\sqrt[3]{2^3a^3 \cdot a} + 6\sqrt[3]{a^6 \cdot a}$ Factor the radicands.

$= a \cdot 2a\sqrt[3]{a} + 6 \cdot a^2\sqrt[3]{a}$ The terms have *like* radicals.

$= 2a^2\sqrt[3]{a} + 6a^2\sqrt[3]{a}$

$= (2 + 6)a^2\sqrt[3]{a}$ Apply the distributive property.

$= 8a^2\sqrt[3]{a}$ Simplify.

3. Recognizing Un*like* Radicals

It is important to realize that only *like* radicals may be added or subtracted. The next example provides extra practice for recognizing un*like* radicals.

example 3

Recognizing Un*like* Radicals

The following radicals cannot be simplified further by adding or subtracting. Why?

a. $2\sqrt{7} - 5\sqrt{3}$ b. $\sqrt{x} + \sqrt[3]{x}$ c. $7 + 4\sqrt{5}$ d. $4\sqrt{20} - 3\sqrt{8}$

Solution:

a. $2\sqrt{7} - 5\sqrt{3}$ Un*like* radicals. Radicands are not the same.

b. $\sqrt{x} + \sqrt[3]{x}$ Un*like* radicals. Indices are not the same.

c. $7 + 4\sqrt{5}$ Un*like* radicals. One term has a radical, one does not.

d. $4\sqrt{20} - 3\sqrt{8}$ Simplify radicals first.

$4 \cdot 2\sqrt{5} - 3 \cdot 2\sqrt{2}$

$8\sqrt{5} - 6\sqrt{2}$ Un*like* radicals. Radicands are not the same.

Calculator Connections

A calculator can be used to evaluate a radical expression and its simplified form. For example, use a calculator to evaluate the expression on the left and its simplified form on the right: $2\sqrt{5} + 6\sqrt{5} = 8\sqrt{5}$

Scientific Calculator

Enter: $\boxed{5}$ $\boxed{\sqrt{\ }}$ $\boxed{\times}$ $\boxed{2}$ $\boxed{=}$ $\boxed{+}$ $\boxed{5}$ $\boxed{\sqrt{\ }}$ $\boxed{\times}$ $\boxed{6}$ $\boxed{=}$ **Result:** $\boxed{17.88854382}$

Enter: $\boxed{5}$ $\boxed{\sqrt{\ }}$ $\boxed{\times}$ $\boxed{8}$ $\boxed{=}$ **Result:** $\boxed{17.88854382}$

Graphing Calculator

```
2√(5)+6√(5)
        17.88854382
8√(5)
        17.88854382
```

A calculator can help you determine when a rule has been applied *incorrectly*. For example, use a calculator to show that $\sqrt{3} + \sqrt{5} \neq \sqrt{8}$

Scientific Calculator

Enter: $\boxed{3}$ $\boxed{\sqrt{\ }}$ $\boxed{+}$ $\boxed{5}$ $\boxed{\sqrt{\ }}$ $\boxed{=}$ **Result:** $\boxed{3.968118785}$ ← Values are

Enter: $\boxed{8}$ $\boxed{\sqrt{\ }}$ **Result:** $\boxed{2.828427125}$ ← not equal.

Graphing Calculator

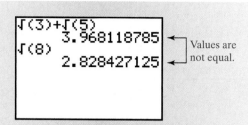

Values are not equal.

Calculator Exercises

Simplify the radical expression algebraically. Then use a calculator to approximate the original expression and its simplified form.

1. $2\sqrt{3} + 4\sqrt{3}$ ❖

2. $-\sqrt{5} - 4\sqrt{5} + 3\sqrt{5}$ ❖

3. $\sqrt{20} + \sqrt{5}$ ❖

4. $4\sqrt[3]{6} - 7\sqrt[3]{6}$ ❖

---

section 8.3   PRACTICE EXERCISES

For Exercises 1–8, simplify the expressions. Assume all variables represent positive real numbers.   8.1–8.2

1. $\sqrt{25w^2}$ $5w$ 2. $\sqrt[3]{8y^3}$ $2y$ 3. $\sqrt[3]{4z^4}$ $z\sqrt[4]{4z}$

4. $\sqrt{36x^3}$ $6x\sqrt{x}$ 5. $\sqrt{\dfrac{3a^5}{b^4}}$ $\dfrac{a^2\sqrt{3a}}{b^2}$ 6. $\dfrac{\sqrt[4]{5c^6}}{\sqrt[4]{16}}$ $\dfrac{c\sqrt[4]{5c^2}}{2}$

7. $\dfrac{\sqrt{2x^3}}{\sqrt{x}}$ $x\sqrt{2}$ 8. $\sqrt[4]{\dfrac{p^3}{q^8}}$ $\dfrac{\sqrt[4]{p^3}}{q^2}$

9. How do you determine whether two radicals are *like* or un*like*?   Two radicals are *like* if they have the same radical factor (that is, same radicand and same index).

10. Write two radicals that are considered un*like*.
    For example: $2\sqrt{3}, 6\sqrt[3]{5}$

11. From the three following pairs of radicals, identify the pair of *like* radicals: ii

    i. $2\sqrt{x}$ and $8\sqrt[3]{x}$ ii. $\sqrt{5}$ and $-3\sqrt{5}$

    iii. $3a\sqrt{3}$ and $3a\sqrt{2}$

12. From the three following pairs of radicals, identify the pair of *like* radicals: iii

    i. $13\sqrt{5b}$ and $13b\sqrt{5}$

    ii. $\sqrt[4]{x^2y}$ and $\sqrt[3]{x^2y}$

    iii. $-2\sqrt[3]{y^2}$ and $6\sqrt[3]{y^2}$

For Exercises 13–26, add or subtract the expressions, if possible. Assume all variables represent positive real numbers.

13. $8\sqrt{6} + 2\sqrt{6}$ $10\sqrt{6}$ 14. $3\sqrt{2} + 5\sqrt{2}$ $8\sqrt{2}$

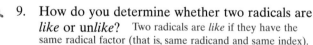

15. $4\sqrt{3} - 2\sqrt{3} + 5\sqrt{3}$ $7\sqrt{3}$

16. $5\sqrt{7} - 3\sqrt{7} + 2\sqrt{7}$ $4\sqrt{7}$

17. $\sqrt{11} + \sqrt{11}$ $2\sqrt{11}$ 18. $\sqrt{10} + \sqrt{10}$ $2\sqrt{10}$

19. $12\sqrt{x} - 3\sqrt{x}$ $9\sqrt{x}$ 20. $15\sqrt{y} - 4\sqrt{y}$ $11\sqrt{y}$

21. $-3\sqrt{a} + 2\sqrt{a} + \sqrt{a}$ 0

22. $5\sqrt{c} - 6\sqrt{c} + \sqrt{c}$ 0

23. $7x\sqrt{11} - 9x\sqrt{11}$
    $-2x\sqrt{11}$
    24. $8y\sqrt{15} - 3y\sqrt{15}$
    $5y\sqrt{15}$

25. $9\sqrt{2} - 9\sqrt{5}$
    $9\sqrt{2} - 9\sqrt{5}$
    26. $x\sqrt{y} - y\sqrt{x}$
    $x\sqrt{y} - y\sqrt{x}$

For Exercises 27–30, translate the English phrases into algebraic expressions. Then simplify the expression.

27. The sum of three times the cube root of six and eight times the cube root of six. $3\sqrt[3]{6} + 8\sqrt[3]{6}; 11\sqrt[3]{6}$

28. The difference of negative two times the cube root of w and five times the cube root of w.
    $-2\sqrt[3]{w} - 5\sqrt[3]{w}; -7\sqrt[3]{w}$

29. Four times the square root of five, minus six times the square root of five. $4\sqrt{5} - 6\sqrt{5}; -2\sqrt{5}$

30. Eight times the square root of two, plus the square root of two. $8\sqrt{2} + \sqrt{2}; 9\sqrt{2}$

 See Additional Answers Appendix   Writing   Translating Expression   Geometry   Scientific Calculator   Video

For Exercises 31–58, add or subtract the expressions, if possible. Assume all variables represent positive real numbers.

 31. $2\sqrt{12} + \sqrt{48}$ $8\sqrt{3}$ 32. $5\sqrt{32} + 2\sqrt{50}$ $30\sqrt{2}$

33. $4\sqrt{45} - 6\sqrt{20}$ 0 34. $8\sqrt{54} - 4\sqrt{24}$ $16\sqrt{6}$

35. $\frac{1}{2}\sqrt{8} + \frac{1}{3}\sqrt{18}$ $2\sqrt{2}$ 36. $\frac{1}{4}\sqrt{32} - \frac{1}{5}\sqrt{50}$ 0

37. $6p\sqrt{20p^2} + p^2\sqrt{80}$ $16p^2\sqrt{5}$ 38. $2q\sqrt{48} + \sqrt{27q^2}$ $11q\sqrt{3}$

39. $-2\sqrt{2k} + 6\sqrt{8k}$ $10\sqrt{2k}$ 40. $5\sqrt{27x} - 4\sqrt{12x}$ $7\sqrt{3x}$

41. $11\sqrt{a^4b} - a^2\sqrt{b} - 9a\sqrt{a^2b}$ $a^2\sqrt{b}$

42. $-7\sqrt{x^4y} + 5x^2\sqrt{y} - 6x\sqrt{x^2y}$ $-8x^2\sqrt{y}$

43. $4\sqrt[3]{5} - \sqrt[3]{5}$ $3\sqrt[3]{5}$ 44. $-3\sqrt[3]{10} - \sqrt[3]{10}$ $-4\sqrt[3]{10}$

45. $\frac{5}{6}z\sqrt[3]{6} + \frac{7}{9}z\sqrt[3]{6}$ $\frac{29}{18}z\sqrt[3]{6}$ 46. $\frac{3}{4}a\sqrt[4]{b} + \frac{1}{6}a\sqrt[4]{b}$ $\frac{11}{12}a\sqrt[4]{b}$

47. $1.1\sqrt[3]{10} - 5.6\sqrt[3]{10} + 2.8\sqrt[3]{10}$ $-1.7\sqrt[3]{10}$

48. $0.25\sqrt[4]{x} + 1.50\sqrt[4]{x} - 0.75\sqrt[4]{x}$ $\sqrt[4]{x}$

49. $4\sqrt[3]{x^4} - 2x\sqrt[3]{x}$ $2x\sqrt[3]{x}$ 50. $8\sqrt[3]{y^{10}} - 2y^2\sqrt[3]{y^4}$ $6y^3\sqrt[3]{y}$

51. $4\sqrt{7} + \sqrt{63} - 2\sqrt{28}$ $3\sqrt{7}$

52. $8\sqrt{3} - 2\sqrt{27} + \sqrt{75}$ $7\sqrt{3}$

53. $\sqrt[3]{16w} + \sqrt[3]{24w} + \sqrt[3]{40w}$ $2\sqrt[3]{2w} + 2\sqrt[3]{3w} + 2\sqrt[3]{5w}$

54. $\sqrt[3]{54y} + \sqrt[3]{81y} - \sqrt[3]{120y}$ $3\sqrt[3]{2y} + 3\sqrt[3]{3y} - 2\sqrt[3]{15y}$

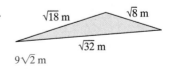 55. $\sqrt[4]{x^6y} + 5x\sqrt[4]{x^2y}$ $6x\sqrt[4]{x^2y}$ 56. $7\sqrt[4]{a^5b^2} - a\sqrt[4]{ab^2}$ $6a\sqrt[4]{ab^2}$

57. $x\sqrt[3]{16} - 2\sqrt[3]{27x} + \sqrt[3]{54x^3}$ $5x\sqrt[3]{2} - 6\sqrt[3]{x}$

58. $5\sqrt[4]{y^5} - 2y\sqrt[4]{y} + \sqrt[4]{16y^7}$ $3y\sqrt[4]{y} + 2y\sqrt[4]{y^3}$

For Exercises 59–60, find the exact perimeter of each figure.

59.

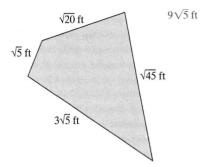

60.

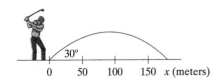

 61. Find the exact perimeter of a rectangle whose width is $2\sqrt{3}$ in. and whose length is $3\sqrt{12}$ in. $16\sqrt{3}$ in.

62. Find the exact perimeter of a square whose side length is $5\sqrt[3]{16}$ cm. $40\sqrt[3]{2}$ cm

EXPANDING YOUR SKILLS

63. Find the slope of the line through the points: $(4, 2\sqrt{3})$ and $(1, \sqrt{3})$. $\frac{\sqrt{3}}{3}$

64. Find the slope of the line through the points: $(7, 4\sqrt{5})$ and $(2, 3\sqrt{5})$. $\frac{\sqrt{5}}{5}$

For Exercises 65–66, add or subtract the expressions, if possible. Assume all variables represent positive real numbers.

65. $\frac{1}{2}ab\sqrt{24a^3} + \frac{4}{3}\sqrt{54a^5b^2} - a^2b\sqrt{150a}$ 0

66. $mn\sqrt{72n} + \frac{3}{2}n\sqrt{8m^2n} - \frac{6}{5}n\sqrt{50m^2n}$ $3mn\sqrt{2n}$

67. A golfer hits a golf ball at an angle of 30° with an initial velocity of 46.0 meters/second (m/s). The horizontal position of the ball, x, (measured in meters) depends on the number of seconds, t, after the ball is struck according to the equation:

$$x = 23\sqrt{3}t$$

Figure for Exercise 67

a. What is the horizontal position of the ball after 2 s? Round the answer to the nearest meter.   80 m

b. What is the horizontal position of the ball after 4 s? Round the answer to the nearest meter.   159 m

c. Convert the answers to parts (a) and (b) to yards (1 m = 1.094 yd).   88 yd, 174 yd

68. A long-jumper leaves the ground at an angle of 30° at a speed of 9 m/s. The horizontal position of the long jumper, x, (measured in meters) depends on the number of seconds, t, after he leaves the ground according to the equation:

$$x = 4.5\sqrt{3}t$$

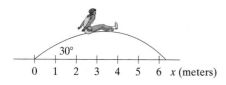

Figure for Exercise 68

a. What is the horizontal position of the long-jumper after 0.5 s? Round the answer to the nearest hundredth of a meter.   3.90 m

b. What is the horizontal position of the long-jumper after 0.75 s? Round the answer to the nearest hundredth of a meter.   5.85 m

c. Convert the answers to parts (a) and (b) to feet (1 m = 3.281 ft)   12.80 ft, 19.19 ft

---

Concepts

1. **Multiplication Property of Radicals**

2. **Multiplying Radical Expressions with One Term**

3. **Multiplying Radical Expressions with Multiple Terms**

4. **Expressions of the Form** $(\sqrt[n]{a})^n$

5. **Special Case Products**

6. **Multiplying Conjugate Radical Expressions**

section

8.4   MULTIPLICATION OF RADICALS

1. Multiplication Property of Radicals

In this section we will learn how to multiply radicals that have the same index. Recall from Section 8.2 the multiplication property of radicals.

Multiplication Property of Radicals

Let a and b represent real numbers such that $\sqrt[n]{a}$ and $\sqrt[n]{b}$ are both real. Then,

$$\sqrt[n]{ab} = \sqrt[n]{a} \cdot \sqrt[n]{b}$$

2. Multiplying Radical Expressions with One Term

To multiply two radical expressions, use the multiplication property of radicals along with the commutative and associative properties of multiplication.

example 1

Multiplying Radical Expressions

Multiply the expressions and simplify the result. Assume all variables represent positive real numbers.

a. $(5\sqrt{3})(4\sqrt{2})$ b. $(2x\sqrt{3})(2\sqrt{15})$ c. $(6a\sqrt{ab})\left(\frac{1}{3}a\sqrt{a}\right)$

Solution:

a. $(5\sqrt{3})(4\sqrt{2})$

$\quad = (5 \cdot 4)(\sqrt{3} \cdot \sqrt{2})$ Commutative and associative properties of multiplication

$\quad = 20\sqrt{3 \cdot 2}$ Multiplication property of radicals

$\quad = 20\sqrt{6}$

b. $(2x\sqrt{3})(2\sqrt{15})$

$\quad = (2x \cdot 2)(\sqrt{3} \cdot \sqrt{15})$ Commutative and associative properties of multiplication

$\quad = 4x\sqrt{45}$ Multiplication property of radicals

$\quad = 4x\sqrt{3^2 \cdot 5}$ Simplify the radical

$\quad = 4x \cdot 3\sqrt{5}$

$\quad = 12x\sqrt{5}$

c. $(6a\sqrt{ab})\left(\frac{1}{3}a\sqrt{a}\right)$

$\quad = \left(6a \cdot \frac{1}{3}a\right)(\sqrt{ab} \cdot \sqrt{a})$ Commutative and associative properties of multiplication

$\quad = (2a^2)\sqrt{ab \cdot a}$ Multiplication property of radicals

$\quad = 2a^2\sqrt{a^2b}$ Multiply.

$\quad = 2a^2 \cdot a\sqrt{b}$ Simplify the radical.

$\quad = 2a^3\sqrt{b}$

3. Multiplying Radical Expressions with Multiple Terms

When multiplying radical expressions with more than one term we use the distributive property.

example 2

Multiplying Radical Expressions with Multiple Terms

Multiply the expressions. Assume all variables represent positive real numbers.

a. $\sqrt{5}(4 + 3\sqrt{5})$ b. $(\sqrt{x} - 10)(\sqrt{y} + 4)$ c. $(2\sqrt{3} - \sqrt{5})(\sqrt{3} + 6\sqrt{5})$

Solution:

a. $\sqrt{5}(4 + 3\sqrt{5})$

$= \sqrt{5}(4) + \sqrt{5}(3\sqrt{5})$ Apply the distributive property.

$= 4\sqrt{5} + 3\sqrt{25}$ Multiplication property of radicals

$= 4\sqrt{5} + 3 \cdot 5$ Simplify the radical.

$= 4\sqrt{5} + 15$

b. $(\sqrt{x} - 10)(\sqrt{y} + 4)$

$= \sqrt{x}(\sqrt{y}) + \sqrt{x}(4) - 10(\sqrt{y}) - 10(4)$ Apply the distributive property.

$= \sqrt{xy} + 4\sqrt{x} - 10\sqrt{y} - 40$ Simplify.

c. $(2\sqrt{3} - \sqrt{5})(\sqrt{3} + 6\sqrt{5})$

$= 2\sqrt{3}(\sqrt{3}) + 2\sqrt{3}(6\sqrt{5}) - \sqrt{5}(\sqrt{3}) - \sqrt{5}(6\sqrt{5})$ Apply the distributive property.

$= 2\sqrt{9} + 12\sqrt{15} - \sqrt{15} - 6\sqrt{25}$ Multiplication property of radicals

$= 2 \cdot 3 + 11\sqrt{15} - 6 \cdot 5$ Simplify radicals. Combine *like* radicals.

$= 6 + 11\sqrt{15} - 30$

$= -24 + 11\sqrt{15}$ Combine *like* terms.

4. Expressions of the Form $(\sqrt[n]{a})^n$

The multiplication property of radicals can be used to simplify an expression of the form $(\sqrt{a})^2$, where $a \geq 0$.

$$(\sqrt{a})^2 = \sqrt{a} \cdot \sqrt{a} = \sqrt{a^2} = a, \text{ where } a \geq 0$$

This logic can be applied to *n*th-roots. If $\sqrt[n]{a}$ is a real number, then, $(\sqrt[n]{a})^n = a$.

example 3   **Simplifying Radical Expressions**

Simplify the expressions. Assume all variables represent positive real numbers.

a. $(\sqrt{7})^2$ b. $(\sqrt[4]{x})^4$ c. $(\sqrt[3]{ab})^3$

Classroom Activity 8.4A

Solution:

a. $(\sqrt{7})^2 = 7$

b. $(\sqrt[4]{x})^4 = x$

c. $(\sqrt[3]{ab})^3 = ab$

5. Special Case Products

From Example 2, you may have noticed a similarity between multiplying radical expressions and multiplying polynomials.

Recall from Section 3.6 that the square of a binomial results in a perfect square trinomial.

$$(a + b)^2 = a^2 + 2ab + b^2$$

$$(a - b)^2 = a^2 - 2ab + b^2$$

The same patterns occur when squaring a radical expression with two terms.

example 4

Squaring a Two-Term Radical Expression

Square the radical expressions as indicated. Assume all variables represent positive real numbers.

a. $(\sqrt{x} + \sqrt{y})^2$ b. $(\sqrt{2} - 4\sqrt{3})^2$

Solution:

a. $(\sqrt{x} + \sqrt{y})^2$

This expression is in the form $(a + b)^2$, where $a = \sqrt{x}$ and $b = \sqrt{y}$.

$$\overbrace{= (\sqrt{x})^2 + 2(\sqrt{x})(\sqrt{y}) + (\sqrt{y})^2}^{a^2 + 2ab + b^2}$$

Apply the formula $(a + b)^2 = a^2 + 2ab + b^2$.

$$= x + 2\sqrt{xy} + y$$

Simplify.

Tip: The product $(\sqrt{x} + \sqrt{y})^2$ can also be found using the distributive property.

$$(\sqrt{x} + \sqrt{y})^2 = (\sqrt{x} + \sqrt{y})(\sqrt{x} + \sqrt{y}) = \sqrt{x} \cdot \sqrt{x} + \sqrt{x} \cdot \sqrt{y} + \sqrt{y} \cdot \sqrt{x} + \sqrt{y} \cdot \sqrt{y}$$

$$= \sqrt{x^2} + \sqrt{xy} + \sqrt{xy} + \sqrt{y^2}$$

$$= x + 2\sqrt{xy} + y$$

b. $(\sqrt{2} - 4\sqrt{3})^2$ This expression is in the form $(a - b)^2$, where $a = \sqrt{2}$ and $b = 4\sqrt{3}$.

$$\overset{\overbrace{a^2 - 2ab + b^2}}{(\sqrt{2})^2 - 2(\sqrt{2})(4\sqrt{3}) + (4\sqrt{3})^2}$$

Apply the formula
$(a - b)^2 = a^2 - 2ab + b^2$.

$= 2 - 8\sqrt{6} + 16 \cdot 3$ Simplify.

$= 2 - 8\sqrt{6} + 48$

$= 50 - 8\sqrt{6}$

6. Multiplying Conjugate Radical Expressions

Recall from Section 3.6 that the product of two conjugate binomials results in a difference of squares.

$$(a + b)(a - b) = a^2 - b^2$$

These same patterns occur when multiplying two conjugate radical expressions.

example 5   **Multiplying Conjugate Radical Expressions**

Multiply the radical expressions. Assume all variables represent positive real numbers.

a. $(\sqrt{5} + 4)(\sqrt{5} - 4)$ b. $(2\sqrt{c} - 3\sqrt{d})(2\sqrt{c} + 3\sqrt{d})$

Classroom Activity 8.4B

Solution:

a. $(\sqrt{5} + 4)(\sqrt{5} - 4)$ This expression is in the form $(a + b)(a - b)$, where $a = \sqrt{5}$ and $b = 4$.

$$= \overset{\overbrace{a^2 - b^2}}{(\sqrt{5})^2 - (4)^2}$$

Apply the formula $(a + b)(a - b) = a^2 - b^2$.

$= 5 - 16$ Simplify.

$= -11$

Tip: The product $(\sqrt{5} + 4)(\sqrt{5} - 4)$ can also be found using the distributive property.

$(\sqrt{5} + 4)(\sqrt{5} - 4) = \sqrt{5} \cdot (\sqrt{5}) + \sqrt{5} \cdot (-4) + 4 \cdot (\sqrt{5}) + 4 \cdot (-4)$
$= 5 - 4\sqrt{5} + 4\sqrt{5} - 16$
$= 5 - 16$
$= -11$

b. $(2\sqrt{c} - 3\sqrt{d})(2\sqrt{c} + 3\sqrt{d})$

This expression is in the form $(a + b)(a - b)$, where $a = 2\sqrt{c}$ and $b = 3\sqrt{d}$

$$= (2\sqrt{c})^2 \overset{a^2 - b^2}{-} (3\sqrt{d})^2$$

Apply the formula $(a + b)(a - b) = a^2 - b^2$.

$$= 4c - 9d$$

Calculator Connections

A calculator can support the multiplication property of radicals. For example, use a calculator to evaluate $(3\sqrt{5})(4\sqrt{2})$ and its simplified form $12\sqrt{10}$.

Scientific Calculator

Tip: The decimal approximations for $(3\sqrt{5})(4\sqrt{2})$ and $12\sqrt{10}$ agree for the first 10 digits. This in itself does not make $(3\sqrt{5})(4\sqrt{2}) = 12\sqrt{10}$. It is the multiplication property of radicals that guarantees that the expressions are equal.

Enter: $\boxed{5}\ \boxed{\sqrt{}}\ \boxed{\times}\ \boxed{3}\ \boxed{=}\ \boxed{\times}\ \boxed{2}\ \boxed{\sqrt{}}\ \boxed{\times}\ \boxed{4}\ \boxed{=}$ **Result:** $\boxed{37.94733192}$

Enter: $\boxed{1}\ \boxed{0}\ \boxed{\sqrt{}}\ \boxed{\times}\ \boxed{1}\ \boxed{2}\ \boxed{=}$ **Result:** $\boxed{37.94733192}$

Graphing Calculator

```
(3√(5))*(4√(2))
        37.94733192
12√(10)
        37.94733192
```

Calculator Exercises

Simplify the radical expressions algebraically. Then use a calculator to approximate the original expression and its simplified form.

1. $(3\sqrt{5})(4\sqrt{10})$ ❖

2. $(4\sqrt{6})(7\sqrt{10})$ ❖

3. $(\sqrt{2} - \sqrt{3})(\sqrt{2} + \sqrt{3})$ ❖

4. $(\sqrt{5} + \sqrt{7})(\sqrt{5} - \sqrt{7})$ ❖

5. $(2 + \sqrt{11})^2$ ❖

6. $(\sqrt{5} - 4)^2$ ❖

---

section 8.4 PRACTICE EXERCISES

For Exercises 1–6, perform the indicated operations and simplify. Assume all variables represent positive real numbers.

1. $\sqrt{25} + \sqrt{16} - \sqrt{36}$ 3

2. $\sqrt{100} - \sqrt{4} + \sqrt{9}$
   11

3. $\dfrac{\sqrt{27}}{2} - \dfrac{\sqrt{3}}{\sqrt{4}}$ $\sqrt{3}$

4. $\dfrac{\sqrt{5}}{\sqrt{9}} + \dfrac{\sqrt{125}}{3}$ $2\sqrt{5}$

5. $6x\sqrt{18} + 2\sqrt{2x^2}$
   $20x\sqrt{2}$

6. $10\sqrt{zw^4} - w^2\sqrt{49z}$
   $3w^2\sqrt{z}$

For Exercises 7–14, use a calculator to approximate the value of the expressions to three decimal places.

7. $3\sqrt{6}$
   7.348

8. $-7\sqrt{21}$
   -32.078

9. $-2 + \sqrt{10}$
   1.162

10. $4 + \sqrt{37}$
    10.083

11. $5 - \sqrt{40}$
    -1.325

12. $3 - \sqrt{63}$
    -4.937

❖ See Additional Answers Appendix   ✎ Writing   ⬌ Translating Expression   ⬤ Geometry   ⬛ Scientific Calculator   ▭ Video

13. $\dfrac{4\sqrt{3}}{3}$ 2.309

14. $\dfrac{-5\sqrt{6}}{6}$ -2.041

For Exercises 15–20, state the property that is used in each exercise. Choose from the associative property of multiplication, the commutative property of multiplication, or the distributive property of multiplication over addition.

15. $\sqrt{3}(4\sqrt{5}) = \sqrt{3}(\sqrt{5} \cdot 4)$ Commutative property of multiplication

16. $3(\sqrt{2} + \sqrt{6}) = 3\sqrt{2} + 3\sqrt{6}$ Distributive property of multiplication over addition

17. $(8\sqrt{7})(\sqrt{2}) = 8(\sqrt{7}\sqrt{2})$ Associative property of multiplication

18. $21 \cdot (2\sqrt{13}) = (21 \cdot 2)\sqrt{13}$ Associative property of multiplication

19. $\sqrt{3}(9 + \sqrt{11}) = \sqrt{3} \cdot 9 + \sqrt{3} \cdot \sqrt{11}$ Distributive property of multiplication over addition

20. $6\sqrt{10} \cdot 4 = 6 \cdot 4\sqrt{10}$ Commutative property of multiplication

For Exercises 21–26, multiply the expressions in parts (a) and (b) and compare the process used. Assume all variables represent positive real numbers.

21. a. $3(x + 2)$ $3x + 6$
    b. $\sqrt{3}(\sqrt{x} + \sqrt{2})$
       $\sqrt{3x} + \sqrt{6}$

22. a. $-5(6 + y)$ $-30 - 5y$
    b. $-\sqrt{5}(\sqrt{6} + \sqrt{y})$
       $-\sqrt{30} - \sqrt{5y}$

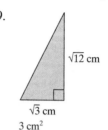 23. a. $(2a + 3)^2$ $4a^2 + 12a + 9$
    b. $(2\sqrt{a} + 3)^2$
       $4a + 12\sqrt{a} + 9$

24. a. $(6 - z)^2$ $36 - 12z + z^2$
    b. $(\sqrt{6} - z)^2$
       $6 - 2z\sqrt{6} + z^2$

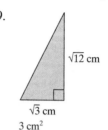 25. a. $(b - 5)(b + 5)$ $b^2 - 25$
    b. $(\sqrt{b} - 5)(\sqrt{b} + 5)$ $b - 25$

26. a. $(3w - 1)(3w + 1)$ $9w^2 - 1$
    b. $(3\sqrt{w} - 1)(3\sqrt{w} + 1)$ $9w - 1$

For Exercises 27–36, multiply the expressions.

27. $\sqrt{5} \cdot \sqrt{3}$ $\sqrt{15}$

28. $\sqrt{7} \cdot \sqrt{6}$ $\sqrt{42}$

29. $\sqrt{11} \cdot \sqrt{11}$ 11

30. $\sqrt{13} \cdot \sqrt{13}$ 13

31. $\sqrt{10} \cdot \sqrt{5}$ $5\sqrt{2}$

32. $\sqrt{2} \cdot \sqrt{10}$ $2\sqrt{5}$

33. $\sqrt{7} \cdot \sqrt{14}$ $7\sqrt{2}$

34. $\sqrt{2} \cdot \sqrt{22}$ $2\sqrt{11}$

35. $\sqrt{2} \cdot \sqrt{18}$ 6

36. $\sqrt{3} \cdot \sqrt{27}$ 9

 For Exercises 37–38, find the exact perimeter and exact area of the rectangles.

37.
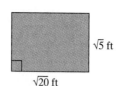
$\sqrt{20}$ ft
$\sqrt{5}$ ft
Perimeter: $6\sqrt{5}$ ft; area: 10 ft^2

38.
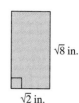
$\sqrt{8}$ in.
$\sqrt{2}$ in.
Perimeter: $6\sqrt{2}$ in.; area: 4 in.^{2}

For Exercises 39–40, find the exact area of the triangles.

39.
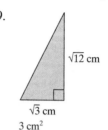
$\sqrt{12}$ cm
$\sqrt{3}$ cm
3 cm^2

40.
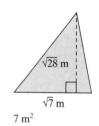
$\sqrt{28}$ m
$\sqrt{7}$ m
7 m^2

For Exercises 41–74, multiply the expressions. Assume all variables represent positive real numbers.

41. $\sqrt{3w} \cdot \sqrt{3w}$ $3w$

42. $\sqrt{6p} \cdot \sqrt{6p}$ $6p$

43. $(8\sqrt{5y})(-2\sqrt{2})$
    $-16\sqrt{10y}$

44. $(4\sqrt{5x})(7\sqrt{3})$
    $28\sqrt{15x}$

45. $\sqrt{2}(\sqrt{6} - \sqrt{3})$
    $2\sqrt{3} - \sqrt{6}$

46. $\sqrt{5}(\sqrt{3} + \sqrt{7})$
    $\sqrt{15} + \sqrt{35}$

47. $4\sqrt{x}(\sqrt{x} + 5)$
    $4x + 20\sqrt{x}$

48. $2\sqrt{y}(3 - \sqrt{y})$
    $6\sqrt{y} - 2y$

49. $(\sqrt{3} + 2\sqrt{10})(4\sqrt{3} - \sqrt{10})$ $-8 + 7\sqrt{30}$

50. $(8\sqrt{7} - \sqrt{5})(\sqrt{7} + 3\sqrt{5})$ $41 + 23\sqrt{35}$

51. $(\sqrt{a} - 3b)(9\sqrt{a} - b)$ $9a - 28b\sqrt{a} + 3b^2$

52. $(11\sqrt{m} + 4n)(\sqrt{m} + n)$ $11m + 15n\sqrt{m} + 4n^2$

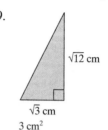 53. $(p + 2\sqrt{p})(8p + 3\sqrt{p} - 4)$ $8p^2 + 19p\sqrt{p} + 2p - 8\sqrt{p}$

54. $(5s - \sqrt{s})(s + 5\sqrt{s} + 6)$ $5s^2 + 24s\sqrt{s} + 25s - 6\sqrt{s}$

55. $(\sqrt{13} + 4)^2$ $29 + 8\sqrt{13}$ 56. $(6 - \sqrt{11})^2$ $47 - 12\sqrt{11}$

57. $(\sqrt{a} - 2)^2$ $a - 4\sqrt{a} + 4$ 58. $(\sqrt{p} + 3)^2$ $p + 6\sqrt{p} + 9$

59. $(2\sqrt{a} - 3)^2$
    $4a - 12\sqrt{a} + 9$

60. $(3\sqrt{w} + 4)^2$
    $9w + 24\sqrt{w} + 16$

61. $(\sqrt{x} - 2\sqrt{y})^2$
    $x - 4\sqrt{xy} + 4y$

62. $(5\sqrt{c} + 2\sqrt{d})^2$
    $25c + 20\sqrt{cd} + 4d$

63. $(\sqrt{5} + 2)(\sqrt{5} - 2)$ 1   64. $(\sqrt{3} - 4)(\sqrt{3} + 4)$ -13

65. $(\sqrt{x} + \sqrt{y})(\sqrt{x} - \sqrt{y})$ $x - y$

66. $(\sqrt{a} + \sqrt{b})(\sqrt{a} - \sqrt{b})$ $a - b$

67. $(\sqrt{10} - \sqrt{11})^2$
    $21 - 2\sqrt{110}$

68. $(\sqrt{3} - \sqrt{2})^2$
    $5 - 2\sqrt{6}$

69. $(\sqrt{10} - \sqrt{11})(\sqrt{10} + \sqrt{11})$ -1

 See Additional Answers Appendix   Writing   Translating Expression   Geometry   Scientific Calculator   Video

70. $(\sqrt{3} - \sqrt{2})(\sqrt{3} + \sqrt{2})$ 1

71. $(\sqrt{6} + \sqrt{2})(\sqrt{6} - \sqrt{2})$ 4

72. $(\sqrt{15} + \sqrt{5})(\sqrt{15} - \sqrt{5})$ 10

73. $(8\sqrt{x} + 2\sqrt{y})(8\sqrt{x} - 2\sqrt{y})$ $64x - 4y$

74. $(4\sqrt{s} + 11\sqrt{t})(4\sqrt{s} - 11\sqrt{t})$ $16s - 121t$

For Exercises 75–76, find the exact volume of the boxes.

75.

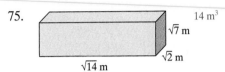

76.

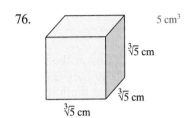

■ **EXPANDING YOUR SKILLS**

For Exercises 77–88, multiply the expressions. Assume all variables represent positive real numbers.

77. $\sqrt[3]{x}(\sqrt[3]{x^2} + 2)$
$x + 2\sqrt[3]{x}$

78. $\sqrt[3]{y}(\sqrt[3]{y^2} - 4)$
$y - 4\sqrt[3]{y}$

79. $\sqrt[3]{9}(\sqrt[3]{3} - 1)$
$3 - \sqrt[3]{9}$

80. $\sqrt[3]{25}(\sqrt[3]{5} + 2)$
$5 + 2\sqrt[3]{25}$

81. $(\sqrt[3]{3} - 2)(\sqrt[3]{9} + 1)$
$1 + \sqrt[3]{3} - 2\sqrt[3]{9}$

82. $(\sqrt[3]{25} + 6)(\sqrt[3]{5} + 2)$
$17 + 2\sqrt[3]{25} + 6\sqrt[3]{5}$

83. $(\sqrt[3]{x} + 7)(4\sqrt[3]{x} - 1)$
$4\sqrt[3]{x^2} + 27\sqrt[3]{x} - 7$

84. $(y + \sqrt[3]{5})(y - \sqrt[3]{25})$
$y^2 - y\sqrt[3]{25} + y\sqrt[3]{5} - 5$

85. $(\sqrt[3]{a} - \sqrt[3]{b})(\sqrt[3]{a^2} + \sqrt[3]{ab} + \sqrt[3]{b^2})$ $a - b$

86. $(\sqrt[3]{a} + \sqrt[3]{b})(\sqrt[3]{a^2} - \sqrt[3]{ab} + \sqrt[3]{b^2})$ $a + b$

87. $(\sqrt[3]{x} + \sqrt[3]{y})(\sqrt[3]{x^2} - \sqrt[3]{xy} + \sqrt[3]{y^2})$ $x + y$

88. $(\sqrt[3]{x} - \sqrt[3]{y})(\sqrt[3]{x^2} + \sqrt[3]{xy} + \sqrt[3]{y^2})$ $x - y$

---

chapter 8 **MIDCHAPTER REVIEW**

For Exercises 1–12, simplify the expression if possible.

1. $\sqrt[3]{-216}$ -6 2. $\sqrt[4]{625}$ 5   3. $\sqrt[4]{\dfrac{256}{16}}$ 2

4. $\sqrt{\dfrac{121}{25}}$ $\dfrac{11}{5}$ 5. $\sqrt{-49}$
Not a real number   6. $\sqrt{-169}$
Not a real number

7. $-\sqrt{49}$ -7 8. $-\sqrt{169}$ -13 9. $\sqrt{m^2}$ $|m|$

10. $\sqrt{p^2}$ $|p|$ 11. $\sqrt[3]{m^3}$ m 12. $\sqrt[3]{p^3}$ p

For Exercises 13–20, simplify the expression. Assume all variables represent positive real numbers.

13. $\sqrt{8q^6}$
$2q^3\sqrt{2}$

14. $\sqrt{27p^8}$
$3p^4\sqrt{3}$

15. $\sqrt[3]{125u^{11}v^{12}}$
$5u^3v^4\sqrt[3]{u^2}$

16. $\sqrt[3]{64r^{15}s^{13}}$
$4r^5s^4\sqrt[3]{s}$

17. $\sqrt{\dfrac{50m^3}{2m}}$ $5m$

18. $\sqrt{\dfrac{18u^5}{2u}}$ $3u^2$

19. $\dfrac{\sqrt{10x^5}}{\sqrt{x}}$

(*Hint*: Write the expression as a single radical and reduce the radicand.) $x^2\sqrt{10}$

20. $\dfrac{\sqrt{15y^3}}{\sqrt{5y}}$

(*Hint*: Write the expression as a single radical and reduce the radicand.) $y\sqrt{3}$

For Exercises 21–34, perform the indicated operations. Assume all variables represent positive real numbers.

21. $2\sqrt{6} - 5\sqrt{6}$ $-3\sqrt{6}$ 22. $5\sqrt{a} + 7\sqrt{a} - \sqrt{a}$ $11\sqrt{a}$

23. $x\sqrt{18} + \sqrt{2x^2}$
$4x\sqrt{2}$

24. $4\sqrt{75} - 20\sqrt{3}$
0

25. $\sqrt{5}(\sqrt{5} + \sqrt{7})$
$5 + \sqrt{35}$

26. $\sqrt{a}(\sqrt{a} + 2)$
$a + 2\sqrt{a}$

27. $(3\sqrt{2} - 4)(5\sqrt{2} + 1)$ $26 - 17\sqrt{2}$

28. $(4\sqrt{x} + \sqrt{y})(\sqrt{x} - 3\sqrt{y})$ $4x - 11\sqrt{xy} - 3y$

29. $(\sqrt{2} + 7)^2$
$51 + 14\sqrt{2}$

30. $(\sqrt{3} + \sqrt{5})^2$
$8 + 2\sqrt{15}$

31. $(\sqrt{5} - \sqrt{11})^2$
$16 - 2\sqrt{55}$

32. $(\sqrt{x} - 6)^2$
$x - 12\sqrt{x} + 36$

33. $(2\sqrt{3} - 10)(2\sqrt{3} + 10)$ -88

❖ See Additional Answers Appendix   ✎ Writing   ⬅ Translating Expression   Geometry   Scientific Calculator   Video

34. $(\sqrt{u} - 3\sqrt{v})(\sqrt{u} + 3\sqrt{v})$ $u - 9v$

35. On a hazy day at the beach, the formula $d = 0.9\sqrt{A}$ relates the distance, d, (in miles) that a person can see to the horizon from an altitude of A feet.

a. How far can a person see to the horizon from a 40-ft lifeguard tower? Use a calculator and round to the nearest tenth of a mile. 5.7 miles

b. How far can a person see to the horizon from a 100-ft high bridge?   9 miles

36. The velocity of an object, v, (in meters per second: m/s) depends on the kinetic energy, E, (in joules: J) and mass, m, (in kilograms) of the object according to the formula:

$$v = \sqrt{\frac{2E}{m}}$$

a. Find the velocity of a 2-kg object whose kinetic energy is 100 J.   10 m/s

b. Find the velocity of a 5-kg object whose kinetic energy is 150 J. Round the answer to one decimal place.   7.7 m/s

The length of a diagonal, D, of a rectangular box is given by $D = \sqrt{L^2 + W^2 + H^2}$, where L, W, and H are the length, width, and height, respectively. For Exercises 37–38, find the length of the diagonal. Use a calculator and round the answer to one decimal place.

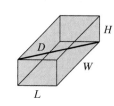

Figure for Exercises 37, 38

37.

$H = 3$ in.
D
$W = 12$ in.
$L = 6$ in.   13.7 in.

38.

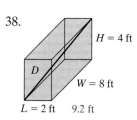

$H = 4$ ft
D
$W = 8$ ft
$L = 2$ ft   9.2 ft

39. The length of the sides of a cube is related to the volume of the cube according to the formula: $x = \sqrt[3]{V}$.

Figure for Exercise 39

a. How long are the sides of a cube if the volume is 100 cm³? Use your calculator and round the answer to the nearest tenth of a centimeter.   4.6 cm

b. How long are the sides of a cube if the volume is 216 m³?   6 m

40. The length of the sides of a square is related to the area according to the formula: $x = \sqrt{A}$.

a. How long are the sides of a square if the area is 90 ft²? Round the answer to the nearest tenth of a foot.   9.5 ft

b. How long are the sides of a square if the area is 600 cm²? Round the answer to the nearest tenth of a centimeter.   24.5 cm

For Exercises 41–44, evaluate the expression using a calculator. Round the answer to three decimal places.

41. $\dfrac{2 + \sqrt{6}}{5}$ 0.890

42. $\dfrac{8 - 3\sqrt{7}}{2}$ 0.031

43. $\dfrac{2 - \sqrt{6}}{5}$ −0.090

44. $\dfrac{8 + 3\sqrt{7}}{2}$ 7.969

section

8.5 RATIONALIZATION

1. Simplified Form of a Radical

Recall that for a radical expression to be in simplified form the following three conditions must be met.

Simplified Form of a Radical

Consider any radical expression where the radicand is written as a product of prime factors. The expression is in simplified form if all of the following conditions are met:

1. The radicand has no factor raised to a power greater than or equal to the index.
2. There are no radicals in the denominator of a fraction.
3. The radicand does not contain a fraction.

The basis of the second and third conditions, which restrict radicals from the denominator of an expression, are largely historical. In some cases, removing a radical from the denominator of a fraction will create an expression that is computationally simpler. For example, we will show that

$$\frac{2}{\sqrt{2}} = \sqrt{2} \quad \text{and} \quad \frac{2}{\sqrt{5} + \sqrt{3}} = \sqrt{5} - \sqrt{3}$$

The process to remove a radical from the denominator is called **rationalizing the denominator**. In this section we will rationalize the denominator in two cases:

1. When the denominator contains a single radical term, and
2. When the denominator contains two terms involving square roots

2. Rationalizing the Denominator: One Term

To begin the first case, recall that the nth-root of a perfect nth power is easily simplified.

$$\sqrt{x^2} = x \quad x \geq 0$$
$$\sqrt[3]{x^3} = x$$
$$\sqrt[4]{x^4} = x \quad x \geq 0$$
$$\sqrt[5]{x^5} = x$$

Therefore, to rationalize a radical term, use the multiplication property of radicals to create an nth-root of an nth power.

example 1

Rationalizing a Radical Expression

Fill in the blanks to rationalize the radical expressions. Assume all variables represent positive real numbers.

a. $\sqrt{a} \cdot \sqrt{?} = \sqrt{a^2} = a$ b. $\sqrt[3]{y} \cdot \sqrt[3]{?} = \sqrt[3]{y^3} = y$

c. $\sqrt[3]{2z^2} \cdot \sqrt[3]{?} = \sqrt[3]{2^3 z^3} = 2z$

Solution:

a. $\sqrt{a} \cdot \sqrt{?} = \sqrt{a^2} = a$ What multiplied by $\sqrt{a}$ will equal $\sqrt{a^2}$?

$\sqrt{a} \cdot \sqrt{a} = \sqrt{a^2} = a$ Notice that $a \cdot a = a^2$.

b. $\sqrt[3]{y} \cdot \sqrt[3]{?} = \sqrt[3]{y^3} = y$ What multiplied by $\sqrt[3]{y}$ will equal $\sqrt[3]{y^3}$?

$\sqrt[3]{y} \cdot \sqrt[3]{y^2} = \sqrt[3]{y^3} = y$ Notice that $y^1 \cdot y^2 = y^3$

c. $\sqrt[3]{2z^2} \cdot \sqrt[3]{?} = \sqrt[3]{2^3 z^3}$ What multiplied by $\sqrt[3]{2z^2}$ will equal $\sqrt[3]{2^3 z^3}$?

$\sqrt[3]{2z^2} \cdot \sqrt[3]{2^2 z} = \sqrt[3]{2^3 z^3}$ Notice that $2^1 \cdot 2^2 = 2^3$ and $z^2 \cdot z^1 = z^3$.

$= 2z$

To rationalize the denominator of a radical expression, multiply the numerator and denominator by an appropriate expression to create the nth-root of an nth power in the denominator.

example 2

Rationalizing the Denominator: One Term

Simplify the expression $\dfrac{4}{\sqrt[3]{x}}$, $(x \neq 0)$

Tip: Notice that for $x \neq 0$:

$$\frac{\sqrt[3]{x^2}}{\sqrt[3]{x^2}} = 1$$

Multiplying the original expression

$$\frac{4}{\sqrt[3]{x}}$$

by this ratio does not change its value. It merely changes its form.

Solution:

A cube root of a perfect cube in the denominator is needed to remove the radical. Multiply numerator and denominator by $\sqrt[3]{x^2}$ because $\sqrt[3]{x} \cdot \sqrt[3]{x^2} = x$.

$$\frac{4}{\sqrt[3]{x}} = \frac{4}{\sqrt[3]{x}} \cdot \frac{\sqrt[3]{x^2}}{\sqrt[3]{x^2}}$$ Multiply by $\sqrt[3]{x^2}$ in the numerator and denominator.

$$= \frac{4\sqrt[3]{x^2}}{\sqrt[3]{x^3}}$$ Multiply the radicals.

$$= \frac{4\sqrt[3]{x^2}}{x}, \quad (x \neq 0)$$ Simplify.

example 3

Rationalizing the Denominator: One Term

Simplify the expressions. Assume all variables represent positive real numbers.

a. $\dfrac{2}{\sqrt{2}}$ b. $\sqrt{\dfrac{x}{5}}$ c. $\dfrac{14\sqrt[3]{w}}{\sqrt[3]{7}}$

Classroom Activity 8.5A

Solution:

a. A square root of a perfect square is needed in the denominator to remove the radical. Multiply numerator and denominator by $\sqrt{2}$ because $\sqrt{2} \cdot \sqrt{2} = \sqrt{2^2}$.

$$\frac{2}{\sqrt{2}} = \frac{2}{\sqrt{2}} \cdot \frac{\sqrt{2}}{\sqrt{2}}$$

$$= \frac{2\sqrt{2}}{\sqrt{2^2}} \qquad \text{Multiply the radicals.}$$

$$= \frac{2\sqrt{2}}{2} \qquad \text{Simplify.}$$

$$= \frac{\cancel{2}\sqrt{2}}{\cancel{2}} \qquad \text{Reduce the fraction.}$$

$$= \sqrt{2}$$

b. $\sqrt{\dfrac{x}{5}}$ The radicand contains an irreducible fraction.

$$= \frac{\sqrt{x}}{\sqrt{5}} \qquad \text{Apply the division property of radicals.}$$

$$= \frac{\sqrt{x}}{\sqrt{5}} \cdot \frac{\sqrt{5}}{\sqrt{5}} \qquad \text{A square root of a perfect square is needed in the denominator to remove the radical. Multiply numerator and denominator } \sqrt{5} \text{ because } \sqrt{5} \cdot \sqrt{5} = \sqrt{5^2}.$$

$$= \frac{\sqrt{5x}}{\sqrt{5^2}} \qquad \text{Multiply the radicals.}$$

 Avoiding Mistakes

In the expression

$$\frac{\sqrt{5x}}{5}$$

do not try to "cancel" the factor of $\sqrt{5}$ from the numerator with the factor of 5 in the denominator.

$$= \frac{\sqrt{5x}}{5} \qquad \text{Simplify the radicals.}$$

c. A cube root of a perfect cube is needed in the denominator to remove the radical. Multiply numerator and denominator by $\sqrt[3]{7^2}$ because $\sqrt[3]{7^2} \cdot \sqrt[3]{7} = \sqrt[3]{7^3}$.

632   **Chapter 8**   Radicals

Tip: In the expression

$$\frac{14\sqrt[3]{49w}}{7}$$

the factor of 14 and the factor of 7 may be reduced because both are outside the radical.

$$\frac{14\sqrt[3]{49w}}{7} = \frac{14}{7} \cdot \sqrt[3]{49w}$$
$$= 2\sqrt[3]{49w}$$

$$\frac{14\sqrt[3]{w}}{\sqrt[3]{7}} = \frac{14\sqrt[3]{w}}{\sqrt[3]{7}} \cdot \frac{\sqrt[3]{7^2}}{\sqrt[3]{7^2}}$$

$$= \frac{14\sqrt[3]{7^2 w}}{\sqrt[3]{7^3}} \qquad \text{Multiply the radicals.}$$

$$= \frac{14\sqrt[3]{49w}}{7} \qquad \text{Simplify.}$$

$$= 2\sqrt[3]{49w}$$

3.   Rationalizing the Denominator: Two Terms

Recall from the multiplication of polynomials that the product of two conjugates results in a difference of squares.

$$(a + b)(a - b) = a^2 - b^2$$

If either a or b has a square root factor, the expression will simplify without a radical. That is, the expression is rationalized. For example,

$$(\sqrt{5} - \sqrt{3})(\sqrt{5} + \sqrt{3}) = (\sqrt{5})^2 - (\sqrt{3})^2$$
$$= 5 - 3$$
$$= 2$$

Multiplying a binomial by its conjugate is the basis for rationalizing a denominator with two terms involving square roots.

example 4   **Rationalizing the Denominator: Two Terms**

Simplify the expression by rationalizing the denominator.

$$\frac{2}{\sqrt{5} + \sqrt{3}}$$

Solution:

$$\frac{2}{\sqrt{5} + \sqrt{3}}$$

To rationalize a denominator with two terms, multiply the numerator and denominator by the conjugate of the denominator.

$$= \frac{2}{(\sqrt{5} + \sqrt{3})} \cdot \frac{(\sqrt{5} - \sqrt{3})}{(\sqrt{5} - \sqrt{3})}$$

Conjugates

The denominator is in the form $(a + b)(a - b)$, where $a = \sqrt{5}$ and $b = \sqrt{3}$.

$$= \frac{2(\sqrt{5} - \sqrt{3})}{(\sqrt{5})^2 - (\sqrt{3})^2}$$

In the denominator, apply the formula $(a + b)(a - b) = a^2 - b^2$.

$$= \frac{2(\sqrt{5} - \sqrt{3})}{5 - 3} \qquad \text{Simplify.}$$

$$= \frac{2(\sqrt{5} - \sqrt{3})}{2}$$

$$= \frac{\cancel{2}(\sqrt{5} - \sqrt{3})}{\cancel{2}} \qquad \text{Reduce.}$$

$$= \sqrt{5} - \sqrt{3}$$

example 5

Rationalizing the Denominator: Two Terms

Simplify the expression by rationalizing the denominator.

$$\frac{\sqrt{x} + \sqrt{2}}{\sqrt{x} - \sqrt{2}}$$

Classroom Activity 8.5B

Solution:

$$\frac{\sqrt{x} + \sqrt{2}}{\sqrt{x} - \sqrt{2}} = \frac{(\sqrt{x} + \sqrt{2})}{(\sqrt{x} - \sqrt{2})} \cdot \frac{(\sqrt{x} + \sqrt{2})}{(\sqrt{x} + \sqrt{2})} \qquad \begin{array}{l}\text{Multiply numerator and de-}\\ \text{nominator by the conjugate}\\ \text{of the denominator.}\end{array}$$

$$\underbrace{\qquad\qquad\qquad}_{\text{Conjugates}}$$

$$= \frac{(\sqrt{x} + \sqrt{2})^2}{(\sqrt{x} - \sqrt{2})(\sqrt{x} + \sqrt{2})}$$

$$= \frac{(\sqrt{x})^2 + 2(\sqrt{x})(\sqrt{2}) + (\sqrt{2})^2}{(\sqrt{x})^2 - (\sqrt{2})^2} \qquad \begin{array}{l}\text{Simplify using special case}\\ \text{products.}\end{array}$$

$$= \frac{x + 2\sqrt{2x} + 2}{x - 2} \qquad \text{Simplify radicals.}$$

4. Reducing a Radical Quotient to Lowest Terms

Sometimes a radical expression within a quotient must be reduced to lowest terms. This is demonstrated in the next example.

example 6

Reducing a Radical Quotient to Lowest Terms

Simplify and reduce the expression $\dfrac{4 - \sqrt{20}}{10}$.

Solution:

$$\frac{4 - \sqrt{20}}{10}$$

First simplify $\sqrt{20}$ by writing the radicand as a product of prime factors.

$$= \frac{4 - \sqrt{2^2 \cdot 5}}{10}$$

$$= \frac{4 - 2\sqrt{5}}{10}$$

Simplify the radical.

The expression $(4 - 2\sqrt{5})/10$ is a rational expression and can be reduced by factoring numerator and denominator and reducing common factors to 1.

$$= \frac{2(2 - \sqrt{5})}{2 \cdot 5}$$

Factor out the GCF.

$$= \frac{\cancel{2}(2 - \sqrt{5})}{\cancel{2} \cdot 5}$$

Reduce.

$$= \frac{2 - \sqrt{5}}{5}$$

⬢ **Avoiding Mistakes**

Remember that it is not correct to reduce terms within a rational expression. In the expression

$$\frac{4 - 2\sqrt{5}}{10}$$

do not try to "cancel" the 4 and the 10.

Calculator Connections

After simplifying a radical, a calculator can be used to support your solution. For example, use a calculator to approximate the right- and left-hand sides of each expression.

$$\frac{2}{\sqrt{2}} = \sqrt{2} \quad \text{and} \quad \frac{2}{\sqrt{5} + \sqrt{3}} = \sqrt{5} - \sqrt{3}$$

Scientific Calculator

Enter: [2] [÷] [2] [√] [=]   Result: [1.414213562]

Enter: [2] [√]   Result: [1.414213562]

To enter $\dfrac{2}{\sqrt{5} + \sqrt{3}}$ use parentheses around the denominator.

Enter: [2] [÷] [(] [5] [√] [+] [3] [√] [)] [=]   Result: [0.50401717]

Enter: [5] [√] [−] [3] [√] [=]   Result: [0.50401717]

Graphing Calculator

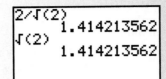

The calculator approximation of each expression and its simplified form agree to 10 decimal places.

Calculator Exercises

Simplify each expression. Then use a calculator to approximate the original expression and its simplified form.

1. $\dfrac{5}{\sqrt{11}}$ ❖ 2. $\dfrac{4}{\sqrt{3}}$ ❖ 3. $\dfrac{6}{\sqrt{2}}$ ❖ 4. $\dfrac{9}{\sqrt[3]{3}}$ ❖

5. $\dfrac{12}{\sqrt[3]{4}}$ ❖ 6. $\dfrac{10}{\sqrt{3}+\sqrt{2}}$ ❖ 7. $\dfrac{4}{\sqrt{5}+1}$ ❖ 8. $\dfrac{\sqrt{2}+\sqrt{7}}{\sqrt{2}-\sqrt{7}}$ ❖

---

section 8.5 PRACTICE EXERCISES

For Exercises 1–10, perform the indicated operations. Assume all variables represent positive real numbers.

8.3–8.4
1. $x\sqrt{45}+4\sqrt{20x^2}$
   $11x\sqrt{5}$

2. $5b\sqrt{72b}-3\sqrt{50b^3}$
   $15b\sqrt{2b}$

3. $(2\sqrt{y}+3)(3\sqrt{y}+7)$ $6y+23\sqrt{y}+21$

4. $(4\sqrt{w}-2)(2\sqrt{w}-4)$ $8w-20\sqrt{w}+8$

5. $4\sqrt{3}+\sqrt{5}\cdot\sqrt{15}$
   $9\sqrt{3}$

6. $\sqrt{7}\cdot\sqrt{21}+2\sqrt{27}$
   $13\sqrt{3}$

7. $(5-\sqrt{a})^2$
   $25-10\sqrt{a}+a$

8. $(\sqrt{z}+3)^2$
   $z+6\sqrt{z}+9$

9. $(\sqrt{2}+\sqrt{7})(\sqrt{2}-\sqrt{7})$ -5

10. $(\sqrt{3}+5)(\sqrt{3}-5)$ -22

 11. a. Use a calculator to approximate $\frac{1}{\sqrt{2}}$ to three decimal places.   0.707

   $\dfrac{\sqrt{2}}{2}$ b. Simplify $\frac{1}{\sqrt{2}}$ by rationalizing the denominator.

   c. Use a calculator to approximate the simplified expression from part (b) and compare the result with part (a).   0.707

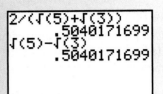 12. a. Use a calculator to approximate $\dfrac{10}{\sqrt{5}}$ to three decimal places.   4.472

   b. Simplify $\dfrac{10}{\sqrt{5}}$ by rationalizing the denominator. $2\sqrt{5}$

   c. Use a calculator to approximate the simplified expression from part (b) and compare the result with part (a).   4.472

13. a. Use a calculator to approximate

   $$\frac{1}{\sqrt{2}-\sqrt{7}}$$

   to three decimal places. -0.812

---

b. Simplify $\dfrac{1}{\sqrt{2} - \sqrt{7}}$
by rationalizing the denominator. $\dfrac{\sqrt{2} + \sqrt{7}}{-5}$

c. Use a calculator to approximate the simplified expression from part (b) and compare the result with part (a). -0.812

14. a. Use a calculator to approximate

$$\dfrac{1}{\sqrt{3} + \sqrt{11}}$$

to three decimal places. 0.198

b. Simplify $\dfrac{1}{\sqrt{3} + \sqrt{11}}$
by rationalizing the denominator. $\dfrac{\sqrt{3} - \sqrt{11}}{-8}$

c. Use a calculator to approximate the simplified expression from part (b) and compare the result with part (a). 0.198

For Exercises 15–30, rationalize the denominators. Assume all variable expressions represent positive real numbers.

15. $\dfrac{1}{\sqrt{6}}$ $\dfrac{\sqrt{6}}{6}$

16. $\dfrac{5}{\sqrt{2}}$ $\dfrac{5\sqrt{2}}{2}$

17. $\dfrac{15}{\sqrt{5}}$ $3\sqrt{5}$

18. $\dfrac{14}{\sqrt{7}}$ $2\sqrt{7}$

19. $\dfrac{6}{\sqrt{x+1}}$ $\dfrac{6\sqrt{x+1}}{x+1}$

20. $\dfrac{8}{\sqrt{y-3}}$ $\dfrac{8\sqrt{y-3}}{y-3}$

21. $\sqrt{\dfrac{6}{x}}$ $\dfrac{\sqrt{6x}}{x}$

22. $\sqrt{\dfrac{8}{y}}$ $\dfrac{2\sqrt{2y}}{y}$

23. $\dfrac{1}{\sqrt[3]{6}}$ $\dfrac{\sqrt[3]{36}}{6}$

24. $\dfrac{6}{\sqrt[3]{3}}$ $2\sqrt[3]{9}$

25. $\dfrac{7}{\sqrt[3]{4}}$ $\dfrac{7\sqrt[3]{2}}{2}$

26. $\dfrac{1}{\sqrt[3]{9}}$ $\dfrac{\sqrt[3]{3}}{3}$

27. $\dfrac{10}{\sqrt{6y}}$ $\dfrac{5\sqrt{6y}}{3y}$

28. $\dfrac{15}{\sqrt{3w}}$ $\dfrac{5\sqrt{3w}}{w}$

29. $\sqrt[4]{\dfrac{16}{3}}$ $\dfrac{2\sqrt[4]{27}}{3}$

30. $\sqrt[4]{\dfrac{81}{8}}$ $\dfrac{3\sqrt[4]{2}}{2}$

For Exercises 31–32, multiply the conjugates.

31. $(\sqrt{2} + 3)(\sqrt{2} - 3)$ -7

32. $(\sqrt{3} + \sqrt{7})(\sqrt{3} - \sqrt{7})$ -4

33. What is the conjugate of $\sqrt{5} - \sqrt{3}$? Multiply $\sqrt{5} - \sqrt{3}$ by its conjugate. $\sqrt{5} + \sqrt{3}$; 2

34. What is the conjugate of $\sqrt{7} + \sqrt{2}$? Multiply $\sqrt{7} + \sqrt{2}$ by its conjugate. $\sqrt{7} - \sqrt{2}$; 5

35. What is the conjugate of $\sqrt{x} + 10$? Multiply $\sqrt{x} + 10$ by its conjugate. $\sqrt{x} - 10$; $x - 100$

36. What is the conjugate of $12 - \sqrt{y}$? Multiply $12 - \sqrt{y}$ by its conjugate. $12 + \sqrt{y}$; $144 - y$

For Exercises 37–46, rationalize the denominators.

37. $\dfrac{4}{\sqrt{2} + 3}$ $\dfrac{4\sqrt{2} - 12}{-7}$ or $\dfrac{12 - 4\sqrt{2}}{7}$

38. $\dfrac{6}{4 - \sqrt{3}}$ $\dfrac{24 + 6\sqrt{3}}{13}$

39. $\dfrac{1}{\sqrt{5} - \sqrt{2}}$ $\dfrac{\sqrt{5} + \sqrt{2}}{3}$

40. $\dfrac{2}{\sqrt{3} + \sqrt{7}}$ $\dfrac{\sqrt{3} - \sqrt{7}}{-2}$ or $\dfrac{\sqrt{7} - \sqrt{3}}{2}$

41. $\dfrac{\sqrt{8}}{\sqrt{3} + 1}$ $\sqrt{6} - \sqrt{2}$

42. $\dfrac{\sqrt{18}}{1 - \sqrt{2}}$ $-3\sqrt{2} - 6$

43. $\dfrac{1}{\sqrt{x} - \sqrt{3}}$ $\dfrac{\sqrt{x} + \sqrt{3}}{x - 3}$

44. $\dfrac{1}{\sqrt{y} + \sqrt{5}}$ $\dfrac{\sqrt{y} - \sqrt{5}}{y - 5}$

45. $\dfrac{\sqrt{5} + 4}{2 - \sqrt{5}}$ $-13 - 6\sqrt{5}$

46. $\dfrac{3 + \sqrt{2}}{\sqrt{2} - 5}$ $\dfrac{17 + 8\sqrt{2}}{-23}$ or $-\dfrac{17 + 8\sqrt{2}}{23}$

For Exercises 47–54, simplify and reduce the expression.

47. $\dfrac{10 - \sqrt{50}}{5}$ $2 - \sqrt{2}$

48. $\dfrac{4 + \sqrt{12}}{2}$ $2 + \sqrt{3}$

49. $\dfrac{21 + \sqrt{98}}{14}$ $\dfrac{3 + \sqrt{2}}{2}$

50. $\dfrac{3 - \sqrt{18}}{6}$ $\dfrac{1 - \sqrt{2}}{2}$

51. $\dfrac{2 - \sqrt{28}}{2}$ $1 - \sqrt{7}$

52. $\dfrac{5 + \sqrt{75}}{5}$ $1 + \sqrt{3}$

53. $\dfrac{14 + \sqrt{72}}{6}$ $\dfrac{7 + 3\sqrt{2}}{3}$

54. $\dfrac{15 - \sqrt{125}}{10}$ $\dfrac{3 - \sqrt{5}}{2}$

55. State the conditions for a radical expression to be in simplified form.

1. The radicand has no factor raised to a power greater than or equal to the index.
2. There are no radicals in the denominator of a fraction.
3. The radicand does not contain a fraction.

❖ See Additional Answers Appendix   Writing   Translating Expression   Geometry   Scientific Calculator   Video

For Exercises 56–67, simplify the radical expressions, if possible. Assume all variables represent positive real numbers.

56. $\sqrt{45}$ $3\sqrt{5}$

57. $-\sqrt{108y^4}$ $-6y^2\sqrt{3}$

58. $-\sqrt{\dfrac{18w^2}{25}}$ $-\dfrac{3w\sqrt{2}}{5}$

59. $\sqrt{\dfrac{8a^2}{7}}$ $\dfrac{2a\sqrt{14}}{7}$

60. $\sqrt{-36}$
    Not a real number

61. $\sqrt[3]{54b^4}$
    $3b\sqrt[3]{2b}$

62. $\sqrt[3]{\dfrac{p^2}{q}}$ $\dfrac{\sqrt[3]{p^2q^2}}{q}$

63. $\sqrt[4]{\dfrac{3x^2y^5}{16z^8}}$ $\dfrac{y\sqrt[4]{3x^2y}}{2z^2}$

64. $\sqrt{\dfrac{81}{t^3}}$ $\dfrac{9\sqrt{t}}{t^2}$

65. $\sqrt[3]{-\dfrac{8}{27}}$ $-\dfrac{2}{3}$

66. $-\sqrt[4]{\dfrac{2m^5}{32m}}$ $-\dfrac{m}{2}$

67. $-\sqrt{a^3bc^6}$ $-ac^3\sqrt{ab}$

EXPANDING YOUR SKILLS

68. Find the slope of the line through the points: $(5\sqrt{2}, 3)$ and $(\sqrt{2}, 6)$. $-\dfrac{3\sqrt{2}}{8}$

69. Find the slope of the line through the points: $(4\sqrt{5}, -1)$ and $(6\sqrt{5}, -5)$. $-\dfrac{2\sqrt{5}}{5}$

70. Find the slope of the line through the points: $(\sqrt{3}, -1)$ and $(4\sqrt{3}, 0)$. $\dfrac{\sqrt{3}}{9}$

71. Find the slope of the line through the points: $(-2\sqrt{7}, -5)$ and $(\sqrt{7}, 2)$ $\dfrac{\sqrt{7}}{3}$

For Exercises 72–75, rationalize the denominators. Assume all variable expressions represent positive real numbers.

72. $\sqrt[3]{\dfrac{8a^3b}{c^2}}$ $\dfrac{2a\sqrt[3]{bc}}{c}$

73. $-\sqrt[3]{\dfrac{16x^6}{y^2z}}$ $-\dfrac{2x^2\sqrt[3]{2yz^2}}{yz}$

74. $-\dfrac{2w}{\sqrt[4]{2w^3}}$ $-\sqrt[4]{8w}$

75. $\dfrac{3p^2}{\sqrt[4]{36p}}$ $\dfrac{p\sqrt[4]{36p^3}}{2}$

RADICAL EQUATIONS

Concepts

1. **Definition of a Radical Equation**
2. **Solutions to Radical Equations**
3. **Solving Radical Equations**
4. **Translations Involving Radical Equations**
5. **Applications of Radical Equations**

1. Definition of a Radical Equation

Definition of a Radical Equation

An equation with one or more radicals containing a variable is called a **radical equation**.

For example, $\sqrt[3]{x} = 2$ is a radical equation. Recall that $(\sqrt[n]{a})^n = a$ provided $\sqrt[n]{a}$ is a real number. The basis to solve a radical equation is to eliminate the radical by raising both sides of the equation to a power equal to the index of the radical.

To solve the equation $\sqrt[3]{x} = 2$, cube both sides of the equation.

$$\sqrt[3]{x} = 2$$
$$(\sqrt[3]{x})^3 = (2)^3$$
$$x = 8$$

2. Solutions to Radical Equations

By raising each side of a radical equation to a power equal to the index of the radical, a new equation is produced. However, it is important to note that the new equation may have extraneous solutions. That is, some or all of the solutions to the new equation may *not* be solutions to the original radical equation. For this reason, it is

necessary to check *all* potential solutions in the original equation. For example, consider the equation $x = 4$. By squaring both sides we produce a quadratic equation.

$$x = 4$$

Square both sides. $$(x)^2 = (4)^2$$ Squaring both sides produces a quadratic equation.

$$x^2 = 16$$ Solving this equation, we find two solutions. However, the solution $x = -4$ does not check in the original equation.

$$x^2 - 16 = 0$$

$$(x - 4)(x + 4) = 0$$

$$x = 4 \quad \text{or} \quad x \neq -4 \quad \text{Does not check.}$$

3.   Solving Radical Equations

Steps to Solving a Radical Equation

1.   Isolate the radical. If an equation has more than one radical, choose one of the radicals to isolate.
2.   Raise each side of the equation to a power equal to the index of the radical.
3.   Solve the resulting equation.
4.   Check the potential solutions in the original equation.*

*Extraneous solutions can only arise when both sides of the equation are raised to an *even power.* Therefore, an equation with odd-index roots will not have an extraneous solution. However, it is still recommended that you check *all* potential solutions regardless of the type of root.

example 1   **Solving Radical Equations**

Solve the equations.

a. $\sqrt{2x + 1} + 5 = 8$ b. $8 + \sqrt[4]{x + 2} = 7$

Solution:

a. $\sqrt{2x + 1} + 5 = 8$

$$\sqrt{2x + 1} = 8 - 5 \qquad \text{Isolate the radical.}$$

$$\sqrt{2x + 1} = 3$$

$$(\sqrt{2x + 1})^2 = (3)^2 \qquad \text{Raise both sides to a power equal to the index of the radical.}$$

$$2x + 1 = 9 \qquad \text{Simplify both sides.}$$

$$2x = 8 \qquad \text{Solve the resulting equation (the equation is linear).}$$

$$x = 4$$

Check: Check $x = 4$ as a potential solution.

$$\sqrt{2x + 1} + 5 = 8$$

$$\sqrt{2(4) + 1} + 5 \stackrel{?}{=} 8$$

$$\sqrt{8+1} + 5 \stackrel{?}{=} 8$$

$$\sqrt{9} + 5 \stackrel{?}{=} 8$$

$$3 + 5 = 8 \checkmark \qquad \text{Answer checks.}$$

$x = 4$ is the solution.

Tip: After isolating the radical in Example 1b, the equation shows a fourth root equated to a negative number.

$$\sqrt[4]{x+2} = -1$$

By definition, a principal fourth-root of any real number must be nonnegative. Therefore, there can be no solution to this equation.

b. $8 + \sqrt[4]{x+2} = 7$

$\sqrt[4]{x+2} = 7 - 8$ Isolate the radical.

$\sqrt[4]{x+2} = -1$

$(\sqrt[4]{x+2})^4 = (-1)^4$ Raise both sides to a power equal to the index of the radical.

$x + 2 = 1$ Simplify.

$x = -1$ Solve the resulting equation.

Check: Check $x = -1$ as a potential solution.

$$8 + \sqrt[4]{x+2} = 7$$

$$8 + \sqrt[4]{(-1)+2} \stackrel{?}{=} 7$$

$$8 + \sqrt[4]{1} \stackrel{?}{=} 7$$

$$8 + 1 \neq 7 \qquad \text{The potential solution does } not \text{ check.}$$

There is no solution to the equation.

---

example 2

Solving Radical Equations

Solve the equations.

a. $p + 4 = \sqrt{p+6}$ b. $\sqrt[3]{2x-3} - \sqrt[3]{x+1} = 0$

Classroom Activity 8.6A

Tip: Recall that

$$(a+b)^2 = a^2 + 2ab + b^2.$$

Hence,

$$(p+4)^2$$
$$= (p)^2 + 2(p)(4) + (4)^2$$
$$= p^2 + 8p + 16$$

Solution:

a. $\qquad p + 4 = \sqrt{p+6}$ The radical is already isolated.

$\qquad (p+4)^2 = (\sqrt{p+6})^2$ Raise both sides to a power equal to the index.

$p^2 + 8p + 16 = p + 6$

$p^2 + 7p + 10 = 0$ Solve the resulting equation (the equation is quadratic).

$$(p + 5)(p + 2) = 0$$ Set the equation equal to zero and factor.

$$p + 5 = 0 \quad \text{or} \quad p + 2 = 0$$ Set each factor equal to zero.

$$p = -5 \quad \text{or} \quad p = -2$$ Solve for p.

Check: $p = -5$

$$p + 4 = \sqrt{p + 6}$$

$$(-5) + 4 \stackrel{?}{=} \sqrt{(-5) + 6}$$

$$-1 \stackrel{?}{=} \sqrt{1}$$

$$-1 \neq 1 \quad \text{Does not check.}$$

Check: $p = -2$

$$p + 4 = \sqrt{p + 6}$$

$$(-2) + 4 \stackrel{?}{=} \sqrt{(-2) + 6}$$

$$2 \stackrel{?}{=} \sqrt{4}$$

$$2 = 2 \; ✔ \quad \text{Solution checks.}$$

The only solution is $p = -2$ ($p = -5$ does not check).

b. $\sqrt[3]{2x - 3} - \sqrt[3]{x + 1} = 0$

$$\sqrt[3]{2x - 3} = \sqrt[3]{x + 1}$$ Isolate the radical.

$$(\sqrt[3]{2x - 3})^3 = (\sqrt[3]{x + 1})^3$$ Raise both sides to a power equal to the index.

$$2x - 3 = x + 1$$ The resulting equation is linear.

$$2x - x = 1 + 3$$ Solve for x.

$$x = 4$$

Check: Check the potential solution $x = 4$.

$$\sqrt[3]{2x - 3} - \sqrt[3]{x + 1} = 0$$

$$\sqrt[3]{2(4) - 3} - \sqrt[3]{(4) + 1} \stackrel{?}{=} 0$$

$$\sqrt[3]{8 - 3} - \sqrt[3]{5} \stackrel{?}{=} 0$$

$$\sqrt[3]{5} - \sqrt[3]{5} = 0 \; ✔ \quad \text{The solution checks.}$$

The solution is $x = 4$.

4. Translations Involving Radical Equations

example 3

Translating English Form into Algebraic Form

The square root of the sum of a number and three is equal to seven. Find the number.

Solution:

Let x represent the number. Label the variable.

$$\sqrt{x + 3} = 7$$ Translate the verbal model into an algebraic equation.

$$(\sqrt{x + 3})^2 = (7)^2$$ The radical is already isolated. Square both sides.

$$x + 3 = 49 \qquad \text{The resulting equation is linear.}$$

$$x = 46 \qquad \text{Solve for } x.$$

Check: $\qquad$ Check $x = 46$ as a potential solution.

$$\sqrt{x + 3} = 7$$

$$\sqrt{46 + 3} \overset{?}{=} 7$$

$$\sqrt{49} \overset{?}{=} 7$$

$$7 = 7 \; \checkmark \qquad \text{The solution checks.}$$

The number is 46.

5. Applications of Radical Equations

example 4

Using a Radical Equation in an Application

For a small company, the weekly sales, y, of its product is related to the money spent on advertising, x, according to the equation:

$$y = 100\sqrt{x} \qquad \text{where } y \text{ is total sales in dollars and } x \text{ is the number of dollars spent on advertising.}$$

a. Find the amount in sales if the company spends $100 in advertising.
b. Find the amount in sales if the company spends $625 in advertising.
c. Find the amount the company spent on advertising if its sales for 1 week totaled $2000.

Classroom Activity 8.6B

Solution:

a. $y = 100\sqrt{x}$

$\quad = 100\sqrt{100} \qquad$ Substitute $x = 100$.

$\quad = 100(10)$

$\quad = 1000$

The amount in sales is $1000.

b. $y = 100\sqrt{x}$

$\quad = 100\sqrt{625} \qquad$ Substitute $x = 625$.

$\quad = 100(25)$

$\quad = 2500$

The amount in sales is $2500.

c. $\quad y = 100\sqrt{x}$

$\quad 2000 = 100\sqrt{x} \qquad$ Substitute $y = 2000$.

$\quad \dfrac{2000}{100} = \dfrac{100\sqrt{x}}{100} \qquad$ Isolate the radical. Divide both sides by 100.

$$20 = \sqrt{x} \qquad \text{Simplify.}$$

$$(20)^2 = (\sqrt{x})^2 \qquad \text{Raise both sides to a power equal to the index.}$$

$$400 = x \qquad \text{Simplify both sides.}$$

$$\underline{\text{Check:}} \qquad \text{Check } x = 400 \text{ as a potential solution.}$$

$$y = 100\sqrt{x}$$

$$2000 \stackrel{?}{=} 100\sqrt{400}$$

$$2000 \stackrel{?}{=} 100(20)$$

$$2000 = 2000 \; ✔ \qquad \text{The solution checks.}$$

The amount spent on advertising was $400.

section 8.6 PRACTICE EXERCISES

For Exercises 1–6, rationalize the denominators. 8.5

1. $\dfrac{1}{\sqrt{3} - \sqrt{7}}$ $\dfrac{\sqrt{3} + \sqrt{7}}{-4}$ or $\dfrac{\sqrt{3} + \sqrt{7}}{4}$

2. $\dfrac{1}{\sqrt{2} + \sqrt{10}}$ $\dfrac{\sqrt{3} + \sqrt{7}}{4}$

3. $\dfrac{6}{\sqrt{6}}$ $\sqrt{6}$

$\dfrac{\sqrt{2} - \sqrt{10}}{-8}$ or $\dfrac{\sqrt{10} - \sqrt{2}}{8}$

4. $\dfrac{2\sqrt{2}}{\sqrt{3}}$ $\dfrac{2\sqrt{6}}{3}$

5. $\dfrac{5}{\sqrt[3]{25}}$ $\sqrt[3]{5}$

6. $\dfrac{4}{\sqrt[4]{8}}$ $2\sqrt[4]{2}$

For Exercises 7–8, simplify the radicals and reduce the expression. 8.5

7. $\dfrac{10 - \sqrt{75}}{5}$ $2 - \sqrt{3}$

8. $\dfrac{6 + \sqrt{32}}{4}$ $\dfrac{3 + 2\sqrt{2}}{2}$

For Exercises 9–12, square the binomials. 3.6

9. $(x + 4)^2$ $x^2 + 8x + 16$

10. $(3 - y)^2$ $9 - 6y + y^2$

11. $(2a - 5)^2$ $4a^2 - 20a + 25$

12. $(7b + 4)^2$ $49b^2 + 56b + 16$

For Exercises 13–18, simplify the expressions. 8.1

13. $(\sqrt{2x - 3})^2$ $2x - 3$

14. $(\sqrt{m + 6})^2$ $m + 6$

15. $(\sqrt[3]{t - 9})^3$ $t - 9$

16. $(\sqrt[4]{5y - 4})^4$ $5y - 4$

17. $(\sqrt[4]{7 - n})^4$ $7 - n$

18. $(\sqrt[3]{8p + 1})^3$ $8p + 1$

For Exercises 19–40, solve the equations. Be sure to check all of the answers.

19. $\sqrt{x + 1} = 4$ $x = 15$

20. $\sqrt{x - 3} = 7$ $x = 52$

21. $\sqrt{y - 4} = -5$ No solution

22. $\sqrt{p + 6} = -1$ No solution

23. $\sqrt{5 - t} = 0$ $t = 5$

24. $\sqrt{13 + m} = 0$ $m = -13$

25. $\sqrt[3]{2n + 10} = 3$ $n = \dfrac{17}{2}$

26. $\sqrt[3]{1 - q} = 15$ $q = -3374$

27. $\sqrt[3]{6w - 8} = -2$ $w = 0$

28. $\sqrt[3]{2z + 1} = -3$ $z = -14$

29. $\sqrt{5a - 4} - 2 = 4$ $a = 8$

30. $\sqrt{3b + 4} - 3 = 2$ $b = 7$

31. $5\sqrt{c} = \sqrt{10c + 15}$ $c = 1$

32. $4\sqrt{x} = \sqrt{10x + 6}$ $x = 1$

33. $\sqrt[3]{3y + 7} = \sqrt[3]{2y - 1}$ $y = -8$

34. $\sqrt[3]{p - 5} - \sqrt[3]{2p + 1} = 0$ $p = -6$

35. $\sqrt[4]{m + 2} + \sqrt[4]{5m - 2} = 0$ No solution; ($m = 1$ does not check.)

36. $\sqrt[4]{7q + 5} = \sqrt[4]{q - 13}$ No solution; ($q = -3$ does not check.)

37. $\sqrt{6t + 7} = t + 2$ $t = 3, t = -1$

38. $\sqrt{y + 1} = y + 1$ $y = 0, y = -1$

39. $\sqrt{3p + 3} + 5 = p$ $p = 11$; ($p = 2$ does not check.)

40. $\sqrt{2m + 1} + 7 = m$ $m = 12$; ($m = 4$ does not check.)

41. Ignoring air resistance, the time, t, (in seconds) required for an object to fall x feet is given by the equation:

$$t = \frac{\sqrt{x}}{4}$$

a. Find the time required for an object to fall 64 ft. 2 s

b. Find the distance an object will fall in 4 s. 256 ft

42. Ignoring air resistance, the velocity, v, (in feet per second: ft/s) of an object in free-fall depends on the distance it has fallen, x, (in feet) according to the equation:

$$v = 8\sqrt{x}$$

a. Find the velocity of an object that has fallen 100 ft.   80 ft/s

b. Find the distance that an object has fallen if its velocity is 136 ft/s.   289 ft

43. The speed of a car, s, (in miles per hour) before the brakes were applied can be approximated by the length of its skid marks, x, (in feet) according to the equation:

$$s = 4\sqrt{x}$$

a. Find the speed of a car before the brakes were applied if its skid marks are 324 ft long.   72 mph

b. How long would you expect the skid marks to be if the car had been traveling the speed limit of 60 mph?   225 ft

44. The height of a certain plant, y, (in inches) can be determined by the time, t, (in weeks) after the seed has germinated according to the equation:

$$y = 8\sqrt{t} \quad 0 \le t \le 40$$

16 in.

a. Find the height of the plant after 4 weeks.

b. In how many weeks will the plant be 40 in. tall?   25 weeks

---

Concepts

1. Evaluating Expressions of the Form $a^{1/n}$

2. Evaluating Expressions of the Form $a^{m/n}$

3. Converting Between Rational Exponents and Radical Notation

4. Properties of Rational Exponents

5. Simplifying Expressions with Rational Exponents

6. Applications of Rational Exponents

section

8.7 RATIONAL EXPONENTS

1. Evaluating Expressions of the Form $a^{1/n}$

In Sections 3.1–3.3, the properties for simplifying expressions with integer exponents were presented. In this section, the properties are expanded to include expressions with rational exponents. We begin by defining expressions of the form $a^{1/n}$.

Definition of $a^{1/n}$

Let a be a real number, and let n be an integer such that $n > 1$. If $\sqrt[n]{a}$ is a real number, then

$$a^{1/n} = \sqrt[n]{a}$$

example 1

Evaluating Expressions of the Form $a^{1/n}$

Evaluate the expressions.

a. $9^{1/2}$ b. $125^{1/3}$ c. $16^{1/4}$ d. $-25^{1/2}$ e. $(-25)^{1/2}$

Classroom Activity 8.7A

Solution:

a. $9^{1/2} = \sqrt{9} = 3$
b. $125^{1/3} = \sqrt[3]{125} = 5$
c. $16^{1/4} = \sqrt[4]{16} = 2$
d. $-25^{1/2}$ is equivalent to $-1(25^{1/2})$
    $\quad = -1 \cdot \sqrt{25}$
    $\quad = -5$
e. $(-25)^{1/2}$ is not a real number because $\sqrt{-25}$ is not a real number

2. Evaluating Expressions of the Form $a^{m/n}$

If $\sqrt[n]{a}$ is a real number, then we can define an expression of the form $a^{m/n}$ in such a way that the multiplication property of exponents holds true. For example,

$$16^{3/4} = \begin{cases} (16^{1/4})^3 = (\sqrt[4]{16})^3 = (2)^3 = 8 \\ (16^3)^{1/4} = \sqrt[4]{16^3} = \sqrt[4]{4096} = 8 \end{cases}$$

Definition of $a^{m/n}$

Let a be a real number, and let m and n be positive integers such that m and n share no common factors and $n > 1$. If $\sqrt[n]{a}$ is a real number, then

$$a^{m/n} = (a^{1/n})^m = (\sqrt[n]{a})^m \qquad \text{and} \qquad a^{m/n} = (a^m)^{1/n} = \sqrt[n]{a^m}$$

The rational exponent in the expression $a^{m/n}$ is essentially performing two operations. The numerator of the exponent raises the base to the mth-power. The denominator takes the nth-root.

example 2

Evaluating Expressions of the Form $a^{m/n}$

Simplify the expressions.

a. $125^{2/3}$ b. $100^{3/2}$ c. $(81)^{3/4}$

Solution:

a. $125^{2/3} = (\sqrt[3]{125})^2$ Take the cube root of 125 and square the result.

    $\qquad\quad = (5)^2$ Simplify.

    $\qquad\quad = 25$

b. $100^{3/2} = (\sqrt{100})^3$ Take the square root of 100 and cube the result.

       $= (10)^3$ Simplify.

       $= 1000$

c. $(81)^{3/4} = (\sqrt[4]{81})^3$ Take the fourth root of 81 and cube the result.

       $= (3)^3$ Simplify.

       $= 27$

3. Converting Between Rational Exponents and Radical Notation

example 3

Using Radical Notation and Rational Exponents

Convert the expressions to radical notation. Assume all variables represent positive real numbers.

a. $x^{3/5}$ b. $(2a^2)^{1/3}$ c. $5y^{1/4}$

Solution:

a. $x^{3/5} = \sqrt[5]{x^3}$ or $(\sqrt[5]{x})^3$

b. $(2a^2)^{1/3} = \sqrt[3]{2a^2}$

c. $5y^{1/4} = 5\sqrt[4]{y}$ The exponent $\frac{1}{4}$ applies only to y.

example 4

Using Radical Notation and Rational Exponents

Convert each expression to an equivalent expression using rational exponents. Assume that all variables represent positive real numbers.

a. $\sqrt[4]{c^3}$ b. $\sqrt{11p}$

Classroom Activity 8.7B

Solution:

a. $\sqrt[4]{c^3} = c^{3/4}$

b. $\sqrt{11p} = (11p)^{1/2}$

4. Properties of Rational Exponents

In Sections 3.1–3.3, several properties and definitions were introduced to simplify expressions with integer exponents. These properties also apply to rational exponents.

Properties of Exponents and Definitions

Let a and b be real numbers. Let m and n be rational numbers such that a^m, a^n, and b^n are real numbers. Then,

	Description	Property	Example
1.	Multiplying like bases	$a^m a^n = a^{m+n}$	$x^{1/3} \cdot x^{4/3} = x^{5/3}$
2.	Dividing like bases	$\dfrac{a^m}{a^n} = a^{m-n}$	$\dfrac{x^{3/5}}{x^{1/5}} = x^{2/5}$
3.	The power rule	$(a^m)^n = a^{mn}$	$(2^{1/3})^{1/2} = 2^{1/6}$
4.	Power of a product	$(ab)^m = a^m b^m$	$(xy)^{1/2} = x^{1/2} y^{1/2}$
5.	Power of a quotient	$\left(\dfrac{a}{b}\right)^m = \dfrac{a^m}{b^m}$ $(b \neq 0)$	$\left(\dfrac{4}{25}\right)^{1/2} = \dfrac{4^{1/2}}{25^{1/2}} = \dfrac{2}{5}$

	Description	Definition	Example
1.	Negative exponents	$a^{-m} = \left(\dfrac{1}{a}\right)^m = \dfrac{1}{a^m}$ $(a \neq 0)$	$(8)^{-1/3} = \left(\dfrac{1}{8}\right)^{1/3} = \dfrac{1}{2}$
2.	Zero exponent	$a^0 = 1$ $(a \neq 0)$	$5^0 = 1$

5. Simplifying Expressions with Rational Exponents

example 5

Simplifying Expressions with Rational Exponents

Use the properties of exponents to simplify the expressions. Write the final answers with positive exponents only. Assume all variables represent positive real numbers.

a. $x^{2/3} x^{1/3}$

b. $\dfrac{y^{4/5}}{y^{1/10}}$

c. $(z^4)^{1/2}$

d. $(s^4 t^8)^{1/4}$

e. $\left(\dfrac{x^{-2/3}}{y^{-1/2}}\right)^6 (x^{-1/5})^{10}$

Classroom Activity 8.7C

Solution:

a. $x^{2/3} x^{1/3} = x^{(2/3)+(1/3)}$ Add exponents.

$= x^{3/3}$ Simplify.

$= x$

b. $\dfrac{y^{4/5}}{y^{1/10}} = y^{(4/5)-(1/10)}$ Subtract exponents.

$= y^{(8/10)-(1/10)}$ Get a common denominator.

$= y^{7/10}$ Simplify.

c. $(z^4)^{1/2} = z^{(4)\cdot(1/2)}$ Multiply exponents.

$= z^2$ Simplify.

d. $(s^4 t^8)^{1/4} = s^{4/4} t^{8/4}$ Multiply exponents.

$\qquad = s t^2$ Reduce fractions.

e. $\left(\dfrac{x^{-2/3}}{y^{-1/2}}\right)^6 (x^{-1/5})^{10} = \dfrac{x^{(-2/3)(6)}}{y^{(-1/2)(6)}} \cdot x^{(-1/5)(10)}$ Multiply exponents.

$\qquad = \left(\dfrac{x^{-4}}{y^{-3}}\right)(x^{-2})$ Simplify exponents.

$\qquad = \dfrac{x^{-4}}{y^{-3}} \cdot \dfrac{x^{-2}}{1}$ Write x^{-2} as $\dfrac{x^{-2}}{1}$

$\qquad = \dfrac{x^{-6}}{y^{-3}}$ Add exponents in the numerator.

$\qquad = \dfrac{y^3}{x^6}$ Simplify negative exponents.

6. Applications of Rational Exponents

example 6

Using Rational Exponents in an Application

Suppose P dollars in principal is invested in an account that earns interest annually. If after t years the investment grows to A dollars, then the annual rate of return, r, on the investment is given by

$$r = \left(\frac{A}{P}\right)^{1/t} - 1$$

Find the annual rate of return on $8000 which grew to $11,220.41 after 5 years (round to the nearest tenth of a percent).

Solution:

$r = \left(\dfrac{A}{P}\right)^{1/t} - 1$ where $A = \$11{,}220.41$, $P = \$8000$ and $t = 5$. Hence

$r = \left(\dfrac{11220.41}{8000}\right)^{1/5} - 1$

$\quad = (1.40255125)^{1/5} - 1$

$\quad \approx 1.070 - 1$

$\quad \approx 0.070 \text{ or } 7.0\%$

There is a 7.0% annual rate of return.

Calculator Connections

Rational exponents provide another method to evaluate nth-roots on a calculator. For example, use rational exponents and the $\boxed{y^x}$ key (or $\boxed{\wedge}$ key) to confirm the solutions to Example 2. That is, evaluate: $125^{2/3}$, $100^{3/2}$, and $81^{3/4}$.

Scientific Calculator

Enter: $\boxed{1}\ \boxed{2}\ \boxed{5}\ \boxed{y^x}\ \boxed{(}\ \boxed{2}\ \boxed{\div}\ \boxed{3}\ \boxed{)}\ \boxed{=}$ *Result: $\boxed{25}$

Enter: $\boxed{1}\ \boxed{0}\ \boxed{0}\ \boxed{y^x}\ \boxed{(}\ \boxed{3}\ \boxed{\div}\ \boxed{2}\ \boxed{)}\ \boxed{=}$ *Result: $\boxed{1000}$

Enter: $\boxed{8}\ \boxed{1}\ \boxed{y^x}\ \boxed{(}\ \boxed{3}\ \boxed{\div}\ \boxed{4}\ \boxed{)}\ \boxed{=}$ *Result: $\boxed{27}$

*Notice that parentheses are used around the rational exponent to ensure that the base is raised to the entire quotient.

Graphing Calculator

```
125^(2/3)
                25
100^(3/2)
              1000
81^(3/4)
                27
```

Calculator Exercises

Simplify the expressions algebraically, and then check your answers on a calculator.

1. $169^{1/2}$ ❖ 2. $729^{1/3}$ ❖ 3. $1024^{1/5}$ ❖ 4. $-10{,}000^{1/4}$ ❖

5. $8^{-2/3}$ ❖ 6. $-27^{2/3}$ ❖ 7. $100{,}000^{4/5}$ ❖ 8. $16^{-3/4}$ ❖

Evaluate the expressions with a calculator.

9. $984^{1/4}$ ❖ 10. $14.8^{1/3}$ ❖ 11. $\sqrt[3]{56^2}$ ❖ 12. $\sqrt[4]{24^3}$ ❖

13. $\sqrt[3]{\dfrac{45}{59}}$ ❖ 14. $\sqrt[3]{\dfrac{3}{104}}$ ❖ 15. $\sqrt[4]{\left(\dfrac{4}{5}\right)^3}$ ❖ 16. $\sqrt[3]{\left(\dfrac{5}{6}\right)^2}$ ❖

section 8.7   PRACTICE EXERCISES

For the exercises in this set, assume that all variables represent positive real numbers unless otherwise stated.

1. Given $\sqrt[3]{125}$ 8.1
    a. Identify the index   3   b. Identify the radicand   125

2. Given $\sqrt{12}$ 8.1
    a. Identify the index   2   b. Identify the radicand   12

For Exercises 3–6, simplify the radicals.   8.1

3. $(\sqrt[4]{81})^3$ 27   4. $(\sqrt[4]{16})^3$ 8

5. $\sqrt[3]{(a + 1)^3}$ $a + 1$ 6. $\sqrt[5]{(x + y)^5}$ $x + y$

7. Explain how to interpret the expression $a^{m/n}$ as a radical. $a^{m/n} = \sqrt[n]{a^m}$ or $(\sqrt[n]{a})^m$, provided the root exists.

8. Explain why $(\sqrt[3]{8})^4$ is easier to evaluate than $\sqrt[3]{8^4}$.   It is easier in this case to evaluate a cube root of a smaller number.

For Exercises 9–20, simplify the expression.

9. $81^{1/2}$ 9   10. $25^{1/2}$ 5   11. $125^{1/3}$ 5

12. $8^{1/3}$ 2   13. $81^{1/4}$ 3   14. $16^{3/4}$ 8

15. $(-8)^{1/3}$ -2 16. $(-9)^{1/2}$ Not a real number   17. $-8^{1/3}$ -2

18. $-9^{1/2}$ -3 19. $\dfrac{1}{36^{-1/2}}$ 6   20. $\dfrac{1}{16^{-1/2}}$ 4

For Exercises 21–32, write the expressions in radical notation.

21. $x^{1/3}$ $\sqrt[3]{x}$ 22. $y^{1/4}$ $\sqrt[4]{y}$ 23. $(4a)^{1/2}$ $\sqrt{4a}$ or $2\sqrt{a}$

24. $(36x)^{1/2}$ $\sqrt{36x}$ or $6\sqrt{x}$ 25. $(yz)^{1/5}$ $\sqrt[5]{yz}$ 26. $(cd)^{1/4}$ $\sqrt[4]{cd}$

27. $(u^2)^{1/3}$ $\sqrt[3]{u^2}$ 28. $(v^3)^{1/4}$ $\sqrt[4]{v^3}$ 29. $5q^{1/2}$ $5\sqrt{q}$

30. $6p^{1/2}$ $6\sqrt{p}$ 31. $\left(\dfrac{x}{9}\right)^{1/2}$ $\sqrt{\dfrac{x}{9}}$ or $\dfrac{\sqrt{x}}{3}$ 32. $\left(\dfrac{y}{8}\right)^{1/3}$ $\sqrt[3]{\dfrac{y}{8}}$ or $\dfrac{\sqrt[3]{y}}{2}$

For Exercises 33–44, write the expressions using rational exponents rather than radical notation.

33. $\sqrt[3]{x}$ $x^{1/3}$ 34. $\sqrt[4]{a}$ $a^{1/4}$ 35. $\sqrt[3]{xy}$ $(xy)^{1/3}$

36. $\sqrt[5]{ab}$ $(ab)^{1/5}$ 37. $5\sqrt{x}$ $5x^{1/2}$ 38. $7\sqrt[3]{z}$ $7z^{1/3}$

39. $\sqrt[3]{y^2}$ $y^{2/3}$ 40. $\sqrt[4]{b^3}$ $b^{3/4}$ 41. $\sqrt[4]{m^3n}$ $(m^3n)^{1/4}$

42. $\sqrt[5]{u^3v^4}$ $(u^3v^4)^{1/5}$ 43. $4\sqrt[3]{k^3}$ $4k^{3/3}$ or $4k$ 44. $6\sqrt[4]{t^4}$ $6t^{4/4}$ or $6t$

For Exercises 45–64, simplify the expressions using the properties of rational exponents. Write the final answer with positive exponents only.

45. $x^{1/4}x^{3/4}$ x 46. $2^{2/3}2^{1/3}$ 2   47. $(y^{1/5})^{10}$ y^2

48. $(x^{1/2})^8$ x^4 49. $6^{-1/5}6^{6/5}$ 6   50. $a^{-1/3}a^{2/3}$ $a^{1/3}$

51. $(a^{1/3}a^{1/4})^{12}$ a^7 52. $(x^{2/3}x^{1/2})^6$ x^7 53. $\dfrac{y^{5/3}}{y^{1/3}}$ $y^{4/3}$

54. $\dfrac{z^{5/2}}{z^{1/2}}$ z^2 55. $\dfrac{2^{4/3}}{2^{1/3}}$ 2   56. $\dfrac{5^{6/5}}{5^{1/5}}$ 5

57. $(5a^2c^{-1/2}d^{1/2})^2$ $\dfrac{25a^4d}{c}$ 58. $(2x^{-1/3}y^2z^{5/3})^3$ $\dfrac{8y^6z^5}{x}$

59. $\left(\dfrac{x^{-2/3}}{y^{-3/4}}\right)^{12}$ $\dfrac{y^9}{x^8}$ 60. $\left(\dfrac{m^{-1/4}}{n^{-1/2}}\right)^{-4}$ $\dfrac{m}{n^2}$

61. $\left(\dfrac{16w^{-2}z}{2wz^{-8}}\right)^{1/3}$ $\dfrac{2z^3}{w}$ 62. $\left(\dfrac{50p^{-1}q}{2pq^{-3}}\right)^{1/2}$ $\dfrac{5q^2}{p}$

63. $(25x^2y^4z^6)^{1/2}$ $5xy^2z^3$ 64. $(8a^6b^3c^9)^{2/3}$ $4a^4b^2c^6$

For Exercises 65–72, use a calculator to approximate the expressions. Round the answers to four decimal places.

65. $19^{1/2}$ 4.3589   66. $150^{1/3}$ 5.3133   67. $50^{-1/4}$ 0.3761

68. $172^{3/5}$ 21.9441   69. $\sqrt[3]{5^2}$ 2.9240   70. $\sqrt[4]{6^3}$ 3.8337

71. $\sqrt{10^3}$ 31.6228   72. $\sqrt[3]{16}$ 2.5198

❖ See Additional Answers Appendix   ✎ Writing   ⬌ Translating Expression   Geometry   Scientific Calculator   Video

73. If the area, A, of a square is known, then the length of its sides, s, can be computed by the formula: $s = A^{1/2}$.

    a. Compute the length of the sides of a square having an area of 100 in.^{2} 10 in.

    b. Compute the length of the sides of a square having an area of 72 in.^{2} Round your answer to the nearest 0.01 in.   8.49 in.

74. The radius, r, of a sphere of volume, V, is given by

$$r = \left(\frac{3V}{4\pi}\right)^{1/3}$$

Find the radius of a spherical ball having a volume of 55 in.^{3} Round your answer to the nearest 0.01 in.   2.36 in.

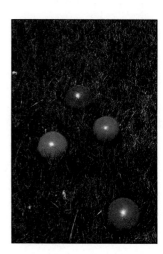

75. If P dollars in principal grows to A dollars after t years with annual interest, then the rate of return is given by

$$r = \left(\frac{A}{P}\right)^{1/t} - 1$$

    a. In one account, \$10,000 grows to \$16,802 after 5 years. Compute the interest rate to the nearest tenth of a percent.   10.9%

    b. In another account \$10,000 grows to \$18,000 after 7 years. Compute the interest rate to the nearest tenth of a percent.   8.8%

    c. Which account produced a higher average yearly return?   The account in part (a)

EXPANDING YOUR SKILLS

76. Is $(a + b)^{1/2}$ the same as $a^{1/2} + b^{1/2}$? Why or why not?   No, for example $(36 + 64)^{1/2} \neq 36^{1/2} + 64^{1/2}$

For Exercises 77–82, simplify the expressions. Write the final answer with positive exponents only.

77. $\left(\frac{1}{8}\right)^{2/3} + \left(\frac{1}{4}\right)^{1/2}$ $\frac{3}{4}$

78. $\left(\frac{1}{8}\right)^{-2/3} + \left(\frac{1}{4}\right)^{-1/2}$ 6

79. $\left(\frac{1}{16}\right)^{-1/4} - \left(\frac{1}{49}\right)^{-1/2}$ -5

80. $\left(\frac{1}{16}\right)^{1/4} - \left(\frac{1}{49}\right)^{1/2}$ $\frac{5}{14}$

81. $\left(\frac{x^2 y^{-1/3} z^{2/3}}{x^{2/3} y^{1/4} z}\right)^{12}$ $\frac{x^{16}}{y^7 z^4}$

82. $\left(\frac{a^2 b^{1/2} c^{-2}}{a^{-3/4} b^0 c^{1/8}}\right)^8$ $\frac{a^{22} b^4}{c^{17}}$

chapter 8   SUMMARY

SECTION 8.1—INTRODUCTION TO ROOTS AND RADICALS

KEY CONCEPTS:

b is a square root of a if $b^2 = a$.

The expression $\sqrt{a}$ represents the principal square root of a.

b is an nth-root of a if $b^n = a$.

1. If n is a positive *even* integer and $a > 0$, then $\sqrt[n]{a}$ is the principal (positive) nth-root of a.
2. If n is a positive *odd* integer, then $\sqrt[n]{a}$ is the nth-root of a.
3. If n is any positive integer, then $\sqrt[n]{0} = 0$

$$\sqrt[n]{a^n} = |a| \text{ if } n \text{ is even.}$$

$$\sqrt[n]{a^n} = a \text{ if } n \text{ is odd.}$$

$\sqrt[n]{a}$ is not a real number if a is *negative* and n is even.

Pythagorean Theorem:

$$a^2 + b^2 = c^2$$

KEY TERMS:

cube root
index
nth root
negative square root
perfect square
positive square root
principal square root
Pythagorean theorem
radicand
square root

EXAMPLES:

The square roots of 16 are 4 and -4 because $(4)^2 = 16$ and $(-4)^2 = 16$.

$\sqrt{16} = 4$ Because $4^2 = 16$

$\sqrt[4]{16} = 2$ Because $2^4 = 16$

$\sqrt[3]{125} = 5$ Because $5^3 = 125$

$\sqrt{y^2} = |y|$ $\sqrt[3]{y^3} = y$ $\sqrt[4]{y^4} = |y|$

$\sqrt[4]{-16}$ is not a real number.

Find the length of the unknown side:

$$a^2 + b^2 = c^2$$
$$(8)^2 + b^2 = (17)^2$$
$$64 + b^2 = 289$$
$$b^2 = 289 - 64$$
$$b^2 = 225$$
$$b = \sqrt{225}$$
$$b = 15$$

The third side is 15 cm.

Because b denotes a length, b must be the positive square root of 225.

SECTION 8.2—PROPERTIES OF RADICALS

KEY CONCEPTS:

Multiplication Property of Radicals:

If $\sqrt[n]{a}$ and $\sqrt[n]{b}$ are both real, then

$$\sqrt[n]{ab} = \sqrt[n]{a} \cdot \sqrt[n]{b}$$

Division Property of Radicals:

$$\sqrt[n]{\frac{a}{b}} = \frac{\sqrt[n]{a}}{\sqrt[n]{b}} \quad b \neq 0$$

Consider a radical expression whose radicand is written as a product of prime factors. Then the radical is in simplified form if each of the following criteria are met:

1. The radicand has no factor raised to a power greater than or equal to the index.
2. There are no radicals in the denominator of a fraction.
3. The radicand does not contain a fraction.

KEY TERMS:

division property of radicals
multiplication property of radicals
simplified form of a radical

EXAMPLES:

$$\sqrt{3} \cdot \sqrt{5} = \sqrt{3 \cdot 5} = \sqrt{15}$$

$$\sqrt{\frac{x}{9}} = \frac{\sqrt{x}}{\sqrt{9}} = \frac{\sqrt{x}}{3}$$

Simplify:

$$\sqrt[3]{16x^5 y^7}$$
$$= \sqrt[3]{2^4 x^5 y^7}$$
$$= \sqrt[3]{2^3 x^3 y^6 \cdot 2x^2 y}$$
$$= \sqrt[3]{2^3 x^3 y^6} \cdot \sqrt[3]{2x^2 y}$$
$$= 2xy^2 \sqrt[3]{2x^2 y}$$

Simplify:

$$\sqrt{\frac{2x^5}{8x}}$$

$$= \sqrt{\frac{x^4}{4}} \qquad \text{Reduce the radicand.}$$

$$= \frac{\sqrt{x^4}}{\sqrt{4}}$$

$$= \frac{x^2}{2}$$

SECTION 8.3—ADDITION AND SUBTRACTION OF RADICALS

KEY CONCEPTS:

Two radical expressions are said to be *like* radicals if their radical factors have the same index and the same radicand.

EXAMPLES:

***Like* radicals:**

$$\sqrt[3]{5z}, 6\sqrt[3]{5z}$$

Un*like* radicals:

$$\sqrt{6p}, \sqrt[3]{6p}$$
$$\sqrt{3c}, \sqrt{5c}$$

Use the distributive property to add or subtract *like* radicals.

KEY TERM:

like radicals

Perform the indicated operations:

$$3x\sqrt{7} - 10x\sqrt{7} + x\sqrt{7}$$
$$= (3 - 10 + 1) \cdot x\sqrt{7}$$
$$= -6x\sqrt{7}$$

SECTION 8.4—MULTIPLICATION OF RADICALS

KEY CONCEPTS:

Multiplication Property of Radicals:

$\sqrt[n]{ab} = \sqrt[n]{a} \cdot \sqrt[n]{b}$ provided $\sqrt[n]{a}$ and $\sqrt[n]{b}$ are both real.

Special Case Products:

$$(a + b)(a - b) = a^2 - b^2$$
$$(a + b)^2 = a^2 + 2ab + b^2$$
$$(a - b)^2 = a^2 - 2ab + b^2$$

KEY TERM:

$(\sqrt[n]{a})^n$

EXAMPLES:

Multiply:

$$(6\sqrt{5})(4\sqrt{3})$$
$$= 6 \cdot 4\sqrt{5 \cdot 3}$$
$$= 24\sqrt{15}$$

Multiply:

$$3\sqrt{2}(\sqrt{2} + 5\sqrt{7} - \sqrt{6})$$
$$= 3\sqrt{4} + 15\sqrt{14} - 3\sqrt{12}$$
$$= 3 \cdot 2 + 15\sqrt{14} - 3 \cdot 2\sqrt{3}$$
$$= 6 + 15\sqrt{14} - 6\sqrt{3}$$

Multiply:

$$(4\sqrt{x} + \sqrt{2})(4\sqrt{x} - \sqrt{2})$$
$$= (4\sqrt{x})^2 - (\sqrt{2})^2$$
$$= 16x - 2$$

Multiply:

$$(\sqrt{x} - \sqrt{5y})^2$$
$$= (\sqrt{x})^2 - 2(\sqrt{x})(\sqrt{5y}) + (\sqrt{5y})^2$$
$$= x - 2\sqrt{5xy} + 5y$$

SECTION 8.5—RATIONALIZATION

KEY CONCEPTS:

Rationalizing the Denominator with One Term:

Multiply numerator and denominator by an appropriate expression to create an nth-root of an nth-power in the denominator.

EXAMPLES:

Simplify:

$$\frac{10}{\sqrt{5}}$$

$$= \frac{10}{\sqrt{5}} \cdot \frac{\sqrt{5}}{\sqrt{5}} = \frac{10\sqrt{5}}{\sqrt{5^2}} = \frac{10\sqrt{5}}{5} = 2\sqrt{5}$$

Simplify:

$$\frac{1}{\sqrt[3]{x}}$$

$$= \frac{1}{\sqrt[3]{x}} \cdot \frac{\sqrt[3]{x^2}}{\sqrt[3]{x^2}} = \frac{\sqrt[3]{x^2}}{\sqrt[3]{x^3}} = \frac{\sqrt[3]{x^2}}{x}$$

Rationalizing a Two-Term Denominator Involving Square Roots:

Multiply numerator and denominator by the conjugate of the denominator.

KEY TERMS:

rationalizing the denominator (one term)
rationalizing the denominator (two terms)

Rationalize the denominator:

$$\frac{\sqrt{2}}{\sqrt{x} - \sqrt{3}}$$

$$= \frac{\sqrt{2}}{(\sqrt{x} - \sqrt{3})} \cdot \frac{(\sqrt{x} + \sqrt{3})}{(\sqrt{x} + \sqrt{3})}$$

$$= \frac{\sqrt{2x} + \sqrt{6}}{x - 3}$$

SECTION 8.6—RADICAL EQUATIONS

KEY CONCEPTS:

An equation with one or more radicals containing a variable is a radical equation.

Steps for Solving a Radical Equation:

1. Isolate the radical. If an equation has more than one radical, choose one of the radicals to isolate.
2. Raise each side of the equation to a power equal to the index of the radical.
3. Solve the resulting equation.
4. Check the potential solutions in the original equation.

EXAMPLES:

Solve: $\sqrt[3]{2x - 4} + 3 = 7$

Step 1: $\sqrt[3]{2x - 4} = 4$

Step 2: $(\sqrt[3]{2x - 4})^3 = (4)^3$

Step 3: $2x - 4 = 64$

$$2x = 68$$

$$x = 34$$

KEY TERM:

radical equations

Step 4:

Check:

$$\sqrt[3]{2x - 4} + 3 = 7$$

$$\sqrt[3]{2(34) - 4} + 3 \stackrel{?}{=} 7$$

$$\sqrt[3]{68 - 4} + 3 \stackrel{?}{=} 7$$

$$\sqrt[3]{64} + 3 \stackrel{?}{=} 7$$

$$4 + 3 = 7 \; \checkmark \qquad \text{The solution checks.}$$

The solution is $x = 34$.

SECTION 8.7—RATIONAL EXPONENTS

KEY CONCEPTS:

If $\sqrt[n]{a}$ and $\sqrt[n]{b}$ are real numbers, then

$$b^{1/n} = \sqrt[n]{b}$$

$$a^{m/n} = (\sqrt[n]{a})^m = \sqrt[n]{a^m}$$

KEY TERM:

$a^{1/n} \qquad a^{m/n}$

EXAMPLES:

$$121^{1/2} = \sqrt{121} = 11$$

$$27^{2/3} = (\sqrt[3]{27})^2 = (3)^2 = 9$$

chapter 8   REVIEW EXERCISES

Section 8.1

For Exercises 1–4, state the principal square root and the negative square root.

1. 196 ❖   2. 1.44 ❖   3. 225 ❖   4. 0.64 ❖

5. Explain why $\sqrt{-64}$ is *not* a real number.
   There is no real number b such that $b^2 = -64$.

6. Explain why $\sqrt[3]{-64}$ *is* a real number.
   $\sqrt[3]{-64} = -4$ because $(-4)^3 = -64$.

For Exercises 7–22, simplify the expressions, if possible.

7. $-\sqrt{144}$
   -12
8. $-\sqrt{25}$
   -5
9. $\sqrt{-144}$
   Not a real number
10. $\sqrt{-25}$
    Not a real number
11. $\sqrt{y^2}$
    $|y|$
12. $\sqrt[3]{y^3}$
    y
13. $\sqrt[4]{y^4}$
    $|y|$
14. $\sqrt[5]{y^5}$
    y
15. $-\sqrt[3]{125}$
    -5
16. $-\sqrt[4]{625}$
    -5
17. $\sqrt[3]{p^{12}}$
    p^4
18. $\sqrt[5]{q^{15}}$
    q^3
19. $\sqrt[4]{\dfrac{81}{t^8}}$
    $\dfrac{3}{t^2}$
20. $\sqrt[3]{\dfrac{-27}{w^3}}$
    $-\dfrac{3}{w}$

21. $\sqrt[5]{-32}$ -2 22. $\sqrt[5]{-1}$ -1

For Exercises 23–30, use a calculator to evaluate the radicals. Round the answer to three decimal places.

23. $\sqrt{10}$ 3.162   24. $\sqrt{31}$ 5.568   25. $\sqrt[3]{15}$ 2.466
26. $\sqrt[3]{63}$ 3.979   27. $\sqrt[4]{8}$ 1.682   28. $\sqrt[4]{25}$ 2.236
29. $\sqrt[5]{82}$ 2.414   30. $\sqrt[5]{100}$ 2.512

31. The radius, r, of a circle can be found from the area of the circle according to the formula:

$$r = \sqrt{\dfrac{A}{\pi}}$$

a. What is the radius of a circular garden whose area is 160 m²? Round to the nearest tenth of a meter.   7.1 m

b. What is the radius of a circular fountain whose area is 1600 ft²? Round to the nearest tenth of a foot.   22.6 ft

❖ See Additional Answers Appendix   🖊 Writing   ↔ Translating Expression   ◆ Geometry   Scientific Calculator   📼 Video

32. Suppose a ball is thrown with an initial velocity of 76 ft/s at an angle of 30° (see figure). Then the horizontal position of the ball, x, (measured in feet) depends on the number of seconds, t, after the ball is thrown according to the equation:

$$x = 38\sqrt{3}t$$

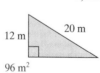

Figure for Exercise 32

a. What is the horizontal position of the ball after 1 s? Round your answer to the nearest tenth of a foot. 65.8 ft

b. What is the horizontal position of the ball after 2 s? Round your answer to the nearest tenth of a foot. 131.6 ft

For Exercises 33–34, translate the English phrases into algebraic expressions.

33. The square of b plus the square root of 5. $b^2 + \sqrt{5}$

34. The difference of the cube root of y and the fourth root of x. $\sqrt[3]{y} - \sqrt[4]{x}$

For Exercises 35–36, translate the algebraic expressions into English phrases. (Answers may vary.)

35. $\dfrac{2}{\sqrt{p}}$ The quotient of 2 and the square root of p

36. $8\sqrt{q}$ The product of 8 and the square root of q

For Exercises 37–38, find the area of the triangle.

37.
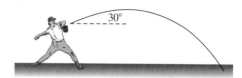

38.
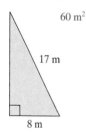

39. A car and truck leave from the same point at 4:00 P.M. The car travels north at 40 mph and the truck travels west at 30 mph. How far apart will the car and truck be at 5:00 P.M.?

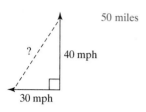

Figure for Exercise 39

40. A hedge extends 5 ft from the wall of a house. A 13-ft ladder is placed on the edge of the hedge. How far up the house is the tip of the ladder?

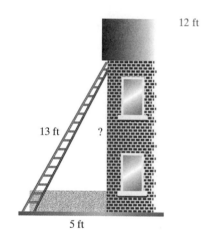

Figure for Exercise 40

41. Nashville, Tennessee, is north of Birmingham, Alabama, a distance of 182 miles. Augusta, Georgia, is east of Birmingham, a distance of 277 miles. How far is it from Augusta to Nashville? Round the answer to the nearest mile.

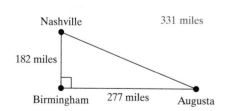

Figure for Exercise 41

Section 8.2

For Exercises 42–49, use the multiplication property of radicals to simplify. Assume all variables represent positive real numbers.

42. $\sqrt{x^{17}}$ $x^8\sqrt{x}$ 43. $\sqrt[5]{a^{11}}$ $a^2\sqrt[5]{a}$ 44. $\sqrt[3]{40}$ $2\sqrt[3]{5}$

45. $\sqrt{28}$ 46. $\sqrt{18x^3}$ 47. $\sqrt{27y^{10}}$
   $2\sqrt{7}$ $3x\sqrt{2x}$ $3y^5\sqrt{3}$

48. $\sqrt[3]{27y^{10}}$ 49. $\sqrt[4]{96x^7}$
   $3y^3\sqrt[3]{y}$ $2x\sqrt[4]{6x^3}$

For Exercises 50–61, use the division property of radicals to simplify. Assume all variables represent positive real numbers.

50. $\dfrac{\sqrt{50}}{\sqrt{49}}$ $\dfrac{5\sqrt{2}}{7}$ 51. $\sqrt{\dfrac{a^8}{25}}$ $\dfrac{a^4}{5}$

52. $\sqrt{\dfrac{2x^3}{y^6}}$ $\dfrac{x\sqrt{2x}}{y^3}$ 53. $\dfrac{\sqrt{18x}}{\sqrt{2x^3}}$ $\dfrac{3}{x}$

54. $\sqrt[5]{\dfrac{t^8}{r^5}}$ $\dfrac{t\sqrt[5]{t^3}}{r}$ 55. $\dfrac{\sqrt{200y^5}}{\sqrt{2y^3}}$ $10y$

56. $\sqrt{\dfrac{w^5}{9z^4}}$ $\dfrac{w^2\sqrt{w}}{3z^2}$ 57. $\sqrt[3]{\dfrac{30p^2}{q^6}}$ $\dfrac{\sqrt[3]{30p^2}}{q^2}$

58. $\dfrac{\sqrt{3n^5}}{\sqrt{48n^7}}$ $\dfrac{1}{4n}$ 59. $\dfrac{\sqrt{5t^6}}{\sqrt{121}}$ $\dfrac{t^3\sqrt{5}}{11}$

60. $\sqrt[4]{\dfrac{32}{m^{12}}}$ $\dfrac{2\sqrt[4]{2}}{m^3}$ 61. $\dfrac{\sqrt{196}}{\sqrt{36}}$ $\dfrac{7}{3}$

Section 8.3

For Exercises 62–71, add or subtract as indicated. Assume all variables represent positive real numbers.

62. $8\sqrt{6} - \sqrt{6}$ $7\sqrt{6}$

63. $1.6\sqrt{y} - 1.4\sqrt{y} + 0.6\sqrt{y}$ $0.8\sqrt{y}$

64. $-5\sqrt[4]{9} + 2\sqrt[4]{9}$ 65. $-3\sqrt[3]{11} + 2\sqrt[3]{11}$
  $-3\sqrt[4]{9}$ $-\sqrt[3]{11}$

66. $x\sqrt{20} - 2\sqrt{45x^2}$ 67. $y\sqrt{64y} + 3\sqrt{y^3}$
  $-4x\sqrt{5}$ $11y\sqrt{y}$

68. $3\sqrt{112} - 4\sqrt{28} + \sqrt{7}$ $5\sqrt{7}$

69. $2\sqrt{50} - 4\sqrt{18} - 6\sqrt{2}$ $-8\sqrt{2}$

70. $2\sqrt[3]{3p^7} + 2p\sqrt[3]{192p^4}$ $10p^2\sqrt[3]{3p}$

71. $5mn\sqrt[4]{m^5n^3} + \sqrt[4]{m^9n^7}$ $6m^2n\sqrt[4]{mn^3}$

 For Exercises 72–73, translate the English phrases into algebraic expressions and simplify.

72. The sum of the square root of the fourth power of x and the square of $5x$. $\sqrt{x^4} + (5x)^2;\ 26x^2$

73. The difference of the cube root of 128 and the cube root of 2. $\sqrt[3]{128} - \sqrt[3]{2};\ 3\sqrt[3]{2}$

74. Find the exact perimeter of the triangle.

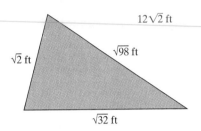

Figure for Exercise 74

75. Find the exact perimeter of a square whose sides are $3\sqrt{48}$ m. $48\sqrt{3}$ m

For Exercises 76–77, determine the slope of the line given two points on the line.

76. $(0, 2\sqrt{5})(-2, 5\sqrt{5})$ $-\dfrac{3\sqrt{5}}{2}$

77. $(-5, -\sqrt{10})(-3, 6\sqrt{10})$ $\dfrac{7\sqrt{10}}{2}$

Section 8.4

For Exercises 78–87, multiply the expressions. Assume all variables represent positive real numbers.

78. $\sqrt{5} \cdot \sqrt{125}$ 25 79. $\sqrt{10p} \cdot \sqrt{6}$ $2\sqrt{15p}$

80. $(5\sqrt{6})(7\sqrt{2x})$ 81. $(3\sqrt{y})(-2z\sqrt{11y})$
  $70\sqrt{3x}$ $-6yz\sqrt{11}$

82. $8\sqrt{m}(\sqrt{m} + 3)$ 83. $\sqrt{2}(\sqrt{7} + 8)$
  $8m + 24\sqrt{m}$ $\sqrt{14} + 8\sqrt{2}$

84. $(5\sqrt{2} + \sqrt{13})(-\sqrt{2} - 3\sqrt{13})$ $-49 - 16\sqrt{26}$

85. $(\sqrt{p} + 2\sqrt{q})(4\sqrt{p} - \sqrt{q})$ $4p + 7\sqrt{pq} - 2q$

86. $(8\sqrt{w} - \sqrt{z})(8\sqrt{w} + \sqrt{z})$ $64w - z$

87. $(2x - \sqrt{y})^2$ $4x^2 - 4x\sqrt{y} + y$

88. Find the exact volume of a cube whose sides are $2\sqrt[3]{4}$ yd. 32 yd^3

89. Find the exact volume of the box.

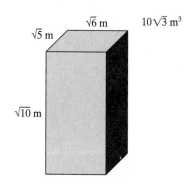

$\sqrt{6}$ m $10\sqrt{3}$ m³

$\sqrt{5}$ m

$\sqrt{10}$ m

Figure for Exercise 89

Section 8.5

90. To rationalize the denominator in the expression

$$\frac{6}{\sqrt{a} + 5}$$

which quantity would you multiply by in the numerator and denominator?   b

a. $\sqrt{a} + 5$ b. $\sqrt{a} - 5$

c. $\sqrt{a}$ d. -5

91. To rationalize the denominator in the expression

$$\frac{w}{\sqrt{w} - 4}$$

which quantity would you multiply by in the numerator and denominator?   b

a. $\sqrt{w} - 4$ b. $\sqrt{w} + 4$

c. $\sqrt{w}$ d. 4

For Exercises 92–99, rationalize the denominators. Assume all variables represent positive real numbers.

92. $\dfrac{11}{\sqrt{7}}$ $\dfrac{11\sqrt{7}}{7}$ 93. $\sqrt{\dfrac{18}{y}}$ $\dfrac{3\sqrt{2y}}{y}$ 94. $\dfrac{\sqrt{24}}{\sqrt{6x^7}}$ $\dfrac{2\sqrt{x}}{x^4}$

95. $\dfrac{5}{\sqrt[3]{9}}$ $\dfrac{5\sqrt[3]{3}}{3}$ 96. $\sqrt[4]{\dfrac{3m}{n}}$ $\dfrac{\sqrt[4]{3mn^3}}{n}$ 97. $\dfrac{10}{\sqrt{7} - \sqrt{2}}$ $\dfrac{2\sqrt{7} + 2\sqrt{2}}{}$

98. $\dfrac{6}{\sqrt{w} + 2}$ 99. $\dfrac{\sqrt{7} + 3}{\sqrt{7} - 3}$

$\dfrac{6\sqrt{w} - 12}{w - 4}$ $-8 - 3\sqrt{7}$

For Exercises 100–103, simplify the radicals. Assume all variables represent positive real numbers.

100. $\sqrt{\dfrac{3}{a^2b}}$ $\dfrac{\sqrt{3b}}{ab}$ 101. $\sqrt[3]{\dfrac{54}{x^2}}$ $\dfrac{3\sqrt[3]{2x}}{x}$

102. $\sqrt[4]{\dfrac{m^8}{n^4}}$ $\dfrac{m^2}{n}$ 103. $\sqrt{\dfrac{6}{y^4}}$ $\dfrac{\sqrt{6}}{y^2}$

104. The velocity of an object, v, (in meters per second: m/s) depends on the kinetic energy, E, (in joules: J) and mass, m, (in kilograms: kg) of the object according to the formula:

$$v = \sqrt{\dfrac{2E}{m}}$$

a. What is the exact velocity of a 3-kg object whose kinetic energy is 100 J? $\dfrac{10\sqrt{6}}{3}$ m/s

b. What is the exact velocity of a 5-kg object whose kinetic energy is 162 J? $\dfrac{18\sqrt{5}}{5}$ m/s

Section 8.6

For Exercises 105–114, solve the equations. Be sure to check the potential solutions.

105. $\sqrt{p + 6} = 12$ $p = 138$ 106. $\sqrt{k + 1} = -7$ No solution

107. $\sqrt[3]{2y + 13} = -5$ $y = -69$

108. $\sqrt{3x - 17} - 10 = 0$ $x = 39$

109. $\sqrt{14n + 10} = 4\sqrt{n}$ $n = 5$

110. $\sqrt[4]{8h + 2} = \sqrt[4]{h - 5}$ No solution; ($h = -1$ does not check)

111. $\sqrt{2z + 2} = \sqrt{3z - 5}$ $z = 7$

112. $\sqrt{5y - 5} - \sqrt{4y + 1} = 0$ $y = 6$

113. $\sqrt{2m + 5} = m + 1$ $m = 2$; ($m = -2$ does not check.)

114. $\sqrt{3n - 8} - n + 2 = 0$ $n = 3, n = 4$

 115. The length of the sides of a cube is related to the volume of the cube according to the formula: $x = \sqrt[3]{V}$.

x

x

x

Figure for Exercise 115

a. What is the volume of the cube if the side length is 21 in.?   9261 in.³

b. What is the volume of the cube if the side length is 15 cm?   3375 cm³

Section 8.7

For Exercises 116–121, simplify the expressions.

116. $(-27)^{1/3}$
     -3

117. $121^{1/2}$
     11

118. $-16^{1/4}$
     -2

119. $(-16)^{1/4}$
     Not a real number

120. $4^{-3/2}$
     $\frac{1}{8}$

121. $\left(\frac{1}{9}\right)^{-3/2}$
     27

For Exercises 122–125, write the expression in radical notation. Assume all variables represent positive real numbers.

122. $z^{1/5}$ $\sqrt[5]{z}$

123. $q^{2/3}$ $\sqrt[3]{q^2}$

124. $(w^3)^{1/4}$ $\sqrt[4]{w^3}$

125. $\left(\frac{b}{121}\right)^{1/2}$ $\sqrt{\frac{b}{121}} = \frac{\sqrt{b}}{11}$

For Exercises 126–129, write the expression using rational exponents rather than radical notation. Assume all variables represent positive real numbers.

126. $\sqrt[5]{a^2}$ $a^{2/5}$

127. $5\sqrt[3]{m^2}$ $5m^{2/3}$

128. $\sqrt[5]{a^2b^4}$ $(a^2b^4)^{1/5}$

129. $\sqrt{6}$ $6^{1/2}$

For Exercises 130–135, simplify using the properties of rational exponents. Write the answer with positive exponents only. Assume all variables represent positive real numbers.

130. $y^{2/3}y^{4/3}$ y^2

131. $\frac{6^{4/5}}{6^{1/5}}$ $6^{3/5}$

132. $a^{1/3}a^{1/2}$ $a^{5/6}$

133. $(5^{1/2})^{3/2}$ $5^{3/4}$

134. $(64a^3b^6)^{1/3}$ $4ab^2$

135. $\left(\frac{b^4b^0}{b^{1/4}}\right)^4$ b^{15}

For Exercises 136–139, use a calculator to approximate the expression and round the answer to four decimal places.

136. $47^{1/2}$ 6.8557

137. $20^{-1/3}$ 0.3684

138. $1744^{2/3}$ 144.8875

139. $\left(\frac{40}{3}\right)^{3/4}$ 6.9776

140. The radius, r, of a right circular cylinder can be found if the volume, V, and height, h, are known. The radius is given by

$$r = \left(\frac{V}{\pi h}\right)^{1/2}$$

Find the radius of a right circular cylinder whose volume is 150.8 cm³ and whose height is 12 cm. Round to the nearest tenth of a centimeter.   2.0 cm

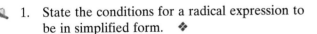

chapter 8   TEST

1. State the conditions for a radical expression to be in simplified form.   ❖

For Exercises 2–7, simplify the radicals, if possible. Assume all variables represent positive real numbers.

2. $\sqrt{242x^2}$
   $11x\sqrt{2}$

3. $\sqrt[3]{48y^4}$
   $2y\sqrt[3]{6y}$

4. $\sqrt{-64}$
   Not a real number

5. $\sqrt[4]{\frac{5a^6}{81}}$ $\frac{a\sqrt[4]{5a^2}}{3}$

6. $\frac{9}{\sqrt{6}}$ $\frac{3\sqrt{6}}{2}$

7. $\frac{2}{\sqrt{5}+6}$ ❖

8. Translate the English phrases into algebraic expressions and simplify.

   a. The sum of the square root of twenty-five and the cube of five. $\sqrt{25} + 5^3$; 130

   b. The difference of the square of four and the fourth root of 16. $4^2 - \sqrt[4]{16}$; 14

9. Estimate the value of the following radicals. Then use your calculator to approximate the value to three decimal places.

   a. $\sqrt{38}$ 6.164   b. $\sqrt[4]{20}$ 2.115

10. A baseball player hits the ball at an angle of 30° with an initial velocity of 112 ft/s. The horizontal

position of the ball, x, (measured in feet) depends on the number of seconds, t, after the ball is struck according to the equation.

$$x = 56\sqrt{3}t$$

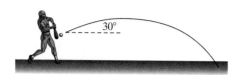

Figure for Exercise 10

a. What is the horizontal position of the ball after 1 s? Round the answer to the nearest foot. **97 ft**

b. What is the horizontal position of the ball after 3.5 s? Round the answer to the nearest foot. **339 ft**

For Exercises 11–16, perform the indicated operations. Assume all variables represent positive real numbers.

11. $6\sqrt{z} - 3\sqrt{z} + 5\sqrt{z}$ **$8\sqrt{z}$**

12. $\sqrt{3}(4\sqrt{2} - 5\sqrt{3})$ **$4\sqrt{6} - 15$**

13. $\sqrt{50t^2} - t\sqrt{288}$ **$-7t\sqrt{2}$**

14. $\sqrt{360} + \sqrt{250} - \sqrt{40}$ **$9\sqrt{10}$**

15. $(6\sqrt{2} - \sqrt{5})(\sqrt{2} + 4\sqrt{5})$ **$-8 + 23\sqrt{10}$**

16. $\dfrac{\sqrt{2m^3n}}{\sqrt{72m^5}}$ **$\dfrac{\sqrt{n}}{6m}$**

17. A triathlon consists of a swim, followed by a bike ride, followed by a run. The swim begins on a beach at point A. The swimmers must swim 50 yd to a buoy at point B and then 200 yd to a buoy at point C and then return to point A on the beach. How far is the distance from point C to point A? (Round to the nearest yard.)

206 yd

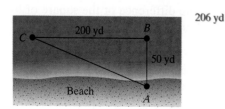

Figure for Exercise 17

For Exercises 18–20, solve the equations.

18. $\sqrt{2x + 7} + 6 = 2$ **No solution**

19. $\sqrt[4]{x + 6} = \sqrt[4]{2x - 8}$ **$x = 14$**

20. $\sqrt{1 - 7x} = 1 - x$ **$x = 0, x = -5$**

21. The height, y, (in inches) of a certain plant can be approximated by the time, t, (in weeks) after the seed has germinated according to the equation

$$y = 6\sqrt{t}$$

The graph of the equation is shown below. The variable t is used in place of x.

a. Use the equation to find the height of the plant after 4 weeks. Verify your answer from the graph. **12 in.**

b. Use the equation to find the height of the plant after 9 weeks. Verify your answer from the graph. **18 in.**

c. Use the equation to find the time required for the plant to reach a height of 30 in. Verify your answer from the graph. **25 weeks**

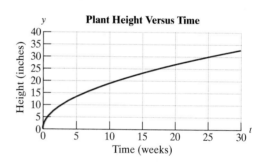

Figure for Exercise 21

For Exercises 22–23, simplify the expression.

22. $10,000^{3/4}$ **1000**   23. $\left(\dfrac{1}{8}\right)^{-1/3}$ **2**

For Exercises 24–25, write the expressions in radical notation. Assume all variables represent positive real numbers.

24. $x^{3/5}$ **$\sqrt[5]{x^3}$ or $(\sqrt[5]{x})^3$**   25. $5y^{1/2}$ **$5\sqrt{y}$**

26. Write the expression using rational exponents: $\sqrt[4]{ab^3}$. (Assume $a \geq 0$ and $b \geq 0$.) **$(ab^3)^{1/4}$**

For Exercises 27–29, simplify using the properties of rational exponents. Write the final answer with positive exponents only. Assume all variables represent positive real numbers.

27. $p^{1/4} \cdot p^{2/3}$ $p^{11/12}$

28. $\dfrac{5^{4/5}}{5^{1/5}}$ $5^{3/5}$

29. $\dfrac{(9m^2n^4)^{1/2}}{3mn^2}$

CUMULATIVE REVIEW EXERCISES, CHAPTERS 1–8

For Exercises 1–2, simplify completely:

1. $\dfrac{|-3 - 12 \div 6 + 2|}{\sqrt{5^2 - 4^2}}$ 1

2. $\left(-\dfrac{4}{5} \div \dfrac{2}{15}\right)^2 + \dfrac{1}{6}$ $\dfrac{217}{6}$

3. Solve for y:
   $2 - 5[2y + 4] - (-3y - 1) = -(y + 5)$ $y = -2$

4. Solve for a: $2a + b + c = A$ $a = \dfrac{A - b - c}{2}$

5. Solve the inequality. Graph the solution set. Then write the solution in set-builder notation and in interval notation: $2x - 5(x + 1) < -x + 3$ ❖

6. The sum of two-thirds of a number and five equals the number. Find the number.
   The number is 15.

7. Two hikers start from two different camps 18 miles apart. One hiker walks twice as fast as the other. How fast does each hiker walk if they meet in 4 hours (h)?

	d	r	t
Hiker 1			
Hiker 2			

Table for Exercise 7
One hiker walks at a rate of 1.5 mph and the other walks 3 mph.

For Exercises 8–9, simplify as much as possible. Write the final answer with positive exponents only.

8. $\dfrac{(x^{-2})^3(x^4)^2}{x^{-2}}$ x^4

9. $\left(\dfrac{1}{3}\right)^0 - \left(\dfrac{1}{4}\right)^{-2}$ -15

10. Perform the indicated operations:
    $2(x - 3) - (3x + 4)(3x - 4)$ $-9x^2 + 2x + 10$

11. Perform the indicated operations:
    $$\left(\dfrac{1}{2}c + 4\right)^2 \quad \tfrac{1}{4}c^2 + 4c + 16$$

12. Divide:
    $$\dfrac{14x^3y - 7x^2y^2 + 28xy^2}{7x^2y^2} \quad \dfrac{2x}{y} - 1 + \dfrac{4}{x}$$

In Exercises 13–15, factor completely:

13. $6ax + 2bx - 3ay - by$ $(3a + b)(2x - y)$

14. $m^4 - 81$ $(m - 3)(m + 3)(m^2 + 9)$

15. $50c^2 + 40c + 8$ $2(5c + 2)^2$

16. Solve for x: $10x^2 = x + 2$ $x = -\dfrac{2}{5}, x = \dfrac{1}{2}$

For Exercises 17–19,

   a. Find the x- and y-intercepts.

   b. Match the equation with the appropriate graph found on page 662.

17. $2x + 3y = 6$ iii
    x-intercept: $(3, 0)$; y-intercept: $(0, 2)$

18. $y = x^2 + 2x - 8$ i
    x-intercepts: $(2, 0)(-4, 0)$; y-intercept: $(0, -8)$

19. $y = x^2 - 9$ ii
    x-intercepts: $(3, 0)(-3, 0)$; y-intercept: $(0, -9)$

❖ See Additional Answers Appendix   ✎ Writing   ⬌ Translating Expression   Geometry   Scientific Calculator   Video

i.

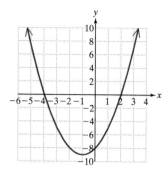

ii.

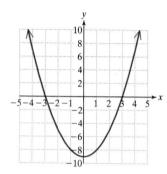

iii.

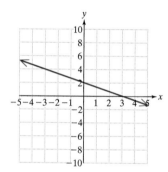

20. What is the domain of the expression?

$$\frac{x + 2}{(x - 5)(x + 1)}$$

Write the answer in set-builder notation.
$\{x \mid x \neq 5, x \neq -1\}$

21. Perform the indicated operations:

$$\frac{5a^2 + 2ab - 3b^2}{10a + 10b} \div \frac{25a^2 - 9b^2}{50a + 30b} \quad 1$$

22. Perform the indicated operations:

$$\frac{p}{2p + 4} - \frac{2}{p^2 + 2p} + \frac{1}{p} \quad \frac{1}{2}$$

23. Solve for z: $\dfrac{1}{5} + \dfrac{z}{z - 5} = \dfrac{5}{z - 5}$ **No solution**

24. $1\frac{1}{4}$ cups of Shredded Wheat and Bran cereal has 200 Calories. How many calories are in 2 cups of Shredded Wheat and Bran? **320 Calories**

25. Simplify:

$$\frac{\dfrac{5}{4} + \dfrac{2}{x}}{\dfrac{4}{x} - \dfrac{4}{x^2}} \quad \frac{x(5x + 8)}{16(x - 1)}$$

For Exercises 26–27, graph the lines.

26. $4x - 2y = 8$ ❖ **27.** $3y = 6$ ❖

28. The equation $y = 210x + 250$ represents the cost, y, (in dollars) of renting office space for x months.

Cost of Renting Office Space Versus Number of Months

Figure for Exercise 28

a. Find y when x is 3. Interpret the result in the context of the problem. $y = 880$; The cost of renting the office space for 3 months is $880.

b. Find x when y is $2770. Interpret the result in the context of the problem. $x = 12$; The cost of renting office space for 12 months is $2770.

c. What is the slope of the line? Interpret the meaning of the slope in the context of the problem. $m = 210$; The increase in cost is $210 per month.

d. What is the y-intercept? Interpret the meaning of the y-intercept in the context of the problem. $(0, 250)$; The down payment of renting the office space is $250.

29. Write an equation of the line passing through the points $(2, -1)$ and $(-3, 4)$. Write the final answer in slope-intercept form. $y = -x + 1$

30. a. What is the slope of a line parallel to the line $x + 2y = 4$? $-\frac{1}{2}$

b. What is the slope of a line perpendicular to the line $x + 2y = 4$? 2

31. Solve the system of equations using the addition method. If the system has no solution or infinitely many solutions, so state:

$$3x - 5y = 23$$
$$2x + 4y = -14$$

$(1, -4)$

32. Solve the system of equations using the substitution method. If the system has no solution or infinitely many solutions, so state:

$$3x - y = 6$$
$$-9x + 3y = -18$$

Infinitely many solutions; $\{(x, y) \mid 3x - y = 6\}$

33. Graph the solution to the inequality: $-2x - y > 3$ ❖

34. Two angles are complementary. One angle is 12° less than 5 times the other angle. Find the two angles.   The angles are 17° and 73°.

35. How many liters (L) of 20% acid solution must be mixed with a 50% acid solution to obtain 12 L of a 30% acid solution?   8 L of 20% solution should be mixed with 4 L of 50% solution.

For Exercises 36–39, simplify the radical expressions (if possible).

36. $\sqrt[3]{-125}$ -5

37. $\sqrt{99}$ $3\sqrt{11}$

38. $\sqrt{-169}$ Not a real number

39. $\sqrt[3]{16c^5d^6}$ $2cd^2\sqrt[3]{2c^2}$

For Exercises 40–41, perform the indicated operations:

40. $5x\sqrt{3} + \sqrt{12x^2}$ $7x\sqrt{3}$

41. $(\sqrt{5} - \sqrt{3})(\sqrt{2} + \sqrt{7})$ $\sqrt{10} + \sqrt{35} - \sqrt{6} - \sqrt{21}$

For Exercises 42–43, rationalize the denominators:

42. $\dfrac{\sqrt{x}}{\sqrt{x} - \sqrt{y}}$ $\dfrac{x + \sqrt{xy}}{x - y}$

43. $\dfrac{5}{\sqrt{a}}$ $\dfrac{5\sqrt{a}}{a}$

44. Solve for y: $\sqrt[3]{2y - 1} - 4 = -1$ $y = 14$

FUNCTIONS, COMPLEX NUMBERS, AND QUADRATIC EQUATIONS

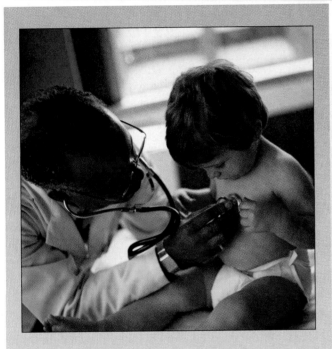

The graphical representation of numerical data can give insight into the relationship between two variables. In this chapter we introduce the concept of a function that gives an equation or rule defining the dependency of one variable on another. For example, the average number of visits to a doctor's office per year, N, depends on the age of the patient, x, according to:

$$N(x) = 0.0014x^2 - 0.0658x + 2.65$$

The function N is a quadratic function and its graph is in the shape of a parabola.

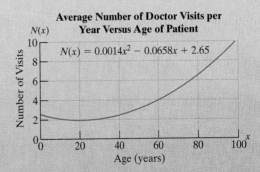

Average Number of Doctor Visits per Year Versus Age of Patient

$N(x) = 0.0014x^2 - 0.0658x + 2.65$

Source: U.S. Center for Health Statistics

In this chapter, we introduce the concept of a function with emphasis on quadratic functions. For more information on graphing quadratic functions, visit xyplot at

www.mhhe.com/miller_oneill

section

9.1 INTRODUCTION TO FUNCTIONS AND FUNCTION NOTATION

1. Definition of a Relation

The number of points scored by Kobe Bryant of the L.A. Lakers during the first six games of the 1999/2000 basketball season is shown in Table 9-1.

Table 9-1

Game, x	Number of Points, y	Ordered Pair
1	35	→ (1, 35)
2	32	→ (2, 32)
3	23	→ (3, 23)
4	23	→ (4, 23)
5	16	→ (5, 16)
6	25	→ (6, 25)

Each ordered pair from Table 9-1 shows a correspondence, or relationship, between the game number and the number of points scored by Kobe Bryant. The set of ordered pairs: {(1, 35), (2, 32), (3, 23), (4, 23), (5, 16), (6, 25)} defines a relation between the game number and the number of points scored.

Definition of a Relation in x and y

Any set of ordered pairs, (x, y), is called a **relation** in x and y. Furthermore:

- The set of first components in the ordered pairs is called the **domain** of the relation.
- The set of second components in the ordered pairs is called the **range** of the relation.

2. Finding the Domain and Range of a Relation

example 1

Finding the Domain and Range of a Relation

Find the domain and range of the relation linking the game number to the number of points scored by Kobe Bryant in the first six games of the 1999/2000 season:

$$\{(1, 35), (2, 32), (3, 23), (4, 23), (5, 16), (6, 25)\}$$

Solution:

Domain: {1, 2, 3, 4, 5, 6}   (Set of first coordinates)

Range: {35, 32, 23, 16, 25}   (Set of second coordinates—Note that 23 is not listed twice)

The domain consists of the game numbers for the first six games of the season. The range represents the corresponding number of points.

---

example 2

Finding the Domain and Range of a Relation

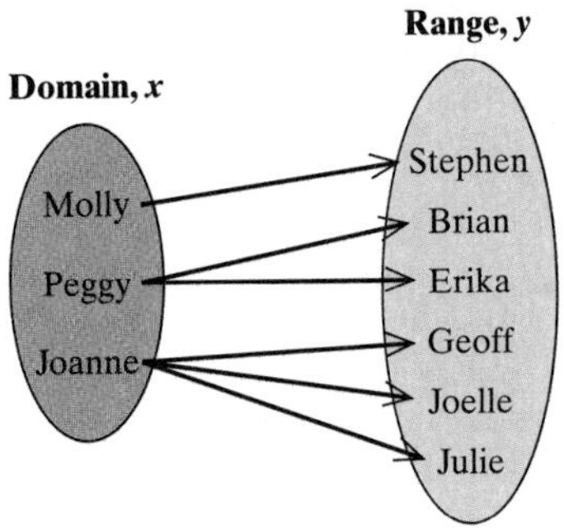

Range, *y*

Domain, *x*

Figure 9-1

The three women represented in Figure 9-1 each have children. Molly has one child, Peggy has two children, and Joanne has three children.

a. If the set of mothers is given as the domain and the set of children is the range, write a set of ordered pairs defining the relation given in Figure 9-1.
b. Write the domain and range of the relation.

Solution:

a. {(Molly, Stephen), (Peggy, Brian), (Peggy, Erika), (Joanne, Geoff), (Joanne, Joelle), (Joanne, Julie)}
b. Domain: {Molly, Peggy, Joanne}
   Range: {Stephen, Brian, Erika, Geoff, Joelle, Julie}

3. Definition of a Function

In mathematics, a special type of relation, called a function, is used extensively.

> **Definition of a Function**
>
> Given a relation in *x* and *y*, we say "*y* is a **function** of *x*" if for every element *x* in the domain, there corresponds exactly one element *y* in the range.

To understand the difference between a relation that is a function and one that is not a function, consider the next example.

---

example 3

Determining Whether a Relation Is a Function

Determine whether the following relations are functions:

a. {(2, −3), (4, 1), (3, −1), (2, 4)}
b. {(−3, 1), (0, 2), (4, −3), (1, 5), (−2, 1)}

Solution:

a.   This relation is defined by the set of ordered pairs.

same x-values

$$\{(2, -3), (4, 1), (3, -1), (2, 4)\}$$

different y-values

When $x = 2$, there are two possibilities for y: $y = -3$ and $y = 4$.

This relation is *not* a function because when $x = 2$, there is more than one corresponding element in the range.

b.   This relation is defined by the set of ordered pairs: $\{(-3, 1), (0, 2), (4, -3), (1, 5), (-2, 1)\}$. Notice that each value in the domain corresponds to only one value in the range. Therefore, this relation *is* a function.

When $x = -3$, there is only one possibility for y: $y = 1$.

When $x = 0$, there is only one possibility for y: $y = 2$.

When $x = 4$, there is only one possibility for y: $y = -3$.

When $x = 1$, there is only one possibility for y: $y = 5$.

When $x = -2$, there is only one possibility for y: $y = 1$.

In Example 2, the relation linking the set of mothers with their respective children is *not* a function. The domain elements, "Peggy" and "Joanne" each have more than one child. Because these x-values in the domain correspond to more than one y-value in the range, the relation is not a function.

The relation relating the game number to the number of points scored by Kobe Bryant *is* a function. That is, for any given game, x, Kobe cannot have more than one point total, y.

4. Vertical Line Test

A relation that is not a function has at least one domain element, x, paired with more than one range element, y. For example, the ordered pairs $(2, 1)$ and $(2, 4)$ do not constitute a function. On a graph, these two points are aligned vertically in the xy-plane, and a vertical line drawn through one point also intersects the other point (Figure 9-2). Thus if a vertical line drawn through a graph of a relation intersects the graph in more than one point, the relation cannot be a function. This idea is stated formally as the **vertical line test**.

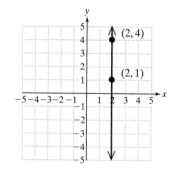

Figure 9-2

Vertical Line Test

Consider a relation defined by a set of points (x, y) on a rectangular coordinate system. Then the graph defines y as a function of x if no vertical line intersects the graph in more than one point.

The vertical line test also implies that if any vertical line drawn through the graph of a relation intersects the relation in more than one point, then the relation does *not* define *y* as a function of *x*.

The vertical line test can be demonstrated by graphing the ordered pairs from the relations in Example 3 (Figure 9-3 and Figure 9-4).

$$\{(2, -3), (4, 1), (3, -1), (2, 4)\} \qquad \{(-3, 1), (0, 2), (4, -3), (1, 5), (-2, 1)\}$$

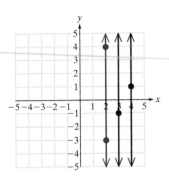

Figure 9-3

Figure 9-4

Not a Function	**Function**
A vertical line intersects in more than one point.	No vertical line intersects more than once.

The relations in Examples 1, 2, and 3 consist of a finite number of ordered pairs. A relation may, however, consist of an *infinite* number of points defined by an equation or by a graph. For example, the equation $y = x + 1$ defines infinitely many ordered pairs whose *y*-coordinate is one more than its *x*-coordinate. These ordered pairs cannot all be listed but can be depicted in a graph.

The vertical line test is especially helpful in determining whether a relation is a function based on its graph.

---

example 4 **Using the Vertical Line Test**

Use the vertical line test to determine whether the following relations are functions.

a. $y = x + 1$ b. $x = y^2$

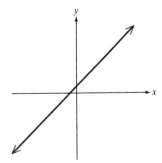

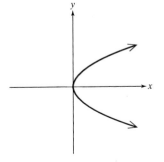

c. $x^2 + y^2 = 1$

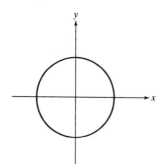

Solution:

a. $y = x + 1$

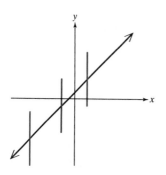

Function
No vertical line intersects
more than once.

b. $x = y^2$

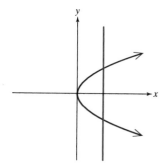

Not a Function
A vertical line intersects
in more than one point.

c. $x^2 + y^2 = 1$

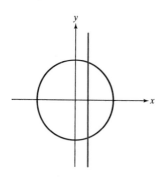

Not a Function
A vertical line intersects
in more than one point.

5. Function Notation

A function is defined as a relation with the added restriction that each value of the domain corresponds to only one value in the range. In mathematics, functions are often given by rules or equations to define the relationship between two or more variables. For example, the equation, $y = x + 1$ defines the set of ordered pairs such that the y-value is one more than the x-value.

When a function is defined by an equation, we often use **function notation**. For example, the equation $y = x + 1$ may be written in function notation as

$$f(x) = x + 1$$

where f is the name of the function, x is an input value from the domain of the function, and $f(x)$ is the function value (or y-value) corresponding to x.

The notation $f(x)$ is read as "f of x" or "the value of the function, f, at x."

A function may be evaluated at different values of x by substituting values of x from the domain into the function. For example, for the function defined by $f(x) = x + 1$ we can evaluate f at $x = 3$ by using substitution.

$$f(x) = x + 1$$

$$f(3) = (3) + 1$$

$$f(3) = 4 \qquad \text{This is read as "}f\text{ of 3 equals 4."}$$

Thus, when $x = 3$, the corresponding function value is 4. This can also be interpreted as an ordered pair: $(3, 4)$

The names of functions are often given by either lowercase letters or uppercase letters such as f, g, h, p, k, M, and so on.

6. Evaluating a Function

example 5

Evaluating a Function

Given the function defined by $h(x) = x^2 - 2$, find the function values.

a. $h(0)$ b. $h(1)$ c. $h(2)$ d. $h(-1)$ e. $h(-2)$

Solution:

a. $h(x) = x^2 - 2$

$\qquad h(0) = (0)^2 - 2 \qquad$ Substitute $x = 0$ into the function.

$\qquad\qquad = 0 - 2$

$\qquad\qquad = -2 \qquad h(0) = -2$ means that when $x = 0, y = -2$, yielding the ordered pair $(0, -2)$.

b. $h(x) = x^2 - 2$

$\qquad h(1) = (1)^2 - 2 \qquad$ Substitute $x = 1$ into the function.

$\qquad\qquad = 1 - 2$

$\qquad\qquad = -1 \qquad h(1) = -1$ means that when $x = 1, y = -1$, yielding the ordered pair $(1, -1)$.

c. $h(x) = x^2 - 2$

$h(2) = (2)^2 - 2$ Substitute $x = 2$ into the function.

$= 4 - 2$

$= 2$ $h(2) = 2$ means that when $x = 2$, $y = 2$, yielding the ordered pair $(2, 2)$.

d. $h(x) = x^2 - 2$

$h(-1) = (-1)^2 - 2$ Substitute $x = -1$ into the function.

$= 1 - 2$

$= -1$ $h(-1) = -1$ means that when $x = -1$, $y = -1$, yielding the ordered pair $(-1, -1)$.

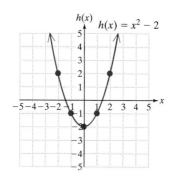

e. $h(x) = x^2 - 2$

$h(-2) = (-2)^2 - 2$ Substitute $x = -2$ into the function.

$= 4 - 2$

$= 2$ $h(-2) = 2$ means that when $x = -2$, $y = 2$, yielding the ordered pair $(-2, 2)$.

The rule $h(x) = x^2 - 2$ is equivalent to the equation $y = x^2 - 2$. The function values $h(0), h(1), h(2), h(-1)$, and $h(-2)$ correspond to the y-values in the ordered pairs $(0, -2), (1, -1), (2, 2), (-1, -1)$, and $(-2, 2)$, respectively. These points can be used to sketch a graph of the function (Figure 9-5).

Figure 9-5

---

example 6   **Evaluating a Function**

Given the functions defined by $f(x) = 2x - 3$ and $g(x) = x^3 - 1$, find the function values.

a. $f(1) + g(3)$ b. $g(t)$ c. $f(a + b)$

Classroom Activity 9.1B

Solution:

a. $f(x) = 2x - 3$ $g(x) = x^3 - 1$

$f(1) = 2(1) - 3$ $g(3) = (3)^3 - 1$

$f(1) = 2 - 3$ $g(3) = 27 - 1$

$f(1) = -1$ $g(3) = 26$

Therefore,

$f(1) + g(3) = -1 + 26$

$f(1) + g(3) = 25$

b. $g(x) = x^3 - 1$

$g(t) = (t)^3 - 1$ Substitute $x = t$ into the function.

$g(t) = t^3 - 1$

c. $f(x) = 2x - 3$

$f(a + b) = 2(a + b) - 3$ Substitute $x = (a + b)$ into the function.

$f(a + b) = 2a + 2b - 3$ Simplify.

7.   Domain and Range of a Function

A function is a relation, and it is often necessary to determine its domain and range. Consider a function defined by the equation $y = f(x)$. The domain of f is the set of all x-values that when substituted into the function produce a real number. The range of f is the set of all y-values corresponding to the values of x in the domain.

For the examples encountered in this text, the domain of a function will be all real numbers unless restricted by the following conditions:

- The domain must exclude values of the variable that make the denominator zero.
- The domain must exclude values of the variable that make the radicand of an even-indexed root negative.

example 7

Finding the Domain of a Function

Find the domain of the functions defined by the following rules. Write the answer in set-builder notation.

a. $f(x) = \dfrac{1}{x - 2}$ b. $h(x) = \sqrt{x - 1}$ c. $g(x) = 2x^2 - 6x$

Solution:

a.   The function defined by $f(x) = \dfrac{1}{x - 2}$

will be a real number as long as the denominator does not equal zero. The denominator will equal zero only if $x = 2$.

Therefore, the domain of f is $\{x \mid x \neq 2\}$.

b.   The function defined by $h(x) = \sqrt{x - 1}$ will be a real number only if the radicand is greater than or equal to zero.

$$x - 1 \geq 0$$

$$x \geq 1$$

The domain of h is $\{x \mid x \geq 1\}$.

c.   The function defined by $g(x) = 2x^2 - 6x$ does not contain a root with an even index, nor does it contain an expression with a variable in the denominator. Therefore, the function has no restrictions on its domain.

The domain of g is all real numbers.

example 8   Finding the Domain and Range of a Function

Find the domain and range of the functions based on the graph of the function. Express the answers in interval notation.

a.

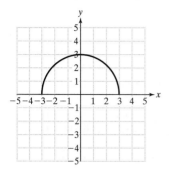

b.

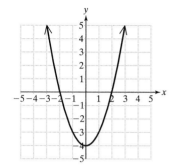

c.

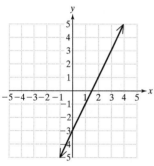

Classroom Activity 9.1C

Solution:

a.

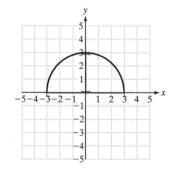

The *x*-values are bounded between -3 and 3.

Domain: $[-3, 3]$

The *y*-values are bounded between 0 and 3.

Range: $[0, 3]$

b.

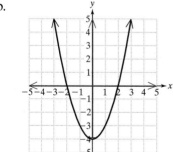

The function extends infinitely far to the left and right.

Domain: $(-\infty, \infty)$

The *y*-values extend infinitely far in the positive direction, but are bounded below at $y = -4$.

Range: $[-4, \infty)$

c.

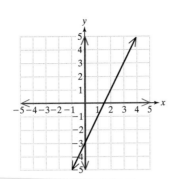

The line extends infinitely far to the left and right, and infinitely far up and down.

Domain: $(-\infty, \infty)$

Range: $(-\infty, \infty)$

8. Applications of Functions

example 9   **Using a Function in an Application**

The score a student receives on an exam is a function of the number of hours the student spends studying. The function defined by

$$P(x) = \frac{100x^2}{40 + x^2} \quad (x \geq 0)$$

indicates that a student will achieve a score of $P\%$ after studying for x hours.

a. Evaluate $P(0)$, $P(10)$, and $P(20)$.
b. Interpret the function values from part (a) in the context of this problem.

Solution:

a. $P(x) = \dfrac{100x^2}{40 + x^2}$

$$P(0) = \frac{100(0)^2}{40 + (0)^2} \qquad P(10) = \frac{100(10)^2}{40 + (10)^2} \qquad P(20) = \frac{100(20)^2}{40 + (20)^2}$$

$$P(0) = \frac{0}{40} \qquad\qquad P(10) = \frac{10{,}000}{140} \qquad\qquad P(20) = \frac{40{,}000}{440}$$

$$P(0) = 0 \qquad\qquad P(10) = \frac{500}{7} \approx 71.4 \qquad P(20) = \frac{1000}{11} \approx 90.9$$

b. $P(0) = 0$ means that if a student spends 0 hours (h) studying, the student will score 0% on the exam.

$$P(10) = \frac{500}{7} \approx 71.4$$

means that if a student spends 10 h studying, the student will score approximately 71.4% on the exam.

$$P(20) = \frac{1000}{11} \approx 90.9$$

means that if a student spends 20 h studying, the student will score approximately 90.9% on the exam.

The graph of the function defined by

$$P(x) = \frac{100x^2}{40 + x^2}$$

is shown in Figure 9-6.

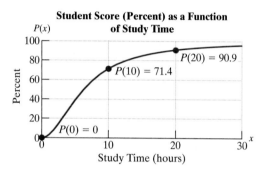

Student Score (Percent) as a Function of Study Time

Figure 9-6

Calculator Connections

A graphing calculator can be used to graph a function defined by $y = f(x)$. For example,

$$f(x) = \frac{1}{4}x^3 - x^2 - x + 4$$

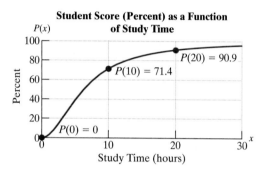

$$g(x) = \sqrt{x + 4}$$

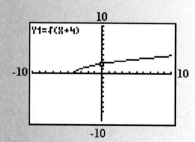

$h(x) = |x - 2| - 6$

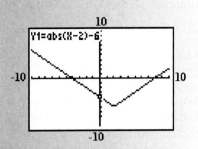

Calculator Exercises

Use a graphing calculator to graph the following functions.

1. $f(x) = x^2 - 5x + 2$ ❖ 2. $g(x) = \sqrt{6 - 2x}$ ❖ 3. $h(x) = |x + 3| - 4$ ❖

4. $y = x^3 - 9x$ ❖ 5. $y = \sqrt{25 - x^2}$ ❖ 6. $y = -|x| + 8$ ❖

section 9.1 PRACTICE EXERCISES

1. Define a relation. **Any set of ordered pairs (x, y) is called a relation in x and y.**

2. What is a function? ❖

3. How can you tell from the graph of a relation if that relation is a function? **By the vertical line test**

4. How can you tell if a set of ordered pairs represents a function? **If each first coordinate is unique, then the set of ordered pairs represents a function.**

For Exercises 5–12, determine which of the relations is a function.

5. $\{(1, 8), (6, 2), (7, 2), (-3, 2)\}$ **Yes**

6. $\{(-6, -1), (8, -1), (0, -1), (4, -1)\}$ **Yes**

7. $\{(3, 1), (3, 4), (3, 0), (3, 3)\}$ **No**

8. $\{(6, 8), (\frac{1}{2}, 3), (6, \frac{1}{4}), (0, 4)\}$ **No**

9.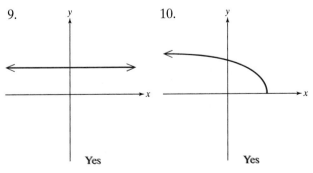

Yes

10.

Yes

11.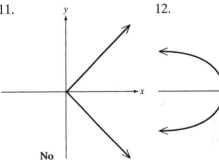

No

12.

No

13. Let $f(x) = 2x - 5$. Find:

$f(0)$ **−5**

$f(2)$ **−1**

$f(-3)$ **−11**

14. Let $g(x) = x^2 + 1$. Find:

$g(0)$ **1**

$g(-1)$ **2**

$g(3)$ **10**

15. Let $h(x) = \dfrac{1}{x + 4}$ Find:

$h(1)$ $\dfrac{1}{5}$

$h(0)$ $\dfrac{1}{4}$

$h(-2)$ $\dfrac{1}{2}$

16. Let $p(x) = \sqrt{x + 4}$ Find:

$p(0)$ **2**

$p(-4)$ **0**

$p(5)$ **3**

For Exercises 17–20, translate the expressions into English phrases.

17. $f(6) = 2$
The function value at $x = 6$ is 2.

18. $f(-2) = -14$
The function value at $x = -2$ is -14.

19. $g\left(\dfrac{1}{2}\right) = \dfrac{1}{4}$
The function value at $x = \frac{1}{2}$ is $\frac{1}{4}$.

20. $h(k) = k^2$
The function value $x = k$ is k^2.

21. Consider a function defined by $y = f(x)$. The function value $f(2) = 7$ corresponds to what ordered pair? $(2, 7)$

22. Consider a function defined by $y = f(x)$. The function value $f(-3) = -4$ corresponds to what ordered pair? $(-3, -4)$

23. Consider a function defined by $y = f(x)$. The function value $f(0) = 8$ corresponds to what ordered pair? $(0, 8)$

24. Consider a function defined by $y = f(x)$. The function value $f(4) = 0$ corresponds to what ordered pair? $(4, 0)$

25. Ice-cream is $1.50 for one scoop plus $0.50 for each topping. The total cost, C, (in dollars) can be found using the function defined by

$$C(x) = 1.50 + 0.50x$$ where x is the number of toppings.

a. Find $C(0)$ and interpret the meaning of this function value in terms of cost and the number of toppings. $C(0) = 1.50$. The cost of a scoop of ice-cream with 0 toppings is $1.50.

b. Find $C(1)$ and interpret the meaning in terms of cost and the number of toppings.
$C(1) = 2.00$. The cost of a scoop of ice-cream with 1 topping is $2.00.

c. Find $C(4)$ and interpret the meaning in terms of cost and the number of toppings. $C(4) = 3.50$.
The cost of a scoop of ice-cream with four toppings is $3.50.

26. Ignoring air resistance, the speed, s, (in feet per second: ft/s) of an object in free-fall is a function of the number of seconds, t, after it was dropped:

$$s(t) = 32t$$

a. Find $s(1)$ and interpret the meaning of this function value in terms of speed and time.
$s(1) = 32$. The speed of an object 1 s after being dropped is 32 ft/s.

b. Find $s(2)$ and interpret the meaning in terms of speed and time. $s(2) = 64$. The speed of an object 2 s after being dropped is 64 ft/s.

c. Find $s(10)$ and interpret the meaning in terms of speed and time. $s(10) = 320$. The speed of an object 10 s after being dropped is 320 ft/s.

d. A ball dropped from the top of the Sears Tower in Chicago falls for approximately 9.2 s. How fast was the ball going the instant before it hit the ground?   294.4 ft/s

27. Ignoring air resistance, the speed, s, (in meters per second: m/s) of an object in free-fall is a function of the number of seconds, t, after it was dropped.

$$s(t) = 9.8t$$

a. Find $s(0)$ and interpret the meaning of this function value in terms of speed and time.
$s(0) = 0$. The initial speed is 0 m/s.

b. Find $s(4)$ and interpret the meaning in terms of speed and time. $s(4) = 39.2$. The speed of an object 4 s after being dropped is 39.2 m/s.

c. Find $s(6)$ and interpret the meaning in terms of speed and time. $s(6) = 58.8$. The speed of an object 6 s after being dropped is 58.8 m/s.

d. A penny dropped from the top of the Texas Commerce Tower in Houston falls for approximately 7.9 s. How fast was it going the instant before it hit the ground?   77.42 m/s

28. A punter kicks a football straight up with an initial velocity of 64 ft/s. The height of the ball, h, (in feet) is a function of the number of seconds, t, after the ball is kicked:

$$h(t) = -16t^2 + 64t + 3$$

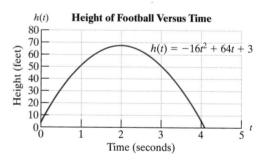

Height of Football Versus Time

$h(t) = -16t^2 + 64t + 3$

Figure for Exercise 28

a. Find $h(0)$ and interpret the meaning of the function value in terms of time and height.
$h(0) = 3$. The initial height of the ball is 3 ft.

b. Find $h(1)$ and interpret the meaning in terms of time and height. $h(1) = 51$. The height of the ball 1 s after being kicked is 51 ft.

c. Find $h(2)$ and interpret the meaning in terms of time and height. $h(2) = 67$. The height of the ball 2 s after being kicked is 67 ft.

d. Find $h(4)$ and interpret the meaning in terms of time and height. $h(4) = 3$. The height of the ball 4 s after being kicked is 3 ft.

29. A punter kicks a football straight up with an initial velocity of 64 ft/s. The velocity of the ball, v,

(in feet per second: ft/s) is a function of the number of seconds, t, after the ball is kicked:

$$v(t) = -32t + 64$$

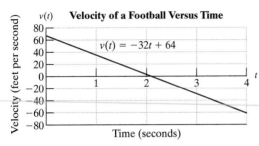

Velocity of a Football Versus Time

$v(t) = -32t + 64$

Time (seconds)

Figure for Exercise 29

a. Find $v(0)$ and interpret the meaning of the function value in terms of time and velocity. $v(0) = 64$. The initial velocity is 64 ft/s.
b. Find $v(1)$ and interpret the meaning in terms of time and velocity. $v(1) = 32$. The velocity after 1 s is 32 ft/s.
c. Find $v(2)$ and interpret the meaning in terms of time and velocity. $v(2) = 0$. The velocity after 2 s is 0 ft/s.
d. Find $v(4)$ and interpret the meaning in terms of time and velocity. $v(4) = -64$. The velocity after 4 s is −64 ft/s, (the football is descending).

30. For the adults, the maximum recommended heart rate, M, (in beats per minute: beats/min) is a function of a person's age, x, (in years).

$$M(x) = 220 - x \quad x \geq 16$$

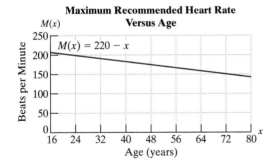

Maximum Recommended Heart Rate Versus Age

$M(x) = 220 - x$

Age (years)

Figure for Exercise 30

a. Find $M(16)$ and interpret the meaning in terms of maximum recommended heart rate and age. $M(16) = 204$. A 16-year-old adult's maximum recommended heart rate is 204 beats/min.
b. Find $M(30)$ and interpret the meaning in terms of maximum recommended heart rate and age. $M(30) = 190$. A 30-year-old adult's maximum recommended heart rate is 190 beats/min.
c. Find $M(60)$ and interpret the meaning in terms of maximum recommended heart rate and age. $M(60) = 160$. A 60-year-old adult's maximum recommended heart rate is 160 beats/min.
d. Find your own maximum recommended heart rate. Answers will vary.

For Exercises 31–38, let the function f and g be defined by $f(x) = x^2$ and $g(x) = 3x$. Find the function values.

31. $f(2) - g(1)$ 1

32. $f(6) + g(-3)$ 27

33. $g(2) \cdot f(-1)$ 6

34. $f(5) \cdot g(-1)$ −75

35. $f(m^2)$ m^4

36. $g(n^2)$ $3n^2$

37. $f(a + b)$ $(a + b)^2$ or $a^2 + 2ab + b^2$

38. $g(4y)$ $12y$

39. What is the domain of a relation? The domain of a relation is the set of first coordinates of the ordered pairs.

40. What is the range of a relation? The range of a relation is the set of second coordinates of the ordered pairs.

For Exercises 41–48, find the domain of the given function. Write the answer in interval notation.

41. $g(x) = 5x - 1$ $(-\infty, \infty)$

42. $g(x) = 2x - 2$ $(-\infty, \infty)$

43. $h(x) = \sqrt{x + 1}$ $[-1, \infty)$

44. $h(x) = \sqrt{x - 2}$ $[2, \infty)$

45. $L(x) = |x|$ $(-\infty, \infty)$

46. $F(x) = x^2 + 2x + 1$ $(-\infty, \infty)$

47. $m(x) = \sqrt{5 - x}$ $(-\infty, 5]$

48. $n(x) = \sqrt{3 - x}$ $(-\infty, 3]$

For Exercises 49–56, find the domain of the given function. Write the answer in set-builder notation.

49. $p(x) = \dfrac{1}{x + 6}$ $\{x | x \neq -6\}$

50. $p(x) = \dfrac{1}{x - 3}$ $\{x | x \neq 3\}$

51. $r(x) = \dfrac{x + 2}{(x + 3)(x - 4)}$ $\{x | x \neq -3, x \neq 4\}$

52. $s(x) = \dfrac{x - 6}{(x + 2)(x - 5)}$ $\{x | x \neq -2, x \neq 5\}$

53. $P(x) = \dfrac{x - 10}{x^2 - 36}$ $\{x | x \neq 6, x \neq -6\}$

54. $A(x) = \dfrac{x + 11}{x^2 - 4}$ $\{x | x \neq 2, x \neq -2\}$

55. $k(x) = \dfrac{4}{2x^2 - 9x - 5}$ $\{x | x \neq -\tfrac{1}{2}, x \neq 5\}$

56. $L(x) = \dfrac{-3}{3x^2 - 2x - 5}$ $\{x | x \neq \tfrac{5}{3}, x \neq -1\}$

For Exercises 57–60, match the domain and range given with a possible graph. For each graph the relation is shown in black, the domain is shown in red, and the range is shown in blue.

57. Domain: All real numbers

Range: $[1, \infty)$ b

58. Domain: $[-4, 4]$

Range: $[-2, 2]$ a

59. Domain: $[-2, \infty)$

Range: All real numbers   c

60. Domain: $\{x | x \neq 2\}$

Range: $\{y | y \neq 3\}$ d

a.

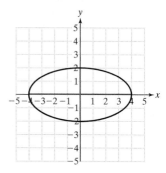

b.

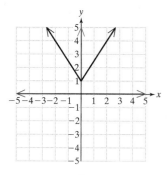

c.

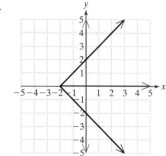

d.

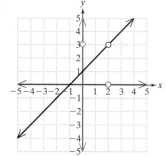

---

section

9.2   COMPLEX NUMBERS

Concepts

1. Definition of *i*
2. Simplifying Expressions in Terms of *i*
3. Simplifying Powers of *i*
4. Definition of a Complex Number
5. Addition, Subtraction, and Multiplication of Complex Numbers
6. Division of Complex Numbers

1. Definition of *i*

In Section 8.1, we learned that there are no real-valued square roots of a negative number. For example, $\sqrt{-9}$ is not a real number because no real number when squared equals -9. However, the square roots of a negative number are defined over another set of numbers called the *imaginary numbers*. The foundation of the set of imaginary numbers is the definition of the imaginary number, *i*, as: $i = \sqrt{-1}$.

Definition of *i*

$$i = \sqrt{-1}$$

Note: From the definition of *i*, it follows that $i^2 = -1$

2. Simplifying Expressions in Terms of *i*

Using the imaginary number *i*, we can define the square root of any negative real number.

Definition of $\sqrt{-b}$, *b* > 0

Let *b* be a real number such that $b > 0$, then $\sqrt{-b} = i\sqrt{b}$

example 1

Simplifying Expressions in Terms of *i*

Simplify the expressions in terms of *i*.

a. $\sqrt{-25}$ b. $\sqrt{-81}$ c. $\sqrt{-13}$

Solution:

a. $\sqrt{-25} = 5i$
b. $\sqrt{-81} = 9i$
c. $\sqrt{-13} = i\sqrt{13}$

◆ **Avoiding Mistakes**

In an expression such as $i\sqrt{13}$ the *i* is usually written in front of the square root. The expression $\sqrt{13}\,i$ is also correct but may be misinterpreted as $\sqrt{13i}$ (with *i* incorrectly placed under the square root).

The multiplication and division properties of radicals were presented in Sections 8.2 and 8.4 as follows.

If *a* and *b* represent real numbers such that $\sqrt[n]{a}$ and $\sqrt[n]{b}$ are both real, then

$$\sqrt[n]{ab} = \sqrt[n]{a} \cdot \sqrt[n]{b} \quad \text{and} \quad \sqrt[n]{\frac{a}{b}} = \frac{\sqrt[n]{a}}{\sqrt[n]{b}} \quad b \neq 0$$

The conditions that $\sqrt[n]{a}$ and $\sqrt[n]{b}$ must both be real numbers prevent us from applying the multiplication and division properties of radicals for square roots with negative radicands. Therefore, to multiply or divide radicals with negative radicands, write the radicals in terms of the imaginary number *i* first. This is demonstrated in the next example.

example 2

Simplifying a Product of Expressions in Terms of *i*

Simplify the expressions.

a. $\dfrac{\sqrt{-100}}{\sqrt{-25}}$ b. $\sqrt{-16} \cdot \sqrt{-4}$

Solution:

a. $\dfrac{\sqrt{-100}}{\sqrt{-25}}$

$= \dfrac{10i}{5i}$ Simplify each radical in terms of i *before* dividing.

$= 2$ Reduce.

b. $\sqrt{-16} \cdot \sqrt{-4}$

$= (4i)(2i)$ Simplify each radical in terms of i first before multiplying.

$= 8i^2$

$= 8(-1)$

$= -8$

3. Simplifying Powers of i

From the definition of $i = \sqrt{-1}$, it follows that

$i = i$

$i^2 = -1$

$i^3 = -i$ because $i^3 = i^2 \cdot i = (-1)i = -i$

$i^4 = 1$ because $i^4 = i^2 \cdot i^2 = (-1)(-1) = 1$

$i^5 = i$ because $i^5 = i^4 \cdot i = (1)i = i$

$i^6 = -1$ because $i^6 = i^4 \cdot i^2 = (1)(-1) = -1$

This pattern of values $i, -1, -i, 1, i, -1, -i, 1, \ldots$ continues for all subsequent powers of i. Here is a list of several powers of i.

Powers of i		
$i^1 = i$	$i^5 = i$	$i^9 = i$
$i^2 = -1$	$i^6 = -1$	$i^{10} = -1$
$i^3 = -i$	$i^7 = -i$	$i^{11} = -i$
$i^4 = 1$	$i^8 = 1$	$i^{12} = 1$

To simplify higher powers of i, we can decompose the expression into multiples of i^4 ($i^4 = 1$) and write the remaining factors as i, i^2 or i^3.

example 3

Simplifying Powers of _i_

Simplify the powers of _i_.

a. i^9 b. i^{14} c. i^{103} d. i^{28}

Classroom Activity 9.2A

Solution:

a. $i^9 = (i^8) \cdot i$

$\quad = (i^4)^2 \cdot (i)$

$\quad = (1)^2(i)$ Recall that $i^4 = 1$.

$\quad = i$ Simplify.

b. $i^{14} = (i^{12}) \cdot i^2$

$\quad = (i^4)^3(i^2)$

$\quad = (1)^3(-1)$ $i^4 = 1$ and $i^2 = -1$.

$\quad = -1$ Simplify.

c. $i^{103} = (i^{100}) \cdot i^3$

$\quad = (i^4)^{25}(i^3)$

$\quad = (1)^{25}(-i)$ $i^4 = 1$ and $i^3 = -i$.

$\quad = -i$ Simplify.

d. $i^{28} = (i^4)^7$

$\quad = (1)^7$ $i^4 = 1$.

$\quad = 1$ Simplify.

4. Definition of a Complex Number

We have already learned the definitions of the integers, rational numbers, irrational numbers, and real numbers. In this section, we define the complex numbers.

Definition of a Complex Number

A **complex number** is a number of the form $a + bi$, where a and b are real numbers and $i = \sqrt{-1}$

Notes:
- If $b = 0$, then the complex number, $a + bi$ is a real number.
- If $b \neq 0$, then we say that $a + bi$ is an **imaginary number**.
- The complex number $a + bi$ is said to be written in **standard form**. The quantities a and b are called the **real** and **imaginary parts**, respectively.
- The complex numbers $(a - bi)$ and $(a + bi)$ are called **conjugates**.

From the definition of a complex number, it follows that all real numbers are complex numbers and all imaginary numbers are complex numbers. Figure 9-7 illustrates the relationship among the sets of numbers we have learned so far.

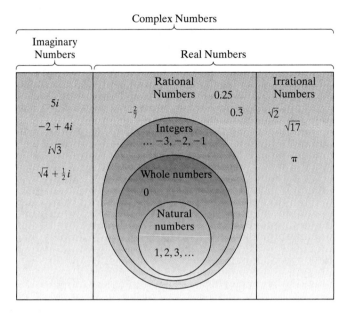

Figure 9-7

---

example 4

Identifying the Real and Imaginary Parts of a Complex Number

Identify the real and imaginary parts of the complex numbers.

a. $7 + 4i$ b. -6 c. $-\dfrac{1}{2}i$

Solution:

a. $7 + 4i$ The real part is 7, and the imaginary part is 4.

b. -6

$= -6 + 0i$ Rewrite -6 in the form $a + bi$.
The real part is -6, and the imaginary part is 0.

Tip: Example 4(b) illustrates that a real number is also a complex number.

c. $-\dfrac{1}{2}i$

$= 0 - \dfrac{1}{2}i$

$= 0 + -\dfrac{1}{2}i$ Rewrite $-\frac{1}{2}i$ in the form $a + bi$.
The real part is 0, and the imaginary part is $-\frac{1}{2}$.

Tip: Example 4(c) illustrates that an imaginary number is also a complex number.

5. Addition, Subtraction, and Multiplication of Complex Numbers

The operations for addition, subtraction, and multiplication of real numbers also apply to imaginary numbers. To add or subtract complex numbers, combine the real parts together and combine the imaginary parts together. The commutative, associative, and distributive properties that apply to real numbers also apply to complex numbers.

example 5   **Adding, Subtracting, and Multiplying Complex Numbers**

a. Add: $(2 - 3i) + (4 + 17i)$
b. Subtract: $\left(-\frac{3}{2} + \frac{1}{3}i\right) - \left(2 - \frac{2}{3}i\right)$
c. Multiply: $(5 - 2i)(3 + 4i)$
d. Multiply: $(2 + 7i)(2 - 7i)$

Solution:

a. $(2 - 3i) + (4 + 17i) = (2 + 4) + (-3 + 17)i$ Add real parts. Add imaginary parts.

$= 6 + 14i$ Simplify

b. $\left(-\frac{3}{2} + \frac{1}{3}i\right) - \left(2 - \frac{2}{3}i\right) = -\frac{3}{2} + \frac{1}{3}i - 2 + \frac{2}{3}i$ Apply the distributive property.

$= \left(-\frac{3}{2} - 2\right) + \left(\frac{1}{3} + \frac{2}{3}\right)i$ Add real parts. Add imaginary parts.

$= \left(-\frac{3}{2} - \frac{4}{2}\right) + \left(\frac{3}{3}\right)i$ Find common denominator and simplify.

$= -\frac{7}{2} + i$

c. $(5 - 2i)(3 + 4i)$

$= (5)(3) + (5)(4i) + (-2i)(3) + (-2i)(4i)$ Apply the distributive property.

$= 15 + 20i - 6i - 8i^2$ Simplify.

$= 15 + 14i - 8(-1)$ Recall $i^2 = -1$

$= 15 + 14i + 8$

$= 23 + 14i$ Write the answer in the form $a + bi$.

d. $(2 + 7i)(2 - 7i)$ The expressions $(2 + 7i)$ and $(2 - 7i)$ are conjugates.

The product is a difference of squares: $(a + b)(a - b) = a^2 - b^2$.

$(a + b)(a - b) = a^2 - b^2$

$(2 + 7i)(2 - 7i) = (2)^2 - (7i)^2$ Apply the formula, where $a = 2$ and $b = 7i$.

$\qquad\qquad = 4 - 49i^2$ Simplify.

$\qquad\qquad = 4 - 49(-1)$ Recall $i^2 = -1$.

$\qquad\qquad = 4 + 49$

$\qquad\qquad = 53$

> **Tip:** The complex numbers $2 + 7i$ and $2 - 7i$ can also be multiplied by using the distributive property.
>
> $$(2 + 7i)(2 - 7i) = 4 - 14i + 14i - 49i^2$$
> $$= 4 - 49(-1)$$
> $$= 4 + 49$$
> $$= 53$$

6. Division of Complex Numbers

Example 5(d) illustrates that the product of a complex number and its conjugate produces a real number. Consider the complex numbers $a + bi$ and $a - bi$, where a and b are real numbers. Then,

$$(a + bi)(a - bi) = (a)^2 - (bi)^2$$
$$= a^2 - b^2 i^2$$
$$= a^2 - b^2(-1)$$
$$= a^2 + b^2 \quad \text{(real number)}$$

To divide by a complex number, multiply the numerator and denominator by the conjugate of the denominator. This produces a real number in the denominator so that the resulting expression can be written in the form $a + bi$.

example 6   **Dividing by a Complex Number**

Divide the complex numbers. Write the answer in the form $a + bi$.

a. $\dfrac{17}{3 + 5i}$ b. $\dfrac{2 + 3i}{4 - 5i}$ c. $\dfrac{-2}{i}$

Solution:

a. $\dfrac{17}{3 + 5i}$

$= \dfrac{17}{(3 + 5i)} \cdot \dfrac{(3 - 5i)}{(3 - 5i)}$ Multiply the numerator and denominator by the conjugate of the denominator.

$= \dfrac{17(3 - 5i)}{(3)^2 - (5i)^2}$ In the denominator apply the formula $(a + b)(a - b) = a^2 - b^2$.

$= \dfrac{17(3 - 5i)}{9 - 25i^2}$ Simplify the denominator.

$= \dfrac{17(3 - 5i)}{9 - 25(-1)}$ Recall $i^2 = -1$.

$= \dfrac{17(3 - 5i)}{9 + 25}$ Simplify.

$= \dfrac{17(3 - 5i)}{34}$

$= \dfrac{\overset{1}{\cancel{17}}(3 - 5i)}{\underset{2}{\cancel{34}}}$ Reduce.

$= \dfrac{3 - 5i}{2}$

$= \dfrac{3}{2} - \dfrac{5}{2}i$ Write in the form $a + bi$.

b. $\dfrac{2 + 3i}{4 - 5i}$

$\dfrac{(2 + 3i)}{(4 - 5i)} \cdot \dfrac{(4 + 5i)}{(4 + 5i)} = \dfrac{(2)(4) + (2)(5i) + (3i)(4) + (3i)(5i)}{(4)^2 - (5i)^2}$ Multiply the numerator and denominator by the conjugate of the denominator.

$= \dfrac{8 + 10i + 12i + 15i^2}{16 - 25i^2}$ Simplify numerator and denominator.

$= \dfrac{8 + 22i + 15(-1)}{16 - 25(-1)}$ Recall $i^2 = -1$

$= \dfrac{8 + 22i - 15}{16 + 25}$

$$= \frac{-7 + 22i}{41} \qquad \text{Simplify.}$$

$$= -\frac{7}{41} + \frac{22}{41}i \qquad \text{Write in the form } a + bi.$$

c. $\dfrac{-2}{i}$ The denominator is equivalent to $0 + i$. The conjugate is $0 - i$, or simply $-i$.

$$= \frac{-2}{i} \cdot \frac{-i}{-i} \qquad \text{Multiply numerator and denominator by the conjugate of the denominator.}$$

$$= \frac{2i}{-i^2} \qquad \text{Multiply.}$$

$$= \frac{2i}{-(-1)} \qquad \text{Recall } i^2 = -1.$$

$$= \frac{2i}{1} \qquad \text{Simplify.}$$

$$= 2i$$

$$= 0 + 2i \qquad \text{Write in the form } a + bi.$$

---

section 9.2   PRACTICE EXERCISES

For Exercises 1–8, simplify the expressions in terms of i.

1. $\sqrt{-49}$ $7i$ 2. $\sqrt{-36}$ $6i$ 3. $\sqrt{-15}$ $i\sqrt{15}$

4. $\sqrt{-21}$ $i\sqrt{21}$ 5. $\sqrt{-12}$ $2i\sqrt{3}$ 6. $\sqrt{-48}$ $4i\sqrt{3}$

7. $\sqrt{-1}$ i 8. $\sqrt{-4}$ $2i$

For Exercises 9–22, perform the indicated operations. Remember to write the radicals in terms of i first.

9. $\sqrt{-100} \cdot \sqrt{-4}$ -20 10. $\sqrt{-9} \cdot \sqrt{-25}$ -15

11. $\sqrt{-3} \cdot \sqrt{-12}$ -6 12. $\sqrt{-8} \cdot \sqrt{-2}$ -4

13. $\dfrac{\sqrt{-81}}{\sqrt{-9}}$ 3 14. $\dfrac{\sqrt{-64}}{\sqrt{-16}}$ 2

15. $\dfrac{\sqrt{-50}}{\sqrt{-2}}$ 5 16. $\dfrac{\sqrt{-45}}{\sqrt{-5}}$ 3

 17. $\sqrt{-9} + \sqrt{-121}$ $14i$ 18. $\sqrt{-36} - \sqrt{-49}$ $-i$

19. $\sqrt{-1} - \sqrt{-144} - \sqrt{-169}$ $-24i$

20. $\sqrt{-4} + \sqrt{-64} + \sqrt{-81}$ $19i$

21. $-\sqrt{25} + \sqrt{-25}$ $-5 + 5i$ 22. $\sqrt{-100} - \sqrt{100}$ $-10 + 10i$

For Exercises 23–30, simplify the powers of i.

23. i^6 -1 24. i^{17} i 25. i^{12} 1 26. i^{27} $-i$

27. i^{101} i 28. i^{92} 1 29. i^{87} $-i$ 30. i^{66} -1

31. Explain how to add or subtract complex numbers.   Add or subtract the real parts. Add or subtract the imaginary parts.

32. Explain how to multiply complex numbers.   Multiply using the distributive property, remembering to replace i^2 with -1.

For Exercises 33–62, perform the indicated operations. Write the answers in standard form, $a + bi$.

33. $(2 + 7i) + (-8 + i)$ $-6 + 8i$ 34. $(6 - i) - (4 + 2i)$ $2 - 3i$

35. $(3 - 4i) + (7 - 6i)$ $10 - 10i$

36. $(-4 - 15i) - (-3 - 17i)$ $-1 + 2i$

37. $4i - (9 + i) + 15$ $6 + 3i$ 38. $10i - (1 - 5i) - 8$ $-9 + 15i$

39. $(5 - 6i) - (9 - 8i) - (3 - i)$ $-7 + 3i$

40. $(1 - i) - (5 - 19i) - (24 + 19i)$ $-28 - i$

41. $(2 - i)(7 - 7i)$ $7 - 21i$ 42. $(1 + i)(8 - i)$ $9 + 7i$

43. $(13 - 5i) - (2 + 4i)$ $11 - 9i$

44. $(1 + 8i) + (-6 + 3i)$ $-5 + 11i$

45. $(5 + 3i)(3 + 2i)$ $9 + 19i$ 46. $(9 + i)(8 + 2i)$ $70 + 26i$

47. $\left(\frac{1}{2} + \frac{1}{5}i\right) - \left(\frac{3}{4} + \frac{2}{5}i\right)$ $-\frac{1}{4} - \frac{1}{5}i$

48. $\left(\frac{5}{6} + \frac{1}{8}i\right) + \left(\frac{1}{3} - \frac{3}{8}i\right)$ $\frac{7}{6} - \frac{1}{4}i$

49. $8.4i - (3.5 - 9.7i)$ $-3.5 + 18.1i$

50. $(4.25 - 3.75i) - (10.5 - 18.25i)$ $-6.25 + 14.5i$

51. $(3 - 2i)(3 + 2i)$ 13 52. $(18 + i)(18 - i)$ 325

53. $(10 - 2i)(10 + 2i)$ 104 54. $(3 - 5i)(3 + 5i)$ 34

55. $\left(\frac{1}{2} - i\right)\left(\frac{1}{2} + i\right)$ $\frac{5}{4}$ 56. $\left(\frac{1}{3} - i\right)\left(\frac{1}{3} + i\right)$ $\frac{10}{9}$

57. $(6 - i)^2$ $35 - 12i$ 58. $(4 + 3i)^2$ $7 + 24i$

59. $(5 + 2i)^2$ $21 + 20i$ 60. $(7 - 6i)^2$ $13 - 84i$

61. $(4 - 7i)^2$ $-33 - 56i$ 62. $(3 - i)^2$ $8 - 6i$

63. What is the conjugate of $7 - 4i$? Multiply $7 - 4i$ by its conjugate. $7 + 4i; 65$

64. What is the conjugate of $-3 - i$? Multiply $-3 - i$ by its conjugate. $-3 + i; 10$

65. What is the conjugate of $\frac{3}{2} + \frac{2}{5}i$? Multiply $\frac{3}{2} + \frac{2}{5}i$ by its conjugate. $\frac{3}{2} - \frac{2}{5}i; \frac{241}{100}$

66. What is the conjugate of $-1.3 + 5.7i$? Multiply $-1.3 + 5.7i$ by its conjugate. $-1.3 - 5.7i; 34.18$

67. What is the conjugate of $4i$? Multiply $4i$ by its conjugate. $-4i; 16$

68. What is the conjugate of $-8i$? Multiply $-8i$ by its conjugate. $8i; 64$

For Exercises 69–84, divide the complex numbers. Write the answers in standard form, $a + bi$.

69. $\dfrac{2}{7 - 4i}$ $\frac{14}{65} + \frac{8}{65}i$ 70. $\dfrac{-3}{-3 - i}$ $\frac{9}{10} - \frac{3}{10}i$ 71. $\dfrac{5}{1 + i}$ $\frac{5}{2} - \frac{5}{2}i$

72. $\dfrac{6}{1 - i}$ $3 + 3i$ 73. $\dfrac{-3i}{2 + i}$ $-\frac{3}{5} - \frac{6}{5}i$ 74. $\dfrac{6i}{3 - 2i}$ $-\frac{12}{13} + \frac{18}{13}i$

75. $\dfrac{4i}{5 - i}$ $-\frac{2}{13} + \frac{10}{13}i$ 76. $\dfrac{6i}{3 + i}$ $\frac{3}{5} + \frac{9}{5}i$ 77. $\dfrac{4 + i}{4 - i}$ $\frac{15}{17} + \frac{8}{17}i$

78. $\dfrac{1 - 5i}{1 + 5i}$ $-\frac{12}{13} - \frac{5}{13}i$ 79. $\dfrac{4 + 3i}{2 + 5i}$ $\frac{23}{29} - \frac{14}{29}i$ 80. $\dfrac{1 + 7i}{3 + 2i}$ $\frac{17}{13} + \frac{19}{13}i$

81. $\dfrac{1}{-18i}$ $\frac{1}{18}i$ 82. $\dfrac{1}{4i}$ $-\frac{1}{4}i$ 83. $\dfrac{-3}{5i}$ $\frac{3}{5}i$

84. $\dfrac{-4}{6i}$ $\frac{2}{3}i$

EXPANDING YOUR SKILLS

For Exercises 85–97, answer true or false. If an answer is false, explain why.

85. Every complex number is a real number.
    False. For example: $2 + 3i$ is not a real number.

86. Every real number is a complex number. True

87. Every imaginary number is a complex number. True

88. $\sqrt{-64}$ is an imaginary number. True

89. $\sqrt[3]{-64}$ is an imaginary number. False. $\sqrt[3]{-64} = -4$.

90. The product $(2 + 3i)(2 - 3i)$ is a real number. True

91. The product $(1 + 4i)(1 - 4i)$ is an imaginary number. False. $(1 + 4i)(1 - 4i) = 17$.

92. The imaginary part of the complex number $2 - 3i$ is 3. False. The imaginary part is -3.

93. The imaginary part of the complex number $4 - 5i$ is -5. True

94. i^2 is a real number. True

95. i^4 is an imaginary number. False. $i^4 = 1$.

96. i^3 is a real number. False. $i^3 = -i$

97. i^4 is a real number. True

section

9.3 COMPLETING THE SQUARE

1. Square Root Property

In Section 4.6, we learned how to solve quadratic equations by factoring and applying the zero product rule. However, the zero product rule can only be used if the equation is factorable. In this section and the next, we will learn two techniques for solving *all* quadratic equations, factorable and nonfactorable.

The first technique will employ the **square root property**.

Square Root Property

For any real number, k, if $x^2 = k$, then $x = \sqrt{k}$ or $x = -\sqrt{k}$.

Note: The solution may also be written as $x = \pm\sqrt{k}$, read "x equals plus or minus the square root of k."

2. Solving Quadratic Equations Using the Square Root Property

example 1   Solving Quadratic Equations Using the Square Root Property

Use the square root property to solve the equations.

a. $x^2 = 25$ b. $2x^2 + 18 = 0$ c. $(t - 4)^2 = 12$

Solution:

a. $x^2 = 25$ The equation is in the form $x^2 = k$.

$\quad\quad = \pm\sqrt{25}$ Apply the square root property.

$\quad\quad = \pm 5$

The solutions are $x = 5$ and $x = -5$.

Tip: The equation $x^2 = 25$ can also be solved by using the zero product rule:

$$x^2 = 25$$
$$x^2 - 25 = 0$$
$$(x - 5)(x + 5) = 0$$
$$x = 5 \quad \text{or} \quad x = -5$$

Avoiding Mistakes

A common mistake is to forget the $\pm$ symbol when solving the equation $x^2 = k$.

$$x = \pm\sqrt{k}$$

b. $2x^2 + 18 = 0$ Rewrite the equation to fit the form $x^2 = k$.

$\quad\quad 2x^2 = -18$

$\quad\quad x^2 = -9$ The equation is now in the form $x^2 = k$.

$\quad\quad x = \pm\sqrt{-9}$ Apply the square root property.

$\quad\quad x = \pm 3i$

The solutions are $x = 3i$ and $x = -3i$.

Classroom Activity 9.3A

Check: $x = 3i$ Check: $x = -3i$

$$2x^2 + 18 = 0$$ $$2x^2 + 18 = 0$$

$$2(3i)^2 + 18 \stackrel{?}{=} 0$$ $$2(-3i)^2 + 18 \stackrel{?}{=} 0$$

$$2(9i^2) + 18 \stackrel{?}{=} 0$$ $$2(9i^2) + 18 \stackrel{?}{=} 0$$

$$2(-9) + 18 \stackrel{?}{=} 0$$ $$2(-9) + 18 \stackrel{?}{=} 0$$

$$-18 + 18 = 0 ✔$$ $$-18 + 18 = 0 ✔$$

c. $(t - 4)^2 = 12$ The equation is in the form $x^2 = k$, where $x = (t - 4)$.

$t - 4 = \pm\sqrt{12}$ Apply the square root property.

$t - 4 = \pm\sqrt{2^2 \cdot 3}$ Simplify the radical.

$t - 4 = \pm 2\sqrt{3}$

$t = 4 \pm 2\sqrt{3}$ Solve for t.

The solutions are: $t = 4 + 2\sqrt{3}$ and $t = 4 - 2\sqrt{3}$.

Check: $t = 4 + 2\sqrt{3}$ Check: $t = 4 - 2\sqrt{3}$

$$(t - 4)^2 = 12$$ $$(t - 4)^2 = 12$$

$$(4 + 2\sqrt{3} - 4)^2 \stackrel{?}{=} 12$$ $$(4 - 2\sqrt{3} - 4)^2 \stackrel{?}{=} 12$$

$$(2\sqrt{3})^2 \stackrel{?}{=} 12$$ $$(-2\sqrt{3})^2 \stackrel{?}{=} 12$$

$$4 \cdot 3 \stackrel{?}{=} 12$$ $$4 \cdot 3 \stackrel{?}{=} 12$$

$$12 = 12 ✔$$ $$12 = 12 ✔$$

3.   Completing the Square

In Example 1(c), we used the square root property to solve an equation in which the square of a binomial was equal to a constant.

$$(t - 4)^2 = 12$$

$$t - 4 = \pm\sqrt{12}$$

$$t = 4 \pm 2\sqrt{3}$$

In general, an equation of the form: $(x - h)^2 = k$ can be solved using the square root property. Furthermore, any equation $ax^2 + bx + c = 0 \, (a \neq 0)$ can be rewritten in the form $(x - h)^2 = k$ by using a process called **completing the square**.

We begin our discussion of completing the square with some vocabulary. For a trinomial $ax^2 + bx + c \, (a \neq 0)$, the term ax^2 is called the **quadratic term**. The term bx is called the **linear term**, and the term, c, is called the **constant term**.

Next, notice that the square of a binomial is the factored form of a perfect square trinomial.

Perfect Square Trinomial	Factored Form
$x^2 + 10x + 25 \longrightarrow$	$(x + 5)^2$
$t^2 - 6t + 9 \longrightarrow$	$(t - 3)^2$
$p^2 - 14p + 49 \longrightarrow$	$(p - 7)^2$

Furthermore, for a perfect square trinomial with a leading coefficient of 1, the constant term is the square of half the coefficient of the linear term. For example:

$x^2 + 10x + 25$ ⟵ $\qquad$ $t^2 - 6t + 9$ ⟵ $\qquad$ $p^2 - 14p + 49$ ⟵ $\qquad$

$\left[\dfrac{1}{2}(10)\right]^2 = [5]^2 = 25$ $\qquad$ $\left[\dfrac{1}{2}(-6)\right]^2 = [-3]^2 = 9$ $\qquad$ $\left[\dfrac{1}{2}(-14)\right]^2 = [-7]^2 = 49$

In general, an expression of the form $x^2 + bx$ will be a perfect square trinomial if the square of half the linear term coefficient, $(\frac{1}{2}b)^2$, is added to the expression.

---

example 2

Completing the Square

Complete the square for each expression. Then factor the expression as the square of a binomial.

a. $x^2 + 12x$ $\qquad$ b. $x^2 - 22x$ $\qquad$ c. $x^2 + 5x$ $\qquad$ d. $x^2 - \dfrac{3}{5}x$

Solution:

The expressions are in the form $x^2 + bx$. Add the square of half the linear term coefficient, $(\frac{1}{2}b)^2$.

a. $x^2 + 12x$

$\quad$ $x^2 + 12x + 36$ $\qquad$ Add $\frac{1}{2}$ of 12, squared. $[\frac{1}{2}(12)]^2 = (6)^2 = 36$.

$\quad$ $(x + 6)^2$ $\qquad$ Factored form

b. $x^2 - 22x$

$\quad$ $x^2 - 22x + 121$ $\qquad$ Add $\frac{1}{2}$ of -22, squared. $[\frac{1}{2}(-22)]^2 = (-11)^2 = 121$.

$\quad$ $(x - 11)^2$ $\qquad$ Factored form

c. $x^2 + 5x$

$\quad$ $x^2 + 5x + \dfrac{25}{4}$ $\qquad$ Add $\frac{1}{2}$ of 5, squared. $[\frac{1}{2}(5)]^2 = (\frac{5}{2})^2 = \frac{25}{4}$.

$\quad$ $\left(x + \dfrac{5}{2}\right)^2$ $\qquad$ Factored form

d. $x^2 - \dfrac{3}{5}x$

$\quad$ $x^2 - \dfrac{3}{5}x + \dfrac{9}{100}$ $\qquad$ Add $\frac{1}{2}$ of $-\frac{3}{5}$, squared.

$\qquad\qquad\qquad\qquad$ $\left[\dfrac{1}{2}\left(-\dfrac{3}{5}\right)\right]^2 = \left(-\dfrac{3}{10}\right)^2 = \dfrac{9}{100}$

$\quad$ $\left(x - \dfrac{3}{10}\right)^2$ $\qquad$ Factored form

4.   Solving Quadratic Equations by Completing the Square

The process of completing the square can be used to write a quadratic equation $ax^2 + bx + c = 0$ ($a \neq 0$) in the form $(x - h)^2 = k$. Then, the square root property can be used to solve the equation. The following steps outline the procedure.

Solving a Quadratic Equation in the Form $ax^2 + bx + c = 0$ ($a \neq 0$) by Completing the Square and Applying the Square Root Property

1.   Divide both sides by a to make the leading coefficient 1.
2.   Isolate the variable terms on one side of the equation.
3.   Complete the square (add the square of one-half the linear term coefficient to both sides of the equation. Then factor the resulting perfect square trinomial).
4.   Apply the square root property and solve for x.

---

example 3

Solving Quadratic Equations by Completing the Square and Applying the Square Root Property

Solve the quadratic equations by completing the square and applying the square root property.

a. $2x^2 - 16x + 40 = 0$ b. $x(2x - 5) = 3$

Classroom Activity 9.3B

Solution:

a. $2x^2 - 16x + 40 = 0$ The equation is in the form $ax^2 + bx + c = 0$.

$\dfrac{2x^2}{2} - \dfrac{16x}{2} + \dfrac{40}{2} = \dfrac{0}{2}$ **Step 1:**   Divide both sides by the coefficient, 2.

$x^2 - 8x + 20 = 0$

$x^2 - 8x = -20$ **Step 2:**   Isolate the variable terms on one side.

$x^2 - 8x + 16 = -20 + 16$ **Step 3:**   To complete the square, add $[\frac{1}{2}(-8)]^2 = 16$ to both sides of the equation.

$(x - 4)^2 = -4$ Factor the perfect square trinomial.

$x - 4 = \pm\sqrt{-4}$ **Step 4:**   Apply the square root property.

$x - 4 = \pm 2i$ Simplify the radical.

$x = 4 \pm 2i$ Solve for x.

The solutions are $x = 4 + 2i$ and $x = 4 - 2i$.

<u>Check:</u> $x = 4 + 2i$

$$2x^2 - 16x + 40 = 0$$

$$2(4 + 2i)^2 - 16(4 + 2i) + 40 \overset{?}{=} 0$$

$$2(16 + 16i + 4i^2) - 64 - 32i + 40 \overset{?}{=} 0$$

$$2(16 + 16i - 4) - 24 - 32i \overset{?}{=} 0$$

$$2(12 + 16i) - 24 - 32i \overset{?}{=} 0$$

$$24 + 32i - 24 - 32i \overset{?}{=} 0$$

$$0 = 0 \checkmark$$

<u>Check:</u> $x = 4 - 2i$

$$2x^2 - 16x + 40 = 0$$

$$2(4 - 2i)^2 - 16(4 - 2i) + 40 \overset{?}{=} 0$$

$$2(16 - 16i + 4i^2) - 64 + 32i + 40 \overset{?}{=} 0$$

$$2(16 - 16i - 4) - 24 + 32i \overset{?}{=} 0$$

$$2(12 - 16i) - 24 + 32i \overset{?}{=} 0$$

$$24 - 32i - 24 + 32i \overset{?}{=} 0$$

$$0 = 0 \checkmark$$

b. $\quad x(2x - 5) = 3$

$\quad 2x^2 - 5x = 3$ Write the equation in the form $ax^2 + bx + c = 0$.

$$2x^2 - 5x - 3 = 0$$

$$\frac{2x^2}{2} - \frac{5x}{2} - \frac{3}{2} = \frac{0}{2}$$

Step 1: Divide both sides by the coefficient, 2.

$$x^2 - \frac{5}{2}x - \frac{3}{2} = 0$$

$$x^2 - \frac{5}{2}x = \frac{3}{2}$$

Step 2: Isolate the variable terms on one side.

$$x^2 - \frac{5}{2}x + \frac{25}{16} = \frac{3}{2} + \frac{25}{16}$$

Step 3: Add $[\frac{1}{2}(-\frac{5}{2})]^2 = (-\frac{5}{4})^2 = \frac{25}{16}$ to both sides.

$$\left(x - \frac{5}{4}\right)^2 = \frac{24}{16} + \frac{25}{16}$$

Factor the perfect square trinomial. Rewrite the right-hand side with a common denominator.

$$\left(x - \frac{5}{4}\right)^2 = \frac{49}{16}$$

$$x - \frac{5}{4} = \pm\sqrt{\frac{49}{16}}$$

Step 4: Apply the square root property.

$$x - \frac{5}{4} = \pm\frac{7}{4}$$

Simplify the radical.

$$x = \frac{5}{4} \pm \frac{7}{4}$$

Solve for x.

The solutions are

$$x = \frac{5}{4} + \frac{7}{4} \quad \text{and} \quad x = \frac{5}{4} - \frac{7}{4}$$

$$x = \frac{12}{4} \quad \text{and} \quad x = -\frac{2}{4}$$

$$x = 3 \quad \text{and} \quad x = -\frac{1}{2}$$

<u>Check:</u> $x = 3$ <u>Check:</u> $x = -\frac{1}{2}$

$$x(2x - 5) = 3 \qquad\qquad x(2x - 5) = 3$$

$$3(2 \cdot 3 - 5) \overset{?}{=} 3 \qquad -\frac{1}{2}\left[2\left(-\frac{1}{2}\right) - 5\right] \overset{?}{=} 3$$

$$3(6 - 5) \overset{?}{=} 3 \qquad\qquad -\frac{1}{2}(-1 - 5) \overset{?}{=} 3$$

$$3(1) \overset{?}{=} 3 \qquad\qquad -\frac{1}{2}(-6) \overset{?}{=} 3$$

$$3 = 3 \checkmark \qquad\qquad 3 = 3 \checkmark$$

Tip: Since the solutions to the equation $x(2x - 5) = 3$ are rational numbers, the equation could have been solved by factoring and using the zero product rule:

$$x(2x - 5) = 3$$

$$2x^2 - 5x - 3 = 0$$

$$(2x + 1)(x - 3) = 0$$

$$2x + 1 = 0 \quad \text{or} \quad x - 3 = 0$$

$$x = -\frac{1}{2} \quad \text{or} \quad x = 3$$

5. Solving Quadratic Equations in Applications

example 4

Solving a Quadratic Equation in an Application

The length of a box is 2 in. more than the width. The height of the box is 4 in. and the volume of the box is 600 in.[3] Find the exact dimensions of the box. Then use a calculator to approximate the dimensions to the nearest tenth of an inch.

Solution:

Label the box as follows (Figure 9-8):

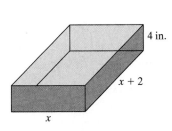

Figure 9-8

Width $= x$

Length $= x + 2$

Height $= 4$

The volume of a box is given by the formula: $V = lwh$

$$V = l \cdot w \cdot h$$

$$600 = (x + 2)(x)(4) \qquad \text{Substitute } V = 600, l = x + 2,$$
$$\qquad\qquad\qquad\qquad w = x, \text{ and } h = 4.$$

$$600 = (x + 2)4x$$

$$600 = 4x^2 + 8x$$

$$0 = 4x^2 + 8x - 600$$

$$4x^2 + 8x - 600 = 0 \qquad \text{The equation is in the form}$$
$$\qquad\qquad\qquad\qquad ax^2 + bx + c = 0.$$

The equation is not factorable. Therefore, complete the square and apply the square root property to solve for x.

$$\frac{4x^2}{4} + \frac{8x}{4} - \frac{600}{4} = \frac{0}{4}$$ **Step 1:** Divide both sides by the coefficient, 4.

$$x^2 + 2x - 150 = 0$$

$$x^2 + 2x = 150$$ **Step 2:** Isolate the variable terms on one side.

$$x^2 + 2x + 1 = 150 + 1$$ **Step 3:** Add $[\frac{1}{2}(2)]^2 = 1$ to both sides.

$$(x + 1)^2 = 151$$ Factor the perfect square trinomial.

$$x + 1 = \pm\sqrt{151}$$ **Step 4:** Apply the square root property.

$$x = -1 \pm \sqrt{151}$$ Solve for x.

Because the width of the box must be positive, use $x = -1 + \sqrt{151}$.

The width is $(-1 + \sqrt{151})$ in. ≈ 11.3 in.

The length is $x + 2$: $(-1 + \sqrt{151} + 2)$ in. or $(1 + \sqrt{151})$ in. ≈ 13.3 in.

The height is 4 in.

---

section 9.3   PRACTICE EXERCISES

1. Identify the equations as linear or quadratic.
   a. $2x - 5 = 3(x + 2) - 1$ Linear
   b. $2x(x - 5) = 3(x + 2) - 1$ Quadratic
   c. $ax^2 + bx + c = 0 \,(a \neq 0)$ Quadratic

2. Identify the equations as linear or quadratic.
   a. $ax + b = 0 \,(a \neq 0)$ Linear
   b. $\frac{1}{2}p - \frac{3}{4}p^2 = 0$ Quadratic
   c. $\frac{1}{2}(p - 3) = 5$ Linear

3. The symbol "$\pm$" is read as ...   Plus or minus

For Exercises 4–23, solve the equations using the square root property.

4. $x^2 = 49$ $x = \pm7$

5. $x^2 = 16$ $x = \pm4$

6. $k^2 - 100 = 0$ $k = \pm10$

7. $m^2 - 64 = 0$ $m = \pm8$

8. $p^2 = -24$ $p = \pm2i\sqrt{6}$

 9. $q^2 = -50$ $q = \pm5i\sqrt{2}$

10. $3w^2 + 9 = 0$ $w = \pm i\sqrt{3}$

11. $4v^2 - 5 = 0$ $v = \frac{\pm\sqrt{5}}{2}$

12. $(a - 5)^2 = 16$ $a = 9, a = 1$

13. $(b + 3)^2 = 1$ $b = -2, b = -4$

14. $(y - 5)^2 = 36$ $y = 11, y = -1$

15. $(y + 4)^2 = 4$ $y = -2, y = -6$

16. $(x - 11)^2 = 5$ $x = 11 \pm \sqrt{5}$

 17. $(z - 2)^2 = 7$ $z = 2 \pm \sqrt{7}$

18. $(a + 1)^2 = -18$ $a = -1 \pm 3i\sqrt{2}$

19. $(b - 1)^2 = -12$ $b = 1 \pm 2i\sqrt{3}$

20. $\left(t - \frac{1}{4}\right)^2 = \frac{7}{16}$ $t = \frac{1}{4} \pm \frac{\sqrt{7}}{4}$

21. $\left(t - \frac{1}{3}\right)^2 = \frac{1}{9}$ $t = \frac{2}{3}, t = 0$

22. $\left(x - \frac{1}{2}\right)^2 + 5 = 20$ $x = \frac{1}{2} \pm \sqrt{15}$

23. $\left(x + \frac{5}{2}\right)^2 - 3 = 18$ $x = -\frac{5}{2} \pm \sqrt{21}$

24. Check the solution $x = -3 + \sqrt{5}$ in the equation $(x + 3)^2 = 5$.   The solution checks.

25. Check the solution $y = 4 + 3i$ in the equation $(y - 4)^2 = -9$.   The solution checks.

26. Check the solution $w = 1 - 4i$ in the equation $(w - 1)^2 = -16$.   The solution checks.

27. Check the solution $p = -5 - \sqrt{7}$ in the equation $(p + 5)^2 = 7$.   The solution checks.

For Exercises 28–33, answer true or false. If a statement is false, explain why.

28. The solution to the equation $x^2 = 64$ is $x = 8$.
   False. −8 is also a solution.

29. The solutions to the equation $x^2 = -25$ are real numbers.   False. The solutions are ±5i.

30. The solutions to the equation $x^2 = \frac{16}{9}$ are rational numbers.   True

31. The solutions to the equation $x^2 = 7$ are irrational numbers.   True

32. There are two solutions to every quadratic equation of the form $x^2 = k$, where k is a real number.   False. If $k = 0$, there is only one solution.

33. The solutions to the equation $x^2 = 100$ are irrational.   False. The solutions are 10 and −10.

34. The area of a circular wading pool is approximately 200 ft². Find the radius to the nearest tenth of a foot.   8.0 ft

35. A sprinkler system covers a circular area of approximately 1700 ft². Find the radius of the area to the nearest tenth of a foot.   23.3 ft

36. An isosceles right triangle has legs of equal length. If the hypotenuse is 10 m long, find the length (in meters) of each leg. Round the answer to the nearest tenth of a meter.   7.1 m

Figure for Exercise 36

37. The diagonal of a square television screen is 24 in. long. Find the length of the sides to the nearest tenth of an inch.   17.0 in.

Figure for Exercise 37

For Exercises 38–49, what constant should be added to the expression to make it a perfect square trinomial?

38. $y^2 + 4y$ 4

39. $w^2 - 6w$ 9

40. $p^2 - 12p$ $\frac{36}{}$

41. $q^2 + 16q$ 64

42. $x^2 - 9x$ $\frac{81}{4}$

43. $a^2 - 5a$ $\frac{25}{4}$

44. $d^2 + \frac{5}{3}d$ $\frac{25}{36}$

45. $t^2 + \frac{1}{4}t$ $\frac{1}{64}$

46. $m^2 - \frac{1}{5}m$ $\frac{1}{100}$

47. $n^2 - \frac{5}{7}n$ $\frac{25}{196}$

48. $u^2 + u$ $\frac{1}{4}$

49. $v^2 - v$ $\frac{1}{4}$

For Exercises 50–63, solve the equations by completing the square and applying the square root property.

50. $x^2 + 4x = 12$
$x = 2, x = -6$

51. $x^2 - 2x = 8$
$x = 4, x = -2$

52. $y^2 + 6y = -5$
$y = -1, y = -5$

53. $t^2 + 10t = 11$
$t = 1, t = -11$

54. $x^2 = 2x + 1$
$x = 1 \pm \sqrt{2}$

55. $x^2 = 6x - 2$
$x = 3 \pm \sqrt{7}$

56. $3x^2 - 6x + 9 = 0$
$x = 1 \pm i\sqrt{2}$

57. $6x^2 + 12x + 6 = 0$
$x = -1$

58. $x^2 + x + 2 = 0$ $x = -\frac{1}{2} \pm \frac{i\sqrt{7}}{2}$

59. $x^2 + 2x + 5 = 0$
$x = -1 \pm 2i$

60. $4x^2 - 6x = -1$ $x = \frac{3}{4} \pm \frac{\sqrt{5}}{4}$ 61. $6x^2 + 3x = -6$ $x = -\frac{1}{4} \pm \frac{i\sqrt{15}}{4}$

62. $3x(x - 1) = -12$ $x = \frac{1}{2} \pm \frac{i\sqrt{15}}{2}$ 63. $2x(x - 2) = -14$
$x = 1 \pm i\sqrt{6}$

The volume of a right circular cylinder is $V = \pi r^2 h$, where r is the radius of the cylinder and h is the height. For Exercises 64–65, find the radius of the cylinder for the given values of volume and height. Round to one decimal place.

64. $V = 1131 \text{ cm}^3, h = 10 \text{ cm}$ $r = 6.0$ cm

65. $V = 37.7 \text{ ft}^3, h = 3 \text{ ft}$ $r = 2.0$ ft

Figure for Exercises 64 and 65

66. Ignoring air resistance, the distance, d, (in feet) that an object drops in t seconds is given by the equation

$$d = 16t^2$$

a. Find the distance traveled in 2 s. 64 ft

b. Find the time required for the object to fall 200 ft. Round to the nearest tenth of a second. 3.5 s

c. Find the time required for an object to fall from the top of the Empire State Building in New York City if the building is 1250 ft. Round to the nearest tenth of a second. 8.8 s

67. Ignoring air resistance, the distance, d (in meters), that an object drops in t seconds is given by the equation

$$d = 4.9t^2$$

a. Find the distance traveled in 5 s. 122.5 m

b. Find the time required for the object to fall 50 m. Round to the nearest tenth of a second. 3.2 s

c. Find the time required for an object to fall from the top of the Canada Trust Tower in Toronto, Canada, if the building is 261 m high. Round to the nearest tenth of a second. 7.3 s

EXPANDING YOUR SKILLS

68. In a rectangle, the length is 4 ft more than the width. The area is 72 ft². Find the exact dimensions of the rectangle. Then use a calculator to approximate the dimensions to the nearest tenth of a foot. The width is $-2 + 2\sqrt{19} \approx 6.7$ ft. The length is $2 + 2\sqrt{19} \approx 10.7$ ft.

69. In a triangle, the base is 4 cm less than twice the height. The area is 60 cm². Find the exact values of the base and height of the triangle. Then use a calculator to approximate the base and height to the nearest tenth of a centimeter. The height is $1 + \sqrt{61} \approx 8.8$ cm. The base is $-2 + 2\sqrt{61} \approx 13.6$ cm.

70. In a right triangle, one leg is 3 m more than the other leg. The hypotenuse is 13 m. Find the exact value of the legs of the triangle. Then use a calculator to approximate the values to the nearest tenth of a meter. The legs are $\frac{3 + \sqrt{329}}{2} \approx 10.6$ m and $\frac{-3 + \sqrt{329}}{2} \approx 7.6$ m.

71. The volume of a rectangular box is 120 in³. The length is 3 in. more than the width, and the height is 2 in. Find the exact dimensions of the box. Then use a calculator to approximate the dimensions to the nearest tenth of an inch.

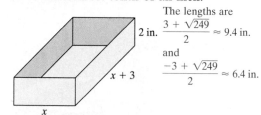

The lengths are $\frac{3 + \sqrt{249}}{2} \approx 9.4$ in. and $\frac{-3 + \sqrt{249}}{2} \approx 6.4$ in.

Figure for Exercise 71

72. The area of a parallelogram is 48 m². If the height is 4 m less than twice the base, find the exact values of the base and height. The base is 6 m, and the height is 8 m.

chapter 9   MIDCHAPTER REVIEW

For Exercises 1–4, evaluate the function values (if possible) for the functions f, g, h, and k. Then write the domain of each function in interval notation.

1. $f(x) = 6x^2 - 3$

   a. $f(0)$ -3

   b. $f(-1)$ 3

   c. $f(1)$ 3

   d. Write the domain of f. $(-\infty, \infty)$

2. $h(x) = \sqrt{x - 3}$

   a. $h(3)$ 0

   b. $h(1)$ Not possible

   c. $h(4)$ 1

   d. Write the domain of h. $[3, \infty)$

3. $g(x) = \dfrac{1}{x + 1}$

   a. $g(0)$ 1

   b. $g(1)$ $\dfrac{1}{2}$

   c. $g(-1)$ Not possible

   d. Write the domain of g.
      $(-\infty, -1) \cup (-1, \infty)$

4. $k(x) = |x - 2|$

   a. $k(0)$ 2

   b. $k(1)$ 1

   c. $k(-1)$ 3

   d. Write the domain of k.
      $(-\infty, \infty)$

For Exercises 5–12, perform the indicated operations. Write the answer in the form $a + bi$.

5. $(8 + 3i) + (2 - i)$
   $10 + 2i$

6. $(8 + 3i) - (2 - i)$
   $6 + 4i$

7. $(2 - i) - (8 + 3i)$
   $-6 - 4i$

8. $(8 + 3i)(2 - i)$
   $19 - 2i$

9. $(8 + 3i)^2$ $55 + 48i$

10. $(2 - i)(2 + i)$ 5

11. $\dfrac{8 + 3i}{2 - i}$ $\dfrac{13}{5} + \dfrac{14}{5}i$

12. $\dfrac{2 - i}{8 + 3i}$ $\dfrac{13}{73} - \dfrac{14}{73}i$

13. Does the graph represent y as a function of x?   Yes

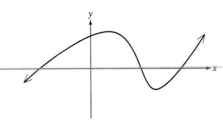

Figure for Exercise 13

14. Is the relation a function?
    $\{(2, 4), (1, 6), (3, 6), (2, -5)\}$ No

15. a. What number should be added to form a perfect square trinomial: $x^2 + 8x$? 16

    b. Solve the equation: $x^2 + 8x - 3 = 0$.
       $x = -4 \pm \sqrt{19}$

16. a. What number should be added to form a perfect square trinomial: $x^2 - \frac{2}{3}x$? $\dfrac{1}{9}$

    b. Solve the equation: $0 = -9x^2 + 6x - 1$. $x = \dfrac{1}{3}$

17. a. What number should be added to form a perfect square trinomial: $x^2 + 2x$? 1

    b. Solve the equation: $2x^2 + 4x + 10 = 0$.
       $x = -1 \pm 2i$

---

section

9.4   QUADRATIC FORMULA

Concepts

1. Derivation of the Quadratic Formula

2. Solving Quadratic Equations Using the Quadratic Formula

3. The Discriminant

4. Review of the Methods for Solving a Quadratic Equation

1.   Derivation of the Quadratic Formula

If we solve a general quadratic equation $ax^2 + bx + c = 0$ ($a \neq 0$) by completing the square and using the square root property, the result is a formula that gives the solutions for x in terms of a, b, and c.

$$ax^2 + bx + c = 0$$
Begin with a quadratic equation in standard form.

$$x^2 + \frac{b}{a}x + \frac{c}{a} = 0$$
Divide by the leading coefficient.

$$x^2 + \frac{b}{a}x = -\frac{c}{a}$$
Isolate the terms containing x.

 See Additional Answers Appendix   Writing   Translating Expression   Geometry   Scientific Calculator   Video

$$x^2 + \frac{b}{a}x + \left(\frac{1}{2} \cdot \frac{b}{a}\right)^2 = \left(\frac{1}{2} \cdot \frac{b}{a}\right)^2 - \frac{c}{a}$$

Add the square of $\frac{1}{2}$ the linear-term coefficient to both sides of the equation.

$$\left(x + \frac{b}{2a}\right)^2 = \frac{b^2}{4a^2} - \frac{c}{a}$$

Factor the left side as a perfect square.

$$\left(x + \frac{b}{2a}\right)^2 = \frac{b^2 - 4ac}{4a^2}$$

Combine fractions on the right side by getting a common denominator.

$$x + \frac{b}{2a} = \pm\sqrt{\frac{b^2 - 4ac}{4a^2}}$$

Apply the square root property.

$$x + \frac{b}{2a} = \frac{\pm\sqrt{b^2 - 4ac}}{2a}$$

Simplify the denominator.

$$x = -\frac{b}{2a} \pm \frac{\sqrt{b^2 - 4ac}}{2a}$$

Subtract $\frac{b}{2a}$ from both sides.

$$= \frac{-b \pm \sqrt{b^2 - 4ac}}{2a}$$

Combine fractions.

Quadratic Formula

For any quadratic equation of the form $ax^2 + bx + c = 0, (a \neq 0)$ the solutions are

$$x = \frac{-b \pm \sqrt{b^2 - 4ac}}{2a}$$

2. Solving Quadratic Equations Using the Quadratic Formula

example 1   **Solving Quadratic Equations Using the Quadratic Formula**

Solve the quadratic equation using the quadratic formula: $3x^2 - 7x = -2$

Solution:

$$3x^2 - 7x = -2$$

$$3x^2 - 7x + 2 = 0$$

Write the equation in the form $ax^2 + bx + c = 0$.

$$a = 3, b = -7, c = 2$$

Identify a, b, and c.

$$x = \frac{-(-7) \pm \sqrt{(-7)^2 - 4(3)(2)}}{2(3)}$$ Apply the quadratic formula.

$$x = \frac{7 \pm \sqrt{49 - 24}}{6}$$ Simplify.

$$= \frac{7 \pm \sqrt{25}}{6}$$

$$= \frac{7 \pm 5}{6}$$

There are two rational solutions.

$$x = \frac{7 + 5}{6} = \frac{12}{6} = 2 \qquad x = 2$$

$$x = \frac{7 - 5}{6} = \frac{2}{6} = \frac{1}{3} \qquad x = \frac{1}{3}$$

Check: $x = 2$ Check: $x = \frac{1}{3}$

$$3x^2 - 7x = -2 \qquad\qquad 3x^2 - 7x = -2$$

$$3(2)^2 - 7(2) \overset{?}{=} -2 \qquad\qquad 3\left(\frac{1}{3}\right)^2 - 7\left(\frac{1}{3}\right) \overset{?}{=} -2$$

$$3(4) - 14 \overset{?}{=} -2 \qquad\qquad 3\left(\frac{1}{9}\right) - \frac{7}{3} \overset{?}{=} -2$$

$$12 - 14 \overset{?}{=} -2 \qquad\qquad \frac{1}{3} - \frac{7}{3} \overset{?}{=} -2$$

$$-2 = -2 \; \checkmark \qquad\qquad -\frac{6}{3} \overset{?}{=} -2$$

$$-2 = -2 \; \checkmark$$

Tip: Because the solutions to the equation $3x^2 - 7x = -2$ are rational numbers, the equation could have been solved by factoring and using the zero product rule.

$$3x^2 - 7x = -2$$

$$3x^2 - 7x + 2 = 0$$

$$(3x - 1)(x - 2) = 0$$

$$3x - 1 = 0 \quad \text{or} \quad x - 2 = 0$$

$$x = \frac{1}{3} \quad \text{or} \quad x = 2$$

example 2

Solving a Quadratic Equation with the Quadratic Formula

Solve the quadratic equation using the quadratic formula: $x(x + 7) + 4 = 0$. Find the exact solution(s), then approximate the solution(s) to three decimal places.

Solution:

$x(x + 7) + 4 = 0$

$x^2 + 7x + 4 = 0$ Write the equation in the form $ax^2 + bx + c = 0$.

$a = 1, b = 7, c = 4$ Identify a, b, and c.

$x = \dfrac{-(7) \pm \sqrt{(7)^2 - 4(1)(4)}}{2(1)}$ Apply the quadratic formula.

$= \dfrac{-7 \pm \sqrt{49 - 16}}{2}$ Simplify.

$= \dfrac{-7 \pm \sqrt{33}}{2}$ The solutions are irrational numbers.

The solutions can be written as

$$x = \frac{-7 + \sqrt{33}}{2} \approx -0.628$$

and

$$x = \frac{-7 - \sqrt{33}}{2} \approx -6.372$$

example 3

Solving a Quadratic Equation Using the Quadratic Formula

Solve the quadratic equation using the quadratic formula: $\frac{1}{4}w^2 - \frac{1}{2}w + \frac{5}{4} = 0$

Classroom Activity 9.4A

Solution:

$\dfrac{1}{4}w^2 - \dfrac{1}{2}w + \dfrac{5}{4} = 0$ To simplify the equation, multiply both sides by 4.

$4\left(\dfrac{1}{4}w^2 - \dfrac{1}{2}w + \dfrac{5}{4}\right) = 4(0)$ Clear fractions.

$w^2 - 2w + 5 = 0$ The equation is in the form $ax^2 + bx + c = 0$.

$a = 1, b = -2, c = 5$ Identify a, b, and c.

$x = \dfrac{-(-2) \pm \sqrt{(-2)^2 - 4(1)(5)}}{2(1)}$ Apply the quadratic formula.

$= \dfrac{2 \pm \sqrt{4 - 20}}{2}$ Simplify.

$$= \frac{2 \pm \sqrt{-16}}{2}$$

$$= \frac{2 \pm 4i}{2}$$ The solutions are imaginary numbers.

$$= \frac{\cancel{2}(1 \pm 2i)}{\cancel{2}}$$ Factor the numerator and reduce.

$$= 1 \pm 2i$$

There are two imaginary solutions: $x = 1 + 2i$ and $x = 1 - 2i$.

3. The Discriminant

From Examples 1–3, we see that the solutions to a quadratic equation may be rational numbers, irrational numbers, or imaginary numbers. The *number* and *type* of solution can be determined by noting the value of the square root term in the quadratic formula. The radicand of the square root, $b^2 - 4ac$, is called the discriminant.

Using the Discriminant to Determine the Number and Type of Solutions of a Quadratic Equation

Consider the equation, $ax^2 + bx + c = 0$, where a, b, and c are rational numbers and $a \neq 0$. The expression $b^2 - 4ac$, is called the **discriminant**. Furthermore,

1. If $b^2 - 4ac > 0$ then there will be two real solutions. Moreover,
   a. If $b^2 - 4ac$ is a perfect square, the solutions will be rational numbers.
   b. If $b^2 - 4ac$ is not a perfect square, the solutions will be irrational numbers.
2. If $b^2 - 4ac < 0$ then there will be two imaginary solutions.
3. If $b^2 - 4ac = 0$ then there will be one rational solution.

example 4

Using the Discriminant

Use the discriminant to determine the type and number of solutions for each equation.

a. $3x^2 - 4x + 7 = 0$ b. $2x^2 + x = 5$
c. $2(x^2 + 10) = 13x$ d. $0.4x^2 - 1.2x + 0.9 = 0$

Classroom Activity 9.4B

Solution:

For each equation, first write the equation in standard form, $ax^2 + bx + c = 0$. Then determine the value of the discriminant.

Equation	Discriminant	Solution Type and Number
a. $3x^2 - 4x + 7 = 0$	$b^2 - 4ac$	Because $-68 < 0$, there will be two imaginary solutions.
$a = 3, b = -4, c = 7$	$(-4)^2 - 4(3)(7)$	
	$16 - 84$	
	-68	

Equation	Discriminant	Solution Type and Number
b. $2x^2 + x = 5$		
$2x^2 + x - 5 = 0$	$b^2 - 4ac$	Because $41 > 0$ but 41 is
$a = 2, b = 1, c = -5$	$(1)^2 - 4(2)(-5)$	not a perfect square, there will be two
	$1 + 40$	irrational solutions.
	41	
c. $2(x^2 + 10) = 13x$		
$2x^2 + 20 = 13x$		
$2x^2 - 13x + 20 = 0$	$b^2 - 4ac$	Because $9 > 0$ and 9 is a
$a = 2, b = -13, c = 20$	$(-13)^2 - 4(2)(20)$	perfect square, there will be two rational solutions.
	$169 - 160$	
	9	
d. $0.4x^2 - 1.2x + 0.9 = 0$		
(clear decimals first)		
$4x^2 - 12x + 9 = 0$	$b^2 - 4ac$	Because the discriminant
$a = 4, b = -12, c = 9$	$(-12)^2 - 4(4)(9)$	equals 0, there will be only one rational
	$144 - 144$	solution.
	0	

4. Review of the Methods for Solving a Quadratic Equation

Three methods have been presented for solving quadratic equations.

Methods for Solving a Quadratic Equation

- Factor and use the zero product rule (Section 4.6)
- Use the square root property. Complete the square if necessary (Section 9.3)
- Use the quadratic formula (Section 9.4)

Using the zero product rule is the simplest method, but it only works if you can factor the equation. The square root property and the quadratic formula can be used to solve any quadratic equation. Before solving a quadratic equation, take a minute to analyze it first. Each problem must be evaluated individually before choosing the most efficient method to find its solutions.

example 5   **Solving Quadratic Equations Using Any Method**

Solve the quadratic equations using any method.

a. $(x + 1)^2 = -4$ b. $t^2 - t - 30 = 0$ c. $2x^2 + 5x + 1 = 0$

Classroom Activity 9.4C

Solution:

a. $(x + 1)^2 = -4$ Because the equation is in the form: $(x + h)^2 = k$, the square root property can be applied easily.

$\qquad x + 1 = \pm\sqrt{-4}$ Apply the square root property.

$\qquad x + 1 = \pm 2i$ Simplify the radical.

$\qquad\qquad x = -1 \pm 2i$ Isolate x.

$\qquad x = -1 + 2i \quad$ or $\quad x = -1 - 2i$

b. $t^2 - t - 30 = 0$ This equation factors easily.

$\qquad (t - 6)(t + 5) = 0$ Factor and apply the zero product rule.

$\qquad t = 6 \quad$ or $\quad t = -5$

c. $2x^2 + 5x + 1 = 0$ The equation does not factor easily. Because it is already in the form $ax^2 + bx + c = 0$, use the quadratic formula.

$\qquad a = 2, b = 5, c = 1$ Identify a, b, and c.

$$x = \frac{-(5) \pm \sqrt{(5)^2 - 4(2)(1)}}{2(2)}$$ Apply the quadratic formula.

$$x = \frac{-5 \pm \sqrt{25 - 8}}{4}$$ Simplify.

$$x = \frac{-5 \pm \sqrt{17}}{4}$$

$$x = \frac{-5 + \sqrt{17}}{4} \quad \text{or} \quad x = \frac{-5 - \sqrt{17}}{4}$$

Calculator Connections

A calculator can be used to obtain decimal approximations for the irrational solutions of a quadratic equation. From Example 2, the solutions to the equation $x(x + 7) + 4 = 0$ are

$$x = \frac{-7 + \sqrt{33}}{2} \quad \text{and} \quad x = \frac{-7 - \sqrt{33}}{2}$$

Scientific Calculator

Enter: 7 +/− + 3 3 √ = ÷ 2 =   **Result:** −0.627718677

Enter: 7 +/− − 3 3 √ = ÷ 2 =   **Result:** −6.372281323

Graphing Calculator

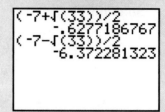

Tip: A graphing calculator can be used to apply the quadratic formula directly. The result of the calculation is stored in memory in a variable such as *Ans*. The solution can be checked by substituting the value of *Ans* into the original equation.

Consider the equation $3x^2 + 6x - 2 = 0$. The solutions are

$$x = \frac{-(6) + \sqrt{(6)^2 - 4(3)(-2)}}{2(3)} \approx 0.2909944487$$

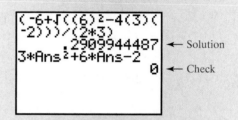

and

$$x = \frac{-(6) - \sqrt{(6)^2 - 4(3)(-2)}}{2(3)} \approx -2.290994449$$

Calculator Exercises

Use a calculator to obtain a decimal approximation of each expression.

1. $\dfrac{-5 + \sqrt{17}}{4}$ and $\dfrac{-5 - \sqrt{17}}{4}$ (Example 5) ❖

2. $\dfrac{-40 + \sqrt{1920}}{-32}$ and $\dfrac{-40 - \sqrt{1920}}{-32}$ ❖

3. $\dfrac{-17 - \sqrt{(17)^2 - 4(4)(-3)}}{2(4)}$ ❖   4. $\dfrac{5.2 + \sqrt{(5.2)^2 - 4(2.1)(1.7)}}{2(2.1)}$ ❖

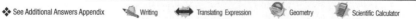

section 9.4   PRACTICE EXERCISES

1. State the quadratic formula from memory. ❖

2. What is $b^2 - 4ac$ called?   The discriminant

For Exercises 3–8, write the equations in the form $ax^2 + bx + c = 0$. Then identify the values of $a, b,$ and c.

3. $2x^2 - x = 5$ ❖

4. $5(x^2 + 2) = -3x$ ❖

5. $-3x(x - 4) = -2x$ ❖

6. $x(x - 2) = 3(x + 1)$ ❖

7. $x^2 - 9 = 0$ ❖

8. $x^2 + 25 = 0$ ❖

For Exercises 9–16, find the discriminant and use the discriminant to determine the number and type of solution to each equation. There may be two rational solutions, two irrational solutions, two imaginary solutions, or one rational solution.

9. $x^2 + 16x + 64 = 0$ ❖

10. $x^2 - 10x + 25 = 0$ ❖

11. $6y^2 - y - 2 = 0$ ❖

12. $3m^2 + 5m - 2 = 0$ ❖

13. $2p^2 + 5p = 1$ ❖

14. $9a^2 + 3a = 1$ ❖

15. $3b^2 = b - 2$ ❖

16. $5q^2 = 2q - 3$ ❖

For Exercises 17–38, solve the equations using the quadratic formula.

17. $t^2 + 16t + 64 = 0$
    $t = -8$

18. $y^2 - 10y + 25 = 0$
    $y = 5$

19. $6k^2 - k - 2 = 0$ ❖

20. $3n^2 + 5n - 2 = 0$ ❖

21. $2a^2 + 5a = 1$ ❖

22. $9b^2 + 3b = 1$ ❖

23. $3c^2 = c - 2$ ❖

24. $5p^2 = 2p - 3$ ❖

25. $k^2 + 2k + 5 = 0$ ❖

26. $p^2 - 7p + 3 = 0$ ❖

27. $-4y^2 - y + 1 = 0$ ❖

28. $-5z^2 - 3z + 4 = 0$ ❖

29. $2x(x + 1) = 3 - x$ ❖

30. $3m(m - 2) = -m + 1$ ❖

31. $0.2y^2 = -1.5y - 1$ ❖

32. $0.2t^2 = t + 0.5$ ❖

33. $-2.5t(t - 4) = 1.5$ ❖

34. $-1.6p(p - 2) = 2.1$ ❖

35. $\frac{2}{3}x^2 + \frac{4}{9}x = \frac{1}{3}$ ❖

36. $\frac{1}{2}x^2 + \frac{1}{6}x = 1$ ❖

37. $(m - 3)(m + 2) = -9$ ❖

38. $(h - 6)(h - 1) = -12$ ❖

39. In a rectangle, the length is 1 m less than twice the width and the area is 100 m². Find the exact dimensions of the rectangle. Then use a calculator to approximate the dimensions to the nearest tenth of a meter. ❖

40. In a triangle, the height is 2 cm more than the base. The area is 72 cm². Find the exact values of the base and height of the triangle. Then use a calculator to approximate the base and height to the nearest tenth of a centimeter. ❖

41. In a right triangle, one leg is 2 ft less than the other leg. The hypotenuse is 12 ft. Find the exact lengths of the legs of the triangle. Then use a calculator to approximate the legs to the nearest tenth of a foot. ❖

42. The volume of a rectangular storage area is 240 ft³. The length is 2 ft more than the width. The height is 6 ft. Find the exact dimensions of the storage area. Then use a calculator to approximate the dimensions to the nearest tenth of an inch. ❖

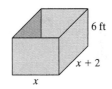

6 ft

$x + 2$

x

Figure for Exercise 42

For Exercises 43–52, choose any method to solve the quadratic equations.

43. $16x^2 - 9 = 0$
    $x = \frac{3}{4}, x = -\frac{3}{4}$

44. $\frac{1}{4}x^2 + 5x + 13 = 0$
    $x = -10 \pm 4\sqrt{3}$

45. $(x - 5)^2 = 21$
    $x = 5 \pm \sqrt{21}$

46. $2x^2 + x + 5 = 0$
    $x = \frac{-1 + i\sqrt{39}}{4}$

47. $\frac{1}{9}x^2 + \frac{8}{3}x + 11 = 0$
    $x = -12 \pm 3\sqrt{5}$

48. $7x^2 = 12x$
    $x = 0, x = \frac{12}{7}$

49. $2x^2 - 6x - 3 = 0$
    $x = \frac{3 \pm \sqrt{15}}{2}$

50. $4(x + 1)^2 = -15$
    $x = -1 \pm \frac{i\sqrt{15}}{2}$

51. $9x^2 = 11x$
    $x = 0, x = \frac{11}{9}$

52. $25x^2 - 4 = 0$
    $x = \frac{2}{5}, -\frac{2}{5}$

❖ See Additional Answers Appendix   🖉 Writing   ⬅ Translating Expression   🌐 Geometry   📋 Scientific Calculator   📺 Video

section

9.5   CONNECTIONS TO GRAPHING: QUADRATIC FUNCTIONS

1. Definition of a Quadratic Function

In Section 3.8 we graphed several nonlinear equations in two variables. In particular, we graphed $y = x^2$ (Table 9-2 and Figure 9-9).

Table 9-2	
x	y
-3	9
-2	4
-1	1
0	0
1	1
2	4
3	9

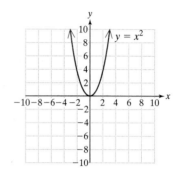

Figure 9-9

The equation $y = x^2$ is a special type of function called a quadratic function, and its graph is in the shape of a **parabola**.

Definition of a Quadratic Function

Let a, b, and c represent real numbers such that $a \neq 0$. Then a function in the form, $y = ax^2 + bx + c$ is called a **quadratic function**.

The graph of a quadratic function is a parabola that opens upward or downward. The leading coefficient, a, determines the direction of the parabola. For the quadratic function defined by $y = ax^2 + bx + c$.

If $a > 0$, the parabola opens *upward*. For example: $y = x^2$.

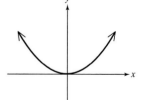

If $a < 0$, the parabola opens *downward*. For example: $y = -x^2$.

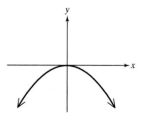

If a parabola opens upward, the **vertex** is the lowest point on the graph. If a parabola opens downward, the **vertex** is the highest point on the graph. For a quadratic function, the **axis of symmetry** is the vertical line that passes through the vertex. Notice that the graph of the parabola is its own mirror image to the left and right of the axis of symmetry.

$y = 0.5x^2 + 2x + 3$
$a > 0$
Vertex $(-2, 1)$
Axis of symmetry: $x = -2$

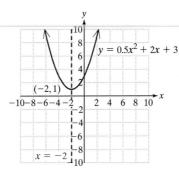

$y = x^2 - 6x + 9$
$a > 0$
Vertex $(3, 0)$
Axis of symmetry: $x = 3$

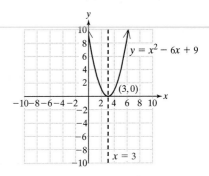

$y = -x^2 + 4x$
$a < 0$
Vertex $(2, 4)$
Axis of symmetry: $x = 2$

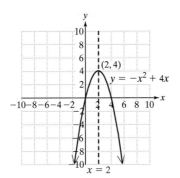

$y = -2x^2 - 4$
$a < 0$
Vertex $(0, -4)$
Axis of symmetry: $x = 0$

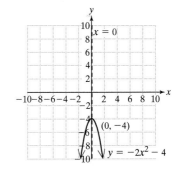

Classroom Activity 9.5A

2. Vertex of a Quadratic Function

Quadratic functions arise in many applications of mathematics and applied sciences. For example, an object thrown through the air follows a parabolic path. The mirror inside a reflecting telescope is parabolic in shape. In applications, it is often advantageous to analyze the graph of a parabola. In particular, we want to find the location of the x- and y-intercepts and the vertex.

To find the vertex of a quadratic function defined by $y = ax^2 + bx + c$ $(a \neq 0)$, we use the following steps:

Vertex of a Quadratic Function

1. The x-coordinate of the vertex of the quadratic function $y = ax^2 + bx + c$ $(a \neq 0)$ is given by

$$x = \frac{-b}{2a}$$

2. To find the corresponding y-coordinate of the vertex, substitute the value of the x-coordinate found in Step 1 and solve for y.

example 1

Analyzing a Quadratic Function

Given the function defined by $y = -x^2 + 4x - 3$.

a.   Determine whether the function opens upward or downward.
b.   Find the vertex of the parabola.
c.   Find the x-intercept(s).
d.   Find the y-intercept.
e.   Sketch the graph of the function.
f.   Write the domain of the function in interval notation.
g.   Write the range of the function in interval notation.

Classroom Activity 9.5B

Solution:

a.   The function $y = -x^2 + 4x - 3$ is written in the form $y = ax^2 + bx + c$, where $a = -1, b = 4$, and $c = -3$. Because the value of a is negative, the parabola opens downward.

b.   The x-coordinate of the vertex is given by $x = \dfrac{-b}{2a}$

$$x = \frac{-b}{2a} = \frac{-(4)}{2(-1)} \qquad \text{Substitute } b = 4 \text{ and } a = -1.$$

$$= \frac{-4}{-2} \qquad \text{Simplify.}$$

$$= 2$$

The y-coordinate of the vertex is found by substituting $x = 2$ into the equation and solving for y.

$$y = -x^2 + 4x - 3$$

$$= -(2)^2 + 4(2) - 3 \qquad \text{Substitute } x = 2.$$

$$= -4 + 8 - 3$$

$$= 1$$

The vertex is at $(2, 1)$.   Because the parabola opens downward, the vertex is the maximum point on the graph of the parabola.

c. To find the x-intercept(s), substitute $y = 0$ and solve for x.

$y = -x^2 + 4x - 3$

$0 = -x^2 + 4x - 3$ Substitute $y = 0$. The resulting equation is quadratic.

$0 = -1(x^2 - 4x + 3)$ Factor out -1.

$0 = -1(x - 3)(x - 1)$ Factor the trinomial.

$x - 3 = 0$ or $x - 1 = 0$ Apply the zero product rule.

$x = 3$ or $x = 1$

The x-intercepts are $(3, 0)$ and $(1, 0)$.

d. To find the y-intercept, substitute $x = 0$ and solve for y.

$y = -x^2 + 4x - 3$

$\quad = -(0)^2 + 4(0) - 3$ Substitute $x = 0$.

$\quad = -3$

The y-intercept is $(0, -3)$.

e. Using the results of parts (a)–(d), we have a parabola that opens downward with vertex at $(2, 1)$, x-intercepts at $(3, 0)$ and $(1, 0)$, and y-intercept at $(0, -3)$ (Figure 9-10).

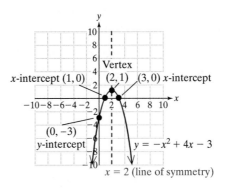

Figure 9-10

f. Because any real number, x, when substituted into the equation $y = -x^2 + 4x - 3$ produces a real number, the domain is all real numbers.

Domain: $(-\infty, \infty)$

g. From the graph, we see that the vertex is the maximum point on the graph. Therefore, the maximum y-value is $y = 1$. The range is restricted to $y \le 1$.

Range: $(-\infty, 1]$

3. Graphing a Quadratic Function

Example 1 illustrates a process to sketch a quadratic function by finding the location of the defining characteristics of the function. These include the vertex and the x- and y-intercepts. Furthermore, notice that the parabola defining the graph of a quadratic function is symmetric with respect to the axis of symmetry.

   To analyze the graph of a parabola, we recommend the following guidelines.

Graphing a Quadratic Function

Given a quadratic function defined by $y = ax^2 + bx + c$ $(a \neq 0)$, consider the following guidelines to graph the function.

1. Determine whether the function opens upward or downward.

    - If $a > 0$, the parabola opens upward.
    - If $a < 0$, the parabola opens downward.

2. Find the vertex.

    - The x-coordinate is given by $x = \dfrac{-b}{2a}$

    - To find the y-coordinate, substitute the x-coordinate of the vertex into the equation and solve for y.

3. Find the x-intercept(s) by substituting $y = 0$ and solving the equation for x.

    - *Note*: If the solutions to the equation in Step 3 are imaginary numbers, then the function has no x-intercepts.

4. Find the y-intercept by substituting $x = 0$ and solving the equation for y.

5. Plot the vertex and x- and y-intercepts. If necessary, find and plot additional points near the vertex. Then use the symmetry of the parabola to sketch the curve through the points.

example 2   Graphing a Quadratic Function

Graph the function defined by $y = x^2 - 6x + 9$.

Solution:

1. The function $y = x^2 - 6x + 9$ is written in the form $y = ax^2 + bx + c$, where $a = 1, b = -6$, and $c = 9$. Because the value of a is positive, the parabola opens upward.

2. The x-coordinate of the vertex is given by

$$x = \frac{-b}{2a} = \frac{-(-6)}{2(1)} = 3$$

Substituting $x = 3$ into the equation, we have

$$y = (3)^2 - 6(3) + 9$$
$$= 9 - 18 + 9$$
$$= 0$$

The vertex is at the point $(3, 0)$.

3. To find the x-intercept(s), substitute $y = 0$ and solve for x.

    Substitute $y = 0$

    $$y = x^2 - 6x + 9 \longrightarrow 0 = x^2 - 6x + 9$$

    $$0 = (x - 3)^2 \qquad \text{Factor.}$$

    $$x = 3 \qquad \text{Apply the zero product rule.}$$

    The x-intercept is $(3, 0)$.

4. To find the y-intercept, substitute $x = 0$ and solve for y.

    Substitute $x = 0$

    $$y = x^2 - 6x + 9 \longrightarrow y = (0)^2 - 6(0) + 9$$
    $$= 9$$

    The y-intercept is $(0, 9)$.

Tip: Using the symmetry of the parabola, we know that the points to the right of the vertex must mirror the points to the left of the vertex.

5. Sketch the parabola through the x- and y-intercepts and vertex (Figure 9-11).

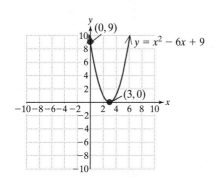

Figure 9-11

---

example 3

Graphing a Quadratic Function

Graph the function defined by $y = -x^2 - 4$.

Classroom Activity 9.5C

Solution:

1. The function $y = -x^2 - 4$ is written in the form $y = ax^2 + bx + c$, where $a = -1, b = 0$, and $c = -4$. Because the value of a is negative, the parabola opens downward.

2. The x-coordinate of the vertex is given by

    $$x = \frac{-b}{2a} = \frac{-(0)}{2(-1)} = 0$$

Substituting $x = 0$ into the equation, we have

$$y = -(0)^2 - 4$$
$$= -4$$

The vertex is at the point $(0, -4)$.

3. Because the vertex is below the x-axis and the parabola opens downward, the function cannot have an x-intercept.

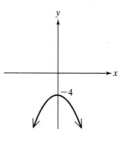

4. The vertex is at $(0, -4)$. This is also the y-intercept.

5. Sketch the parabola through the y-intercept and vertex (Figure 9-12).

To verify the proper shape of the graph, find additional points to the right or left of the vertex and use the symmetry of the parabola to sketch the curve.

x	y
1	−5
2	−8
3	−13

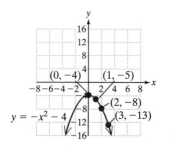

Figure 9-12

Tip: Substituting $y = 0$ into the equation $y = -x^2 - 4$ results in an equation with no real solutions. Therefore, the function $y = -x^2 - 4$ has no x-intercepts.

$$y = -x^2 - 4$$

$$0 = -x^2 - 4$$

$$x^2 = -4$$

$$x = \pm\sqrt{-4}$$

$$x = \pm 2i \qquad \text{Not a real number}$$

4. Applications of Quadratic Functions

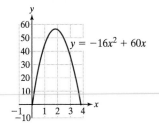

Figure 9-13

Classroom Activity 9.5D

Using a Quadratic Function in an Application

A golfer hits a ball at an angle of 30°. The height of the ball, y, (in feet) can be represented by

$$y = -16x^2 + 60x$$

where x is the time in seconds after the ball was hit (Figure 9-13).

Find the maximum height of the ball. In how many seconds will the ball reach its maximum height?

Solution:

The function is written in the form $y = ax^2 + bx + c$ where $a = -16$, $b = 60$, and $c = 0$. Because a is negative, the function opens downward. Therefore, the maximum height of the ball occurs at the vertex of the parabola.

The x-coordinate of the vertex is given by

$$x = \frac{-b}{2a} = \frac{-(60)}{2(-16)} = \frac{-60}{-32} = \frac{15}{8}$$

Substituting $x = \frac{15}{8}$ into the equation, we have

$$y = -16\left(\frac{15}{8}\right)^2 + 60\left(\frac{15}{8}\right)$$

$$= -16\left(\frac{225}{64}\right) + \frac{900}{8}$$

$$= -\frac{225}{4} + \frac{450}{4}$$

$$= \frac{225}{4}$$

The vertex is at the point $\left(\frac{15}{8}, \frac{225}{4}\right)$ or equivalently $(1.875, 56.25)$.

The ball reaches its maximum height of 56.25 ft after 1.875 s.

Calculator Connections

Some graphing calculators have *Minimum* and *Maximum* features that enable the user to approximate the minimum and maximum values of a function. Otherwise, *Zoom* and *Trace* can be used.

For example, the maximum value of the function from Example 4, $y = -16x^2 + 60x$, can be found using the *Maximum* feature.

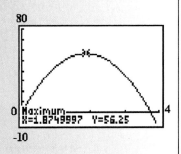

The minimum value of the function from Example 2, $y = x^2 - 6x + 9$, can be found using the *Minimum* feature.

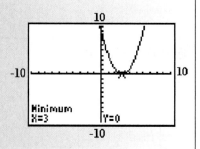

Calculator Exercises

Without using a calculator, find the location of the vertex for each function. Then verify the answer using a graphing calculator.

1. $y = x^2 + 4x + 7$ ❖

2. $y = x^2 - 20x + 105$ ❖

3. $y = -x^2 - 3x - 4.85$ ❖

4. $y = -x^2 + 3.5x - 0.5625$ ❖

5. $y = 2x^2 - 10x + \dfrac{25}{2}$ ❖

6. $y = 3x^2 + 16x + \dfrac{64}{3}$ ❖

7. Use a graphing calculator to graph the functions on the same screen. Describe the relationship between the graph of $y = x^2$ and the graphs in parts (b) and (c). ❖

    a. $y = x^2$
    b. $y = x^2 + 4$
    c. $y = x^2 - 3$

8. Use a graphing calculator to graph the functions on the same screen. Describe the relationship between the graph of $y = x^2$ and the graphs in parts (b) and (c). ❖

    a. $y = x^2$
    b. $y = (x - 3)^2$
    c. $y = (x + 2)^2$

9. Use a graphing calculator to graph the functions on the same screen. Describe the relationship between the graph of $y = x^2$ and the graphs in parts (b) and (c). ❖

    a. $y = x^2$
    b. $y = 2x^2$
    c. $y = \frac{1}{2}x^2$

10. Use a graphing calculator to graph the functions on the same screen. Describe the relationship between the graph of $y = x^2$ and the graphs in parts (b) and (c). ❖

    a. $y = x^2$
    b. $y = -2x^2$
    c. $y = -\frac{1}{2}x^2$

❖ See Additional Answers Appendix   Writing   Translating Expression   Geometry   Scientific Calculator   Video

section 9.5   PRACTICE EXERCISES

For Exercises 1–8, solve the quadratic equations using any one of the following methods: factoring, the square root property, or the quadratic formula.   9.3, 9.4

1. $3(y^2 + 1) = 10y$
   $y = \frac{1}{3}, y = 3$

2. $3 + a(a + 2) = 18$
   $a = -5, a = 3$

3. $4t^2 - 7 = 0$
   $t = \pm\frac{\sqrt{7}}{2}$

4. $9p^2 = 5$
   $p = \pm\frac{\sqrt{5}}{3}$

5. $(b + 1)^2 = 6$
   $b = -1 \pm \sqrt{6}$

6. $(x - 2)^2 = 8$
   $x = 2 \pm 2\sqrt{2}$

7. $w^2 - w + 5 = 0$
   $w = \frac{1 \pm i\sqrt{19}}{2}$

8. $z^2 + 2z + 7 = 0$
   $z = -1 \pm i\sqrt{6}$

For Exercises 9–18, identify the equations as linear, quadratic, or neither.

9. $y = -8x + 3$ Linear

10. $y = 5x - 12$ Linear

11. $y = 4x^2 - 8x + 22$
    Quadratic

12. $y = x^2 + 10x - 3$
    Quadratic

13. $y = -5x^3 - 8x + 14$ Neither

14. $y = -3x^4 + 7x - 11$ Neither

15. $y = 15x$ Linear

16. $y = -9x$ Linear

17. $y = -21x^2$ Quadratic

18. $y = 3x^2$ Quadratic

19. How do you determine whether the graph of a function $y = ax^2 + bx + c$ ($a \neq 0$) opens upward or downward?   If $a > 0$ the graph opens upward; if $a < 0$ the graph opens downward.

For Exercises 20–25, identify a and determine if the parabola opens upward or downward.

20. $y = -5x^2 - x + 10$
    $a = -5$; downward

21. $y = -7x^2 + 3x - 1$
    $a = -7$; downward

22. $y = x^2 - 15$
    $a = 1$; upward

23. $y = 2x^2 + 23$
    $a = 2$; upward

24. $y = -3x^2 + x - 18$ $a = -3$; downward

25. $y = -10x^2 - 6x - 20$ $a = -10$; downward

26. How do you find the vertex of a parabola?   ❖

For Exercises 27–34, find the vertex of the parabola, and write the equation of the axis of symmetry for each parabola.

27. $y = 2x^2 + 4x - 6$
    Vertex: $(-1, -8)$; axis of symmetry: $x = -1$

28. $y = x^2 - 4x - 4$
    Vertex: $(2, -8)$; axis of symmetry: $x = 2$

29. $y = -x^2 + 2x - 5$
    Vertex: $(1, -4)$; axis of symmetry: $x = 1$

30. $y = 2x^2 - 4x - 6$
    Vertex: $(1, -8)$; axis of symmetry: $x = 1$

31. $y = x^2 - 2x + 3$
    Vertex: $(1, 2)$; axis of symmetry: $x = 1$

32. $y = -x^2 + 4x - 2$
    Vertex: $(2, 2)$; axis of symmetry: $x = 2$

33. $y = x^2 - 4$
    Vertex: $(0, -4)$; axis of symmetry: $x = 0$

34. $y = x^2 - 1$
    Vertex: $(0, -1)$; axis of symmetry: $x = 0$

For Exercises 35–38, find the x- and y-intercepts of the function. Then match the function with the graph.

35. $y = x^2 - 7$ ❖

36. $y = x^2 - 9$ ❖

37. $y = (x + 3)^2 - 4$ ❖

38. $y = (x - 2)^2 - 1$ ❖

a.
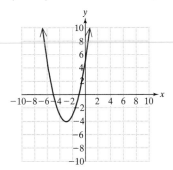

b.
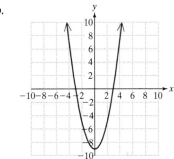

c.
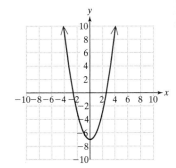

d.
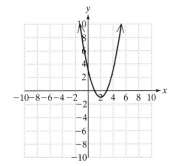

 ❖ See Additional Answers Appendix   Writing   Translating Expression   Geometry   Scientific Calculator   Video

39. What is the y-intercept of the function defined by $y = ax^2 + bx + c$? $(0, c)$

40. What is the y-intercept of the function defined by $y = 4x^2 - 3x + 5$? $(0, 5)$

41. What is the domain of the quadratic function defined by $y = x^2 - x - 12$? $(-\infty, \infty)$

42. What is the domain of the quadratic function defined by $y = x^2 + 10x + 9$? $(-\infty, \infty)$

For Exercises 43–54,

    a. Determine whether the graph of the parabola opens upward or downward.

    b. Find the vertex.

    c. Find the x-intercept(s) if possible.

    d. Find the y-intercept.

    e. Sketch the function.

    f. Identify the domain of the function.

    g. Identify the range of the function.

43. $y = x^2 - 9$ ❖   44. $y = x^2 - 4$ ❖

45. $y = x^2 - 2x - 8$ ❖   46. $y = x^2 + 2x - 24$ ❖

47. $y = -x^2 + 6x - 9$ ❖

48. $y = -x^2 + 10x - 25$ ❖

49. $y = -x^2 + 8x - 15$ ❖   50. $y = -x^2 - 4x + 5$ ❖

51. $y = x^2 + 6x + 10$ ❖   52. $y = x^2 + 4x + 5$ ❖

53. $y = -2x^2 - 2$ ❖   54. $y = -x^2 - 5$ ❖

55. True or False: The function $y = -5x^2$ has a maximum value but no minimum value.   True

56. True or False: The graph of $y = -4x^2 + 9x - 6$ opens upward.   False

57. True or False: The graph of $y = 1.5x^2 - 6x - 3$ opens downward.   False

58. True or False: The function $y = 2x^2 - 5x + 4$ has a maximum value but no minimum value.   False

59. A concession stand at the Arthur Ashe Tennis Center sells a hamburger/drink combination dinner for $5. The profit, y, (in dollars) can be approximated by

$$y = -0.001x^2 + 3.6x - 400$$ where x is the number of dinners prepared.

    a. Find the number of dinners that should be prepared to maximize profit.   1800 dinners

    b. What is the maximum profit? $2840

60. For a fund-raising activity, a charitable organization produces calendars to sell in the community. The profit, y, (in dollars) can be approximated by

$$y = -\frac{1}{40}x^2 + 10x - 500$$ where x is the number of calendars produced.

    a. Find the number of calendars that should be produced to maximize profit.   200 calenders

    b. What is the maximum profit? $500

61. The pressure, x, in an automobile tire can affect its wear. Both over-inflated and under-inflated tires can lead to poor performance and poor mileage. For one particular tire, the number of miles that a tire lasts, y, (in thousands) is given by

$$y = -0.875x^2 + 57.25x - 900$$ where x is the tire pressure in pounds per square inch (psi).

    a. Find the tire pressure that will yield the maximum number of miles that a tire will last. Round to the nearest whole unit.   33 psi

    b. Find the maximum number of miles that a tire will last if the proper tire pressure is maintained. Round to the nearest whole unit.   36,000 miles

❖ See Additional Answers Appendix   ✎ Writing   ⬌ Translating Expression   Geometry   Scientific Calculator   Video

62. A child kicks a ball into the air, and the height of the ball, y, (in feet) can be approximated by

$$y = -16t^2 + 40t + 3$$ where t is the number of seconds after the ball was kicked.

a. Find the maximum height of the ball.   28 ft

b. How long will it take the ball to reach its maximum height?   1.25 s

---

chapter 9   SUMMARY

SECTION 9.1—INTRODUCTION TO FUNCTIONS AND FUNCTION NOTATION

KEY CONCEPTS:

Any set of ordered pairs, (x, y), is called a relation in x and y.

The domain of a relation is the set of first components in the ordered pairs in the relation. The range of a relation is the set of second components in the ordered pairs.

Given a relation in x and y, we say "y is a function of x" if for every element x in the domain there corresponds exactly one element y in the range.

Vertical Line Test for Functions

Consider any relation defined by a set of points (x, y) on a rectangular coordinate system. Then the graph defines y as a function of x if no vertical line intersects the graph in more than one point.

EXAMPLES:

Find the domain and range of the relation:

$$\{(0, 0), (1, 1), (2, 4), (3, 9), (-1, 1), (-2, 4), (-3, 9)\}$$

Domain: $\{0, 1, 2, 3, -1, -2, -3\}$

Range: $\{0, 1, 4, 9\}$

Function: $\{(1, 3), (2, 5), (6, 3)\}$

Nonfunction: $\{(1, 3), (2, 5), (1, -2)\}$

different y-values for the same x-value

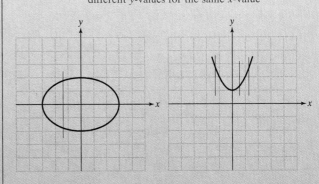

Not a Function
Vertical line intersects more than once.

Function
No vertical line intersects more than once.

Function Notation

$f(x)$ is the value of the function, f, at x.

Given $f(x) = -3x^2 + 5x$, find $f(-2)$:

$$f(-2) = -3(-2)^2 + 5(-2)$$
$$= -12 - 10$$
$$= -22$$

The domain of a function defined by $y = f(x)$ is the set of x-values that make the function a real number. In particular,

- Exclude values of x that make the denominator of a fraction zero.
- Exclude values of x that make the radicand of an even-indexed root negative.

Find the domain of the functions:

1. $f(x) = \dfrac{x + 4}{(x - 5)}$; $\{x | x \neq 5\}$

2. $f(x) = \sqrt{x - 3}$; $\{x | x \geq 3\}$

3. $f(x) = \sqrt[3]{x - 6}$;   All real numbers

KEY TERMS:

domain of a relation	range of a relation
function	relation
function notation	vertical line test

SECTION 9.2—COMPLEX NUMBERS

KEY CONCEPTS:

The imaginary number i is defined as

$$i = \sqrt{-1} \quad \text{and} \quad i^2 = -1$$

For a real number

$$b > 0, \sqrt{-b} = i\sqrt{b}$$

EXAMPLES:

Simplify: $\sqrt{-4} \cdot \sqrt{-9}$

$$= (2i)(3i)$$
$$= 6i^2$$
$$= -6$$

Powers of i Follow a Pattern

$i = i$	$i^5 = i$	and so on . . .
$i^2 = -1$	$i^6 = -1$	
$i^3 = -i$	$i^7 = -i$	
$i^4 = 1$	$i^8 = 1$	

A complex number is in the form $a + bi$, where a and b are real numbers. a is called the real part, and b is called the imaginary part.

Simplify the powers of i:

$$i^{13} \quad \text{and} \quad i^{22}$$

$$
\begin{aligned}
i^{13} &= i^{12} \cdot i \\
&= (i^4)^3 \cdot i \\
&= (1)^3 \cdot i \\
&= i
\end{aligned}
\qquad
\begin{aligned}
i^{22} &= i^{20} \cdot i^2 \\
&= (i^4)^5 \cdot i^2 \\
&= (1)^5 \cdot (-1) \\
&= -1
\end{aligned}
$$

Adding or Subtracting Complex Numbers

Combine the real parts and combine the imaginary parts.

Perform the indicated operations:

$$(3 - 5i) - (2 + i) + (3 - 2i)$$
$$= 3 - 5i - 2 - i + 3 - 2i$$
$$= 4 - 8i$$

Multiplying Complex Numbers

Use the distributive property.

Multiply: $(1 + 6i)(2 + 4i)$

$$= 2 + 4i + 12i + 24i^2$$
$$= 2 + 16i + 24(-1)$$
$$= -22 + 16i$$

Dividing Complex Numbers

Multiply the numerator and denominator by the conjugate of the denominator.

KEY TERMS:

$\sqrt{-b}$ for $b > 0$
complex number
conjugates
i
imaginary number
imaginary part of a complex number
real part of a complex number
standard form of a complex number

Divide: $\dfrac{3}{2 - 5i}$

$$= \frac{3}{(2 - 5i)} \cdot \frac{(2 + 5i)}{(2 + 5i)}$$

$$= \frac{6 + 15i}{4 - 25i^2}$$

$$= \frac{6 + 15i}{4 - 25(-1)}$$

$$= \frac{6 + 15i}{4 + 25}$$

$$= \frac{6 + 15i}{29} \quad \text{or} \quad \frac{6}{29} + \frac{15}{29}i$$

SECTION 9.3—COMPLETING THE SQUARE

KEY CONCEPTS:

Square Root Property

If $x^2 = k$, then $x = \pm\sqrt{k}$.

The square root property can be used to solve quadratic equations in the form: $(x + h)^2 = k$

Solving a Quadratic Equation of the Form $ax^2 + bx + c = 0$ ($a \neq 0$) by Completing the Square and Applying the Square Root Property

1. Divide both sides by a to make the leading coefficient 1.
2. Isolate the variable terms on one side of the equation.
3. Complete the square (add the square of $\frac{1}{2}$ the linear term coefficient to both sides of the equation. Then factor the resulting perfect square trinomial).
4. Apply the square root property and solve for x.

KEY TERMS:

completing the square
constant term
linear term
quadratic term
square root property

EXAMPLES:

Solve using the square root property:

$$(x - 5)^2 = -13$$

$x - 5 = \pm\sqrt{-13}$ Square root property
$x = 5 \pm i\sqrt{13}$

Solve by completing the square and applying the square root property:

$$2x^2 - 6x - 5 = 0$$

Step 1: $\dfrac{2x^2}{2} - \dfrac{6x}{2} - \dfrac{5}{2} = \dfrac{0}{2}$

Step 2: $x^2 - 3x = \dfrac{5}{2}$ *Note:* $\left[\frac{1}{2}(-3)\right]^2 = \dfrac{9}{4}$

Step 3: $x^2 - 3x + \dfrac{9}{4} = \dfrac{5}{2} + \dfrac{9}{4}$

$$\left(x - \frac{3}{2}\right)^2 = \frac{19}{4}$$

Step 4: $x - \dfrac{3}{2} = \pm\sqrt{\dfrac{19}{4}}$

$$x = \dfrac{3}{2} \pm \dfrac{\sqrt{19}}{2} \quad \text{or} \quad x = \dfrac{3 \pm \sqrt{19}}{2}$$

SECTION 9.4—QUADRATIC FORMULA

KEY CONCEPTS:

The solutions to a quadratic equation of the form $ax^2 + bx + c = 0\ (a \neq 0)$ is given by the quadratic formula:

$$x = \dfrac{-b \pm \sqrt{b^2 - 4ac}}{2a}$$

The discriminant of a quadratic equation $ax^2 + bx + c = 0\ (a \neq 0)$ is $b^2 - 4ac$. If a, b, and c are rational numbers, then

1. If $b^2 - 4ac > 0$, then there will be two real solutions. Furthermore,

    a. If $b^2 - 4ac$ is a perfect square, the solutions will be rational numbers.

    b. If $b^2 - 4ac$ is not a perfect square, the solutions will be irrational numbers.

2. If $b^2 - 4ac < 0$, then there will be two imaginary solutions.

3. If $b^2 - 4ac = 0$, then there will be one rational solution.

Three Methods for Solving a Quadratic Equation

1. Factoring.
2. Completing the square and applying the square root property.
3. Using the quadratic formula.

KEY TERMS:

discriminant
methods to solve quadratic equations
quadratic formula

EXAMPLES:

Solve the equation by using the quadratic formula:

$$3x^2 = 2x - 4$$

$$3x^2 - 2x + 4 = 0 \qquad a = 3, b = -2, c = 4$$

$$x = \dfrac{-(-2) \pm \sqrt{(-2)^2 - 4(3)(4)}}{2(3)}$$

$$= \dfrac{2 \pm \sqrt{4 - 48}}{6}$$

$$= \dfrac{2 \pm \sqrt{-44}}{6}$$

$$= \dfrac{2 \pm 2i\sqrt{11}}{6} \qquad \text{Simplify radical.}$$

$$= \dfrac{2(1 \pm i\sqrt{11})}{6} \qquad \text{Factor.}$$

$$= \dfrac{1 \pm i\sqrt{11}}{3} \qquad \text{Reduce.}$$

SECTION 9.5—CONNECTIONS TO GRAPHING: QUADRATIC FUNCTIONS

KEY CONCEPTS:

Let a, b, and c represent real numbers such that $a \neq 0$. Then a function in the form, $y = ax^2 + bx + c$ is called a quadratic function.

The graph of a quadratic function is called a parabola.

The leading coefficient, a, of a quadratic function, $y = ax^2 + bx + c$, determines if the parabola will open upward or downward. If $a > 0$, then the parabola opens upward. If $a < 0$, then the parabola opens downward.

Finding the Vertex of a Quadratic Function

1. The x-coordinate of the vertex of the quadratic function $y = ax^2 + bx + c$ ($a \neq 0$) is given by

$$x = \frac{-b}{2a}$$

2. To find the corresponding y-coordinate of the vertex, substitute the value of the x-coordinate found in Step 1 and solve for y.

If a parabola opens upward, the vertex is the lowest point on the graph. If a parabola opens downward, the vertex is the highest point on the graph.

KEY TERMS:

axis of symmetry
parabola
quadratic function
vertex

EXAMPLES:

$y = x^2 - 4x - 3$ is a quadratic function.

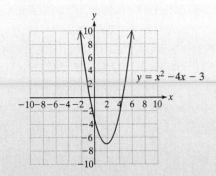

$y = x^2 - 4x - 3$

Find the vertex of the parabola defined by $y = 3x^2 + 6x - 1$.

$$x = \frac{-b}{2a} = \frac{-6}{2 \cdot 3} = -1$$

$$y = 3(-1)^2 + 6(-1) - 1 = -4$$

The vertex is the point $(-1, -4)$.

Because $a = 3 > 0$, the parabola opens upward so the vertex $(-1, -4)$ represents the lowest point of the graph.

chapter 9   REVIEW EXERCISES

Section 9.1

For Exercises 1–6, state the domain and range of each relation. Then determine whether the relation is a function.

1. $\{(6, 3), (10, 3), (-1, 3), (0, 3)\}$
   Domain: $\{6, 10, -1, 0\}$; range: $\{3\}$; function
2. $\{(2, 0), (2, 1), (2, -5), (2, 2)\}$
   Domain: $\{2\}$; range: $\{0, 1, -5, 2\}$; not a function
3.

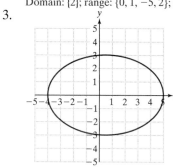

   Domain: $[-4, 5]$
   Range: $[-3, 3]$
   not a function

4.

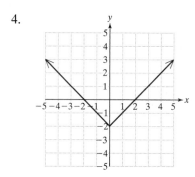

   Domain: $(-\infty, \infty)$
   Range: $[-2, \infty)$
   function

5. $\{(4, 23), (3, -2), (-6, 5), (4, 6)\}$
   Domain: $\{4, 3, -6\}$; range: $\{23, -2, 5, 6\}$; not a function
6. $\{(3, 0), (-4, \frac{1}{2}), (0, 3), (2, -12)\}$
   Domain: $\{3, -4, 0, 2\}$; range: $\{0, \frac{1}{2}, 3, -12\}$; function
7. Given the function defined by $f(x) = x^3$ find:
   a. $f(0)$ 0
   b. $f(2)$ 8
   c. $f(-3)$ -27
   d. $f(-1)$ -1
   e. $f(4)$ 64
   f. Write the domain in interval notation. $(-\infty, \infty)$

8. Given the function defined by $g(x) = \dfrac{x}{5 - x}$
   find:
   a. $g(0)$ 0

b. $g(4)$ 4
c. $g(-1)$ $-\dfrac{1}{6}$
d. $g(3)$ $\dfrac{3}{2}$
e. $g(-5)$ $-\dfrac{1}{2}$
f. Write the domain in interval notation.
   $(-\infty, 5) \cup (5, \infty)$

For Exercises 9–14, let h and k be defined by $h(x) = x^3 + 2x - 1$ and $k(x) = -4x$. Evaluate the following.

9. $h(-2) + k(8)$ -45
10. $h(1) \cdot k(0)$ 0
11. $k(m)$ $-4m$
12. $h(m)$ $m^3 + 2m - 1$
13. $h(2y)$ $8y^3 + 4y - 1$
14. $k(-3a)$ $12a$

15. The landing distance that a certain plane will travel on a runway is determined by the initial landing speed at the instant the plane touches down. The following function relates landing distance, $D(x)$, to initial landing speed, x:

$$D(x) = \frac{1}{10}x^2 - 3x + 22 \qquad \text{where } D \text{ is in feet and } x \text{ is in feet per second.}$$

Distance of Runway Versus Speed of Plane

$D(x) = \frac{1}{10}x^2 - 3x + 22$

(graph: Distance (feet) on vertical axis from 0 to 1200; Speed (feet/second) on horizontal axis from 0 to 125)

$D(90) = 562$. A plane traveling 90 ft/s when it touches down will require 562 ft of runway.

a. Find $D(90)$ and interpret the meaning of the function value in terms of landing speed and length of the runway.

b. Find $D(110)$ and interpret the meaning in terms of landing speed and length of the runway.

$D(110) = 902$. A plane traveling 110 ft/s when it touches down will require 902 ft of runway.

For Exercises 16–21, write the domain of each function in set-builder notation.

16. $f(x) = \sqrt{x - 8}$
   $\{x | x \geq 8\}$

17. $p(x) = x^4 - 16$
   $\{x | x \text{ is a real number}\}$

18. $h(x) = \dfrac{1}{x - 4}$
   $\{x | x \neq 4\}$

19. $k(x) = \sqrt{2 + x}$
   $\{x | x \geq -2\}$

20. $g(x) = x^2 - 3$
   $\{x | x \text{ is a real number}\}$

21. $q(x) = \dfrac{2}{x + 1}$
   $\{x | x \neq -1\}$

Section 9.2

For Exercises 22–27, write the expression in terms of i and simplify.

22. $\sqrt{-36}$ $6i$

23. $\sqrt{-5}$ $i\sqrt{5}$

24. $\sqrt{-32}$ $4i\sqrt{2}$

25. $\sqrt{-10} \cdot \sqrt{10}$ $10i$

26. $\sqrt{-25} \cdot \sqrt{-1}$ -5

27. $\dfrac{\sqrt{-64}}{\sqrt{-4}}$ 4

For Exercises 28–31, simplify the powers of i.

28. i^7 $-i$

29. i^2 -1

30. i^{20} 1

31. i^{41} i

For Exercises 32–43, perform the indicated operations. Write the answers in standard form, $a + bi$.

32. $(8 - 3i) - 9$ $-1 - 3i$

33. $(-2 - 4i) + (6 - 8i)$ $4 - 12i$

34. $7i + (6 - 2i)$ $6 + 5i$

35. $-2i \cdot 11i$ 22

36. $-3i(-1 + 9i)$ $27 + 3i$

37. $(3 + i)(2 + 4i)$ $2 + 14i$

38. $(4 - 2i)^2$ $12 - 16i$

39. $(5 - i)(5 + i)$ 26

40. $\dfrac{9}{1 - 2i}$ $\dfrac{9}{5} + \dfrac{18}{5}i$

41. $\dfrac{6i}{1 + 2i}$ $\dfrac{12}{5} + \dfrac{6}{5}i$

42. $\dfrac{3 + 5i}{1 + i}$ $4 + i$

43. $\dfrac{5 + 8i}{2i}$ $4 - \dfrac{5}{2}i$

Section 9.3

For Exercises 44–47, identify the equations as linear or quadratic.

44. $5x - 10 = 3x - 6$ Linear

45. $(x + 6)^2 = 6$ Quadratic

46. $x(x - 4) = 5x - 2$ Quadratic

47. $3(x + 6) = 18(x - 1)$ Linear

For Exercises 48–55, solve the equations using the square root property.

48. $x^2 = 25$ $x = \pm 5$

49. $x^2 - 19 = 0$ $x = \pm\sqrt{19}$

50. $x^2 + 49 = 0$ $x = \pm 7i$

51. $x^2 = -48$ $x = \pm 4i\sqrt{3}$

52. $(x + 1)^2 = 14$ $x = -1 \pm \sqrt{14}$

53. $(x - 2)^2 = -64$ $x = 2 \pm 8i$

54. $\left(x - \dfrac{1}{8}\right)^2 = -\dfrac{3}{64}$ $x = \dfrac{1}{8} \pm \dfrac{i\sqrt{3}}{8}$

55. $(2x - 3)^2 = 20$ $x = \dfrac{3 \pm 2\sqrt{5}}{2}$

For Exercises 56–59, find the constant that should be added to each expression to make it a perfect square trinomial.

56. $x^2 + 12x$ 36

57. $x^2 - 18x$ 81

58. $x^2 - \dfrac{2}{3}x$ $\dfrac{1}{9}$

59. $x^2 + \dfrac{1}{7}x$ $\dfrac{1}{196}$

For Exercises 60–63, solve the quadratic equations by completing the square and applying the square root property.

60. $x^2 + 8x + 3 = 0$ $x = -4 \pm \sqrt{13}$

61. $x^2 - 2x - 4 = 0$ $x = 1 \pm \sqrt{5}$

62. $2x^2 - 6x + 5 = 0$ $\dfrac{3}{2} \pm \dfrac{1}{2}i$

63. $3x^2 - 7x - 3 = 0$ $x = \dfrac{7 \pm \sqrt{85}}{6}$

64. An isosceles right triangle has legs of equal length. If the hypotenuse is 15 ft long, find the length of each leg. Round the answer to the nearest tenth of a foot.   10.6 ft

65. A can in the shape of a right circular cylinder holds approximately 362 cm^3 of liquid. If the height of the can is 12.1 cm, find the radius of the can. Round to the nearest tenth of a centimeter. (*Hint*: The volume of a right circular cylinder is given by: $V = \pi r^2 h$)   3.1 cm

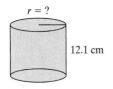

$r = ?$

12.1 cm

Figure for Exercise 65

Section 9.4

For $ax^2 + bx + c = 0$, $x = \dfrac{-b \pm \sqrt{b^2 - 4ac}}{2a}$

66. Write the quadratic formula from memory.

67. Write the discriminant from memory. $b^2 - 4ac$

For Exercises 68–71, find the discriminant for each equation. Then match the equation with the number and type of solutions listed on the right.

68. $5x^2 + x - 7 = 0$ b
    Discriminant is 141.

69. $x^2 + 4x + 4 = 0$ c
    Discriminant is 0.

70. $3x^2 - 2x + 2 = 0$ d
    Discriminant is -20.

71. $2x^2 - x - 3 = 0$ a
    Discriminant is 25.

a. Two rational solutions

b. Two irrational solutions

c. One rational solution

d. Two imaginary solutions

Figure for Exercise 84

For Exercises 72–81, solve the quadratic equations using the quadratic formula.

72. $5x^2 + x - 7 = 0$
$$x = \frac{-1 \pm \sqrt{141}}{10}$$

73. $x^2 + 4x + 4 = 0$
   $x = -2$

74. $3x^2 - 2x + 2 = 0$
$$x = \frac{1 \pm i\sqrt{5}}{3}$$

75. $2x^2 - x - 3 = 0$ $x = \frac{3}{2}, x = -1$

76. $\frac{1}{8}x^2 + x = \frac{5}{2}$ $x = -10, x = 2$

77. $\frac{1}{6}x^2 + x + \frac{1}{3} = 0$ $x = -3 \pm \sqrt{7}$

78. $0.01x^2 - 0.06x + 0.09 = 0$ $x = 3$

79. $1.2x^2 + 6x = 7.2$ $x = 1, x = -6$

80. $(x + 6)(x + 2) = 10$ $x = -4 \pm \sqrt{14}$

81. $(x - 1)(x - 7) = -18$ $x = 4 \pm 3i$

82. One number is two more than another number. Their product is 11.25. Find the numbers.
   The numbers are -2.5 and -4.5 or 2.5 and 4.5.

83. The base of a parallelogram is 1 cm more than the height, and the area is 24 cm². Find the exact values of the base and height of the parallelogram. Then use a calculator to approximate the values to the nearest tenth of a centimeter. ❖

84. An astronaut on the moon tosses a rock upward with an initial velocity of 25 ft/s. The height of the rock, $h(t)$, (in feet) is determined by the number of seconds, t, after the rock is released according to the equation:

$$h(t) = -2.7t^2 + 25t + 5$$

Find the time required for the rock to hit the ground. (*Hint*: At ground level, $h(t) = 0$.) Round to the nearest tenth of a second.   9.5 s

Section 9.5

For Exercises 85–90, find the *x*- and *y*-intercepts and match the equation with the graph. Round to the nearest hundredth where necessary.

85. $y = x^2 - 12x + 25$ ❖

86. $y = x^2 + 6x + 4$ ❖

87. $y = -5x^2 + 5$ ❖

88. $y = -3x^2 + 3$ ❖

89. $y = 2x^2 + 8x$ ❖

90. $y = x^2 - 2x + 4$ ❖

a.

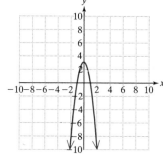

b.

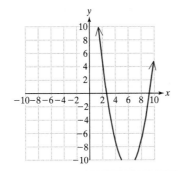

f.

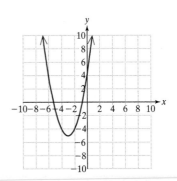

c.

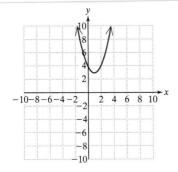

For Exercises 91–94, identify a and determine if the parabola opens upward or downward.

91. $y = x^2 - 3x + 1$
   $a = 1$; Upward

92. $y = -x^2 + 8x + 2$
   $a = -1$; Downward

93. $y = -2x^2 + x - 12$
   $a = -2$; Downward

94. $y = 5x^2 - 2x - 6$
   $a = 5$; Upward

For Exercises 95–98, find the vertex and the axis of symmetry for each parabola.

95. $y = 3x^2 + 6x + 4$ ❖

96. $y = -x^2 + 8x + 3$ ❖

97. $y = -2x^2 + 12x - 5$ ❖

98. $y = 2x^2 + 2x - 1$ ❖

d.

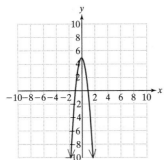

For Exercises 99–102,

   a. Determine whether the graph of the parabola opens upward or downward.

   b. Find the vertex.

   c. Find the x-intercept(s) if possible.

   d. Find the y-intercept.

   e. Sketch the function.

   f. Identify the domain of the function.

   g. Identify the range of the function.

99. $y = 2x^2 + 4x - 1$ ❖

100. $y = -3x^2 + 12x - 10$ ❖

101. $y = -8x^2 - 16x - 12$ ❖

102. $y = 3x^2 + 12x + 9$ ❖

e.

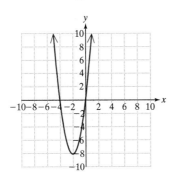

103. An object is launched into the air from ground level with an initial velocity of 256 ft/s. The height of the object, y, (in feet) can be approximated by the function

$$y = -16t^2 + 256t$$

where t is the number of seconds after launch.

a. Find the maximum height of the object.   1024 ft
b. Find the time required for the object to reach its maximum height.   8 s

---

chapter 9   TEST

1. a. Draw a graph of a relation that is *not* a function. (Answers may vary.) ❖
    b. Draw a graph of a relation that *is* a function. (Answers may vary.) ❖

2. For the function defined by $f(x) = \sqrt{x + 3}$
    a. Find the function values: $f(0), f(-2), f(6)$
        $f(0) = \sqrt{3}, f(-2) = 1, f(6) = 3$
    b. What is the domain of f? Write the domain in interval notation. $[-3, \infty)$

3. The number of diagonals, D, of a polygon is a function of the number of sides, x, of the polygon according to the equation:

$$D(x) = \frac{1}{2}x(x - 3)$$

a. Find $D(5)$ and interpret the meaning of the function value. Verify your answer by counting the number of diagonals in the pentagon in the figure below.
    $D(5) = 5$; a five-sided polygon has five diagonals.
b. Find $D(10)$ and interpret its meaning.
    $D(10) = 35$; a 10-sided polygon has 35 diagonals.
c. If a polygon has 20 diagonals, how many sides does it have? (*Hint*: Substitute $D(x) = 20$ and solve for x. Try clearing fractions first.)   8 sides

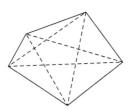

Figure for Exercise 3

For Exercises 7–8, simplify the powers of i.

7. i^{13} i 8. i^{35} $-i$

For Exercises 9–12, perform the indicated operation. Write the answer in standard form, $a + bi$.

9. $(2 - 7i) - (-3 - 4i)$
    $5 - 3i$
10. $(8 + i)(-2 - 3i)$
    $-13 - 26i$
11. $(10 - 11i)(10 + 11i)$
    221
12. $\dfrac{1}{10 - 11i}$ $\dfrac{10}{221} + \dfrac{11}{221}i$

13. Solve the equation by applying the square root property.
$$(3x + 1)^2 = -14 \qquad x = \frac{-1 \pm i\sqrt{14}}{3}$$

14. Solve the equation by completing the square and applying the square root property.
$$x^2 - 8x - 5 = 0 \qquad x = 4 \pm \sqrt{21}$$

15. Solve the equation by using the quadratic formula.
$$3x^2 - 5x = -1 \qquad x = \frac{5 \pm \sqrt{13}}{6}$$

16. The surface area, S, of a sphere is given by the formula $S = 4\pi r^2$, where r is the radius of the sphere. Find the radius of a sphere whose surface area is 201 in.^{2} Round to the nearest tenth of an inch.   4.0 in.

$S = 201$ in.^{2}

Figure for Exercise 16

For Exercises 4–6, simplify the expressions in terms of i.

4. $\sqrt{-100}$ $10i$ 5. $\sqrt{-23}$ $i\sqrt{23}$ 6. $\sqrt{-9} \cdot \sqrt{-49}$ -21

17. The height of a triangle is 2 m more than twice the base, and the area is 24 m². Find the exact values of the base and height. Then use a calculator to approximate the base and height to the nearest tenth of a meter. ❖

18. Explain how to determine if a parabola opens upward or downward. ❖

19. Graph the parabola and label the vertex, x-intercepts, and y-intercept. ❖

$$y = -x^2 + 25$$

20. A certain sports arena holds 20,000 seats. If the sports franchise charges x dollars per ticket for a game, then the total revenue, y, (in dollars) can be approximated by

$$y = -400x^2 + 20,000x$$ where x is the price per ticket.

a. Find the ticket price that will produce the maximum revenue. $25 per ticket

b. What is the maximum revenue? $250,000

CUMULATIVE REVIEW EXERCISES, CHAPTERS 1–9

1. Solve for x: $3x - 5 = 2(x - 2)$ $x = 1$

2. Solve for h: $A = \frac{1}{2}bh$ $h = \frac{2A}{b}$

3. Solve for y: $\frac{1}{2}y - \frac{5}{6} = \frac{1}{4}y + 2$ $y = \frac{34}{3}$

4. Determine whether $x = 2$ is a solution to the inequality: $-3x + 4 < x + 8$ Yes, x = 2 is a solution.

5. Graph the solutions to the inequality: $-3x + 4 < x + 8$. Then write the solution in set-builder notation and in interval notation. ❖

6. The graph depicts the death rate from 60 U.S. cities versus the median education level of the people living in that city. The death rate, y, is measured in number of deaths per 100,000 people. The median education level, x, is a type of "average" and is measured by grade level. (*Source:* U.S. Bureau of the Census)

    The death rate can be predicted from the median education level according to the equation:

    $$y = -37.6x + 1353$$ where $8 \le x \le 13$.

    a. From the graph, does it appear that the death rate increases or decreases as the median education level increases?   Decreases

    b. What is the slope of the line? Interpret the slope in the context of death rate and education level.   m = −37.6. For each additional increase in education level, the death rate decreases by approximately 38 deaths per 100,000 people.

    c. For a city in the United States with a median education level of 12, what would be the expected death rate?   901.8 per 100,000

d. If the death rate of a certain city is 977 per 100,000 people, what would be the approximate median education level?   10th grade

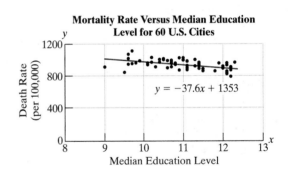

Mortality Rate Versus Median Education Level for 60 U.S. Cities

$$y = -37.6x + 1353$$

Figure for Exercise 6

7. Simplify completely. Write the final answer with positive exponents only:

$$\left(\frac{2a^2b^{-3}}{c}\right)^{-1} \cdot \left(\frac{4a^{-1}}{b^2}\right)^2$$ $\frac{8c}{a^4b}$

8. Approximately 5.2×10^7 disposable diapers are thrown into the trash each day in the United States and Canada. How many diapers are thrown away each year? 1.898×10^{10} diapers

9. The distance between the earth and the star, Polaris, is approximately 3.83×10^{15} miles. If 1 light-year is approximately 5.88×10^{12} miles, how many light-years is Polaris from the earth?   Approximately 651 light-years

10. Perform the indicated operation: $(2x - 3)^2 - 4(x - 1)$ $4x^2 - 16x + 13$

❖ See Additional Answers Appendix   ✎ Writing   ⬄ Translating Expression   ⬙ Geometry   ⌨ Scientific Calculator   ▣ Video

11. Divide using long division:
$(2y^4 - 4y^3 + y - 5) \div (y - 2)$ $2y^3 + 1 - \dfrac{3}{y-2}$

12. Which of the equations is linear?   a
    a. $2x - 4y = 8$ b. $x^2 + y = 4$
    c. $y = \sqrt{x}$

13. Factor completely: $2xy + 8xa - 3by - 12ab$
    $(y + 4a)(2x - 3b)$

14. The base of a triangle is 1 m more than the height. If the area is 36 m^2, find the base and height.   The base is 9 m and the height is 8 m.

15. Multiply.
$$\frac{x^2 + 10x + 9}{x^2 - 81} \cdot \frac{18 - 2x}{x^2 + 2x + 1} \qquad -\frac{2}{x+1}$$

16. What is the domain of the expression.
$$\frac{5}{x^2 - 4}? \quad \{x \mid x \neq 2, x \neq -2\}$$

17. Perform the indicated operations.
$$\frac{x^2}{x - 5} - \frac{10x - 25}{x - 5} \qquad x - 5$$

18. Simplify completely.
$$\frac{\dfrac{1}{x+1} - \dfrac{1}{x-1}}{\dfrac{x}{x^2 - 1}} \qquad -\frac{2}{x}$$

19. Solve for y.
$$1 - \frac{1}{y} = \frac{12}{y^2} \qquad y = 4, y = -3$$

20. At a certain lake, a fishing enthusiast must release any fish that is less than 8 in. in length. If a fish is caught and measured at 20 cm, must the fish be released back to the lake? (*Hint:* 1 in. = 2.54 cm)   20 cm ≈ 7.9 in. The fish must be released.

21. Write an equation of the line passing through the point $(-2, 3)$ and having a slope of $\frac{1}{2}$. Write the final answer in slope-intercept form. $y = \frac{1}{2}x + 4$

For Exercises 22–23,

   a. Find the x-intercept (if it exists).
   b. Find the y-intercept (if it exists).
   c. Find the slope (if it exists).
   d. Graph the line.

22. $2x - 4y = 12$ ❖   23. $4x + 12 = 0$ ❖

24. Solve the system by using the addition method. If the system has no solution or infinitely many solutions, so state
$$\frac{1}{2}x - \frac{1}{4}y = \frac{1}{6} \qquad \left(1, \frac{4}{3}\right)$$
$$12x - 3y = 8$$

25. Solve the system by using the substitution method. If the system has no solution or infinitely many solutions, so state
$$2x - y = 8 \qquad (5, 2)$$
$$4x - 4y = 3x - 3$$

26. In a right triangle, one acute angle is 2° more than three times the other acute angle. Find the measure of each angle.   The angles are 22° and 68°.

27. A bank of 27 coins contains only dimes and quarters. The total value of the coins is $4.80. Find the number of dimes and the number of quarters.   There are 13 dimes and 14 quarters.

28. Sketch the inequality, $x - y \leq 4$.   ❖

29. Which of the following are irrational numbers? $\{0, -\frac{2}{3}, \pi, \sqrt{7}, 1.2, \sqrt{25}\}$ $\pi, \sqrt{7}$

For Exercises 30–31, simplify the radicals.

30. $\sqrt{\dfrac{1}{7}}$ $\dfrac{\sqrt{7}}{7}$

31. $\dfrac{\sqrt[3]{16x^4}}{\sqrt[3]{2x}}$ $2x$

32. Perform the indicated operation: $(4\sqrt{3} + \sqrt{x})^2$
    $48 + 8\sqrt{3x} + x$

33. Add the radicals: $-3\sqrt{2x} + \sqrt{50x}$ $2\sqrt{2x}$

34. Rationalize the denominator.
$$\frac{4}{2 - \sqrt{a}} \qquad \frac{8 + 4\sqrt{a}}{4 - a}$$

35. Solve for x: $\sqrt{x + 11} = x + 5$
    $x = -2$ ($x = -7$ does not check.)

36. Factor completely: $8c^3 - y^3$ $(2c - y)(4c^2 + 2cy + y^2)$

37. Which of the following graphs define y as a function of x?   c, d

a.

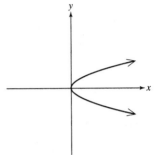

b.

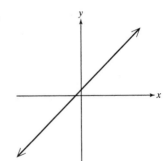

c.

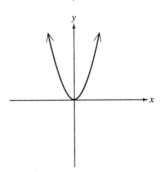

d.

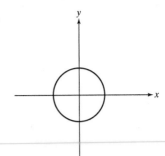

38. Given the functions defined by $f(x) = -\frac{1}{2}x + 4$ and $g(x) = x^2$, find

a. $f(6)$ 1   b. $g(-2)$ 4

c. $f(0) + g(3)$ 13   d. $g(x + h)$ $x^2 + 2xh + h^2$

39. Find the domain and range of the function:
    $\{(2, 4), (-1, 3), (9, 2), (-6, 8)\}$

    Domain: $\{2, -1, 9, -6\}$; range: $\{4, 3, 2, 8\}$

For Exercises 40–41, perform the indicated operations. Write the answer in the standard form of a complex number, $a + bi$.

40. $(3 + 2i) - (4 - 7i) + (1 - 6i)$
    $0 + 3i$

41. $\dfrac{4 - 2i}{3 + i}$
    $1 - i$

42. What value of k would make the expression a perfect square trinomial?   25

$$x^2 + 10x + k$$

43. Solve the quadratic equation by completing the square and applying the square root property:
    $2x^2 + 12x + 6 = 0$. $x = -3 \pm \sqrt{6}$

44. Solve the quadratic equation by using the quadratic formula: $2x^2 + 12x + 6 = 0$. $x = -3 \pm \sqrt{6}$

45. Graph the parabola defined by the equation.   ❖

$$y = -\frac{8}{9}x^2 + \frac{8}{3}x$$

Label the vertex, x-intercepts, and y-intercept.

STUDENT ANSWER APPENDIX

CHAPTER R

Section R.1 Practice Exercises, pp. 13–16

1. Numerator: 7; denominator: 8; proper **3.** Numerator: 9; denominator: 5; improper **5.** Numerator: 6; denominator: 6; improper

7. Numerator: 12; denominator: 1; improper **9.** $\frac{3}{4}$ **11.** $\frac{4}{3}$ **13.** $\frac{1}{6}$

15. $\frac{2}{2}$ **17.** $\frac{5}{2}$ or $2\frac{1}{2}$ **19.** $\frac{6}{2}$ or 3 **21.** The set of whole numbers includes the number 0 and the set of natural numbers does not.

23. For example: $\frac{2}{4}$ **25.** Prime **27.** Composite **29.** Composite

31. Prime **33.** $2 \times 2 \times 3 \times 3$ **35.** $2 \times 3 \times 7$ **37.** $2 \times 5 \times 11$

39. $3 \times 3 \times 3 \times 5$ **41.** $\frac{1}{5}$ **43.** $\frac{3}{8}$ **45.** $\frac{7}{8}$ **47.** $\frac{3}{4}$ **49.** $\frac{5}{8}$ **51.** $\frac{3}{4}$

53. False: When adding or subtracting fractions it is necessary to have a common denominator. **55.** $\frac{4}{3}$ or $1\frac{1}{3}$ **57.** $\frac{2}{3}$ **59.** $\frac{9}{2}$ or $4\frac{1}{2}$ **61.** 46

63. $\frac{14}{5}$ or $2\frac{4}{5}$ **65.** $\frac{11}{54}$ **67.** \$300 **69.** Four eggs **71.** Eight jars

73. 12 servings **75.** 12 batches **77.** 24 **79.** 40 **81.** 90 **83.** $\frac{3}{7}$

85. $\frac{1}{2}$ **87.** $\frac{7}{8}$ **89.** $\frac{3}{40}$ **91.** $\frac{3}{26}$ **93.** $\frac{29}{36}$ **95.** $\frac{7}{10}$ **97.** $\frac{35}{48}$

99. $\frac{7}{2}$ or $3\frac{1}{2}$ **101.** $\frac{23}{24}$ **103.** $\frac{59}{12}$ or $4\frac{11}{12}$ **105.** $\frac{1}{8}$ **107.** $5\frac{7}{12}$ ft

109. $1\frac{7}{12}$ cups **111.** $\frac{2}{3}$ miles **113.** $\frac{7}{8}$ yd **115.** $244\frac{1}{2}$ yd

117. Approximately 2 **119.** Approximately 8

Calculator Connections R.2, p. 24

1. $0.\overline{4}$
2. $0.\overline{63}$

3. $0.1\overline{36}$
4. 0.384615

Section R.2 Practice Exercises, pp. 24–26

1. Tens **3.** Hundreds **5.** Tenths **7.** Hundredths **9.** Ones **11.** Ten-thousandths **13.** No, the symbols I, V, X, and so on each represent certain values but the values are not dependent on the position of the symbol within the number. **15.** 0.9 **17.** 0.12 **19.** $1.\overline{7}$

21. $0.\overline{18}$ **23.** $\frac{13}{20}$ **25.** $\frac{273}{1000}$ **27.** $\frac{301}{50}$ or $6\frac{1}{50}$ **29.** $\frac{12003}{1000}$ or $12\frac{3}{1000}$

31. $\frac{8}{9}$ **33.** $\frac{7}{3}$ or $2\frac{1}{3}$ **35.** $0.8, \frac{4}{5}$ **37.** $0.25, \frac{1}{4}$ **39.** $0.045, \frac{9}{200}$

41. $0.058, \frac{29}{500}$ **43.** Remove the % sign and divide by 100. **45.** 6%

47. 70% **49.** 480% **51.** 930% **53.** 53.6% **55.** 0.2% **57.** 46%

59. 175% **61.** 12.5% **63.** 43.75% **65.** $26.\overline{6}\%$ **67.** $27.\overline{7}\%$
69. 12,096 students **71.** \$5.95 **73. a.** \$792 **b.** \$408 **c.** \$192
d. \$1488 **75.** \$960

Section R.3 Practice Exercises, pp. 34–40

1. b, e, i **3.** 108 cm; 704 cm^2 **5.** 1 ft; 0.0625 ft^2 **7.** $11\frac{1}{2}$ in.
9. 31.4 ft **11.** a, f, g **13.** 0.0004 m^2 **15.** 40 miles^2
17. 132.665 cm^2 **19.** 66 in.^{2} **21.** 6 km^2 **23.** 75.36 ft^3
25. 39 in.^{3} **27.** 3052.08 in.^{3} **29.** 113.04 cm^3 **31.** 32.768 ft^3
33. 82 ft **35.** 36 in.^{2} **37.** 15.2464 cm^2 **39. a.** \$0.31/ft^2 **b.** \$129
41. Perimeter **43. a.** 50.24 in.^{2} **b.** 113.04 in.^{2} **c.** One 12-in. pizza
45. 289.3824 cm^3 **47.** True **49.** True **51.** True **53.** True
55. 45° **57.** Not possible **59.** For example: 100°, 80°
61. a. $\angle 1$ and $\angle 3$, $\angle 2$ and $\angle 4$ **b.** $\angle 1$ and $\angle 2$, $\angle 2$ and $\angle 3$, $\angle 3$ and $\angle 4$, $\angle 1$ and $\angle 4$ **c.** $m(\angle 1) = 100°, m(\angle 2) = 80°$, $m(\angle 3) = 100°$. **63.** 57° **65.** 78° **67.** 60° **69.** 20° **71.** 147°
73. 58° **75.** 135° **77.** 45° **79.** 7 **81.** 1 **83.** 1 **85.** 5
87. $m(\angle a) = 45°, m(\angle b) = 135°, m(\angle c) = 45°, m(\angle d) = 135°$, $m(\angle e) = 45°, m(\angle f) = 135°, m(\angle g) = 45°$ **89.** Scalene
91. Isosceles **93.** No, a 90° angle plus an angle greater than 90° would make the sum of the angles greater than 180°. **95.** 40°
97. 37° **99.** $m(\angle a) = 80°, m(\angle b) = 80°, m(\angle c) = 100°$, $m(\angle d) = 100°, m(\angle e) = 65°, m(\angle f) = 115°, m(\angle g) = 115°$, $m(\angle h) = 35°, m(\angle i) = 145°, m(\angle j) = 145°$ **101.** $m(\angle a) = 70°$, $m(\angle b) = 65°, m(\angle c) = 65°, m(\angle d) = 110°, m(\angle e) = 70°$, $m(\angle f) = 110°, m(\angle g) = 115°, m(\angle h) = 115°, m(\angle i) = 65°$, $m(\angle j) = 70°, m(\angle k) = 65°$

CHAPTER 1

Section 1.1 Practice Exercises, pp. 51–53

1.

3. a **5.** b **7.** a **9.** a **11.** c **13.** a **15.** a **17.** a **19.** b **21.** c
23. $0.29, 3.8, \frac{1}{9}, \frac{1}{3}, \frac{1}{8}, \frac{1}{5}, 5, 2, -0.125, -3.24, -3, -6, \frac{7}{20}, \frac{5}{8}, 0.\overline{2}, 0.\overline{6}$
25. $\{1, 2, 3, 4, \ldots\}$ **27.** The set of all real numbers that are not rational **29.** The set of all numbers that includes all rational and irrational numbers **31.** For example: $\pi, -\sqrt{2}, \sqrt{3}$ **33.** For example: $-4, -1, 0$ **35.** For example: $-2, \frac{1}{2}, 0$ **37.** No, $\frac{3}{0}$ is undefined.

39. $\frac{0}{5}, 1$ **41.** $\sqrt{11}, \sqrt{7}$

43.

45.

47.

49. $-2, 2$ **51.** $2.5, 2.5$ **53.** $\frac{1}{3}, \frac{1}{3}$ **55.** $-\frac{1}{9}, \frac{1}{9}$ **57.** 5.1 **59.** 7

61. False, $|m|$ is never negative. **63.** True **65.** False **67.** True **69.** False **71.** False **73.** False **75.** False **77.** True **79.** False **81.** True **83.** True **85.** True **87.** For all $a \geq 0$

Calculator Connections 1.2, p. 60

1–3.

4–6.

7–9.

Section 1.2 Practice Exercises, pp. 61–63

1.

3.

5. a. $<$ **b.** $>$ **c.** $>$ **d.** $<$ **7.** 15 **9.** 3 **11.** 5 **13.** 9 **15.** $\frac{2}{3}$

17. 17 **19.** $57,600 \text{ ft}^2$ **21.** 21 ft^2 **23. a.** x **b.** Yes, 1 **25.** $\left(\frac{1}{6}\right)^4$

27. $a^3 b^2$ **29.** $(5c)^5$ **31.** $8yx^6$ **33.** $x \cdot x \cdot x$ **35.** $2b \cdot 2b \cdot 2b$

37. $10 \cdot y \cdot y \cdot y \cdot y \cdot y$ **39.** $2 \cdot w \cdot z \cdot z$ **41.** 25 **43.** $\frac{1}{49}$

45. 0.015625 **47.** 64 **49.** 9 **51.** 2 **53.** 10 **55.** 4 **57.** 20

59. 60 **61.** 8 **63.** 0 **65.** $\frac{7}{8}$ **67.** 45 **69.** 16 **71.** 15 **73.** 19

75. 3 **77.** $\frac{5}{12}$ **79.** $\frac{5}{2}$ **81.** 39 **83.** 26 **85.** $3x$ **87.** $\frac{x}{7}$ or $x \div 7$

89. $2 - a$ **91.** $2y + x$ **93.** $4(x + 12)$ **95.** $21 - 2x$ **97.** $t - 14$

99. The sum of 5 and r **101.** The difference of s and 14 **103.** The quotient of 5 and the product of 2 and p **105.** One more than the product of 7 and x **107.** 5, squared **109.** The square root of 5 **111.** 7, cubed **113.** The sum of 2 and the square of x **115.** The sum of 3 and the square root of r

117. a. $36 \div 4 \cdot 3$ Division must be performed before
    $= 9 \cdot 3$ multiplication.
    $= 27$

b. $36 - 4 + 3$ Subtraction must be performed before
    $= 32 + 3$ addition.
    $= 35$

119. This is acceptable, provided division and multiplication are performed in order from left to right, and subtraction and addition are performed in order from left to right. **121.** 2 **123.** 1

Section 1.3 Practice Exercises, pp. 67–69

1. Rational **3.** Rational **5.** Irrational **7.** Rational **9.** $>$ **11.** $>$ **13.** $>$ **15.** 3 **17.** -3 **19.** -17 **21.** 7 **23.** -19 **25.** -23 **27.** -5 **29.** -3 **31.** 0 **33.** 0 **35.** -5 **37.** -3 **39.** 0 **41.** -23 **43.** -6 **45.** -3 **47.** $-5 + 13 + (-11)$, $-3°$ **49.** $3 + (-5) + 14$, 12-yd gain **51.** To add two numbers with different signs, subtract the smaller absolute value from the larger absolute value and apply the sign of the number with the larger absolute value.

53. 21.3 **55.** $-\frac{3}{14}$ **57.** -2.4 or $-\frac{12}{5}$ **59.** $\frac{1}{4}$ or 0.25 **61.** 0

63. $-\frac{7}{8}$ **65.** $\frac{11}{9}$ **67.** -23.08 **69.** 494.686 **71.** -0.002117

73. a. $52.23 + (-52.95)$ **b.** Yes **75. a.** $100 + 200 + (-500) + 300 + 100 + (-200)$ **b.** \$0 **77.** -1 **79.** 10 **81.** 5 **83.** $-6 + (-10) = -16$ **85.** $-3 + 8 = 5$ **87.** $-21 + 17 = -4$ **89.** $3(-14 + 20) = 18$ **91.** $(-7 + (-2)) + 5 = -4$

Calculator Connections 1.4, p. 74

1–3.

4–6.

7, 8.

Section 1.4 Practice Exercises, pp. 74–77

1. For example: $0, 1, 2, 3, 4$ **3.** For example: $-\sqrt{2}, \sqrt{3}, \sqrt{5}, -2\pi, \pi$ **5.** For example: $1, 2, 3, 4, 5$ **7.** $\sqrt{6}$ **9.** $-7 + 10$ **11.** -3 **13.** -12 **15.** 4 **17.** 3 **19.** -2 **21.** 8 **23.** -8 **25.** 2 **27.** 6 **29.** 40 **31.** -40 **33.** -6 **35.** -20 **37.** -24 **39.** 25

41. -5 **43.** $\frac{2}{5}$ **45.** $-\frac{2}{3}$ **47.** 9.2 **49.** -5.72 **51.** -10 **53.** -14

55. -51 **57.** -173.188 **59.** 3.243 **61.** $6 - (-7) = 13$ **63.** $3 - 18 = -15$ **65.** $-5 - (-11) = 6$ **67.** $-1 - (-13) = 12$ **69.** $-32 - 20 = -52$ **71.** 13 **73.** -9 **75.** 5 **77.** -25 **79.** -25 **81.** $19,881 \text{ m}$ **83.** $10 - 10 + 7 + 7 = 14$ **85.** $152°$ **87.** -7 **89.** 5 **91.** 5 **93.** 3 **95.** -2 **97.** -11 **99.** 2

Midchapter Review, p. 77

1. Add their absolute values and apply a negative sign. **2.** Subtract the smaller absolute value from the larger absolute value. Apply the sign of the number with the larger absolute value. **3.** Add their absolute values. **4.** Add their absolute values and apply a negative sign. **5.** 41 **6.** 13 **7.** 31 **8.** 46 **9.** -1.3 **10.** -3.6 **11.** -16

12. -7 **13.** $-\frac{1}{12}$ **14.** $\frac{7}{24}$ **15.** -36 **16.** -59 **17.** -12 **18.** -50

19. $-\frac{19}{6}$ **20.** $-\frac{8}{5}$ **21.** -5 **22.** -32 **23.** 0 **24.** 0 **25.** -7.7

26. -10.5 **27.** -114 **28.** -56 **29.** -32 **30.** -46 **31.** -60 **32.** -70 **33.** -30 **34.** -400

Calculator Connections 1.5, p. 86

1–3.

4–6.

7, 8.

9, 10.

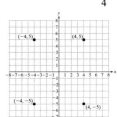

Section 1.5 Practice Exercises, pp. 87–89

1. True **3.** False **5.** True **7.** False **9.** $4 + 4 + 4 + 4 + 4$
11. $(-2) + (-2) + (-2)$ **13.** $(-2)(-7) = 14$ **15.** $-5 \cdot 0 = 0$
17. No number multiplied by zero equals 6. **19.** 6 **21.** -6 **23.** -6
25. 6 **27.** 8 **29.** -8 **31.** -8 **33.** 8 **35.** 0 **37.** Undefined
39. 0 **41.** 0 **43.** 2 **45.** $-\dfrac{1}{5}$ **47.** $-\dfrac{3}{2}$ **49.** $\dfrac{3}{10}$ **51.** -2

53. -7.912 **55.** 0.092 **57.** -6 **59.** 2.1 **61.** 9 **63.** -9 **65.** $-\dfrac{8}{27}$

67. 0.0016 **69.** -0.0016 **71.** -3 **73.** 3 **75.** -7 **77.** 7
79. -30 **81.** 96 **83.** 2 **85.** -1 **87.** $-2(3) + 3 = -3$, loss of \$3
89. -29 **91.** 48 **93.** -14.28 **95.** 340 **97.** $-\dfrac{10}{9}$ **99.** $\dfrac{14}{9}$

101. No, parentheses are required around the quantity $5x$; $10/(5x)$

103. $-3.75(0.3) = -1.125$ **105.** $\dfrac{16}{5} \div \left(-\dfrac{8}{9}\right) = -\dfrac{18}{5}$

107. $-0.4 + 6(-0.42) = -2.92$ **109.** $-\dfrac{1}{4} - 6\left(-\dfrac{1}{3}\right) = \dfrac{7}{4}$

111. a. -10 **b.** 24 **c.** In part (a) we subtract; in part (b) we multiply.

113. -23 **115.** 12 **117.** $\dfrac{9}{7}$ **119.** Undefined **121.** -2 **123.** -6

125. -1 **127.** 12 **129.** -40 **131.** $\dfrac{7}{2}$ **133.** $-\dfrac{5}{3}$ **135.** For

$x = 2, 10$; for $x = -2, 10$

Section 1.6 Practice Exercises, pp. 98–100

1. 8 **3.** -8 **5.** $-\dfrac{9}{2}$, or -4.5 **7.** 0 **9.** $\dfrac{7}{8}$ **11.** $\dfrac{1}{3}$ **13.** $-\dfrac{4}{45}$

15. $\dfrac{11}{15}$ **17.** Reciprocal **19.** 0 **21.** b **23.** i **25.** g **27.** d

29. h **31.** Identity property of addition **33.** Inverse property of
multiplication **35.** Inverse property of addition **37.** Commutative
property of multiplication **39.** Associative property of addition
41. Distributive property of multiplication over addition **43.** Commutative property of addition **45.** Commutative property of addition **47.** Equivalent **49.** Not equivalent. The terms are not *like*
terms and cannot be combined. **51.** Not equivalent; subtraction is
not commutative. **53.** Equivalent **55.** $5\frac{1}{8} + (18\frac{2}{5} + 1\frac{3}{5})$ is easier.
57. $2x + 14$ **59.** $-6z - 27$ **61.** $4w - 52z$ **63.** $-8x + 2$

65. $-\dfrac{1}{2}b + 2$ **67.** $-7q - 1$ **69.** $7a + b$ **71.** 243 **73.** 3995

75. $2x, 2; -y, -1; 18xy, 18; 5, 5$ **77.** $-x, -1; 8y, 8; -9x^2y, -9; -3, -3$
79. The variable factors are different. **81.** The variables are the
same and raised to the same power. **83.** For example: $5y, -2x, 6$

85. $2p - 12$ **87.** $-13z$ **89.** $3t - \dfrac{7}{5}$ **91.** $-10.4w + 3.3$

93. $-8a - 20$ **95.** $10r$ **97.** $-6x - 12$ **99.** $-16y - 27$ **101.** 23

103. $-3q + 1$ **105.** $-314p + 107$ **107.** $-7b + 8$ **109.** $-\dfrac{9}{5}p + \dfrac{5}{2}$

111. $3k + 11$ **113.** $-y + 25$ **115.** $-6.12q + 29.72$

Section 1.7 Practice Exercises, pp. 106–110

1. $2x + 6$ **3.** $-5x + 2$ **5.** $10x - 10$ **7.** $-18x + 11$ **9.** $\dfrac{5}{4}x + 0.3$

11.

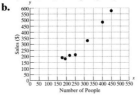

13.

15.

17. IV **19.** II **21.** $(0, -5)$ lies on the y-axis. **23.** $\left(\frac{7}{8}, 0\right)$ is located
on the x-axis.
25. a. $(250, 225)$ 250 people produce \$225 in popcorn sales.
$(175, 193)$ 175 people produce \$193 in popcorn sales.
$(315, 330)$ 315 people produce \$330 in popcorn sales.
$(220, 209)$ 220 people produce \$209 in popcorn sales.
$(450, 570)$ 450 people produce \$570 in popcorn sales.
$(400, 480)$ 400 people produce \$480 in popcorn sales.
$(190, 185)$ 190 people produce \$185 in popcorn sales.

b.

27. a. $(0, 251000)$ In 1700 the population of the U.S. colonies was
251,000.
$(10, 332000)$ In 1710 the population of the U.S. colonies was
332,000.
$(20, 466000)$ In 1720 the population of the U.S. colonies was
466,000.
$(30, 629000)$ In 1730 the population of the U.S. colonies was
629,000.
$(40, 906000)$ In 1740 the population of the U.S. colonies was
906,000.
$(50, 1171000)$ In 1750 the population of the U.S. colonies was
1,171,000.
$(60, 1594000)$ In 1760 the population of the U.S. colonies was
1,594,000.
$(70, 2148000)$ In 1770 the population of the U.S. colonies was
2,148,000.

b.

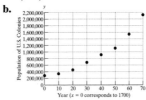

29. a. $(1, -10.2)$ $(2, -9.0)$ $(3, -2.5)$ $(4, 5.7)$ $(5, 13.0)$ $(6, 18.3)$ $(7, 20.9)$
$(8, 19.6)$ $(9, 14.8)$ $(10, 8.7)$ $(11, 2.0)$ $(12, -6.9)$

b.

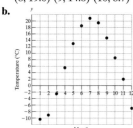

31. a. $(1, 89.25)$ On day 1 the closing price per share was $89.25.
$(2, 92.50)$ On day 2 the closing price per share was $92.50.
$(3, 91.25)$ On day 3 the closing price per share was $91.25.
$(4, 93.00)$ On day 4 the closing price per share was $93.00.
$(5, 90.25)$ On day 5 the closing price per share was $90.25.
$(6, 91.50)$ On day 6 the closing price per share was $91.50.
$(7, 92.25)$ On day 7 the closing price per share was $92.25.
$(8, 94.50)$ On day 8 the closing price per share was $94.50.

b.

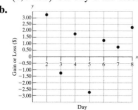

33. a. $(1, 42)$ In year 1 the gross income was $42,000,000.
$(2, 57)$ In year 2 the gross income was $57,000,000.
$(3, 39)$ In year 3 the gross income was $39,000,000.
$(4, 46)$ In year 4 the gross income was $46,000,000.
$(5, 58)$ In year 5 the gross income was $58,000,000.
$(6, 62)$ In year 6 the gross income was $62,000,000.

b. $(1, 47)$ In year 1 the cost was $47,000,000.
$(2, 50)$ In year 2 the cost was $50,000,000.
$(3, 49)$ In year 3 the cost was $49,000,000.
$(4, 40)$ In year 4 the cost was $40,000,000.
$(5, 50)$ In year 5 the cost was $50,000,000.
$(6, 62)$ In year 6 the cost was $62,000,000.

c.

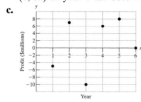

Chapter 1 Review Exercises, pp. 115–118

1. Real numbers: $7, \frac{1}{3}, -4, 0, -\sqrt{3}, -0.\overline{2}, \pi, 1$;
Integers: $7, -4, 0, 1$; Rational numbers: $7, \frac{1}{3}, -4, 0, -0.\overline{2}, 1$;
Whole numbers: $7, 0, 1$; Irrational numbers: $-\sqrt{3}, \pi$;
Natural numbers: $7, 1$ **3.** 6   **5.** 0   **7.** False   **9.** True   **11.** True
13. True   **15.** $\frac{7}{y}$, or $7 \div y$ **17.** $a - 5$ **19.** $13z - 7$ **21.** $3y + 12$
23. $2p - 5$ **25.** 225   **27.** $\frac{1}{16}$ **29.** $\frac{27}{8}$ **31.** 11   **33.** 10   **35.** 26
37. 20   **39.** 4   **41.** -17 **43.** $-\frac{5}{22}$ **45.** $-\frac{27}{10}$ **47.** -4.28

49. When a and b are both negative or when a and b have different signs and the number with the larger absolute value is negative
51. -12 **53.** -1 **55.** $-\frac{29}{18}$ **57.** -1.2 **59.** -10.2 **61.** -1
63. $-7 - (-18) = 11$ **65.** $7 - 13 = -6$
67. $(6 + (-12)) - 21 = -27$ **69.** -170 **71.** -2 **73.** $-\frac{1}{6}$ **75.** 0
77. 0   **79.** 2.25   **81.** $-\frac{3}{2}$ **83.** 17   **85.** $-\frac{7}{120}$ **87.** 70.6   **89.** True
91. True   **93.** True   **95.** For example: $2 + 3 = 3 + 2$ **97.** For example: $5 + (-5) = 0$ **99.** For example: $5 \cdot 2 = 2 \cdot 5$ **101.** For example: $3 \cdot \frac{1}{3} = 1$ **103.** $5x - 2y = 5x + (-2y)$, then use the commutative property of addition.   **105.** $3y, 10x, -12, xy$
107. a. $8a - b - 10$ **b.** $-7p - 11q + 16$ **109.** $p - 2$
111. $-14q - 1$ **113.** $4x + 24$
115–122.

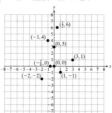

123. a. $(1, 26.25)$ On day 1 the price per share was $26.25.
$(2, 28.50)$ On day 2 the price per share was $28.50.
$(3, 28.00)$ On day 3 the price per share was $28.00.
$(4, 27.00)$ On day 4 the price per share was $27.00.
$(5, 24.75)$ On day 5 the price per share was $24.75.
$(6, 24.50)$ On day 6 the price per share was $24.50.
$(7, 24.50)$ On day 7 the price per share was $24.50.
$(8, 26.25)$ On day 8 the price per share was $26.25.

b.

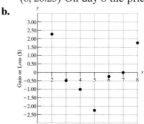

Chapter 1 Test, p. 119

1. Rational, all repeating decimals are rational numbers.
2.

3. a. False **b.** True **c.** True **d.** True   **4. a.** $(4x)(4x)(4x)$
b. $4 \cdot x \cdot x \cdot x$ **5. a.** Twice the difference of a and b **b.** The difference of twice a and b **6.** $\frac{\sqrt{c}}{d^2}$ or $\sqrt{c} \div d^2$ **7.** 6   **8.** 28   **9.** $-\frac{7}{8}$
10. 4.66   **11.** -12 **12.** Undefined   **13.** -28 **14.** 0   **15.** -8
16. 9   **17.** $\frac{1}{3}$ **18.** $-\frac{3}{5}$ **19. a.** Commutative property of multiplication **b.** Identity property of addition **c.** Associative property of addition **d.** Inverse property of multiplication **e.** Associative property of multiplication   **20.** $-6k - 8$ **21.** $-4p - 23$

22. a. $(5, 46)$ At age 5 the boy's height was 46 in.
$(7, 50)$ At age 7 the boy's height was 50 in.
$(9, 55)$ At age 9 the boy's height was 55 in.
$(11, 60)$ At age 11 the boy's height was 60 in.

b.

c. 57.5 in. **d.** 62.5 in. **e.** No, his height will maximize in his teen years. No.

CHAPTER 2

Section 2.1 Practice Exercises, pp. 130–132

1. Expression **3.** Equation **5.** Equation **7.** Expression
9. Substitute the value into the equation and determine if the right-hand side is equal to the left-hand side. **11.** No **13.** Yes
15. a. No **b.** Yes **c.** No **d.** Yes **17.** $x = -1$ **19.** $q = 20$
21. $m = -17$ **23.** $y = -16$ **25.** $z = \frac{11}{2}$ or $5\frac{1}{2}$ **27.** $x = -2$
29. $a = 1.3$ **31.** $c = 0$ **33.** $c = -2.13$ **35.** $t = -3.2675$
37. $-8 + x = 42, x = 50$ **39.** $x - (-6) = 18, x = 12$
41. $x + \frac{5}{8} = \frac{13}{8}, x = 1$ **43.** $x = 9$ **45.** $p = -4$ **47.** $y = 0$
49. $y = -15$ **51.** $t = -\frac{4}{5}$ **53.** $a = -10$ **55.** $b = 4$ **57.** $x = 41$
59. $p = -127$ **61.** $y = -2.6$ **63.** $x \cdot 7 = -63$ or $7x = -63, x = -9$
65. $\frac{x}{12} = \frac{1}{3}, x = 4$ **67.** $1\frac{1}{3}$ h **69.** 8 mph **71.** 14 cm **73.** 4.04 ft
75. $a = 10$ **77.** $x = -\frac{1}{9}$ **79.** $h = -12$ **81.** $t = \frac{22}{3}$ or $7\frac{1}{3}$
83. $r = -36$ **85.** $k = 16$ **87.** $k = 2$ **89.** $q = -\frac{7}{4}$ or $-1\frac{3}{4}$
91. $q = 11$ **93.** $d = -36$ **95.** $z = \frac{7}{2}$ or $3\frac{1}{2}$ **97.** $y = 4$
99. $y = 3.6$ **101.** $y = 0.4084$

Section 2.2 Practice Exercises, pp. 139–140

1. $-7t - 6$ **3.** $-5z + 2$ **5.** $10p - 10$ **7.** $20a + 15$ **9.** To simplify an expression, clear parentheses and combine *like* terms. To solve an equation, use the addition, subtraction, multiplication, and division properties of equality to isolate the variable. **11.** $y = -3$
13. $z = -5$ **15.** $b = \frac{59}{5}$ or $11\frac{4}{5}$ **17.** $h = \frac{1}{20}$ **19.** First use the addition property, then the division property. **21.** $x = -3$ **23.** $w = 2$
25. $q = -\frac{3}{4}$ **27.** $n = -2$ **29.** $b = \frac{21}{8}$ or $2\frac{5}{8}$ **31.** $a = -2$
33. $r = \frac{3}{7}$ **35.** $v = 16$ **37.** $u = 1$ **39.** $b = 34$ **41.** $x = 2$
43. $t = 4$ **45.** $x = -1$ **47.** $t = -10$ **49.** $k = \frac{1}{4}$ **51.** $w = 22$
53. $p = 8$ **55.** $y = 1$ **57.** $k = \frac{3}{7}$ **59.** $z = 5$ **61.** $m = -0.65$

63. No solution **65.** Contradiction; no solution **67.** Conditional equation; $x = -15$ **69.** Identity; all real numbers **71.** Identity; all real numbers **73.** Conditional equation; $x = 0$ **75.** $a = 15$ **77.** $a = 4$
79. Contradiction

Section 2.3 Practice Exercises, pp. 147–150

1. $x + 16 = -31, x = -47$ **3.** $x - 6 = -3, x = 3$ **5.** $x - 16 = -1$, $x = 15$ **7.** Contradiction; no solution **9.** Conditional equation; $y = -7$ **11.** Identity; all real numbers **13.** The number is -3.
15. The number is 11. **17.** The number is 5. **19.** The number is -5.
21. The number is 10. **23.** The number is 9. **25.** There were 165 Republicans and 269 Democrats. **27.** The lengths of the pieces are 33 cm and 53 cm. **29.** The Congo River is 4370 km long, and the Nile River is 6825 km. **31. a.** $x + 1, x + 2$ **b.** $x - 1, x - 2$
33. a. $x + 2, x + 4$ **b.** $x - 2, x - 4$ **35.** The integers are -34 and -33. **37.** The page numbers are 470 and 471. **39.** The numbers are 17, 19, and 21. **41.** The sides are 13 in., 14 in., and 15 in. **43.** The sides are 14 in., 15 in., 16 in., 17 in., and 18 in. **45.** The area of New Guinea is 792,500 km^2. **47.** The area of Africa is 30,065,000 km^2. The area of Asia is 44,579,000 km^2. **49.** Kariya and Lindros earned $8.5 million, and Fedorov earned $14 million.

Section 2.4 Practice Exercises, pp. 157–160

1. $x = -\frac{3}{5}$ **3.** $m = 46$ **5.** $x = -2$ **7.** $b = 22$ **9.** $y = -5$
11. No solution **13.** 18, 36 **15.** 100; 1000; 10,000 **17.** $x = 4$
19. $y = -\frac{19}{2}$ **21.** $q = -\frac{15}{4}$ **23.** $w = 8$ **25.** No solution
27. All real numbers **29.** The number is -16. **31.** The number is $\frac{9}{5}$.
33. The number is $\frac{7}{16}$. **35.** $y = 6$ **37.** $w = 3$ **39.** All real numbers
41. 12.5% **43.** 85% **45.** 0.75 **47.** 310.8 **49.** 885 **51.** 2200
53. 46.2% **55.** 34.5% **57.** $80.20 **59.** $16.00 **61.** 5% **63.** $29.96
65. $4562.50 **67.** $3879 **69.** 10% **71.** $420 **73.** $1200 **75.** 6%
77. $a = -1$ **79.** $b = 2$ **81.** $645 **83.** $60

Midchapter Review, p. 160

1. $b = 7$ **2.** $x = -7$ **3.** $y = -8$ **4.** $p = -\frac{3}{2}$, or $-1\frac{1}{2}$ **5.** $a = 20$
6. $w = \frac{1}{2}$ **7.** $x = 0$ **8.** $q = -10$ **9.** $h = 3$ **10.** No solution
11. $x = \frac{8}{3}$, or $2\frac{2}{3}$ **12.** $y = \frac{2}{3}$ **13.** No solution **14.** $x = -3.5$
15. $w = \frac{25}{2}$, or $12\frac{1}{2}$ **16.** $c = -1$ **17.** All real numbers
18. $z = \frac{9}{2}$, or $4\frac{1}{2}$ **19.** $a = -6$ **20.** $x = \frac{5}{4}$, or $1\frac{1}{4}$ **21.** $h = \frac{1}{3}$
22. $w = 0.25$ **23.** $t = -6$ **24.** All real numbers **25.** The sum of a number and 18 is equal to three times the sum of the number and two. **26.** The difference of a number and five is the same as twice the sum of the number and one. **27.** Twice the difference of a number and five is the same as one third of the number. **28.** One half the sum of a number and four is the same as twice the number.

Calculator Connections 2.5, p. 167

1, 2. 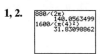   **3, 4.** [box: 20/((-0.05)*(5)) = -80; 10/(0.5(6+4)) = 2]

1. 140.056   **2.** 31.831   **3.** −80   **4.** 2

Section 2.5 Practice Exercises, pp. 168–171

1. $z = 7$ **3.** $k = 35$ **5.** $a = -3$ **7.** No solution   **9.** $y = -5$
11. $x = 0$ **13.** All real numbers   **15.** $y = -2$ **17.** $a = P - b - c$
19. $y = x + z$ **21.** $q = p - 250$ **23.** $t = \dfrac{d}{r}$ **25.** $t = \dfrac{PV}{nr}$
27. $x = 5 + y$ **29.** $y = -19 - 3x$ **31.** $x = \dfrac{6 - 3y}{2}$
33. $x = \dfrac{9 + y}{-2}$ or $-\dfrac{9 + y}{2}$ **35.** $y = \dfrac{12 - 4x}{-3}$ or $-\dfrac{12 - 4x}{3}$
37. $x = \dfrac{c - by}{a}$ **39. a.** $A = lw$ **b.** $w = \dfrac{A}{l}$ **c.** The width is 29.5 ft.
41. a. $P = 2l + 2w$ **b.** $l = \dfrac{P - 2w}{2}$ **c.** The length is 103 m.
43. a. $A = \dfrac{1}{2}bh$ **b.** $h = \dfrac{2A}{b}$ **c.** The height is 4 km.
45. a. $C = 2\pi r$ **b.** $r = \dfrac{C}{2\pi}$ **c.** The radius is approximately 140 ft.
47. $h = \dfrac{V}{\pi r^2}$ **49.** $B = \dfrac{2A}{h} - b$ or $\dfrac{2A - bh}{h}$ **51.** The length is 7 ft
and the width is 5 ft.   **53.** The length is 195 m and the width is 100 m.
55. The measures of the angles are 30°, 60°, and 90°.   **57.** The measures of the angles are 42°, 54°, and 84°.   **59.** $x = 17$; the measures of the angles are 34° and 56°.   **61.** Adjacent **S**upplementary angles form a **S**traight angle. The words *supplementary* and *straight* both begin with the same letter.   **63.** The angles are 30° and 60°.   **65.** The angles are 45° and 135°.   **67.** The angles are 34.8° and 145.2°.   **69.** $y = 40$; the vertical angles measure 146°.   **71. a.** 33.18 ft² **b.** 265.46 ft³
73. a. 28 m² **b.** 14 m² **c.** The area of the triangle is one half the area of the parallelogram.

Section 2.6 Practice Exercises, pp. 180–183

1. $x = 2$ **3.** $y = \dfrac{20 - 4x}{5}$ **5.** $a = 0.95$ **7.** $p = 36$ **9.** $h = \dfrac{3V}{lw}$
11. a. $1200 **b.** $6x$ **c.** $6(750 - x)$ or $4500 - 6x$ **13.** 53 tickets were sold at $3 and 28 tickets were sold at $2.   **15. a.** $300 **b.** $0.06x$
c. $0.06(20{,}000 - x)$ or $1200 - 0.06x$ **17.** There is $400 in the 8% account and $1200 in the 6% account.   **19.** $5000 should be invested in the 6% account and $3750 should be invested in the 8% account.
21. a. 2 lb **b.** $0.10x$ **c.** $0.10(x + 3) = 0.10x + 0.30$ **23.** Ten pounds of coffee sold at $12 per pound and 40 lb of coffee sold at $8 per pound.   **25.** 2 lb of raisins and 4 lb of granola   **27. a.** 300 miles
b. $5x$ **c.** $5(x + 12)$ or $5x + 60$ **29.** The car travels 5.5 h.   **31.** The slower car travels at 46 mph and the faster car travels at 50 mph.
33. The Cessna's speed is 110 mph and the Piper Cub's speed is 120 mph.
35. The rates of the boats are 20 mph and 40 mph.   **37.** $x = 15$
39. $z = 13$ **41.** $a = 57$ **43.** $k = 378$ **45.** $p = 206.4$ **47.** 30 tbsp of fertilizer is needed.   **49.** 30 red M&Ms   **51.** 4160 incorrectly marked ballots are expected.   **53.** $x = 3.75$ cm; $y = 4.5$ cm   **55.** The height of the pole is 7 m.   **57. a.** $3.50 **b.** $0.05x$ **c.** $0.05(30 + x)$ or $1.5 + 0.05x$ **59.** Jean-Paul has three quarters and nine dimes.   **61.** She receives four $10 bills and eight $20 bills.

Section 2.7 Practice Exercises, pp. 193–197

1. a. $x + 13$ **b.** $-7x - 11$ **c.** $x = -3$ **3.** The number is 25.
5. The United States scored $11\frac{1}{2}$ points, and Europe scored $16\frac{1}{2}$ points.
7. ⟶ ; $[6, \infty)$ **9.** ⟵ ; $(-\infty, 2.1]$
11. ⟶ ; $(-2, 7]$ **13.** $\{x \mid x > \frac{3}{4}\}$; $(\frac{3}{4}, \infty)$
15. $\{x \mid -1 < x < 8\}$; $(-1, 8)$ **17.** $\{x \mid x < -14\}$; $(-\infty, -14)$
19. $\{x \mid x \geq 18\}$; ⟶   **21.** $\{x \mid x < -0.6\}$; ⟵
23. $\{x \mid -3.5 \leq x < 7.1\}$; ⟶   **25. a.** A $[93, 100]$;
B+ $[89, 93)$; B $[84, 89)$; C+ $[80, 84)$; C $[75, 80)$; F $[0, 75)$ **b.** B **c.** C
27. $s \geq 110$ **29.** $t > 90$ **31.** $h \leq 2$ **33.** $t \geq 100$ **35.** $d \leq 10$
37. a. $x = 3$ **39. a.** $p = 13$ **41. a.** $c = -3$
b. $x > 3$ **b.** $p \leq 13$ **b.** $c < -3$
⟶ (3)   ⟵ (13)   ⟵ (−3)
43. a. $z = -\dfrac{3}{2}$ **45.** No   **47.** Yes
b. $z \geq -\dfrac{3}{2}$
⟶ $-\frac{3}{2}$
49. $(-\infty, 1]$ ⟵ (1)   **51.** $(10, \infty)$ ⟶ (10)
53. $(3, \infty)$ ⟶ (3)   **55.** $(-\infty, 8]$ ⟵ (8)
57. $(2, \infty)$ ⟶ (2)   **59.** $(-\infty, -2)$ ⟵ (−2)
61. $[14, \infty)$ ⟶ (14)   **63.** $[-24, \infty)$ ⟶ (−24)
65. $[-3, 3)$ ⟶ (−3, 3)   **67.** $\left(0, \dfrac{5}{2}\right)$ ⟶ (0, $\frac{5}{2}$)
69. $[90, \infty)$ ⟶ (90)   **71.** $(-9, \infty)$ ⟶ (−9)
73. $\left[-\dfrac{15}{2}, \infty\right)$ ⟶ ($-\frac{15}{2}$)   **75.** $\left[-\dfrac{1}{3}, \infty\right)$ ⟶ ($-\frac{1}{3}$)
77. $(-3, \infty)$ ⟶ (−3)   **79.** $(-\infty, 7)$ ⟵ (7)
81. $\left(-\infty, \dfrac{15}{4}\right]$ ⟵ ($\frac{15}{4}$)   **83.** $(-3, \infty)$ ⟶ (−3)
85. More than 10.2 in. of rain is needed.   **87. a.** $1539 **b.** 200 birdhouses cost $1440. It is cheaper to purchase 200 birdhouses because the discount is higher.   **89. a.** $2.00x > 75 + 0.17x$ **b.** $x > 41$; profit occurs when more than 41 lemonades are sold.
91. $[13, \infty)$ ⟶ (13)   **93.** $[-4, \infty)$ ⟶ (−4)
95. $(14.5, \infty)$ ⟶ (14.5)

Calculator Connections 2.8, pp. 203–205

1. a. [graph]   **b.** [graph]

2. a.

b.

3. a.

b.

4. a.

b.

5. a.

b.

Section 2.8 Practice Exercises, pp. 206–208

1. 3 is a solution. **3.** −4 is a solution. **5.** $y = \dfrac{8}{5}$ **7.** $x = \dfrac{5}{2}$

9. $y = 10$ **11.** $x = -2$ **13.** Yes **15.** Yes **17.** No

19.
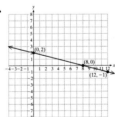
21.

23.
25.
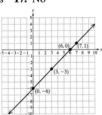

27.
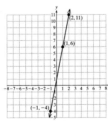
29.
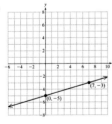

31.
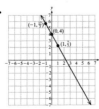
33.
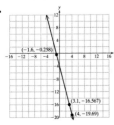

35. a. $y = 17.95$ **b.** $x = 145$ **c.** $(55, 17.95)$ Collecting 55 lb of cans yields \$17.95. $(145, 80.05)$ Collecting 145 lb of cans yields \$80.05.
d.
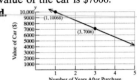

37. a. $y = 10{,}068$ **b.** $x = 3$ **c.** $(1, 10068)$ One year after purchase the value of the car is \$10,068. $(3, 7006)$ Three years after purchase the value of the car is \$7006.
d.

39. a. \$9926 million **b.** \$13,334 million **c.** 1997 **d.** 1999

Chapter 2 Review Exercises, pp. 217–221

1. a. Equation **b.** Expression **c.** Equation **d.** Equation
3. a. Nonlinear **b.** Linear **c.** Nonlinear **d.** Linear

5. $a = -8$ **7.** $k = \dfrac{21}{4}$ or $5\frac{1}{4}$ **9.** $x = -\dfrac{21}{5}$ or $-4\frac{1}{5}$

11. $k = -\dfrac{10}{7}$ or $-1\frac{3}{7}$ **13.** The number is 60. **15.** The number is −8.

17. $d = 1$ **19.** $c = \dfrac{9}{4}$ or $2\frac{1}{4}$ **21.** $b = -3$ **23.** $p = \dfrac{3}{4}$ **25.** $a = 0$

27. $b = \dfrac{3}{8}$ **29.** A contradiction has no solution and an identity is true

for all real numbers. **31.** The number is 18. **33.** The number is 30.
35. The number is −7. **37.** The minimum salary was \$30,000 in 1980.
39. The sides are 36 in., 37 in., 38 in., 39 in., and 40 in. **41.** No solution **43.** $q = 2.5$ **45.** $a = -4.2$ **47.** All real numbers

49. $r = \dfrac{10}{3}$, or $3\frac{1}{3}$ **51.** 28.8 **53.** 95% **55.** 1750 **57.** The total

price is \$28.24. **59. a.** \$840 **b.** \$3840 **61.** $K = C + 273$

63. $s = \dfrac{P}{4}$ **65.** $x = \dfrac{y - b}{m}$ **67.** $y = \dfrac{-2 - 2x}{5}$

69. The height is 5.1 in. **71.** The width is 5 ft. **73.** The angles are 32° and 58°. **75.** Peggy invested \$3000 in savings and \$1500 in stocks.

77. It will take $\frac{3}{4}$ h. **79.** $m = \dfrac{10}{3}$ or $3\frac{1}{3}$ **81.** Five bags will be needed.

83. a. $(-2, \infty)$ **b.** $\left(-\infty, \dfrac{1}{2}\right]$

c. $(-1, 4]$ **85.** $(-\infty, 17)$

87. $(-\infty, -6]$ **89.** $[49, \infty)$

91. $(18, \infty)$

93. More than 2.5 in. is required.

95.

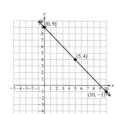

97. It rained 10.8 h.

99. a. \$19.10 **b.** (5.0, 7.70), (7.5, 11.55), (10.0, 15.40), (12.5, 19.25), (15.0, 23.10) **c.**

d. 6.5 gal **e.** 17 gal

Chapter 2 Test, pp. 222–223

1. $t = -16$ **2.** $p = 12$ **3.** $t = -\dfrac{16}{9}$ **4.** $x = \dfrac{7}{3}$ **5.** $p = 15$

6. $d = \dfrac{13}{4}$ **7.** All real numbers   **8.** No solution   **9.** No solution

10. $c = -47$ **11.** $y = -3x - 4$ **12.** $r = \dfrac{C}{2\pi}$ **13.** The sides are 61 in., 62 in., 63 in., 64 in., and 65 in.   **14.** The basketball tickets were \$36.32, and the hockey tickets were \$40.64.   **15.** The cost was \$82.00. **16.** They invested \$3200 in the 8% account and \$4000 in the 10% account.   **17.** The angles are 32° and 58°.   **18.** Their speeds were 50 mph and 55 mph.   **19.** There are 4.5 g of fat.
20. a. $(-\infty, 0)$ ⟵————)   **b.** $[-2, 5)$ ————[———)
                                    0   −2   5
21. $(-2, \infty)$ ⟵(————→   **22.** $(-\infty, -4]$ ⟵————]—→
                        −2   −4
23. More than 26.5 in. is required.
24. a. 202 beats per minute.   **c.** (20, 200) A 20-year-old has a maximum recommended heart rate of 200 beats per minute.   (30, 190) A 30-year-old has a maximum recommended heart rate of 190 beats per minute.   (40, 180) A 40-year-old has a maximum recommended heart rate of 180 beats per minute.   (50, 170) A 50-year-old has a maximum recommended heart rate of 170 beats per minute.   (60, 160) A 60-year-old has a maximum recommended heart rate of 160 beats per minute.

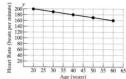

d. 28 years old

Cumulative Review Exercises, Chapters 1–2, pp. 223–224

1. -4 **2.** $\dfrac{1}{2}$ **3.** -2.75 **4.** -7 **5.** $-\dfrac{5}{12}$ **6.** 16 **7.** 4 **8.** -3.5
9. Whole numbers 2; Integers 2, -2; Rational numbers 2, -2, $0.\overline{7}$, 0.25, $\dfrac{3}{5}$; Irrational numbers $\sqrt{7}$, π
10. a. $\dfrac{1}{3} \cdot \dfrac{3}{4} = \dfrac{1}{4}$ **b.** $\sqrt{5^2 - 9} = 4$

11.

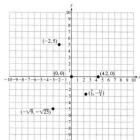

a. Quadrant II
b. Quadrant IV
c. Quadrant III
d. x-axis
e. Origin

12. $-7x^2y, 4xy, -6$ **13.** 1   **14.** $9x + 13$ **15.** $x = -7.2$ **16.** All real numbers   **17.** The numbers are 77 and 79.   **18.** The cost before tax was \$350.00.   **19.** The height is $6\frac{5}{6}$ cm.   **20.** Rachael invests \$3600 in the 4% account and \$7200 in the 9% account.   **21.** Her average running rate is 8 mph and her average biking rate is 20 mph.
22. $(-2, \infty)$ ⟵(————→   **23.** There are 28 candies in the box.
                    −2

24. a.

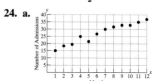

b. Third and fourth months
c. 147%

25. It rained 8 h.

CHAPTER 3

Calculator Connections 3.1, pp. 232–233

1, 2, 3.
```
(1.06)^5
       1.338225578
(1.02)^40
       2.208039664
5000(1.06)^5
       6691.127888
```

4, 5.
```
2000(1.02)^40
       4416.079327
3000(1+0.06/12)^
(12*2)
       3381.479329
```

6.
```
1000(1+0.05/4)^(
4*3)
       1160.754518
```

Section 3.1 Practice Exercises, pp. 233–235

1. Base: r; exponent: 4   **3.** Base: 5; exponent: 2   **5.** Base: -4; exponent: 8   **7.** Base: x; exponent: 1   **9.** y **11.** x **13.** No; $-5^2 = -25$ and $(-5)^2 = 25$ **15.** Yes; $-2^5 = -32$ and $(-2)^5 = -32$
17. $\left(\dfrac{1}{2}\right)^3 = \dfrac{1}{8}$ and $\dfrac{1}{2^3} = \dfrac{1}{8}$ **19.** $\left(\dfrac{3}{10}\right)^2 = \dfrac{9}{100}$ and $(0.3)^2 = 0.09$
21. 201 in.²   **23.** 113 cm³   **25.** 34,560 in.², or 240 ft²   **27.** \$5744.41
29. \$4776.21   **31. a.** $(x \cdot x \cdot x \cdot x)(x \cdot x \cdot x) = x^7$
b. $(5 \cdot 5 \cdot 5 \cdot 5)(5 \cdot 5 \cdot 5) = 5^7$ **33.** z^8 **35.** a^9 **37.** 4^{14} **39.** 9^5
41. c^{14} **43. a.** $\dfrac{p \cdot p \cdot p \cdot p \cdot p \cdot p \cdot p \cdot p}{p \cdot p \cdot p} = p^5$
b. $\dfrac{8 \cdot 8 \cdot 8 \cdot 8 \cdot 8 \cdot 8 \cdot 8 \cdot 8}{8 \cdot 8 \cdot 8} = 8^5$ **45.** x^2 **47.** a^9 **49.** 7^7 **51.** 5^7
53. y **55.** h^4 **57.** x^5 **59.** 7^7 **61.** x^6 **63.** 10^9 **65.** 6^{10} **67.** z^4

69. $40a^5b^5$ **71.** $13r^8s^5$ **73.** $16m^{20}n^{10}$ **75.** $2cd^4$ **77.** $\dfrac{x^2y^2}{4}$

79. $\dfrac{25hjk^4}{12}$ **81.** x^{2n+1} **83.** p^{2m+3} **85.** z **87.** r^3

Section 3.2 Practice Exercises, p. 239

1. 4^9 **3.** a^{20} **5.** d^9 **7.** 7^6 **9.** When multiplying expressions with the same base, add the exponents. When raising an expression with an exponent to a power, multiply the exponents.   **11.** 5^{12}
13. 12^6 **15.** y^{14} **17.** w^{25} **19.** a^{36} **21.** y^{14} **23.** They are both equal to 2^6.   **25.** $2^{(2^2)} = 2^{16}$ $(2^2)^4 = 2^8$; $2^{(2^2)}$ is greater than $(2^2)^4$.

27. $\dfrac{8}{27}$ **29.** $\dfrac{1}{16}$ **31.** $\dfrac{x^5}{y^5}$ **33.** $\dfrac{1}{t^4}$ **35.** $81a^4$ **37.** $-27a^3b^3c^3$

39. $216\,u^6v^{12}$ **41.** $5x^8y^4$ **43.** $\dfrac{1024}{r^5s^{20}}$ **45.** $\dfrac{343p^5}{q^{15}}$ **47.** y^{14}

49. $25a^{18}b^6$ **51.** $3x^{12}y^4$ **53.** $12c^{14}d^{12}$ **55.** $\dfrac{c^{27}d^{31}}{2}$ **57.** x^{2m}

59. $125a^{6n}$ **61.** $\dfrac{m^{2b}}{n^{3b}}$ **63.** $\dfrac{3^n a^{3n}}{5^n b^{4n}}$

Section 3.3 Practice Exercises, pp. 246–247

1. b^{11} **3.** x^4 **5.** 9^{11} **7.** $7776a^5b^{15}c^{10}$ **9.** $\dfrac{s^6t^{15}}{64}$ **11. a.** 1 **b.** 1

13. 1   **15.** 1   **17.** -1 **19.** 1   **21.** 1   **23.** -7 **25.** a **27. a.** $\dfrac{1}{t^5}$
b. $\dfrac{1}{t^5}$ **29.** $\dfrac{x^4}{x^{-6}} = x^{4-(-6)} = x^{10}$ **31.** $2a^{-3} = 2 \cdot \dfrac{1}{a^3} = \dfrac{2}{a^3}$ **33.** $\dfrac{343}{8}$
35. 25   **37.** $\dfrac{1}{a^3}$ **39.** $\dfrac{1}{12}$ **41.** $\dfrac{1}{16b^2}$ **43.** $\dfrac{6}{x^2}$ **45.** $\dfrac{1}{w^6}$ **47.** $\dfrac{1}{x^4}$ **49.** 1
51. y^4 **53.** $\dfrac{n^{27}}{m^{18}}$ **55.** $\dfrac{81k^{24}}{j^{20}}$ **57.** $\dfrac{1}{p^6}$ **59.** $\dfrac{1}{r^3}$ **61.** $\dfrac{1}{7^7}$ **63.** $\dfrac{1}{a^4b^6}$
65. $\dfrac{1}{w^{21}}$ **67.** $\dfrac{-16y^4}{z^2}$ **69.** $-\dfrac{a^{12}}{6}$ **71.** $80c^{21}d^{24}$ **73.** $\dfrac{p^{27}}{8}$ **75.** $\dfrac{2d^8}{c}$
77. $\dfrac{9}{20}$ **79.** $\dfrac{9}{10}$ **81.** $\dfrac{26}{81}$

Calculator Connections 3.4, pp. 252–253

1, 2.
```
(5.2E6)*(4.6E-3)
           2.392E4
(2.19E-8)*(7.84E
-4)
       1.71696E-11
```

3, 4.
```
(4.76E-5)/(2.38E
9)
            2E-14
(8.5E4)/(4.0E-1)
          2.125E5
```

5.
```
((9.6E7)*(4.0E-3
))/(2.0E-2)
          1.92E7
```

6.
```
((5.0E-12)*(6.4E
-5))/((1.6E-8)*(
4.0E2))
           5E-11
```

7, 8.
```
(7.8E-3)+(9.4E-3
)
          1.72E-2
(4.3E8)+(8.2E9)
          1.25E9
```

9, 10.
```
(6.43E5)-(1.91E5
)
          4.52E5
(2.01E-6)-(3.16E
-6)
         -1.15E-6
```

Section 3.4 Practice Exercises, pp. 254–256

1. $\dfrac{1}{a}$ **3.** $\dfrac{1}{10}$ **5.** $\dfrac{1}{x^3}$ **7.** $\dfrac{1}{10^3}$ **9.** z^{10} **11.** 10^{10} **13.** Move the decimal point between 2 and 3 and multiply by 10^{-10}; 2.3×10^{-10}.
15. 6.8×10^7 gal; 1.0×10^2 miles   **17.** 1.7×10^{-24} g
19. 1.4115999×10^8 shares   **21.** Move the decimal nine places to the left; 0.000 000 0031   **23.** 0. 000 000 000 001 g   **25.** 1600 calories and 2800 calories   **27.** 5.0×10^4 **29.** 2.25×10^{-13} **31.** 3.2×10^{14}
33. 2.432×10^{-10} **35.** 3.0×10^{13} **37.** 6.0×10^5 **39.** 1.38×10^1
41. 5.0×10^{-14} **43.** 3.75 in.   **45.** Approximately $1714 per commercial   **47. a.** 6.5×10^7 **b.** 2.3725×10^{10} days **c.** 5.694×10^{11} h
d. 2.04984×10^{15} s   **49.** 6.0×10^7 **51.** 3.5×10^{-12} **53.** 1.4×10^7
55. 3.0×10^{-4} **57.** 2.8×10^4 employees   **59.** 1.6423×10^{10}
61. a. U.S. 6.82×10^{11}; Germany 5.40×10^{11}; Japan 3.88×10^{11}; France 3.05×10^{11}; Britain 2.72×10^{11}; Italy 2.42×10^{11}; Canada 2.14×10^{11}; Netherlands 1.99×10^{11}; China 1.84×10^{11}
b. 3.026×10^{12} **c.** 2.04×10^{11} **d.** 7%   **63.** Approximately 3.19×10^9

Midchapter Review, pp. 256–257

1. 3   **2.** -4 **3.** $\dfrac{b^9}{2^{15}}$ **4.** $\dfrac{81}{y^6}$ **5.** $\dfrac{16y^4}{81x^4}$ **6.** $\dfrac{25d^6}{36c^2}$ **7.** $3a^7b^5$
8. $64x^7y^{11}$ **9.** $\dfrac{y^4}{x^8}$ **10.** $\dfrac{1}{a^{10}b^{10}}$ **11.** $\dfrac{1}{t^2}$ **12.** $\dfrac{1}{p^7}$ **13.** $\dfrac{8w^6x^9}{27}$
14. $\dfrac{25b^8}{16c^6}$ **15.** $\dfrac{q^3s}{r^2t^5}$ **16.** $\dfrac{m^2p^3q}{n^3}$ **17.** $\dfrac{1}{y^{13}}$ **18.** w^{10} **19.** $-\dfrac{1}{8a^{18}b^6}$
20. $\dfrac{4x^{18}}{9y^{10}}$ **21.** $\dfrac{k^8}{5h^6}$ **22.** $\dfrac{6n^{10}}{m^{12}}$ **23.** 1, 10, 100, 1000, 10,000
24. 1, 0.1, 0.01, 0.001, 0.0001   **25.** 2.7233×10^8 **26.** 2.8×10^7
27. 0.000001   **28.** 0.001   **29.** $2 \times 10^3 = 2000$ mutations

Section 3.5 Practice Exercises, pp. 265–267

1. $4p^2$ **3.** $12y^6$ **5.** $\dfrac{1}{8^5}$ **7.** 3.0×10^7 is scientific notation in which 10 is raised to the seventh power. 3^7 is not scientific notation and 3 is being raised to the seventh power.   **9.** 1.312×10^2 **11.** For example: $x^3 + 6$ **13.** For example: $8x^5$ **15.** For example: $-6x^2 + 2x + 5$
17. $-7x^4 + 7x^2 + 9x + 6$ **19.** Binomial   **21.** Monomial
23. Trinomial   **25.** Trinomial   **27.** Monomial   **29.** Binomial
31. Exercise 21: degree 2; Exercise 22: degree 0; Exercise 26: degree 3; Exercise 27: degree 0; Exercise 30: degree 3   **33.** Exercise 19: degree 2; Exercise 24: degree 1; Exercise 28: degree 1; Exercise 29: degree 4   **35.** Exercise 20: 13; Exercise 23: -1; Exercise 25: 12
37. a. 134 ft, 114 ft, 86 ft **b.** 150 ft   **39.** The exponents on the x-factors are different.   **41.** $35x^2y$ **43.** $10y$ **45.** $8b^2 - 9$
47. $4y^2 + y - 9$ **49.** $\dfrac{4}{3}z^2 - \dfrac{5}{3}$ **51.** $7.9t^3 - 3.4t^2 + 6t - 4.2$
53. $4y^3 + 2y^2 + 2$ **55.** $-4h + 5$ **57.** $2m^2 - 3m + 15$
59. $-3v^3 - 5v^2 - 10v - 22$ **61.** $9t^4 + 8t + 39$ **63.** $-8a^3b^2$
65. $-53x^3$ **67.** $-5a - 3$ **69.** $16k + 9$ **71.** $2s + 14$
73. $3t^2 - 4t - 3$ **75.** $-\dfrac{2}{3}h^2 + \dfrac{3}{5}h - \dfrac{5}{2}$ **77.** $2.4x^4 - 3.1x^2 - 4.4x - 6.7$
79. $3a^2 - 3a + 5$ **81.** $-3x^3 - 2x^2 + 11x - 31$
83. $4b^3 + 12b^2 - 5b - 12$ **85.** $9ab^2 - 3ab + 16a^2b$
87. $4z^5 + z^4 + 9z^3 - 3z - 2$ **89.** $2x^4 + 11x^3 - 3x^2 + 8x - 4$

Section 3.6 Practice Exercises, pp. 274–276

1. $9x$ **3.** $20x^2$ **5.** $-7a^3b$ **7.** $10a^6b^2$ **9.** $4c^2 - c$ **11.** $-4c^3$
13. 6.6×10^6 **15.** 9.89×10^{12} **17.** -1.14×10^{-11}
19. -1.953×10^{-23} **21.** $32x$ **23.** $-50z$ **25.** $4x^{13}$ **27.** $-12m^9n^8$
29. $16p^2q^2 - 24p^2q + 40pq^2$ **31.** $-4k^3 + 52k^2 + 24k$
33. $-45p^3q - 15p^4q^3 + 30pq^2$ **35.** $y^2 - y - 90$
37. $m^2 - 14m + 24$ **39.** $3x^2 + 28x + 32$ **41.** $5s^3 + 8s^2 - 7s - 6$
43. $27w^3 - 8$ **45.** $9a^2 - 16b^2$ **47.** $81k^2 - 36$ **49.** $\frac{1}{4} - t^2$
51. $u^6 - 25v^2$ **53.** $a^2 + 2ab + b^2$ **55.** $x^2 - 2xy + y^2$
57. $4c^2 + 20c + 25$ **59.** $9t^4 - 24st^2 + 16s^2$ **61. a.** 36 **b.** 20
c. $(a + b)^2 \neq a^2 + b^2$ in general   **63.** $4x^2 - 25$ **65.** $16p^2 + 40p + 25$
67. $49x^2 - y^2$ **69.** $25s^2 + 30st + 9t^2$ **71.** $21x^2 - 65xy + 24y^2$
73. $2t^2 + \frac{26}{3}t + 8$ **75.** $5z^3 + 23z^2 + 7z - 3$ **77.** $\frac{1}{9}m^2 - \frac{2}{3}mn + n^2$
79. $42w^3 - 84w^2$ **81.** $16y^2 - 65.61$ **83.** $21c^4 + 4c^2 - 32$
85. $9.61x^2 + 27.9x + 20.25$ **87.** $k^3 - 12k^2 + 48k - 64$
89. $15a^5 - 6a^2$ **91.** $27p^3 - 135p^2 + 225p - 125$ **93.** $2x - 7$

Section 3.7 Practice Exercises, pp. 282–284

1. $6z^5 - 10z^4 - 4z^3 - z^2 - 6$ **3.** $10x^2 - 29xy - 3y^2$
5. $4w^6 + 20w^3 + 25$ **7.** $\frac{49}{64}w^2 - 1$ **9.** $a^3 + 27$ **11.** Use long
division when the divisor is a polynomial with two or more terms.
13. $5t^2 + 6t$ **15.** $3a^2 + 2a - 7$ **17.** $3p^2 - p$ **19.** $1 + \frac{2}{m}$
21. $-2y^2 + y - 3$ **23.** $x^2 - 6x - \frac{1}{4} + \frac{2}{x}$ **25.** $a - 1 + \frac{b}{a}$
27. $3t - 1 + \frac{3}{2t} - \frac{1}{2t^2} + \frac{2}{t^3}$ **29.** $z + 2 + \frac{1}{z + 5}$ **31.** $t + 3$
33. $7b + 4$ **35.** $k - 6$ **37.** $2p^2 + 3p - 4$ **39.** $k - 2 + \frac{-4}{k + 1}$
41. $2x^2 - x + 6 + \frac{2}{2x - 3}$ **43.** $a - 3 + \frac{18}{a + 3}$
45. $w^2 + 5w - 2 + \frac{1}{w^2 - 3}$ **47.** $n^2 + n - 6$
49. $x - 1 + \frac{-8}{5x^2 + 5x + 1}$ **51.** To check, multiply
$(x - 2)(x^2 + 4) = x^3 - 2x^2 + 4x - 8$, which does not equal $x^3 - 8$.
53. Monomial division; $3a^2 + 4a$ **55.** Long division; $p + 2$
57. Long division; $t^3 - 2t^2 + 5t - 10 + \frac{4}{t + 2}$ **59.** Long division;
$w^2 + 3 + \frac{1}{w^2 - 2}$ **61.** Long division; $n^2 + 4n + 16$
63. Monomial division; $-3r + 4 - \frac{3}{r^2}$ **65.** $x + 1$
67. $x^3 + x^2 + x + 1$ **69.** $x + 1 + \frac{1}{x - 1}$ **71.** $x^3 + x^2 + x + 1 + \frac{1}{x - 1}$

Calculator Connections 3.8, pp. 290–291

1. 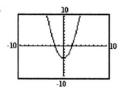   **2.**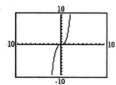

3. 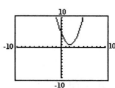   **4.**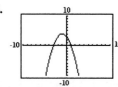

5. 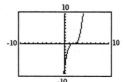   **6.**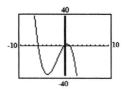

7. 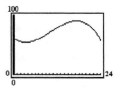   **8.**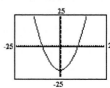

9.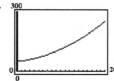

Section 3.8 Practice Exercises, pp. 292–298

1. 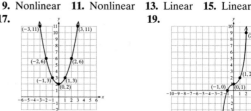   **3.** 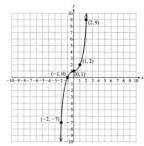   **5.** b   **7.** Linear

9. Nonlinear   **11.** Nonlinear   **13.** Linear   **15.** Linear
17.   **19.**

21. a, b. $(0, 0)$ When the plane is 0 miles from the airport its altitude
is 0 ft.
$(0.25, 156.25)$ When the plane is 0.25 miles from the airport
its altitude is 156.25 ft.
$(0.50, 500)$ When the plane is 0.50 miles from the airport its
altitude is 500 ft.
$(0.75, 843.75)$ When the plane is 0.75 miles from the airport
its altitude is 843.75 ft.
$(1.00, 1000)$ When the plane is 1 mile from the airport its
altitude is 1000 ft.
c. A nonlinear pattern is preferred so that the flight path
levels off at ground level.

23. a. Nonlinear
b, c. (0, 0) In 0 s the object travels 0 ft.
(1, 16) In 1 s the object travels 16 ft.
(2, 64) In 2 s the object travels 64 ft.
(3, 144) In 3 s the object travels 144 ft.
(4, 256) In 4 s the object travels 256 ft.
(5, 400) In 5 s the object travels 400 ft.

25. a. Nonlinear
b, c. (0, 0) If \$0 is charged per ticket the revenue will be \$0.
(10, 160000) If \$10 is charged per ticket the revenue will be \$160,000.
(20, 240000) If \$20 is charged per ticket the revenue will be \$240,000.
(30, 240000) If \$30 is charged per ticket the revenue will be \$240,000.
(40, 160000) If \$40 is charged per ticket the revenue will be \$160,000.
(50, 0) If \$50 is charged per ticket the revenue will be \$0.

27. a. Nonlinear
b. $(0, 0)$ $\left(1, \frac{1}{8}\right)$ $(2, 1)$ $\left(3, \frac{27}{8}\right)$ $(4, 8)$ $\left(5, \frac{125}{8}\right)$
c.

29. a. Nonlinear
b, c. (0, 4) At 0 s the height of the ball is 4 ft.
(0.5, 15) At 0.5 s the height of the ball is 15 ft.
(1.0, 18) At 1.0 s the height of the ball is 18 ft.
(1.5, 13) At 1.5 s the height of the ball is 13 ft.
(2.0, 0) At 2 s the height of the ball is 0 ft.

Chapter 3 Review Exercises, pp. 304–308

1. Base: 5; exponent: 3   **3.** Base: -2; exponent: 0   **5. a.** 36 **b.** 36
c. -36 **7.** 5^{13} **9.** x^9 **11.** 10^3 **13.** b^8 **15.** k **17.** 2^8
19. Exponents are added only when multiplying factors with the same base. In such a case, the base does not change.
21. \$7164.31   **23.** 7^{12} **25.** p^{18} **27.** $\dfrac{a^2}{b^2}$ **29.** $\dfrac{5^2}{c^4 d^{10}}$ **31.** $2^4 a^4 b^8$
33. $\dfrac{-3^3 x^9}{5^3 y^6 z^3}$ **35.** a^{11} **37.** $4h^{14}$ **39.** $54x^{12}y^{14}$ **41.** 1   **43.** 1   **45.** 2
47. $\dfrac{1}{z^5}$ **49.** $\dfrac{1}{36a^2}$ **51.** $\dfrac{17}{16}$ **53.** $\dfrac{1}{t^8}$ **55.** $\dfrac{2y^7}{x^6}$ **57.** $\dfrac{n^{16}}{16m^8}$ **59.** $\dfrac{k^{21}}{5}$
61. 6   **63. a.** $\$5.9148 \times 10^{12}$ **b.** 4.2×10^{-3} in. **c.** 1.66241×10^8 km^2
65. 9.43×10^5 **67.** 2.5×10^8 **69.** 9.5367×10^{13}. This number is too big to fit on most calculator displays.   **71. a.** 5.84×10^8 miles
b. 6.67×10^4 mph   **73. a.** Trinomial **b.** 4 **c.** 7   **75.** $7x - 3$
77. $14a^2 - 2a - 6$ **79.** $\dfrac{15}{2}x^3 + \dfrac{1}{4}x^2 + \dfrac{1}{2}x + 2$ **81.** $-2x^2 - 9x - 6$
83. For example: $-5x^2 + 2x - 4$ **85.** $-75x^6 y^4$
87. $15c^4 - 35c^2 + 25c$ **89.** $5k^2 + k - 4$ **91.** $6q^2 + 47q - 8$
93. $49a^2 + 7a + \dfrac{1}{4}$ **95.** $8p^3 - 27$ **97.** $b^2 - 16$
99. $49z^4 - 84z^2 + 36$ **101.** $4y^2 - 2y$ **103.** $-3x^2 + 2x - 1$
105. $x + 2$ **107.** $p - 3 + \dfrac{5}{2p + 7}$ **109.** $b^2 + 5b + 25$

111. $y^2 - 4y + 2 + \dfrac{9y - 4}{y^2 + 3}$ **113.** $2x^2 - 3x + 2 + \dfrac{1}{3x + 2}$
115. $t^2 - 3t + 1 + \dfrac{-2t - 6}{3t^2 + t + 1}$ **117.** a, c
119. a, b.
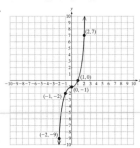

Chapter 3 Test, pp. 308–309

1. $\dfrac{(3 \cdot 3 \cdot 3 \cdot 3) \cdot (3 \cdot 3 \cdot 3)}{3 \cdot 3 \cdot 3 \cdot 3 \cdot 3 \cdot 3} = 3$ **2.** 9^6 **3.** q^8 **4.** $27a^6 b^3$
5. $\dfrac{16x^4}{y^{12}}$ **6.** 1   **7.** $\dfrac{1}{c^3}$ **8.** 14   **9.** $49s^{18}t$ **10.** $\dfrac{a^6}{2b^{24}}$ **11.** $\dfrac{8a^{15}}{9b^7}$
12. a. 4.3×10^{10} **b.** 0.000 0056   **13. a.** 2.4192×10^8 m^3
b. 8.83008×10^{10} m^3 **14.** $5x^3 - 7x^2 + 4x + 11$ **a.** 3 **b.** 5
15. $24w^2 - 3w - 4$ **16.** $-10x^5 - 2x^4 + 30x^3$ **17.** $8a^2 - 10a + 3$
18. $4y^3 - 25y^2 + 37y - 15$ **19.** $4 - 9b^2$ **20.** $25z^2 - 60z + 36$
21. Perimeter: $12x - 2$; area: $5x^2 - 13x - 6$
22. a. $-3x^6 + \dfrac{x^4}{4} - 2x$ **b.** $2y - 7$ **23. a.** Nonlinear **b.** 55 **c.** 55
d. First 20: 210; first 100: 5050
24. a, b.

Cumulative Review Exercises, Chapters 1–3, pp. 309–311

1. $-\dfrac{35}{2}$ **2.** 4   **3.** 4   **4.** $5^2 - \sqrt{4}, 23$ **5.** $-7, \dfrac{0}{4}, 2, 0.8, \sqrt{100}$
6. $x = \dfrac{28}{3}$ **7.** No solution   **8.** Quadrant III   **9.** y-axis
10. The measures are $31°, 54°, 95°$.   **11.** He sold \$12,000 worth of merchandise.   **12.** The dimensions are 24 ft $\times$ 120 ft.
13. a. 12 in. **b.** 19.5 in. **c.** 5.5 h **d.**
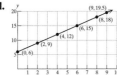
14. $[-5, \infty)$;   **15.** $5x^2 - 9x - 15$
16. $-2y^2 - 13yz - 15z^2$ **17.** $16t^2 - 24t + 9$ **18.** $\dfrac{4}{25}a^2 - \dfrac{1}{9}$
19. $-x^2 + x - 17$ **20.** $-4a^3 b^2 + 2ab - 1$
21. $4m^2 + 8m + 11 + \dfrac{24}{m - 2}$ **22.** $\dfrac{1}{x}$ **23.** $\dfrac{c^2}{16d^4}$ **24.** $\dfrac{2b^3}{a^2}$
25. a. 4.071×10^8 **b.** 4.071×10^{-6} **26. a.** 0.000 389
b. 4,500,000,000,000   **27.** 2.788×10^{-2}

CHAPTER 4

Section 4.1 Practice Exercises, pp. 321–322

1. 7 **3.** 6 **5.** ab **7.** $4w^2z$ **9.** $(x - y)$ **11.** $7(3x + 1)$
13. a. $3x - 6y$ **b.** $3(x - 2y)$ **15.** $4(p + 3)$ **17.** $5c(c - 2)$
19. $x^3(x^2 + 1)$ **21.** $t(t^3 - 4)$ **23.** $2ab(1 + 2a^2)$ **25.** $19x^2y(2 - y^3)$
27. $7pq^2(6p^2 + 2 - p^3q^2)$ **29.** $t^2(t^3 + 2rt - 3t^2 + 4r^2)$
31. $(a + 6)(13 - 4b)$ **33.** $(w^2 - 2)(8v + 1)$ **35.** $7x(x + 3)^2$
37. $(z - 1)(13z^2 - 19z + 5)$ **39.** $P = 2(l + w)$
41. $S = 2\pi r(r + h)$ **43. a.** $-2x(x^2 + 2x - 4)$ **b.** $2x(-x^2 - 2x + 4)$
45. $-1(8t^2 + 9t + 2)$ **47.** $-1(4y^3 - 5y + 7)$ **49.** $-15p^2(p + 2)$
51. $-q(q^3 - 2q + 9)$ **53.** $-1(7x + 6y + 2z)$
55. $-1(2c + 5)(3 + 4c)$ **57.** $(2a - b)(4a + 3c)$ **59.** $(q + p)(3 + r)$
61. $(2x + 1)(3x + 2)$ **63.** $(t + 3)(2t - 5)$ **65.** $(3y - 1)(2y - 3)$
67. $(b + 1)(b^3 - 4)$ **69.** $(j^2 + 5)(3k + 1)$ **71.** $(2x^6 + 1)(7w^6 - 1)$
73. $5x(x^2 + y^2)(3x + 2y)$ **75.** $4b(a - b)(x - 1)$
77. $6t(t - 3)(s - t^2)$ **79.** For example: $6x^2 + 9x$
81. For example: $16p^4q^2 + 8p^3q - 4p^2q$

Section 4.2 Practice Exercises, pp. 328–329

1. $8p^3(p^6 + 3)$ **3.** $6u(2u - 1)$ **5.** $3xy(3x + 4y - 5xy)$
7. $(x - 2)(5x - 2)$ **9.** $(3p - q)(4p + 5)$ **11.** $(c + d)(2c + 3d)$
13. $(u - 5)(2 - v)$ **15.** $6(a + 4)(a - 2)$ **17.** A polynomial that
cannot be factored is prime. **19.** 12, 1 **21.** $-8, -1$ **23.** $-1, 6$
25. $-12, 6$ **27.** $(x + 4)(3x + 1)$ **29.** $(w - 2)(4w - 1)$
31. $(m + 3)(2m - 1)$ **33.** $(4k + 3)(2k - 3)$ **35.** $(2k - 5)^2$
37. Prime **39.** $(p + 2q)(4p - 3q)$ **41.** $(5m + 2n)(3m - n)$
43. $(r + 2s)(3r - 7s)$ **45.** Prime **47.** $(q - 10)(q - 1)$
49. $(r + 4)(r - 10)$ **51.** $(x + 7)(x - 1)$ **53.** $(m - 6)(m - 7)$
55. $(a + 5)(a + 4)$ **57.** Prime **59.** $(p + 10q)^2$
61. $(x - 7y)(x + 6y)$ **63.** $(r + 5s)(r + 3s)$ **65.** No, $(2x + 4)$
contains a common factor of 2. **67.** $2(3x + 7)(x + 1)$
69. $2p(p - 15)(p - 4)$ **71.** $x^2(y + 3)(y + 11)$
73. $-1(k + 2)(k + 5)$ **75.** $-3(n + 6)(n - 5)$ **77.** $(2z - 1)(8z - 3)$
79. $(b - 4)^2$ **81.** $-5(x - 2)(x - 3)$ **83.** $(t - 3)(t + 2)$

Section 4.3 Practice Exercises, pp. 337–338

1. $7a^3(a^6 + 4)$ **3.** $4w(3w - 1)$ **5.** $3ab(7ab + 4b - 5a)$
7. $(y - 1)(4y - 5)$ **9.** $(4p - 9)(p + 4)$ **11.** $(s - 2)(t + 3)$
13. $(x - 2)(5 - y)$ **15.** $5(t - 2)(t + 6)$ **17.** Different **19.** Both
negative **21.** $(x + 8)$ **23.** $(x - 8)$ **25.** A polynomial that cannot
be factored is prime. **27.** $(2y + 1)(y - 2)$ **29.** $(3x - 2)^2$
31. $(2a + 3)(a + 2)$ **33.** $(2t + 3)(3t - 1)$ **35.** $(2m - 5)^2$
37. Prime **39.** $(2x - 5y)(3x - 2y)$ **41.** $(4m + 5n)(3m - n)$
43. $(3r + 2s)(2r - s)$ **45.** Prime **47.** $(x + 9)(x - 2)$
49. $(a - 12)(a + 2)$ **51.** $(r + 8)(r - 3)$ **53.** $(w - 7)^2$
55. $(k + 4)(k + 1)$ **57.** Prime **59.** $(m - 8n)(m - 5n)$
61. $(a + 8b)(a + b)$ **63.** $(x + 5y)(x + 4y)$ **65.** No, $(3x + 6)$ has a
common factor of 3. **67.** $d(5d^2 + 3d - 10)$ **69.** $4b(b - 5)(b + 4)$
71. $y^2(x - 3)(x - 10)$ **73.** $-1(a + 17)(a - 2)$
75. $-2(u - 5)(u - 9)$ **77.** $(8x + 1)(2x + 1)$ **79.** $(c - 1)^2$
81. $-2(z - 9)(z - 1)$ **83.** $(q - 7)(q - 6)$

Section 4.4 Practice Exercises, pp. 343–345

1. $(3x - 5)(x + 2)$ **3.** $yz(x^2z + 6y + 1)$ **5.** $2(3x - 1)(2x - 5)$
7. $(x + b)(a - 6)$ **9.** $(x + 3)^2$ **11.** $4x^2 + 12x + 9$
13. $36h^2 - 12h + 1$ **15. a.** $x^2 + 4x + 4$ **b.** $(x + 2)^2$; $(x + 4)(x + 1)$
17. a. $4x^2 - 20x + 25$ **b.** $(x - 5)(4x - 5)$; $(2x - 5)^2$ **19.** $(y - 5)^2$
21. $(m + 3)^2$ **23.** Prime **25.** $(7q - 2)^2$ **27.** $(3p + 7)^2$
29. $(5x + 2)(5x + 8)$ **31.** $(4a + b)^2$ **33.** $(4q + 5r)^2$ **35.** $(a + b)^2$
37. $\left(k - \dfrac{1}{2}\right)^2$ **39.** $\left(3x + \dfrac{1}{6}\right)^2$ **41.** $x^2 - 25$ **43.** $4w^2 - 9$
45. $(x - 6)(x + 6)$ **47.** $(w - 10)(w + 10)$
49. $(2a - 11b)(2a + 11b)$ **51.** $(7m - 4n)(7m + 4n)$
53. Prime **55.** $(c^3 - 5)(c^3 + 5)$ **57.** $(5 - 4t)(5 + 4t)$
59. $\left(p - \dfrac{1}{3}\right)\left(p + \dfrac{1}{3}\right)$ **61.** Prime **63.** $\left(\dfrac{2}{3} - w\right)\left(\dfrac{2}{3} + w\right)$
65. a. $a^2 - b^2$ **b.** $(a - b)(a + b)$ **67.** $3(w - 3)(w + 3)$
69. $2(5p^2 - 1)(5p^2 + 1)$ **71.** $2(x + 6)^2$ **73.** $2(t - 5)(t - 1)(t + 1)$
75. $25(4y^4 + x^2)$ **77.** $4b(a - 5b)^2$ **79.** $(2x + 3)(x - 1)(x + 1)$
81. $(3y - 2)(3y + 2)(9y^2 + 4)$ **83.** $(27k + 1)(3k + 1)$
85. $(k + 4)(k - 3)(k + 3)$ **87.** $(2m^7 - 5)^2$
89. $(0.6x - 0.1)(0.6x + 0.1)$ **91.** $\left(\dfrac{1}{2}w - \dfrac{1}{3}v\right)\left(\dfrac{1}{2}w + \dfrac{1}{3}v\right)$
93. $y(y - 6)$ **95.** $(2p - 5)(2p + 7)$
97. $(-t + 2)(t + 6)$ or $-1(t - 2)(t + 6)$
99. $(-2b + 15)(2b + 5)$ or $-1(2b - 15)(2b + 5)$

Midchapter Review, pp. 345–346

1. A prime factor cannot be factored further. **2.** Factor out the
GCF. **3.** Look for the difference of squares: $a^2 - b^2$. **4.** Look for
a perfect square trinomial: $a^2 + 2ab + b^2$ or $a^2 - 2ab + b^2$. **5.** Not
all polynomials factor completely in one step. For example: $2x^4 - 32$.
6. $3(2x + 3)(x - 5)$ **7.** $2(5y - 1)(2y - 1)$ **8.** $abc^2(5ac - 7)$
9. $2(2a - 5)(2a + 5)$ **10.** $(t + 9)(t - 7)$ **11.** $(b + 10)(b - 8)$
12. $(b + y)(a - b)$ **13.** $3x^2y^4(2x + y)$ **14.** $(7u - 2v)(2u - v)$
15. Prime **16.** $2(2q^2 - 4q - 3)$ **17.** $3(3w^2 + w - 5)$ **18.** Prime
19. $5(b - 3)^2$ **20.** $(3r + 1)(2r + 3)$ **21.** $(2s - 3)(2s + 5)$
22. $(2a - 1)(2a + 1)(4a^2 + 1)$ **23.** $(p + c)(p - 3)(p + 3)$
24. $(9u - 5v)^2$ **25.** $4(x^2 + 4)$ **26.** $2(x - 3y)(a + 2b)$
27. $m(4m + 1)(2m - 3)$ **28.** $x^2y(3x + 5)(7x + 2)$
29. $2(m^2 - 8)(m^2 + 8)$ **30.** $(4v - 3)(2u + 3)$ **31.** $(t - 5)(4t + s)$
32. $3(2x - 1)^2$ **33.** $(p + q)^2$ **34.** $n(2n - 1)(3n + 4)$
35. $k(2k - 1)(2k + 3)$ **36.** $(8 - y)(8 + y)$ **37.** $b(6 - b)(6 + b)$

Section 4.5 Practice Exercises, pp. 353–354

1. $x^3 - y^3$ **3.** $x^2 + 2x + 4$ **5.** $2x + 1$ **7.** $x^3, 8, y^6, 27q^3, w^{12}, r^3s^6$
9. If the binomial is of the form $a^3 + b^3$.
11. $(a + b)(a^2 - ab + b^2)$ **13.** $(y - 2)(y^2 + 2y + 4)$
15. $(1 - p)(1 + p + p^2)$ **17.** $(w + 4)(w^2 - 4w + 16)$
19. $(10a + 3)(100a^2 - 30a + 9)$ **21.** $\left(n - \dfrac{1}{2}\right)\left(n^2 + \dfrac{1}{2}n + \dfrac{1}{4}\right)$
23. $(a + b^2)(a^2 - ab^2 + b^4)$ **25.** $(x^3 + 4y)(x^6 - 4x^3y + 16y^2)$
27. Prime **29.** $(a - b)(a + b)$ **31.** $(x^2 - 2)(x^2 + 2)$ **33.** Prime
35. $(t + 4)(t^2 - 4t + 16)$ **37.** Prime **39.** $4(b + 3)(b^2 - 3b + 9)$
41. $5(p - 5)(p + 5)$ **43.** $\left(\dfrac{1}{4} - 2h\right)\left(\dfrac{1}{16} + \dfrac{1}{2}h + 4h^2\right)$
45. $(x - 2)(x + 2)(x^2 + 4)$
47. $(q - 2)(q^2 + 2q + 4)(q + 2)(q^2 - 2q + 4)$
49. $4(b + 4)$ **51.** $(y + 3)(y + 1)$ **53.** $(2z + 3)(2z - 3)(4z^2 + 9)$
55. $5(r + 1)(r^2 - r + 1)$ **57.** $(7p - 1)(p - 4)$ **59.** $-2(x - 2)^2$
61. $2(3 - y)(9 + 3y + y^2)$ **63.** $(4t + 1)(t - 8)$

65. $(w - 5)(2x + 3y)$ **67.** $(2p - 3)(2p + 3)$
69. $\left(\frac{4}{5}p - \frac{1}{2}q\right)\left(\frac{16}{25}p^2 + \frac{2}{5}pq + \frac{1}{4}q^2\right)$ **71.** $(a^4 + b^4)(a^8 - a^4b^4 + b^8)$
73. a. The quotient is $x^2 + 2x + 4$. **b.** $(x - 2)(x^2 + 2x + 4)$
75. a. The quotient is $m^2 - m + 1$. **b.** $(m + 1)(m^2 - m + 1)$

Section 4.6 Practice Exercises, pp. 363–367

1. $(2x - 1)(2 + b)$ **3.** $4(b - 5)(b - 6)$ **5.** $(4w - 1)(4w + 1)$
7. $4(3k + 4)$ **9.** $(2y + 11)(y - 4)$ **11.** Linear   **13.** Quadratic
15. Neither   **17.** Quadratic   **19.** If $ab = 0$, then $a = 0$ or $b = 0$.
21. $x = -3, x = 1$ **23.** $x = \frac{7}{2}, x = -\frac{7}{2}$ **25.** $x = -5$
27. $x = 0, x = -\frac{1}{3}, x = -1$ **29.** $p = 5, p = -3$
31. $z = -12, z = 2$ **33.** $q = 4, q = -\frac{1}{2}$ **35.** $x = \frac{2}{3}, x = -\frac{2}{3}$
37. $k = 6, k = 8$ **39.** $m = 0, m = -\frac{3}{2}, m = 4$ **41.** The equation
must have one side equal to zero and the other side factored completely.
43. $x = 8, x = 2$ **45.** $p = \frac{7}{2}, p = -\frac{7}{2}$ **47.** $q = -5$
49. $x = 0, x = -\frac{1}{3}, x = -1$ **51.** $k = \frac{3}{4}, k = -3$
53. $p = 3, p = -2$ **55.** $w = 0, w = \frac{2}{3}$ **57.** $d = 0, d = -\frac{1}{4}$
59. $t = -2, t = 4, t = -4$ **61.** $w = -7, w = 5$ **63.** $k = 4, k = 2$
65. The numbers are $-\frac{9}{2}$ and 4.   **67.** The numbers are 5 and -4.
69. The numbers are 6 and 8 or -8 and -6.   **71.** The numbers are 0
and 1 or 9 and 10.   **73.** The painting has length 12 in. and width 10 in.
75. a. The picture is 13 in. by 6 in. **b.** 38 in.   **77.** The base is 10 cm
and the height is 25 cm.   **79.** 2 s   **81.** 0 s and 4 s   **83.** Given a right
triangle with legs a and b and hypotenuse c, then $a^2 + b^2 = c^2$.
85. Yes   **87.** No   **89.** The bottom of the ladder is 8 ft from the house.
The distance from the top of the ladder to the ground is 15 ft.
91. 10 m   **93. a.** Two diagonals **b.** Five diagonals **c.** Ten sides

Calculator Connections 4.7, pp. 373–374

1. x-intercept: $(2, 0)$; y-intercept: $(0, -4)$

2. x-intercept: $(2, 0)$; y-intercept: $(0, 6)$

3. x-intercept: $(2, 0)$; y-intercept: $\left(0, \frac{3}{2}\right)$

4. x-intercept: $(2, 0)$; y-intercept: $\left(0, -\frac{4}{5}\right)$

5. x-intercepts: $(-7, 0)$, $(1, 0)$; y-intercept: $(0, -7)$

6. x-intercepts: $(-1, 0)$, $(1, 0)$; y-intercept: $(0, -1)$

7. x-intercepts: $(-2, 0)$, $(3, 0)$; y-intercept: $(0, -12)$

8. x-intercepts: $(-2, 0)$, $(9, 0)$; y-intercept: $(0, -18)$

9. x-intercept: $(15, 0)$; y-intercept: $(0, -15)$

Section 4.7 Practice Exercises, pp. 374–378

1. $2(x - 9)(x + 8)$ **3.** $x = 9, x = -8$ **5.** The base is 16 ft, and the
height is 9 ft.   **7.** An x-intercept is a point $(a, 0)$ where a graph intersects the x-axis.   **9.** Substitute $x = 0$, and solve for y.
11. x-intercept: $(0, 0)$; y-intercepts: $(0, 0)(0, -3)$
13. x-intercept: $(-2, 0)$; y-intercepts: $(0, 4)(0, -3)$
15. x-intercepts: $(2, 0)(-2, 0)$; y-intercept: $(0, 2)$
17. x-intercept: $(4, 0)$; y-intercept: $(0, -2)$
19. x-intercept: $(-2, 0)$; y-intercept: $(0, 10)$
21. x-intercepts: $(4, 0)(-4, 0)$; y-intercept: $(0, -16)$
23. x-intercepts: $(5, 0)(2, 0)$; y-intercept: $(0, 10)$
25. x-intercept: $(-2, 0)$; y-intercept: $(0, 4)$
27. x-intercepts: $(-5, 0)(3, 0)(0, 0)$; y-intercept: $(0, 0)$
29. x-intercepts: $(4, 0)(-4, 0)(1, 0)$; y-intercept: $(0, 16)$
31. a. $(0, 0)$ **b.** 0 s after release, the ball is at ground level.
c. $(0, 0)(2, 0)$ **d.** At 0 s and 2 s after release, the ball is at ground level.
33. a. \$1000 **b.** \$500 **c.** $(0, 1500)$ If the air-conditioner lasts 0 years
(it breaks close to the time of purchase) the dealer will provide a full
refund of \$1500. **d.** $(3, 0)$ If the air-conditioner breaks after 3 years,
the dealer pays nothing. **35. a.** 65,000 tickets **b.** 50,000 tickets
c. $(0, 75000)$ If tickets cost \$0 (free), the promoter will sell 75,000
tickets. **d.** $(150, 0)$ If tickets cost \$150, the promoter will sell 0 tickets.

37. **39.**

Chapter 4 Review Exercises, pp. 382–385

1. 6 **3.** ab^4 **5.** $2c(3c - 5)$ **7.** $2x(3x + x^2 - 4)$ **9.** $16(2y^2 - 3)$
11. $t(-t + 5)$ or $-t(t - 5)$ **13.** $(b + 2)(3b - 7)$
15. $(w + 2)(7w + b)$ **17.** $(x - 6)(x - 4)$ **19.** $3(4y - 3)(5y - 1)$
21. $-6, 1$ **23.** $8, 3$ **25.** $5, -1$ **27.** $(c - 2)(3c + 1)$
29. $(t + 4s)(2t + 3s)$ **31.** $w(w + 5)(w - 1)$ **33.** $2(4v + 3)(5v - 1)$
35. $(x - 2)(x + 11)$ **37.** $ab(a - 6b)(a - 4b)$
39. $(3m - 1)(3m + 2)$ **41.** Different **43.** Both positive
45. $(2y + 3)(y - 4)$ **47.** $2(p - 6)(p + 4)$ **49.** $(2z + 5)(5z + 2)$
51. Prime **53.** $10(w - 9)(w + 3)$ **55.** $(3c - 5d)^2$
57. $(v^2 + 1)(v^2 - 3)$ **59.** $(2x - 5)^2$ **61.** $(c - 3)^2$ **63.** Not a
perfect square trinomial. The middle term does not equal $2(t)(7)$.
65. $(a - 7)(a + 7)$ **67.** Not a difference of squares. h is not a
perfect square. **69.** $(10 - 9t)(10 + 9t)$ **71.** This is a sum of squares.
73. $2(c^2 - 3)(c^2 + 3)$ **75.** $2(2x + 3)^2$ **77.** $(p + 3)(p - 4)(p + 4)$
79. $(a + b)(a^2 - ab + b^2)$ **81.** $(z - w)(z^2 + zw + w^2)$
83. $(4 + a)(16 - 4a + a^2)$ **85.** $(p^2 + 2)(p^4 - 2p^2 + 4)$
87. $6(x - 2)(x^2 + 2x + 4)$ **89.** x. **91.** vi.
93. $(6w - 1)(36w^2 + 6w + 1)$ **95.** $2(4 + v^2)(16 - 4v^2 + v^4)$
97. $(q - 1)(q^2 + q + 1)(q + 1)(q^2 - q + 1)$ **99.** $3(p - 1)^2$
101. $(k - 7)(k - 6)$ **103.** $q(q - 4)(q^2 + 4q + 16)$ **105.** Prime
107. $(x + 4)(x + 1)(x - 1)$ **109.** $x = \frac{1}{4}, x = -\frac{2}{3}$

111. $w = 0, w = -3, w = -\frac{2}{5}$ **113.** $k = -\frac{5}{7}, k = 2$

115. $q = 12, q = -12$ **117.** $v = 0, v = \frac{1}{5}$ **119.** $t = -\frac{5}{6}$

121. $y = \frac{2}{3}, y = 6$ **123.** The height is 6 ft, and the base is 13 ft.

125. Yes **127.** The numbers are -8 and 8. **129.** The height is 4 m
and the base is 9 m. **131.** x-intercept: $(8, 0)$; y-intercept $(0, -2)$
133. x-intercept: $(8, 0)$; y-intercept: $(0, 4)$
135. x-intercepts: $(4, 0)(-3, 0)$; y-intercept: $(0, -12)$
137. x-intercept: $(-1, 0)$; y-intercept: $(0, 1)$
139. a. x-intercept: $(0, 0)$; y-intercept: $(0, 0)$
b.

Chapter 4 Test, pp. 385–386

1. $3x(5x^3 - 1 + 2x^2)$ **2.** $(a - 5)(7 - a)$ **3.** $(6w - 1)(w - 7)$
4. $(13 - p)(13 + p)$ **5.** $(q - 8)^2$ **6.** $(2 + t)(4 - 2t + t^2)$
7. $3(a + 6b)(a + 3b)$ **8.** $(c - 1)(c + 1)(c^2 + 1)$
9. $(y - 7)(x + 3)$ **10.** Prime **11.** $-10(u - 2)(u - 1)$
12. $3(2t - 5)(2t + 5)$ **13.** $5(y - 5)^2$ **14.** $7q(3q + 2)$
15. $(2x + 1)(x - 2)(x + 2)$ **16.** $(y - 5)(y^2 + 5y + 25)$
17. $x = \frac{3}{2}, x = -5$ **18.** $x = 0, x = 7$ **19.** $x = 8, x = -2$

20. $x = \frac{1}{5}, x = -1$ **21.** The tennis court is 12 yd by 26 yd.
22. The shorter leg is 5 ft. **23. a.** x-intercept: $(9, 0)$; y-intercept:
$(0, -6)$ ii **b.** x-intercepts: $(5, 0)(-1, 0)$; y-intercept: $(0, -5)$ i
24. a. $(0, 0)(260, 0)$ **b.** The baker makes no profit if 0 loaves or 260
loaves of bread are produced. **c.** If the baker produces too many
loaves, the baker floods the market and cannot sell them all.

Cumulative Review Exercises, Chapters 1–4, pp. 387–388

1. $\frac{7}{5}$ **2.** $-\frac{20}{3}$ **3.** $x = 0.25$ **4.** $t = -3$ **5.** $\frac{8 - 3x}{-2}$ or $\frac{3x - 8}{2}$

6. The radius is 8.0 ft. **7.** There are 10 quarters, 12 nickels, and
7 dimes. **8.** ⟵—E———⟶ $[-4, \infty)$ **9.** $-\frac{7}{2}y^2 - 5y - 14$
    $\qquad\qquad\quad -4$

10. $8p^3 - 22p^2 + 13p + 3$ **11.** $4w^2 - 28w + 49$
12. $r^3 + 5r^2 + 15r + 40 + \frac{121}{r - 3}$ **13.** c^4 **14.** $\frac{b^6c^2}{4a^4}$ **15.** -15
16. 1.6×10^3 **17.** $(w - 2)(w + 2)(w^2 + 4)$ **18.** $(a + 5b)(2x - 3y)$
19. $(2a - 3)^2$ **20.** $(2x - 5)(2x + 1)$ **21.** $(y - 3)(y^2 + 3y + 9)$
22. $(p^2 + q^2)(p^4 - p^2q^2 + q^4)$ **23.** $x = 0, x = \frac{1}{2}, x = -5$

24. $x = -7, x = 5$
25. a. Quadratic **b.** $(-3, -5), (-2, 0), (-1, 3), (0, 4), (1, 3), (2, 0),$
$(3, -5)$ **c.**   **d.** $(-2, 0)(2, 0)$ **e.** $(0, 4)$

26. a. \$10,000 **b.** \$5000 **c.** $(0, 24000)$ The y-intercept represents the
initial value of the truck \$24,000. **d.** $(10, 0)$ The x-intercept indicates
that after 10 years the value of the truck is \$0.

CHAPTER 5

Section 5.1 Practice Exercises, pp. 397–399

1. a. A number $\frac{p}{q}$, where p and q are integers and $q \neq 0$.

b. An expression $\frac{p}{q}$, where p and q are polynomials and $q \neq 0$.

3. $-\frac{1}{8}$ **5.** $-\frac{5}{3}$ **7.** 0 **9.** Undefined **11. a.** $3\frac{1}{5}$ h or 3.2 h
b. $1\frac{3}{4}$ h or 1.75 h **13.** $\{k | k \neq -2\}$ **15.** $\{x | x \neq 5, x \neq -8\}$
17. $\left\{b \middle| b \neq -2, b \neq -3, b \neq \frac{5}{2}\right\}$ **19.** For example: $\frac{1}{x - 2}$
21. For example: $\frac{1}{(x + 3)(x - 7)}$ **23. a.** $\frac{5}{3}$ **b.** $\frac{5}{3}$ **25. a.** $\frac{2}{5}$ **b.** $\frac{2}{5}$
27. $\frac{b}{3}$ **29.** $\frac{3}{2}t^2$ **31.** $-\frac{3xy}{z^2}$ **33.** $\frac{1}{2}$ **35.** $\frac{p - 3}{p + 4}$ **37.** $\frac{1}{4(m - 11)}$
39. a. $\frac{3(y + 2)}{6(y + 2)}$ **b.** $\{y | y \neq -2\}$ **c.** $\frac{1}{2}$ **41. a.** $\frac{(t - 1)(t + 1)}{t + 1}$
b. $\{t | t \neq -1\}$ **c.** $t - 1$ **43. a.** $\frac{7w}{7w(3w - 5)}$ **b.** $\left\{w | w \neq 0, w \neq \frac{5}{3}\right\}$

c. $\dfrac{1}{3w-5}$ **45. a.** $\dfrac{(3x-2)(3x+2)}{2(3x+2)}$ **b.** $\left\{x\,|\,x\neq-\dfrac{2}{3}\right\}$ **c.** $\dfrac{3x-2}{2}$

47. a. $\dfrac{(a+5)(a-2)}{(a+3)(a-2)}$ **b.** $\{a\,|\,a\neq-3,a\neq2\}$ **c.** $\dfrac{a+5}{a+3}$ **49.** $\dfrac{y+3}{2y-5}$

51. $\dfrac{3x-2}{x+4}$ **53.** $\dfrac{5}{(q+1)(q-1)}$ **55.** $\dfrac{c-d}{2c+d}$ **57.** $\dfrac{7p-2q}{2}$

59. $5x+4$ **61.** $\dfrac{x+y}{x-4y}$ **63.** They are opposites.   **65.** -1 **67.** -1

69. $-\dfrac{1}{2}$ **71.** $-\dfrac{x+3}{4+x}$ **73.** $w-2$ **75.** $\dfrac{z+4}{z^2+4z+16}$

Section 5.2 Practice Exercises, pp. 407–409

1. $\{x\,|\,x\neq3,x=-2\}$, $\dfrac{x-1}{x-3}$ **3.** $\{a\,|\,a\neq2\}$, $\dfrac{a+2}{a-2}$

5. $\left\{t\,|\,t\neq\dfrac{1}{2}\right\}$, -2 **7.** $\dfrac{3}{10}$ **9.** 2 **11.** $\dfrac{5}{2}$ **13.** $\dfrac{15}{4}$ **15.** $\dfrac{x-6}{8}$

17. $\dfrac{2}{y}$ **19.** $-\dfrac{5}{8}$ **21.** $-\dfrac{b+a}{a-b}$ **23.** $\dfrac{10}{9}$ **25.** $\dfrac{6}{7}$ **27.** $\dfrac{y+9}{y-6}$

29. $\dfrac{t+4}{t+2}$ **31.** $-m(m+n)$ **33.** $\dfrac{3p+4q}{4(p+2q)}$ **35.** $\dfrac{w}{2w-1}$ **37.** $\dfrac{5}{6}$

39. $\dfrac{q+1}{q-6}$ **41.** $\dfrac{1}{4}$ **43.** $\dfrac{3t+8}{t+2}$ **45.** $\dfrac{x+4}{x+1}$ **47.** $-\dfrac{w-3}{2}$

49. $\dfrac{k+6}{k+3}$ **51.** 6.8 miles   **53.** 1.5 m by 0.76 m   **55.** 59,970 gal

57. Approximately 5.88×10^{12} miles   **59.** 41.8 kg   **61.** $0.61/ft^2

63. 2500 mL   **65.** 1.55 km   **67.** 2 **69.** $\dfrac{1}{a-2}$ **71.** $\dfrac{p+q}{2}$

Section 5.3 Practice Exercises, pp. 414–416

1. $\{x\,|\,x\neq1,x\neq-1\}$; $\dfrac{3}{5(x-1)}$ **3.** $\{t\,|\,t\neq-3,t\neq-1\}$; $\dfrac{t-4}{t+3}$

5. $\dfrac{a+5}{a+7}$ **7.** $\dfrac{4(a+3b)}{3(a-2b)}$ **9.** 56.8 L   **11.** 36   **13.** 6   **15.** $15p$

17. $12xyz^3$ **19.** $w^2+8w+12$ **21.** -6 **23.** $5t^2+13t-6$

25. $p^2-10p-24$ **27.** a, b, c, d   **29.** x^5 is the lowest power of x that has x^3, x^5, x^4 as factors.   **31.** The product of unique factors is $(x+3)(x-2)$.   **33.** Because $(b-1)$ and $(1-b)$ are opposites, they differ by a factor of -1.   **35.** 45   **37.** 16   **39.** 63   **41.** $9x^2y^3$

43. w^2y **45.** $(p+3)(p-1)(p+2)$ **47.** $9t(t+1)^2$

49. $(y-2)(y+2)(y+3)$ **51.** $12x(2x-1)$ or $12(1-2x)$

53. $\dfrac{10}{12a^2b}$, $\dfrac{a^3}{12a^2b}$ **55.** $\dfrac{6m-6}{(m+4)(m-1)}$, $\dfrac{3m+12}{(m+4)(m-1)}$

57. $\dfrac{6w+6}{(w+3)(w-8)(w+1)}$, $\dfrac{w^2+3w}{(w+3)(w-8)(w+1)}$

59. $\dfrac{6p^2+12p}{(p-2)(p+2)^2}$, $\dfrac{3p-6}{(p-2)(p+2)^2}$

61. $\dfrac{1}{a-4}$, $\dfrac{-a}{a-4}$ or $\dfrac{-1}{4-a}$, $\dfrac{a}{4-a}$ **63.** $\dfrac{z^2+3z}{(z+2)(z+7)(z+3)}$,
$\dfrac{-3z^2-6z}{(z+2)(z+7)(z+3)}$, $\dfrac{5z+35}{(z+2)(z+7)(z+3)}$

65. $\dfrac{3p+6}{(p-2)(p^2+2p+4)(p+2)}$, $\dfrac{p^3+2p^2+4p}{(p-2)(p^2+2p+4)(p+2)}$,
$\dfrac{5p^3-20p}{(p-2)(p^2+2p+4)(p+2)}$

Section 5.4 Practice Exercises, pp. 424–426

1. a. $-\dfrac{1}{2}$, -2, 0, undefined, undefined   **b.** $(x-5)(x-2)$;
$\{x\,|\,x\neq5,x\neq2\}$ **c.** $\dfrac{x+1}{x-2}$ **3.** $\dfrac{8}{b-1}$ **5.** 2 cups   **7.** $\dfrac{5}{4}$ **9.** $\dfrac{3}{8}$

11. 2 **13.** 5 **15.** $\dfrac{-2(t-2)}{t-8}$ **17.** $\dfrac{5}{3x-7}$ **19.** $m+5$ **21.** $\dfrac{15x}{y}$

23. $\dfrac{12}{15xy^3}$, $\dfrac{2x^2y}{15xy^3}$ **25.** $\dfrac{z+7}{3(z-3)}$, $\dfrac{3z-18}{3(z-3)}$

27. $\dfrac{6a^2}{(a-b)(a+b)a^2}$, $\dfrac{2a^2-2ab}{(a-b)(a+b)a^2}$ **29.** $\dfrac{19}{3(a+1)}$

31. $\dfrac{-3(k+4)}{(k-3)(k+3)}$ **33.** $\dfrac{a-4}{2a}$ **35.** $\dfrac{5}{3x-7}$ or $\dfrac{-5}{7-3x}$

37. $\dfrac{6n-1}{n-8}$ or $\dfrac{-6n+1}{8-n}$ **39.** $\dfrac{2(4x+5)}{x(x+2)}$ **41.** $\dfrac{2(w-3)}{(w+3)(w-1)}$

43. $\dfrac{4a-13}{(a-3)(a-4)}$ **45.** $\dfrac{2(3x+7)}{(x+3)(x+2)}$ **47.** $\dfrac{1}{n}$ **49.** $\dfrac{12}{p}$

51. $n+\left(7\cdot\dfrac{1}{n}\right)=\dfrac{n^2+7}{n}$ **53.** $\dfrac{1}{n}-\dfrac{2}{n}=-\dfrac{1}{n}$ **55.** $\dfrac{3k+5}{4k+7}$ **57.** $\dfrac{1}{a}$

59. $\dfrac{-w^2}{(w+3)(w-3)(w^2-3w+9)}$ **61.** $\dfrac{p^2-2p+7}{(p+2)(p+3)(p-1)}$

63. $\dfrac{-m-21}{2(m+5)(m-2)}$ or $\dfrac{m+21}{2(m+5)(2-m)}$

Midchapter Review, pp. 426–427

1. a. Factor the numerator and denominator completely, then reduce factors whose ratio is 1 or -1.   **b.** $\dfrac{a}{a-7}$ **2. a.** 1. Factor the denominator of each rational expression. 2. Identify the LCD. 3. Rewrite each rational expression as an equivalent expression with the LCD as the denominator. 4. Add or subtract the numerators and write the result over the common denominator. 5. Simplify and reduce.   **b.** $\dfrac{1}{3}$

3. a. Factor the numerators and denominators completely. Multiply across. Then reduce factors whose ratio is 1 or -1.   **b.** $\dfrac{x+3}{5}$

4. a. Factor the numerators and denominators completely. Multiply the first expression by the reciprocal of the second expression. Then reduce factors whose ratio is 1 or -1.   **b.** $\dfrac{c+3}{c}$ **5.** Adding and subtracting rational expressions   **6.** Multiplying, dividing, and reducing rational expressions   **7.** Multiplying, dividing, and reducing rational expressions.   **8.** Adding and subtracting rational expressions.

9. $\dfrac{a}{12b^4c}$ **10.** $\dfrac{2a-b}{a-b}$ **11.** $\dfrac{p-q}{5}$ **12.** 4 **13.** $\dfrac{10}{2x+1}$

14. $\dfrac{w+2z}{w+z}$ **15.** $(h-7)(h-1)$ **16.** $\dfrac{y+7}{x+a}$ **17.** $\dfrac{1}{2(a+3)}$

18. $\dfrac{2(3y+10)}{(y-6)(y+6)(y+2)}$ **19.** $(t+8)^2$ **20.** $b+7$

Section 5.5 Practice Exercises, pp. 433–434

1. $\{c\,|\,c\neq-1,c\neq2\}$; $\dfrac{c+3}{c+1}$ **3.** $\{x\,|\,x\neq2,x\neq-2\}$; $\dfrac{2}{x-2}$

5. $\dfrac{5w-6}{w(w-2)}$ **7.** $\dfrac{5}{12}$ **9.** $\dfrac{5(t-4)}{t+2}$ **11.** $\dfrac{1}{z+1}$ **13.** 368 square miles

15. $\dfrac{\frac{1}{2}+\frac{2}{3}}{5}=\dfrac{7}{30}$ **17.** $\dfrac{3}{\frac{2}{3}+\frac{3}{4}}=\dfrac{36}{17}$ **19.** $\dfrac{35}{2}$ **21.** $k+h$ **23.** $\dfrac{n+1}{2(n-3)}$

25. $\dfrac{2x+1}{4x+1}$ **27.** $m-7$ **29.** $\dfrac{2y(y-5)}{7y^2+10}$ **31.** $-\dfrac{a+8}{a-2}$ or $\dfrac{a+8}{2-a}$

33. $\dfrac{t-2}{t-4}$ **35.** $\dfrac{2z-5}{3(z+3)}$ **37.** $-\dfrac{x+1}{x-1}$ or $\dfrac{x+1}{1-x}$ **39. a.** $\dfrac{6}{5}\,\Omega$ **b.** $6\,\Omega$

41. $\dfrac{3}{2}$ **43.** $\dfrac{8}{5}$

Section 5.6 Practice Exercises, pp. 441–442

1. $\dfrac{2x+1}{(x-3)(x+2)}$ **3.** $\dfrac{(t-3)(t+1)}{(t-6)(t+2)}$ **5.** $5(h+1)$ **7.** 247.8 km

9. $z=4$ **11.** $p=-\dfrac{13}{12}$ **13.** $x=-\dfrac{1}{2}$ **15. a.** $4w$ **b.** $w=4$

17. a. $(x+3)(x-1)$ **b.** $x=-5$ **19.** $y=3, y=-1$ **21.** $a=4$
23. $w=5$; ($w=0$ does not check.)   **25.** $m=-5$ **27.** No solution;
($p=4$ does not check.)   **29.** $t=4$ **31.** No solution; ($x=-4$ does
not check.)   **33.** $x=4$; ($x=-6$ does not check.)   **35.** The number

is 8.   **37.** The number is -26.   **39.** $m=\dfrac{FK}{a}$ **41.** $E=\dfrac{IR}{K}$

43. $R=\dfrac{E-Ir}{I}$ or $R=\dfrac{E}{I}-r$

45. $B=\dfrac{2A-hb}{h}$ or $B=\dfrac{2A}{h}-b$ **47.** $h=\dfrac{V}{r^2\pi}$

Section 5.7 Practice Exercises, pp. 449–451

1. Equation; $b=30$ **3.** Expression; $\dfrac{2a-5}{(a+5)(a-5)}$

5. Expression; $\dfrac{3}{10}$ **7.** Equation; $p=2$ **9.** $a=\dfrac{40}{3}$ **11.** $x=40$

13. $y=3$ **15.** $z=-1$ **17.** $a=1$ **19. a.** $V_f=\dfrac{V_iT_f}{T_i}$ **b.** $T_f=\dfrac{T_iV_f}{V_i}$

21. 102   **23.** 4.8 ft   **25.** 225 miles   **27.** The speed of the current is
2 mph.   **29.** The wind is 35 mph.   **31.** The speeds are 45 mph and

60 mph.   **33.** $\dfrac{1}{2}$ of the room   **35.** $5\frac{5}{11}\ (5.\overline{45})$ minutes   **37.** $22\frac{2}{9}\ (22.\overline{2})$ min

39. $3\frac{1}{3}\ (3.\overline{3})$ days   **41. a.** 4 cm **b.** 5 cm   **43.** The light post is 24 ft high.

Chapter 5 Review Exercises, pp. 458–460

1. a. $-\dfrac{2}{9}, -\dfrac{1}{10}, 0, -\dfrac{5}{6}$, undefined **b.** $\{t\,|\,t\neq -9\}$ **3.** a, c, d

5. $\left\{h\,\Big|\,h\neq -\dfrac{1}{3}, h\neq -7\right\}$; $\dfrac{1}{3h+1}$ **7.** $\{w\,|\,w\neq 4, w\neq -4\}$; $\dfrac{2w+3}{w-4}$

9. $\{k\,|\,k\neq 0, k\neq 5\}$; $-\dfrac{3}{2k}$ **11.** $\{m\,|\,m\neq -1\}$; $\dfrac{m-5}{3}$

13. $\{p\,|\,p\neq -7\}$; $\dfrac{1}{p+7}$ **15.** $\dfrac{u^2}{2}$ **17.** $\dfrac{3}{2(x-5)}$ **19.** $\dfrac{q-2}{4}$

21. $4s(s-4)$ **23.** $\dfrac{1}{n-2}$ **25.** $\dfrac{1}{m+3}$ **27.** 250 cc   **29.** $2y+4$

31. $2r+6$ **33.** u^2-5u-6 **35.** xy^2z^4 **37.** $q(q+8)$
39. $(n-3)(n+3)(n+2)$ **41.** $3k-1$ or $1-3k$

43. $3-x$ or $x-3$ **45.** 2   **47.** $x-7$ **49.** $\dfrac{t^2+2t+3}{(2-t)(2+t)}$

51. $\dfrac{3(r-4)}{2r(r+6)}$ **53.** $\dfrac{q}{(q+5)(q+4)}$ **55.** $\dfrac{1}{3}$ **57.** $\dfrac{3(z+5)}{z(z-5)}$ **59.** $\dfrac{8}{y}$

61. $-(b+a)$ **63.** $-\dfrac{k+10}{k+4}$ **65.** $y=-2$ **67.** $w=2$ **69.** $p=3$

71. $y=-11, y=1$ **73.** $h=\dfrac{3V}{\pi r^2}$ **75.** $m=\dfrac{6}{5}$ **77.** 12 g

79. Together the pumps would fill the pool in 16.8 min.

Chapter 5 Test, pp. 460–461

1. a. $\{x\,|\,x\neq 2\}$ **b.** $-\dfrac{x+1}{6}$ **2. a.** $\{a\,|\,a\neq 0, a\neq 6, a\neq -2\}$

b. $\dfrac{7}{a+2}$ **3.** b, c, d   **4.** $\dfrac{y+7}{3(y+3)(y+1)}$ **5.** $-\dfrac{b+3}{5}$ **6.** $\dfrac{1}{w+1}$

7. $\dfrac{t+4}{t+2}$ **8.** $\dfrac{x(x+5)}{(x+4)(x-2)}$ **9.** $\dfrac{1}{m+4}$ **10.** 212.9 ft/s   **11.** $p=2$

12. $c=1$ **13.** $r=\dfrac{2A}{C}$ **14.** The number is $-\dfrac{2}{5}$.   **15.** $y=-8$

16. $1\frac{1}{4}$ (1.25) cups of carrots   **17.** The speed of the current is 5 mph.
18. It would take the second printer 3 h to do the job working alone.

Cumulative Review Exercises, Chapters 1–5, pp. 461–463

1. 32   **2.** 7   **3.** Rational: $\sqrt{4}, \sqrt{9}, \sqrt{16}, \sqrt{49}$; irrational: $\sqrt{5}, \sqrt{20}$

4. $y=\dfrac{10}{9}$ **5.** $(8,\infty)$ ⟵———⟶
                                                    8

6. ⟵—[———⟶ $[-1,\infty)$; $\{x\,|\,x<5\}$ ⟵———)—⟶
        -1 5

7. The width is 17 m and the length is 35 m.   **8.** The base is 10 in.
and the height is 8 in.   **9.** $x=10$; the angles are 37°.
10. 9.9932×10^{-1} g or 0.99932 g.   **11.** $12^2+16^2=20^2$ **12.** 1
                                                                              $144+256=400$
                                                                              $400=400$ ✓

13. $\dfrac{x^2yz^{17}}{2}$ **14. a.** $6x+4$ **b.** $2x^2+x-3$ **15.** $25x^2-30x+9$

16. $(5x-3)^2$ **17.** $4ab^2-b+\dfrac{a^2}{2}$ **18.** $3(3x-5y)(3x+5y)$

19. $(2c+1)(5d-3)$ **20.** $(x-5)(x+4)$ **21.** $\left\{x\,\Big|\,x\neq 5, x\neq -\dfrac{1}{2}\right\}$

22. $\dfrac{x-3}{x+5}$ **23.** $\dfrac{1}{5(x+4)}$ **24.** $x-3$ **25.** -3 **26.** $y=1$

27. $b=-\dfrac{7}{2}$ **28.** The speed of the current is 3 mph.   **29. a.** Linear

b.

x	y
0	-10
2	0
1	-5

c.

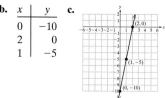

30. a. Nonlinear **b.** $(0,-25)$ **c.** $(5,0)(-5,0)$

CHAPTER 6

Calculator Connections 6.1, p. 474

1. **2.** **3.**

4. **5.** **6.**

Section 6.1 Practice Exercises, pp. 474–477

1. a. Yes **b.** No **c.** Yes **d.** Yes **3.** Linear **5.** Nonlinear
7. Linear **9.** The third point can be used as a check point.
11. $x - y = 4$ **13.** $2x - 5y = 10$ **15.** $y = -2x$

17. $y = \frac{1}{4}x - 2$ **19.** $-x + y = 0$ **21.** $4x - 5y = 15$

23. $y = -0.8x - 1.2$ **25.** $-30x - 20y = 60$

27. a. 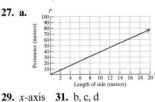   **b.** 8.8 m **c.** 66 m

29. x-axis   **31.** b, c, d

33. x-intercept: $(3, 0)$;
y-intercept: $(0, 15)$

35. x-intercept: $\left(\frac{3}{2}, 0\right)$;
y-intercept: $(0, -1)$

37. x-intercept: $(3, 0)$;
y-intercept: $(0, -3)$

39. x-intercept: $(0, 0)$;
y-intercept: $(0, 0)$

41. x-intercept: none;
y-intercept: $\left(0, \frac{8}{5}\right)$

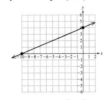

43. x-intercept: $(1, 0)$;
y-intercept: none

45. x-intercept: $(-10, 0)$;
y-intercept: $(0, 4)$

47. x-intercept: $(3, 0)$;
y-intercept: $(0, -1.5)$

49. x-intercept: $(0, 0)$;
y-intercept: $(0, 0)$

51. x-intercept: $(-2, 0)$;
y-intercept: none

53. x-intercept: none;
y-intercept: $(0, 2)$

55. a. $(0, 30{,}000)$ **b.** At the time of purchase ($n = 0$), the value of the car is \$30,000. **c.** \$15,000

57.

They have the same slope but different y-intercepts.
59. a. False, $x = 3$ is vertical. **b.** True
61. Vertical   **63.** Horizontal

65. Vertical

67. Horizontal

69. Vertical

71. Horizontal

73. a. 1555 h **b.** $y = 1555$ **75.** $y = 0$ **77.** $y = 2$ **79.** $x = 3$
81. When $c = 0$, the x- and y-intercepts are at the origin.

Section 6.2 Practice Exercises, pp. 485–489

1. x-intercept: $(6, 0)$;
y-intercept: $(0, -2)$

3. x-intercept: $(7, 0)$;
y-intercept: none

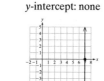

5. x-intercept: none;
y-intercept: $(0, \frac{3}{2})$

7. x-intercept: $(0, 0)$;
y-intercept: $(0, 0)$

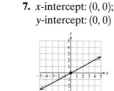

9. Undefined   **11.** Positive   **13.** Negative   **15.** Zero
17. Undefined   **19.** Positive   **21.** $\frac{3}{4}$ **23.** $-\frac{1}{2}$ **25.** -1 **27.** Zero
29. $\frac{5}{3}$ **31.** -3 **33.** Zero   **35.** Undefined   **37.** $\frac{28}{5}$ **39.** $-\frac{31}{35}$
41. -0.45 **43.** -1.6375 **45.** -1833.3 or $-\frac{5500}{3}$

47.

49.

51.

53. a. $\frac{765}{16}$ **b.** The number of male inmates increased by 765 thousand in 16 years, or approximately 47.8 thousand inmates per year.
55. a. Males had a larger rate of increase. **b.** No, the females will not catch up because their rate of increase is less than the males' rate of increase.

57.

59.

61.

63. a. -2 **b.** $\frac{1}{2}$ **65. a.** 0 **b.** Undefined   **67.** Parallel
69. Perpendicular   **71.** Neither   **73.** Parallel   **75.** $\pm\frac{1}{3}$
77. a. 1 mile **b.** 2 miles **c.** 3 miles **d.** $m = 0.2$; The distance between a lightning strike and an observer increases by 0.2 miles for every additional second between seeing lightning and hearing thunder.
79. $\frac{3m - 3n}{2b}$ **81.** $\left(\frac{c}{a}, 0\right)$ **83.** For example: $(7, 1)$

Calculator Connections 6.3, pp. 497–498

1. Perpendicular

2. Parallel

3. Neither

4. The lines may appear parallel; however, they are not parallel because the slopes are different.   **5.** The lines may appear to coincide on a graph; however, they are not the same line because the y-intercepts are different.

Section 6.3 Practice Exercises, pp. 498–501

1. a. x-intercept: $(10, 0)$;
y-intercept: $(0, -2)$
b.

c. $\frac{1}{5}$

3. a. x-intercept: none;
y-intercept: $(0, -3)$
b.

c. 0

5. a. x-intercept: $(0, 0)$;
y-intercept: $(0, 0)$
b.

c. $-\frac{2}{3}$

7. a. x-intercept: $(-5, 0)$;
y-intercept: none
b.

c. Undefined

9. $y = \frac{2}{5}x - \frac{4}{5}$; $m = \frac{2}{5}$; y-intercept: $\left(0, -\frac{4}{5}\right)$

11. $y = 3x - 5$; $m = 3$; y-intercept: $(0, -5)$
13. $y = -x + 6$; $m = -1$; y-intercept: $(0, 6)$ **15.** $x = 2$; Cannot be written in slope-intercept form; undefined slope; no y-intercept
17. $y = -\dfrac{1}{4}$; $m = 0$; y-intercept: $\left(0, -\dfrac{1}{4}\right)$
19. $y = \dfrac{2}{3}x$; $m = \dfrac{2}{3}$; y-intercept: $(0, 0)$

21.
23.

25.
27.

29.
31.

33. $y = \dfrac{1}{2}x - 3$
35. $y = -2x + 9$

37. $y = -2x - \dfrac{1}{3}$
39. $y = -\dfrac{1}{2}x + \dfrac{3}{2}$

41. $y = -x$
43. $y = 0.4x - 0.2$

45. $y = \dfrac{9}{5}x$
47. $y = -\dfrac{2}{3}$

49. $x = 2$
51. $y = -2x + 2$

53. Perpendicular   **55.** Parallel   **57.** Neither   **59.** $y = -\dfrac{1}{3}x + 2$
61. $y = 2x + 10$ **63.** $y = -1$ **65.** $y = 2x - 3$ **67.** $x = -3$
69. $y = \dfrac{1}{3}x - 1$ **71.** $x = 2$ **73.** $y = 4$ **75.** $y = 2$ **77.** $x = 3$
79. a. $m = 49.95$. The cost increases \$49.95 per day. **b.** $(0, 31.95)$. The cost to rent the car for 0 days is \$31.95. **c.** \$381.60
81. $y = -3x - 14$ **83.** $y = \dfrac{1}{3}x + 2$ **85.** $y = -\dfrac{1}{5}x - 2$
87. $y = 2x + 8.7$

Midchapter Review, pp. 501–502

1. Slope-intercept form   **2.** Standard form   **3.** Standard form
4. Slope-intercept form   **5.** Standard form   **6.** Slope-intercept form   **7.** Horizontal   **8.** Vertical   **9.** Vertical   **10.** Horizontal
11. Horizontal   **12.** Vertical

13. a.

x	y
2	$-\dfrac{2}{3}$
3	-2
-3	6

b. x-intercept: $\left(\dfrac{3}{2}, 0\right)$; y-intercept: $(0, 2)$

c. $y = -\dfrac{4}{3}x + 2$
d. Answers may vary.

14. a.

x	y
-3	4
-2	$\dfrac{13}{3}$
-6	3

b. x-intercept: $(-15, 0)$; y-intercept: $(0, 5)$

c. $y = \dfrac{1}{3}x + 5$
d. Answers may vary.

15. a.

x	y
1	-4
2	-8
-4	16

b. x-intercept: $(0, 0)$; y-intercept: $(0, 0)$

c. $y = -4x$
d. Answers may vary.

16. a.

x	y
0	$\dfrac{3}{4}$
1	$\dfrac{3}{4}$
2	$\dfrac{3}{4}$

b. x-intercept: none; y-intercept: $\left(0, \dfrac{3}{4}\right)$

c. $y = \dfrac{3}{4}$
d. Answers may vary.

17. a.

x	y
2	0
2	1
2	2

18.

b. x-intercept: $(2, 0)$;
   y-intercept: none

c. $x = 2$; not possible to write
   in slope-intercept form
d. Answers may vary.

19.   **20.**

21.

Section 6.4 Practice Exercises, pp. 507–509

1. $4x + 3y = 9$ **3.** $y = -2x$

5. $3x - 1 = 5$ **7.** $y = -4$

9. iv   **11.** vi   **13.** iii   **15.** 9   **17.** 0   **19.** Undefined

21. $y = 3x + 7$ **23.** $y = \dfrac{1}{4}x + 8$ **25.** $y = 4.1x - 23.93$

27. $y = -2$ **29.** $y = 2x - 2$ **31.** $y = -\dfrac{5}{8}x - \dfrac{19}{8}$

33. $y = -0.84x + 4.948$ **35.** $y = 1$ **37.** $x = 2$ **39.** $x = \dfrac{5}{2}$

41. $y = 2$ **43.** $x = -6$ **45.** $x = -4$ **47. a.** 5 **b.** $y = 5x + 67$
c. The median price of a one-family house in 2005 would be $192,000.
49. a. $y = 2x - 4$ **b.** $y = 2x - 4$ **c.** The equations are the same.

51. a. $y = -\dfrac{5}{2}x + 3$ **b.** $y = -\dfrac{5}{2}x + 3$ **c.** The equations are the same.

53. $y = -2x - 8$ **55.** $y = 3x - 8$ **57.** $y = -\dfrac{1}{5}x - 6$

Calculator Connections 6.5, pp. 515–516

1. 13.3   **2.** -42.3

3. 345   **4.** 95

Section 6.5 Practice Exercises, pp. 516–521

1. a. \$95 **b.** \$190 **c.** $(0, 0)$. For 0 kilowatt-hours used, the cost is \$0.
d. $m = 0.095$. The cost increases by \$0.095 for each kilowatt-hour
used. **e.**

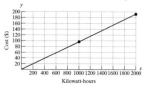

3. a. The women with less than 12 years of education **b.** The line
depicting the women with less than 12 years of education **c.** The line
depicting the women with more than 12 years of education **d.** The
women with less than 12 years of education **e.** The women with more
than 12 years of education **5. a.** y, temperature **b.** x, latitude
c. $30.7°$ **d.** $13.4°$ **e.** $m = -2.333$. The average temperature in January decreases $2.333°$ per $1°$ of latitude. **f.** $(53.2, 0)$. At $53.2°$ latitude,
the average temperature in January is $0°$.   **7. a.** $y = -\dfrac{13}{5}x + 169$

b. 153.4 h   **9. a.** $y = 3.5x - 1.75$ **b.** $m = 3.5$. For each additional
inch in length of a person's arm, the person's height increases by
3.5 in. **c.** 73.5 in. or 6 ft $1\frac{1}{2}$ in.   **11. a.** $y = 0.25x + 20$ **b.** \$84.50
13. a. $y = 25x + 20$ **b.** \$520.00   **15. a.** $y = 35x + 1200$
b. \$4700.00   **17. a.** $y = 0.8x + 100$ **b.** \$260.00

Chapter 6 Review Exercises, pp. 525–528

1. Yes   **3.** No   **5. a.** For example: $3x + 2y = 6$ **b.** For example:
$5x^2 + 8y^2 = 16$ **7.** x-intercept: $(0, 0)$; y-intercept: $(0, 0)$
9. x-intercept: $(2, 0)$; y-intercept: none

11. $y = -\dfrac{3}{2}$ **13.** $y = \dfrac{1}{4}x - 2$ **15.** $6x + y = 0$

17. For example: $y = -2$
19.   **21.**

23. $m = \dfrac{12}{5}$ **25.** $-\dfrac{2}{3}$ **27.** Undefined   **29. a.** -5 **b.** $\dfrac{1}{5}$

31. $m_1 = \dfrac{2}{3}; m_2 = \dfrac{2}{3};$ parallel   **33.** $m_1 = -\dfrac{5}{12}; m_2 = \dfrac{12}{5};$

perpendicular

35. $y = \dfrac{5}{2}x - 5; m = \dfrac{5}{2};$
y-intercept: $(0, -5)$

37. $y = -2x - 3; m = -2;$
y-intercept: $(0, -3)$

39. $y = \dfrac{1}{3}x; m = \dfrac{1}{3};$
y-intercept: $(0, 0)$

41. $y = -\dfrac{5}{2}; m = 0;$
y-intercept: $\left(0, -\dfrac{5}{2}\right)$

43. $y = -\dfrac{4}{3}x - 1$ **45.** $y = -3x - 16$ **47.** $y = 17$

49. $y = 4x - 20$ **51.** For example: $y = 3x + 2$ **53.** $m = \dfrac{y_2 - y_1}{x_2 - x_1}$

55. For example: $x = 6$ **57.** $y = -6x + 2$ **59.** $y = \dfrac{1}{4}x - 4$

61. $y = \dfrac{6}{5}x + 6$ **63. a.** y, the number of crimes **b.** x, the year where $x = 0$ corresponds to 1994 **c.** 22,100,000 **d.** $m = -1.8$. Between 1994 and 1997 the number of reported property crimes dropped by an average of 1.8 million per year. **e.** $(0, 32.9)$. In the year 1994 32.9 million property crimes were reported.   **65. a.** $y = 30.6x + 264$ **b.** \$508.8 billion   **67. a.** $y = 8x + 700$ **b.** \$1340

Chapter 6 Test, pp. 528–529

1. a. No **b.** No **c.** Yes   **2.** x-intercept: $\left(-\dfrac{3}{2}, 0\right)$; y-intercept: $(0, 2)$

3. $\dfrac{2}{5}$ **4. a.** $\dfrac{1}{3}$ **b.** $\dfrac{4}{3}$ **5. a.** $-\dfrac{1}{4}$ **b.** 4   **6. a.** Undefined **b.** 0

7. x-intercept: $\left(-\dfrac{1}{4}, 0\right)$;
y-intercept: $(0, 2)$

8. x-intercept: $(0, 0)$;
y-intercept: $(0, 0)$

9. x-intercept: $(3, 0)$;
y-intercept: none

10. x-intercept: none;
y-intercept: $\left(0, -\dfrac{1}{3}\right)$

11. Perpendicular   **12.** $y = \dfrac{1}{4}x + \dfrac{1}{2}$ **13.** $y = -\dfrac{7}{2}x + 15$

14. $y = -6$ **15.** $y = -\dfrac{1}{3}x + 1$ **16.** $y = 3x + 8$ **17. a.** $\dfrac{3}{4}$ **b.** $\dfrac{1}{2}$

18. a. $y = 1.5x + 10$ **b.** \$25   **19. a.** $m = -0.33$. The winning time has dropped by an average of 0.33 s/year between 1912 and 1996. **b.** $y = -0.33x + 82.2$ **c.** 70.32 s

Cumulative Review Exercises, Chapters 1–6, pp. 529–530

1. $\dfrac{1}{2}$ **2. a.** Rational **b.** Irrational   **3.** $\dfrac{7}{5}$ **4.** $x = -4$

5. $(-\infty, -4)$ **6.** All real numbers   **7.** Two angles are complementary if their sum is 90°. Two angles are supplementary if their sum is 180°.   **8.** The angles are 33° and 57°   **9.** 550,136 people

10. $\dfrac{a^8}{4b^2}$ **11.** 15   **12.** 5.0×10^{-5} **13. a.** $x^2 + 4x + 4$ **b.** $-12x - 3$

c. $x^2 - 8x + 1$ **14.** $3x^3 + 4x^2 + 8x + 21 + \dfrac{35}{x - 2}$

15. $2y(2x - 3)(4x - 1)$ **16.** $(2b - 3a)(2b + 3a)(4b^2 + 9a^2)$

17. $t = \dfrac{5}{3}$ **18.** The width is 3 yd and the length is 11 yd.

19. $\dfrac{4a^2b^8c^3}{5}$ **20.** $\{x \mid x \neq 3, x \neq -3\}$ **21.** $c + 3$ **22.** $\dfrac{c}{d}$

23. The number is 4.   **24.** $y = \dfrac{1}{2}$ **25. a.** Linear **b.** $\left(\dfrac{3}{2}, 0\right)$ **c.** $(0, 2)$

d. $y = -\dfrac{4}{3}x + 2$ **e.** $m = -\dfrac{4}{3}$ **f.**

26. $y = \dfrac{1}{2}x - 4$

CHAPTER 7

Calculator Connections 7.1, pp. 537–538

1. $(2, 1)$ **2.** $(6, -1)$ **3.** $(3, 1)$

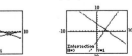

4. $(-2, 0)$ **5.** No solution   **6.** Dependent system

Section 7.1 Practice Exercises, pp. 539–541

1. Yes   **3.** No   **5.** Yes

7. a.   **b.**   **c.**

 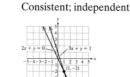

9. c   **11.** a   **13.** a   **15.** b   **17.** c   **19.** b   **21.** d

23. $(3, 1)$
Consistent; independent

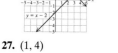

25. $(1, -2)$
Consistent; independent

27. $(1, 4)$
Consistent; independent

29. No solution
Inconsistent; independent

31. Infinitely many solutions
$\{(x, y) | y = 2x + 3\}$;
Consistent; dependent

33. $(-3, 6)$
Consistent; independent

35. Infinitely many solutions
$\{(x, y) | y = \frac{5}{3}x - 3\}$;
Consistent; dependent

37. $(2, -2)$
Consistent; independent

39. No solution
Inconsistent; independent

41. $(4, 2)$
Consistent; independent

43. No solution
Inconsistent; independent

45. $\left(\frac{1}{2}, 3\right)$
Consistent; independent

47. Infinitely many solutions
$\{(x, y) | y = 0.5x + 2\}$; Consistent; dependent

49. 4 lessons will cost $120 for each instructor.   **51.** 1997   **53.** For example: $4x + y = 9; -2x - y = -5$ **55.** For example: $2x + 2y = 1$

Section 7.2 Practice Exercises, pp. 548–551

1. $y = 2x - 4$; $y = 2x - 4$; coinciding lines   **3.** $y = -\frac{2}{3}x + 2$; $y = x - 5$; intersecting lines   **5.** $y = 4x - 4$; $y = 4x - 13$; parallel lines   **7.** $(3, -6)$ **9.** $(0, 4)$ **11.** No solution   **13.** Infinitely many solutions; $\{(x, y) | x = \frac{6}{5}y + 3\}$ **15. a.** y in the second equation is easiest to solve for because its coefficient is 1.   **b.** $(1, 5)$ **17.** $\left(\frac{1}{2}, 3\right)$
19. $(5, 2)$ **21.** $(10, 5)$ **23.** No solution   **25.** Infinitely many solutions; $\{(x, y) | y = \frac{1}{3}x + \frac{1}{3}\}$ **27.** $(5, -7)$ **29.** $\left(-5, \frac{3}{2}\right)$ **31.** $(2, -5)$
33. No solution   **35.** $(0, 2)$ **37.** Infinitely many solutions; $\{(x, y) | y = 0.25x + 1\}$ **39.** $(1, 1)$ **41.** No solution   **43.** $(-1, 5)$
45. $(-6, -4)$ **47.** The numbers are 48 and 58.   **49.** The angles are 165° and 15°.   **51.** The angles are 70° and 20°.   **53.** The angles are 42° and 48°.   **55.** A hot dog costs $1.75 and a drink costs $0.75.
57. a. 22 months   **b.** (22, 495); After renting a system for 22 months, both Company A and Company B will charge $495.   **c.** Company B is more expensive than Company A for more than 22 months. Company A is more expensive than Company B for less than 22 months.   **59.** For example: $(0, 3)(1, 5)(-1, 1)$

Section 7.3 Practice Exercises, pp. 558–561

1. Yes   **3.** No   **5. a.** False, multiply by -2.   **b.** True   **7. a.** y would be easier.   **b.** $(2, 1)$ **9.** The system will have no solution. The lines are parallel.   **11.** There are infinitely many solutions. The lines coincide.   **13.** $(2, 3)$ **15.** $(5, -2)$ **17.** $(0, 1)$ **19.** $(1, -2)$ **21.** No solution   **23.** Infinitely many solutions; $\{(x, y) | y = -\frac{1}{2}x + 1\}$
25. $(1, 4)$ **27.** $(-1, -2)$ **29.** $(2, 1)$ **31.** No solution   **33.** $(2, 3)$
35. $\left(\frac{7}{2}, \frac{5}{2}\right)$ **37.** $\left(\frac{1}{3}, 2\right)$ **39.** Infinitely many solutions;
$\{(x, y) | y = -5x + 1\}$ **41.** $\left(\frac{2}{3}, \frac{5}{3}\right)$ **43.** No solution   **45.** $(1, 4)$
47. $(4, 0)$ **49.** Infinitely many solutions; $\{(a, b) | b = \frac{9}{2}a - 4\}$
51. The numbers are 20 and 6.   **53.** The numbers are 4 and 6.
55. The cake has 340 Calories, and the ice cream has 120 Calories.

57. $\left(\frac{21}{5}, -\frac{4}{5}\right)$ **59.** $\left(\frac{8}{9}, -\frac{4}{9}\right)$ **61.** A line would have to "bend" to have two points of intersection. This is not possible. **63.** $A = -5$, $B = 2$

Midchapter Review, pp. 561–562

1. b. $(5, 0)$ **2. b.** $(4, -5)$ **3. b.** $(-2, -6)$ **4. b.** Infinitely many solutions; $\{(x, y)|y = 2x - 14\}$ **5. b.** No solution **6.** $\left(\frac{1}{2}, -\frac{5}{2}\right)$

7. $(-1, -1)$ **8.** $\left(-5, \frac{3}{4}\right)$ **9.** $\left(3, \frac{2}{5}\right)$ **10.** $(5, -7)$ **11.** $(2, -1)$

12. No solution **13.** $(2, 0)$ **14.** $(2200, 1000)$ **15.** $(3300, 1200)$
16. The numbers are 30 and 26. **17.** The numbers are 49 and 26.
18. Denisha is 26 and her sister is 22. **19.** The angles are 70° and 20°. **20.** The angles are 52° and 38°. **21.** The angles are 165° and 15°.

Section 7.4 Practice Exercises, pp. 568–570

1. $(-1, 4)$ **3.** $\left(\frac{5}{2}, 1\right)$ **5.** The angles are 80° and 10°. **7.** Dallas scored 30 points, and Buffalo scored 13 points. **9.** Tapes are $10.50 each, and CDs are $15.50 each. **11.** Technology stock costs $16 per share, and the mutual fund costs $11 per share. **13.** Shanelle invested $3500 in the 10% account and $6500 in the 7% account. **15.** Janise invested $8000 in the 7% account and $4000 in the 4% account. **17.** 15 gallons of the 50% mixture should be mixed with 10 gal of the 40% mixture. **19.** 12 gal of the 45% disinfectant solution should be mixed with 8 gal of the 30% disinfectant solution. **21.** The speed of the boat in still water is 6 mph, and the speed of the current is 2 mph. **23.** The speed of the plane in still air is 300 mph, and the wind is 20 mph. **25.** There are 17 dimes and 22 nickels. **27. a.** 835 free-throws and 1597 field goals **b.** 4029 points **c.** Approximately 50 points per game **29.** The speed of the plane in still air is 160 mph, and the wind is 40 mph. **31.** 12 pounds of candy should be mixed with 8 lb of nuts. **33.** $15,000 is invested in the 5.5% account, and $45,000 is invested in the 6.5% account. **35.** 20 oz of Miracle-Gro should be mixed with 40 oz of Green Light. **37.** There were 300 women and 200 men in the survey. **39.** There are six 15-s commercials and sixteen 30-s commercials.

Section 7.5 Practice Exercises, pp. 577–579

1. 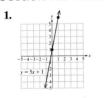 **3.** **5.**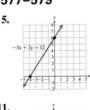

7. 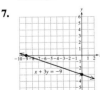 **9.** **11.**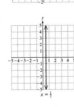

13. When the inequality symbol is $\leq$ or $\geq$ **15.** Test a point in each region. Shade the region that yields a true statement. **17.** a

19. For example: $(0, 5)(2, 7)(-1, 8)$

21. For example: $(1, -1)(3, 0)(-2, -9)$

23. For example: $(0, 0)(0, 2)(-1, -3)$

25. **27.** **29.**

31. **33.** **35.**

37.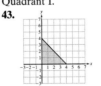

39. a. The set of ordered pairs above the line $x + y = 4$; for example: $(6, 3)(-2, 8)(0, 5)$. **b.** The set of ordered pairs on the line $x + y = 4$; for example: $(0, 4)(4, 0)(2, 2)$. **c.** The set of ordered pairs below the line $x + y = 4$; for example: $(0, 0)(-2, 1)(3, 0)$. **41.** This is Quadrant I.

43. **45.** **47.**

49. 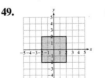 **51.**

Chapter 7 Review Exercises, pp. 585–589

1. Yes **3.** No **5.** Intersecting lines **7.** Parallel lines
9. Coinciding lines

11. $(0, -2)$
Consistent; independent

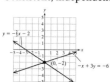

13. Infinitely many solutions
$\{(x, y) | y = -2x + 5\}$;
Consistent; dependent

15. $(1, -1)$
Consistent; independent

17. No solution
Inconsistent; independent

19. 1994 **21.** $(-4, 1)$ **23.** Infinitely many solutions
$\{(x, y) | y = -2x + 2\}$ **25. a.** y in the second equation is easiest to solve for because its coefficient is 1. **b.** $\left(\frac{9}{2}, 3\right)$ **27.** $(0, 4)$ **29.** No solution **31.** The angles are 42° and 48°. **33.** The angles are $115\frac{1}{3}°$ and $64\frac{2}{3}°$. **35.** $(-3, -2)$ **37.** $(2, -1)$ **39.** $\left(-\frac{1}{2}, \frac{1}{3}\right)$ **41.** Infinitely many solutions $\{(x, y) | y = -\frac{2}{3}x + \frac{1}{3}\}$ **43.** $(-4, -2)$ **45.** $(5, -3)$
47. There were 8 adult tickets and 52 childrens tickets sold.
49. 5 liters of the 20% solution should be mixed with 10 L of the 14% solution. **51.** Suzanne has seven quarters and three dollar coins.
53. a
55. For example:
$(1, -1)(0, -4)(2, 0)$

57. For example:
$(0, 0)(0, -5)(-1, 1)$

59.

61.

63.

65.

67.

69.

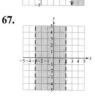

Chapter 7 Test, pp. 589–591

1. Parallel lines **2. a.** $(-3, 3)$

3. a. $15 **b.** 5,000,000 items **4.** $(-2, 0)$ **5.** Cooper had 680

points, and Swoopes had 453 points. **6.** $\left(2, -\frac{1}{3}\right)$ **7.** 12 milliliters of the 50% acid solution should be mixed with 24 mL of the 20% solution. **8. a.** No solution **b.** Infinitely many solutions **c.** One solution **9.** $(-5, 4)$ **10.** No solution **11.** Infinitely many solutions $\{(x, y) | y = -5x - 4\}$ **12.** $(1, -2)$ **13.** The angles are 57° and 33°.
14. a. 13 dimes, 7 quarters, 10 nickels **b.** $3.55 **15.** $7000 was invested at 9%, $12,000 was invested at 11%, totaling $19,000.
16. a. 15 months **b.** For 15 months of service the cost is $510 for either cable or a satellite dish. **17.** The plane is traveling at 470 mph, and the wind is 30 mph.
18.

19.

Cumulative Review Exercises, Chapters 1–7, pp. 591–592

1. $\frac{11}{6}$ **2.** Natural numbers $\sqrt{9}$; Whole numbers $\sqrt{9}, 0$; Integers $\sqrt{9}, 0, -4$; Rational numbers $\sqrt{9}, 0, -4, 6.7, \frac{1}{3}$; Irrational numbers $\pi, \sqrt{2}$
3. $x = -\frac{21}{2}$ **4.** No solution **5.** $y = \frac{3}{2}x - 3$
6. $\left[\frac{3}{11}, \infty\right)$ **7.** 180° **8.** The angles are 37°, 33°, 110°. **9.** The rates of the hikers are 2 mph and 4 mph.
10. a. $\frac{10d^5}{c^2}$ **b.** x^{18} **11.** 1.8×10^{13} **12.** $3y^3 - 4y^2 - 8y + 8$
13. $-w^2 - \frac{7}{10}w + \frac{5}{2}$ **14.** $\frac{2p}{q} - 1 + \frac{4q}{p}$ **15.** $5(x - 5)(x + 5)$
16. $(a - 2b)(5x - 2y)$ **17.** $y = 0, y = 3, y = -\frac{1}{2}$ **18.** $t = 5$
19. a. Nonlinear **b.** x-intercepts: $(4, 0)(-3, 0)$; y-intercept: $(0, -12)$
20. $\frac{5(x + 3)}{2x + 3}$ **21.** $\frac{2(a + 5)(a + 1)}{(a + 3)(a - 3)}$ **22.** $a = -5, a = -1$
23. In Problem 21 you must change the fractions to equivalent fractions with a common denominator. In Problem 22 you must clear the denominators. **24.** It would take them $1\frac{5}{7}$ h.
25. a. $-\frac{2}{3}$ **b.** $\frac{3}{2}$ **26.** $y = -3x + 3$
27. a. b.   **c.** $(0, 2)$ **28.** $(0, 2)$

29. a.

b.

c. Part (a) represents the solutions to an equation. Part (b) represents the solutions to a strict inequality. **30.** 20 gal of the 15% solution should be mixed with 40 gal of the 60% solution. **31.** x is 27°; y is 63° **32. a.** 1.4 **b.** Each year between 1920 and 1990 the winning speed in the Indianapolis 500 increased on average by 1.4 mph.

CHAPTER 8

Calculator Connections 8.1, pp. 601–602

1. 2. 3.

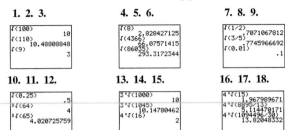

4. 5. 6.

7. 8. 9.

10. 11. 12.

13. 14. 15.

16. 17. 18.

19. 20.

Section 8.1 Practice Exercises, pp. 602–605

1. a. 8 **b.** −8 **3.** No, only positive numbers have two square roots. Zero has only one square root, and negative numbers have no real-valued square roots. **5.** 0, 1, 4, 16, 25, 36, 49 **7.** 0, 1, 8, 27 **9.** There is no real value of b for which $b^2 = -16$. **11.** Yes, −3 **13.** 13 **15.** 15 **17.** 3 **19.** 1 **21.** 0 **23.** −2 **25.** Not a real number **27.** Not a real number **29.** −2 **31.** Not a real number **33.** −4 **35.** $\dfrac{1}{3}$ **37.** 0.5 **39.** Not a real number **41.** 2.236 **43.** 7.071 **45.** 5.745 **47.** 8.944 **49.** 1.913 **51.** 4.021 **53.** 2.031 **55.** 1.968 **57.** $x^2, y^4, (ab)^6, w^8x^8, m^{10}$. The expression is a perfect square if the exponent is even. **59.** $p^4, t^8, (cd)^{12}$. The expression is a perfect fourth power if the exponent is a multiple of 4. **61.** 4 **63.** 4 **65.** 5 **67.** −5 **69.** 2 **71.** 2 **73.** 1 **75.** −1 **77.** 5 **79.** 5 **81.** $|a|$ **83.** y **85.** $|w|$ **87.** x **89.** $|m|$ **91.** c **93.** $5|x|$ **95.** $5|x|$ **97.** $5p^2$ **99.** $5p^2$ **101.** $\sqrt{q} + p^2$ **103.** $\dfrac{6}{\sqrt[4]{x}}$ **105.** The difference of 5 and the square root of 6 **107.** The product of 4 and the cube root of x **109.** 9 cm **111.** 5 ft **113.** 6.9 cm **115.** 17.0 in. **117.** 219 miles **119.** 268 km **121. a.** 3.6 ft **b.** 7.2 ft **c.** 144 ft **123.** $x \le 0$

Calculator Connections 8.2, p. 611

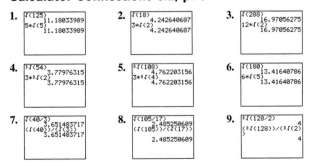

Section 8.2 Practice Exercises, pp. 612–614

1. 4, 16, 25, x^2, x^{20} **3.** 16, 81, w^4, w^{16}, w^{20} **5.** −5 **7.** −3 **9.** a^2 **11.** $2xy^2$ **13.** 446 km **15. a.** 8.944 **b.** $4\sqrt{5}$ **c.** 8.944 **d.** They are

the same. **17. a.** 9.899 **b.** $7\sqrt{2}$ **c.** 9.899 **d.** They are the same. **19.** $3\sqrt{2}$ **21.** $2\sqrt{7}$ **23.** $a^2\sqrt{a}$ **25.** w^{11} **27.** $m^2n^2\sqrt{n}$ **29.** $x^6y^5\sqrt{x}$ **31.** $2x\sqrt{2x}$ **33.** $4z\sqrt{z}$ **35.** $-3w^3\sqrt{5}$ **37.** $-z^5\sqrt{15z}$ **39.** $2ab^3\sqrt{26b}$ **41.** $\sqrt{26pq}$ **43.** $a^2\sqrt[3]{a^2}$ **45.** $2z\sqrt[3]{2}$ **47.** $2ab\sqrt[4]{ab^2}$ **49.** $u^2v\sqrt[5]{v^2}$ **51.** $2k\sqrt[5]{2}$ **53.** $m^2n^4\sqrt[4]{m^3}$ **55.** $2bc\sqrt[4]{3a^3b}$ **57.** $5u^2v^2\sqrt{3v}$ **59.** $\dfrac{\sqrt{3}}{4}$ **61.** $\dfrac{a^2}{b^2}$ **63.** a^4 **65.** $\dfrac{1}{2}$ **67.** $\dfrac{c\sqrt{c}}{2}$ **69.** $\dfrac{a^4\sqrt{a}}{b^2}$ **71.** $\dfrac{10\sqrt{2}}{9}$ **73.** $\dfrac{2}{5}$ **75.** $\dfrac{1}{2p}$ **77.** z **79.** $\dfrac{\sqrt[3]{x^2}}{3}$ **81.** $\dfrac{\sqrt[3]{y^2}}{3}$ **83.** $\dfrac{w}{2}$ **85.** $pq\sqrt[4]{q}$ **87.** $11\sqrt{2}$ ft **89.** $2\sqrt{66}$ cm **91.** $2a\sqrt[3]{2}$ **93.** $4a\sqrt{a}$ **95.** $\dfrac{2x\sqrt{x}}{y}$ **97.** $\dfrac{b}{3}$ **99.** $2\sqrt[4]{2}$ **101.** $2u^2v^3\sqrt{13v}$ **103.** $6\sqrt{6}$ **105.** 6 **107.** $\dfrac{1}{3}$ **109.** x **111.** $mn^7\sqrt[3]{15mn}$ **113.** $\sqrt[4]{8p^2q}$ **115.** $5\sqrt{2}$ **117.** $x + 5$

Calculator Connections 8.3, pp. 617–618

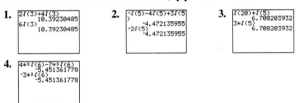

Section 8.3 Practice Exercises, pp. 618–620

1. $5w$ **3.** $z\sqrt[4]{4z}$ **5.** $\dfrac{a^2\sqrt{3a}}{b^2}$ **7.** $x\sqrt{2}$ **9.** Two radicals are *like* if they have the same radical factor (that is, same radicand and same index). **11.** ii **13.** $10\sqrt{6}$ **15.** $7\sqrt{3}$ **17.** $2\sqrt{11}$ **19.** $9\sqrt{x}$ **21.** 0 **23.** $-2x\sqrt{11}$ **25.** $9\sqrt{2} - 9\sqrt{5}$ **27.** $3\sqrt[3]{6} + 8\sqrt[3]{6}$; $11\sqrt[3]{6}$ **29.** $4\sqrt{5} - 6\sqrt{5}$; $-2\sqrt{5}$ **31.** $8\sqrt{3}$ **33.** 0 **35.** $2\sqrt{2}$ **37.** $16p^2\sqrt{5}$ **39.** $10\sqrt{2k}$ **41.** $a^2\sqrt{b}$ **43.** $3\sqrt[3]{5}$ **45.** $\dfrac{29}{18}z\sqrt[3]{6}$ **47.** $-1.7\sqrt[3]{10}$ **49.** $2x\sqrt[3]{x}$ **51.** $3\sqrt{7}$ **53.** $2\sqrt[3]{2w} + 2\sqrt[3]{3w} + 2\sqrt[3]{5w}$ **55.** $6x\sqrt[4]{x^2y}$ **57.** $5x\sqrt[3]{2} - 6\sqrt[3]{x}$ **59.** $9\sqrt{2}$ m **61.** $16\sqrt{3}$ in. **63.** $\dfrac{\sqrt{3}}{3}$ **65.** 0 **67. a.** 80 m **b.** 159 m **c.** 88 yd, 174 yd

Calculator Connections 8.4, p. 625

Section 8.4 Practice Exercises, pp. 625–627

1. 3 **3.** $\sqrt{3}$ **5.** $20x\sqrt{2}$ **7.** 7.348 **9.** 1.162 **11.** −1.325 **13.** 2.309 **15.** Commutative property of multiplication **17.** Associative property of multiplication **19.** Distributive property of multiplication over addition **21. a.** $3x + 6$ **b.** $\sqrt{3x} + \sqrt{6}$ **23. a.** $4a^2 + 12a + 9$ **b.** $4a + 12\sqrt{a} + 9$ **25. a.** $b^2 - 25$ **b.** $b - 25$

27. $\sqrt{15}$ **29.** 11   **31.** $5\sqrt{2}$ **33.** $7\sqrt{2}$ **35.** 6   **37.** Perimeter: $6\sqrt{5}$ ft; area: 10 ft^2 **39.** 3 cm^2 **41.** $3w$ **43.** $-16\sqrt{10y}$
45. $2\sqrt{3} - \sqrt{6}$ **47.** $4x + 20\sqrt{x}$ **49.** $-8 + 7\sqrt{30}$
51. $9a - 28b\sqrt{a} + 3b^2$ **53.** $8p^2 + 19p\sqrt{p} + 2p - 8\sqrt{p}$
55. $29 + 8\sqrt{13}$ **57.** $a - 4\sqrt{a} + 4$ **59.** $4a - 12\sqrt{a} + 9$
61. $x - 4\sqrt{xy} + 4y$ **63.** 1   **65.** $x - y$ **67.** $21 - 2\sqrt{110}$ **69.** -1
71. 4   **73.** $64x - 4y$ **75.** 14 m^3 **77.** $x + 2\sqrt[3]{x}$ **79.** $3 - \sqrt[3]{9}$
81. $1 + \sqrt[3]{3} - 2\sqrt[3]{9}$ **83.** $4\sqrt[3]{x^2} + 27\sqrt[3]{x} - 7$ **85.** $a - b$
87. $x + y$

Midchapter Review, pp. 627–628

1. -6 **2.** 5   **3.** 2   **4.** $\dfrac{11}{5}$ **5.** Not a real number   **6.** Not a real number   **7.** -7 **8.** -13 **9.** $|m|$ **10.** $|p|$ **11.** m **12.** p
13. $2q^3\sqrt{2}$ **14.** $3p^4\sqrt{3}$ **15.** $5u^3v^4\sqrt[3]{u^2}$ **16.** $4r^5s^4\sqrt[3]{s}$ **17.** $5m$
18. $3u^2$ **19.** $x^2\sqrt{10}$ **20.** $y\sqrt{3}$ **21.** $-3\sqrt{6}$ **22.** $11\sqrt{a}$ **23.** $4x\sqrt{2}$
24. 0   **25.** $5 + \sqrt{35}$ **26.** $a + 2\sqrt{a}$ **27.** $26 - 17\sqrt{2}$
28. $4x - 11\sqrt{xy} - 3y$ **29.** $51 + 14\sqrt{2}$ **30.** $8 + 2\sqrt{15}$
31. $16 - 2\sqrt{55}$ **32.** $x - 12\sqrt{x} + 36$ **33.** -88 **34.** $u - 9v$
35. a. 5.7 miles **b.** 9 miles   **36. a.** 10 m/s **b.** 7.7 m/s   **37.** 13.7 in.
38. 9.2 ft   **39. a.** 4.6 cm **b.** 6 m   **40. a.** 9.5 ft **b.** 24.5 cm
41. 0.890   **42.** 0.031   **43.** -0.090 **44.** 7.969

Calculator Connections 8.5, pp. 634–635

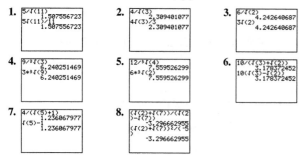

Section 8.5 Practice Exercises, pp. 635–637

1. $11x\sqrt{5}$ **3.** $6y + 23\sqrt{y} + 21$ **5.** $9\sqrt{3}$ **7.** $25 - 10\sqrt{a} + a$
9. -5 **11. a.** 0.707 **b.** $\dfrac{\sqrt{2}}{2}$ **c.** 0.707   **13. a.** -0.812
b. $\dfrac{\sqrt{2} + \sqrt{7}}{-5}$ **c.** -0.812 **15.** $\dfrac{\sqrt{6}}{6}$ **17.** $3\sqrt{5}$ **19.** $\dfrac{6\sqrt{x+1}}{x+1}$
21. $\dfrac{\sqrt{6x}}{x}$ **23.** $\dfrac{\sqrt[3]{36}}{6}$ **25.** $\dfrac{7\sqrt[3]{2}}{2}$ **27.** $\dfrac{5\sqrt[3]{6y}}{3y}$ **29.** $\dfrac{2\sqrt[4]{27}}{3}$ **31.** -7
33. $\sqrt{5} + \sqrt{3}$; 2   **35.** $\sqrt{x} - 10$; $x - 100$ **37.** $\dfrac{4\sqrt{2} - 12}{-7}$ or
$\dfrac{12 - 4\sqrt{2}}{7}$ **39.** $\dfrac{\sqrt{5} + \sqrt{2}}{3}$ **41.** $\sqrt{6} - \sqrt{2}$ **43.** $\dfrac{\sqrt{x} + \sqrt{3}}{x - 3}$
45. $-13 - 6\sqrt{5}$ **47.** $2 - \sqrt{2}$ **49.** $\dfrac{3 + \sqrt{2}}{2}$ **51.** $1 - \sqrt{7}$
53. $\dfrac{7 + 3\sqrt{2}}{3}$ **55.** 1. The radicand has no factor raised to a power greater than or equal to the index. 2. There are no radicals in the denominator of a fraction. 3. The radicand does not contain a fraction.
57. $-6y^2\sqrt{3}$ **59.** $\dfrac{2a\sqrt{14}}{7}$ **61.** $3b\sqrt[3]{2b}$ **63.** $\dfrac{y\sqrt[4]{3x^2y}}{2z^2}$ **65.** $-\dfrac{2}{3}$

67. $-ac^3\sqrt{ab}$ **69.** $-\dfrac{2\sqrt{5}}{5}$ **71.** $\dfrac{\sqrt{7}}{3}$ **73.** $-\dfrac{2x^2\sqrt[3]{2yz^2}}{yz}$
75. $\dfrac{p\sqrt[4]{36p^3}}{2}$

Section 8.6 Practice Exercises, pp. 642–643

1. $\dfrac{\sqrt{3} + \sqrt{7}}{-4}$ or $-\dfrac{\sqrt{3} + \sqrt{7}}{4}$ **3.** $\sqrt{6}$ **5.** $\sqrt[3]{5}$ **7.** $2 - \sqrt{3}$
9. $x^2 + 8x + 16$ **11.** $4a^2 - 20a + 25$ **13.** $2x - 3$ **15.** $t - 9$
17. $7 - n$ **19.** $x = 15$ **21.** No solution   **23.** $t = 5$ **25.** $n = \dfrac{17}{2}$
27. $w = 0$ **29.** $a = 8$ **31.** $c = 1$ **33.** $y = -8$ **35.** No solution; ($m = 1$ does not check.)   **37.** $t = 3, t = -1$ **39.** $p = 11$; ($p = 2$ does not check.)   **41. a.** 2 s **b.** 256 ft   **43. a.** 72 mph **b.** 225 ft

Calculator Connections 8.7, p. 648

Section 8.7 Practice Exercises, pp. 649–650

1. a. 3 **b.** 125   **3.** 27   **5.** $a + 1$ **7.** $a^{m/n} = \sqrt[n]{a^m}$ or $(\sqrt[n]{a})^m$, provided the root exists.   **9.** 9   **11.** 5   **13.** 3   **15.** -2 **17.** -2 **19.** 6
21. $\sqrt[3]{x}$ **23.** $\sqrt{4a}$ or $2\sqrt{a}$ **25.** $\sqrt[5]{yz}$ **27.** $\sqrt[3]{u^2}$ **29.** $5\sqrt{q}$
31. $\sqrt{\dfrac{x}{9}}$ or $\dfrac{\sqrt{x}}{3}$ **33.** $x^{1/3}$ **35.** $(xy)^{1/3}$ **37.** $5x^{1/2}$ **39.** $y^{2/3}$
41. $(m^3n)^{1/4}$ **43.** $4k^{3/3}$ or $4k$ **45.** x **47.** y^2 **49.** 6   **51.** a^7 **53.** $y^{4/3}$
55. 2   **57.** $\dfrac{25a^4d}{c}$ **59.** $\dfrac{y^9}{x^8}$ **61.** $\dfrac{2z^3}{w}$ **63.** $5xy^2z^3$ **65.** 4.3589
67. 0.3761   **69.** 2.9240   **71.** 31.6228   **73. a.** 10 in. **b.** 8.49 in.
75. a. 10.9% **b.** 8.8% **c.** The account in part (a)
77. $\dfrac{3}{4}$ **79.** -5 **81.** $\dfrac{x^{16}}{y^7z^4}$

Chapter 8 Review Exercises, pp. 655–659

1. Principal square root: 14; negative square root: -14 **3.** Principal square root: 15; negative square root: -15 **5.** There is no real number b such that $b^2 = -64$.   **7.** -12 **9.** Not a real number   **11.** $|y|$
13. $|y|$ **15.** -5 **17.** p^4 **19.** $\dfrac{3}{t^2}$ **21.** -2 **23.** 3.162   **25.** 2.466
27. 1.682   **29.** 2.414   **31. a.** 7.1 m **b.** 22.6 ft   **33.** $b^2 + \sqrt{5}$
35. The quotient of 2 and the square root of p **37.** 96 m^2
39. 50 miles   **41.** 331 miles   **43.** $a^2\sqrt[5]{a}$ **45.** $2\sqrt{7}$ **47.** $3y^5\sqrt{3}$
49. $2x\sqrt[4]{6x^3}$ **51.** $\dfrac{a^4}{5}$ **53.** $\dfrac{3}{x}$ **55.** $10y$ **57.** $\dfrac{\sqrt[3]{30p^2}}{q^2}$ **59.** $\dfrac{t^3\sqrt{5}}{11}$
61. $\dfrac{7}{3}$ **63.** $0.8\sqrt{y}$ **65.** $-\sqrt[3]{11}$ **67.** $11y\sqrt{y}$ **69.** $-8\sqrt{2}$
71. $6m^2n\sqrt[4]{mn^3}$ **73.** $\sqrt[3]{128} - \sqrt[3]{2}$; $3\sqrt[3]{2}$ **75.** $48\sqrt{3}$ m   **77.** $\dfrac{7\sqrt{10}}{2}$

79. $2\sqrt{15p}$ **81.** $-6yz\sqrt{11}$ **83.** $\sqrt{14} + 8\sqrt{2}$
85. $4p + 7\sqrt{pq} - 2q$ **87.** $4x^2 - 4x\sqrt{y} + y$ **89.** $10\sqrt{3}$ m³
91. b **93.** $\dfrac{3\sqrt{2y}}{y}$ **95.** $\dfrac{5\sqrt{3}}{3}$ **97.** $2\sqrt{7} + 2\sqrt{2}$ **99.** $-8 - 3\sqrt{7}$
101. $\dfrac{3\sqrt[3]{2x}}{x}$ **103.** $\dfrac{\sqrt{6}}{y^2}$ **105.** $p = 138$ **107.** $y = -69$ **109.** $n = 5$
111. $z = 7$ **113.** $m = 2; (m = -2 \text{ does not check.})$
115. a. 9261 in.³ **b.** 3375 cm³ **117.** 11 **119.** Not a real number
121. 27 **123.** $\sqrt[3]{q^2}$ **125.** $\sqrt{\dfrac{b}{121}} = \dfrac{\sqrt{b}}{11}$ **127.** $5m^{2/3}$ **129.** $6^{1/2}$
131. $6^{3/5}$ **133.** $5^{3/4}$ **135.** b^{15} **137.** 0.3684 **139.** 6.9776

Chapter 8 Test, pp. 659–661

1. 1. The radicand has no factor raised to a power greater than or equal to the index. 2. There are no radicals in the denominator of a fraction. 3. The radicand does not contain a fraction. **2.** $11x\sqrt{2}$
3. $2y\sqrt[3]{6y}$ **4.** Not a real number **5.** $\dfrac{a\sqrt[5]{5a^2}}{3}$ **6.** $\dfrac{3\sqrt{6}}{2}$
7. $\dfrac{2\sqrt{5} - 12}{-31}$ or $\dfrac{12 - 2\sqrt{5}}{31}$ **8. a.** $\sqrt{25} + 5^3; 130$
b. $4^2 - \sqrt[4]{16}; 14$ **9. a.** 6.164 **b.** 2.115 **10. a.** 97 ft **b.** 339 ft
11. $8\sqrt{z}$ **12.** $4\sqrt{6} - 15$ **13.** $-7t\sqrt{2}$ **14.** $9\sqrt{10}$
15. $-8 + 23\sqrt{10}$ **16.** $\dfrac{\sqrt{n}}{6m}$ **17.** 206 yd **18.** No solution
19. $x = 14$ **20.** $x = 0, x = -5$ **21. a.** 12 in. **b.** 18 in. **c.** 25 weeks
22. 1000 **23.** 2 **24.** $\sqrt[5]{x^3}$ or $(\sqrt[5]{x})^3$ **25.** $5\sqrt{y}$ **26.** $(ab^3)^{1/4}$
27. $p^{11/12}$ **28.** $5^{3/5}$ **29.** $3mn^2$

Cumulative Review Exercises, Chapters 1–8, pp. 661–663

1. 1 **2.** $\dfrac{217}{6}$ **3.** $y = -2$ **4.** $a = \dfrac{A - b - c}{2}$
5. $\{x \mid x > -4\}; (-4, \infty)$ ⟵————→ **6.** The number is 15.
                                    -4
7. One hiker walks at a rate of 1.5 mph and the other walks 3 mph.
8. x^4 **9.** -15 **10.** $-9x^2 + 2x + 10$ **11.** $\dfrac{1}{4}c^2 + 4c + 16$
12. $\dfrac{2x}{y} - 1 + \dfrac{4}{x}$ **13.** $(3a + b)(2x - y)$
14. $(m - 3)(m + 3)(m^2 + 9)$ **15.** $2(5c + 2)^2$ **16.** $x = -\dfrac{2}{5}, x = \dfrac{1}{2}$
17. iii; x-intercept: $(3, 0)$; y-intercept: $(0, 2)$
18. i; x-intercepts: $(2, 0)(-4, 0)$; y-intercept: $(0, -8)$
19. ii; x-intercepts: $(3, 0)(-3, 0)$; y-intercept: $(0, -9)$
20. $\{x \mid x \neq 5, x \neq -1\}$ **21.** 1 **22.** $\dfrac{1}{2}$ **23.** No solution
24. 320 Calories **25.** $\dfrac{x(5x + 8)}{16(x - 1)}$
26. **27.**

28. a. $y = 880$; The cost of renting the office space for 3 months is $880. **b.** $x = 12$; The cost of renting office space for 12 months is $2770. **c.** $m = 210$; The increase in cost is $210 per month.
d. $(0, 250)$; The down payment of renting the office space is $250.

29. $y = -x + 1$ **30. a.** $-\dfrac{1}{2}$ **b.** 2 **31.** $(1, -4)$
32. Infinitely many solutions; $\{(x, y) \mid 3x - y = 6\}$
33.
34. The angles are 17° and 73°. **35.** 8 L of 20% solution should be mixed with 4 L of 50% solution. **36.** -5 **37.** $3\sqrt{11}$ **38.** Not a real number **39.** $2cd^2\sqrt[3]{2c^2}$ **40.** $7x\sqrt{3}$ **41.** $\sqrt{10} + \sqrt{35} - \sqrt{6} - \sqrt{21}$
42. $\dfrac{x + \sqrt{xy}}{x - y}$ **43.** $\dfrac{5\sqrt{a}}{a}$ **44.** $y = 14$

CHAPTER 9

Calculator Connections 9.1, pp. 676–677

1. **2.** **3.**

4. **5.** **6.**

Section 9.1 Practice Exercises, pp. 677–680

1. Any set of ordered pairs (x, y) is called a relation in x and y.
3. By the vertical line test **5.** Yes **7.** No **9.** Yes **11.** No
13. $-5; -1; -11$ **15.** $\dfrac{1}{5}, \dfrac{1}{4}, \dfrac{1}{2}$ **17.** The function value at $x = 6$ is 2.
19. The function value at $x = \frac{1}{2}$ is $\frac{1}{4}$. **21.** $(2, 7)$ **23.** $(0, 8)$
25. a. $C(0) = 1.50$. The cost of a scoop of ice-cream with 0 toppings is $1.50. **b.** $C(1) = 2.00$. The cost of a scoop of ice-cream with 1 topping is $2.00. **c.** $C(4) = 3.50$. The cost of a scoop of ice-cream with four toppings is $3.50. **27. a.** $s(0) = 0$. The initial speed is 0 m/s.
b. $s(4) = 39.2$. The speed of an object 4 s after being dropped is 39.2 m/s. **c.** $s(6) = 58.8$. The speed of an object 6 s after being dropped is 58.8 m/s. **d.** 77.42 m/s **29. a.** $v(0) = 64$. The initial velocity is 64 ft/s. **b.** $v(1) = 32$. The velocity after 1 s is 32 ft/s. **c.** $v(2) = 0$. The velocity after 2 s is 0 ft/s. **d.** $v(4) = -64$. The velocity after 4 s is -64 ft/s, (the football is descending). **31.** 1 **33.** 6 **35.** m^4
37. $(a + b)^2$ or $a^2 + 2ab + b^2$ **39.** The domain of a relation is the set of first coordinates of the ordered pairs. **41.** $(-\infty, \infty)$
43. $[-1, \infty)$ **45.** $(-\infty, \infty)$ **47.** $(-\infty, 5]$ **49.** $\{x \mid x \neq -6\}$
51. $\{x \mid x \neq -3, x \neq 4\}$ **53.** $\{x \mid x \neq 6, x \neq -6\}$
55. $\{x \mid x \neq -\frac{1}{2}, x \neq 5\}$ **57.** b **59.** c

Section 9.2 Practice Exercises, pp. 688–689

1. $7i$ **3.** $i\sqrt{15}$ **5.** $2i\sqrt{3}$ **7.** i **9.** -20 **11.** -6 **13.** 3 **15.** 5
17. $14i$ **19.** $-24i$ **21.** $-5 + 5i$ **23.** -1 **25.** 1 **27.** i **29.** $-i$
31. Add or subtract the real parts. Add or subtract the imaginary parts.
33. $-6 + 8i$ **35.** $10 - 10i$ **37.** $6 + 3i$ **39.** $-7 + 3i$ **41.** $7 - 21i$

43. $11 - 9i$ **45.** $9 + 19i$ **47.** $-\frac{1}{4} - \frac{1}{5}i$ **49.** $-3.5 + 18.1i$ **51.** 13

53. 104 **55.** $\frac{5}{4}$ **57.** $35 - 12i$ **59.** $21 + 20i$ **61.** $-33 - 56i$

63. $7 + 4i; 65$ **65.** $\frac{3}{2} - \frac{2}{5}i; \frac{241}{100}$ **67.** $-4i; 16$ **69.** $\frac{14}{65} + \frac{8}{65}i$

71. $\frac{5}{2} - \frac{5}{2}i$ **73.** $-\frac{3}{5} - \frac{6}{5}i$ **75.** $-\frac{2}{13} + \frac{10}{13}i$ **77.** $\frac{15}{17} + \frac{8}{17}i$

79. $\frac{23}{29} - \frac{14}{29}i$ **81.** $\frac{1}{18}i$ **83.** $\frac{3}{5}i$ **85.** False. For example: $2 + 3i$ is not a real number.   **87.** True   **89.** False. $\sqrt[3]{-64} = -4$.   **91.** False. $(1 + 4i)(1 - 4i) = 17$.   **93.** True   **95.** False. $i^4 = 1$.   **97.** True

Section 9.3 Practice Exercises, pp. 696–698

1. a. Linear   **b.** Quadratic   **c.** Quadratic   **3.** Plus or minus

5. $x = \pm 4$ **7.** $m = \pm 8$ **9.** $q = \pm 5i\sqrt{2}$ **11.** $v = \frac{\pm\sqrt{5}}{2}$

13. $b = -2, b = -4$ **15.** $y = -2, y = -6$ **17.** $z = 2 \pm \sqrt{7}$

19. $b = 1 \pm 2i\sqrt{3}$ **21.** $t = \frac{2}{3}, t = 0$ **23.** $x = -\frac{5}{2} \pm \sqrt{21}$

25. The solution checks.   **27.** The solution checks.   **29.** False. The solutions are $\pm 5i$.   **31.** True   **33.** False. The solutions are 10 and -10.   **35.** 23.3 ft   **37.** 17.0 in.   **39.** 9   **41.** 64   **43.** $\frac{25}{4}$ **45.** $\frac{1}{64}$

47. $\frac{25}{196}$ **49.** $\frac{1}{4}$ **51.** $x = 4, x = -2$ **53.** $t = 1, t = -11$

55. $x = 3 \pm \sqrt{7}$ **57.** $x = -1$ **59.** $x = -1 \pm 2i$

61. $x = -\frac{1}{4} \pm \frac{i\sqrt{15}}{4}$ **63.** $x = 1 \pm i\sqrt{6}$ **65.** $r = 2.0$ ft

67. a. 122.5 m   **b.** 3.2 s   **c.** 7.3 s   **69.** The height is $1 + \sqrt{61} \approx 8.8$ cm. The base is $-2 + 2\sqrt{61} \approx 13.6$ cm.   **71.** The lengths are $\frac{3 + \sqrt{249}}{2} \approx 9.4$ in. and $\frac{-3 + \sqrt{249}}{2} \approx 6.4$ in.

Midchapter Review, p. 699

1. a. -3 **b.** 3   **c.** 3   **d.** $(-\infty, \infty)$ **2. a.** 0   **b.** Not possible   **c.** 1
d. $[3, \infty)$ **3. a.** 1   **b.** $\frac{1}{2}$ **c.** Not possible   **d.** $(-\infty, -1) \cup (-1, \infty)$
4. a. 2   **b.** 1   **c.** 3   **d.** $(-\infty, \infty)$ **5.** $10 + 2i$ **6.** $6 + 4i$ **7.** $-6 - 4i$
8. $19 - 2i$ **9.** $55 + 48i$ **10.** 5   **11.** $\frac{13}{5} + \frac{14}{5}i$ **12.** $\frac{13}{73} - \frac{14}{73}i$
13. Yes   **14.** No   **15. a.** 16   **b.** $x = -4 \pm \sqrt{19}$
16. a. $\frac{1}{9}$ **b.** $x = \frac{1}{3}$ **17. a.** 1   **b.** $x = -1 \pm 2i$

Calculator Connections 9.4, pp. 705–706

1.
```
(-5+√(17))/4
      .2192235936
(-5-√(17))/4
      -2.280776406
```

2.
```
(-40+√(1920))/-3
2
      -11.93063938
(-40-√(1920))/-3
2
      2.619306394
```

3.
```
(-17-√((17)²-4*(
4)(-3)))/(2*4)
      -4.419694969
```

4.
```
(5.2+√((5.2)²-4(
2.1)(1.7)))/(2*2
.1)
      2.088598624
```

Section 9.4 Practice Exercises, p. 707

1. For $ax^2 + bx + c = 0, x = \dfrac{-b \pm \sqrt{b^2 - 4ac}}{2a}$

3. $2x^2 - x - 5 = 0; a = 2, b = -1, c = -5$

5. $-3x^2 + 14x + 0 = 0; a = -3, b = 14, c = 0$

7. $x^2 + 0x - 9 = 0; a = 1, b = 0, c = -9$

9. Discriminant is 0; one rational solution   **11.** Discriminant is 49; two rational solutions   **13.** Discriminant is 33; two irrational solutions   **15.** Discriminant is -23; two imaginary solutions   **17.** $t = -8$

19. $k = \frac{2}{3}, k = -\frac{1}{2}$ **21.** $a = \dfrac{-5 \pm \sqrt{33}}{4}$ **23.** $c = \dfrac{1 \pm i\sqrt{23}}{6}$

25. $k = -1 \pm 2i$ **27.** $y = \dfrac{1 \pm \sqrt{17}}{-8}$ or $\dfrac{-1 \pm \sqrt{17}}{8}$

29. $x = \dfrac{-3 \pm \sqrt{33}}{4}$ **31.** $y = \dfrac{-15 \pm \sqrt{145}}{4}$

33. $t = \dfrac{-10 \pm \sqrt{85}}{-5}$ or $\dfrac{10 \pm \sqrt{85}}{5}$ **35.** $x = \dfrac{-2 \pm \sqrt{22}}{6}$

37. $m = \dfrac{1 \pm i\sqrt{11}}{2}$ **39.** The width is $\dfrac{1 + 3\sqrt{89}}{4} \approx 7.3$ m

The length is $\dfrac{-1 + 3\sqrt{89}}{2} \approx 13.7$ m

41. The legs are $1 + \sqrt{71} \approx 9.4$ ft and $-1 + \sqrt{71} \approx 7.4$ ft.

43. $x = \frac{3}{4}, x = -\frac{3}{4}$ **45.** $x = 5 \pm \sqrt{21}$ **47.** $x = -12 \pm 3\sqrt{5}$

49. $x = \dfrac{3 \pm \sqrt{15}}{2}$ **51.** $x = 0, x = \frac{11}{9}$

Calculator Connections 9.5, pp. 715–716

1. $(-2, 3)$ **2.** $(10, 5)$ **3.** $(-1.5, -2.6)$

4. $(1.75, 2.5)$ **5.** $\left(\frac{5}{2}, 0\right)$ **6.** $\left(-\frac{8}{3}, 0\right)$

7.

The graph in part (b) is shifted up 4 units. The graph in part (c) is shifted down 3 units.

8.

In part (b), the graph is shifted to the right 3 units. In part (c) the graph is shifted to the left 2 units.

9.

In part (b) the graph is stretched vertically by a factor of 2. In part (c) the graph is shrunk vertically by a factor of $\frac{1}{2}$.

10.

In part (b) the graph has been stretched vertically and reflected across the x-axis. In part (c) the graph has been shrunk vertically and reflected across the x-axis.

Section 9.5 Practice Exercises, pp. 717–719

1. $y = \frac{1}{3}, y = 3$ **3.** $t = \pm\frac{\sqrt{7}}{2}$ **5.** $b = -1 \pm \sqrt{6}$

7. $w = \frac{1 \pm i\sqrt{19}}{2}$ **9.** Linear   **11.** Quadratic   **13.** Neither

15. Linear   **17.** Quadratic   **19.** If $a > 0$ the graph opens upward; if $a < 0$ the graph opens downward.   **21.** $a = -7$; downward
23. $a = 2$; upward   **25.** $a = -10$; downward   **27.** Vertex: $(-1, -8)$; axis of symmetry: $x = -1$ **29.** Vertex: $(1, -4)$; axis of symmetry: $x = 1$ **31.** Vertex: $(1, 2)$; axis of symmetry: $x = 1$ **33.** Vertex: $(0, -4)$; axis of symmetry: $x = 0$
35. c x-intercepts: $(\sqrt{7}, 0)(-\sqrt{7}, 0)$; y-intercept: $(0, -7)$
37. a x-intercepts: $(-1, 0)(-5, 0)$; y-intercept: $(0, 5)$
39. $(0, c)$ **41.** $(-\infty, \infty)$

43. a. Upward
b. $(0, -9)$
c. $(3, 0)(-3, 0)$
d. $(0, -9)$
e.

f. $(-\infty, \infty)$
g. $[-9, \infty)$

45. a. Upward
b. $(1, -9)$
c. $(4, 0)(-2, 0)$
d. $(0, -8)$
e.

f. $(-\infty, \infty)$
g. $[-9, \infty)$

47. a. Downward
b. $(3, 0)$
c. $(3, 0)$
d. $(0, -9)$
e.

f. $(-\infty, \infty)$
g. $(-\infty, 0]$

49. a. Downward
b. $(4, 1)$
c. $(3, 0)(5, 0)$
d. $(0, -15)$
e.

f. $(-\infty, \infty)$
g. $(-\infty, 1]$

51. a. Upward
b. $(-3, 1)$
c. none
d. $(0, 10)$
e.

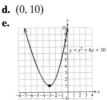

f. $(-\infty, \infty)$
g. $[1, \infty)$

53. a. Downward
b. $(0, -2)$
c. none
d. $(0, -2)$
e.

f. $(-\infty, \infty)$
g. $(-\infty, -2]$

55. True   **57.** False   **59. a.** 1800 dinners **b.** $2840
61. a. 33 psi **b.** 36,000 miles

Chapter 9 Review Exercises, pp. 724–728

1. Domain: $\{6, 10, -1, 0\}$; range: $\{3\}$; function
3. Domain: $[-4, 5]$; range: $[-3, 3]$; not a function
5. Domain: $\{4, 3, -6\}$; range: $\{23, -2, 5, 6\}$; not a function
7. a. 0 **b.** 8 **c.** -27 **d.** -1 **e.** 64 **f.** $(-\infty, \infty)$ **9.** -45 **11.** $-4m$
13. $8y^3 + 4y - 1$ **15. a.** $D(90) = 562$. A plane traveling 90 ft/s when it touches down will require 562 ft of runway. **b.** $D(110) = 902$. A plane traveling 110 ft/s when it touches down will require 902 ft of runway.   **17.** $\{x | x \text{ is a real number}\}$ **19.** $\{x | x \geq -2\}$
21. $\{x | x \neq -1\}$ **23.** $i\sqrt{5}$ **25.** $10i$ **27.** 4   **29.** -1 **31.** i
33. $4 - 12i$ **35.** 22   **37.** $2 + 14i$ **39.** 26   **41.** $\frac{12}{5} + \frac{6}{5}i$

43. $4 - \frac{5}{2}i$ **45.** Quadratic   **47.** Linear   **49.** $x = \pm\sqrt{19}$

51. $x = \pm 4i\sqrt{3}$ **53.** $x = 2 \pm 8i$ **55.** $x = \frac{3 \pm 2\sqrt{5}}{2}$ **57.** 81

59. $\frac{1}{196}$ **61.** $x = 1 \pm \sqrt{5}$ **63.** $x = \frac{7 \pm \sqrt{85}}{6}$ **65.** 3.1 cm
67. $b^2 - 4ac$ **69.** c; Discriminant is 0.   **71.** a; Discriminant is 25.

73. $x = -2$ **75.** $x = \frac{3}{2}, x = -1$ **77.** $x = -3 \pm \sqrt{7}$

79. $x = 1, x = -6$ **81.** $x = 4 \pm 3i$
83. The height is $\frac{-1 + \sqrt{97}}{2} \approx 4.4$ cm.

The base is $\frac{1 + \sqrt{97}}{2} \approx 5.4$ cm.

85. b; x-intercepts: $(9.32, 0)(2.68, 0)$; y-intercept: $(0, 25)$
87. d; x-intercepts: $(1, 0)(-1, 0)$; y-intercept: $(0, 5)$
89. e; x-intercepts: $(0, 0)(-4, 0)$; y-intercept: $(0, 0)$
91. $a = 1$; Upward   **93.** $a = -2$; Downward
95. Vertex: $(-1, 1)$; axis of symmetry: $x = -1$
97. Vertex: $(3, 13)$; axis of symmetry: $x = 3$

99. a. Upward
   b. $(-1, -3)$
   c. Approximately
     $(0.22, 0)(-2.22, 0)$
   d. $(0, -1)$
   e.

   f. $(-\infty, \infty)$
   g. $[-3, \infty)$
103. a. 1024 ft **b.** 8 s

101. a. Downward
   b. $(-1, -4)$
   c. No x-intercepts
   d. $(0, -12)$
   e.

   f. $(-\infty, \infty)$
   g. $(-\infty, -4]$

Chapter 9 Test, pp. 728–729

1. a.

b.

2. a. $f(0) = \sqrt{3}, f(-2) = 1, f(6) = 3$ **b.** $[-3, \infty)$ **3. a.** $D(5) = 5$; a five-sided polygon has five diagonals. **b.** $D(10) = 35$; a 10-sided polygon has 35 diagonals. **c.** 8 sides **4.** $10i$ **5.** $i\sqrt{23}$ **6.** -21

7. i **8.** $-i$ **9.** $5 - 3i$ **10.** $-13 - 26i$ **11.** 221 **12.** $\dfrac{10}{221} + \dfrac{11}{221}i$

13. $x = \dfrac{-1 \pm i\sqrt{14}}{3}$ **14.** $x = 4 \pm \sqrt{21}$ **15.** $x = \dfrac{5 \pm \sqrt{13}}{6}$

16. 4.0 in. **17.** The base is $\dfrac{-1 + \sqrt{97}}{2} \approx 4.4$ m.

The height is $1 + \sqrt{97} \approx 10.8$ m.
18. For $y = ax^2 + bx + c$, if $a > 0$ the parabola opens upward, if $a < 0$ the parabola opens downward.
19. Vertex: $(0, 25)$; x-intercepts: $(-5, 0)(5, 0)$; y-intercept: $(0, 25)$

20. a. \$25 per ticket **b.** \$250,000

Cumulative Review Exercises, Chapters 1–9, pp. 729–731

1. $x = 1$ **2.** $h = \dfrac{2A}{b}$ **3.** $y = \dfrac{34}{3}$ **4.** Yes, $x = 2$ is a solution.

5. $\{x \mid x > -1\}; (-1, \infty)$

6. a. Decreases **b.** $m = -37.6$. For each additional increase in education level, the death rate decreases by approximately 38 deaths per 100,000 people. **c.** 901.8 per 100,000 **d.** 10th grade **7.** $\dfrac{8c}{a^4 b}$

8. 1.898×10^{10} diapers **9.** Approximately 651 light-years

10. $4x^2 - 16x + 13$ **11.** $2y^3 + 1 - \dfrac{3}{y - 2}$ **12.** a

13. $(y + 4a)(2x - 3b)$ **14.** The base is 9 m and the height is 8 m.

15. $-\dfrac{2}{x + 1}$ **16.** $\{x \mid x \neq 2, x \neq -2\}$ **17.** $x - 5$ **18.** $-\dfrac{2}{x}$

19. $y = 4, y = -3$ **20.** 20 cm $\approx$ 7.9 in. The fish must be released.

21. $y = \dfrac{1}{2}x + 4$

22. a. $(6, 0)$
   b. $(0, -3)$
   c. $\dfrac{1}{2}$
   d.

23. a. $(-3, 0)$
   b. No y-intercept
   c. Slope is undefined.
   d.

24. $\left(1, \dfrac{4}{3}\right)$ **25.** $(5, 2)$ **26.** The angles are 22° and 68°.
27. There are 13 dimes and 14 quarters.
28.

29. $\pi, \sqrt{7}$ **30.** $\dfrac{\sqrt{7}}{7}$ **31.** $2x$ **32.** $48 + 8\sqrt{3x} + x$ **33.** $2\sqrt{2x}$

34. $\dfrac{8 + 4\sqrt{a}}{4 - a}$ **35.** $x = -2$ ($x = -7$ does not check.)

36. $(2c - y)(4c^2 + 2cy + y^2)$ **37.** c, d **38. a.** 1 **b.** 4 **c.** 13
d. $x^2 + 2xh + h^2$ **39.** Domain: $\{2, -1, 9, -6\}$; range: $\{4, 3, 2, 8\}$
40. $0 + 3i$ **41.** $1 - i$ **42.** 25 **43.** $x = -3 \pm \sqrt{6}$
44. $x = -3 \pm \sqrt{6}$

45. Vertex: $\left(\dfrac{3}{2}, 2\right)$; x-intercepts: $(0, 0)(3, 0)$; y-intercept: $(0, 0)$

ADDITIONAL ANSWER APPENDIX

CHAPTER R

Calculator Connections R.2, p. 24

1. $0.\overline{4}$ **2.** $0.\overline{63}$

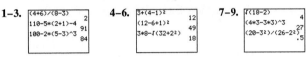

3. $0.1\overline{36}$ **4.** $0.\overline{384615}$

Section R.2 Practice Exercises, pp. 24–26

13. No, the symbols I, V, X, and so on each represent certain values but the values are not dependent on the position of the symbol within the number.

Section R.3 Practice Exercises, pp. 34–40

54. False; an isosceles triangle has two sides that are the same length. A scalene triangle has three sides of different length.

CHAPTER 1

Section 1.1 Practice Exercises, pp. 51–53

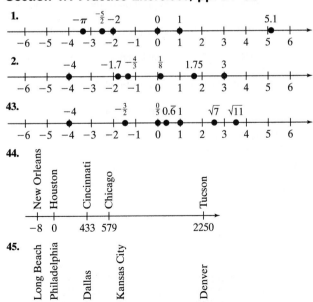

44.

45.

46.

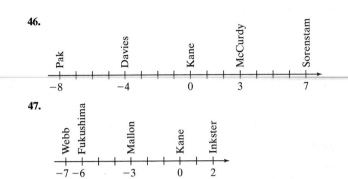

47.

Calculator Connections 1.2, p. 60

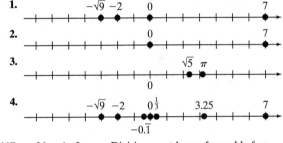

Section 1.2 Practice Exercises, pp. 61–63

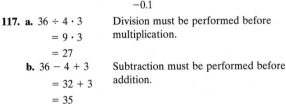

117. a. $36 \div 4 \cdot 3$ Division must be performed before
$= 9 \cdot 3$ multiplication.
$= 27$
b. $36 - 4 + 3$ Subtraction must be performed before
$= 32 + 3$ addition.
$= 35$

118. Multiplication and division are performed in order from left to right. Addition and subtraction are performed in order from left to right.   **119.** This is acceptable, provided division and multiplication are performed in order from left to right, and subtraction and addition are performed in order from left to right.

Section 1.3 Practice Exercises, pp. 67–69

51. To add two numbers with different signs, subtract the smaller absolute value from the larger absolute value and apply the sign of the number with the larger absolute value.

Calculator Connections 1.4, p. 74

1–3.

-8+(-5)
4+(-5)+(-1)   -13
627-(-84)   -2
711

4–6.

-0.06-0.12
-3.2+(-14.5)   -.18
-472+(-518)   -17.7
-990

7, 8.

-12-9+4
209-108+(-63)   -17
38

Calculator Connections 1.5, p. 86

1–3.

-6(5)
-5.2/2.6   -30
(-5)(-5)(-5)(-5)   -2
625

4–6.

(-5)^4
-5^4   625
-2.4²   -625
-5.76

7, 8.

(-2.4)²
(-1)(-1)(-1)   5.76
-1

9, 10.

-8.4/-2.1
90/(-5)(2)   4
-36

Section 1.6 Practice Exercises, pp. 98–100

48. Not equivalent. The terms are not *like* terms and cannot be combined. **49.** Not equivalent. The terms are not *like* terms and cannot be combined.

Section 1.7 Practice Exercises, pp. 106–110

11.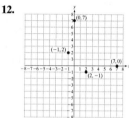

12.

13.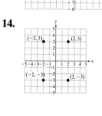

14.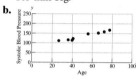

15.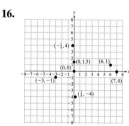

16.

25. a. (250, 225) 250 people produce $225 in popcorn sales.
(175, 193) 175 people produce $193 in popcorn sales.
(315, 330) 315 people produce $330 in popcorn sales.
(220, 209) 220 people produce $209 in popcorn sales.
(450, 570) 450 people produce $570 in popcorn sales.
(400, 480) 400 people produce $480 in popcorn sales.
(190, 185) 190 people produce $185 in popcorn sales.

b.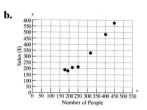

26. a. (57, 149) A 57-year-old woman has a systolic blood pressure of 149 mm Hg.
(41, 120) A 41-year-old woman has a systolic blood pressure of 120 mm Hg.
(71, 158) A 71-year-old woman has a systolic blood pressure of 158 mm Hg.
(36, 115) A 36-year-old woman has a systolic blood pressure of 115 mm Hg.
(64, 151) A 64-year-old woman has a systolic blood pressure of 151 mm Hg.
(25, 110) A 25-year-old woman has a systolic blood pressure of 110 mm Hg.
(40, 118) A 40-year-old woman has a systolic blood pressure of 118 mm Hg.
(77, 165) A 77-year-old woman has a systolic blood pressure of 165 mm Hg.

b.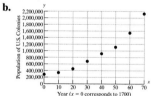

27. a. (0, 251000) In 1700 the population of the U.S. colonies was 251,000.
(10, 332000) In 1710 the population of the U.S. colonies was 332,000.
(20, 466000) In 1720 the population of the U.S. colonies was 466,000.
(30, 629000) In 1730 the population of the U.S. colonies was 629,000.
(40, 906000) In 1740 the population of the U.S. colonies was 906,000.
(50, 1171000) In 1750 the population of the U.S. colonies was 1,171,000.
(60, 1594000) In 1760 the population of the U.S. colonies was 1,594,000.
(70, 2148000) In 1770 the population of the U.S. colonies was 2,148,000.

b.

28. a. (0, 1490) In 1960 the poverty threshold was $1490.
(5, 1582) In 1965 the poverty threshold was $1582.
(10, 1954) In 1970 the poverty threshold was $1954.
(20, 4190) In 1980 the poverty threshold was $4190.
(30, 6652) In 1990 the poverty threshold was $6652.
(35, 7763) In 1995 the poverty threshold was $7763.

b.

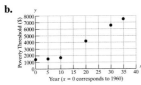

29. a. (1, −10.2) (2, −9.0) (3, −2.5) (4, 5.7) (5, 13.0) (6, 18.3) (7, 20.9)
(8, 19.6) (9, 14.8) (10, 8.7) (11, 2.0) (12, −6.9)

b.

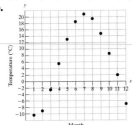

30. a. (1, −12.8) (2, −4.0) (3, 8.4) (4, 30.2) (5, 48.2) (6, 59.4) (7, 61.5)
(8, 56.7) (9, 45.0) (10, 25.0) (11, 6.1) (12, −10.1)

b.

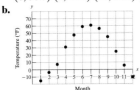

31. a. (1, 89.25) On day 1 the closing price per share was $89.25.
(2, 92.50) On day 2 the closing price per share was $92.50.
(3, 91.25) On day 3 the closing price per share was $91.25.
(4, 93.00) On day 4 the closing price per share was $93.00.
(5, 90.25) On day 5 the closing price per share was $90.25.
(6, 91.50) On day 6 the closing price per share was $91.50.
(7, 92.25) On day 7 the closing price per share was $92.25.
(8, 94.50) On day 8 the closing price per share was $94.50.

b.

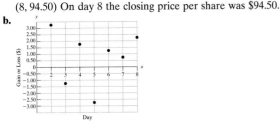

32. a. (1, 10.125) On day 1 the closing price per share was $10.125.
(2, 9.875) On day 2 the closing price per share was $9.875.
(3, 9.500) On day 3 the closing price per share was $9.500.
(4, 9.500) On day 4 the closing price per share was $9.500.
(5, 8.625) On day 5 the closing price per share was $8.625.
(6, 8.250) On day 6 the closing price per share was $8.250.
(7, 8.125) On day 7 the closing price per share was $8.125.
(8, 7.875) On day 8 the closing price per share was $7.875.

b.

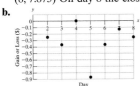

33. a. (1, 42) In year 1 the gross income was $42,000,000.
(2, 57) In year 2 the gross income was $57,000,000.
(3, 39) In year 3 the gross income was $39,000,000.
(4, 46) In year 4 the gross income was $46,000,000.
(5, 58) In year 5 the gross income was $58,000,000.
(6, 62) In year 6 the gross income was $62,000,000.

b. (1, 47) In year 1 the cost was $47,000,000.
(2, 50) In year 2 the cost was $50,000,000.
(3, 49) In year 3 the cost was $49,000,000.
(4, 40) In year 4 the cost was $40,000,000.
(5, 50) In year 5 the cost was $50,000,000.
(6, 62) In year 6 the cost was $62,000,000.

c.

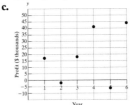

34. a. (1, 102) In year 1 the income was $102,000.
(2, 84) In year 2 the income was $84,000.
(3, 100) In year 3 the income was $100,000.
(4, 137) In year 4 the income was $137,000.
(5, 104) In year 5 the income was $104,000.
(6, 152) In year 6 the income was $152,000.

b. (1, 85) In year 1 the total cost was $85,000.
(2, 86) In year 2 the total cost was $86,000.
(3, 82) In year 3 the total cost was $82,000.
(4, 96) In year 4 the total cost was $96,000.
(5, 110) In year 5 the total cost was $110,000.
(6, 108) In year 6 the total cost was $108,000.

c.

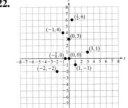

Chapter 1 Review Exercises, pp. 115–118

49. When a and b are both negative or when a and b have different signs and the number with the larger absolute value is negative
115–122.

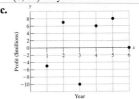

123. a. (1, 26.25) On day 1 the price per share was $26.25.
(2, 28.50) On day 2 the price per share was $28.50.
(3, 28.00) On day 3 the price per share was $28.00.
(4, 27.00) On day 4 the price per share was $27.00.
(5, 24.75) On day 5 the price per share was $24.75.
(6, 24.50) On day 6 the price per share was $24.50.
(7, 24.50) On day 7 the price per share was $24.50.
(8, 26.25) On day 8 the price per share was $26.25.

b.

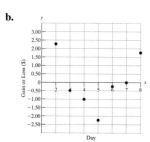

124. a. (0, 25.9) In 1965 25,900,000 people in the United States were living below the poverty level.

(5, 29.3) In 1970 29,300,000 people in the United States were living below the poverty level.

(10, 33.1) In 1975 33,100,000 people in the United States were living below the poverty level.

(25, 33.6) In 1990 33,600,000 people in the United States were living below the poverty level.

(30, 36.4) In 1995 36,400,000 people in the United States were living below the poverty level.

b.

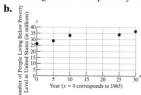

Chapter 1 Test, p. 119

2.

22. a. (5, 46) At age 5 the boy's height was 46 in.

(7, 50) At age 7 the boy's height was 50 in.

(9, 55) At age 9 the boy's height was 55 in.

(11, 60) At age 11 the boy's height was 60 in.

b.

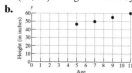

c. 57.5 in. **d.** 62.5 in. **e.** No, his height will maximize in his teen years. No.

CHAPTER 2

Section 2.2 Practice Exercises, pp. 139–140

9. To simplify an expression, clear parentheses and combine *like* terms. To solve an equation, use the addition, subtraction, multiplication, and division properties of equality to isolate the variable.

Midchapter Review, p. 160

25. The sum of a number and 18 is equal to three times the sum of the number and two. **26.** The difference of a number and five is the same as twice the sum of the number and one.

Calculator Connections 2.5, p. 167

1, 2.
```
880/(2π)
        140.0563499
1600/(π(4)²)
        31.83098862
```

3, 4.
```
20/((-0.05)*(5))
        -80
10/(0.5(6+4))
        2
```

Section 2.5 Practice Exercises, pp. 168–171

31. $x = \dfrac{6 - 3y}{2}$ **32.** $y = \dfrac{10 - 5x}{2}$

Section 2.7 Practice Exercises, pp. 193–197

25. a. A [93, 100]; B+ [89, 93); B [84, 89); C+ [80, 84); C [75, 80); F [0, 75) **b.** B **c.** C **26. a.** A [90, 100]; B+ [86, 90); B [80, 86); C+ [76, 80); C [70, 76); D+ [66, 70); D [60, 66); F [0, 60) **b.** B+ **c.** D+

37. a. $x = 3$
   b. $x > 3$

38. a. $y = 18$
   b. $y \geq 18$

39. a. $p = 13$
   b. $p \leq 13$

40. a. $k = 2$
   b. $k < 2$

41. a. $c = -3$
   b. $c < -3$

42. a. $d = -7$
   b. $d > -7$

43. a. $z = -\dfrac{3}{2}$
   b. $z \geq -\dfrac{3}{2}$

44. a. $w = -7$
   b. $w > -7$

49. $(-\infty, 1]$

50. $(-\infty, 13)$

51. $(10, \infty)$

52. $[-5, \infty)$

53. $(3, \infty)$

54. $(-\infty, 9)$

55. $(-\infty, 8]$

56. $[-5, \infty)$

57. $(2, \infty)$

58. $(-\infty, 3]$

59. $(-\infty, -2)$

60. $[-3, \infty)$

61. $[14, \infty)$

62. $(-\infty, 7)$

63. $[-24, \infty)$

64. $(-\infty, -18)$

65. $[-3, 3)$

66. $(-7, -1)$

67. $\left(0, \dfrac{5}{2}\right)$

68. $\left[-1, \dfrac{7}{2}\right]$

69. $[90, \infty)$

70. $(-\infty, -40)$

71. $(-9, \infty)$

72. $(-\infty, -16]$

73. $\left[-\dfrac{15}{2}, \infty\right)$

74. $\left(-\infty, \dfrac{1}{3}\right)$

75. $\left[-\dfrac{1}{3}, \infty\right)$

76. $\left(-\infty, \dfrac{35}{4}\right]$

77. $(-3, \infty)$
78. $[-3, \infty)$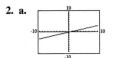
79. $(-\infty, 7)$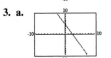
80. $(-5, \infty)$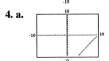
81. $\left(-\infty, \frac{15}{4}\right]$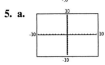
82. $[-2, \infty)$
83. $(-3, \infty)$
84. $(-1, \infty)$

87. a. $1539 **b.** 200 birdhouses cost $1440. It is cheaper to purchase 200 birdhouses because the discount is higher.

91. $[13, \infty)$
92. $(-\infty, -6)$
93. $[-4, \infty)$
94. $(-\infty, -7)$
95. $(14.5, \infty)$
96. $(-\infty, -1.\overline{3}]$

Calculator Connections 2.8, p. 205

1. a. **b.**
2. a. **b.**
3. a. **b.**
4. a. **b.**
5. a. **b.**

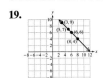

Section 2.8 Practice Exercises, pp. 206–208

19.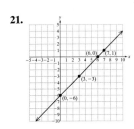
20.
21.
22.

23.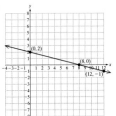
24.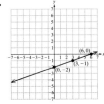
25.
26.
27.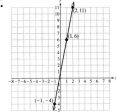
28.
29.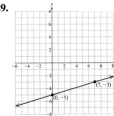
30.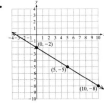
31.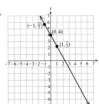
32.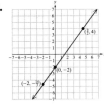
33.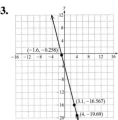
34.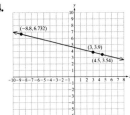

35. a. $y = 17.95$ **b.** $x = 145$ **c.** $(55, 17.95)$ Collecting 55 lb of cans yields $17.95. $(145, 80.05)$ Collecting 145 lb of cans yields $80.05.
d.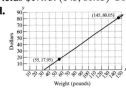

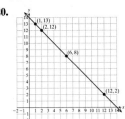

36. a. $y = 181.87$ **b.** $x = 20$ **c.** $(13, 181.87)$ Selling 13 compact discs yields \$181.87 in revenue. $(20, 279.80)$ Selling 20 compact discs yields \$279.80 in revenue.

d.

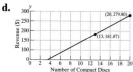

37. a. $y = 10{,}068$ **b.** $x = 3$ **c.** $(1, 10068)$ One year after purchase the value of the car is \$10,068. $(3, 7006)$ Three years after purchase the value of the car is \$7006.

d.

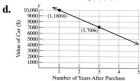

38. a. $y = 3{,}891{,}110$ **b.** $x = 13$ **c.** $(5, 3891110)$ In 1975 the total enrollment was approximately 3,891,110. $(13, 3136086)$ In 1983, the total enrollment was 3,136,086.

d.

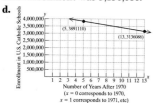

Chapter 2 Review Exercises, pp. 217–221

83. a. $(-2, \infty)$

b. $\left(-\infty, \dfrac{1}{2}\right]$

c. $(-1, 4]$

85. $(-\infty, 17)$

86. $\left(-\dfrac{1}{3}, \infty\right)$

87. $(-\infty, -6]$

88. $\left(-\infty, -\dfrac{14}{5}\right]$

89. $[49, \infty)$

90. $(-\infty, 34.5)$

91. $(18, \infty)$

92. $\left(-\infty, \dfrac{5}{2}\right]$

95.

96.

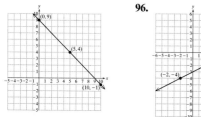

98. a. £82.32 **c.** $(10, 6.10)$ \$10 is equivalent to £6.10. $(25, 15.25)$ \$25 is equivalent to £15.25. $(100, 60.98)$ \$100 is equivalent to £60.98. $(500, 304.90)$ \$500 is equivalent to £304.90. **d.** \$714.99

99. a. \$19.10 **b.** $(5.0, 7.70)$, $(7.5, 11.55)$, $(10.0, 15.40)$, $(12.5, 19.25)$, $(15.0, 23.10)$ **c.**

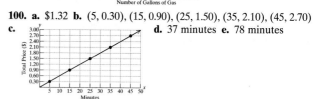

d. 6.5 gal **e.** 17 gal

100. a. \$1.32 **b.** $(5, 0.30)$, $(15, 0.90)$, $(25, 1.50)$, $(35, 2.10)$, $(45, 2.70)$

c.

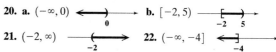

d. 37 minutes **e.** 78 minutes

Chapter 2 Test, pp. 222–223

20. a. $(-\infty, 0)$

b. $[-2, 5)$

21. $(-2, \infty)$

22. $(-\infty, -4]$

24. a. 202 beats per minute. **c.** $(20, 200)$ A 20-year-old has a maximum recommended heart rate of 200 beats per minute. $(30, 190)$ A 30-year-old has a maximum recommended heart rate of 190 beats per minute. $(40, 180)$ A 40-year-old has a maximum recommended heart rate of 180 beats per minute. $(50, 170)$ A 50-year-old has a maximum recommended heart rate of 170 beats per minute. $(60, 160)$ A 60-year-old has a maximum recommended heart rate of 160 beats per minute. **d.** 28 years old

Cumulative Review Exercises, Chapters 1–2, pp. 223–224

11.

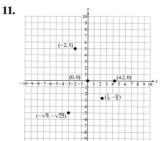

a. Quadrant II
b. Quadrant IV
c. Quadrant III
d. x-axis
e. Origin

22. $(-2, \infty)$

24. a.

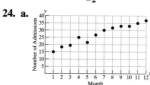

b. Third and fourth months
c. 147%

CHAPTER 3

Calculator Connections 3.1, pp. 232–233

1, 2, 3.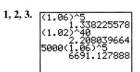

4, 5.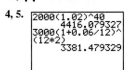

6.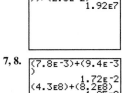

Section 3.2 Practice Exercises, p. 239

9. When multiplying expressions with the same base, add the exponents. When raising an expression with an exponent to a power, multiply the exponents. **10.** When multiplying expressions with the same base, add the exponents. When dividing expressions with the same base, subtract the exponents.

Calculator Connections 3.4, pp. 252–253

1, 2.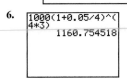

3, 4.

5.

6.

7, 8.

9, 10.

Section 3.4 Practice Exercises, pp. 254–256

61. a. U.S. $\$6.82 \times 10^{11}$; Germany $\$5.40 \times 10^{11}$; Japan $\$3.88 \times 10^{11}$; France $\$3.05 \times 10^{11}$; Britain $\$2.72 \times 10^{11}$; Italy $\$2.42 \times 10^{11}$; Canada $\$2.14 \times 10^{11}$; Netherlands $\$1.99 \times 10^{11}$; China $\$1.84 \times 10^{11}$
b. $\$3.026 \times 10^{12}$ **c.** $\$2.04 \times 10^{11}$ **d.** 7%

Section 3.5 Practice Exercises, pp. 265–267

7. 3.0×10^7 is scientific notation in which 10 is raised to the seventh power. 3^7 is not scientific notation and 3 is being raised to the seventh power. **8.** 4.0×10^{-2} is scientific notation in which 10 is being raised to the -2 power. 4^{-2} is not scientific notation and 4 is being raised to the -2 power.

Section 3.7 Practice Exercises, pp. 282–284

52. To check, multiply $(y + 3)(y^2 + 9) = y^3 + 3y^2 + 9y + 27$, which does not equal $y^3 + 27$. **61.** Long division; $n^2 + 4n + 16$

62. Long division; $3s + 5 + \dfrac{-43}{5s + 3}$ **69.** $x + 1 + \dfrac{1}{x - 1}$

70. $x^2 + x + 1 + \dfrac{1}{x - 1}$

Calculator Connections 3.8, pp. 290–291

1.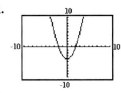

2.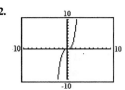

3.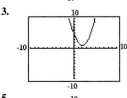

4.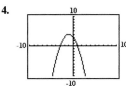

5.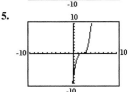

6.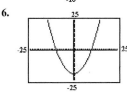

7.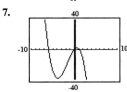

8.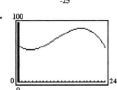

9.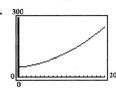

Section 3.8 Practice Exercises, pp. 294–298

1.

2.

3.
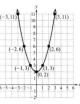

4.
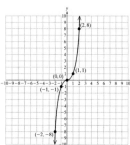

17.
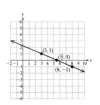

18.
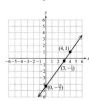

19.
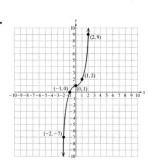

20.
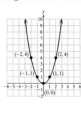

21. b. (0, 0) When the plane is 0 miles from the airport its altitude is 0 ft.
(0.25, 156.25) When the plane is 0.25 miles from the airport its altitude is 156.25 ft.
(0.50, 500) When the plane is 0.50 miles from the airport its altitude is 500 ft.
(0.75, 843.75) When the plane is 0.75 miles from the airport its altitude is 843.75 ft.
(1.00, 1000) When the plane is 1 mile from the airport its altitude is 1000 ft.
c. A nonlinear pattern is preferred so that the flight path levels off at ground level.

22. a.
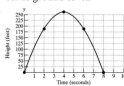

b. (0, 0) After 0 s the height is 0 ft.
(2, 192) After 2 s the height is 192 ft.

(4, 256) After 4 s the height is 256 ft.
(6, 192) After 6 s the height is 192 ft.
(8, 0) After 8 s the height is 0 ft.
c. The rocket reaches a height of 192 ft on the way up and again on the way down.

23. a. Nonlinear
c. (0, 0) In 0 s the object travels 0 ft.
(1, 16) In 1 s the object travels 16 ft.
(2, 64) In 2 s the object travels 64 ft.
(3, 144) In 3 s the object travels 144 ft.
(4, 256) In 4 s the object travels 256 ft.
(5, 400) In 5 s the object travels 400 ft.

24. a. Nonlinear
c. (0, 0) In 0 s the object travels 0 m.
(1, 4.9) In 1 s the object travels 4.9 m.
(2, 19.6) In 2 s the object travels 19.6 m.
(3, 44.1) In 3 s the object travels 44.1 m.
(4, 78.4) In 4 s the object travels 78.4 m.

25. a. Nonlinear
c. (0, 0) If \$0 is charged per ticket the revenue will be \$0.
(10, 160000) If \$10 is charged per ticket the revenue will be \$160,000.
(20, 240000) If \$20 is charged per ticket the revenue will be \$240,000.
(30, 240000) If \$30 is charged per ticket the revenue will be \$240,000.
(40, 160000) If \$40 is charged per ticket the revenue will be \$160,000.
(50, 0) If \$50 is charged per ticket the revenue will be \$0.

26. a. Nonlinear
c. (0, 0) If \$0 is charged per ticket the revenue will be \$0.
(25, 1687500) If \$25 is charged per ticket the revenue will be \$1,687,500.
(50, 3000000) If \$50 is charged per ticket the revenue will be \$3,000,000.
(100, 4500000) If \$100 is charged per ticket the revenue will be \$4,500,000.
(250, 0) If \$250 is charged per ticket the revenue will be \$0.

27. a. Nonlinear
b. $(0, 0)$ $\left(1, \dfrac{1}{8}\right)$ $(2, 1)$ $\left(3, \dfrac{27}{8}\right)$ $(4, 8)$ $\left(5, \dfrac{125}{8}\right)$
c.

28. a. Nonlinear
b. $(0, 0)$ $(1, 1)$ $(2, 4)$ $(3, 9)$ $(4, 16)$ $(5, 25)$
c.

d. 36 ft²; area is increased by four times.

29. a. Nonlinear
  c. (0, 4) At 0 s the height of the ball is 4 ft.
    (0.5, 15) At 0.5 s the height of the ball is 15 ft.
    (1.0, 18) At 1.0 s the height of the ball is 18 ft.
    (1.5, 13) At 1.5 s the height of the ball is 13 ft.
    (2.0, 0) At 2 s the height of the ball is 0 ft.

30. a. Nonlinear
  c. (0, 5) At 0 s the height of the ball is 5 ft.
    (0.5, 13.5) At 0.5 s the height of the ball is 13.5 ft.
    (1.0, 14) At 1.0 s the height of the ball is 14 ft.
    (1.5, 6.5) At 1.5 s the height of the ball is 6.5 ft.
    (1.7, 1.26) At 1.7 s the height of the ball is 1.26 ft.

Chapter 3 Review Exercises, pp. 304–308

1. Base: 5; exponent: 3   **2.** Base: x; exponent: 4   **3.** Base: -2; exponent: 0   **4.** Base: y; exponent: 1   **19.** Exponents are added only when multiplying factors with the same base. In such a case, the base does not change.   **20.** Exponents are subtracted only when dividing factors with the same base. In such a case, the base does not change.

79. $\dfrac{15}{2}x^3 + \dfrac{1}{4}x^2 + \dfrac{1}{2}x + 2$ **118. a, b.**

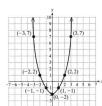

119. a, b.

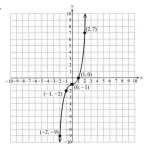

120. a. Nonlinear
  c. (0, 400) At 0 s the height is 400 ft.
    (1.0, 384) At 1 s the height is 384 ft.
    (2.0, 336) At 2 s the height is 336 ft.
    (2.5, 300) At 2.5 s the height is 300 ft.
    (3.0, 256) At 3.0 s the height is 256 ft.
    (4.0, 144) At 4.0 s the height is 144 ft.
    (5.0, 0) At 5.0 s the height is 0 ft.

Chapter 3 Test, pp. 308–309

24. b.

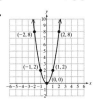

Cumulative Review Exercises, Chapters 1–3, pp. 309–311

13. a. 12 in. **b.** 19.5 in. **c.** 5.5 h **d.**

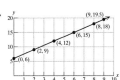

CHAPTER 4

Section 4.3 Practice Exercises, pp. 337–338

25. A polynomial that cannot be factored is prime.   **26.** A trinomial $ax^2 + bx + c$ is prime if no combination of binomials (using the factors of a and c as the coefficients of its terms) produces the correct product.   **65.** No, $(3x + 6)$ has a common factor of 3.   **66.** No, $(4x - 12)$ has a common factor of 4.

Section 4.5 Practice Exercises, pp. 353–354

30. Factor out the GCF. If possible, identify and factor the binomial as a difference of squares, a difference of cubes, or a sum of cubes.
69. $\left(\dfrac{4}{5}p - \dfrac{1}{2}q\right)\left(\dfrac{16}{25}p^2 + \dfrac{2}{5}pq + \dfrac{1}{4}q^2\right)$
70. $\left(\dfrac{1}{10}r + \dfrac{2}{3}s\right)\left(\dfrac{1}{100}r^2 - \dfrac{1}{15}rs + \dfrac{4}{9}s^2\right)$

Section 4.6 Practice Exercises, pp. 363–367

33. $q = 4, q = -\dfrac{1}{2}$ **34.** $x = -\dfrac{1}{4}, x = 3$ **35.** $x = \dfrac{2}{3}, x = -\dfrac{2}{3}$

36. $a = \dfrac{7}{2}, a = -\dfrac{7}{2}$ **41.** The equation must have one side equal to zero and the other side factored completely.

48. $x = 0, x = -\dfrac{3}{2}, x = 4$ **49.** $x = 0, x = -\dfrac{1}{3}, x = -1$

50. $a = -3$ **51.** $k = \dfrac{3}{4}, k = -3$ **65.** The numbers are $-\dfrac{9}{2}$ and 4.

74. a. The slab is 7 m by 4 m. **b.** 22 m   **75. a.** The picture is 13 in. by 6 in. **b.** 38 in.   **82.**

90. The first boat traveled 3 miles; the second boat traveled 4 miles.

Calculator Connections 4.7, pp. 373–374

1. x-intercept: (2, 0); y-intercept: $(0, -4)$

2. *x*-intercept: $(2, 0)$; *y*-intercept: $(0, 6)$

3. *x*-intercept: $(2, 0)$; *y*-intercept: $\left(0, \frac{3}{2}\right)$

4. *x*-intercept: $(2, 0)$; *y*-intercept: $\left(0, -\frac{4}{5}\right)$

5. *x*-intercepts: $(-7, 0)$, $(1, 0)$; *y*-intercept: $(0, -7)$

6. *x*-intercepts: $(-1, 0)$, $(1, 0)$; *y*-intercept: $(0, -1)$

7. *x*-intercepts: $(-2, 0)$, $(3, 0)$; *y*-intercept: $(0, -12)$

8. *x*-intercepts: $(-2, 0)$, $(9, 0)$; *y*-intercept: $(0, -18)$

9. *x*-intercept: $(15, 0)$; *y*-intercept: $(0, -15)$

Section 4.7 Practice Exercises, pp. 374–378

24. *x*-intercepts: $(-6, 0)(-1, 0)$; *y*-intercept: $(0, 6)$
27. *x*-intercepts: $(-5, 0)(3, 0)(0, 0)$; *y*-intercept: $(0, 0)$
28. *x*-intercepts: $(0, 0)(8, 0)(2, 0)$; *y*-intercept: $(0, 0)$ **33. a.** $1000
b. $500 **c.** $(0, 1500)$ If the air-conditioner lasts 0 years (it breaks close
to the time of purchase) the dealer will provide a full refund of $1500.
d. $(3, 0)$ If the air-conditioner breaks after 3 years, the dealer pays
nothing. **34. a.** $400 **b.** $200 **c.** $(0, 800)$ If the stereo lasts 0 years
(it breaks near the time of purchase) the store will provide a full
refund of $800. **d.** $(2, 0)$ If the stereo breaks after 2 years the store
pays nothing.

37.

38.

39.

40.

Chapter 4 Review Exercises, pp. 382–385

63. Not a perfect square trinomial. The middle term does not equal
$2(t)(7)$. **64.** Not a perfect square trinomial. The middle term does
not equal $-2(k)(8)$.
138. a. *x*-intercept: $(5, 0)$; *y*-intercept: $(0, -2)$
b.

139. a. *x*-intercept: $(0, 0)$; *y*-intercept: $(0, 0)$
b.

Cumulative Review Exercises, Chapters 1–4, pp. 387–388

8. $[-4, \infty)$

25. a. Quadratic **b.** $(-3, -5)$, $(-2, 0)$, $(-1, 3)$, $(0, 4)$, $(1, 3)$, $(2, 0)$,
$(3, -5)$ **c.** **d.** $(-2, 0)(2, 0)$ **e.** $(0, 4)$

CHAPTER 5

Section 5.1 Practice Exercises, pp. 397–399

39. a. $\dfrac{3(y + 2)}{6(y + 2)}$ **b.** $\{y | y \neq -2\}$ **c.** $\dfrac{1}{2}$ **40. a.** $\dfrac{8(x - 1)}{4(x - 1)}$ **b.** $\{x | x \neq 1\}$

c. 2 **41. a.** $\dfrac{(t - 1)(t + 1)}{t + 1}$ **b.** $\{t | t \neq -1\}$ **c.** $t - 1$

42. a. $\dfrac{(r - 2)(r + 2)}{r - 2}$ **b.** $\{r | r \neq 2\}$ **c.** $r + 2$ **43. a.** $\dfrac{7w}{7w(3w - 5)}$

b. $\left\{w | w \neq 0, w \neq \dfrac{5}{3}\right\}$ **c.** $\dfrac{1}{3w - 5}$ **44. a.** $\dfrac{12a^2}{6a(4a - 3)}$

b. $\left\{a \mid a \neq 0, a \neq \dfrac{3}{4}\right\}$ **c.** $\dfrac{2a}{4a - 3}$ **45. a.** $\dfrac{(3x - 2)(3x + 2)}{2(3x + 2)}$

b. $\left\{x \mid x \neq -\dfrac{2}{3}\right\}$ **c.** $\dfrac{3x - 2}{2}$ **46. a.** $\dfrac{4(2b - 5)}{(2b - 5)(2b + 5)}$

b. $\left\{b \mid b \neq \dfrac{5}{2}, b \neq -\dfrac{5}{2}\right\}$ **c.** $\dfrac{4}{2b + 5}$ **47. a.** $\dfrac{(a + 5)(a - 2)}{(a + 3)(a - 2)}$

b. $\{a \mid a \neq -3, a \neq 2\}$ **c.** $\dfrac{a + 5}{a + 3}$ **48. a.** $\dfrac{(t + 5)(t - 2)}{(t + 5)(t - 4)}$

b. $\{t \mid t \neq -5, t \neq 4\}$ **c.** $\dfrac{t - 2}{t - 4}$

Section 5.3 Practice Exercises, pp. 414–416

55. $\dfrac{6m - 6}{(m + 4)(m - 1)}, \dfrac{3m + 12}{(m + 4)(m - 1)}$

56. $\dfrac{3n + 6}{(n - 5)(n + 2)}, \dfrac{7n - 35}{(n - 5)(n + 2)}$

57. $\dfrac{6w + 6}{(w + 3)(w - 8)(w + 1)}, \dfrac{w^2 + 3w}{(w + 3)(w - 8)(w + 1)}$

58. $\dfrac{t^2 - 2t}{(t + 2)(t - 2)(t + 12)}, \dfrac{18t + 216}{(t + 2)(t - 2)(t + 12)}$

63. $\dfrac{z^2 + 3z}{(z + 2)(z + 7)(z + 3)}, \dfrac{-3z^2 - 6z}{(z + 2)(z + 7)(z + 3)},$
$\dfrac{5z + 35}{(z + 2)(z + 7)(z + 3)}$

64. $\dfrac{6w + 30}{(w - 4)(w + 1)(w + 5)}, \dfrac{w - 4}{(w - 4)(w + 1)(w + 5)},$
$\dfrac{-9w^2 - 9w}{(w - 4)(w + 1)(w + 5)}$

65. $\dfrac{3p + 6}{(p - 2)(p^2 + 2p + 4)(p + 2)}, \dfrac{p^3 + 2p^2 + 4p}{(p - 2)(p^2 + 2p + 4)(p + 2)},$
$\dfrac{5p^3 - 20p}{(p - 2)(p^2 + 2p + 4)(p + 2)}$

66. $\dfrac{7q - 35}{(q + 5)(q^2 - 5q + 25)(q - 5)}, \dfrac{q^3 - 5q^2 + 25q}{(q + 5)(q^2 - 5q + 25)(q - 5)},$
$\dfrac{12q^2 - 300}{(q + 5)(q^2 - 5q + 25)(q - 5)}$

Section 5.4 Practice Exercises, pp. 424–426

25. $\dfrac{z + 7}{3(z - 3)}, \dfrac{3z - 18}{3(z - 3)}$ **26.** $\dfrac{6w + 8}{2(w - 2)}, \dfrac{w - 3}{2(w - 2)}$ **29.** $\dfrac{19}{3(a + 1)}$

30. $\dfrac{11}{5(c - 4)}$ **31.** $\dfrac{-3(k + 4)}{(k - 3)(k + 3)}$ **32.** $\dfrac{5h - 32}{(h - 5)(h + 5)}$ **33.** $\dfrac{a - 4}{2a}$

34. $\dfrac{5}{24}$ **35.** $\dfrac{5}{3x - 7}$ or $\dfrac{-5}{7 - 3x}$ **36.** $\dfrac{4}{2w - 1}$ or $\dfrac{-4}{1 - 2w}$

37. $\dfrac{6n - 1}{n - 8}$ or $\dfrac{-6n + 1}{8 - n}$ **38.** $\dfrac{4m + 1}{m - 2}$ or $\dfrac{-4m - 1}{2 - m}$

59. $\dfrac{-w^2}{(w + 3)(w - 3)(w^2 - 3w + 9)}$ **60.** $\dfrac{2m^2 + 1}{(m - 1)^2(m^2 + m + 1)}$

61. $\dfrac{p^2 - 2p + 7}{(p + 2)(p + 3)(p - 1)}$ **62.** $\dfrac{-17t^2 - 6t + 2}{(2t + 1)(4t - 1)(t - 5)}$

63. $\dfrac{-m - 21}{2(m + 5)(m - 2)}$ or $\dfrac{m + 21}{2(m + 5)(2 - m)}$

64. $\dfrac{-5n + 17}{2(3n + 1)(n - 3)}$ or $\dfrac{5n - 17}{2(3n + 1)(3 - n)}$

Midchapter Review, pp. 426–427

2. a. 1. Factor the denominator of each rational expression. 2. Identify the LCD. 3. Rewrite each rational expression as an equivalent expression with the LCD as the denominator. 4. Add or subtract the numerators and write the result over the common denominator. 5. Simplify and reduce. **b.** $\dfrac{1}{3}$ **3. a.** Factor the numerators and denominators completely. Multiply across. Then reduce factors whose ratio is 1 or -1. **b.** $\dfrac{x + 3}{5}$ **4. a.** Factor the numerators and denominators completely. Multiply the first expression by the reciprocal of the second expression. Then reduce factors whose ratio is 1 or -1. **b.** $\dfrac{c + 3}{c}$

Section 5.5 Practice Exercises, pp. 433–434

17. $\dfrac{3}{\dfrac{2}{3} + \dfrac{3}{4}} = \dfrac{36}{17}$

Section 5.7 Practice Exercises, pp. 449–451

3. Expression; $\dfrac{2a - 5}{(a + 5)(a - 5)}$ **4.** Expression; $\dfrac{2n + 5}{4(n + 1)}$

Chapter 5 Review Exercises, pp. 458–460

4. $\left\{x \mid x \neq \dfrac{5}{2}, x \neq 3\right\}; \dfrac{1}{2x - 5}$ **5.** $\left\{h \mid h \neq -\dfrac{1}{3}, h \neq -7\right\}; \dfrac{1}{3h + 1}$

6. $\{a \mid a \neq 2, a \neq -2\}; \dfrac{4a - 1}{a - 2}$ **7.** $\{w \mid w \neq 4, w \neq -4\}; \dfrac{2w + 3}{w - 4}$

8. $\{z \mid z \neq 4\}; -\dfrac{z}{2}$ **9.** $\{k \mid k \neq 0, k \neq 5\}; -\dfrac{3}{2k}$

10. $\{b \mid b \neq -3\}; \dfrac{b - 1}{2}$ **11.** $\{m \mid m \neq -1\}; \dfrac{m - 5}{3}$

12. $\{n \mid n \neq -3\}; \dfrac{1}{n + 3}$ **13.** $\{p \mid p \neq -7\}; \dfrac{1}{p + 7}$

48. $\dfrac{-y - 18}{(y - 9)(y + 9)}$ or $\dfrac{y + 18}{(9 - y)(y + 9)}$ **49.** $\dfrac{t^2 + 2t + 3}{(2 - t)(2 + t)}$

50. $\dfrac{m + 8}{3m(m + 2)}$ **51.** $\dfrac{3(r - 4)}{2r(r + 6)}$

Cumulative Review Exercises, Chapters 1–5, pp. 461–463

3. Rational: $\sqrt{4}, \sqrt{9}, \sqrt{16}, \sqrt{49}$; irrational: $\sqrt{5}, \sqrt{20}$
5. $(8, \infty)$ ⟵————→
         8

6. ⟵—[————————→ $[-1, \infty)$; $\{x \mid x < 5\}$ ⟵————————→
   -1 5

7. The width is 17 m and the length is 35 m.   **8.** The base is 10 in. and the height is 8 in.   **11.** $12^2 + 16^2 = 20^2$
$$144 + 256 = 400$$
$$400 = 400 \checkmark$$

18. $3(3x - 5y)(3x + 5y)$ **19.** $(2c + 1)(5d - 3)$ **20.** $(x - 5)(x + 4)$
28. The speed of the current is 3 mph.   **29. a.** Linear   **b.**

x	y
0	-10
2	0
1	-5

c.

CHAPTER 6

Calculator Connections 6.1, p. 474

1.   **2.**   **3.**

4.   **5.**   **6.**

Section 6.1 Practice Exercises, pp. 474–477

10. $x = y + 2$ **11.** $x - y = 4$ **12.** $-3x + y = -6$

13. $2x - 5y = 10$ **14.** $y = 4x$ **15.** $y = -2x$

16. $y = -\dfrac{1}{2}x + 3$ **17.** $y = \dfrac{1}{4}x - 2$ **18.** $x + y = 0$

   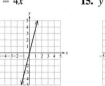

19. $-x + y = 0$ **20.** $2x + 3y = 8$ **21.** $4x - 5y = 15$

22. $y = 0.75x + 0.25$ **23.** $y = -0.8x - 1.2$ **24.** $50x - 40y = 200$

      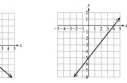

25. $-30x - 20y = 60$

26. a.   **b.** 124 miles   **c.** 201.5 miles

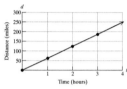

27. a.   **b.** 8.8 m   **c.** 66 m

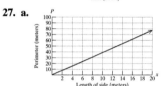

33. x-intercept: $(3, 0)$;   **34.** x-intercept: $(-9, 0)$;
      y-intercept: $(0, 15)$ y-intercept: $(0, 3)$

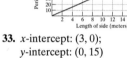

35. x-intercept: $\left(\dfrac{3}{2}, 0\right)$;   **36.** x-intercept: $\left(\dfrac{8}{3}, 0\right)$;
      y-intercept: $(0, -1)$ y-intercept: $(0, 2)$

37. x-intercept: $(3, 0)$;   **38.** x-intercept: $(-4, 0)$;
      y-intercept: $(0, -3)$ y-intercept: $(0, 8)$

39. *x*-intercept: (0, 0);
    y-intercept: (0, 0)

40. *x*-intercept: (1, 0);
    y-intercept: none

53. *x*-intercept: none;
    y-intercept: (0, 2)

54. *x*-intercept: (1, 0);
    y-intercept: none

41. *x*-intercept: none;

    y-intercept: $\left(0, \dfrac{8}{5}\right)$

42. *x*-intercept: none;
    y-intercept: (0, −2)

57.

They have the same slope but different *y*-intercepts.

58.

They have the same slope but different *y*-intercepts.

43. *x*-intercept: (1, 0);
    y-intercept: none

44. *x*-intercept: $\left(\dfrac{5}{4}, 0\right)$;

    y-intercept: none

61. Vertical

62. Horizontal

45. *x*-intercept: (−10, 0);
    y-intercept: (0, 4)

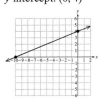

46. *x*-intercept: (10, 0);
    y-intercept: (0, 5)

63. Horizontal

64. Vertical

47. *x*-intercept: (3, 0);
    y-intercept: (0, −1.5)

48. *x*-intercept: (−2, 0);
    y-intercept: (0, −1.5)

65. Vertical

66. Horizontal

49. *x*-intercept: (0, 0);
    y-intercept: (0, 0)

50. *x*-intercept: (0, 0);
    y-intercept: (0, 0)

67. Horizontal

68. Vertical

51. *x*-intercept: (−2, 0);
    y-intercept: none

52. *x*-intercept: none;
    y-intercept: (0, −4)

69. Vertical

70. Horizontal

71. Horizontal

72. Vertical

Section 6.2 Practice Exercises, pp. 485–489

1. x-intercept: $(6, 0)$;
y-intercept: $(0, -2)$

2. x-intercept: none;
y-intercept: $(0, -3)$

3. x-intercept: $(7, 0)$;
y-intercept: none

4. x-intercept: $(0, 0)$;
y-intercept: $(0, 0)$

5. x-intercept: none;
y-intercept: $(0, \frac{3}{2})$

6. x-intercept: $(2, 0)$;
y-intercept: $(0, 8)$

7. x-intercept: $(0, 0)$;
y-intercept: $(0, 0)$

8. x-intercept: $(-2, 0)$;
y-intercept: none

47.

48.

49.

50.

51.

52.

57.

58.

59.

60.

61.

62.

Calculator Connections 6.3, pp. 497–498

1. Perpendicular   **2.** Parallel   **3.** Neither

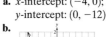

Section 6.3 Practice Exercises, pp. 498–501

1. a. x-intercept: $(10, 0)$;
y-intercept: $(0, -2)$
b.

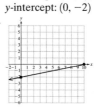

c. $\frac{1}{5}$

2. a. x-intercept: $(-4, 0)$;
y-intercept: $(0, -12)$
b.

c. -3

3. a. x-intercept: none;
y-intercept: $(0, -3)$
b.

c. 0

4. a. x-intercept: none;
y-intercept: $(0, 3)$
b.

c. 0

5. a. x-intercept: $(0, 0)$;
     y-intercept: $(0, 0)$

b.

c. $-\dfrac{2}{3}$

6. a. x-intercept: $\left(\dfrac{3}{2}, 0\right)$;
     y-intercept: $\left(0, -\dfrac{3}{2}\right)$

b.

c. 1

7. a. x-intercept: $(-5, 0)$;
     y-intercept: none

b.

c. Undefined

8. a. x-intercept: $(4, 0)$;
     y-intercept: none

b.

c. Undefined

9. $y = \dfrac{2}{5}x - \dfrac{4}{5}$; $m = \dfrac{2}{5}$; y-intercept: $\left(0, -\dfrac{4}{5}\right)$

10. $y = -\dfrac{3}{2}x + \dfrac{9}{2}$; $m = -\dfrac{3}{2}$; y-intercept: $\left(0, \dfrac{9}{2}\right)$

12. $y = \dfrac{7}{3}x + 2$; $m = \dfrac{7}{3}$; y-intercept: $(0, 2)$

15. $x = 2$; Cannot be written in slope-intercept form; undefined slope; no y-intercept   **16.** $x = 5$; Cannot be written in slope-intercept form; undefined slope; no y-intercept

19. $y = \dfrac{2}{3}x$; $m = \dfrac{2}{3}$; y-intercept: $(0, 0)$

20. $y = \dfrac{5}{6}x$; $m = \dfrac{5}{6}$; y-intercept: $(0, 0)$

21.

22.

23.

24.

25.

26.

27.

28.

29.

30.

31.

32.

33. $y = \dfrac{1}{2}x - 3$

34. $y = \dfrac{5}{2}x - 1$

35. $y = -2x + 9$
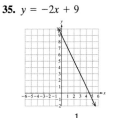

36. $y = 6x + 8$
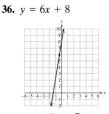

37. $y = -2x - \dfrac{1}{3}$

38. $y = -\dfrac{1}{2}x + \dfrac{7}{4}$

39. $y = -\dfrac{1}{2}x + \dfrac{3}{2}$

40. $y = 3x + 7$

41. $y = -x$

42. $y = x$

43. $y = 0.4x - 0.2$

44. $y = -0.6x + 2$

45. $y = \dfrac{9}{5}x$

46. $y = -\dfrac{2}{5}x$

47. $y = -\dfrac{2}{3}$

48. $y = 1$

49. $x = 2$

50. $x = -3$

51. $y = -2x + 2$

52. $y = 2x - 3$

Midchapter Review, pp. 501–502

13. a.

x	y
2	$-\dfrac{2}{3}$
3	-2
-3	6

b. x-intercept: $\left(\dfrac{3}{2}, 0\right)$;
y-intercept: $(0, 2)$

c. $y = -\dfrac{4}{3}x + 2$
d. Answers may vary.

14. a.

x	y
-3	4
-2	$\dfrac{13}{3}$
-6	3

b. x-intercept: $(-15, 0)$;
y-intercept: $(0, 5)$

c. $y = \dfrac{1}{3}x + 5$
d. Answers may vary.

15. a.

x	y
1	-4
2	-8
-4	16

b. x-intercept: $(0, 0)$;
y-intercept: $(0, 0)$

c. $y = -4x$
d. Answers may vary.

16. a.

x	y
0	$\dfrac{3}{4}$
1	$\dfrac{3}{4}$
2	$\dfrac{3}{4}$

b. x-intercept: none;
y-intercept: $\left(0, \dfrac{3}{4}\right)$

c. $y = \dfrac{3}{4}$
d. Answers may vary.

17. a.

x	y
2	0
2	1
2	2

b. x-intercept: $(2, 0)$;
y-intercept: none

c. $x = 2$; not possible to write
in slope-intercept form
d. Answers may vary.

18.

19.

20.

21.

Section 6.4 Practice Exercises, pp. 507–509

1. $4x + 3y = 9$

2. $2x - 3y = -3$

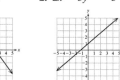

3. $y = -2x$

4. $3 - y = 9$

5. $3x - 1 = 5$

6. $x = -1$

7. $y = -4$

8. $x = \dfrac{4}{5}x$

Calculator Connections 6.5, pp. 515–516

1. 13.3

2. -42.3

3. 345

4. 95

Section 6.5 Practice Exercises, pp. 516–521

1. a. \$95   **b.** \$190   **c.** $(0, 0)$. For 0 kilowatt-hours used, the cost is \$0.
d. $m = 0.095$. The cost increases by \$0.095 for each kilowatt-hour
used.   **e.**

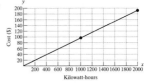

6. a. \$12.60   **b.** \$21.00   **c.** $(0, 0)$. If 0 gallons of water are used, the
water bill will be \$0.   **d.** $m = 4.20$. For every additional 1000 gallons
of water used, the bill increases by \$4.20.
e.

9. a. $y = 3.5x - 1.75$ **b.** $m = 3.5$. For each additional inch in length
of a person's arm, the person's height increases by 3.5 in.
c. 73.5 in. or 6 ft $1\frac{1}{2}$ in.

Chapter 6 Review Exercises, pp. 525–528

10. $x - 5y = 10$

11. $y = -\dfrac{3}{2}$

12. $y = 3x$

13. $y = \dfrac{1}{4}x - 2$

14. $x - 7 = 0$

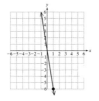

15. $6x + y = 0$

19.

20.

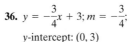

21.

22.

35. $y = \dfrac{5}{2}x - 5;\ m = \dfrac{5}{2};$
y-intercept: $(0, -5)$

36. $y = -\dfrac{3}{4}x + 3;\ m = -\dfrac{3}{4};$
y-intercept: $(0, 3)$

37. $y = -2x - 3;\ m = -2;$
y-intercept: $(0, -3)$

38. $y = 3x - 4;\ m = 3;$
y-intercept: $(0, -4)$

39. $y = \frac{1}{3}x; m = \frac{1}{3}$;
y-intercept: (0, 0)

40. $y = \frac{12}{5}; m = 0$;
y-intercept: $\left(0, \frac{12}{5}\right)$

41. $y = -\frac{5}{2}; m = 0$;
y-intercept: $\left(0, -\frac{5}{2}\right)$

42. $y = x; m = 1$;
y-intercept: (0, 0)

53. $m = \dfrac{y_2 - y_1}{x_2 - x_1}$

Chapter 6 Test, pp. 528–529

7. x-intercept: $\left(-\frac{1}{4}, 0\right)$;
y-intercept: (0, 2)

8. x-intercept: (0, 0);
y-intercept: (0, 0)

9. x-intercept: (3, 0);
y-intercept: none

10. x-intercept: none;
y-intercept: $\left(0, -\frac{1}{3}\right)$

Cumulative Review Exercises, Chapters 1–6, pp. 529–530

5. $(-\infty, -4)$

25. a. Linear **b.** $\left(\frac{3}{2}, 0\right)$ **c.** (0, 2)
d. $y = -\frac{4}{3}x + 2$ **e.** $m = -\frac{4}{3}$ **f.**

CHAPTER 7

Calculator Connections 7.1, pp. 537–538

1. (2, 1)

2. (6, −1)

3. (3, 1)

4. (−2, 0)

5. No solution

6. Dependent system

Section 7.1 Practice Exercises, pp. 539–541

7. a.

b.

c.

23. (3, 1)
Consistent; independent

24. (−2, −4)
Consistent; independent

25. (1, −2)
Consistent; independent

26. (−2, 1)
Consistent; independent

27. (1, 4)
Consistent; independent

28. (3, −1)
Consistent; independent

29. No solution
Inconsistent; independent

30. No solution
Inconsistent; independent

43. No solution
Inconsistent; independent

44. No solution
Inconsistent; independent

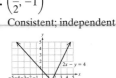

31. Infinitely many solutions
$\{(x, y)|y = 2x + 3\}$;
Consistent; dependent

32. Infinitely many solutions
$\{(x, y)|y = -\frac{1}{3}x\}$;
Consistent; dependent

45. $\left(\frac{1}{2}, 3\right)$
Consistent; independent

46. $\left(\frac{3}{2}, -1\right)$
Consistent; independent

33. $(-3, 6)$
Consistent; independent

34. $(6, -2)$
Consistent; independent

47. Infinitely many solutions
$\{(x, y)|y = 0.5x + 2\}$;
Consistent; dependent

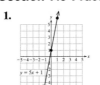

48. Infinitely many solutions
$\{(x, y)|y = \frac{3}{4}x - \frac{3}{2}\}$;
Consistent; dependent

35. Infinitely many solutions
$\{(x, y)|y = \frac{5}{3}x - 3\}$;
Consistent; dependent

36. Infinitely many solutions
$\{(x, y)|y = -2x + 3\}$;
Consistent; dependent

Section 7.2 Practice Exercises, pp. 548–551

57. a. 22 months **b.** $(22, 495)$; After renting a system for 22 months, both Company A and Company B will charge $495. **c.** Company B is more expensive than Company A for more than 22 months. Company A is more expensive than Company B for less than 22 months.

Section 7.4 Practice Exercises, pp. 568–570

19. 12 gal of the 45% disinfectant solution should be mixed with 8 gal of the 30% disinfectant solution.   **20.** 12 gal of the 25% antifreeze solution should be mixed with 3 gal of the 15% antifreeze solution.

37. $(2, -2)$
Consistent; independent

38. $(3, 4)$
Consistent; independent

Section 7.5 Practice Exercises, pp. 577–579

1.

2.

3.

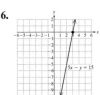

39. No solution
Inconsistent; independent

40. No solution
Inconsistent; independent

4.

5.

6.

41. $(4, 2)$
Consistent; independent

42. $(-2, 2)$
Consistent; independent

7.

8.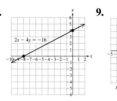

9.

34.

35.

36.

10.

11.

12.

37.

38.

19. For example:
(0, 5)(2, 7)(−1, 8)

20. For example:
(0, −1)(2, −2)(−1, −5)

39. a. The set of ordered pairs above the line $x + y = 4$; for example: $(6, 3)(−2, 8)(0, 5)$. **b.** The set of ordered pairs on the line $x + y = 4$; for example: $(0, 4)(4, 0)(2, 2)$. **c.** The set of ordered pairs below the line $x + y = 4$; for example: $(0, 0)(−2, 1)(3, 0)$. **40. a.** The set of ordered pairs below the line $x + y = 3$; for example: $(0, 2)(−1, −1)(3, −2)$. **b.** The set of ordered pairs on the line $x + y = 3$; for example: $(0, 3)(3, 0)(1, 2)$. **c.** The set of ordered pairs above the line $x + y = 3$; for example: $(4, 0)(−1, 6)(2, 2)$.

21. For example:
(1, −1)(3, 0)(−2, −9)

22. For example:
(2, 1)(0, 4)(−1, 8)

43.

44.

45.

23. For example:
(0, 0)(0, 2)(−1, −3)

24. For example:
(0, −3)(4, −4)(−1, −6)

46.

47.

48.

25.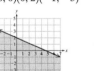

26.

27.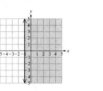

49.

50.

51.

28.

29.

30.

52.

31.

32.

33.

Chapter 7 Review Exercises, pp. 585–589

11. $(0, −2)$
Consistent; independent

12. $(−2, 3)$
Consistent; independent

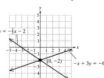

13. Infinitely many solutions
$\{(x, y)|y = -2x + 5\}$;
Consistent; dependent

14. Infinitely many solutions
$\{(x, y)|y = \frac{1}{5}x - 1\}$;
Consistent; dependent

15. $(1, -1)$
Consistent; independent

16. $(-1, -6)$
Consistent; independent

17. No solution
Inconsistent; independent

18. No solution
Inconsistent; independent

34. 1. Write both equations in standard form.
2. Multiply one or both equations by a constant to create opposite coefficients for one of the variables.
3. Add equations to eliminate the variable.
4. Solve for the remaining variable.
5. Substitute the known variable into an original equation to solve for the other variable.

55. For example:
$(1, -1)(0, -4)(2, 0)$

56. For example:
$(5, 5)(4, 0)(0, 7)$

57. For example:
$(0, 0)(0, -5)(-1, 1)$

58. For example:
$(-4, 0)(-2, -2)(1, -4)$

59.

60.

61.

62.

63.

64.

65.

66.

67.

68.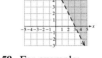

69.

70.

Chapter 7 Test, pp. 589–591

2. a. $(-3, 3)$

18.

19.

Cumulative Review Exercises, Chapters 1–7, pp. 591–592

6. $\left[\frac{3}{11}, \infty\right)$

23. In Problem 21 you must change the fractions to equivalent fractions with a common denominator. In Problem 22 you must clear the denominators.

27. a. b. 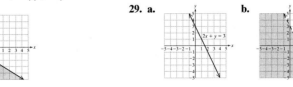 **c.** $(0, 2)$

29. a. **b.**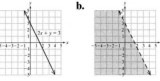

c. Part (a) represents the solutions to an equation. Part (b) represents the solutions to a strict inequality.

30. 20 gal of the 15% solution should be mixed with 40 gal of the 60% solution.

CHAPTER 8

Calculator Connections 8.1, pp. 601–602

1. 2. 3.

4. 5. 6.

7. 8. 9.

10. 11. 12.

13. 14. 15.

16. 17. 18.

19. 20.

Section 8.1 Practice Exercises, pp. 602–605

3. No, only positive numbers have two square roots. Zero has only one square root, and negative numbers have no real-valued square roots. **57.** x^2, y^4, $(ab)^6$, w^8x^8, m^{10}. The expression is a perfect square if the exponent is even. **58.** b^3, e^6, $(pq)^9$, $u^{12}v^{12}$. The expression is a perfect cube if the exponent is a multiple of 3. **59.** p^4, t^8, $(cd)^{12}$. The expression is a perfect fourth power if the exponent is a multiple of 4. **60.** d^5, y^{10}. The expression is a perfect fifth power if the exponent is a multiple of 5.

Calculator Connections 8.2, p. 611

1.   **2.**   **3.**

4.   **5.**   **6.**

7.   **8.**   **9.**

Calculator Connections 8.3, pp. 617–618

1.   **2.**   **3.**

4.

Calculator Connections 8.4, p. 625

1.   **2.**   **3.**

4.   **5.**   **6.**

Calculator Connections 8.5, pp. 634–635

1.   **2.**   **3.**

4.   **5.**   **6.**

7.   **8.**

Calculator Connections 8.7, p. 648

1. 2. 3.

4. 5. 6.

7. 8. 9.

10. 11. 12.

13. 14. 15.

16.

Chapter 8 Review Exercises, pp. 655–659

1. Principal square root: 14; negative square root: -14 **2.** Principal square root: 1.2; negative square root: -1.2 **3.** Principal square root: 15; negative square root: -15 **4.** Principal square root: 0.8; negative square root: -0.8

Chapter 8 Test, pp. 659–661

1. 1. The radicand has no factor raised to a power greater than or equal to the index. 2. There are no radicals in the denominator of a fraction. 3. The radicand does not contain a fraction.
7. $\dfrac{2\sqrt{5} - 12}{-31}$ or $\dfrac{12 - 2\sqrt{5}}{31}$

Cumulative Review Exercises, Chapters 1–8, pp. 661–663

5. $\{x | x > -4\}; (-4, \infty)$

26.

27.

33.

CHAPTER 9

Calculator Connections 9.1, pp. 676–677

1.
2.
3.

4.
5.
6.

Section 9.1 Practice Exercises, pp. 677–680

2. Given a relation in x and y, y is a function of x if for every element x in the domain, there corresponds exactly one element y in the range.

Calculator Connections 9.4, pp. 705–706

1.
2.
3.

4.

Section 9.4 Practice Exercises, p. 707

1. For $ax^2 + bx + c = 0$, $x = \dfrac{-b \pm \sqrt{b^2 - 4ac}}{2a}$

3. $2x^2 - x - 5 = 0; a = 2, b = -1, c = -5$
4. $5x^2 + 3x + 10 = 0; a = 5, b = 3, c = 10$
5. $-3x^2 + 14x + 0 = 0; a = -3, b = 14, c = 0$
6. $x^2 - 5x - 3 = 0; a = 1, b = -5, c = -3$
7. $x^2 + 0x - 9 = 0; a = 1, b = 0, c = -9$
8. $x^2 + 0x + 25 = 0; a = 1, b = 0, c = 25$
9. Discriminant is 0; one rational solution **10.** Discriminant is 0; one rational solution **11.** Discriminant is 49; two rational solutions **12.** Discriminant is 49; two rational solutions **13.** Discriminant is 33; two irrational solutions **14.** Discriminant is 45; two irrational solutions **15.** Discriminant is -23; two imaginary solutions **16.** Discriminant is -56; two imaginary solutions

19. $k = \dfrac{2}{3}, k = -\dfrac{1}{2}$ **20.** $n = -2, n = \dfrac{1}{3}$ **21.** $a = \dfrac{-5 \pm \sqrt{33}}{4}$

22. $b = \dfrac{-1 \pm \sqrt{5}}{6}$ **23.** $c = \dfrac{1 \pm i\sqrt{23}}{6}$ **24.** $p = \dfrac{1 \pm i\sqrt{14}}{5}$

25. $k = -1 \pm 2i$ **26.** $p = \dfrac{7 \pm \sqrt{37}}{2}$

27. $y = \dfrac{1 \pm \sqrt{17}}{-8}$ or $\dfrac{-1 \pm \sqrt{17}}{8}$

28. $z = \dfrac{3 \pm \sqrt{89}}{-10}$ or $\dfrac{-3 \pm \sqrt{89}}{10}$

29. $x = \dfrac{-3 \pm \sqrt{33}}{4}$ **30.** $m = \dfrac{5 \pm \sqrt{37}}{6}$ **31.** $y = \dfrac{-15 \pm \sqrt{145}}{4}$

32. $t = \dfrac{5 \pm \sqrt{35}}{2}$ **33.** $t = \dfrac{-10 \pm \sqrt{85}}{-5}$ or $\dfrac{10 \pm \sqrt{85}}{5}$

34. $p = \dfrac{-4 \pm i\sqrt{5}}{-4}$ or $\dfrac{4 \pm i\sqrt{5}}{4}$ **35.** $x = \dfrac{-2 \pm \sqrt{22}}{6}$

36. $x = \dfrac{-1 \pm \sqrt{73}}{6}$ **37.** $m = \dfrac{1 \pm i\sqrt{11}}{2}$ **38.** $h = \dfrac{7 \pm i\sqrt{23}}{2}$

39. The width is $\dfrac{1 + 3\sqrt{89}}{4} \approx 7.3$ m

The length is $\dfrac{-1 + 3\sqrt{89}}{2} \approx 13.7$ m

40. The base is $-1 + \sqrt{145} \approx 11.0$ cm. The height is $1 + \sqrt{145} \approx 13.0$ cm.
41. The legs are $1 + \sqrt{71} \approx 9.4$ ft and $-1 + \sqrt{71} \approx 7.4$ ft.
42. The length is $1 + \sqrt{41} \approx 7.4$ ft. The width is $-1 + \sqrt{41} \approx 5.4$ ft. The height is 6 ft.

Calculator Connections 9.5, pp. 715–716

1. $(-2, 3)$ **2.** $(10, 5)$ **3.** $(-1.5, -2.6)$

4. $(1.75, 2.5)$ **5.** $\left(\dfrac{5}{2}, 0\right)$ **6.** $\left(-\dfrac{8}{3}, 0\right)$

7.

The graph in part (b) is shifted up 4 units. The graph in part (c) is shifted down 3 units.

8.

In part (b), the graph is shifted to the right 3 units. In part (c) the graph is shifted to the left 2 units.

9.

In part (b) the graph is stretched vertically by a factor of 2. In part (c) the graph is shrunk vertically by a factor of $\frac{1}{2}$.

10.

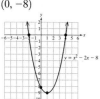

In part (b) the graph has been stretched vertically and reflected across the x-axis. In part (c) the graph has been shrunk vertically and reflected across the x-axis.

Section 9.5 Practice Exercises, pp. 717–719

26. Find the x-coordinate by $\frac{-b}{2a}$. Then substitute the value of x into the equation and solve for y.   **35.** c; x-intercepts: $(\sqrt{7}, 0)(-\sqrt{7}, 0)$; y-intercept: $(0, -7)$ **36.** b; x-intercepts: $(3, 0)(-3, 0)$; y-intercept: $(0, -9)$ **37.** a; x-intercepts: $(-1, 0)(-5, 0)$; y-intercept: $(0, 5)$ **38.** d; x-intercepts: $(3, 0)(1, 0)$; y-intercept: $(0, 3)$

43. a. Upward
   b. $(0, -9)$
   c. $(3, 0)(-3, 0)$
   d. $(0, -9)$
   e.

   f. $(-\infty, \infty)$
   g. $[-9, \infty)$

44. a. Upward
   b. $(0, -4)$
   c. $(2, 0)(-2, 0)$
   d. $(0, -4)$
   e.

   f. $(-\infty, \infty)$
   g. $[-4, \infty)$

45. a. Upward
   b. $(1, -9)$
   c. $(4, 0)(-2, 0)$
   d. $(0, -8)$
   e.

   f. $(-\infty, \infty)$
   g. $[-9, \infty)$

46. a. Upward
   b. $(-1, -25)$
   c. $(4, 0)(-6, 0)$
   d. $(0, -24)$
   e.

   f. $(-\infty, \infty)$
   g. $[-25, \infty)$

47. a. Downward
   b. $(3, 0)$
   c. $(3, 0)$
   d. $(0, -9)$
   e.

   f. $(-\infty, \infty)$
   g. $(-\infty, 0]$

48. a. Downward
   b. $(5, 0)$
   c. $(5, 0)$
   d. $(0, -25)$
   e.

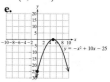

   f. $(-\infty, \infty)$
   g. $(-\infty, 0]$

49. a. Downward
   b. $(4, 1)$
   c. $(3, 0)(5, 0)$
   d. $(0, -15)$
   e.

   f. $(-\infty, \infty)$
   g. $(-\infty, 1]$

50. a. Downward
   b. $(-2, 9)$
   c. $(-5, 0)(1, 0)$
   d. $(0, 5)$
   e.

   f. $(-\infty, \infty)$
   g. $(-\infty, 9]$

51. a. Upward
   b. $(-3, 1)$
   c. none
   d. $(0, 10)$
   e.

   f. $(-\infty, \infty)$
   g. $[1, \infty)$

52. a. Upward
   b. $(-2, 1)$
   c. none
   d. $(0, 5)$
   e.

   f. $(-\infty, \infty)$
   g. $[1, \infty)$

53. a. Downward
   b. $(0, -2)$
   c. none
   d. $(0, -2)$
   e.

   f. $(-\infty, \infty)$
   g. $(-\infty, -2]$

54. a. Downward
   b. $(0, -5)$
   c. none
   d. $(0, -5)$
   e.

   f. $(-\infty, \infty)$
   g. $(-\infty, -5]$

Chapter 9 Review Exercises, pp. 724–728

83. The height is $\dfrac{-1 + \sqrt{97}}{2} \approx 4.4$ cm.

The base is $\dfrac{1 + \sqrt{97}}{2} \approx 5.4$ cm.

85. b; x-intercepts: $(9.32, 0)(2.68, 0)$; y-intercept: $(0, 25)$
86. f; x-intercepts: $(-0.76, 0)(-5.23, 0)$; y-intercept: $(0, 4)$
87. d; x-intercepts: $(1, 0)(-1, 0)$; y-intercept: $(0, 5)$
88. a; x-intercepts: $(1, 0)(-1, 0)$; y-intercept: $(0, 3)$
89. e; x-intercepts: $(0, 0)(-4, 0)$; y-intercept: $(0, 0)$
90. c; x-intercepts: none; y-intercept: $(0, 4)$
95. Vertex: $(-1, 1)$; axis of symmetry: $x = -1$
96. Vertex: $(4, 19)$; axis of symmetry: $x = 4$
97. Vertex: $(3, 13)$; axis of symmetry: $x = 3$
98. Vertex: $\left(-\dfrac{1}{2}, -\dfrac{3}{2}\right)$; axis of symmetry: $x = -\dfrac{1}{2}$

99. a. Upward
   b. $(-1, -3)$
   c. Approximately $(0.22, 0)(-2.22, 0)$
   d. $(0, -1)$
   e.
   f. $(-\infty, \infty)$
   g. $[-3, \infty)$

100. a. Downward
   b. $(2, 2)$
   c. Approximately $(2.82, 0)(1.18, 0)$
   d. $(0, -10)$
   e.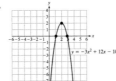
   f. $(-\infty, \infty)$
   g. $(-\infty, 2]$

101. a. Downward
   b. $(-1, -4)$
   c. No x-intercepts
   d. $(0, -12)$
   e.
   f. $(-\infty, \infty)$
   g. $(-\infty, -4]$

102. a. Upward
   b. $(-2, -3)$
   c. $(-3, 0)(-1, 0)$
   d. $(0, 9)$
   e.
   f. $(-\infty, \infty)$
   g. $[-3, \infty)$

19. Vertex: $(0, 25)$; x-intercepts: $(-5, 0)(5, 0)$; y-intercept: $(0, 25)$

Cumulative Review Exercises, Chapters 1–9, pp. 729–731

5. $\{x \mid x > -1\}$; $(-1, \infty)$

22. a. $(6, 0)$
   b. $(0, -3)$
   c. $\frac{1}{2}$
   d.

23. a. $(-3, 0)$
   b. No y-intercept
   c. Slope is undefined.
   d.

28.

45. Vertex: $\left(\dfrac{3}{2}, 2\right)$; x-intercepts: $(0, 0)(3, 0)$; y-intercept: $(0, 0)$

Chapter 9 Test, pp. 728–729

1. a.   **b.**

17. The base is $\dfrac{-1 + \sqrt{97}}{2} \approx 4.4$ m.

The height is $1 + \sqrt{97} \approx 10.8$ m.

18. For $y = ax^2 + bx + c$, if $a > 0$ the parabola opens upward, if $a < 0$ the parabola opens downward.

APPLICATION INDEX

Making holiday aprons, 14
Mixed nuts, 181, 182
Mortgage payments, 26
Ordering shrimp, 15
Phone bill, 500, 520
Raffle tickets, 149, 180
Raise, 23
Sales tax on textbook, 25
Saving for vacation, 25
Saving from monthly pay, 14
Servings of oatmeal per box, 14
Sewing dress, 15
Take-home pay, 14
Tip at restaurant, 23, 25
Total cost to rent water purification system, 550–551
TV commercials, 570
Water bill, 519
Water in orange juice, 177–178
Width of bed, 15

DISTANCE/SPEED

Airplane altitude *versus* horizontal ground distance, 293–294
Distance, 131, 174–176, 214
Distance a person can see to the horizon, 628
Distance between Augusta and Nashville, 656
Distance between Beijing and London, 441
Distance between boats, 366
Distance between Frankville and Clayton, 605
Distance between Greensboro and Asheville, 605
Distance between Jackson and Tupelo, 605
Distance between Reno and Salinas, 604
Distance between Spokane and Portland, 612
Distance between St. Johns and Gander, 441
Distance between Washington DC and Louisville, 604
Distance of runway *versus* speed of plane, 724
Distance of skydiver falling, 313
Distance traveled by car and bus, 182, 193
Distance traveled by car and truck, 656
Distance traveled by cars, 181, 219
Distance traveled by kayak, 601
Distance traveled by plane, 181
Speed of bikers, 222
Speed of boat, 182
Speed of boat and speed of current, 450, 569, 583–584, 588
Speed of canoe, 182
Speed of cars, 182, 456–457, 643
Speed of motorist, 446–447, 450, 460
Speed of plane, 182, 445–446, 450, 566–567, 569, 570

Speed of train, 450
Time for bicycle ride, 398
Time required to commute to work, 520

ENVIRONMENT

Altitude and temperature, 471–472
Amoco Cadiz tanker disaster, 254
Area of Greenland, 149
Average daily temperature in January for cities, 518–519
Average depth of Gulf of Mexico, 148
Average temperature, 85
Death Valley, 76
Deepest point in Pacific Ocean, 149
Distance between lightning strike and observer, 488–489
Elevation of US cities, 52–53
Hurricane Floyd, 528–529
Hurricane Irene, 220
Land area of Asia and Africa, 150
Land area of Texas, 254
Longest river in Africa, 148
Masking tape over windows before hurricane, 605
Mount Rainier National Park, 433
Mt. Everest, 76, 150
Rainfall for Bermuda, 220
Rainfall in Miami, 195
Rainfall in Tampa, 15
Snow depth *versus* time, 465
Snowfall in Burlington, 196
Snowfall in Syracuse, 222
Snowstorm, 310
Speed of current, 449, 461, 462, 569, 583–584, 588
Speed of wind, 450, 569, 570
Temperatures for cities, 61, 68, 72, 76–77, 103–104, 107–108, 116, 117
Tropical storm in Texas, 224
Water flowing over Niagara Falls, 308
Yellowstone National Park, 433

INVESTMENT

Annual rate of return on investment, 647, 650
Compound interest, 232, 234, 304
Piggy bank, 183, 387
Price per share of stock, 104–105, 108–109, 118, 568
Price per share of stock for Hershey Foods, 1
Principal, 173–174, 180, 181, 213–214, 219, 222, 224, 563–565, 568–569, 570, 583–584, 588
Simple interest, 156–157, 159, 173–174, 180, 213–214, 218

POLITICS

Composition of the House of Representatives by political party, 148
Election, 182, 588
Incorrectly marked ballots, 182

SCHOOL

Applicants accepted at Union College, 23
Cost of advertising school play, 514
Earning A in math class, 193
Enrollment at community college, 25
Enrollment in Catholic schools, 207
Lighting at college, 570
Memory devices, 63, 170
Printing out book report, 14
Ratio of men to women in chemistry class, 445
Scores in math class, 194
Scores in science class, 194
Students earning money from recycling, 207
Test score as a function of study time, 675–676

SCIENCE

Charles' law, 449
Dinosaurs, 255
Distance an object drops, 698
Distance between Earth and Polaris, 729
Distance between Earth and Sun, 255
Distance between Mercury and Sun, 305
Distance to Mars, 251
Earth's orbit around the Sun, 305
Height of ball, 259–260, 265, 288–289, 297, 365, 376, 384
Height of object, 727
Height of rock, 308, 372, 375
Height of rock on Moon, 726
Height of rocket, 294, 365, 376
Height of stone, 360, 365
Height of tree, 178–179, 451
Object in free fall, 294–295, 678
Speed of Space Shuttle, 406
Total resistance, 434
Velocity of object, 628, 643, 658

SPORTS

Area of volleyball court, 168
Baskets by Cynthia Cooper, 590
Baskets by Kareem Abdul-Jabbar, 570
Baskets by Wilt Chamberlain, 569–570
Batting average, 477

STATISTICS/ DEMOGRAPHICS

INDEX

Difference of Squares

$$a^2 - b^2 = (a + b)(a - b)$$

Difference of Cubes

$$a^3 - b^3 = (a - b)(a^2 + ab + b^2)$$

Sum of Cubes

$$a^3 + b^3 = (a + b)(a^2 - ab + b^2)$$

Perfect Square Trinomials

$$a^2 + 2ab + b^2 = (a + b)^2$$
$$a^2 - 2ab + b^2 = (a - b)^2$$

Measure Abbreviations

Length

in.	inch
ft	foot
yd	yard
mm	millimeter
cm	centimeter
m	meter
km	kilometer

Volume/Capacity

pt	pint
gal	gallon
mL	milliliter
L	liter
cc	cubic centimeter (also written as cm^3)

Mass

mg	milligram
g	gram
kg	kilogram

Force

lb	pound

Sets of Real Numbers

Natural numbers: $\{1, 2, 3, \ldots\}$

Whole numbers: $\{0, 1, 2, 3, \ldots\}$

Integers: $\{\ldots -3, -2, -1, 0, 1, 2, 3, \ldots\}$

Rational numbers: $\{\frac{p}{q} | p$ and q are integers and q does not equal $0\}$

Irrational numbers: $\{x | x$ is a real number that is not rational$\}$

Quadratic Formula

The solutions to $ax^2 + bx + c = 0$ $(a \neq 0)$ are given by

$$x = \frac{-b \pm \sqrt{b^2 - 4ac}}{2a}$$

Proportions

An equation that equates two ratios is called a proportion:

$$\frac{a}{b} = \frac{c}{d} \quad (b \neq 0, d \neq 0)$$

The cross products are equal: $ad = bc$.

Measure Conversion

Length	Area	Volume	Mass/Force
1 mile = 5280 ft	1 in.^{2} = 6.452 cm^2	1 pt = 2 c	1 g = 1000 mg
1 mile = 1.609 km	1 yd^2 = 9 ft^2	1 qt = 2 pt	1 kg = 1000 g
1 in. = 2.54 cm	1 m^2 = 10.76 ft^2	1 gal = 4 qt	1 kg = 2.2 lb*
1 m = 3.281 ft	1 mile^2 = 640 acres	1 L = 1000 cm^3	1 lb = 16 oz
1 lightyear = 9.46 × 10^{15} m		1 gal = 3.785 L	
1 m = 1000 mm		1 ft^3 = 7.481 gal	
1 m = 100 cm		1 L = 1000 mL	
1 km = 1000 m		1 mL = 1 cc	

*1 kg = 2.2 lbs on Earth

GLOSSARY

4th dimension (4D) roller coaster a roller coaster where the riders rotate independently of the track

A-frame a support structure shaped like the letter *A*

axle a rod in the center of a wheel around which the wheel turns

ballast any heavy material that adds weight to an object, in this case, crushed stones used to hold the track in place

canopy a decorative cover or rooflike structure

catapult a large weapon, similar to a slingshot, used in the past for firing objects over castle walls

circuit a loop that electricity flows around

conductor a material that lets heat, electricity, or sound travel easily through it; metal is a good conductor

cranking rod a bar connecting and passing movement from a rotating gear to another part of a machine

electromagnet a magnet that is temporarily magnetized by an electric current

engineer someone trained to design and build machines, vehicles, bridges, roads, or other structures

force the push or pull on an object that results from its interaction with another object

friction a force produced when two objects rub against each other; friction slows down objects

gear a toothed wheel that fits into another toothed wheel; gears can change the direction of a force or can transfer power

gravity a force that pulls objects with mass together; gravity pulls objects down toward the center of Earth

horizontal flat and parallel to the ground

hydraulic powered by fluid forced through pipes or chambers

iconic widely viewed as perfectly capturing the meaning or spirit of something or someone

inertia tendency of an object to remain either at rest or in motion unless affected by an outside force

innovative advanced or unlike anything done before

kinetic energy the energy of a moving object

laws of motion the three main laws of physics

magnetic field the area around a magnet that can attract other metals

mass the amount of material in an object

mechanism a system of moving parts inside a machine

momentum the force or speed created by movement

pendulum a weight that swings back and forth from a fixed point

perpendicular straight up and down relative to another surface; the two lines that form the letter *T* are perpendicular to each other

piston a disc or short cylinder within an engine that moves inside a closed tube and pushes against a liquid or a gas

pneumatic operated by compressed air

potential energy energy stored within an object, waiting to be released

pulley a grooved wheel turned by a rope, belt, or chain that is often used to move heavy objects

ratchet a device made up of a bar or wheel with slanted teeth that allow movement only in one direction

sleeper a piece of timber, stone, or steel on or near the ground to keep railroad rails in place

spring a device that can be pulled or compressed but always returns to its original shape

steel a strong, hard metal formed from iron, carbon, and other materials

stress the physical pressure, pull, or other force on an object

traction engine a heavy engine that burns coal to produce steam that is used to power machines or move heavy loads

winch a lifting device that winds in or lets out wire, cable, or a chain

INDEX